DISCRETE STRUCTURES

By

Kalika Patrai

Deptt. of Computer Applications

(Galgotia Institute of Management & Technology)

Greater Noida

S.K. KATARIA & SONS®

Publisher of Engineering Books

4885/109, Prakash Mahal, Dr. Subhash Bhargav Lane,

Opposite Delhi Medical Association, Daryaganj, New Delhi–110002 (INDIA)

Phone: +91-11-43551243, Mobile: +91-9871775858, 7678425209

e-mail: katariabook@gmail.com, katariabooks@yahoo.com

Website: www.skkatariaandsons.com

(e-books, subscriptions and bulk buying also available on our website)

Published by:

S.K. Kataria & Sons®

4885/109, Prakash Mahal, Dr. Subhash Bhargav Lane,
Opposite Delhi Medical Association, Daryaganj, New Delhi-110002 (INDIA)
Phone: +91-11-43551243, Mobile: +91-9871775858, 7678425209
e-mail: katariabook@gmail.com, katariabooks@yahoo.com
Website: www.skkatariaandsons.com

First Edition : 2008-09, Third Edition: 2010-11, Reprint: 2026

Price : ₹ 595/-

Laser Typesetting :
QUICK MEDIA, Delhi

Printed at :
NISHA ENTERPRISES

Preface

The book addresses a full spectrum of discrete structures used in computer science. To introduce the formal aspects of Computer Science, the author has presented discrete structure as foundation. For student who are ambitious to achieve unparalleled goals in the field of computer science, and want to develop their logical skills, it becomes necessary that their mathematical concepts are very clear.

The book is written as a text, with examples and exercise. In every chapter of this book, due consideration has been given to the pattern and standard of various universities.

To facilitate easy understanding, the text and problems have been put in simple form. The more and more emphasis is on simpler language and set of examples relating to every concepts in each chapter. This will help readers to understand the subject in better way. Suggestions are alwyas welcome from readers.

Kalika Patrai

Acknowledgement

I take this opportunity to express my deep sense of gratitude to Dr. J.C. Bhatia (Advisor, GIMT) for his constant inspiration and encouragement. Also, Prof. Unni Krishnan A. (Head, MCA, GIMT) for his kind support.

I also want to express my heartiest thanks to Mr. Vivek Keshari, Dr. Indu Uprety, Mrs. Jaya Sinha, Mrs. Varsha Gautam (GCET), Ms. Garima, Mr. S.K. Singh and Ms. Deepika for their valuable suggestions, support, guidance and encouragement.

Last but not the least, I would also like to thank Mr. Sanjeev Kataria of M/s S.K. Kataria and Sons for publishing this book. I acknowledge my indebtness for various authors on graph theory whose books have been freely consulted.

Kalika Patrai

Acknowledgement

[illegible]

[illegible]

[illegible]

[illegible]

Chapter

1 SET THEORY

1.1 WHAT IS DISCRETE MATHEMATICS

Discrete mathematics deals with discrete objects, which are separated from (not connected to/distinct from) each other. Integers (whole numbers), rational numbers (ones that can be expressed as the quotient of two integers), automobiles, houses, people etc. are all discrete objects. On the other hand real numbers, which include irrational as well as rational numbers are not discrete. As you know between any two real numbers there is another real number different from either of them. So they are packed without any gaps and cannot be separated from their immediate neighbours. In that sense they are not discrete. In this course we will be concerned with objects such as integers, propositions, sets, relations and functions, which are all discrete. We are going to learn concepts associated with them, their properties, and relationships among them and among others.

1.2 INTRODUCTION

The concept of set theory was developed in the year 1895 by the German mathematician George Cantor. It is the basic to computer science and mathematics and proves to be a useful tool to solve the problems which seems to be complicated in classical mathematics.

In this chapter, we will cover some basic concepts, definitions and notations of set theory. Also we will define and understand the operations on sets, with the help of vien diagrams too.

1.2.1 Set

George Cantor defined a set as "a collection of well defined objects or things" Objects in a set are called **elements** or **members** of a set. The term 'well defined' indicates that there must be common property between the elements of the set.

Usually a set is denoted by capital letters $A, B, C, D, \ldots$ and elements of set by lower case letters a, b, c, d, e, x, y, z.

For example $A = \{a, b, c, d\}$ denotes a set whose elements are a, b, c and d.

1.2.2 Set Membership

Consider a set A, then a statement "x is an element of A" or "x is a member of A" or "x belongs to A" is written as

$$x \in A$$

If x is not an element of A, or x does not belongs to A, then we write

$$x \notin A$$

Two sets A and B are said to be equal if and only if they have same elements here order of placement is immaterial. It is denoted as $A = B$. And if they are not equal, we write $A \neq B$.

For example $A = \{a, c, d\}$ and $B = \{c, a, d\}$ are same or equal sets, with elements a, c and d, and $X = \{1, 2, 3, 4\}$, $Y = \{a, b, c, d\}$ are not equal as they do not have same elements in them.

1.3 SPECIFYING SET

Essentially there are two ways to specify a set

(*a*) Tabular form of a set (list its members).

(*b*) Set builder form (state the common property of the elements).

1.3.1 Tabular Form of A set

Under this form, set can be represented by listing all its members if possible. *For examples*

(*i*) $A = \{1, 2, 3, 4, 5, 6, 7, 8\}$

(*ii*) A set containing all even numbers from 1 to 10 can be defined as

$A = \{2, 4, 6, 8, 10\}$

(*iii*) A set containing all the vowels in English alphabet can be written as :

$A = \{a, c, o, i, u\}$

In this representation the elements are separated by commas (,) and enclosed in braces { }.

1.3.2 Set Builder Form

Here we specify a set with the help of the common propery of its elements. *For examples*

(*i*) $Z = \{x : x \text{ is integer}\}$

We read it as Z is the set of all those elements x, which is an integer.

(*ii*) $A \{x : x \text{ is odd integer} > 0\}$

Which reads "A is a set of all those elements x which are odd integers and greater them 0".

Usually, we specify a set with the help of set builder form when it is difficult or impossible to list all its elements. *For example* if we want to list all the members of set of children born in the world during a year. To write the set in tabular form is practically not possible though theoretically it is possible to compile such a list. So we say that tabular form do specify a set if it has few elements, otherwise we describe a set by the propery which characterizes its elements.

This fact of describing or specifying a set in terms of property common to its elements is formally stated as principle of abstraction.

1.4 PRINCIPLE OF ABSTRACTION

Consider any set X and property P, there is a set say A such that the members of A are exactly those elements of X which have the property P.

1.5 DIFFERENT TYPES OF SET

1.5.1 Finite and Infinite Set

Finite set is one which has finite or countable number of elements otherwise the set is said to be infinite. *For example*

$A = \{x : x \text{ is born in } 2008\}$ is a finite set whereas

$B = \{x : x \text{ is rational}\}$ is infinite number.

The number of different elements in a finite set is called **cardinality** of set and is denoted as $n(A)$. Sometimes it is also called as the **order** of a set.

For example, consider set

$$A = \{1, 2, 3, 4, 5, 6, 7, 8\}$$

clearly A is finite and thus $n(A) = 8$. Again

$$B = \{x, y, a, c, d, x\}$$

is also finite set with cardinality or order $n(B) = 5$ (as x is appearing twice).

1.5.2 Universal Set

In any application of sets, the elements of all the sets under investigation usually belongs to some fixed large set called the **universal set**, usually denoted by U.

For example, the set consisting all possible numbers is the universal set in number system.

1.5.3 Null (Empty) Set

A set is called null or empty set if it's cardinality is zero. That is, a null set contains no element in it. Usually a null set is denoted by ϕ or { }. Note that, there is only one null set that is if, A and B are both empty then $A = B$ since they have exactly the same members, namely, none.

1.5.4 Singleton Set

A set whose cardinality is one that is set containing single element is called a **singleton set.**

For example : $A = \{a\}$, $B = \{3\}$, both are singleton sets.

1.6 SUBSET

Consider two sets A and B. If every element of A also belongs to B, *i.e.*, if every element of set A is also an element of set B, them A is called a subset of B and denoted as $A \subseteq B$ or $B \supseteq A$.

If at least one element of set A does not belongs to B then A is not a subset of B and is denoted as $A \not\subseteq B$ or $B \not\supseteq A$.

1.6.1 Proper Subset

If $A \subseteq B$ but $A \neq B$ *i.e.*, there is at least one element in B which is not a member of set A, then A is called a proper subset of B and is denoted as $A \subset B$.

For example, let $A = \{a, c\}$ $B = \{a, b, c\}$, $C = \{b, a, c\}$ be three sets, then clearly $A \subset B$ and $A \subset C$ as both the elements of set A are from set B as cells as from set C. That is A is a proper subset of set B as well as set C whereas B is not a proper subset of C since $B = C$.

Notes : (*i*) Every set is a subset of itself.

(*ii*) Null set ϕ is the subset of every set.

(*iii*) Null set ϕ has no proper subset.

(*iv*) Every set has at least two subsets, empty set or null set ϕ and the set itself. They are called improper subsets.

(*v*) If A is a set with n-elements then it has a total of 2^n subsets.

1.6.2 Comparable Sets

Two sets A and B are said to be comparable if either of following happens :

(*a*) $A \subset B$ (*b*) $B \subset A$ (*c*) $A = B$

Otherwise they are said to be incomparable. *For example,* let $A = \{a, b\}$ and $B = \{a, c, d, b\}$. Here $A \subset B$, hence A and B are comparable. But $A = \{1, 3, 5, 7\}$ and $B = \{2, 4, 6, 8\}$ are incomparable as they do not satisfy any of the three properties for being comparable.

1.6.3 Set of Sets

A set whose elements themselves are some sets, is known as **set of sets** or **class of sets** or **family of sets.**

For example if ϕ, $\{1\}$, $\{2, 3\}$, $\{4\}$ are the sets then the set A of these sets is the set of sets

$$A = \{\phi, \{1\}, \{2, 3\}, \{4\}\}$$

1.6.4 Disjoint Sets

Consider two sets A and B. If there is no element common between A and B then they are said to be disjoint. That is

$$A \cap B = \phi \text{ or } \{\ \}$$

For example $A = \{a, b, c\}$ and $B = \{1, 2, 3\}$ are disjoint sets.

1.6.5 Power Set

The collection of all the subsets of a finite set A is called the power set of A and is denoted as $P(A)$. The cardinality of $P(A)$ is 2^n as there are a total of 2^n subsets of a set A containing n-elements.

For example, consider a set $A = \{1, 2\}$. Then there will be $2^2 = 4$ subsets of A as ϕ, $\{1\}$, $\{2\}$ and $\{1, 2\}$. The power set of A will be

$$P(A) = \{\phi, \{1\}, \{2\}, \{1, 2, 3\}\}$$

If $A = \{a, b, c\}$ then

$$P(A) = \{\phi, \{a\}, \{b\}, \{c\} \{a, b\}, \{b, c\}, \{a, c\}, \{a, b, c\}\}$$

1.6.6 Equivalent Sets

If two sets A and B have a one-to-one correspondence between their elements then A and B are said to be equivalent sets.

For example, consider Two sets $A = \{a, b, c, d\}$ and $B = \{1, 2, 3, 4\}$. Then $a \leftrightarrow 1$, $b \leftrightarrow 2$, $c \leftrightarrow 3$ and $d \leftrightarrow 4$.

Since the elements of A and B have one-to-one correspondence *i.e.,* number of elements in both the sets are equal hence A and B are equivalent sets and is denoted as $A \sim B$.

Note : Equal sets and equivalent sets are altogether different terminologies.

1.6.7 Complements

Consider the universal set U and a set $A \subset U$. Then a set A' defined as $A' = \{x : x \in U \text{ and } x \notin A\}$ is called the complement of A. That is a set of all those elements of U which do not belong to set A, is called complement of A. Clearly A and A' are disjoint set.

For Example : Consider the universal set $U = \{1, 2, 3, 4, 5, 6, 7, 8, 9, 10\}$ and a subset A of U as $A = \{2, 4, 6, 8, 10\}$ then complement A' of A is

$$A' = \{1, 3, 5, 7, 9\}$$

EXAMPLES

Example 1.1. Which of the following are empty sets ?

(*i*) $A = \{x : x \in I, 6x^2 - 11x + 3 = 0\}$

(*ii*) $B = \{x : x \in N, 4 + x = 3\}$

(*iii*) $C = \{0\}$.

Solution. (*i*) Solving $6x^2 - 11x + 3 = 0$, we get $x = \frac{3}{2}, \frac{1}{3}$.

Since, no value of $x \in I$.

$\therefore$ $x \notin A$. Hence A is an empty set.

(*ii*) No natural number satisfies the condition $4 + x = 3$. Hence $B = \phi$ *i.e.,* null set.

(*iii*) Since the set C has only one element *i.e.,* 0, therefore, it is not an empty set.

Example 1.2. Let P = **Set of all equilateral triangles.**

Q = **Set of all isosceles triangles.**

R = **Set of all triangles, and**

S = **Set of all equiangular triangles.**

(*i*) State symbolically, that all equilateral triangles are equiangular.

Solution. $P \subseteq S.$

(*ii*) Make the correct symbolic statement about sets R and Q.

Solution. $Q \subseteq R.$

Example 1.3. Categorize the following into finite and infinite sets :

(*i*) $A = \{1, 1, 1, ...\}$.

Solution. In a set, an element is to be counted once only,

$\therefore$ The set $A = \{1\}$, which is a singleton set.

Hence it is a finite set.

(*ii*) $B = \{x : x \in \mathrm{I}$ **and** x **is even}.**

Solution. Since elements of B are all even integers, and there are infinite number of distinct even integers, therefore, the set B is an infinite set.

Example 1.4. Show that $P = \{1, 2, 3\}$ **is not a subset of** $Q = \{1, 4, 5, 6\}$**.**

Solution. If at least one element of P is not in Q, then P is not subset of Q.

Here, $2 \in P$ but $2 \notin Q$.

Hence, $P \subseteq Q.$

Example 1.5. Find the power set of the following sets :

(*i*) $S = \{a, b, c, d\}$.

(*ii*) $A = \{\phi \{1, 2\}, 3\}$

(*iii*) $B = \{ \}$ ***i.e.,*** B **is an empty set.**

Solution. (*i*) $P(S) = \begin{Bmatrix} S, \phi, \{a\}, \{b\}, \{c\}, \{d\}, \{a, b\}, \{a, c\}, \{a, d\} \{b, c\}, \{b, d\}, \\ \{c, d\}, \{a, b, c\}, \{b, c, d\}, \{a, c, d\}, \{a, b, d\} \end{Bmatrix}$

Notes. (*i*) The power set $P(S)$ has $2^4 = 16$ elements.

(*ii*) $P(A) = \{\phi, \{\phi\}, \{1, 2\}, \{3\}, \{\phi, \{1, 2\}\}, \{1, 2\}, 3\}, \{\phi, 3\}, A\}$.

(*iii*) $P(B) = \{\phi\}$.

The power set of an empty set contains one element.

Example 1.6. What is the cardinality of each of the following sets ?

(*i*) $\{1\}$ **Ans.** 1

(*ii*) $\{(1)\}$ **Ans.** 1

(*iii*) $\{\phi\}$ **Ans.** 1

(*iv*) $\{\phi, \{\phi\}\}$ **Ans.** 2

(*v*) $\{\phi, \{\phi\}, \{\phi, \{\phi\}\}\}$ **Ans.** 3

(*vi*) ϕ **Ans.** 0

Example 1.7. Are the following sets equal :

$A = \{x : x$ **is a letter in the word TEA}**

$B = \{x : x$ **is a letter in the word ATE}**

$C = \{x : x$ **is a letter in the word EAT}**

and $D = \{x : x$ **is a letter in the word LATE}.**

Solution. $A = \{T, E, A\}, B = \{A, T, E\}, C = \{E, A, T\}, D = \{L, A, T, E\}$,

Clearly $A = B = C \neq D$.

Example 1.8. Is every set comparable to empty set ϕ and the universal set U.

Solution. Yes, every set is comparable to ϕ and U.

Let A be any set, then $\phi \subseteq A$ and $A \subseteq U$.

Example 1.9. Consider the following sets :

$A = \{a, b\}, B = \{a, b, c, d\}, C = \{a, e\}, D = \{c, d, e\}, E = \{d, e\}$

Which of these sets are disjoint ?

Solution. Sets A & D and sets A & E are disjoint as $A \cap D = \phi$ and $A \cap E = \phi$. Other sets have one or more elements in common *e.g.*, $A \cap B = \{a, b\}$

1.7. VENN DIAGRAM

John Venn gave diagrammatic representations of sets called diagrams. The universal set U is represented by the points of a rectangle and the interior of a rectangle and subset of U, such as A and B, are represented by the point of circles and their interior points.

For Examples :

Set | *Venn Diagram*

(*i*) $A \subset U$

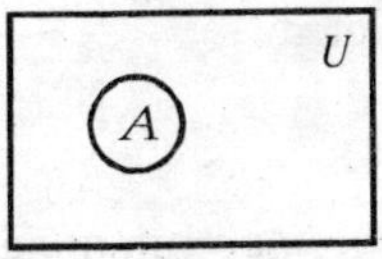

Fig. 1.1.

(*ii*) $A \subseteq B$

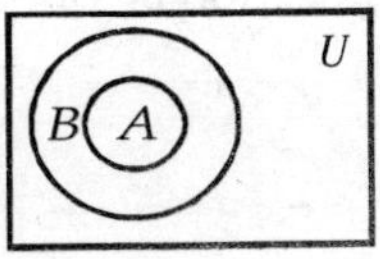

Fig. 1.2.

(*iii*) U = set of real number

A = set of positive odd integers

B = set of positive even integers.

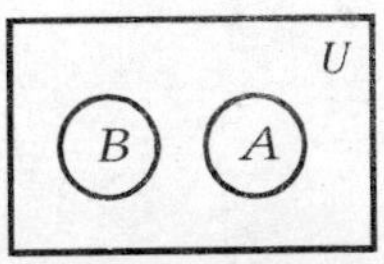

Fig. 1.3.

(*iv*) $U = \{a, b, c, d, ..., x, y, z\}$

$A = \{b, c, d, e\}$

$B = \{c, d, f, g\}$.

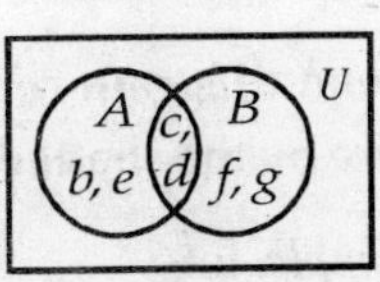

Fig. 1.4.

1.7.1. Set Operations and Venn Diagram

Two or more different sets can be joined in many different ways. *For example,* we have set of students studying English and set of students studying Discrete Mathematics. Then we can form the set of students studying English or Discrete Mathematics, the set of students studying both, set of students studying none, and so on. In this section, we shall define certain operations with sets, yielding new sets.

(*a*) **Union of sets.** The union or join of two sets A and B is the set of all elements, which are either in A or in B, or in both. It is denoted by $A \cup B$, and it is read as "A union B".

i.e., $$A \cup B = \{x \mid x \in A \text{ or } x \in B\}$$

The Venn diagram for $A \cup B$ is given by the shaded region in the figure below :

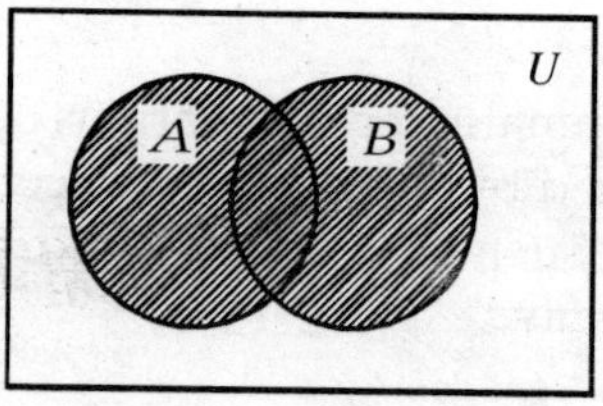

Fig. 1.5.

e.g., let $$A = \{a, b, c\}, B = \{b, c, d, e\}$$

then $$A \cup B = \{a, b, c\} \cup \{b, c, d, e\} = \{a, b, c, d, e\}.$$

(*b*) **Intersection of sets.** The intersection or meet of the two sets A and B is the set of all elements which belong to both A and B. It is denoted by $A \cap B$ and it is read as "A intersection B" or "A meet B".

i.e. $$A \cap B = \{x \mid x \in A \text{ and } x \in B\}$$

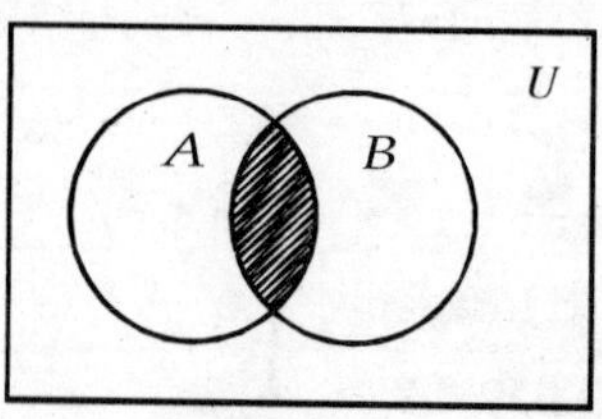

Fig. 1.6.

The Venn diagram representing the intersection of A and B is given in the above figure by the shaded area.

For example, let $A = \{1, 3, 5\}, B = \{1, 3, 7\}$ then $A \cap B = \{1, 3\}$.

Note. The two sets are disjoint if, their intersection is the empty set.

Example 1.10. If $A = \{a, e, i\}$ $B = \{0, 4\}$ then $A \cap B = \phi$

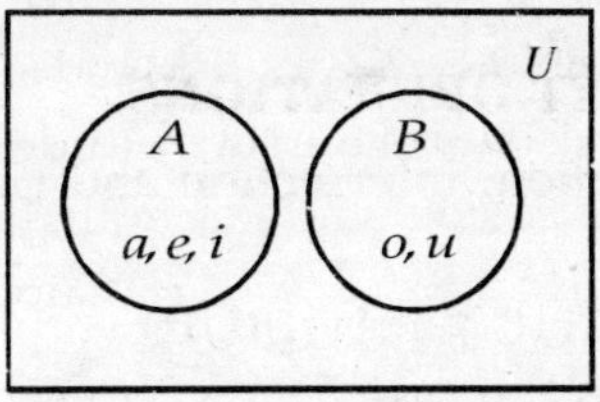

Fig. 1.7. Venn diagram representing disjoint sets *A* and *B*.

(*c*) **Complement of a set.** Let A be a subset of the universal set U. Then the set of element which are not in A is called the complement of A w.r.t. U. It is denoted by

$$A' \text{ or } A^c \text{ or } \overline{A}$$

i.e. $$A^c = \{x \mid x \in \text{U} \text{ but } x \notin A\}$$

The shaded area in the Venn diagram given below represents complement of A.

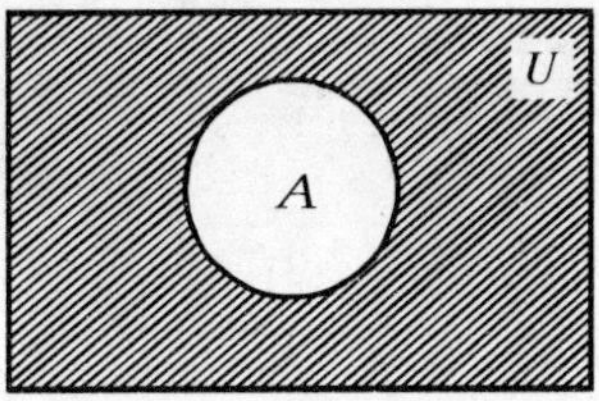

Fig. 1.8.

For example, let $U = \{2, 4, 6, 8, 10, 12, 14, 16\}$

$A = \{2, 6, 10\}$

then $A^c = \{4, 8, 12, 14, 16\}$.

(*d*) **Difference of two sets :** The difference of two sets A and B is the set of all elements of A which are not in B. Symbolically, it is represented as A/B or $A - B$ or $A \sim B$ and is read as "A difference B"

i.e., $A - B = \{x \mid x \in A, x \notin B\}$

Similarly, $B - A = \{x \mid x \in B, x \notin A\}$

For example, let $A = \{1, 3, 5\}, B = \{1, 2, 3\}$

then $A - B = \{5\}, B - A = \{2\}$.

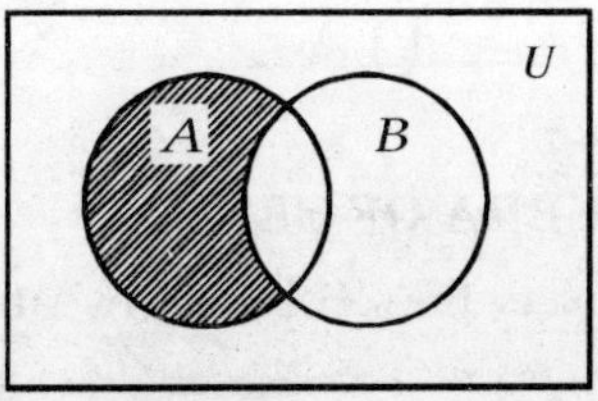

Fig. 1.9. Venn diagram representing $A - B$ is the shaded area.

Notes. (*i*) $A - B \neq B - A$

(*ii*) The complement of the set A w.r.t the universal set U is also defined as $U - A$.

1.8 GENERALIZED SET OPERATIONS

As we saw earlier, union, intersection and Cartesian product of sets are **associative**. *For example*

$$(A \cup B) \cup C = A \cup (B \cup C).$$

To denote either of these we often use $A \cup B \cup C$.

This can be generalized for the union of any finite number of sets as $A_i \cup A_2 \cup \ldots \cup A_n$, which we write as $\bigcup_{i=1}^{n} A_i$

This **generalized union** of sets can be rigorously defined as follows :

➤ **Definition** $\left(\bigcup_{i=1}^{n} A_i\right)$ **:**

Basic Clause : For $n = 1$, $\bigcup_{i=1}^{n} A_i = A_1$.

Inductive Clause : $\bigcup_{i=1}^{n+1} A_i = \left(\bigcup_{i=1}^{n} A_i\right) \cup A_{n+1}$

Similarly the **generalized intersection** $\bigcap_{i=1}^{n} A_i$ and **generalized Cartesian product we** $\bigtimes_{i=1}^{n} A_i$ can be defined.

Based on these definitions, De Morgan's law on set union and intersection can also be generalized as follows :

Theorem (Generalized De Morgan)

$$\overline{\bigcup_{i=1}^{n} A_i} = \bigcap_{i=1}^{n} \overline{A}_i,$$

and

$$\overline{\bigcap_{i=1}^{n} A_i} = \bigcup_{i=1}^{n} \overline{A}_i$$

1.9. LAWS OF THE ALGEBRA OF SETS

If A, B and C are any three sets, the set operations satisfy the following properties or laws :

1. Commutative Laws :

(*a*) $A \cup B = B \cup A$

Proof. Let $x \in A \cup B$, then $x \in A$ or $x \in B$

$\Rightarrow \quad x \in B$ or $x \in A$

$\Rightarrow \quad x \in B \cup A$

So $\quad A \cup B \subseteq B \cup A.$ ...(1)

Similarly, let $y \in B \cup A$

$\Rightarrow \quad y \in B$ or $y \in A$

$\Rightarrow \quad y \in A$ or $y \in B$

$\Rightarrow \quad y \in A \cup B.$

So $\quad B \cup A \subseteq A \cup B$ ...(2)

From (1) and (2) we have

$$A \cup B = A \cup B$$

(*b*) $A \cap B = B \cap A$

Proof. Let $x \in A \cap B \Rightarrow x \in A$ and $x \in B$

$\Rightarrow \quad x \in B$ and $x \in A$

$\Rightarrow \quad x \in B \cap A$

So $\quad A \cap B \subseteq B \cap A$ ...(1)

Let $\quad y \in B \cap A \Rightarrow y \in B$ and $y \in A$

$\Rightarrow y \in A$ and $y \in B$

$\Rightarrow y \in A \cap B$

So $B \cap A \subseteq A \cap B$ ...(2)

From (1) and (2), we have

$$A \cap B = B \cap A$$

2. Idempotent Laws :

(*a*) $A \cup A = A$

Proof. Let $x \in A \cup A$

$\Rightarrow \quad x \in A$ or $x \in A$

$\Rightarrow \quad x \in A$

So $\quad A \cup A \subseteq A$ ...(1)

Let $y \in A$

$\Rightarrow \quad y \in A$ or $y \in A$

$\Rightarrow \quad y \in (A \cup A)$

So $\quad A \subseteq (A \cup A)$ ...(2)

From (1) and (2), we have

$$A \cup A = A.$$

(*b*) $A \cap A = A$

Proof. Let $x \in A \cap A$

$\Rightarrow \quad x \in A$ and $x \in A$

$\Rightarrow \quad x \in A$

So, $A \cap A \subseteq A$ ...(1)

Let $y \in A \Rightarrow y \in A$ and $y \in A$

$\Rightarrow$ $y \in A \cap A$

So $A \subseteq A \cap A$ ...(2)

From (1) and (2),

$$A \cap A = A$$

3. Identity Laws :

(*a*) $A \cup \phi = A$

Proof. Let $x \in A \cup \phi$

$\Rightarrow$ $x \in A$ or $x \in \phi$

$\Rightarrow$ $x \in A$

So $A \cup \phi \subseteq A$ ...(1)

Let $y \in A$

$\Rightarrow$ $y \in A$ or $y \in \phi$

$\Rightarrow$ $y \in A \cup \phi$

So $A \subseteq A \cup \phi$ ...(2)

From (1) and (2),

$$A = A \cup \phi$$

(*b*) $A \cap U = A$

Proof. Let $x \in A \cap U$

$\Rightarrow$ $x \in A$ and $x \in U$

$\Rightarrow$ $x \in A$ $(A \subset U)$

So $A \cap U \subseteq A$ ...(1)

Let $y \in A$, since $A \subset U$.

$y \in A$ and $y \in U$

$\Rightarrow$ $y \in A \cap U$

So $A \subseteq A \cap U$ ...(2)

From (1) and (2), we have $A \cap U = A$.

4. Domination Laws :

(*a*) $A \cup U = U$

Proof. Let $x \in A \cup U$

$\Rightarrow$ $x \in A$ or $x \in U$

$\Rightarrow$ $x \in U$

So $A \cup U \subseteq U$ ...(1)

Also, let $y \in U$ then

$y \in$ union of A and U also.

i.e., $y \in A$ or $y \in U$

$\Rightarrow$ $y \in A \cup U$.

So $$U \subseteq A \cup U \quad ...(2)$$

From (1) and (2) we have,

$$A \cup U = U.$$

(*b*) $A \cap \phi = \phi$

Proof. The empty set ϕ contains no element. Therefore, the set $A \cap \phi$ contains no element since there is no common element between A and ϕ.

So $$A \cap \phi = \phi.$$

5. Associative Laws :

(*a*) $$A \cup (B \cup C) = (A \cup B) \cup C$$

Proof. Let $x \in A \cup (B \cup C)$

$\Rightarrow \quad x \in A \quad$ or $x \in (B \cup C)$

$\Rightarrow \quad x \in A \quad$ or $(x \in B$ or $x \in C)$

$\Rightarrow \quad (x \in A$ or $x \in B) \quad$ or $x \in C$

$\Rightarrow \quad x \in (A \cup B) \quad$ or $x \in C$

$\Rightarrow \quad x \in (A \cup B) \cup C$

So $$A \cup (B \cup C) \subseteq (A \cup B) \cup C \quad ...(1)$$

Let $y \in (A \cup B) \cup C$

$\Rightarrow \; y \in (A \cup B) \quad$ or $y \in C$

$\Rightarrow \; (y \in A$ or $y \in B) \quad$ or $y \in C$

$\Rightarrow \; y \in A \quad$ or $(y \in B$ or $y \in C)$

$\Rightarrow \; y \in A \quad$ or $y \in (B \cup C)$

$\Rightarrow \; y \in A \cup (B \cup C)$

So $$(A \cup B) \cup C \subseteq A \cup (B \cup C) \quad ...(2)$$

From (1) and (2), we have

$$A \cup (B \cup C) = (A \cup B) \cup C.$$

(*b*) $(A \cap B) \cap C = A \cap (B \cap C)$

Proof. Let $x \in (A \cap B) \cap C$

$\Rightarrow \; x \in (A \cap B)$ and $x \in C$

$\Rightarrow \; (x \in A$ and $x \in B)$ and $x \in C$

$\Rightarrow \; x \in A$ and $(x \in B$ and $x \in C)$

$\Rightarrow \; x \in A$ and $x \in (B \cap C)$

$\Rightarrow \; x \in A \cap (B \cap C)$

So $(A \cap B) \cap C \subseteq A \cap (B \cap C)$ ...(1)

Let $y \in A \cap (B \cap C)$

$\Rightarrow$ $y \in A$ and $y \in (B \cap C)$

$\Rightarrow$ $y \in A$ and ($y \in B$ and $y \in C$)

$\Rightarrow$ ($y \in A$ and $y \in B$) and $y \in C$

$\Rightarrow$ $y \in (A \cap B)$ and $y \in C$

$\Rightarrow$ $y \in (A \cap B) \cap C$

So $A \cap (B \cap C) \subseteq (A \cap B) \cap C$ ...(2)

From (1) and (2), we have

$$(A \cap B) \cap C = A \cap (B \cap C).$$

6. Distributive Laws :

(*a*) $A \cap (B \cup C) = (A \cap B) \cup (A \cap C)$

i.e., the intersection of sets distributes over the union of sets.

Proof. Let $x \in A \cap (B \cup C)$

$\Rightarrow$ $x \in A$ and $x \in (B \cup C)$

$\Rightarrow$ $x \in A$ and ($x \in B$ or $x \in C$)

$\Rightarrow$ ($x \in A$ and $x \in B$) or ($x \in A$ and $x \in C$)

$\Rightarrow$ $x \in (A \cap B)$ or $x \in (A \cap C)$

$\Rightarrow$ $x \in (A \cap B) \cup (A \cap C)$

So $A \cap (B \cup C) \subseteq (A \cap B) \cup (A \cap C)$ ...(1)

Let $y \in (A \cap B) \cup (A \cap C)$

$\Rightarrow$ $y \in (A \cap B)$ or $y \in (A \cap C)$

$\Rightarrow$ ($y \in A$ and $y \in B$) or ($y \in A$ and $y \in C$)

$\Rightarrow$ $y \in A$ and ($y \in B$ or $y \in C$)

$\Rightarrow$ $y \in A$ and $y \in (B \cup C)$

$\Rightarrow$ $y \in A \cap (B \cup C)$

So $(A \cap B) \cup (A \cap C) \subseteq A \cap (B \cup C)$ ...(2)

From (1) and (2), we have

$$A \cap (B \cup C) = (A \cap B) \cup (A \cap C)$$

➤ **Venn Diagram and Distributed Law :**

The distributive law is,

$$A \cap (B \cup C) = (A \cap B) \cup (A \cap C)$$

is shown by shaded area in the following Venn diagrams :

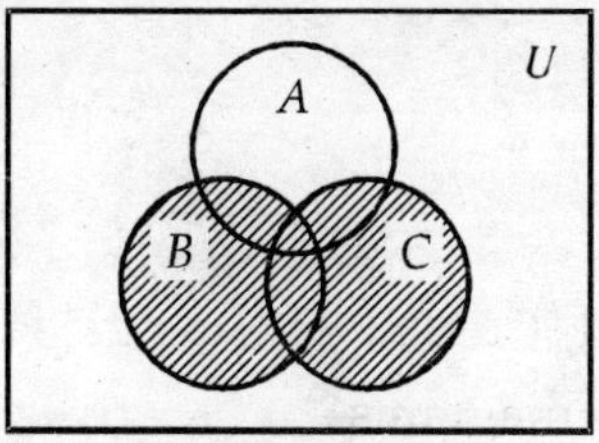

Fig. 1.10. $B \cup C$

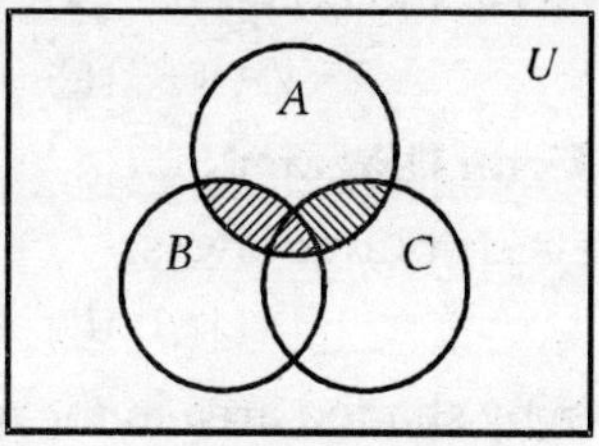

Fig. 1.11. $A \cap (B \cup C)$

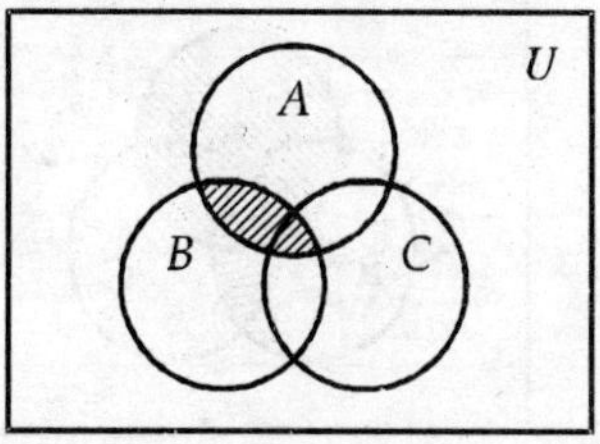

Fig. 1.12. $A \cap B$

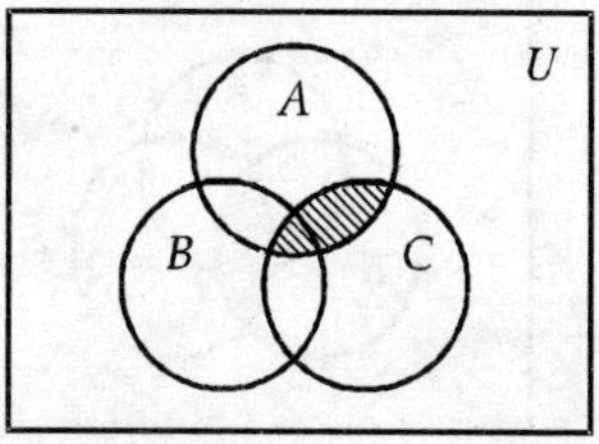

Fig. 1.13. $A \cap C$

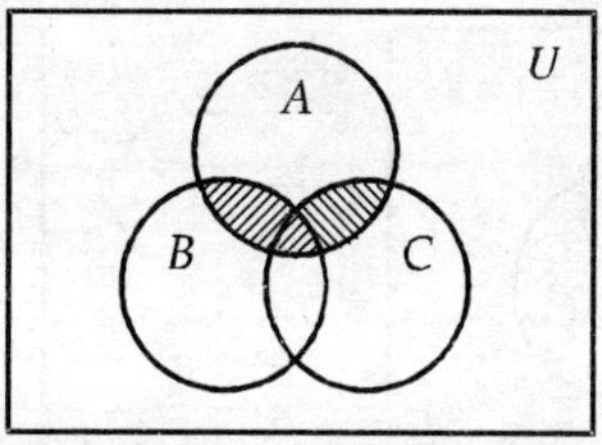

Fig. 1.14. $(A \cap B) \cup (A \cap C)$

(*b*) $$A \cup (B \cap C) = (A \cup B) \cap (A \cup C)$$

i.e., the union of sets distributes over the intersection of sets.

Proof. Let $x \in A \cup (B \cap C)$

$\Rightarrow \quad x \in A \text{ or } x \in (B \cap C)$

$\Rightarrow \quad x \in A \text{ or } (x \in B \text{ and } x \in C)$

$\Rightarrow \quad (x \in A \text{ or } x \in B) \text{ and } (x \in A \text{ or } x \in C)$

$\Rightarrow \quad x \in (A \cup B) \text{ and } x \in (A \cup C)$

$\Rightarrow \quad x \in (A \cup B) \cap (A \cup C)$

So $A \cup (B \cap C) \subseteq (A \cup B) \cap (A \cup C)$ ...(1)

Let $y \in (A \cup B) \cap (A \cup C)$

$\Rightarrow \quad y \in (A \cup B) \text{ and } y \in (A \cup C)$

$\Rightarrow \quad y \in A \text{ or } y \in B \text{ and } y \in A \text{ or } y \in C$

$\Rightarrow \quad y \in A \text{ or } (y \in B \text{ and } y \in C)$

$\Rightarrow \quad y \in A \text{ or } y \in (B \cap C)$

$\Rightarrow \quad y \in A \cup (B \cap C)$

So $(A \cup B) \cap (A \cup C) \subseteq A \cup (B \cap C)$ ...(2)

From (1) and (2) we have,

$$A \cup (B \cap C) = (A \cup B) \cap (A \cup C)$$

Use of Venn Diagram

The distributive law is,

$$A \cup (B \cap C) = (A \cup B) \cap (A \cup C)$$

is shown by shaded area in the following Venn diagrams :

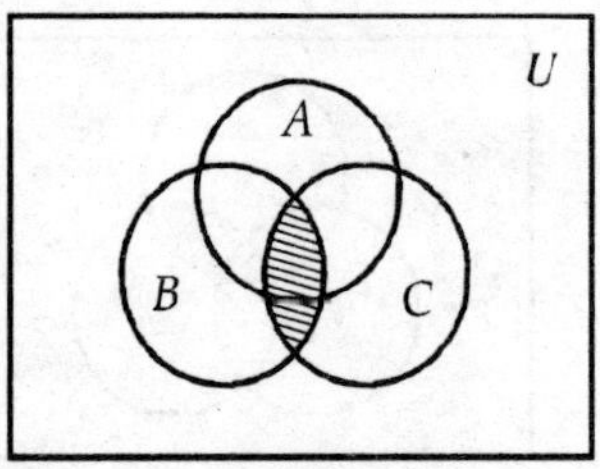

Fig. 1.15. $B \cap C$

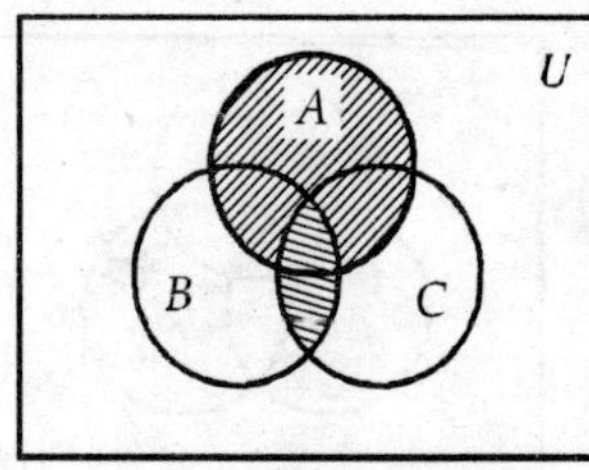

Fig. 1.16. $A \cup (B \cap C)$

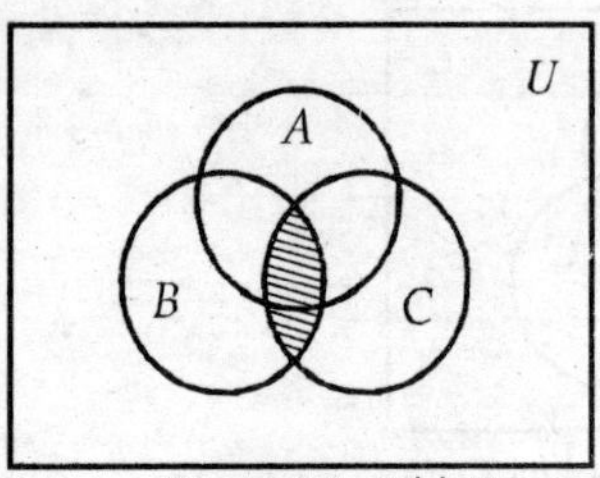

Fig. 1.17. $A \cup B$

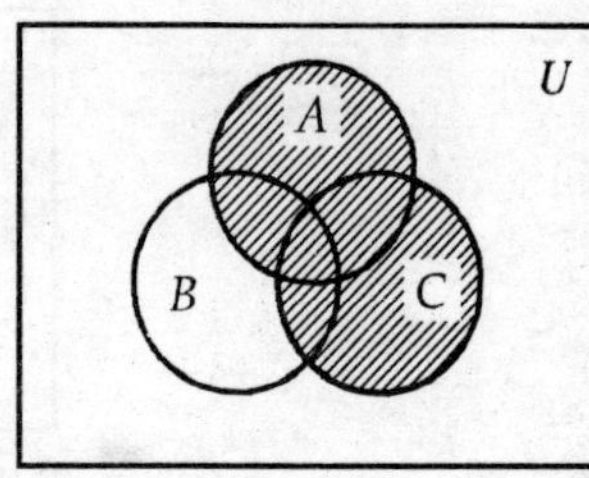

Fig. 1.18. $A \cup C$

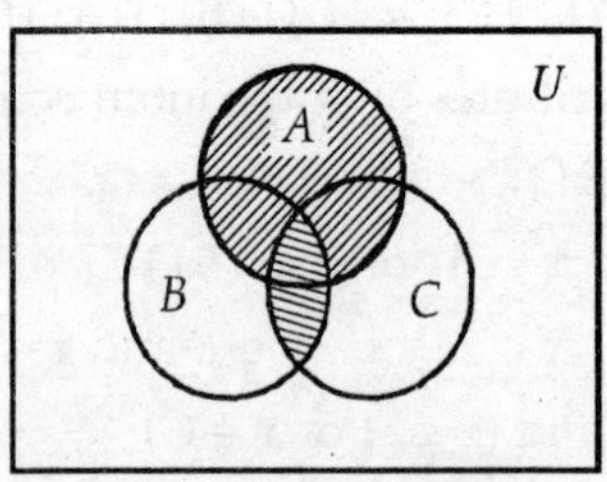

Fig. 1.19. $(A \cup B) \cap (A \cap C)$

7. Complementation Laws :

(*a*) $(A^c)^c = A$

Proofs. Let $x \in (A^c)^c$

$\Rightarrow$ $x \notin A^c$

$\Rightarrow$ $x \in A$

So $(A^c)^c \subseteq A$ ...(1)

Let $y \in A \Rightarrow y \notin A^c$

$\Rightarrow y \in (A^c)^c$

So $A \subseteq (A^c)^c$ ...(2)

From (1) and (2), we have, $(A^c)^c = A$.

This is also known as *involution law*.

(*b*) $A \cup A^c = U$

Proof. Every set is a subset of universal set.

So $\qquad A \cup A^c \subseteq U$ ...(1)

Let $\qquad x \in U \Rightarrow x \in A$ or $x \in A^c$

$\Rightarrow \qquad x \in A \cup A^c$

So $\qquad U \subseteq A \cup A^c$ ...(2)

From (1) and (2), we have

$$A \cup A^c = U$$

(*c*) $\qquad A \cap A^c = \phi$

Proof. Let $x \in A \cap A^c$

$\Rightarrow \quad x \in A$ and $x \in A^c$

$\Rightarrow \quad x \in \phi$, since there is no common element in A and A^c.

So $A \cap A^c \subseteq \phi$ ...(1)

Also, the empty set ϕ is the subset of every set, therefore, $\phi \subseteq A \cap A^c$ ...(2)

From (1) and (2), we have

$$A \cap A^c = \phi$$

(*d*) $U^c = \phi$

Proof. Let $x \in U^c$

$\Rightarrow \qquad x \notin U$

Since there is no element which does not belong to the universal set.

So $\qquad U^c = \phi$

(*e*) $\phi^c = U$

Proof. The null set ϕ has no element. So, the complement of null set *i.e.*, ϕ^c will be set consisting of all elements in the universal set.

Hence, $\qquad \phi^c = U$.

8. De-Mogran's Law :

(*a*) $(A \cup B)^c = A^c \cap B^c$

Proof. Let $x \in (A \cup B)^c$

$\Rightarrow \qquad x \notin A \cup B$

$\Rightarrow \qquad x \notin A$ and $x \notin B$

$\Rightarrow \qquad x \in A^c$ and $x \in B^c$

$\Rightarrow \qquad x \in A^c \cap B^c$

So $\qquad (A \cup B)^c \subseteq A^c \cap B^c$ ...(1)

Let $y \in A^c \cap B^c$

$\Rightarrow \qquad y \in A^c$ and $y \in B^c$

$\Rightarrow \qquad y \notin A$ and $y \notin B$

$\Rightarrow \quad y \notin A \cup B$

$\Rightarrow \quad y \in (A \cup B)^c$

So $\quad A^c \cap B^c \subseteq (A \cup B)^c$ ...(2)

From (1) and (2), we have

$$(A \cup B) = A^c \cap B^c$$

(*b*) $(A \cap B)^c = A^c \cup B^c$.

Proof. Let $x \in (A \cap B)^c$

$\Rightarrow \quad x \notin A \cap B$

$\Rightarrow \quad x \notin A$ or $x \notin B$

$\Rightarrow \quad x \in A^c$ or $x \in B^c$

$\Rightarrow \quad x \in A^c \cup B^c$

So $\quad (A \cap B)^c \subseteq A^c \cup B^c$ ...(1)

Let $y \in A^c \cup B^c$

$\Rightarrow \quad y \in A^c$ or $y \in B^c$

$\Rightarrow \quad y \notin A$ or $y \notin B$

$\Rightarrow \quad y \notin (A \cap B)$

$\Rightarrow \quad y \in (A \cap B)^c$

So $\quad A^c \cup B^c \subseteq (A \cap B)^c$ ...(2)

From (1) and (2), we have

$$(A \cap B)^c = A^c \cup B^c$$

(*c*) $A - (B \cup C) = (A - B) \cap (A - C)$

Proof. Let $x \in A - (B \cup C)$

$\Rightarrow \quad x \in A$ and $x \notin (B \cup C)$

$\Rightarrow \quad x \in A$ and $(x \notin B$ and $x \notin C)$

$\Rightarrow \quad x \in A$ and $x \notin B$ and $x \in A$ and $x \notin C$

$\Rightarrow \quad x \in (A - B)$ and $x \in (A - C)$

$\Rightarrow \quad x \in (A - B) \cap (A - C)$

So $\quad A - (B \cup C) \subseteq (A - B) \cap (A - C)$ ...(1)

Let $y \in (A - B) \cap (A - C)$

$\Rightarrow \quad y \in (A - B)$ and $y \in (A - C)$

$\Rightarrow \quad y \in A$ and $y \notin B$ and $y \in A$ and $y \notin C$

$\Rightarrow \quad y \in A$ and $y \notin B$ and $y \notin C$

$\Rightarrow \quad y \in A$ and $y \notin (B \cup C)$

$\Rightarrow \quad y \in A - (B \cup C)$

So $\quad (A - B) \cap (A - C) \subseteq A - (B \cup C)$ ...(2)

From (1) and (2), we have

$$A - (B \cup C) = (A - B) \cap (A - C)$$

(*d*) $A - (B \cap C) = (A - B) \cup (A - C)$.

Proof. Let $x \in A - (B \cap C)$

$\Rightarrow \quad x \in A$ and $x \notin (B \cap C)$

$\Rightarrow \quad x \in A$ and $(x \notin B$ and $x \notin C)$

$\Rightarrow \quad (x \in A \text{ and } x \notin B) \text{ and } (x \in A \text{ and } x \notin C)$

$\Rightarrow \quad x \in (A - B) \text{ and } x \in (A - C)$

$\Rightarrow \quad x \in ((A - B) \cup (A - C))$

So $\quad A - (B \cap C) \subseteq (A - B) \cup (A - C)$ ...(1)

Let $y \in (A - B) \cup (A - C)$

$\Rightarrow \quad y \in (A - B) \text{ or } y \in (A - C)$

$\Rightarrow \quad y \in A \text{ and } y \notin B \text{ or } y \in A \text{ and } y \notin C$

$\Rightarrow \quad y \in A \text{ and } (y \notin B \text{ and } y \notin C)$

$\Rightarrow \quad y \in A \text{ and } y \notin (B \cap C)$

$\Rightarrow \quad y \in A - (B \cap C)$

So $(A - B) \cup (A - C) \subseteq A - (B \cap C)$ ...(2)

From (1) and (2), we have

$$A - (B \cap C) = (A - B) \cup (A - C)$$

1.10. SYMMETRIC DIFFERENCE OF TWO SETS

If A and B be two sets, then the set

$$(A - B) \cup (B - A)$$

is called the symmetric difference of A and B and symbolically denoted by $A \Delta B$.

i.e., $\quad A \Delta B = \{x : x \in A \text{ and } x \notin B\} \text{ or } \{x : x \in B \text{ and } x \notin A\} = (A \cup B) - (A \cap B)$

We can also say that the symmetric difference of two sets is the set of all those elements which are not common to both the given sets.

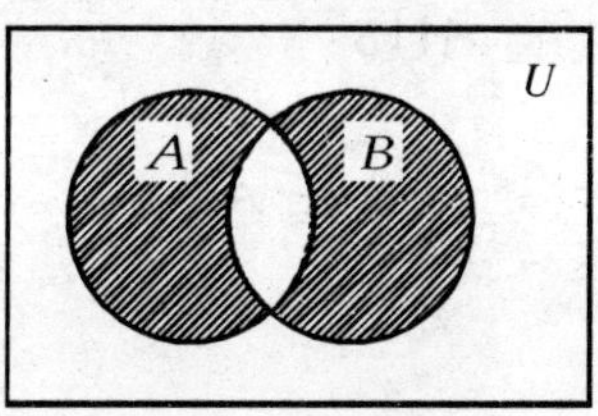

Fig. 1.20. Shaded portion of Venn diagram represents $A \Delta B$.

Example 1.10. Let $A = \{a, b, c\}$, $B = \{a, c, d, e\}$ then $A - B = \{b\}$, $B - A = \{d, e\}$.

Solution: $\quad A \Delta B = \{b\} \cup \{d, e\} = \{b, d, e\}$.

The symmetric difference of two sets A and B is also denoted by $A \oplus B$.

1.11. DUALITY OF SETS

The dual D^* of an equation D involving sets is the equation obtained by interchanging $\cup$, $\cap$, U and ϕ in D by $\cap$, $\cup$, ϕ and U respectively.

e.g.,: Dual of an equation

$$(A \cap U) \cap (\phi \cup A^c) = \phi$$

is

$$(A \cup \phi) \cup (U \cap A^c) = U$$

EXAMPLES

Example 1.11. Let $A = \{a, b, c, d, e\}$ and $B = \{a, b, c, d, e, f, g, h\}$. Find (*i*) $A \cup B$; (*ii*) $A \cap B$; (*iii*) $A - B$; (*iv*) $B - A$; (*v*) $A \Delta B$.

Solution. $A = \{a, b, c, d, e\},\ B = \{a, b, c, d, e, f, g, h\}$

(*i*) $A \cup B = \{a, b, c, d, e, f, g, h\}$

(*ii*) $A \cap B = \{a, b, c, d, e\}$

(*iii*) $A - B = \phi$

(*iv*) $B - A = \{f, g, h\}$

(*v*) $A \Delta B = (A - B) \cup (B - A)$

$= \{f, g, h\}$

Example 1.12. Prove that (*i*) $A \cup B = A$ iff $B \subseteq A$

(*ii*) $A \cap B = A$ iff $A \subseteq B$.

Solution. (*i*) **Proof.** Let $B \subseteq A$

Let $x \in B$

$\Rightarrow \quad x \in A$ ...(1)

Let $x \in A \cup B$

$\Rightarrow \quad x \in A$ or $x \in B$

$\Rightarrow \quad x \in A$ or $x \in A$ [by (1)]

$\Rightarrow \quad x \in A$

So $\quad A \cup B \subseteq A$ ...(2)

Conversely, let $x \in A$

$\Rightarrow \quad x \in A$ or $x \in B$ $(B \subseteq A)$

$\Rightarrow \quad x \in A \cup B$

$\therefore \quad A \subseteq A \cup B$ ...(3)

From (2) and (3), we have

$$A \cup B = A, \text{ iff } B \subseteq A.$$

(*ii*) **Proof.** Let $A \subseteq B$.

Let $x \in A \Rightarrow x \in B$ ...(1)

Let $x \in A \cap B$

$\Rightarrow \quad x \in A$ and $x \in B$

$\Rightarrow \quad x \in A$

So $\quad A \cap B \subseteq A$ ...(2)

Conversely, let $x \in A$

$\Rightarrow \quad x \in A$ and $x \in B$ [by (1)]

$\Rightarrow \quad x \in A \cap B$

So $\quad A \subseteq A \cap B$ ...(3)

From (2) and (3), we have

$$A \cap B = A, \text{ iff } A \subseteq B.$$

Example 1.13. If $A = \{a, b, c, d, e\}$, $B = \{a, c, e, g\}$, $C = \{b, e, f, g\}$ then prove the following:

(i) $A \cup B = B \cup A$

(ii) $A \cap B = B \cap A$

(iii) $(A \cup B) \cup C = A \cup (B \cup C)$

(iv) $(A \cup B) \cap C \neq A \cup (B \cap C)$

(v) $(A \cup B) \cap C = (A \cap C) \cup (B \cap C)$

(vi) $(A - B) \cap (A - C) = A - (B \cup C)$.

Solution. (i) $A \cup B = \{a, b, c, d, e, g\}$

$$B \cup A = \{a, b, c, d, e, g\}$$

So $$A \cup B = B \cup A$$

(ii) $A \cap B = \{a, c, e\}$, $B \cap A = \{a, c, e\}$

So $$A \cap B = B \cap A$$

(iii) $(A \cup B) \cup C = \{a, b, c, d, e, g\} \cup \{b, e, f, g\}$

$$= \{a, b, c, d, e, f, g\}$$

$$A \cup (B \cup C) = \{a, b, c, d, e\} \cup \{\{a, c, e, g\} \cup \{b, e, f, g\}\}$$

$$= \{a, b, c, d, e\} \cup \{a, b, c, e, f, g\}$$

$$= \{a, b, c, d, e, f, g\}$$

So $$(A \cup B) \cup C = A \cup (B \cup C).$$

(iv) $(A \cup B) \cap C = \{a, b, c, d, e, g\} \cap \{b, e, f, g\}$

$$= \{b, e, g\}$$

$$A \cup (B \cap C) = \{a, b, c, d, e\} \cup \{\{a, c, e, g\} \cap \{b, e, f, g\}\}$$

$$= \{a, b, c, d, e\} \cup \{e, g\}$$

$$= \{a, b, c, d, e, g\}$$

So $$(A \cup B) \cap C \neq A \cup (B \cap C)$$

(v) $(A \cup B) \cap C = \{b, e, g\}$

$$(A \cap C) \cup (B \cap C) = \{\{a, b, c, d, e\} \cap \{b, e, f, g\}\} \cup \{e, g\}$$

$$= \{b, e\} \cup \{e, g\}$$

$$= \{b, e, g\}$$

So $$(A \cup B) \cap C = (A \cap C) \cup (B \cap C)$$

(vi) $A - B = \{b, d\}$

$$A - C = \{a, c, d\}$$

$$(A - B) \cap (A - C) = \{d\}$$

$$A - (B \cup C) = \{a, b, c, d, e\} - \{a, b, c, e, f, g\} = \{d\}$$

So $$(A - B) \cap (A - C) = A - (B \cup C).$$

Example 1.14. Draw Venn diagrams for each of the following:

(i) $A^C \cap B^C$

(ii) $A \cap B^C \cap C^C$

(iii) $A^C \cap (B \cup C)$

(iv) $(A - B) \cap (A - C)$.

Solution. (i) $A^C \cap B^C$ is the unshaded area.

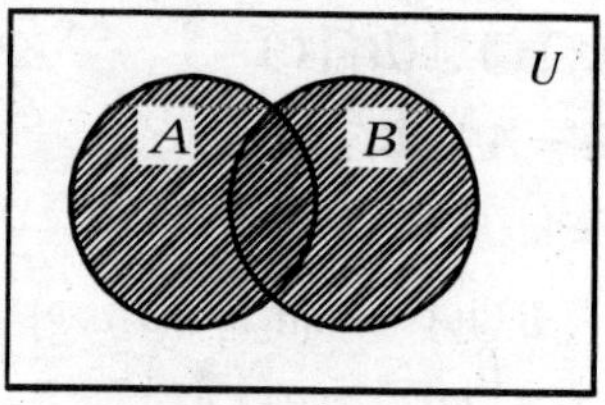

Fig. 1.21.

(ii) $A \cap B^C \cap C^C$

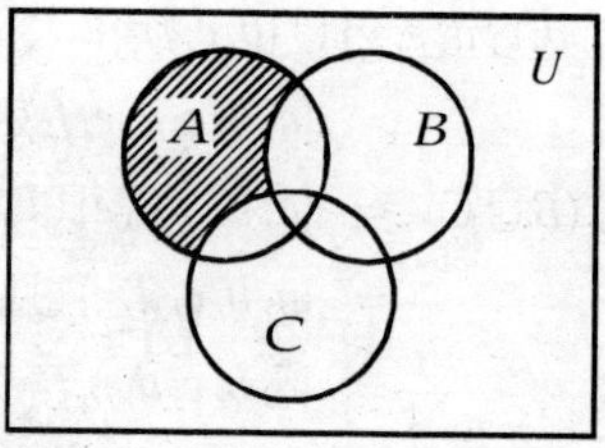

Fig. 1.22.

The shaded area represents $A \cap B^C \cap C^C$ in the above Venn diagram.

(iii) $A^C \cap (B \cup C)$ is represented by shaded area in the Venn diagram below :

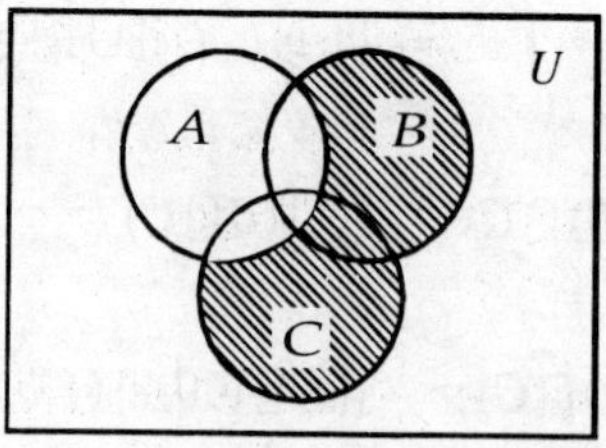

Fig. 1.23.

(iv) $(A - B) \cap (A - C)$ is represented by shaded area in the Venn diagram below :

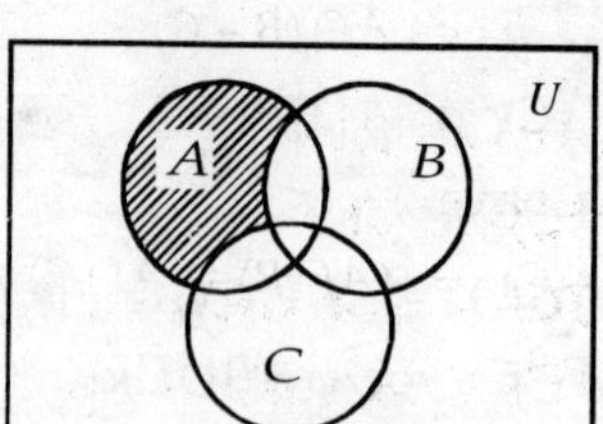

Fig. 1.24.

Example 1.15. Prove that $A - B = A \cap B^C = B^C - A^C$.

Solution. Let $x \in A - B$

$\Rightarrow \quad x \in A$ and $x \notin B$.

$\Rightarrow \quad x \in A$ and $x \in B^C$

$\Rightarrow \quad x \in A \cap B^C$

So $\quad (A - B) \subseteq A \cap B^C$ ...(1)

Conversely, let $x \in A \cap B^C$

$\Rightarrow \quad x \in A$ and $x \in B^C$

$\Rightarrow \quad x \in A$ and $x \notin B$

$\Rightarrow \quad x \in A - B.$

So $\quad A \cap B^C \subseteq A - B$ ...(2)

From (1) and (2), we have

$$A - B = A \cap B^C.$$

Also, $\quad B^C - A^C = B^C \cap (A^C)^C$ (Since $A - B = A \cap B^C$)

$= B^C \cap A$ (Since $(A^C)^C = A$)

$= A \cap B^C$

Hence, $\quad A - B = A \cap B^C = B^C - A^C.$

Example 1.16. Prove that (*i*) $A \cap (B - C) = (A \cap B) - (A \cap C)$

(*ii*) $(A - B) \cup B = A$ iff $B \subseteq A$.

Solution. (*i*) **Proof.** Let $x \in A \cap (B - C)$

$\Rightarrow \quad x \in A$ and $x \in B - C$

$\Rightarrow \quad x \in A$ and $x \in B$ and $x \notin C$

$\Rightarrow \quad (x \in A$ and $x \in B)$ and $x \in A$ and $x \notin C.$

$\Rightarrow \quad x \in A \cap B$ and $x \notin A \cap C$

$\Rightarrow x \in (A \cap B) - (A \cap C)$

So $A \cap (B - C) \subseteq (A \cap B) - (A \cap C)$ ...(1)

Let $y \in (A \cap B) - (A \cap C)$

$\Rightarrow \quad y \in A \cap B$ and $y \notin A \cap C$

$\Rightarrow \quad y \in A$ and $y \in B$ and $y \in A$ and $y \notin C$

$\Rightarrow \quad y \in A$ and $y \in B$ and $y \notin C$

$\Rightarrow \quad y \in A$ and $y \in (B - C)$

$\Rightarrow \quad y \in A \cap (B - C)$

So $\quad (A \cap B) - (A \cap C) \subseteq A \cap (B - C)$ ...(2)

So From (1) and (2), we have

$$A \cap (B - C) = (A \cap B) - (A \cap C).$$

(*ii*) **Proof.** $(A - B) \cup B = A$

$\Leftrightarrow \quad (A \cup B^C) \cup B = A$ (Since $A - B = A \cap B^C$)

$\Leftrightarrow \quad (A \cup B) \cap (B^C \cup B) = A$ (By distributive law)

$\Leftrightarrow \quad (A \cup B) \cap (U) = A$ (Since $B^C \cup B = U$)

$\Leftrightarrow \quad A \cup B = A$ (Since $A \cap U = A$)

But, $A \cup B = A$ iff all elements of A or B are elements of A.

$\Leftrightarrow$ All elements of B are elements of A

$\Leftrightarrow B \subseteq A$

So $(A - B) \cup B = A$ iff $B \subseteq A$.

Example 1.17. Prove that $A \cap (B \Delta C) = (A \cap B) \Delta (A \cap C)$.

Solution. $A \cap (B \Delta C) = A \cap [(B - C) \cup (C - B)]$

$= [A \cap (B - C)] \cup [A \cap (C - B)]$ (By distributive law)

$= [(A \cap B) - (A \cap C)] \cup [(A \cap C) - (A \cap B)]$

[Since $A \cap (B - C) = (A \cap B) - (A \cap C)$]

$= (A \cap B) \Delta (A \cap C)$ (By definition of symmetric difference)

Example 1.18. For any two sets A and B, prove that

$$(A - B) \cup (B - A) = (A \cup B) - (A \cap B).$$

Solution. Let $x \in (A - B) \cup (B - A)$

$\Rightarrow \quad x \in (A - B)$ or $x \in (B - A)$

$\Rightarrow \quad x \in A$ and $x \notin B$ or $x \in B$ and $x \notin A$

$\Rightarrow \quad x \in A$ or $x \in B$ and $x \notin A, x \notin B$

$\Rightarrow \quad x \in A \cup B$ and $x \notin A \cap B$

$\Rightarrow \quad x \in (A \cup B) - (A \cap B)$

So $(A - B) \cup (B - A) \subseteq (A \cup B) - (A \cap B)$ ...(1)

Conversely, let $y \in (A \cup B) - (A \cap B)$

$\Rightarrow \quad y \in A \cup B$ and $y \notin A \cap B$

$\Rightarrow \quad y \in A$ or $y \in B$ and $y \notin A$ and $y \notin B$

$\Rightarrow \quad y \in A$ and $y \notin B$ or $y \in B$ and $y \notin A$

$\Rightarrow \quad y \in (A - B)$ or $y \in (B - A)$

$\Rightarrow \quad y \in (A - B) \cup (B - A)$

So $(A \cup B) - (A \cap B) \subseteq (A - B) \cup (B - A)$ ...(2)

From (1) and (2), we have

$$(A - B) \cup (B - A) = (A \cup B) - (A \cap B).$$

Example 1.19. Write the dual of each of the following set equation :

(*i*) $(A \cup B \cup C)^C = (A \cup C)^C \cap (A \cup B)^C$

(*ii*) $(A \cap U)^C \cap A = \phi$

(*iii*) $(A \cap B) \cup (A^C \cap B) \cup (A \cap B^C) \cup (A^C \cap B^C) = U$.

Solution. (*i*) Dual is $(A \cap B \cap C)^C = (A \cap C)^C \cup (A \cap B)^C$

(*ii*) Dual is $(A \cup \phi)^C \cup A = U$

$\Rightarrow \quad A^C \cup A = U$

(*iii*) Dual is $(A \cup B) \cap (A^C \cup B) \cap (A \cup B^C) \cap (A^C \cup B^C) = \phi$.

Example 1.20. Let $U = \{a : a < 12, a \in N\}$

$$A = \{a : a = 1, 2, 3, 4\}$$
$$B = \{a : a < 11 \text{ and } a \text{ is an even natural number}\}$$
$$C = \{a : a \in N, 2 < a < 7\}$$

Find (*i*) A^C, (*ii*) B^C and (*iii*) $(B - C)^C$.

Solution. (*i*) $A^C = \{5, 6, 7, 8, 9, 10, 11\}$

(*ii*) $B = \{2, 4, 6, 8, 10\}$

So $B^C = \{1, 3, 5, 7, 9, 11\}$

(*iii*) $C = \{3, 4, 5, 6\}$

$$B - C = \{2, 4, 6, 8, 10\} - \{3, 4, 5, 6\}$$
$$= \{2, 8, 10\}$$

So $\{B - C\}^C = \{1, 3, 4, 5, 6, 7, 9, 11\}$

1.12 APPLICATION OF SET THEORY

In this section we present some applications of set theory which will make us more clear about it.

1.12.1 Counting Principle (Principle of Intension-Exclusion)

Consider two finite sets A and B with $n(A)$ and $n(B)$ elements. Then the cardinality of the union of two sets A and B is given by

$$n(A \cup B) = n(A) + n(B) - n(A \cap B)$$

Proof. We will prove this with the help of Venn diagram.

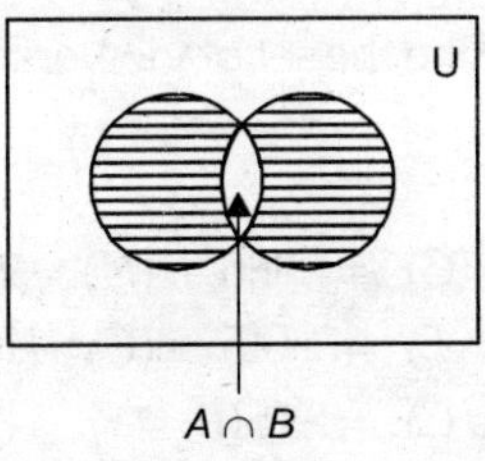

Fig. 1.25.

Here we can conclude that the number of elements in $(A \cup B)$ includes the number of elements of $(A \cap B)$ twice, once in $n(A)$ and another in $n(B)$.

Therefore if we substract $n(A \cap B)$ from the sum of $n(A)$ and $n(B)$ we get the elements in $(A \cup B)$ *i.e.*, $n(A \cup B) = n(A) + n(B) - n(A \cap B)$.

Notes :

(*i*) If A and B are disjoint *i.e.*, $(A \cap B) = \phi$ then $n(A \cup B) = n(A) + n(B)$

(*ii*) $n(A) = n(A - B) + n(A \cap B)$

(*iii*) $n(B) = n(B - A) + n(A \cap B)$

(*iv*) $n(A \cup B) = n(A - B) + n(A \cap B) + n(B - A)$

(*v*) If A, B and C are three finite sets, then

$$n(A \cup B \cup C) = n(A) + n(B) + n(C) - n(A \cap B) - n(A \cap C) - n(B \cap C) + n(A \cap B \cap C)$$

EXAMPLES

Example 1.21. A survey of 1000 customers reading newspaper conducted and reported that 720 customers read *Times of India* and 450 liked *Hindustan times*. What is the least number that reads both the newspapers?

Solution. Let x be the number of persons which reads *Times of India* and y be the number of persons which reads *Hindustan Times*.

i.e., $n(x) = 720$ and $n(y) = 450$

Given $n(x \cup y) = 1000$

We know that

$$n(x \cup y) = n(x) + n(y) - n(x \cap y)$$

$$n(x \cap y) = n(x) + n(y) - n(x \cup y)$$

$$= 720 + 450 - 1000 = 170$$

170 persons read both *Times of India* and *Hindustan Times*.

Example 1.22. In a survey of 600 television viewers given the following information : 385 watch cricket matches, 295 watch hockey matches, 215 watch football matches, 145 watch cricket and football matches, 170 watch cricket and hockey matches, 150 watch hockey and football matches and 150 do not watch any of the three kind of matches.

(*i*) How many people in survey watch all three kinds of matches?

(*ii*) How many people watch exactly one of the spot?

Solution. Let C, H and F be the set of viewers watch cricket, hockey and football matches respectively.

Given that

$$n(C) = 385, n(H) = 295, n(F) = 215$$

$$n(C \cap F) = 145, n(C \cap H) = 170, n(F \cap H) = 150$$

$$n(A \cup B \cup C) = 150$$

$$\Rightarrow \quad n(C \cup H \cup F) = 600 - 150 = 450$$

We know that

$$n(C \cup H \cup F) = n(C) + n(H) + n(F) - n(C \cap F) - n(C \cup H) - n(F \cap H) + n(C \cup H \cap F)$$

Thus

$$n(C \cap H \cap I) = 450 - 385 - 295 - 215 + 145 + 170 + 150$$

$$= 915 - 895 = 20$$

Thus 20 people watch all the three matches.

(*ii*) Now we can show this with the help of Venn diagram as :

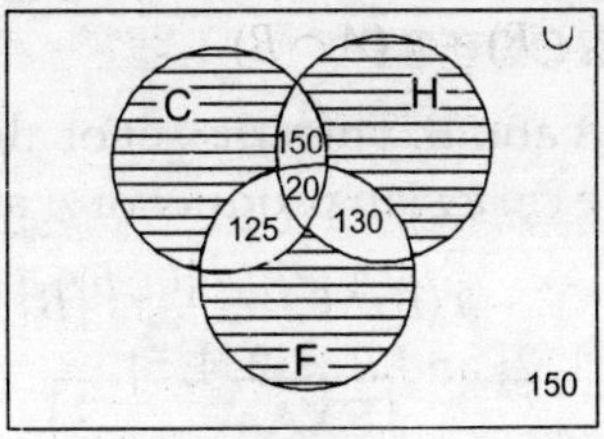

Fig. 1.26.

Number of viewers who watch exactly one of the matches (shaded region)

= Number of people watch cricket only + Number of people watch only hockey matches + Number of people watch only football matches

$= n(C) - n(C \cap H) - n(C \cap F) + n(C \cap F \cap H)$

$+ n(F) - n(C \cap F) - n(F \cap H) + n(C \cap F \cap H)$

$+ n(H) - n(C \cap H) - n(F \cap H) + n(C \cap F \cap H)$

$+ n(C) - n(F) - n(H) - 2n(C \cap F) - 2n(C \cap H)$

$- 2n(H \cap F) + 3n(C \cap H \cap F)$

$= 385 + 295 + 215 - 2.145 - 2.170 - 2.150 + 3.20$

$= 895 + 60 - 2.(465) = 955 - 930 = 25$

25 people watch exactly one of the matches.

1.13 ORDERED PAIRS

Let A and B be two sets and let $x \in A$ and $y \in B$, then a set of two elements whose elements have been listed in a specific order is called an ordered pair and denoted as (x, y).

Clearly $(x, y) \neq (y, x)$ and if $(x_1, y_1) = (x_2, y_2)$

then $x_1 = x_2$ and $y_1 = y_2$

This idea of ordered pair can be extended to define an ordered triple and generally an n-tuple. An ordered triple can be defined as an ordered pair whose first element is itself an ordered pair. Thus we can unite an ordered triple as (x, y, z). How from the definition of equality of an ordered pair, we can write the quality of $((x, y), z)$ and $((a, b), c)$ as

$(x, y) = (a, b)$ and $z = c$

But $(x, y) = (a, b)$ if $x = a, y = b$

Hence, we say

$$((x, y), z) = ((a, b), c) \Leftrightarrow x = a, y = b \text{ and } z = c$$

Hence we can also write an ordered triple as (x, y, z) with an understanding that (x, y, z) actually stands for $((x, y), z)$. Here also the order of placing maters so $(x, y, z) \neq (x, z, y)$ continuing the same process, an ordered n-type can be written as $(x_1, x_2, ..., x_n)$ with an understanding that $(x_1, x_2 ... x_n)$ actually means

$((x_1, x_2, ... x_{n-1}), x_n)$.

1.14 CARTESIAN PRODUCT

Consider two sets A and B. Then the set of all ordered pair (a, b) such that $a \in A$ and $b \in B$ is called the cartesian product of A and B and denoted as $A \times B$.

i.e., $$A \times B = \{(a, b) : a \in A, b \in B\}$$

For example, if $A = \{1, 2\}$ and $B = \{3, 4, 5\}$.

Then

$A \times B = \{(1, 3), (1, 4), (1, 5) (2, 3) (2, 4) (2, 5)\}$

and $B \times A = \{(3, 1), (3, 2), (4, 1) (4, 2), (5, 1), (5, 2)\}$

Clearly $\quad A \times B \neq B \times A$.

Notes :

(*i*) If $n(A) = m$ and $n(B) = n$ then

$n(A \times B) = m \times n$

(*ii*) If $A = \phi$ or $B = \phi$ then $A \times B = \phi$

(*iii*) If $A \times B = B \times A$, then $A = B$.

1.15 PARTITIONS

Let X be a non-empty set. Then a partition of X is a sub-division of X into non-overlaping, non-empty subsets. In other words, a **partition** of X is a collection of non-empty subsets $\{Ai\}$ of X such that

(*i*) Each $a \in X$, belongs to one of the Ai

(*ii*) If $Ai \neq Aj$ then $Ai \cap Aj = \phi$

That is the sets of $\{Ai\}$ are mutually disjoint.

The subsets in a partition are called **Cells.**

The partition of any set s is shown in the venn diagram given below

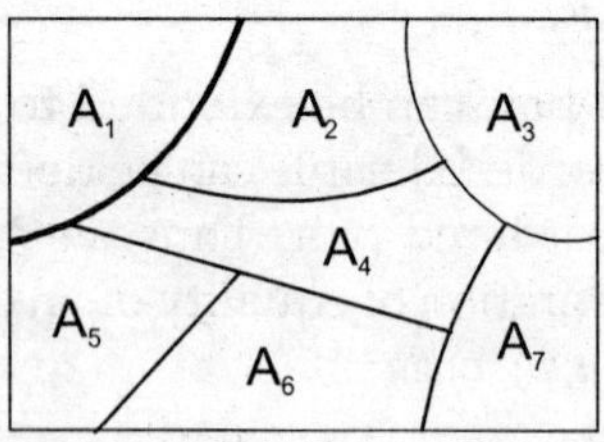

Fig. 1.27.

The partition of s contains 7 cells $A_1, A_2, A_3, A_4, A_5, A_6$ and A_7.

EXAMPLES

Example 1.23 : Consider a set $S = \{1, 2, 3, 4, 5, 6, 7\}$. Then the collection of subsets of S say $A_1 = \{1, 2\}, \{3, 4, 5\} \{6, 7\}$.

Solution. Terms a partition of S whereas another collection $A_2 = \{1, 2, 3, 4\}, \{4, 5, 6, 7\}$ is not a partition of S as element 4 is in both subsets. That is the two subsets $\{1, 2, 3, 4\}$ and $\{4, 5, 6, 7\}$ of S area not disjoint.

Another collection of $A_3 = \{\{1, 2\}, \{4, 5\}, \{6, 7\}$ is also not a partition of S as

element 2 of S does not belong to any of the three subsets of collection A_3.

Example 1.24 : Given three sets $A = \{a, b\}$ $B = \{1, 2, 3\}$ and $C = \{x, y, z\}$. Find $A \times B \times C$.

Solution. $A \times B \times C$ consists of all triplets (p, q, r) where $p \in A, q \in B$ and $r \in C$.

With the help of a tree diagram we can obtain all the triplets of $A \times B \times C$ as follows :

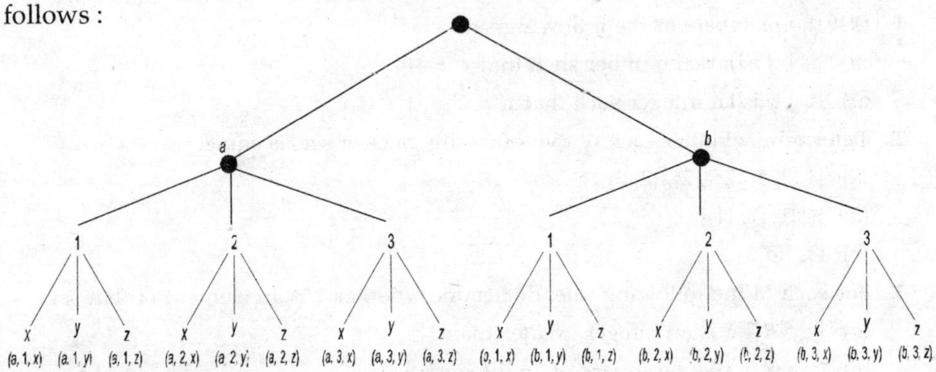

Fig. 1.28.

There are 18 ordered triplets in $A \times B \times C$. One can observe that $n\ (A) = 2$, $n\ (B) = 3$ and $n\ (C) = 3$, and thus $n\ (A \times B \times C) = n\ (A).\ n(B).\ n\ (C) = 18$

Example. 1.25. Prove that (*i*) $A \times (B \cup C) = (A \times B) \cup (A \times C)$

(*ii*) $A \times (B \cap C) = (A \times B) \cap (A \times C)$

Solution. (*i*) Let $(a, b) \in A \times (B \cup C)$

$\Rightarrow \quad a \in A$ and $b \in (B \cup C)$

$\Rightarrow \quad a \in A$ and $(b \in B$ or $b \in C)$

$\Rightarrow \quad a \in A$ and $b \in B$ or $a \in A$ and $b \in C$

$\Rightarrow \quad (a, b) \in (A \times B)$ or $(a, b) \in (B \times C)$

$\Rightarrow \quad (a, b) \in (A \times B) \cup (A \times C)$

So $A \times (B \cup C) \subseteq (A \times B) \cup (A \times C)$ ...(1)

Again let $(x, y) \in (A \times B) \cup (A \times C)$

$\Rightarrow \quad (x, y) \in A \times B$ or $(x, y) \in (A \times C)$

$\Rightarrow \quad (x \in A$ and $y \in B)$ or $(x \in A$ and $y \in C)$

$\Rightarrow \quad (x \in A$ and $(y \in B$ or $y \in C)$

$\Rightarrow \quad x \in A$ and $y \in (B \cup C)$

$\Rightarrow \quad (x, y) \in A \times (B \cup C)$

So $(A \times B) \cup (A \times C) \subseteq A \times (B \cup C)$ ...(2)

By (1) and (2) we conclude that

$$A \times (B \cup C) = (A \times B) \cup (A \times C)$$

(*ii*) $(A \times B) \cap (A \times C) = \{(x, y) : (x, y) \in A \times B$ and $(x, y) \in A \times C\}$

$\Leftrightarrow \quad \{(x, y) : x \in A, y \in B$ and $x \in A, y \in C\}$

$$\Leftrightarrow \quad \{(x, y) : x \in A, y \in B \text{ and } y \in C\}$$
$$\Leftrightarrow \quad \{(x, y) : x \in A, y \in B \cap C\} \Leftrightarrow \{(x, y) : x \in A \times (B \cap C)\}$$
$$= A \times (B \cap C)$$

EXERCISES

1. List the members of the following sets :
   (*a*) $\{x \mid x \text{ is a real number such that } x^2 = 4\}$
   (*b*) $\{x \mid x \text{ is an integer such that } x^2 = 2\}$.
2. Determine whether each of the following pairs of sets is equal.
   (*a*) $\{1, 2, 1, 3, 1, 2\}, \{2, 3, 1\}$
   (*b*) $\{\{1\}\}, \{1, \{1\}\}$
   (*c*) $\varnothing, \{\varnothing\}$.
3. For each of the following sets, determine whether 1 is an element of that set.
   (*a*) $\{x \in R \mid x \text{ is an integer greater than } 1\}$
   (*b*) $\{x \in R \mid x \text{ is the square of an integer}\}$
   (*c*) $\{1, 2, \{1\}\}$
   (*d*) $\{\{1\}, \{\{1\}\}\}$
   (*e*) $\{\{1, 2\}, \{1, \{\{1\}\}\}$
   (*f*) $\{\{\{1\}\}\}$
4. Suppose that A, B and C are sets such that $A \subseteq B$ and $B \subseteq C$. Show that $A \subseteq C$.
5. Find the power set of each of the following sets.
   (*a*) $\{a\}$
   (*b*) $\{a, \{a\}\}$
   (*c*) $\{\varnothing, \{\varnothing\}\}$.
6. Let A be a set. Show that $\varnothing \times A = A \times \varnothing = \varnothing$.
7. How many different elements does $A \times B$ have if A has m elements and B has n elements ?
8. Let $A = \{1, 2, 3, 4\}$ and $b = \{0, 3, 5\}$. Find
   (*a*) $A \cup B$ (*b*) $A \cap B$
   (*c*) $A - B$ (*d*) $B - A$.
9. What can you say about the sets A and B if the following are true ?
   (*a*) $A \cup B = A$ (*b*) $A - B = A$.
10. Let A and B be subsets of a universal set U. Show that $A \subseteq B$ if and only if $\bar{B} \subseteq \bar{A}$.
11. Let A and B be sets. Show that
    (*a*) $A \cup B = B \cup A$ (*b*) $A \cap B = B \cap A$.
12. Show that if A and B are sets, then $\overline{(A \cup B)} = \bar{A} \cap \bar{B}$ by showing each side is a subset of the other side.
13. Let A, B, and C be sets. Show that $A \cup (B \cup C) = (A \cup B) \cup C$.
14. Let $A = \{a, b, c\ d, e\}$ and $B = \{c, g, f\}$. Find
    (*a*) $A \cup B$ (*b*) $A \cap B$
    (*c*) $A - B$ (*d*) $B - A$.

15. Let U = Set of all alphabets in English. If $A = \{a, b, c\}$, $B = \{a, c, m, n\}$, $C = \{c, l, m, n, q\}$ then prove the following :

(*a*) $(A \cap B) \cup C = (A \cup C) \cap (B \cup C)$.

(*b*) $(A - B) \neq B - A$.

(*c*) $A - (B \cap C) = (A - B) \cup (A - C)$.

(*d*) $(A \cup B)^C = A^C \cap B^C$.

16. Determine the sets P, Q, and R, if given that

$$P \cup Q = \{1, 2, 3, 4\}, P \cup R = \{2, 3, 4, 5\}$$
$$P \cap Q = \{2, 3\}, P \cap R = \{2, 4\}.$$

17. Write the dual of each of the following set equation :

(*a*) $A = (B^C \cap A) \cup (C^C \cap A) \cup (A \cap B \cap C)$.

(*b*) $(A \cup U) \cap (A \cap \phi) = \phi$.

18. Prove the following :

(*a*) $(A - B) \cap (B - A) = \phi$.

(*b*) $(A \cup B) \cap B^C = A \Leftrightarrow A \cap B = \phi$.

(*c*) $A - (A - B) = A \cap B$.

(*d*) $A \cup B = A \cap B \Leftrightarrow A = B$.

(*e*) $A \cup B = (A - B) \cup B$.

ANSWERS

1. (*a*) $\{-2, 2\}$ (*b*) Ø.

2. (*a*) Yes (*b*) No (*c*) No

3. (*a*) No (*b*) Yes (*c*) Yes (*d*) No

(*e*) No (*f*) No

4. Let x be an arbitrary element of A. We are going to show that $x \in C$. Since $A \subseteq B$, this implies that $x \in B$. Since $B \subseteq C$, we see that $x \in C$. Since $x \in A$ implies that $x \in C$, it follows that $A \subseteq C$.

5. (*a*) $\{\emptyset, \{a\}\}$ (*b*) $\{\emptyset, \{a\}, \{\{a\}\}, \{a, \{a\}\}\}$

(*c*) $\{\emptyset, \{\emptyset\}, \{\{\emptyset\}\}, \{\emptyset, \{\emptyset\}\}\}$.

6. $\emptyset \times A = \{(x, y) \mid x \in \emptyset \text{ and } y \in A\} = \emptyset = \{(x, y) \mid x \in A \text{ and } y \in \emptyset\} = A \times \emptyset$.

7. mn

8. (*a*) $\{0, 1, 2, 3, 4, 5\}$ (*b*) $\{3\}$ (*c*) $\{1, 2, 4\}$ (*d*) $\{0, 5\}$

9. (*a*) $B \subseteq A$. For since $A \cup B = A$ every element of B is also in A. But A may have more elements than B.

(*b*) $A \cap B = \emptyset$. For $A - B = A$ means that subtracting elements of B do not change A. Hence no elements of B are in A. Hence A and B have no elements in common.

10. $A \subseteq B \Leftrightarrow \forall x\, (x \in A \rightarrow x \in B) \Leftrightarrow \forall x\, (x \notin B \rightarrow x \notin A) \Leftrightarrow \forall x\, (x \in \overline{B} \rightarrow x \in \overline{A}) \Leftrightarrow \overline{B} \subseteq \overline{A}$.

11. (*a*) $A \cup B = \{x \mid x \in A \vee x \in B\} = \{x \mid x \in B \vee x \in A\} = B \cup A$

(*b*) $A \cap B = \{x \mid x \in A \wedge x \in B\} = \{x \mid x \in B \wedge x \in A\} = B \cap A$.

12. For an arbitrary element x in the universe,

$x \in (\overline{A \cup B}) \Leftrightarrow x \notin A \cup B \Leftrightarrow \neg(x \in A \vee x \in B) \Leftrightarrow (\neg x \in A \wedge \neg x \in B) \Leftrightarrow (x \notin A \wedge x \notin B)$

$\Leftrightarrow (x \in \overline{A} \wedge x \in \overline{B}) \Leftrightarrow x \in \overline{A} \cap \overline{B}.$

13. For an arbitrary element x in the universe,

$x \in A \cup (B \cup C) \Leftrightarrow (x \in A \,\Delta\, x \in B \cup C) \Leftrightarrow (x \in A \,\Delta\, (x \in B \,\Delta\, x \in C)) \Leftrightarrow ((x \in A \,\Delta\, x \subset B)$ $\Delta\, x \in C) \Leftrightarrow x \in (A \cup B) \cup C.$

14. (*a*) $\{a, b, c, d, e, g, f\}$ (*b*) $\{c\}$

(*c*) $\{a, b, d, e\}$ (*d*) $\{g, f\}$.

16. $P = \{2, 4, 3\}$, $Q = \{1, 2, 3\}$, $R = \{2, 4, 5\}$

17. (*a*) $A = \{B^C \cup A\} \cap \{C^C \cup A) \cap (A \cup B \cup C)$

(*b*) $(A \cap \emptyset) \cup (A \cup U) = U.$

■■■

Chapter

2 RELATION

2.1. INTRODUCTION TO RELATION

The **relation** we are going to study here is an abstraction of relations we see in our everyday life such as those between parent and child, between car and owner, among name, social security number, address and telephone number etc. We are going to focus our attention on one key property which all the everyday relations have in common, define everything that has that property as a relation, and study properties of those relations.

2.2 RELATION

Let A and B be two non-empty sets. The set of ordered pairs (a, b) such that $a \in A$ and $b \in B$ and the first co-ordinate of every ordered pair of the set is related to that of its second co-ordinate according to a given rule is called **a relation** from A to B.

Note that the set of such ordered pairs is called **relation** not the "Verbal phrase" defining the relationship between the elements.

For example, let A be the set of well defined persons and

$$R = \{(a, b) : a, b \in A \text{ and } a \text{ is the brother of } b\}.$$

Then the set R is the relation not the phrase "is the brother of".

Definition : Let A and B be two non-empty sets. Then a binary relation or simply, relation from A to B is a subset of $A \times B$ *i.e.,* $R \subseteq A \times B$.

Symbolically,

$R = \{(a, b) : a \in A, b \in B \text{ and } a \, R \, b\}$.

Here aRb indicates that a is R related to b

For example, consider a set A of well defined persons.

Then,

$R = \{(a, b) : a, b \in A$ and a is a friend of $b\}$ is a relation.

Again consider two sets $A = \{1, 2, 3\}$ and $B = \{x, y, z, w\}$ then a relation R from A to B may be defined as

$R = \{(1, x) (2, y) (2, z) (3, x) (3, y) (3, w)\}$

Clearly $\quad R \subseteq A \times B = \{(1, x) (1, y) (1, z) (1, w) (2, x) (2, y), (2, z) (2, w) (3, x) (3, y) (3, z) (3, w)\}$

Notes : (*i*) Let A and B be two non-empty sets such that $n(A) = m$ and $n(B) = n$. Then $n(A \times B) = m \times n$.

So the power set of $A \times B$ *i.e.*; $P(A \times B)$ will have $2^{m \times n}$ elements and since every subset of $(A \times B)$ can be a relation from A to B, hence there may be $2^{m \times n}$ different relations from set A to set B.

(*ii*) As we know that null set ϕ is a subset of $A \times B$ hence ϕ is a relation from A to B and is called **null relation** or **void relation.**

(*iii*) We know that every set is a subset of itself. So $A \times B \subseteq A \times B$. Hence $A \times B$ is also a relation from A to B, called **universal relation.**

(*iv*) If $a \in A$ and $b \in B$ such that $(a, b) \notin R$, then we say that a is not R-related to b. Symbolically we can write $a \text{ R } b$.

(*v*) Let A be any non-empty set. Then an important relation from A to itself is that of equality.

$$R = \{(a, a) : a \in A \text{ and } a \text{ R } a\}$$

This relation is called **identity relation** or **diagonal relation** and denoted by I_A or Δ_A or simply Δ.

2.3 DOMAIN AND RANGE OF RELATION

Let R be a relation defined from a set A to B, then $R = \{(a, b) : a \in A, b \in B$ and $a R b\}$.

The Domain of relation R is defined as the set of all first elements of the ordered pair which belongs to R.

That is, $D = \{a : a \in A \text{ and } (a, b) \in R\}$.

The Range, usually denoted by Ran (R) is defined as the set of all second elements of the ordered pair which belongs to R.

That is Ran $(R) = \{b : b \in B \text{ and } (a, b) \in R\}$

Clearly, one can observe that Dom (R) or $D \subseteq A$ and Ran $(R) \subseteq B$. Any other elements of A and B which are not in Dom (R) or Ran (R) respectively are not in the relation.

EXAMPLES

Example 2.1. Let $A = \{4, 5, 6, 7\}$ and $B = \{2, 4, 7\}$, then find out the range and domain of the relation R defined from A to B by > "is greater than".

Solution. Let us find out the relation R first by writing the set $A \times B$.

$A \times B = \{(4, 2), (4, 4), (4, 7) (5, 2) (5, 4) (5, 7) (6, 2) (6, 4) (6, 7) (7, 2) (7, 4) (7, 7)\}$.

The given relation is "is greater than".

i.e., $R = \{((a, b) : a \in A, b \in B \text{ and } a \text{ is greater than } b\}$.

Since

$$4 > 2 \Rightarrow (4, 2) \in R$$

$$4 \not> 4 \Rightarrow (4, 4) \notin R$$

$$4 \not> 7 \Rightarrow (4, 7) \notin R$$

$$5 > 2 \Rightarrow (5, 2) \in R$$

$$5 > 4 \Rightarrow (5, 4) \in R$$

and hence

$$R = \{(4, 2) (5, 2) (5, 4) (6, 2) (6, 4) (7, 2) (7, 4)\}$$

We know that

$$\text{Dom}(R) = D = \{a : a \in A \text{ and } aRb\}$$

So $D = \{4, 5, 6, 7\}$

and $\text{Ran}(R) = \{b : b \in B \text{ and } (a, b) \in R\}$

Thus $R = \{2, 4\}$

Example 2.2. Let $A = \{1, 2, 3, 4\}$, write down the relation R from A to A defined by

$$R = \{(a, b) : a, b \in A \text{ and a divides } b\}.$$

Find out the domain and range also.

Solution. R is defined as

$$R = \{(a, b) : a, b \in A \text{ and } a \text{ divides } b\}$$

Thus the ordered pair $(1, 2) \in R$ as 1 divides 2.

Similarly $(1, 3) \in R$ $(1, 4) \in R$ and $(1, 1) \in R$ as 1 divides 1 itself. By using the same logic R can be written as

$$R = \{(1, 1) (1, 2) (1, 3) (1, 4) (2, 2) (2, 4) (3, 3) (4, 4)\}$$

As 3 does not divide any of 1, 2 or 4.

So $(3, 1) \notin R$, $(3, 2) \notin R$ and $(3, 4) \notin R$.

Similarly $(2, 3) \notin R$ as 2 does not divide 3.

Similar arguments can be given for the ordered pairs (4, 1) (4, 2) (4, 3).

Now $\text{Dom}(R) = \{a : a \in A \text{ and } (a, b) \in R\}$

So $\text{Dom}(R) = \{1, 2, 3, 4\}$

and $\text{Ran}(R) = \{b : b \in A \text{ and } (a, b) \in R\}$

So $\text{Ran}(R) = \{1, 2, 3, 4\}$

2.4 INVERSE RELATION

Let R be a relation from A to B, then an inverse relation R^{-1} from B to A is defined by.

$$R^{-1} = \{(b, a) : b \in B, a \in A \text{ and } (a, b) \in R\}$$

That is, inverse of relation R is obtained by reversing the order of the ordered pairs which belongs to R.

Then clearly $\text{Dom}(R^{-1}) = \text{Ran}(R)$

and $\text{Ran}(R^{-1}) = \text{Dom}(R)$

For example, If $A = \{1, 2, 3\}$, $B = \{(2, 4)\}$ and R is defined as

$$R = \{(1, 2) (2, 2) (2, 4) (3, 4)\}$$

Then $R^{-1} = \{(2, 1) (2, 2) (4, 2) (4, 3)\}$

and $\text{dom}(R^{-1}) = \{2, 4\} = \text{ran}(R)$

$\text{ran}(R^{-1}) = \{1, 2, 3\} = \text{dom}(R)$

2.5 SOME RESULTS

If R is a relation from A to B and A_1, A_2 are two subsets of A, then (*i*) If $A_1 \subseteq A_2$, then $R(A_1) \subseteq R(A_2)$

If $b \in R(A_1)$, then aRb for some $a \in A_1$

Given $A_1 \subseteq A_2 \Rightarrow a \in A_2 \Rightarrow b \in R(A_2)$

Hence $R(A_1) \subseteq R(A_2)$

(*ii*) $R(A_1 \cup A_2) = R(A_1) \cup R(A_2)$

Let an element $b \in R(A_1 \cup A_2)$. Then for some a $\in (A_1 \cup A_2)$, $(a, b) \in R$ or aRb.

Now $a \in A_1 \cup A_2 \Rightarrow a \in A_1$ or $a \in A_2$

If $a \in A_1$ and we have $aRb \Rightarrow b \in R(A_1)$

and if $a \in A_2$ and we have $aRb \Rightarrow b \in R(A_2)$

$\Rightarrow b \in R(A_1)$ or $b \in R(A_2)$

$\Rightarrow b \in R(A_1) \cup R(A_2)$

Hence $R(A_1 \cup A_2) \subseteq R(A_1) \cup R(A_2)$ ...(1)

For the converse part, we know that

$$\boxed{A_1 \subseteq A_1 \cup A_2 \Rightarrow R(A_1) \subseteq R(A_1 \cup A_2)}$$

and $R(A_2) \subseteq R(A_1 \cup A_2)$ [from result (*a*)]

$\Rightarrow R(A_1) \cup R(A_2) \subseteq R(A_1 \cup A_2)$ ...(2)

By (1) and (2) we conclude that

$$R(A_1 \cup A_2) = R(A_1) \cup R(A_2)$$

(*iii*) $R(A_1 \cap A_2) = R(A_1) \cap R(A_2)$

Let be the any element, such that

$b \in R(A_1) \cap R(A_2)$

$\Rightarrow b \in R(A_1)$ and $b \in R(A_2)$

If $b \cup R(A_1) \Rightarrow$ for some $a \in A_1$, aRb exists.

and $b \in R(A_2) \Rightarrow$ for some $a \in A_2$, aRb exists.

$\Rightarrow a \in (A_1 \cap A_2)$ and as aRb exists, we have $b \in R(A_1 \cap A_2)$

and hence $R(A_1) \cap R(A_2) \subseteq R(A_1 \cap A_2)$ ...(1)

Conversely, as $(A_1 \cap A_2) \subseteq A_1$ and $(A_1 \cap A_2) \subseteq A_2$

$\Rightarrow R(A_1 \cup R_2) \subseteq R(A_1)$ (by result (*a*)}

and $R(A_1 \cap A_2) \cap R(A_2)$

$\Rightarrow R(A_1 \cap A_2) \subseteq R(A_1) \cap R(A_2)$ ...(2)

By (*i*) and (*ii*) we conclude that

$$\boxed{R(A_1 \cap A_2) = R(A_1) \cap R(A_2)}$$

EXAMPLES

Example 2.3. Let *R* and *S* be two relations from set *A* to set *B*. Then

(*a*) If $R \subseteq S$, *then* $R^{-1} \subseteq S^{-1}$ **(*b*) $(R \cup S)^{-1} = R^{-1} \cup S^{-1}$**

(*c*) $(R \cap S)^{-1} = R^{-1} \cap S^{-1}$

Solution. (*a*) Let $R \subseteq S$ and $(a, b) \in R^{-1}$

then $(b, a) \in R$ and $(b, a) \in S$ as $R \subseteq S$.

Now if $(b, a) \in S \Rightarrow (a, b) \in S^{-1}$

and hence $R^{-1} \subseteq S^{-1}$

(*b*) Let $(a, b) \in (R \cup S)^{-1} \Rightarrow (b, a) \in (R \cup S)$

$\Rightarrow (b, a) \in R$ or $(b, a) \in S$

or $(a, b) \in R^{-1}$ or $(a, b) \in S^{-1}$

$\in (a, b) \in R^{-1} \cup S^{-1}$

Hence $(R \cup S)^{-1} \subseteq (R^{-1} \cup S^{-1})$ ...(1)

Conversely, let $(a, b) \in R^{-1} \cup S^{-1}$

$\Rightarrow (a, b) \in R^{-1}$ or $(a, b) \in S^{-1}$

$\Rightarrow (b, a) \in R$ or $(b, a) \in S$

$\Rightarrow (a, b) \in (R \cup S)$

$\Rightarrow (a, b) \in (R \cup S)^{-1}$

Hence $(R^{-1} \cup S^{-1}) \subseteq (R \cup S)^{-1}$ ...(2)

By (*i*) and (*ii*) we conclude that

$$\boxed{(R \cup S)^{-1} = R^{-1} \cup S^{-1}}$$

(*c*) Let $(a, b) \in (R \cap S)^{-1}$

$\Rightarrow (b, a) \in R \cap S$

$\Rightarrow (b, a) \in R$ and $(b, a) \in S$

$\Rightarrow (a, b) \in R^{-1}$ and $(a, b) \in S^{-1}$

$\Rightarrow (a, b) \in R^{-1} \cap S^{-1}$

Hence $(R \cap S)^{-1} \subseteq R^{-1} \cap S^{-1}$ ...(1)

Conversely, let $(a, b) \in R^{-1} \cap S^{-1}$

$\Rightarrow (a, b) \in R^{-1}$ and $(a, b) \in S^{-1}$

$\Rightarrow (b, a) \in R$ and $(b, a) \in S$

$\Rightarrow (b, a) \in (R \cap S)$

$\Rightarrow (b, a) \in (R \cap S)^{-1}$

Hence $(R^{-1} \cap S^{-1}) \subseteq (R \cap S)^{-1}$ ...(2)

By (1) and (2), we conclude that

$$\boxed{(R \cap S)^{-1} = R^{-1} \cap S^{-1}}$$

Example 2.4. Let $A = \{a, b, c\}$ and $B = \{a, b, c, d\}$. The relation R and S are defined as

$$R = \{(a, a)\ (b, b)\ (b, c)\ (c, c)\} \text{ and}$$

$$S = \{(a, a)\ (a, b)\ (b, c)\ (b, d)\ (c, c)\ (c, d)\}$$

Find (*i*) $R \cup S$ (*ii*) $R \cap S$ (*iii*) $R^{-1} \cap S^{-1}$ and (*iv*) $R^{-1} \cup S^{-1}$

Solution. (*i*) $R \cup S = \{(a,a)\ (b,b)\ (b,c)\ (c,c)\ (a,b)\ (b,d)\ (c,d)\}$

(*ii*) $R \cap S = \{(b,c)\ (a,a)\ (c,c)\}$

(*iii*) Ass we know that $R^{-1} \cap S^{-1} = (R \cap S)^{-1}$

So $R^{-1} \cap S^{-1} = \{(a,a)\ (c,c)\ (c,b)\}$

(*iv*) We know that $(R \cup S)^{-1} = R^{-1} \cup S^{-1}$

So $R^{-1} \cup S^{-1} = \{(a,a)\ (b,a)\ (b,b)\ (c,b)\ (d,b), (c,c), (d,c)\}$

2.6 PICTORIAL REPRESENTATION OF RELATION

Pictorial representation of relation is sometimes also called graph of the relation.

For example consider the relation R on the set of Real number defined by the equation $x^2 + y^2 = 16$.

That is R consists of all ordered pairs (x, y) which satisfies the given equation. The pictorial representation or graph of the relation is given as

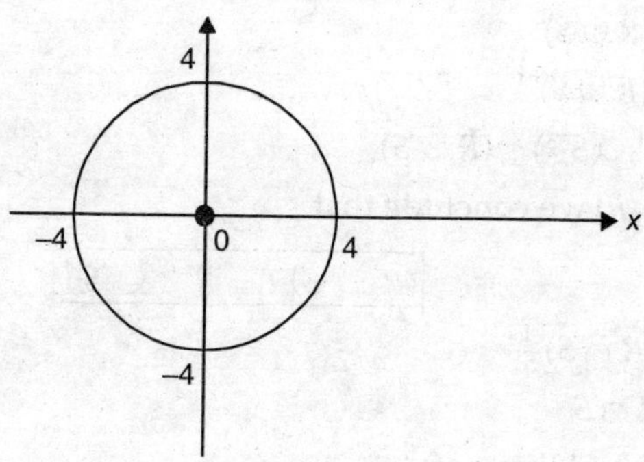

Fig. 2.1. Circle with radius 4 and center (0, 0)

Let A and B be two finite sets. Then we can represent a relation R from A to B in three ways (*i*) Matrix representation (*ii*) Arrow Diagram and (*iii*) By Directed Graphs.

2.6.1 Matrix Representation

Here we form a matrix $M(R)$ of relation R, whose rows are labelled by the elements of set A and columns are labeled by the elements of set B.

$$M(R) = [m_{ij}] = \begin{cases} 1, \text{ if } a_i R\, b_j \\ 0, \text{ otherwise} \end{cases}$$

For example, if $A = \{1, 2, 3, 4\}$, $B = \{a, b, c\}$

and $R = \{(1,a)\ (1,b)\ (2,b)\ (2,c)\ (3,a)\ (3,b)\ (3,c)\ (4,b)\}$

then

$$M(R) = \begin{array}{c} \\ 1 \\ 2 \\ 3 \\ 4 \end{array} \begin{array}{c} \begin{array}{ccc} a & b & c \end{array} \\ \begin{bmatrix} 1 & 1 & 0 \\ 0 & 1 & 1 \\ 1 & 1 & 1 \\ 0 & 1 & 0 \end{bmatrix} \end{array}$$

2.6.2 Arrow Diagram

Write the elements of A and B in circular disks. Whenever aRb, draw an arrow from disk A to disk B.

For example, If $A = \{a, b, c\}$ $B = \{1, 2, 3\}$ and relation R is defined as :

$$R = \{(a, 1)\ (a, 2), (b, 3), (c, 1)\}$$

The arrow diagram of R is given in **Fig. 2.2**

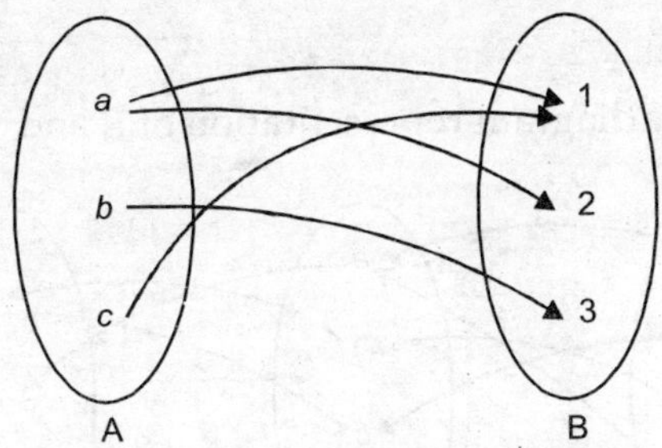

Fig. 2.2. Arrow diagram of R

2.6.3 Directed graph of Relation

If R is a relation from a set A to itself, then R can be represented with the help of diagraphs or directed graphs. In this representation the nodes or vertices of diagraph are labelled by the elements of the set A and the directed edges represents the relation

$$R \subseteq A \times A$$

For example, consider a set $A = \{a, b, c\}$ and R from A to A is defined as

$R = \{(a, a)\ (a, b)\ (a, c)\ (b, b)\ (b, c)\ (c, b)\}$

The diagraphical representation of R is given in **Fig. 2.3.**

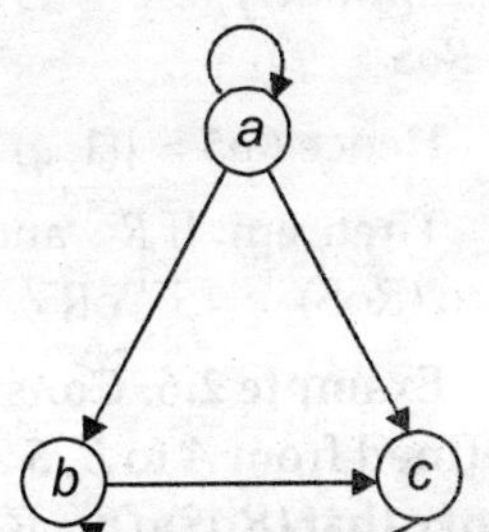

Fig. 2.3. Diagraph of R

Observations. (*i*) Self loop on node a indicates that $(a, a) \in R$ *i.e.,* a is related to itself.

(*ii*) Two edges between nodes b and c, one directed from b to c, another directed from c to b indicates that $(b, c) \in R$ as well as $(c, b) \in R$.

(*iii*) If there is no directed edge between a pair (a_i, a_j), this simply indicates that a_i is not related to a_j i.e., $(a_i, a_j) \notin R$.

2.7 COMPOSITION RELATION OR COMPOSITION OF RELATIONS

Consider three sets A, B and C. Let R and S be the two relations defined on set A and B, and B and C respectively *i.e.,* $R \subseteq A \times B$ and $S \subseteq B \times C$.

Then *a* relation RoS can be defined from set A to set C as

RoS = $\{(a, c) : \}\ b \in B$ such that $(a, b) \in R$ and $(b, c) \in S\}$

is called composite relation of relation R and S.

Note : The composition of a relation R with itself is denoted by a power of relation R. That is, if R is a relation defined on the set A, then $R^2 = RoR$, $R^3 = RoRoR$ and so on.

EXAMPLES

Example 2.5. Let $A = \{1, 2, 3, 4\}$, $B = \{a, b, c, d\}$ and $C = \{p, q, r\}$ and

$R = \{(1, a)\ (2, c)\ (3, b)\ (4, a)\ (4, d)\}$ and

$S = \{(a, q)\ (b, p)\ (d, r)\ (c, p)\}$

Find RoS.

Solution. The arrow diagram representation of R and S is given in **Fig. 2.4.**

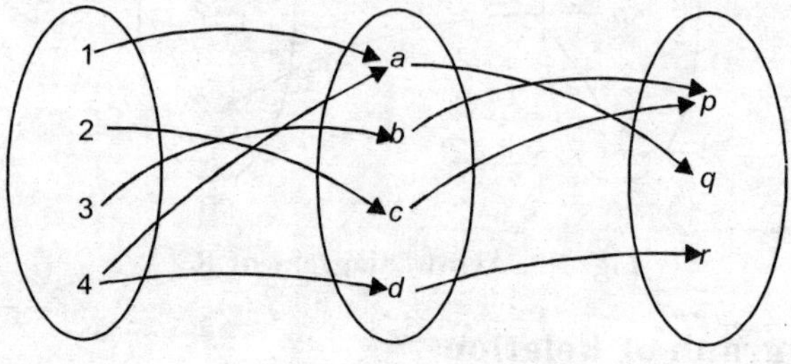

Fig. 2.4.

One can observe that, there is an arrow from 1 to a and is followed by an arrow from a to q. These two arrows can be viwed as a path from 1 to q, clearly 1 is an element of set A and q is the element of set C and they are connected with an element a of set B.

Thus $(1, q) \in RoS$

Similarly, $2 \in A$ is connected with $p \in C$ through an element $c \in B$, so $(2, p) \in RoS$

Hence $RoS = \{(1, q)\ (2, p)\ (4, r)\ (4, q)\ (3, p)\}$

Theorem. If R^{-1} and S^{-1} are the inverse of the relations R and S respectively, then $(\text{RoS})^{-1} = S^{-1}\ \text{oR}^{-1}$

Example 2.6. Consider three non-empty sets *A*, *B*, *C* and *D*. *A* relation *R* is defined from *A* to *B*, *S* is defined from *B* to *C* and *T* is defined from *C* to *D*. Then prove that (*RoS*)*oT* = *Ro*(*SoT*) *i.e.*, the composition of relation is associative.

Solution. To prove $(RoS)\ \text{oT} = Ro\ (SoT)$ we need to show that each ordered pair belongs to $(RoS)\ oT$ also belongs to $Ro(SoT)$ and *vice-versa*.

Let an orderd pair $(a, d) \in (RoS)oT$

$\Rightarrow$ an element $c \in C$ such that

$(a, c) \in RoS$ and $(c, d) \in \text{T}$.

Now $(a, c) \in RoS \Rightarrow \exists\ b \in B$ such that $(a, b) \in R$ and $(b, c) \in S$

Since $(b, c) \in s$ and $(c, d) \in T$

$\Rightarrow (b, d) \in (SoT)$ (by definition of composite relation)

and we have $(a, b) \in R$

So $(a, b) \in R, (b, d) \in (SoT) \Rightarrow (a, d) \in \text{Ro}\ (SoT)$

Thus $(RoS)\ oT \subseteq Ro\ (SoT)$ ...(1)

Similarly, we can show that

$Ro(SoT) \subseteq (RoS)oT$ ...(2)

By (1) and (2) we conclude that

$$\boxed{Ro(SoT) = (RoS)oT}$$

Example 2.7. Let $A = \{a, b, c\}$, $B = \{1, 2, 3\}$ and $C = \{x, y, z\}$

Given relation $R = \{(a, 2), (b, 1), (b, 3)\}$ and

$S = \{(1, y), (2, x), (3, y), (3, 2)\}$

(*a*) Find RoS (*b*) Find M_R, M_S, M_{RoS} (*c*) Find $M_R.M_s$ and compare it with M_{RoS} (*c*) Find $M_R.M_S$ and compare it with M_{RoS}.

Solution. By arrow diagram we can obtain RoS.

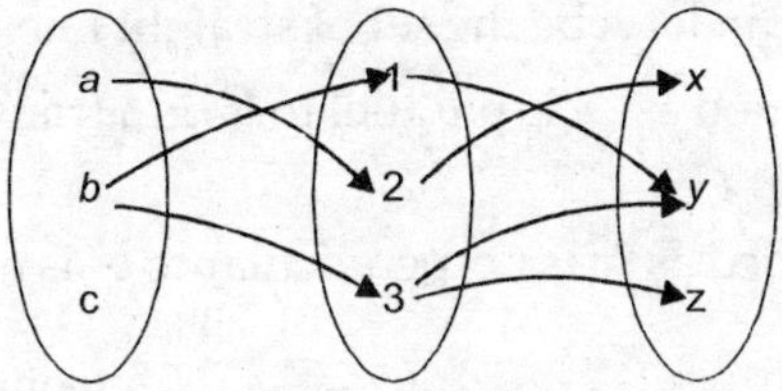

Fig. 2.5.

$RoS = \{(a, x), (b, y), (b, z)\}$

$$(b)\quad M_R = \begin{array}{c} \\ a \\ b \\ c \end{array}\begin{array}{c} \begin{array}{ccc} 1 & 2 & 3 \end{array} \\ \begin{bmatrix} 0 & 1 & 0 \\ 1 & 0 & 1 \\ 0 & 0 & 0 \end{bmatrix} \end{array} \qquad M_S = \begin{array}{c} \\ 1 \\ 2 \\ 3 \end{array}\begin{array}{c} \begin{array}{ccc} x & y & z \end{array} \\ \begin{bmatrix} 0 & 1 & 0 \\ 1 & 0 & 0 \\ 0 & 1 & 1 \end{bmatrix} \end{array} \text{ and } \quad M_{RoS} = \begin{array}{c} \\ a \\ b \\ c \end{array}\begin{array}{c} \begin{array}{ccc} x & y & z \end{array} \\ \begin{bmatrix} 1 & 0 & 0 \\ 0 & 1 & 1 \\ 0 & 0 & 0 \end{bmatrix} \end{array}$$

$$(c)\quad M_R.M_S = \begin{bmatrix} 0 & 1 & 0 \\ 1 & 0 & 1 \\ 0 & 0 & 0 \end{bmatrix}\begin{bmatrix} 0 & 1 & 0 \\ 1 & 0 & 0 \\ 0 & 1 & 1 \end{bmatrix} = \begin{bmatrix} 1 & 0 & 0 \\ 0 & 2 & 1 \\ 0 & 0 & 0 \end{bmatrix}$$

Clearly here $M_R.M_S \neq M_{RoS}$. But both the matrices have same zero entries and same non-zero enteries.

2.8 BINARY RELATION

Let A be a non-empty set. Then a subset of $A \times A$ is called a binary relation on A. That is a relation R defined on set A to itself is called a binary relation.

$$R = \{(a, b) : a, b \in A \text{ and } aRb\}$$

2.8.1 Properties of Binary Relation

(*i*) **Reflexive Relation. A binary relation R defined on set A is said to be reflexive if every element of set A is related to itself. That is**

$$\forall\, a \in A,\ aRa \text{ or } (a, a) \in R$$

(ii) Symmetric Relation. A binary relation R is said to be symmetric if

$$(a, b) \in R \Rightarrow (b, a) \in R, a, b \in A$$

That is, R is symmetric if a related to b implies that b is also R-related to a.

For example, let A be the set of all straight lines in a plane. A relation R is defined by "a is parallel to b" is a symmetric relation because if a is parallel to b then b is also parallel to a *i.e.,* $a \parallel b \Rightarrow b \parallel$

(iii) Transitive Relation. A binary relation R defined on set A is said to be transitive if

$$(a, b) \in R, (b, c) \in R \Rightarrow (a, c) \in R, a, b, c \in A$$

or if $aRb, bRc \Rightarrow aRc$

For example, again let A be the set of straight lines.

(*i*) Relation R defined as "a is parallel to b" is a transitive relation as $a \parallel b$, $b \parallel c \Rightarrow a \parallel c$.

(*ii*) Relation R defined as "a is perpendicular to b" is not a transitive relation, because

$$\text{If } a \perp b, b \perp c \Rightarrow a \perp c \not\Rightarrow a \perp c$$

(iii) Asymmetric Relation. A binary relation R on a set A is said to be asymmetric if $(a, b) \in R \Rightarrow (b, a) \notin R$

(iv) Antisymmetric Relation. A binary relation R defined on a set A is said to be Antisymmetric if

$(a, b) \in R$ and $(b, a) \in R \Rightarrow a = b$

For example, let I be the set of integers. Relation R defined as "a divides b" on set I is an antisymmetric relation as

If a divides b i.e., a/b and b/a

$$\Rightarrow a = b$$

2.9 CLOSURE PROPERTIES

Let A be a set. Consider the collection of all relations on A. Let P be the property of such relations, such as being symmetric or being transitive. Then a relation with property P is called a P-relation.

The P-closure of relation R on A, denoted as $P(R)$, is a P-relation such that

$$R \subseteq P(R) \subseteq S$$

for every P-relation S containing R, we will write reflexive (R), symmetric (R) and transitive (R) for the reflexive, symmetric and transitive closures of R.

2.9.1 Reflexive Closure

Let R be a relation defined on set A, then $(R \cup \Delta A)$, is called the reflexive closure of R, where $\Delta A = \{(a, a) : a \in A\}$ is the diagonal or equality relation on A.

For example: for a given set $A = \{1, 2, 3, 4\}$ relation R is defined as $R = \{(1, 2)$ $(3, 2)$ $(4, 3)$ $(3, 1)\}$.

Then reflexive $(R) = R \cup \Delta A$

$\Delta A = \{(a, a) : a \in A\} = \{(1, 1) (2, 2) (3, 3) (4, 4)\}$

then reflexive $(R) = \{(1, 2) (3, 2) (4, 3) (3, 1) (1, 1) (2, 2) (3, 3) (4, 4)\}$

2.9.2 Symmetric Closure

Let R be a relation defined on a set A, then symmetric closure of R is defined as

Symmetric $(R) = R \cup R^{-1}$

where R^{-1} is the inverse of relation R on A.

For example, let $A = \{1, 2, 3, 4\}$. Then relation R on A is $R = \{(1, 2) (1, 1) (3, 2) (4, 1)\}$

Then $R^{-1} = \{(2, 1) (1, 1) (3, 3) (1, 4)\}$

and Symmetric $(R) = R \cup R^{-1}$

$= \{(1, 2) (2, 1) (1, 1) (3, 2) (2, 3) (4, 1) (1, 4)\}$

2.9.3 Transitive Closure

Let R be a relation defined on set A. We define

$$R^* = \bigcup_{i=1}^{n} R^i = R \cup R^2 \cup ... \cup R^n$$

then R^* is called the transitive closure of a relation R.

For example, let $A = \{a, b, c\}$ and relation R is defined as $R = \{(a, b) (b, c) (c, c)\}$

then $R^2 = RoR = \{(a, c) (b, c) (c, c)\}$

and $R^3 = RoRoR = \{(a, c) (b, c) (c, c)\}$

Now transitive $(R) = = \bigcup_{i=1}^{3} R^i = ROR^2OR^3$

$= \{(a, b) (b, c) (c, c)\} \cup \{(a, c) (b, c) (c, c)\} \cup \{(a, c) (b, c) (c, c)\}$

$= \{(a, b) (a, c) (b, c) (c, c)\}$

Notes : (*i*) One can observe that reflexive closure of R *i.e.,* reflexive (R) is a obtained by simply adding to R those ordered pairs (a, a) in the diagonal which do not already belong to R.

(*ii*) Symmetric closure *i.e.,* symmetric (R) is obtained by adding to R all pairs (b, a) whenever (a, b) belongs to R.

2.10 EQUIVALENCE RELATION

A binary relation R defined on set A is called an equivalence. Relation if

(*a*) R is reflexive *i.e.,* $\forall\, a \in A, aRa$.

(*b*) R is symmetric *i.e.,* if $(a, b) \in R \Rightarrow (b, a) \in R$ and

(*c*) R is transitive *i.e.,* if $(a, b) \in R \Rightarrow (a, c) \in R$

For example, the relation R defined as "a is parallel to b" on a set A of straight lines, is an equivalence relation. Because

(*a*) R is reflexive as every line is parallel to itself.

(*b*) R is symmetric and (*c*) R is transitive also

EXAMPLES

Example 2.8. Prove that in the set of integers, the relation defined by the statement "xRy

$\Rightarrow x - y$ is divisible by 5" is an equivalence relation.

Solution. Defined R as $xRy \Rightarrow x–y$ is divisible by 5

$\Rightarrow (x - y) = 5$ m, where m is an integer.

To prove that R is an equivalence relation, we need to show that R is (*i*) reflexive (*ii*) symmetric and (*iii*) transitive.

(*i*) **Reflexive.** Clearly xRx

as $x - x = 5.0$ (*i.e.*, $m = 0$)

i.e., $x - x$ is divisible by 5.

here R is reflexive.

(*ii*) **Symmetric :** If $xRy \Rightarrow x - y$ is divisible by 5, *i.e.*, $x - y = 5$ m

or $- x + y = - 5\,m$

or $y - x = 5\,(- m) = 5$ m; where m' is also an integer

$\Rightarrow (y - x)$ is also divisible by 5

$\Rightarrow yRx$

So, if, $xRy \Rightarrow yRx$

hence R is a symmetric relation.

(*iii*) **Transitive.** To show that R is transitive we need to show that if xRy and $yRz \Rightarrow xRz$.

If xRy and yRz, *i.e.*, $(x - y)$ is divisible by 5 and $(y - z)$ is also divisible by 5.

$\Rightarrow x - y = 5\,m$, and $y - z = 5\,m_2$

By adding these two, we obtain

$(x - y) + (y - z) = 5\,(m_1 + m_2)$

or $x - z = 5\,(k)$ where $k = m_1 + m_2$ is also an integer

$\Rightarrow (x - z)$ is also divisible by 5

Hence, R is transitive also and thus equivalence.

Example 2.9. If R and S are two equivalence relations on set A, then show that $R \cap S$ is also an equivalence relation.

Solution. To show that $R \cap S$ is an equivalence relation we need to prove that $R \cap S$ is reflexive, symmetric and transitive.

(*i*) **Reflexive.** Since R and S are equivalence relations defined on set A.

$\Rightarrow \quad \forall a \in A, (a, a) \in R$ and $(a, a) \in S$

$\Rightarrow (a,a) \in R \cap S$

i.e., $\forall a \in A, (a,a) \in R \cap S$

Thus $R \cap S$ is reflexive.

(*ii*) **Symmetric.** Let $(a,b) \in R \cap S$

$\Rightarrow (a,b) \in R$ and $(a,b) \in S$

Since, R and S both are equivalence relations.

If $(a,b) \in R \Rightarrow (b,a) \in R$

and if $(a,b) \in S \Rightarrow (b,a) \in S$

Now, $(b,a) \in R$ and $(b,a) \in S$

$\Rightarrow (b,a) \in R \cap S$

So, if $(a,b) \in R \cap S \Rightarrow (b,a) \in R \cap S$

Thus R is symmetric.

(*iii*) **Transitive.** Let $(a,b) \in R \cap S$ and $(b,c) \in (R \cap S)$

$\Rightarrow (a,b) \in R$ and $(a,b) \in S$

and $(b,c) \in R, (b,c) \in S$

Since, both R and S are equivalence relations.

So, if $(a,b) \in R, (b,c) \in R \Rightarrow (a,c) \in R$

and if $(a,b) \in S, (b,c) \in S \Rightarrow (a,c) \in S$

$\Rightarrow (a,c) \in R \cap S$

So, if $(a,b) \in R \cap S$ and $(b,c) \in R \cap S$

$\Rightarrow (a,c) \in R \cap S$

Hence, $(R \cap S)$ is transitive

Since, $(R \cap S)$ is reflexive, symmetric and transitive, thus equivalence.

Example 2.10. Let R be an equivalence relation on set A, then prove that R^{-1} is also an equivalence relation on set A.

Solution. To prove that R^{-1} is an equivalence relation, we need to prove that R^{-1} is reflexive, symmetric and transitive.

(*i*) **Reflexive**. We know that R is an equivalence relation, then

$\forall a \in A, (a,a) \in R$

if $(a,a) \in R \Rightarrow (a,a) \in R^{-1}$ (by the definition of inverse)

$\Rightarrow \forall a \in A, (a,a) \in R^{-1}$

Hence, R^{-1} is a reflexive relation.

(*ii*) **Symmetric.** Let $(a,b) \in R$, for $a, b \in A$

Now $(a,b) \in R \Rightarrow (b,a) \in R$ as R is symmetric

By the definition of inverse

If $(a,b) \in R \Rightarrow (b,a) \in R^{-1}$

also $(b,a) \in R \Rightarrow (a,b) \in R^{-1}$

Thus, if $(b,a) \in R^{-1} \Rightarrow (a,b) \in R^{-1}$, for $a, b \in A$

Hence, R is a symmetric relation.

(*iii*) **Transitive.** Let $(a, b) \in R$ and $(b, c) \in R$

$\Rightarrow (b, a) \in R^{-1}$ and $(c, b) \in R^{-1}$ (by definition of inverse)

By the transitive property of R.

If $(a, b) \in R$ and $(b, c) \in R \Rightarrow (a, c) \in R$

If $(a, c) \in R \Rightarrow (c, a) \in R^{-1}$

Thus $(c, b) \in R^{-1}$ and $(b, a) \in R^{-1} \Rightarrow (c, a) \in R^{-1}$

Hence, R^{-1} is transitive.

So, R^{-1} is and so an equivalence relation as R^{-1} is reflexive, symmetric and transitive.

Example 2.11. If R be a relation, in the set $N \times N$ where, N is a set of positive integers, defined by $(a, b)\ R\ (c, d) \Leftrightarrow ad = ac$, $\forall\ a, b, c, d \in N$ prove that R is an equivalence relation.

Solution : We need to show that R is reflective, symmetric and transitive to prove R an equivalence relation.

(*i*) **Reflective :** Since $a \cdot b = b \cdot a$. Thus by definition of the relation (a, b) $R(a, b)$. So R is reflective.

(*ii*) **Symmetric :** Let $(a, b)\ R\ (c, d)$

$\Rightarrow \quad ad = b \cdot c$

or $\quad d \cdot a = c \cdot b$

or $\quad c \cdot b = d \cdot a$

$\Rightarrow \quad (c, d) = (a, b)$

So, if $(a, b)\ R(c, d) \Rightarrow (c, d)\ R(a, b)$ hence R is symmetric

(*iii*) **Transitive :** Let $(a, b)\ R\ (c, d)$ and $(c, d)\ R(e, f)$

$\Rightarrow \quad ad = bc \quad$ and $cf = de$

or $\quad ad = b \cdot \left(\dfrac{de}{f}\right) \quad \Rightarrow a \cdot f = b \cdot e$

$\Rightarrow (a, b)\ R\ (e, f)$

So, if $(a, b)\ R(c, d)$ and $(c, d)\ R\ (e, f) \Rightarrow (a, b)\ R\ (e, f)$. Hence R is transitive and thus equivalence.

Example 2.12. Give an example of the relation R which is

(*a*) **Symmetric and transitive but not reflexive.**

(*b*) **R is both symmetric and antisymmetric.**

(*c*) **R is transitive but $R \cup R^{-1}$ is not transitive.**

Solution : Let $A = \{1, 2, 3\}$ be any set and a relation R is defined on A, *i.e.*, $R \subseteq A \times A$.

(*a*) Define R as $R = \{(1, 2)\ (2, 1)\ (2, 2)\ (3, 3)\}$ clearly as $(1, 1) \notin R$, so R is reflexive.

Since (1, 2) and (2, 1) both belongs to R. So R is symmetric

Again $\quad (1, 2) \in R \Rightarrow 1R_2$

$\quad (2, 1) \in R \Rightarrow 2R_1$

Now $2R_1$ and $1R_2 \Rightarrow 2R_2$ *i.e.*, $(2, 2) \in R$ so R is transitive.

Note : Three can be many other examples of such type

(*b*) For both symmetric and antisymmetric, R can be defined as

$$R = \{(1, 1), (2, 2), (3, 2)\}$$

(*c*) Define R as $R = \{(1, 2)\ (2, 3)\ (1, 3)\}$ is transitive and $R^{-1} = \{(2, 1)\ (3, 2)\ (3, 1)\}$ is also transitive. $R \cup R^{-1} = \{(1, 2)\ (2, 3)\ (1, 3)\ (2, 1)\ (3, 2)\ (3, 1)\}$ is not transitive as $(2, 3) \in R \cup R^{-1}$ and $(3, 2) \in R \cup R^{-1}$ but $(2, 2) \notin R \cup R^{-1}$.

2.11 EQUIVALENCE SET OR EQUIVALENCE CLASS

Definition : Let A be a non-empty set and R be an equivalence relation defined on A. Consider an element $a \in A$. Then the set of all those elements of set A which are R-related to $a \in A$, is called the equivalence set or equivalence class of a with respect to R. The equivalence class of a is symbollicaly denoted as $[a]$ and defined as

$$[a] = \{x : x \in A \text{ and } x\, R\, a\}$$

2.11.1 Some Properties of Equivalence Classes

Let A be a non-empty set and R be an equivalence relation defined on set A. Then for any element $a \in A$, $[a]$ is the equivalence class of a with respect to the relation R, then

(*i*) $a \in [a]$

(*ii*) $a \in [b] \Leftrightarrow [a] = [b]$

(*iii*) If a and b be two distinct elements of A, then either $[a] \cap [b] = \phi$ or $[a] = [b]$.

Proof: (*i*) Since relation R defined on A is an equivalence relation, so R is reflexive

$$\Rightarrow \quad \forall\, a \in A,\ (a, a) \in R$$

So by the definition of equivalence class

$$\forall\, a \in A, a \in [a]$$

i.e., every element of set A belongs to its equivalence class

(*ii*) Let $\quad a \in [b] \Rightarrow a\, R\, b$

(by the definition of equivalence class)

To show that $\quad [a] = [b]$

We observe that for an arbitrary element

$x \in [a]$, we have $x \in [a] \Leftrightarrow xRa$

$$\Leftrightarrow \ xRa \text{ and } aRb$$

$\Leftrightarrow\ xRb$ (since $a \in [b]$)

$\Leftrightarrow\ x \in [b]$ (by pensitive property of R)

Hence $[a] = [b]$

Again let $[a] = [b]$

then from property (i) $a \in [a]$

$\Rightarrow a \in [b]$

So if $[a] = [b] \Rightarrow a \in [b]$

Hence $a \in [b] \Leftrightarrow [a] = [b]$

(iii) Let $[a] \cap [b] = \phi$

Consider on arbitrary element x of $[a] \cap [b]$

$\Rightarrow x \in [a] \cap [b]$

$\Rightarrow x \in [a]$ and $x \in [b]$

$\Rightarrow xRa$ and xRb

or $\quad aRx$ and xRb

$\Rightarrow aRb$ (by transitive property of R)

$\Rightarrow a \in [b]$

How $a \in [b] \Leftrightarrow [a] = [b]$ (by property (2))

Similarly we can prove if $[a] \neq [b]$ then

$$[a] \cap [b] = \phi$$

Therefore, either $[a] \cap [b] = \phi$ or $[a] = [b]$

Example 2.13. Let $A = \{0, 1, 2, 3, ...\}$ and $R = \{(a, b) : a - b = 4m, m$ is any integer$\}$ is an equivalence relation defined on set A. Find out the equivalence classes of set A.

Solution : By the definition of equivalence class of $a \in A$

$$[a] = \{x : x \in A \text{ and } x\,R\,a\}$$

so
$$[0] = [x : x \in A \text{ and } x\,R\,0\}$$
$$= \{x : x \in A \text{ and } x - 0 = 4 \cdot m, m \text{ is any integer}\}$$
$$= \{x : x \in A \text{ and } x = 4m, m \text{ is any integer}\}$$
$$= \{x : x \in A \text{ and } x = 4m, m \in \{... -3 -1, 0, 1, 2 ...\}\}$$
$$= \{0, 4, 8, 12, ...\}$$

Similarly

$$[1] = \{x : x \in A \text{ and } x - 1 = 4m, m \in (... -1, 0, 1, 2, ...)$$

So
$$[1] = \{x : x \in A \text{ and } x = 4m + 1, m \in (..., -1, 0, 1, 2, ...)$$
$$= \{1, 5, 9, 13, ...\}$$

and
$$[2] = \{2, 6, 10, 14, ...\}$$
$$[3] = \{3, 7, 11, 15, ...\}$$
$$[4\} = \{4, 8, 12, 16, ...\}$$

(Since m may be negative also to satisfy the condition $x \in A$ and $x - A = 4m$, therefore

$$m = -1, 0 \in [4]$$

Thus we find that $[4] = [0]$

Similarly we can show that

$$[5] = [1]$$

$$[6] = [2]$$

$$[7] = [3]$$

$$[8] = [4] \text{ and so on.}$$

Which clearly indicates that, for the above defined equivalence relation there are only 4 *i.e.*, [0], [1], [2] and [3] distinct equivalent sets.

Example 2.14. Let $A = \{1, 2, 3\}$ be a set. Define a relation R on $A \times A$ such that $(a, b)\, R\, (c, d) \Leftrightarrow a + d = b + c$.

(*i*) Show that R is an equivalence relation.

(*ii*) Find the equivalence classes of $A \times A$ for the given relation.

Solution : Lets prove that R is an equivalence relation by taking set A as a set of positive integers.

(*i*) **Reflexive :** Since $a + b = a + b$. So by the definition of R $(a, b)\, R\, (a, b)$. Hence R is reflective.

(*ii*) **Symmetric :** Let $(a, b)\, R(c, d) \Rightarrow a + b = b + c$

or $\quad c + b = d + a$

$\Rightarrow (c, d)\, R\, (a, b)$

So if $(a, b)\, R(c, d) \Rightarrow (c, d)\, R(a, b)$

Thus R is symmetric.

(*iii*) Let $(a, b)\, R(c, d)$ and $(c, d)\, R\, (e, f)$

$\Rightarrow a + b = b + c$ and $c + f = d + e$

or $\quad a + d = b + d\ (d + e - f)$

$\Rightarrow a + f = b + e$

$\Rightarrow (a, b)\, R\, (e, f)$

Thus R is transitive

Since R is reflective, symmetric and transitive. So R is an equivalence relation.

(*b*) We know that for $a \in A$ an equivalence class of s is defined as

$$[a] = \{x : x \in A \text{ and } xRa\}$$

where R is an equivalence relation. How we need to find out the equivalence classes of the elements of set $A \times A$. Given

$$A = \{1, 2, 3\}$$

So $A \times A = \{(1, 1) (1, 2) (1, 3) (2, 1) (2, 2) (2, 3) (3, 1) (3, 2) (3, 3)\}$

So for any $(a, b) \in A \times A$

$$[a, b] = \{(x, y) : (x, y) \in A \times A \text{ and } (x, y) R(a, b)\}$$

$$[1, 1] = \{(x, y) : (x, y) \in A \times A \text{ and } (x, y) R(1, 1)\}$$

$$= \{(x, y) : x + 1 = y + 1\}$$

$$= \{(x, y) : x = y; x, y \in A\}$$

$$\boxed{[1,1] = \{(1,1), (2,2), (3,3)\}}$$

Similarly, $[1, 2] = \{(x, y) : (x, y) \in A \times A \text{ and } (x, y) R(1, 2)\}$

$$= \{(x, y) : x + 2 = y + 1 \Rightarrow x + 1 = y\}$$

So $\boxed{[1,1] = \{(1,2), (2,3)\}}$

$$[1, 3] = \{(x, y) : (x, y) \in A \times A \text{ and } (x, y) R(1, 3)\}$$

$$= \{(x, y) : (x, y) \in A \times A \text{ and } x + 3 = y + 1\}$$

So

$\boxed{[1,3] = \{(1, 3)\}}$ $\boxed{[2,2] = [3,3] = [1,1]}$

$\boxed{[2,1] = \{[2,1](3,2]\}}$ $\boxed{[2,3] = \{[1,2][2,3]\} = [1,2]}$

$\boxed{[3,1] = \{[3,1]\}}$

$\boxed{[3,2] = \{[3,2][2,1]\} = [2,1]}$

So total distinct equivalence values are 5 as

(*i*) $[1, 1] = [2, 2] = [3, 3]$

(*ii*) $[1, 2] = [2, 3]$

(*iii*) $[1, 3]$

(*iv*) $[3, 1]$

(*v*) $[3, 2] = [2, 1]$

2.12 PARTITION OF A SET

Definition : Let A be a non-empty set. Then $P = \{A_1, A_2, A_3, ... A_n\}$ is called a position of set A if

(*i*) Every element of set P *i.e.,* $A_1, A_2, ... A_n$ is a non-empty subset of A.

(*ii*) $A = A_1 \cup A_2 \cup A_3 \cup ... \cup A_n$

(*iii*) The elements of set P are pairwise disjoint *i.e.;* $A_i, A_j \in P \Rightarrow$ either $A_i \cap A_j = \phi$ or $A_i = A_j$

In other words we can say that a position of a set A is a collection of mutually disjoint, non-empty subsets of A, whose union is A. These subsets are called parts of the position. *For example;* for a set $A = \{1, 2, 3, 4 ..., 10\}$

$$P = \{\{1, 3, 5\}, \{2, 4\}, \{6, 7, 9\} \{8, 10\}\}$$

is a position of set A as

(i) All elements of P belongs to same subset of A *i.e.*, there is no element of P which is not the element of any subset in P.

(ii) Clearly $A = \{1, 3, 5\} \cup \{2, 4\} \cup \{6, 7, 9\} \cup \{8, 10\}$.

(iii) All the subsets of A are pair wise disjoint *i.e.*, There is no element common in any pair of subsets.

2.13 PARTIAL ORDER RELATION

A binary relation R defined on a set A is called Partial Order Relation (P.O.R.) if R is

(a) **Reflexive *i.e.*; $\forall\, a \in A, (a, a) \in R$**

(b) **Antisymmetric *i.e.*, if $(a, b) \in R$ and $(b, a) \in R \Rightarrow a = b$, for $a, b \in A$**

(c) **Transitive *i.e.*, if $(a, b) \in R$ and $(b, c) \in R \Rightarrow (a, c) \in R$ for $a, b, c \in A$**

A set S together with a partial order relation R is called a **Partial Order Set** or **Poset.**

Example 2.15. Consider a set of integers Z. A relation R defined as "a is $\leq b$". Show that the relation is a partial order relation and set Z is a Poset.

Solution : To show that Z with defined relation R is a Poset, we need to show that R is

(i) reflexive (ii) antisymmetric and (iii) transitive

(i) **Reflexive :** Consider any element $a \in Z$ clearly $a \leq a$ (as a number is equal to itself).

So $\forall\, a \in Z, a \leq a$ exists.

So R is reflexive.

(ii) **Antisymmetric :** Let any $a, b \in Z$ $(a, b) \in R$ and $(b, a) \in R$.

$\Rightarrow a \leq b$ and $b \leq a$

$\Rightarrow a = b$

Hence R is antisymmetric

(iii) **Transitive :** Let for any $a, b, c \in Z$ if $(a, b) \in R$ and $(b, c) \in R$.

$\Rightarrow a \leq b$ and $b \leq c$

or $a \leq b \leq c$

$\Rightarrow a \leq c$

$\Rightarrow (a, c) \in R$

Hence R is transitive.

Since, R defined as "a is $\leq b$" is reflexive, antisymmetric and transitive. Thus partial order relation and set Z together with this partial order relation is a poset.

2.14. TOTAL ORDER

A binary relation R on a set A is a **total order** if and only if it is

(i) a partial order, and

(ii) for any pair of elements a and b of A, $\langle a, b\rangle \in R$ or $\langle b, a\rangle \in R$.

That is, every element is related with every element one way or the other.

A total order is also called a ***linear order.***

Example 2.16. The less-than-or-equal-to relation on the set of integers *I* is a total order.

The strictly-less-than and proper-subset relations are not partial order because they are not reflexive. They are examples of some relation called quasi order.

2.15. QUASI ORDER

Definition : A binary relation *R* on a set *A* is a quasi order if and only if it is

(*i*) irreflexive, and

(*ii*) transitive.

A quasi order is necessarily antisymmetric as one can easily verify.

Example 2.17. The less-than relation on the set of integers *I* is a quasi order.

Example 2.18. The proper subset relation on the power set of a set, say {1, 2}, is also a quasi order.

2.16 MINIMAL/MAXIMAL ELEMENT

"The concept of *least/greatest number* in a set of integers can be generalized for a general poset. Here we discuss the concept of minimal/maximal elements."

Let $<A, \leq>$ be a poset, where "$\leq$" represents an arbitrary partial order. Then an element $b \in A$ is a minimal element of *A* if there is no element $a \in A$ that satisfies $a \leq b$. Similarly an element $b \in A$ is a maximal element of *A* if there is no element $a \in A$ that satisfies $b \leq a$.

Example 2.19. The set of {{1}, {2}, {1, 2}} with $\subseteq$ has two minimal elements {1} and {2}. Note that {1}, and {2} are not related to each other in $\subseteq$. Hence we can not say which is "smaller than" which, that is, they are not comparable.

2.17 LEAST/GREATEST ELEMENT

Let $<A, \leq>$ be a poset. Then an element $b \in A$ is the least element of *A* if for every element $a \in A$, $b \leq a$. Similarly an element $b \in$ A is called greater if for every $a \in A$, $a \leq b$.

Note that the least/greatest element of a poset is unique if one exists because of the antisymmetry of $\leq$.

Example 2.20. The poset of the set of natural numbers with the less-than-or-equal-to relation has the least element 0.

Example 2.21. The poset of the powerset of {1, 2} with $\subseteq$ has the least element Ø.

2.18 WELL ORDER

A total order *R* on a set *A* is a well order if every non-empty subset of *A* has the least element.

Example 2.22. The poset of the set of natural numbers with the less-than-or-equal-to relation is a well order, because every set of natural numbers has the least element.

The poset of the set of positive real numbers with the less-than-or-equal-to relation is not a well order, because the set itself does not have any least element (0 is not in the set).

A digraph of a binary relation on a set can be simplified if the relation is a partial order. Hasse diagrams defined as follows are such graphs.

2.19 HASSE DIAGRAM

A Hasse diagram is a graph for a poset which does not have loops and arcs implied by the transitivity. Further, it is drawn so that all arcs point upward eliminating arrowheads.

To obtain the Hasse diagram of a poset, first remove the loops, then remove arcs *<a, b>* if and only if there is an element *c* that *<a, c>* and *<c, b>* exist in the given relation.

Example 2.23. For the relation {*<a, a>*, *<a, b>*, *<a, c>* *<b, b>*, *<b, c>*, *<c, c>*} on set {*a*, *b*, *c*}, the Hasse diagram has the arcs {*<a, b>*, *<b, c>*} as shown below.

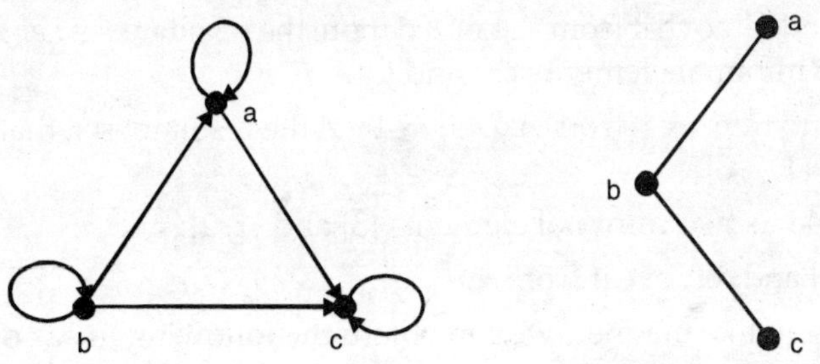

(*a*) Digraph for partial order. (*b*) Hasse Diagram.

Fig. 2.6.

Topological Sorting

The elements in a finite poset can be ordered linearly in a number of ways while preserving the partial order. *For example* {Ø, {1}, {2}, {1, 2}} with the partial order ⊆ can be ordered linearly as Ø, {1}, {2}, {1, 2}, or Ø, {2}, {1}, {1, 2}. In these orders a set appears before (to left of) another set if it is a subset of the other. In real life, tasks for manufacturing goods in general can be partially ordered based on the prerequisite relation, that is certain tasks must be completed before certain other tasks can be started. *For example* the arms of a chair must be carved before the chair is assembled. Scheduling those tasks is essentially the same as arranging them with a linear order (ignoring here some possible concurrent processing for simplicity's sake).

The topological sorting is a procedure to find from a partial order on a finite set a linear order that does not violate the partial order. It is based on the fact that a finite poset has at least one minimal element. The basic idea of the topological sorting is to first remove a minimal element from the given poset, and then repeat that for the resulting set until no more elements are left. The order of removal of the minimal elements gives a linear order. The following algorithm formally describes the topological sorting.

Algorithm Topological Sort

Input. A finite poset *<A, R>*.

Output. A sequence of the elements of A preserving the order R.

```
integer i;
i : = 1;
while (A ¹ Ø) {
    pick a minimal element b from A;
    A : =A - {b};
    i : = i + 1;
    output b
}
```

EXAMPLES

Example 2.24. Let $A = \{\{1\}, \{2\}, \{3\}, \{1, 2\}, \{1, 3\}, \{2, 3\}, \{1, 2, 3\}\}$ with the partial order $\subseteq$.

This given A has three minimal elements $\{1\}$, $\{2\}$, and $\{3\}$.

Select $\{2\}$ and remove it from A. Let A denote the resultant set *i.e.*, $A : A - \{2\}$. The new A has two minimal elements $\{1\}$, and $\{3\}$.

Select $\{1\}$ and remove it from A. Denote by A the resultant set, that is $A = \{\{3\}, \{1, 2\}, \{1, 3\}, \{2, 3\}, \{1, 2, 3\}\}$.

This new A has two minimal elements $\{3\}$ and $\{1, 2\}$.

Select $\{1, 2\}$ and remove it from A.

Proceeding in like manner, we can obtain the following linear order :

$\{\{2\}, \{1\}, \{1, 2\}, \{3\}, \{1, 3\}, \{2, 3\}, \{1, 2, 3\}\}$.

Example 2.25. Let A be the set of all coplanar triangles then the relation, R "is similar to" is an equivalence relation. Since,

(*i*) Let $t \in A$, then tRt, $\because$ every triangle t is similar to itself. Thus, R is reflexive.

(*ii*) Let $t_1, t_2 \in A$, then $t_1Rt_2 \Rightarrow t_2\,Rt_1$ for if triangle t_1 is similar to t_2 then triangle t_2 is also similar to t_1. Thus, R is symmetric.

(*iii*) **Transitivity :** Let $t_1, t_2, t_3 \in T$. If t_1 "is similar to" t_2 and t_2 "is similar to" t_3, then we have t_1 "Similar to" t_3. Hence, $t_1\,Rt_2$, $t_2\,Rt_3 \Rightarrow t_1\,Rt_3$ *i.e.* R is transitive. Thus, R is an equivalence relation.

Example 2.26. Let $A = \{1, 2, 3\}$ and $R = \{(1, 1), (1, 2), (2, 2), (2, 3)\}$ is a relation in A. Is the relation 'R' a reflexive relation.

Solution. No, R is not reflexive since, $3 \in A$ but 3 is not R-related to 3, *i.e.*, $(3, 3) \notin R$.

Example 2.27. Is the relation "is parallel to" in a set A of all straight lines in a plane is a transitive relation.

Solution. Yes, the relation "is parallel to" is a transitive relation since, if $x, y, z \in A$ and $(x, y) \in R$, $(y, z) \in R$ then $(x, z) \in R$, *i.e.*, if x is parallel to y and y is parallel to z then x is parallel to z.

Example 2.28. If A be the set of all triangles in a plane and $R = \{(a, b) : \Delta a = \Delta b\}$, *i.e.*, $aRb \Rightarrow$ area of the triangle a = area of the triangle b, then prove that R is an equivalence relation.

Solution. (*a*) **Reflexive:** Since $\forall a \in A$,

$$\Delta a = \Delta a$$

Hence, aRa, *i.e.*, $(a, a) \in R$

$\Rightarrow$ R is *a* reflexive relation.

(*b*) **Symmetric:** For any $a, b \in A$

$$(a, b) \in R \Rightarrow \Delta a = \Delta b$$
$$\Rightarrow \Delta b = \Delta a$$
$$\Rightarrow (b, a) \in R$$

Hence R is *a* symmetric relation in A.

(*c*) **Transitive:** $\forall a, b, c \in A$, we have

$(a, b) \in R, (b, c) \in R$

$\Rightarrow$ $\Delta a = \Delta b$ and $\Delta b = \Delta c$

$\Rightarrow$ $\Delta a = \Delta c$

$\Rightarrow$ $(a, c) \in R$

Hence, R is transitive.

Since, R is a reflexive, symmetric and transitive relation in A. Hence it is an equivalence relation in A.

Example 2.29. Let R and R' be relations on a set A. Prove that if R is reflexive, symmetric and R' is also so correspondingly then $R \cup R'$ is reflexive and symmetric.

Solution. Let $X = \{(x, y\} : xRy$ and $x = y : x \in A\}$ and $R = \{(x, y) : xRy\}$

R is reflexive (given).

$\therefore$ $X \subset R$, But $R \subset (R \cup R')$

$\therefore$ $X \subset (R \cup R')$

Hence $R \cup R'$ is also reflexive.

Since R is symmetric, $\therefore$ $(x, y), (y, x) \in R$ when $x, y \in A$.

Also R' is symmetric, $\therefore$ $(x, y), (y, x) \in R'$.

Hence $(x, y), (y, x) \in R \cup R'$, where $x, y \in A$.

Hence $R \cup R'$ is symmetric.

Example 2.30. Prove that if R is transitive, inverse relation R^{-1} is also transitive.

Solution. Let R be defined on the set A. Also let $a, b, c \in A$. Hence, R transitive on $A \Rightarrow$ if aRb, bRc then aRc.

$\Rightarrow$ $(b, a), (c, b), (c, a) \in R^{-1}$

$\Rightarrow$ $(c, b), (b, a), (c, a) \in R^{-1}$

$\Rightarrow$ if $cR^{-1}b$, $bR^{-1}a$ then $cR^{-1}a$

$\Rightarrow$ R^{-1} is transitive.

Example 2.31. Are the following relations symmetric or antisymmetric (or asymmetric) for the set $A = \{a, b, c\}$

(i) $R = \{(a, b), (b, c), (a, c), (b, a), (c, b), (c, a)\}$

(ii) $R = \{(a, b), (b, c), (a, c)\}$

(iii) $R = \{(a, b), (b, c), (b, a), (a, c)\}$.

Solution. (i) R is symmetric as

$$aRb \Rightarrow bRa$$
$$bRc \Rightarrow cRb$$
$$aRc \Rightarrow cRa$$

(ii) R is asymmetric as $\forall (a, b) \in R$, we have $(a, b) \notin R$.

(iii) R is neither symmetric nor asymmetric.

Example 2.32. Let $A = \{0, 1, 2, 3, ...\}$ and $R = \{(x, y) : x - y = 3k, k$ is an integer$\}$ *i.e.* xRy if '$x - y$' is divisble by 3, then prove that R is an equivalence relation.

Solution. (i) **Reflexive :** For any $x \in A, x - x = 0$

$\Rightarrow$ $x - x$ is divisible by 3

So $\forall x \in A, x\,R\,x$

Hence R is reflexive.

(ii) **Symmetric :** For any $x, y \in A$

$$(x, y) \in R \Rightarrow x - y = 3k$$
$$\Rightarrow y - x = -3k \qquad (k \text{ is an integer})$$
$$\Rightarrow y - x = 3(-k) \qquad (\text{since } k \text{ is also an integer})$$
$$\Rightarrow y - x \in R$$

Hence, R is symmetric relation.

(iii) **Transitive :** For all $x, y, z \in A$

$(x, y) \in R, (y, z) \in R$

$$\Rightarrow \quad x - y = 3k_1 \text{ and } y - z = 3k_2$$
$$\Rightarrow \quad (x - y) + (y - z) = 3(k_1 + k_2) \qquad (k_1, k_2 \text{ are integers})$$
$$\Rightarrow \quad x - z = 3k \qquad (k = k_1 + k_2 \text{ which is an integer})$$

$\Rightarrow$ $(x, z) \in R$

Hence, R is transitive.

Since, R is reflexive, symmetric and transitive relation, hence R is an equivalence relation.

Example 2.33. If R be *a* relation in the set $N \times N$ where, N is a set of positive integers, defined by

$$(a, b)\ R\ (c, d) \Leftrightarrow a + d = b + c,$$

$a, b, c, d \in N$, then prove R is an equivalence relation.

Solution. Since the ordered pairs (a, b) and (c, d) are R related to each other only when

$$a + d = c + d.$$

(*i*) R is reflexive : Since $a + b = b + a$.

So $(a, b) R (a, b)$, Clearly R is reflexive.

(*ii*) R is symmetric. Since $(a, b) R (c, d)$

$$\Leftrightarrow \quad a + d = b + c$$

$$\Leftrightarrow \quad b + c = a + d$$

$$\Leftrightarrow \quad c + d = d + a$$

$\Leftrightarrow$ $(c, d) R (a, b)$

Hence R is *a* symmetric relation.

(*iii*) R is transitive : Let $(a, b), (c, d), (e, f) \in N \times N$

s.t. $(a, b) R (c, d)$ and $(c, d) R (e, f)$

then $a + d = b + c$ and $c + f = d + e$.

$$\Rightarrow \quad (a + d) + (c + f) = (b + c) + (d + e)$$

$$\Rightarrow \quad (a + f) + (c + d) = (b + e) + (c + d)$$

$$\Rightarrow \quad a + f = b + e$$

$\Rightarrow$ $(a, b) R (e, f)$

Thus, R is transitive.

Hence R is an equivalence relation.

Example 2.34. If R and S be two equivalence relation in the set X, then prove, that $R \cap S$ is also an equivalence relation in X.

Solution. (*a*) Reflexive : Since R and S are equivalence relation in X, hence $\forall\, x \in X$.

$(x, x) \in R, (x, x) \in S$

$$\Rightarrow \quad (x, x) \in R \cap S \quad \forall x \in X.$$

$\Rightarrow$ $R \cap S$ is *a* reflexive relation in X.

(*b*) **Symmetric :** Since R and S are equivalence relation in X, hence for $x, y \in X$, we have

$$(x, y) \in R \Rightarrow (y, x) \in R$$

and

$$(x, y) \in S \Rightarrow (y, x) \in S$$

Hence

$$\begin{aligned} (x, y) \in R \cap S &\Rightarrow (x, y) \in R \text{ and } (x, y) \in S \\ &\Rightarrow (y, x) \in R \quad \text{and} \quad (y, x) \in S \\ &\Rightarrow (y, x) \in R \cap S. \end{aligned}$$

So $R \cap S$ is a symmetric relation.

(*c*) **Transitive :** Since for any three elements $x, y, z \in X$.

$$(x, y), (y, z) \in R \Rightarrow (x, z) \in R$$

and

$$(x, y), (y, z) \in S \Rightarrow (x, z) \in S$$

$$\begin{aligned} \Rightarrow \quad (x, y), (y, z) \in R \cap S &\Rightarrow (x, y), (y, z) \in R \text{ and } (x, y), (y, z) \in S \\ &\Rightarrow (x, z) \in R \text{ and } (x, z) \in S \\ &\Rightarrow (x, z) \in R \cap S. \end{aligned}$$

Hence $R \cap S$ is a transitive relation.

Thus, $R \cap S$ is an equivalence relation.

Example 2.35. Prove that the relation R on the set N defined by xRy if $x \neq y$, $x, y \in N$ is symmetric but, neither reflexive nor transitive.

Solution. Reflexive : xRx is not true, as x is always equal to x.

$\Rightarrow$ R is not reflexive.

Symmetric : If $x \neq y$ then $y \neq x$.

$$\Rightarrow \qquad xRy \Rightarrow yRx$$

$\Rightarrow$ R is symmetric.

Transitive : If $x \neq y, y \neq z$ then $x \neq z$ is not always true.

For example, $1R2, 2R1$; then $1R1$ is not true.

Hence, R is not transitive.

Example 2.36. Let the relation R on Q_0 be defined by xRy if $x = \dfrac{1}{y}, x\ y \in Q_0$ where Q_0 is the set of non-zero rational numbers. Prove that R is not an equivalence relation.

Solution. Let $x \in Q_0, x \neq \dfrac{1}{x}, \forall x \in Q_0$

So xRx is not true.

$\Rightarrow$ R is not reflexive.

$\Rightarrow$ R is not an equivalence relation.

Example 2.37. If R is an equivalence relation in a set A, then prove that R^{-1} is also an equivalence relation in A.

Solution. (*i*) **Reflexive :** Since R is an equivalence relation in A,

$(a, a) \in R\ \forall\ a \in R$

Since $\qquad (a, a) \in R \Rightarrow (a, a) \in R^{-1}$

$\therefore\ \forall a \in A, (a, a) \in R^{-1}$

Hence, R^{-1} is a reflexive relation in A.

(*ii*) **Symmetric :** Let $(a, b) \in R$ where $a, b \in A$

$$(a, b) \in R \Rightarrow (b, a) \in R^{-1}$$

Also, $\qquad (a, b) \in R \Rightarrow (b, a) \in R$

$$\Rightarrow (a, b) \in R^{-1}$$

$$\therefore \qquad (b, a) \in R^{-1} \Rightarrow (a, b) \in R^{-1}, \forall\ a, b \in A$$

Hence, R^{-1} is a symmetric relation in A.

(*iii*) **Transitive :** Let $(a, b), (b, c) \in R$, then $(b, a), (c, b) \in R^{-1}$.

Also $\qquad (a, b), (b, c) \in R \Rightarrow (a, c) \in R \Rightarrow (c, a) \in R^{-1}$.

Thus, $\qquad (c, b) \in R^{-1}$ and $(b, a) \in R^{-1}$

$\Rightarrow\ (c, a) \in R^{-1}$.

Hence, R^{-1} is *a* transitive relation.

Since R^{-1} is reflexive, symmetric and transitive, therefore, R^{-1} is an equivalence relation in A.

Example 2.38. (*a*) Let R be the relation on $A = \{1, 2, 3\}$ defined by $R\{(1, 1), (1, 2), (2, 3)\}$. Then the reflexive closure of R is

$$R \cup \Delta = \{(1,1), (1,2), (2,3), (2,2), (3,3)\}$$

(*b*) Let $A = \{a, b, c\}$ defined by $R = \{(a, a), (a, b), (b, c), (c, c)\}$

Then the reflexive closure of A is

$$R \cup \Delta = \{(a,a), (a,b), (b,c), (c,c), (b,b)\}$$

Example 2.39. Let R be the relation on set $A = \{a, b, c, d\}$ defined by

$$R = \{(a, b), (b, c), (d, c), (d, a), (a, d), (d, d)\}.$$

Determine reflexive closure, symmetric closure and transitive closure of R.

Solution. $R = \{(a, b), (b, c), (d, c), (d, a), (a, d), (d, d)\}$

Reflexive closure is

$$R \cup \Delta = \{(a, a), (b, b), (c, c), (d, d), (a, b), (b, c), (d, c), (d, a), (a, d)\}$$

Symmetric closure is

$$R \cup R^{-1} = \{(a, b), (b, a), (b, c), (c, b), (d, c), (c, d), (d, a), (a, d), (d, d)\}$$

Transitive closure is

$$\{(a, a), (a, b), (a, c), (a, d), (b, c), (d, a), (d, b), (d, c), (d, d)\}.$$

Since, digraph of R is

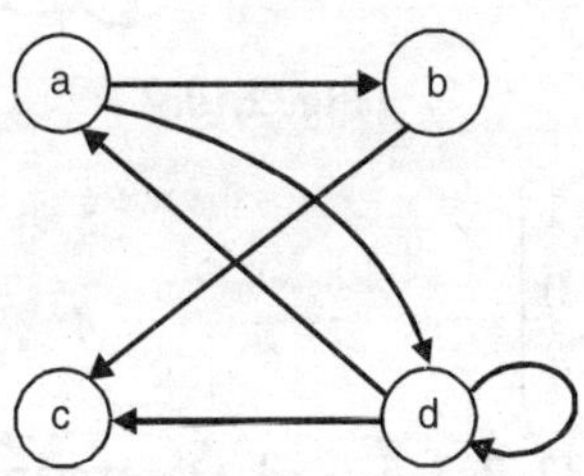

Fig. 2.7.

Example 2.40. Let $A = \{1, 2, 3, 4\}$ $R = \{(1, 1), (1, 2), (2, 1), (2, 2), (2, 3), (2, 4), (3, 4), (4, 1)\}$. Then the digraph of R is as shown in **Fig. 2.8.**

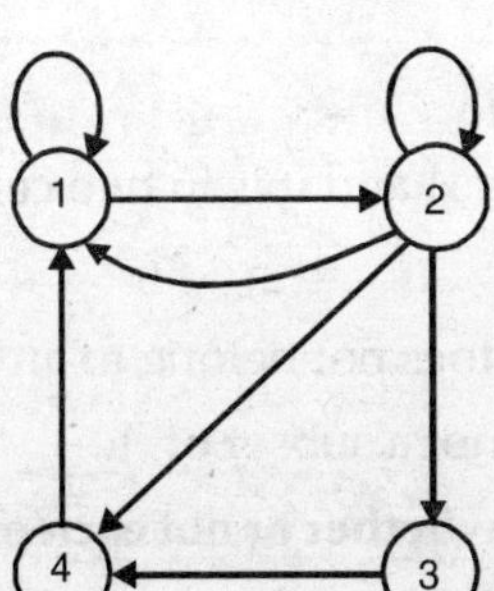

Fig. 2.8.

Example 2.41. Find the relation determined by following figure :

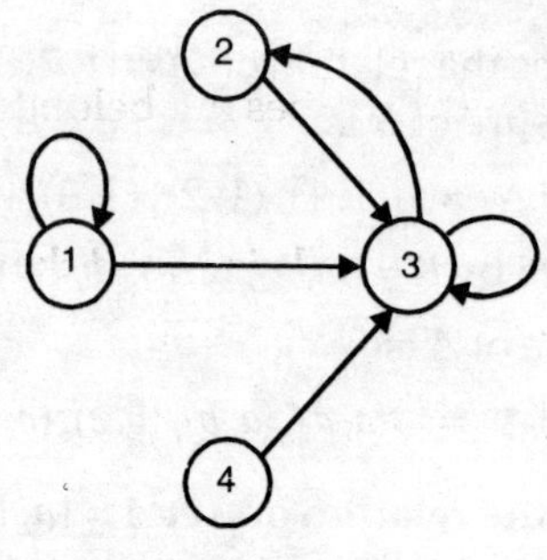

Fig. 2.9.

$$R = \{(1, 1), (1, 3), (3, 2), (2, 3), (3, 3), (4, 3)\}.$$

Example 2.42. Let $A = \{1, 4, 5\}$. Let R be given by the diagraph shown in Fig. 2.20. Find M_R.

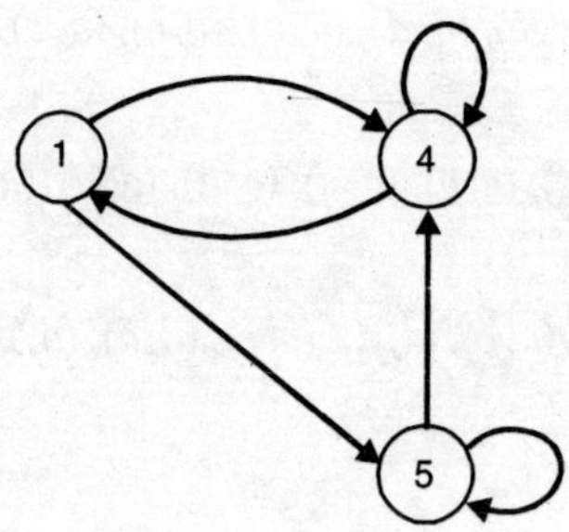

Fig. 2.10.

Solution. $M_R = \begin{bmatrix} 0 & 1 & 1 \\ 1 & 1 & 0 \\ 0 & 1 & 1 \end{bmatrix}$.

Example 2.43. Let $A = \{1, 2, 3, 4, 5, 6\}$. Determine whether or not each of the following is a partition of A:

(*i*) $A_1 = \{(1, 2, 3), (1, 4, 5, 6)\}$

(*ii*) $A_2 = \{(1, 3, 5), (2, 4), (6)\}$

(*iii*) $A_3 = \{(1, 2), (3, 5, 6)\}$

(*iv*) $A_4 = \{(1, 3, 5), (2, 4, 6, 7)\}$.

Solution. (*i*) No, since $1 \in A$ and it is in two cells.

(*ii*) Yes, it is a partition.

(*iii*) No, since $4 \in A$ but it does not belong to any cell.

(*iv*) No, since $\{2, 4, 6, 7\}$ is not a subset of A.

Example 2.44. Determine whether or not each of the following is a partition of the set N of positive integers.

(*i*) $[\{n : n > 5\}], \{n : n < 5\}$

(ii) $[\{n : n > 5\}], \{0\} \{(1, 2, 3, 4, 5)\}$.

(iii) $[(n : n^2 > 11) ; (n : n^2 < 11)]$.

Solution. (*i*) No, since $5 \in N$ but does not belong to any cell.

(*ii*) No, since $0 \notin N$.

(*iii*) Yes, the union of two given cells is N and they are disjoint.

Example 2.45. If $A = \{2, 3, 4\}$, $B = \{p, q\}$, then find $A \times B$ and $B \times A$ and represent them by tree diagram.

Solution.
$$A \times B = \{(2, p), (2, q), (3, p), (3, q), (4, p), (4, q)\}$$
$$B \times A = \{(p, 2), (p, 3), (p, 4), (q, 2), (q, 3), (q, 4)\}$$

Tree diagram :

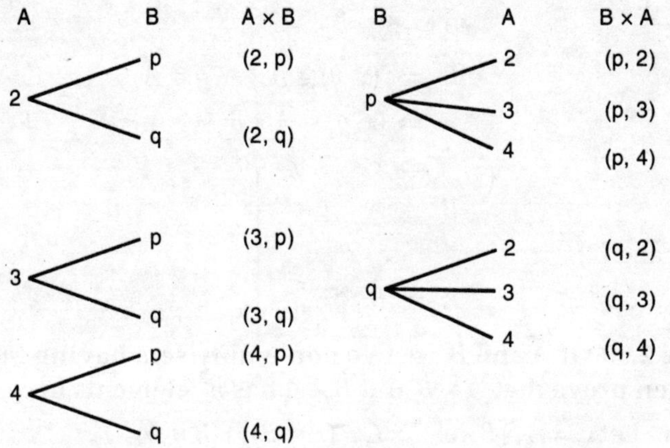

Fig. 2.11.

Example 2.46. If $A \times B = \{(3, 2), (3, 4), (5, 2), (5, 4)\}$, then determine the sets A and B.

Solution. $A = \{3, 5\}, B = \{2, 4\}$.

Example 2.47. If $A \subseteq B$ and $C \subseteq D$, then prove that $(A \times C) \subseteq (B \times D)$.

Solution. Let (a, c) be an arbitrary element of $A \times C$, then

$$\begin{aligned} (a, c) \in A \times C &\Rightarrow a \in A \text{ and } c \in C \\ &\Rightarrow a \in B \text{ and } c \in D \\ &\Rightarrow (a, c) \in B \times D \end{aligned}$$

So $(A \times C) \subseteq (B \times D)$.

Example 2.48. Prove that $A \times (B \cup C) = (A \times B) \cup (A \times C)$.

Solution. Let (x, y) be an arbitrary element of $A \times (B \cup C)$.

$$\begin{aligned} (x, y) \in A \times (B \cup C) &\Leftrightarrow x \in A \text{ and } y \in (B \cup C) \\ &\Leftrightarrow x \in A \text{ and } (y \in B \text{ or } y \in C) \\ &\Leftrightarrow (x \in A \text{ and } y \in B) \text{ or } (x \in A \text{ and } y \in C) \\ &\Leftrightarrow (x, y) \in A \times B \text{ or } (x, y) \in A \times C \end{aligned}$$

$$\Leftrightarrow (x, y) \in (A \times B) \cup (A \times C)$$

So $\quad A \times (B \cup C) = (A \times B) \cup (A \times C).$

Example 2.49. For non-empty sets A and B, prove that

$$\boldsymbol{A \times B = B \times A \Leftrightarrow A = B.}$$

Solution. Let $A = B$, then $A \times B = A \times A$

and $\quad B \times A = A \times A$

$\Rightarrow \quad A \times B = B \times A.$

Also, let $A \times B = B \times A$.

Let $\quad x \in A \Rightarrow (x, y) \in A \times B, y \in B$

$\Rightarrow (x, y) \in B \times A$

$\Rightarrow x \in B, y \in A$

So $\quad A \subseteq B.$ ...(1)

Similarly, $\quad x \in B \Rightarrow (x, y) \in B \times A, y \in A$

$\Rightarrow (x, y) \in A \times B$

$\Rightarrow x \in A, y \in B$

So $\quad B \subseteq A$ ...(2)

From (1) and (2),

$$A = B.$$

Example 2.50. If A and B are two non-empty sets having 'n' elements in common, then prove that $A \times B$ and $B \times A$ has n^2 elements in common.

Solution. Let $C = A \cap B$, let $C \times C = (A \times B) \cap (B \times A)$

Let $\quad (x, y) \in C \times C \Leftrightarrow x \in C$ and $y \in C$

$\Leftrightarrow x \in A \cap B$, and $y \in A \cap B$ [Since $C = A \cap B$]

$\Leftrightarrow (x \in A$ and $x \in B)$ and $(y \in A$ and $y \in B)$

$\Leftrightarrow x \in A$ and $y \in B$ and $x \in B$ and $y \in A$

$\Leftrightarrow (x, y) \in A \times B$ and $(x, y) \in B \times A$

$\Leftrightarrow (x, y) \in (A \times B) \cap (B \times A)$

So $\quad C \times C = (A \times B) \cap (B \times A).$

As $C \times C$ has n^2 elements. So $(A \times B) \cap (B \times A)$ also has n^2 elements.

$\Rightarrow \quad (A \times B)$ and $(B \times A)$ has n^2 elements in common.

Example 2.51. Let $A = \{1, 2, 3\}$, $B = \{2, 3, 4\}$, $S = \{1, 3, 4\}$, $T = \{2, 4, 5\}$ verify that

$$\boldsymbol{(A \times B) \cap (S \times T) = (A \cap S) \times (B \cap T).}$$

Solution.

$$A \times B = \{(1,2), (1,3), (1,4), (2,2), (2,3), (2,4), (3,2), (3,3), (3,4)\}$$

$$S \times T = \{(1,2), (1,4), (1,5), (3,2), (3,4), (3,5), (4,2), (4,4), (4,5)\}$$

$$A \cap S = \{1, 3\}$$

$$B \cap T = \{2, 4\}$$

$$(A \times B) \cap (S \times T) = \{(1, 2), (1, 4), (3, 2), (3, 4)\}$$

$$(A \cap S) \times (B \cap T) = \{(1, 2), (1, 4), (3, 2), (3, 4)\}$$

Hence, $(A \times B) \cap (S \times T) = (A \cap S) \times (B \cap T)$.

Example 2.52. Prove that $(A \cap B) \times (C \cap D) = (A \times C) \cap (B \times D)$.

Solution. Let, (x, y) be any element of $(A \cap B) \times (C \cap D)$.

Then $(x, y) \in (A \cap B) \times (C \cap D)$

$$\Leftrightarrow x \in A \cap B, y \in C \cap D$$
$$\Leftrightarrow x \in A \text{ and } x \in B \text{ and } y \in C \text{ and } y \in D$$
$$\Leftrightarrow x \in A \text{ and } y \in C \text{ and } x \in B \text{ and } y \in D$$
$$\Leftrightarrow (x, y) \in A \times C \text{ and } (x, y) \in (B \times D)$$
$$\Leftrightarrow (x, y) \in (A \times C) \cap (B \times D).$$

So $\quad (A \cap B) \times (C \cap D) = (A \times C) \cap (B \times D).$

Example 2.53. Prove that $(A - B) \times C = (A \times C) - (B \times C)$.

Solution. Let (x, y) be any arbitrary element of $(A - B) \times C$.

So $\quad (x, y) \in (A - B) \times C \Rightarrow x \in (A - B) \text{ and } y \in C$

$$\Rightarrow (x \in A \text{ and } x \notin B) \text{ and } y \in C$$
$$\Rightarrow x \in A \text{ and } y \in C \text{ and } x \notin B \text{ and } y \in C$$
$$\Rightarrow (x, y) \in (A \times C) \text{ and } (x, y) \notin (B \times C)$$
$$\Rightarrow (x, y) \in (A \times C) - (B \times C)$$

So $(A - B) \times C \subseteq (A \times C) - (B \times C)$ ...(1)

Let $(x, y) \in (A \times C) - (B \times C)$

$$\Rightarrow (x, y) \in (A \times C) \text{ and } (x, y) \notin (B \times C)$$
$$\Rightarrow x \in A \text{ and } y \in C \text{ and } (x \notin B \text{ and } y \in C)$$
$$\Rightarrow x \in A \text{ and } x \notin B \text{ and } y \in C$$
$$\Rightarrow x \in (A - B) \text{ and } y \in C$$
$$\Rightarrow (x, y) \in (A - B) \times C$$

So $\quad (A \times C) - (B \times C) \subseteq (A - B) \times C$ ...(2)

Clearly from (1) and (2), we have

$$(A - B) \times C = (A \times C) - (B \times C).$$

Example 2.54. For any two sets A and B, show that $A \times B$ and $B \times A$ have one element in common if A and B have one element in common.

Solution. Let $S = A \cap B$,

then $\quad S \times S = (A \times B) \cap (B \times A).$

Since A and B have one element in common. Therefore, S has only one element.

So $S \times S$ will have $1.1 = 1^2 = 1$ element.

$\Rightarrow (A \times B) \cap (B \times A)$ will have only one element.

$\Rightarrow (A \times B)$ and $(B \times A)$ will have only one element in common.

Conversely, let $(A \times B)$ and $(B \times A)$ have one element in common.

Hence, $n[(A \times B) \cap (B \times A)] = 1$

$\Rightarrow \quad S \times S = 1 = 1^2$

$\Rightarrow S = A \cap B$ has only one element

$\Rightarrow A$ and B have only one element in common.

Hence proved.

Example 2.55. Find the element of the set $A \times B \times C$, where $A = \{a, b, c\}$, $B = \{d, e\}$, $C = \{f, g, h\}$.

Solution. $A \times B \times C$ will consist of $3 \times 2 \times 3 = 18$ ordered triplets.

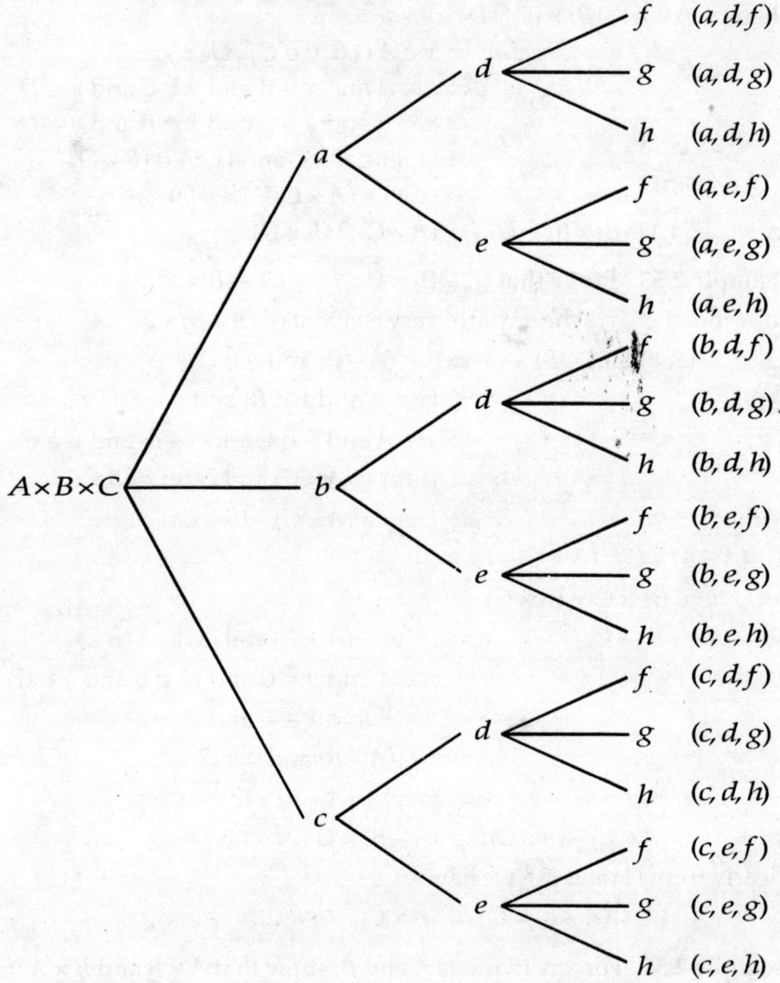

Fig. 2.12.

Tree diagram representing 18 ordered triplets of $A \times B \times C$.

Example 2.56. In $N \times N$, show that the relation defined by $(a, b) R (c, d)$ if $ad = bc$ is an equivalence relation.

Solution. (*i*) Given, $(a, b) R (c, d)$ if $ad = bc$

So $(a, b) R (a, b)$ if $ab = ba$

But in set of natural number, $ab = ba$

i.e., commutative under multiplication is true.

Hence R is reflexive.

(*ii*) Let $(a, b) R (c, d) \Rightarrow ad = bc$

$$\Rightarrow \qquad cb = da$$

$\Rightarrow$ $(c, d)\, R\, (a, b)$.

Hence R is symmetric.

(*iii*) $(a, b)\, R\, (c, d)$ and $(c, d)\, R\, (e, f)$

$$\Rightarrow ad = bc \quad \text{and} \quad cf = de$$
$$\Rightarrow (ad)\,(cf) = (bc)\,(de)$$
$$\Rightarrow adcf = bcde$$
$$\Rightarrow af\,(dc) = be\,(dc)$$
$$\Rightarrow af = be$$
$$\Rightarrow (a, b)\, R\, (e, f)$$

This R is transitive.

Since R is reflexive, symmetric and transitive.

So R is an equivalence relation.

Example 2.57. Show that the relation $R = \{(a, b) : a - b = \text{even integer and } a, b \in I\}$, i.e., $aRb \Rightarrow a - b = \text{even integer}$, is an equivalence relation.

Solution. (*i*) Let $a \in I$, then $a - a = 0$, which is an even integer, therefore,

$$(a, a) \in R \text{ i.e., } aRa.$$

Hence, R is reflexive.

(*ii*) Let $a, b \in I$. If $a - b$ is an even integer, then $b - a$ is also an even integer, therefore, $aRb \Rightarrow bRa$.

i.e., $$(a, b) \in R \;\Rightarrow (b, a) \in R$$

Hence, R is symmetric relation.

(*iii*) Let $a, b, c \in I$ if aRb, bRc then $a - b$ and $b - c$ both are even integers.

Now $$a - c \;=\; a - b + b - c.$$

Sum of two even integers is also an even integer, therefore, $(a - c)$ is also an even integer. Hence aRc *i.e.,* $(a, c) \in R$.

Since R is reflexive, symmetric and transitive.

So R is an equivalence relation.

Example 2.58. In a set L of all straight lines in a plane discuss which of the following two relations are equivalence relations ?

(*i*) $R_1 = \{(x, y) : x, y \in L$ and x is parallel to $y\}$

(*ii*) $R_2 = \{(x, y) : x, y \in L$ and x is perpendicular to $y\}$.

Solution. (*i*) (*a*) Reflexive : $\forall\, x \in L$, x is parallel to itself, hence $(x, x) \in R$ and hence R_1 is reflexive.

(*b*) **Symmetric :** For $x, y \in L$

$$(x, y) \in R_1 \;\Rightarrow x \parallel y$$
$$\Rightarrow y \parallel x$$
$$\Rightarrow (y, x) \in R_1$$

Hence R_1 is symmetric.

(c) **Transitive :** For $x, y_1 \in L$

$(x, y) \in R_1, (y, z) \in R_1$

$$\Rightarrow x \parallel y \text{ and } y \parallel z$$
$$\Rightarrow x \parallel z$$
$$\Rightarrow (x, z) \in R.$$

Hence, R_1 is transitive.

Since R_1 is reflexive, symmetric and transitive hence it is an equivalence relation.

(ii) $\forall\, x \in L$, we have, x is not perpendicular to itself, hence $(x, x) \notin R$.

So R_2 is not a reflexive relation.

Hence, R_2 is not an equivalence relation.

Example 2.59. In the set of all integers, a relation is defined by $a \equiv b \pmod{m}$ if $m \mid (a - b)$, i.e., m divides $(a - b)$, m being a positive integer. Show that this relation is an equivalence relation.

Solution. Let I be the set of all integers.

(i) **Reflexive :** $a \equiv a \pmod{m}$

Since $m \mid (a - a)$ i.e., $m \mid 0$ [m divides 0]

So the relation $\equiv$ is reflexive.

(ii) **Symmetric :** Let $a, b \in I$, s.t.

$a \equiv b \pmod{m}$ i.e., $m \mid a - b$ [m divides $a - b$]

Now $b - a = -(a - b)$.

Also m divides $a - b$, So m divides $-(a - b)$, i.e., m divides $b - a$.

Hence $b \equiv a \pmod{m}$

So Relation $\equiv$ is symmetric.

(iii) **Transitive :** Let $a \equiv b \pmod{m}$, $b \equiv c \pmod{m}$ i..e, m [$(a - b)$ and $m \mid (a - c)$.

Now, $(a - c) = a - b + b - c$

Hence, m divides $a - c$ or $a \equiv c \pmod{m}$

So Relation '$\equiv$' is transitive.

Since in the set I, the relation is reflexive, symmetric and transitive and therefore '$\equiv$' is an equivalence relation.

Example 2.60. If I be a set of non-zero integers, and a relation R is $xRy \Rightarrow x^y = y^x$, $x, y \Leftrightarrow I$, then show that R is not an equivalence relation in I.

Solution. Reflexive : Let $x \in I$, then

$$x^x = x^x \Rightarrow xRx \;\forall\, x \in I$$

$\Rightarrow$ R is reflexive.

Symmetric : Let $x, y \in I$ s.t. xRy i.e., $x^y = y^x$

Since $x^y = y^x \Rightarrow y^x = x^y$.

So $xRy \Rightarrow yRx \;\forall\, x, y \in I$

$$\Rightarrow R \text{ is symmetric.}$$

Transitive : Let x, y, z be three elements of I s.t.

$$xRy,\ yRz \quad i.e.,\quad x^y = y^x,\ y^z = z^y$$

which will not establish,

$$x^z = z^x$$

i.e., $xRy, yRz \neq xRz$.

$\Rightarrow$ R is not transitive. Therefore, R is not an equivalence relation.

Example 2.61. The relation R on the set S of all real numbers is defined as aRb iff $(1 + ab) > 0$. Show that this relation is reflexive and symmetric but not transitive.

Solution. (*i*) **Reflexive :** Let 'a' be any real number.

Then $\quad 1 + aa = 1 + a^2 > 0 \qquad [\geq a^2 \geq 0]$

So aRa for all $a \in S$.

Thus R is reflexive.

(*ii*) **Symmetric :** Let $a, b \in S$, then

$$aRb \Rightarrow 1 + ab > 0$$
$$\Rightarrow 1 + ba > 0$$
$$\Rightarrow bRa.$$

Thus, R is symmetric.

(*iii*) **Transitive :** Consider three real numbers as

$$1, -\frac{1}{4} \text{ and } -4.$$

$$1 + 1\left(-\frac{1}{4}\right) = \frac{3}{4} > 0 \qquad \text{so } 1R\left(-\frac{1}{4}\right)$$

$$1 + \left(-\frac{1}{4}\right)(-4) = 2 > 0 \qquad \text{so } \left(-\frac{1}{4}\right)R(-4)$$

But $1 + 1\,(-4) = -3 < 0$. So 1 is not related to -4.

Hence R is not transitive.

Example 2.62. Let $A = \{2, 4, 6, 8\}$ and $R = \{(2, 2), (2, 8), (4, 4), (4, 6), (6, 8)\}$ is a relation on set A. Determine transitive closure of R.

Solution. $R^* = \{(2, 2), (2, 8), (4, 4), (4, 6), (6, 8)\}$ is the transitive closure.

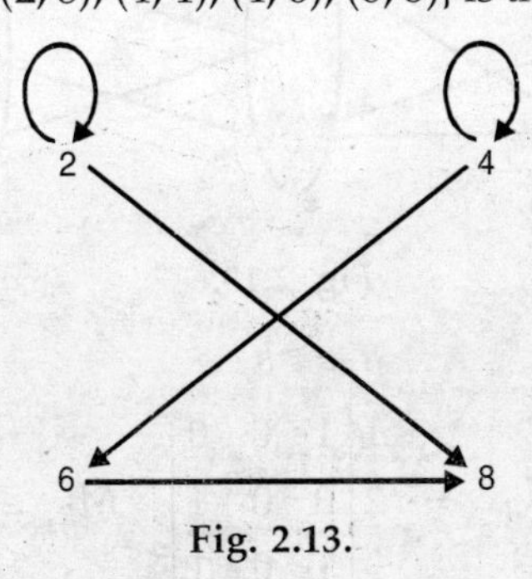

Fig. 2.13.

Example 2.63. Let R be a relation on set $A = \{a, b, c, d, e\}$ whose matrix is given below :

$$M_R = \begin{bmatrix} 1 & 1 & 1 & 1 & 1 \\ 0 & 0 & 0 & 0 & 0 \\ 1 & 1 & 0 & 0 & 0 \\ 0 & 0 & 0 & 0 & 0 \\ 1 & 1 & 1 & 1 & 1 \end{bmatrix}$$

Find the directed graph of relation R.

Solution. Directed graph of given relation R is :

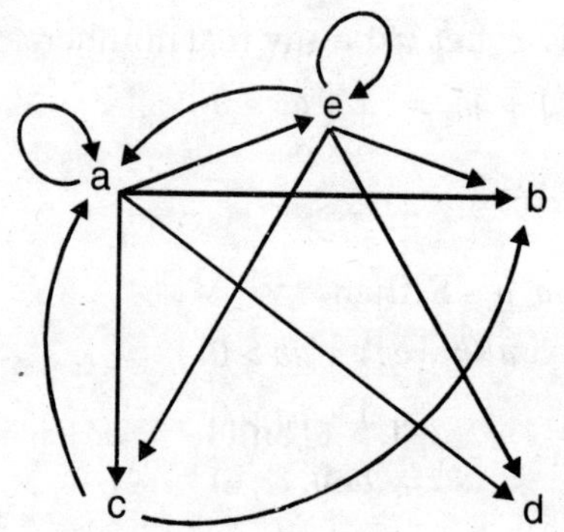

Fig. 2.14.

Example 2.64. Let $A = \{1, 2, 3\}$, $B = \{a, b, c\}$ and $C = \{x, y, z\}$ and let

$R_1 = \{(1, a), (2, c), (3, a), (3, c)\}$ and $R_2 = \{(b, x), (b, z), (c, y)\}$

(a) Find the composition relation $R_1 \, 0 \, R_2$.

(b) Find the matrices M_{R_1}, M_{R_2}, $M_{(R_1 0 R_2)}$ of respective relations R_1, R_2 and $R_1 \, 0 \, R_2$.

Solution. (a) $R_1 \, 0 \, R_2 = \{(2, y), (3, y)\}$.

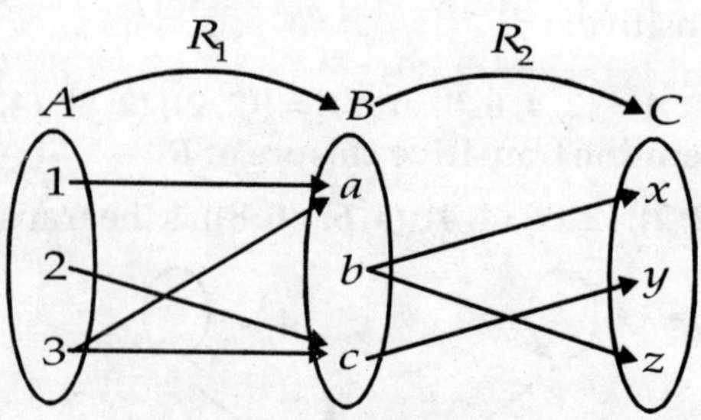

Fig. 2.15.

(b)
$$M_{R_1} = \begin{array}{c} \\ 1 \\ 2 \\ 3 \end{array}\begin{array}{c} \begin{array}{ccc} a & b & c \end{array} \\ \begin{bmatrix} 1 & 0 & 0 \\ 0 & 0 & 1 \\ 1 & 0 & 1 \end{bmatrix} \end{array}, \quad M_{R_2} = \begin{array}{c} \\ a \\ b \\ c \end{array}\begin{array}{c} \begin{array}{ccc} x & y & z \end{array} \\ \begin{bmatrix} 0 & 0 & 0 \\ 1 & 0 & 1 \\ 0 & 1 & 0 \end{bmatrix} \end{array}$$

$$M_{R_1 0 R_2} = \begin{array}{c} \\ 1 \\ 2 \\ 3 \end{array}\begin{array}{c} x \quad y \quad z \\ \begin{bmatrix} 0 & 0 & 0 \\ 0 & 1 & 0 \\ 0 & 1 & 0 \end{bmatrix} \end{array}$$

Example 2.65. Let I be the set of all integers and $R = \{(a, b); a, b \in I$ and $(a - b)$ is divisible by 3$\}$ is an equivalence relation. Find equivalence classes of I.

Solution. Given $I = \{-4, -3, -2, -1, 0, 1, 2, 3, 4, ...\}$ is the set of integers.

Also $\forall\, x, y \in I, xRy$ if $(x - y)$ is divisible by 3.

i.e., $x\,R\,y$ if $(x - y)/3 = m, m \in I$

or $x \equiv y \pmod 3$, then equivalence classes will be,

$$I_0 = \{..., -9, -6, -3, 0, 3, 6, 9, ...\}$$

$$I_1 = \{..., -8, -5, -2, 1, 4, 7, 10, ...\}$$

$$I_2 = \{...., 4, -1, 2, 5, 8, ...\}$$

Hence, equivalence classes are [0], [1] and [2].

Example 2.66. Let $A = \{1, 2, 3, 4, 5, 6\}$ and let R be the equivalence relation on A defined by $R = \{(1, 1)\ (1, 5), (2, 2), (2, 3), (2, 6), (3, 2), (3, 3), (3, 6), (4, 4), (5, 1), (5, 5), (6, 2), (6, 3), (6, 6)\}$.

Find the partition of A induced by R. *i.e.*, find the equivalence classes of R.

Solution.

$$[1] = \{1, 5\} = [5]$$

$$[2] = \{2, 3, 6\} = [3] = [6]$$

$$[4] = \{4\}$$

So $[\{1, 5\}, (2, 3, 6), \{4\}]$ is the partition of A induced by R.

Example 2.67. Let $A = \{1, 2, 3, ..., 14, 15\}$. Consider the equivalence relation on $A \times A$ defined by

$$(a, b) \sim (c, d) \text{ if } a + d = b + c.$$

Find the equivalence class of (2, 11).

Solution. To find equivalence class of (2, 11), we have to find all (m, n) pairs *s.t.* $(2, 11) \sim (m, n)$

and $\quad 2 + n = 11 + m$

$\Rightarrow \quad n = 9 + m, \quad (m = 1, 2, ...)$

So $\quad [(2, 11)] = \{(1, 10, (2, 11), (3, 12), (4, 13), (5, 14), (6, 15)\}$

Example 2.68. Consider a relation R from a set A to B. The complement of relation R denoted by $\overline{R}$ is a relation from A to B *s.t.*

$$\overline{R} = \{(a, b) : (a, b) \notin R\}.$$

Let $\quad A = \{1, 2, 3\}, B = \{4, 5\}.$

$$R = \{(1, 4), (2, 4), (1, 5), (3, 5)\}.$$

Find the complement of relation *R*.

Solution. Universal relation $A \times B$ is $A \times B = \{(1, 4), (1, 5), (2, 4), (2, 5), (3, 4), (3, 5)\}$.

So $\overline{R}$ w.r.t. $A \times B$ is given by

$$\overline{R} = \{(2, 5), (3, 4)\}.$$

Example 2.69. Consider *a* relation *R* from a set *A* to *B* given by following matrix.

$$M_R = \begin{array}{c} \\ 1 \\ 2 \\ 3 \\ 4 \end{array}\!\!\begin{array}{c} \begin{array}{cccc} 1 & 2 & 3 & 4 \end{array} \\ \begin{bmatrix} 1 & 0 & 0 & 1 \\ 1 & 1 & 1 & 1 \\ 0 & 0 & 1 & 1 \\ 1 & 1 & 0 & 0 \end{bmatrix} \end{array}$$

Find complement of *R*.

Solution. The complement of *R i.e.*,

$$\overline{R} = \begin{array}{c} \\ 1 \\ 2 \\ 3 \\ 4 \end{array}\!\!\begin{array}{c} \begin{array}{cccc} 1 & 2 & 3 & 4 \end{array} \\ \begin{bmatrix} 0 & 1 & 1 & 0 \\ 0 & 0 & 0 & 0 \\ 1 & 1 & 0 & 0 \\ 0 & 0 & 1 & 1 \end{bmatrix} \end{array}$$

It is obtained by interchanging 0′s with 1′s and 1′s with 0′s.

So $$\overline{R} = \{(1, 2), (1, 3), (3, 1), (3, 2), (4, 3), (4, 4)\}.$$

EXERCISES

1. List the ordered pairs in the relation R from $A = \{0, 1, 2, 3\}$ to $B = \{0, 1, 2, 3, 4\}$ where $(a, b) \in R$ if and only if
   (*a*) $a > b$. (*b*) $a + b = 3$.
   (*c*) a divides b. (*d*) $a - b = 0$
   (*e*) gcd $(a, b) = 1$ (*f*) lcm $(a, b) = 6$.
2. Recursively define the relation $\{(a, b) \mid a = 2b\}$, where a and b are natural numbers.
3. List unary relation on $\{1, 2, 3\}$.
4. Prove that there are $2n^2$ binary relations on a set of cardinality n.
5. For each of the following relations on the set $\{1, 2, 3, 4\}$, decide whether it is reflexive, symmetric, antisymmetric and/or transitive.
   (*a*) $\{(1, 1), (1, 2), (2, 1), (2, 2), (3, 3), (4, 4)\}$
   (*b*) $\{(1, 3), (1, 4), (2, 3), (3, 4)\}$
   (*c*) $\{(1, 1), (1, 3), (1, 4), (2, 1), (2, 3), (2, 4), (3, 1), (3, 3), (3, 4)\}$.
6. Determine whether the relation R on the set of all integers is reflexive, symmetric, antisymmetric, and/or transitive,
   where $(x, y) \in R$ if and only if

(*a*) x is divisible by y. (*b*) $x \neq y$

(*c*) $y = x + 2$ or $y = x - 2$. (*d*) $x = y^2 + 1$.

7. Let A be the set of people in your town. Let $R1$ be the unary relation representing the people in your town who were registered in the last election and $R2$ be the unary relation representing the people in your town who voted in the last election. Describe the 1-tuples in each of the following relations.

(*a*) $R1 \cup R2$.

(*b*) $R1 \cap R2$.

8. Draw the directed graph that represents the relation $\{(a, b), (a, c), (b, c), (c, b), (c, c), (c, d), (d, a), (d, b)\}$.

9. Let R be the parent-child relation on the set of people that is, $R = \{(a, b) \mid a$ is a parent of $b\}$. Let S be the sibling relation on the set of people that is, $R = \{(a, b) \mid a$ and b are siblings (brothers or sisters)$\}$. What are $S°R$ and $R°S$?

10. Let R be a reflexive relation on a set A. Show that R^n is reflexive for all positive integers n.

11. Let R be the relation on the set $\{1, 2, 3, 4\}$ containing the ordered pairs (1, 1), (1, 2), (2, 2), (2, 4), (3, 4) and (4, 1). Find.

(*a*) the reflexive closure of R,

(*b*) symmetric closure of R and

(*c*) transitive closure of R.

12. Let R be the relation $\{(a, b) \mid a$ is a (integer) multiple of $b\}$ on the set of integers. What is the symmetric closure of R ?

13. Suppose that a binary relation R on a set A is reflexive. Show that R^* is reflexive, where $R^* = \bigcup_{i=1}^{n} R^i$.

14. Suppose that A is a non-empty set, and f is a function that has A as its domain. Let R be the relation on A consisting of all ordered pairs (x, y) where $f(x) = f(y)$.

(*a*) Show that R is an equivalence relation on A.

(*b*) What are the equivalence classes of R ?

15. Which of the following collections of subsets are partitions of $\{1, 2, 3, 4, 5, 6\}$?

(*a*) $\{1, 2, 3\}, \{3, 4\}, \{4, 5, 6\}$

(*b*) $\{1, 2, 6\}, \{3, 5\}, \{4\}$

(*c*) $\{2, 4, 6\}, \{1, 5\}$

(*d*) $\{1, 4, 5\}, \{2, 3, 6\}$.

16. Consider the equivalence relation on the set of integers $R = \{(x, y) \mid x - y$ is an integer$\}$.

(*a*) What is the equivalence class of 1 for this equivalence relation ?

(*b*) What is the equivalence class of 0.3 for this equivalence relation ?

17. Which of the following are posets ?

(*a*) (Z, =)

(*b*) (Z, ≠)

(*c*) A collection of sets, $\subseteq$).

18. Draw the Hasse diagram for the divisibility relation on the following sets
    (*a*) {1, 2, 3, 4, 5, 6, 7, 8, 9, 10}
    (*b*) {1, 2, 5, 8, 16, 32}.
19. Answer the following questions concerning the poset ({1}, {2}, {3}, {4}, {1, 3}, {1, 4}, {2, 4}, {3, 4}, {1, 2, 4}, {2, 3, 4}, ⊆).
    (*a*) Find the maximal elements.
    (*b*) Find the minimal elements.
    (*c*) Is there *a* greatest element ?
    (*d*) Is there a least element ?
    (*e*) Find all upper bounds of {{2}, {4}}.
    (*f*) Find the least upper bound of {{2}, {4}}, if it exists.
    (*g*) Find all lower bounds of {{1, 2, 4}, {2, 3, 4}}.
    (*h*) Find the greatest lower bound of {{1, 2, 4}, {2, 3, 4}}, if it exists.

ANSWERS

1. (*a*) {(1, 0,), (2, 0), (2, 1), (3, 0), (3, 1),(3, 2)}
   (*b*) {(0, 3), (1, 2), (2, 1), (3, 0)}
   (*c*) {(1, 0), (1, 1), (1, 2), (1, 3), (1, 4), (2, 0),(2, 2),(2, 4), (3, 0), (3, 3)}.
   (*d*) {(0, 0), (1, 1), (2, 2), (3, 3)}
   (*e*) {(0, 1), (1, 0), (1, 1), (1, 2), (1, 3), (1, 4), (2, 1), (2, 3), (3, 1), (3, 2), (3, 4)}.
   (*f*) {(2, 3), (3, 2)}.
2. Let *R* denote the relation.

   Basis Clause. $(0, 0) \in R$.

   Inductive Clause. For any natural numbers x and y, if $(x, y) \in R$, then $(x + 2, y + 1) \in R$.

   Extremal Clause. Nothing is in *R* unless it is obtained by the Basis and Inductive clauses.
3. A unary relation is a set of 1-tuples. Hence there are eight unary relations on {1, 2, 3}. The are Ø, {(1)}, {(2)}, {(3)}, {(1), (2)}, {(1), (3)}, {(2), (3)}, {(1), (2), (3)}.
4. A binary relation on a set is a subset of the Cartesian product of the set with itself. If the cardinality of the set is n, then the cardinality of the Cartesian product of the set with itself is n^2. Thus the power set of the Cartesian product has 2^{n2} elements. Hence there are 2^{n2} binary relations on a set of cardinality n.
5. (*a*) Reflexive, symmetric, transitive
   (*b*) Antisymmetric
   (*c*) Transitive.
6. (*a*) Reflexive, antisymmetric, transitive
   (*b*) Symmetric
   (*c*) Symmetric
   (*d*) Antisymmetric.
7. (*a*) {(*a*) | *a* lives in your town, and was registered for vote (or voted) in the last election.}

   (*b*) {(*a*) | a lives in your town, and (was registered for vote) and voted in the last election.}

8. $S^\circ R = \{(a, b) \mid a \text{ is a parent of } b \text{ and } b \text{ has a sibling}\}$, $R^\circ S = \{(a, b) \mid a \text{ is an aunt or uncle of } b\}$.

9. Proof by mathematical induction on n.

Basis Step. Since $R' = R$, R' is reflexive.

Inductive Step. Assume R'' is reflexive. Then for any $a \in A$, $(a, a) \in R^n$ and $(a, a) \in R$. Since $R^{n+1} = R^n \circ R$, $(a, a) \in R^{n\circ} R = R^{n+1}$. Hence R^{n+1} is reflexive.

10. (*a*) $\{(1, 1), (1, 2), (2, 2), (2, 4), (3, 3), (3, 4), (4, 1), (4, 4)\}$

(*b*) $\{(1, 1), (1, 2), (1, 4), (2, 1), (2, 2), (2, 4), (3, 4), (4, 1), (4, 2), (4, 3)\}$

(*c*) $\{(1, 1), (1, 2), (1, 4), (2, 1), (2, 2), (2, 4), (3, 1), (3, 2), (3, 4), (4, 1), (4, 2), (4, 4)\}$.

11. $\{(a, b) \mid a \text{ is a multiple of } b \text{ or } b \text{ is a multiple of } a\}$.

12. Since R' hence R is a subset of R^* and since R is reflexive, all the ordered pairs of form (a, a) are in R^*. Hence R^* is reflexive.

13. (*a*) Equivalence relation.

(*b*) Equivalence relation.

(*c*) Not symmetric, not transitive.

14. (*a*) $(x, x) \in R$ since $f(x) = f(x)$. Hence R is reflexive.

$(x, y) \in R$ if and only if $f(x) = f(y)$, which holds if and only if $f(y) = f(x)$. Hence $(x, y) \in R$ if and only if $(y, x) \in R$. Hence R is symmetric. If $(x, y) \in R$ and $(y, z) \in R$, then $f(x) = f(y)$ and $f(y) = f(z)$. Hence $f(x) = f(z)$. Thus $(x, z) \in R$. Hence R is transitive.

(*b*) The set $f^{-1}(b)$ for b in the range of f. That is the set of all elements of A that are mapped to the same element of the range by f.

15. (*a*) No

(*b*) Yes

(*c*) No

(*d*) Yes.

16. (*a*) The set of integers

(*b*) $\{n + 0.3 \mid n \in s$

17. (*a*) yes

(*b*) No

(*c*) Yes.

18. (*a*)

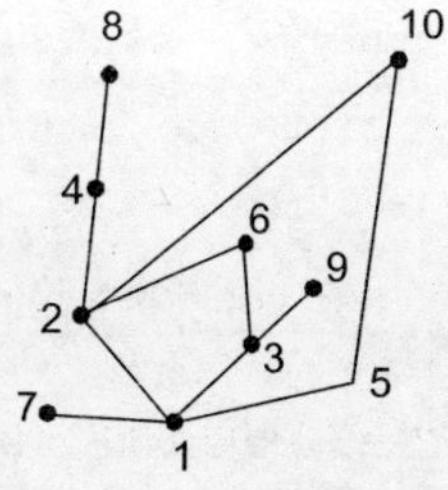

(*b*)

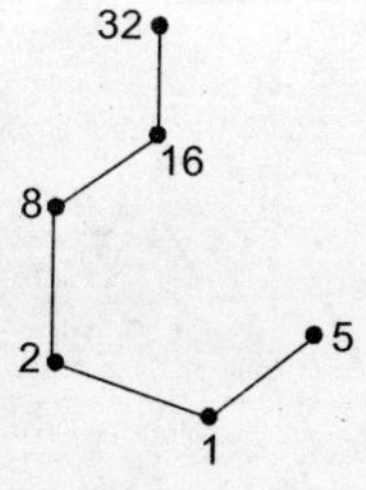

19. (*a*) {1, 3}, {1, 2, 4}, {2, 3, 4}

(*b*) {1}, {2}, {3}, {4}

(*c*) No

(*d*) No

(*e*) {2, 4}, {1, 2, 4}, {2, 3, 4}

(*f*) {2, 4}

(*g*) {2, 4}, {2}, {4}

(*h*) {2, 4}.

■ ■ ■

Chapter

3 FUNCTION

3.1. FUNCTIONS

Function or mapping is a special type of relation. It is an important concept in the field of mathematics and have numerous applications in camputer science. In this section we will discuss some basic properties of functions, their types and applications

3.2 DEFINITION

Consider two non-empty sets A and B. Then a rule 'f' which associates each element of set A to a unique, well defined element of set B is called a function.

Symbolically function is denoted by

$$f : A \longrightarrow B$$

which reads as "f is a function from A to B". Let $a \in A$, then rule f associates 'a' to a unique element $b \in B$ such that

$$f(a) = b$$

element b is called the **image** of a under f.

3.3 DOMAIN AND RANGE

Set A is called the *domain* of function 'f' *i.e.*,

$$Dom(f) \text{ or } D = \{a : a \in A\} = A$$

The range or rangest set R of a function f is the collection of all image values. That is

$$R = \{b : b \in B \text{ and } b = f(a),\ \forall\ a \in A\}$$

Clearly $R \subseteq B$

Note : Each element of set A should have an image in B. Whereas each element of B need not to be the image of an element of set A.

Example 3.1. Consider a function $f : Z \rightarrow Z$ such that

$$F(x) = 2x^2, \qquad \forall\ x \in Z$$

Find out the demain and range of f.

Solution : Clearly $D = Z = \{0, \pm 1, \pm 2, \pm 3, ...\}$

Range $R = \{y : y \in z \text{ such that } y = f(x),\ \forall\ x \in z\}$

$$= \{y : y = f(x) = 2x^2, \forall\ x \in \{0, \pm 1, \pm 2, ...\}\}$$
$$= \{y : y = 2x^2, \forall\ x \in (0, \pm 1, \pm 2, ...)\}$$
$$= \{0, 2(\pm 1)^2, 2(\pm 2)^2, 2(\pm 3)^2, ...\}$$
$$= \{0, 2, 8, 18, 32, ...\}$$

3.4 ANOTHER DEFINITION OF FUNCTION (AS A SET OF ORDERED PAIR)

Let A and B be two non-empty sets than a function or mapping f defined from A to B is a subset of $A \times B$ which safisfied the following conditions.

(a) $\forall\ a \in A, (a, b) \in f$ for some $b \in B$.

(b) If $(a, b) \in f$ and $(a, c) \in f$ then $b = c$

i.e., there is one and only one image possible for any element of domain set or every element of domain set A has a unique image in co-domain set B.

3.5 FUNCTIONS AS RELATION

By the definition of function, we know that $f: A \to B$ is a subset of $A \times B$ such that.

(*i*) $\forall\ a \in A, f(a, b) \in f$ for some $b \in B$

(*ii*) If $(a, b) \in f$ and $(a, c) \in f \Rightarrow b = c$.

Clearly in function the entire domain set is mapped to some elements of co-domain set uniquely. Whereas in case of relation we may have an element $a \in$ A which is not related to any element of set B. And also in relation an element say $x \in A$ can be related with more than one element of set B *i.e.*, in relation $(x, y_1) \in R$ is possible as well as $(x, y_2) \in R$ is also possible.

Thus mean say that every function is a relation but the reverse in not true.

Example 3.2. Check which of the followings are functions.

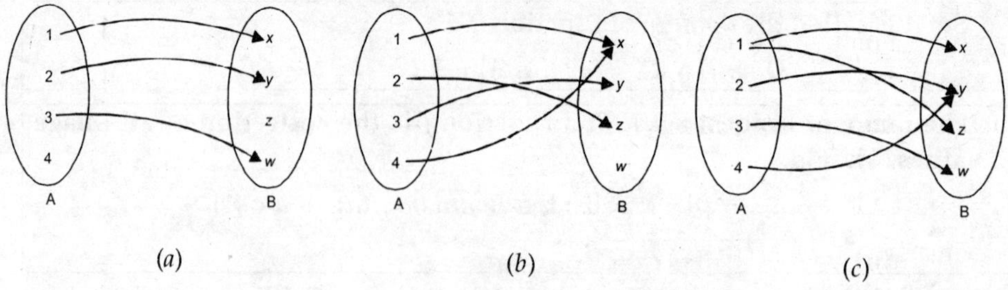

Fig. 3.1

Solution : (*a*) It is not a function as the image of element 4 does not exist. (*b*) It is a function. (*c*) It is not a function as element 1 has two image x and z. Here one can observe that Fig. (*a*) and (*c*) represents a relation but not function whereas **Fig. 3.1** (*b*) represents both relation and function.

3.6 TRANSFORMATION (OPERATOR)

If f is a function from set A to itself *i.e.*, $f: A \to A$, then f is called transformation or operator on A. If A is a finite set, then f is called a permutation on A.

3.7 EQUALITY OF TWO FUNCTIONS

Two functions say f_1 and f_2 defined on sets A and B *i.e.*, $A \to B$ and $f_2 : A \to B$ are said to be equal if and only if

$$\forall\ a \in A, f_1(a) = f_2(a)$$

and then we write $\qquad f_1 = f_2$

If there exist at least one element say $a_i \in A$ such that

$$f_1(a_i) \neq f_2(a_i)$$

then we say $\qquad f_1 \neq f_2$

and the functions are said to be unequal functions.

3.8 DIFFERENT TYPES OF FUNCTION

3.8.1 Into and Onto Function

Consider two non-empty sets A and B. Then a function $f : A \to B$ is called an Into function if there exist at least one element in set B (or co-demain) which is not the f-image of any element of set A (or domain).

In other words, f is said to be **Into** if at least one element $b \in B$ does not have any pre-image $a \in A$. In this case we can say that the range set R of f is a proper subset of co-domain set *i.e.*, $R \subset B$.

For example, $f : Z \to Z$ such that $f(x) = x^2$ is an Into function, where Z is the set of integers because the range set of f will be

$$R(f) = \{x^2 : x \in (0, \pm 1, \pm 2, ...)\} = \{0, 1, 4, 9, ...\} \cdot \subset Z$$

A function $f : A \to B$ is said to be onto if all element of set B (co-domain) have f-image in set A (Domain). That is, for every $b \in B$, $\exists$ same $a \in A$ such that $f(a) = b$

In this case the range of f is equal to the co-domain of f *i.e.*, $f(A) = B$.

For example, the function $f : A \to B$, as $f(x) = x^2$, where $A = \{-2, -1, 0, 1, 2\}$ and $B = \{0, 1, 4\}$ is an onto function. The pictorial representation (arrow diagram) of onto function is given below.

Clearly $\qquad R(f) = B$

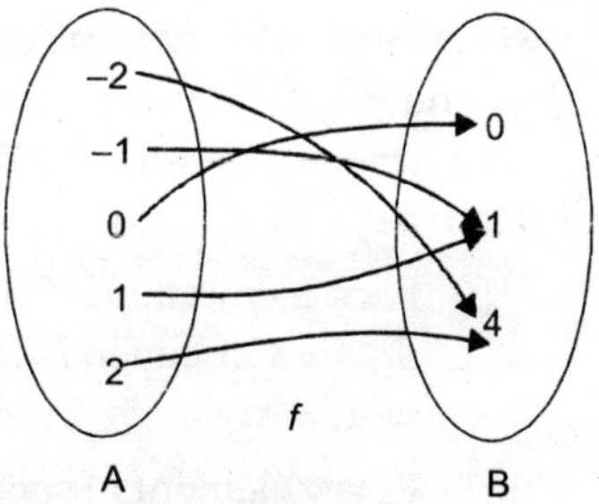

Fig. 3.2

3.8.2 One-to-One and Many-one Functions

Consider two non-empty sets A and B.

(*a*) A function $f : A \to B$ is called one-to-one or simply one-one function if every inique element in set A has only one unique image in set B, *i.e.*, $f : A \to B$ is one-one iff $f(a_1) = f(a_2)$

$$\Rightarrow \quad a_1 = a_2 \qquad \forall\ a_1, a_2 \in A$$

or $\qquad a_1 \neq a_2 \qquad \Rightarrow f(a_1) \neq f(a_2)$

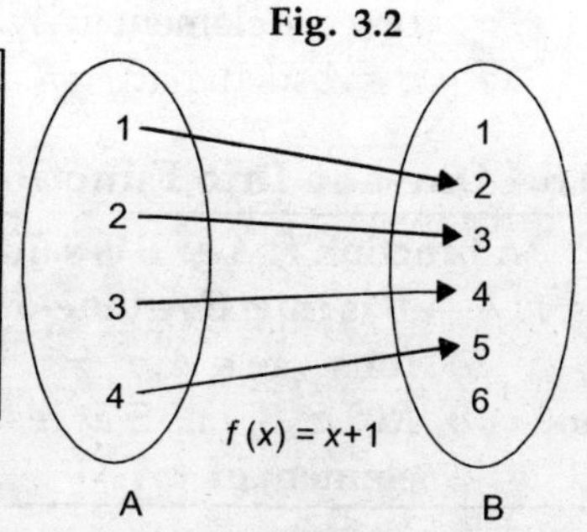

Fig. 3.3

For example, $f : A \to B$ defined as $f(x) = x + 1$ where $A = \{1, 2, 3, 4\}$ and $B = \{1, 2, 3, 4, 5, 6\}$ is a one-to-one function. The arrow diagram of f is given in **Fig. 3.3**

(b) A Function $f: A \to B$ is said to many one if two or more than two elements of set A has same image in set B. In other words. $f: A \to B$ is many one if one element of set B has two or more than two pre images in set A.

i.e., $f: A \to B$ **is many one if**

$$f(a_1) = f(a_2) \text{ where } a_1 \neq a_2$$

For example, $f: Z \to Z$ defined as $f(x) = x^2$ is a many-one function. (Here Z is the set of integers).

As $f(\pm a) = a^2; \quad \pm a \in Z$

Pictorial representation in given in the **Fig. 3.4.**

$$f(\pm 1) = 1$$
$$f(\pm 2) = 4$$
$$f(\pm 3) = 9$$

and so on.

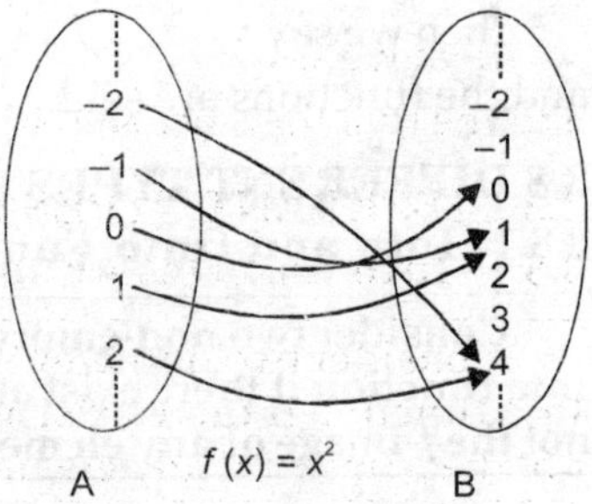

Fig. 3.4.

3.8.3 One-one Onto Function (Bijective)

If a function $f: A \to B$ is one-one and onto, then f is said to be one-one onto or **bijective function,**

i.e., $F: A \xrightarrow[1\text{-}1]{\text{onto}} B$

Function f is also called a one-one correspondence between A and B. Thus A and B have same number of elements in them. Sets A and B once said to be the **cardinality equivalent** sets.

For example, A function f defined on $A = \{a, b, c\}$ and $B = \{1, 2, 3\}$ as $f = \{a, 2\}$ $(b, 1)$ $(c, 3)\}$ is an example of a one-to-one onto (bijective function) as

(*i*) Each element of A has a single and distinct image in B, *i.e.*, $f(x_1) = f(x_2) \Rightarrow x_1 = x_2$ so f is one-one.

(*ii*) Every element of set B is the image of some element in A *i.e.*, $\forall\, y \in B, \exists\, x \in A$ such that $y = f(x)$ so f is onto.

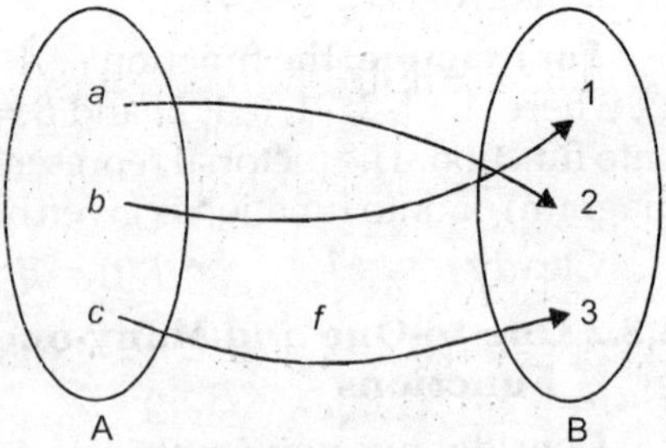

Fig. 3.5.

3.8.4 One-one Into Function (Injective)

A function $f: A \to B$ is said to be injective if f is one-one and into both. That is, $f: A \to B$ is injective (one-one into) if

(*i*) for $x_1, x_2 \in A, x_1 \neq x_2 \Rightarrow f(x_1) \neq f(x_2)$

and (*ii*) $R(f) \subset B$, *i.e.*, $\exists$ at least one element in B which is not f-image of any element of A.

For example, a function $f: R \to R$, where R is a real line defined by $f(x) = 2x(x \in R)$ is one-one and into or injective function as (*i*) **F is One-one :** Let $f(x_1) = f(x)$ for $x_1, x_2 \in R$

$\Rightarrow \quad 2x_1 = 2x_2$ (by definition of f)

$\Rightarrow \quad x_1 = x_2$

Thus each element of domain set has a unique image in co-domain.

(*ii*) **f is Into :** Give $f(x) = 2x \in R$ (co-domain) *i.e.*, each element $x \in R$ mapped to on even number (as $f(x) = 2x$).

So Range (f) = $\{2x : x \in R$ such that $f(x) = 2x\}$ which is clearly $R(f) \subset R$ (real line or (co-domain) so f is into.

3.8.5 Many-one Onto Function.

A function $f : A \to B$ is said to be many-one onto if

(*i*) $f(x_1) = f(x_2) \Rightarrow x_1 \neq x_2, x_1, x_2 \in A$

and (*ii*) $\forall\ y \in B, \exists$ some $x \in A$ S to $f(x) = y$ or $R_f = B$, where R_f is the range of function f.

For example, A function F defined on sets

$A = \{1, 2, 3\}$ and $B = \{a, b\}$ as

$F = \{(1, a)\ (2, b)\ (3, a)\}$ is many-one

onto as (*i*) $f(1) = f(3) = a$ but $1 \neq 3$

and (*ii*) Clearly the $R_f = \{a, b\} = B$ so onto.

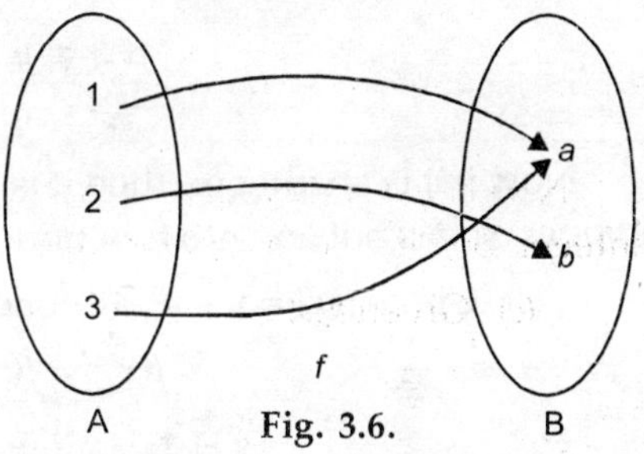

Fig. 3.6.

3.8.6 Many-one Into.

A function $f : A \to B$ is many-one into if

(*i*) $f(x_1) = f(x_2) \Rightarrow x_1 \neq x_2, x_1, x_2 \in A$

and (*ii*) $R_f \subset B$, where R_f is the range set of function f.

For examples, $F : Z \to Z$ as $f(x) = x^2$ is many-one into function as

(*i*) $f(-2) = f(2) = 4$ but $-2 \neq 2$. In general $f(-a) = f(a) = a^2$ but $-a \neq a$. So f is many-one function.

and (*ii*) $R_f = \{x^2 : x \in 2$ such that $f(x) = x^2\}$

$= \{0, 1, 4, 9, 16, ...\} \subset Z$ (set integers) so f is into.

EXAMPLES

Example 3.3. Let Z be the set of integers, them which of the following functions $f : Z \to Z$ are one-one, onto or both.

(*a*) $f(x) = 5x + 3$ (*b*) $f(x) = x^4$ (*c*) $f(x) = x + x^3$

Solution : (*a*) Let $x_1, x_2 \in Z$ such that

$$f(x_1) = f(x_2)$$

$\Rightarrow \quad 5x_1 + 3 = 5x_2 + 3$ (by given definition)

$\Rightarrow \quad x_1 = x_2$

So f is **one-one**

Now let $y \in B$ such that

$$f(x) = y$$

$$\Rightarrow \quad 5x + 3 = y$$

$$\Rightarrow \quad 5x = y - 3 \quad \text{or} \quad \boxed{x = \frac{y-3}{5}}$$

If $y \in Z$, *i.e.*, y is an integer, then $\frac{y-3}{5}$ is not necessarily an integer as $y = 6 \in Z$ by $\frac{6-3}{5} = \frac{3}{5} \notin Z$. Thus f is not onto $\Rightarrow f$ is into function. Hence $f(x) = 5x + 3$, on a set of integers is one-one but not onto.

(*b*) $f(x) = x^4$ let $x_1, x_2 \in A$ such that

$$f(x_1) = f(x_2)$$

$$\Rightarrow \quad x_1^4 = x_2^4$$

$$\Rightarrow \quad (x_1^4)^{1/4} = (x_2^4)^{1/4} \quad \Rightarrow \quad \boxed{x_1 = x_2}$$

Thus f is **one-one.** For onto, let $y \in Z$ such that

$$f(x) = y$$

$$\Rightarrow \quad x^4 = y \quad \Rightarrow \quad x = (y)^{1/4}$$

Now if y is an integer, then it is not necessarily true that $(y)^{1/3}$ will also beam integer. So f is not an onto function.

(*c*) Given $f(x) = x + x^3$. For one-one, let $x_1, x_2 \in Z$ such that

$$f(x_1) = f(x_2)$$

$$\Rightarrow \quad x_1 + x_1^3 = x_2 + x_2^3$$

$$\Rightarrow \quad x_1(1 + x_1^2) = x_2(1 + x_2^2)$$

Clearly, one can observe that $x_1 \neq x_2$ so f is not one-one. Clearly f is many-one. Further, it is also not onto.

Example 3.4. Show that $f : R \rightarrow R$ defined by $f(x) = \sin x$ is not one-one and onto.

Solution : For one-one, let $x_1, x_2 \in R$ such that

$$f(x_1) = f(x_2)$$

$$\Rightarrow \quad \sin x_1 = \sin x_2$$

$$\Rightarrow \quad x_1 = x_2$$

as $\quad \sin 0 = \sin \Pi = 0$ but $0 \neq \pi$

So f is not **one-one**, it is many one. Now, we know that $-1 \leq \sin x \leq 1$, *i.e.*, $\forall\ x \in R \sin x$ takes all values between -1 and 1.

Thus, the Range set $\quad R_f = \{y : y \in (-1, 1)$ such that $y = f(x)\}$

$\Rightarrow \quad R_f = [-1, 1] \subset R$ (real line)

As the range set R_f is a subset of co-domain *i.e.*, real line, the function f is not onto. So f is into function.

Hence $f : R \rightarrow R$, defined by $f(x) = \sin x$ is not one-one onto, it is many one into function.

Example 3.5. Check whether the following functions $f: R \to R$ are one-one onto.

(a) $f(x) = e^x$ **(b) $f(x) = 3x + 5$** **(c) $f(x) = |x|$**

Solution : (*a*) Let $x_1, x_2 \in R$ such that

$$f(x_1) = f(x_2)$$

$$\Rightarrow \quad e^{x_1} = e^{x_2}$$

taking log both the sides, we have

$$\log_e e^{x_1} = \log_e e^{x_2}$$

$$\Rightarrow \quad \boxed{x_1 = x_2}$$

So f is one-one

Let $y \in R$ such that $y = f(x) = e^x \Rightarrow x = \log_e y$. f is not onto as logarithm of negative real number in not real number. Thus there exists negative real numbers in co-domain which are not the images of any element of domain *i.e.*, $R_f \subset R$. Thus $f: R \to R$ defined by $f(x) = e^x$ is one one but not onto. So f is one-one into function.

(*b*) Let $x_1, x_2 \in R$ such that

$$f(x_1) = f(x_2)$$

$$\Rightarrow \quad 3x_1 + 5 = 3x_2 + 5$$

$$\Rightarrow \quad \boxed{x_1 = x_2}$$

So f is one-one. Let $y \in R$ such that $y = f(x) = 3x + 5$

$$\Rightarrow \quad x = \frac{y-5}{3} \in R$$

Now $$f(x) = f\left(\frac{y-5}{3}\right) = 3\cdot\left(\frac{y-5}{3}\right) + 5 = y$$

i.e., $\forall\ y \in R, \exists$ same $x \in R$ such that $f(x) = y$. So f is onto.

Hence $f: R \to R$ defined as $f(x) = 3x + 5$ is one-one onto function.

(*c*) Let $x_1, x_2 \in R$ such that

$$f(x_1) = f(x_2)$$

$$|x_1| = |x_2|$$

but $x_1 \neq x_2$ | as $|-1| = |1| = 1$ but $-1 \neq 1$. So f is not one-one.

Clearly, an element $\in R$ negative or positive mapped into its modulus which is always positive. So f is not onto because the R_f, range set of f will contain only positive real numbers and hence

$$R_f \subset R$$

Thus $f: R \to R$ defined by $f(x) = |x|$ is not one-one onto.

3.9 COMPOSITE FUNCTIONS

Let A, B and C be three non-empty sets. Let f be a function from A to B and g be a function from B to C.

i.e., $$f : A \to B \quad \text{and} \quad g : B \to C$$

Then the function $g\,o\,f : A \to C$ defined as

$$g\,o\,f(x) = g(f(x)),\ \forall\, x \in A$$

is called the composite (or product) function of the two functions f and g. Diagrammatically, this function is shown as :

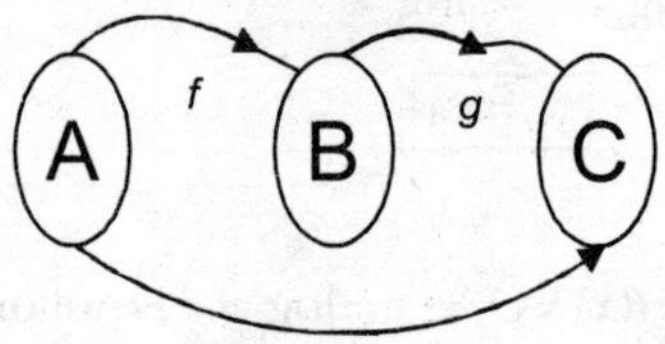

Fig. 3.7.

EXAMPLES

Example 3.6. Let $f : A \to B$ and $g : B \to C$ be two functions, defined on set $A = \{a, b, c\}$, $B = \{1, 2, 3\}$ and $C = \{x, y, z\}$ as

$$f(a) = 1, f(b) = 3\ f(c) = 2 \text{ and } g(1) = y$$
$$g(2) = z \text{ and } g(3) = x$$

Then find out the images of a, b, c under the mapping $g\,o\,f : A \to C$.

Solution :
$$g\,o\,f(a) = g\,(f(a)) = g(1) = y$$
$$g\,o\,f(b) = g\,(f(b)) = g(3) = x$$
$$g\,o\,f(c) = g\,(f(c)) = g\,(2) = z$$

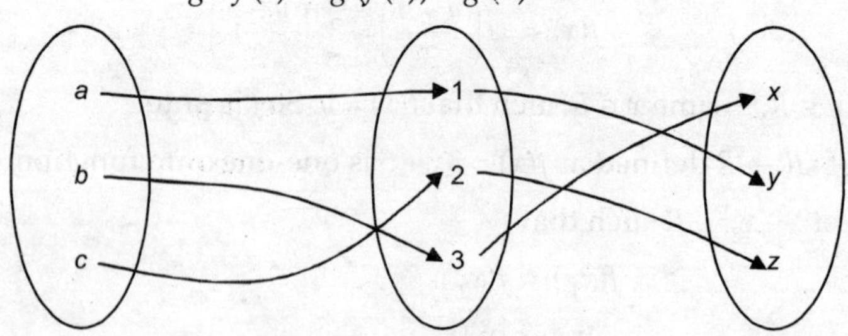

Fig. 3.8.

Example 3.7. Let $f : R \to R$ be defined as

$$f(x) = 2x + 3$$

and $g : R \to R$ be defined as $g(x) = x^2$ then find out (*i*) $g\,o\,f\,(x)$ and (*ii*) $f\,o\,g\,(x)$

Solution : (*a*)
$$\begin{aligned} g\,o\,f(x) &= g\,(f(x), \quad \forall\ x \in R \\ &= g(2x+3) \\ &= (2x+3)^2 \\ &= 4x^2 + 12x + 9 \in R \end{aligned}$$

So $g\,o\,f: R \to R$ as $g\,o\,f(x) = 4x^2 + 12x + 9$

(b) $$f\,o\,g = f(g(x)),\ \forall\ x \in R$$
$$= f(x^2)$$
$$= 2x^2 + 3 \in R$$

Hence, we conclude that $(g\,o\,f)$ and $(f\,o\,g)$ both are functions from $R \to R$ but $f\,o\,g \neq g\,o\,f$.

Example 3.8. If $f: A \to B$ and $g: B \to C$ both are one-one onto mappings then show that $g\,o\,f: A \to C$ is also one-one and onto.

Solution : To show $g\,o\,f: A \to C$ is one-one onto we shall prove that
(*i*) $g\,o\,f$ is one-one and (*ii*) $g\,o\,f$ is onto

(*a*) Let $x_1, x_2 \in A$ such that
$$g\,o\,f(x_1) = g\,o\,f(x_2)$$
$$\Rightarrow \quad g(f(x_1)) = g(f(x_2))$$
$$\Rightarrow \quad f(x_1) = f(x_2) \qquad \text{(as } g \text{ is one-one)}$$
$$\Rightarrow \quad x_1 = x_2 \qquad \text{(as } f \text{ is one-one)}$$

So $gof: A \to C$ *is one-one.*

(*ii*) Let $z \in C$

Since g is onto, so $\forall\, z \in C, \exists$ same $y \in B$ such that
$$g(y) = z$$
and f is also onto, so $\forall\, y \in B, \exists$ same x such that
$$f(x) = y$$
Now $$g(f(x) = g(y)$$
$$\Rightarrow \quad g\,o\,f(x) = z$$

Thus, $\forall\, z, \exists$ same $x \in A$ such that $g\,o\,f(x) = z$ so $g\,o\,f$ is onto. Hence $g\,o\,f$ is also one-one onto function.

Example 3.9. Prove that the composition of functions satisfies the associative property. That is; if f_1, f_2 and f_3 are 3 functions.

then $$(f_1\,o\,f_2)\,o\,f_3 = f_1\,o\,(f_2\,o\,f_3)$$

Solution : Let A, B, C and D be four non-empty sets. Let $f_3: A \to B, f_2: B \to C$ and $f_1: C \to D$. Then clearly $(f_1\,o\,f_2)\,o\,f_3$ is a function from A to D and $f_1\,o\,(f_2\,o\,f_3)$ is also a function of from A to D. Now we need to show that
$$[(f_1\,o\,f_2)\,o\,f_3](x) = [f_1\,o\,(f_2\,o\,f_3)](x)$$
So let $x \in A, y \in B, z \in C$ and $w \in D$, such that
$$f_3(x) = y, f_2(y) = z, f_1(z) = w$$
them $$[(f_1\,o\,f_2)\,o\,f_3](x) = (f_1\,o\,f_2)(f_3(x))$$
$$= (f_1\,o\,f_2)(y) \qquad (\text{as } f_3(x) = y)$$
$$= f_1(f_2(y))$$
$$= f_1(z) = w \qquad ...(1)$$

Again $\quad [f_1 o (f_2 o f_3)](x) = f_1[(f_2 o f_3)(x)]$

$$= f_1[\{f_2(f_3(x))\}$$

$$= f_1[f_2(y)] = f_1[z]$$

$$= w \quad ...(2)$$

By (1) and (2), we conclude that

$$(f_1 o f_2) o f_3 = f_1(f_2 o f_3)$$

3.10 INVERSE FUNCTION

Let $f : A \to B$ is one-one onto function. Then a function $f^{-1} : B \to A$, associates each element $b \in B$ to the element $a \in A$, such that $f(a) = b$ is called an Inverse Function of $f : A \to B$.

For example, $f : A \to B$ defined as $f(a) = 1, f(b) = 2$ and $f(c) = 3$ on set $A = \{a, b, c\}$ and $B = \{1, 2, 3\}$ is one-one onto function. Then $f^{-1} : B \to A$ is defined as $f^{-1}(1) = a, f^{-1}(2) = b$ and $f^{-1}(3) = c$.

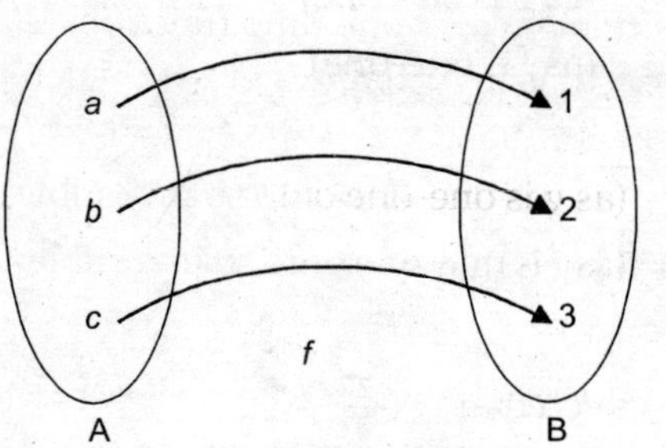

Fig. 3.9.

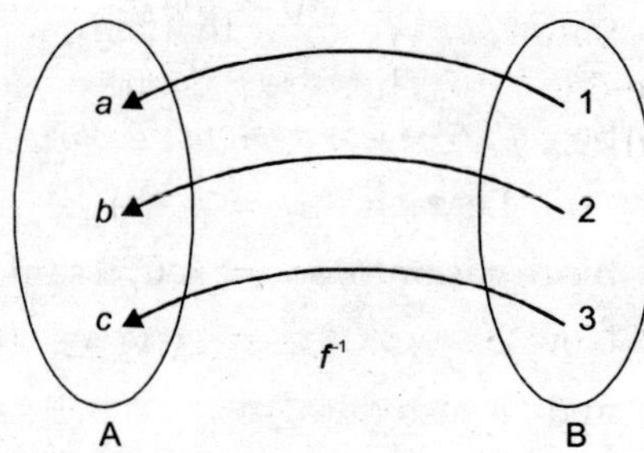

Fig. 3.10.

Theorem 3.1. If $f : A \to B$ is one-one onto function, then $f^{-1} : B \to A$ is also one-one onto function.

Proof : f^{-1} is One-one :

Let $\quad f^{-1}(b_1) = a_1$ and $f^{-1}(b_2) = a_2$

$\Rightarrow \quad b_1 = f(a_1)$ and $b_2 = f(a_2)$

Since f is one-one, so $\quad f(a_1) = f(a_2) \Rightarrow b_1 = b_2$

So $\quad f^{-1}(b_1) = f^{-1}(b_2) \Rightarrow b_1 = b_2$

Hence f^{-1} is one-one.

(*ii*) **f^{-1} is Onto :** Let $a \in A$ 1 then $\exists\, b \in B$ such that $f(a) = b \Rightarrow \exists\, b \in B$ such that $a = f^{-1}(b)$. Hence f^{-1} is onto.

Thus if $f : A \to B$ is one-one onto, then $f^{-1} : B \to A$ is also one-one onto.

Theorem 3.2. If $f : A \to B$ is one-one onto, then there exists a unique inverse of f.

Proof : Let $f_1 : B \to A$ and $f_2 : B \to A$ be the two inverses of $f : A \to B$. Then to prove that there is only one inverse possible we need to show that $f_1 = f_2$. Let $b \in B$, then $\exists\, a_1, a_2 \in A$ such that $f_1(b) = a_1$ and $f_2(b) = a_2$. Since f_1 and f_2 are the inverse mappings of f, so

$$f_1(b) = a_1 \Rightarrow f(a_1) = b$$

and $f_2(b) = a_2 \Rightarrow f(a_2) = b$

i.e., $f(a_1) = f(a_2) = b$

$\Rightarrow$ Two distinct elements of set A has over unique image in B, but as we know that f is one-one. So if $f(a_1) = f(a_2)$

$\Rightarrow$ $a_1 = a_2$

and if $a_1 = a_2$

$\Rightarrow$ $f_1(b) = f_2(b)$

$\Rightarrow$ $\boxed{f_1 = f_2}$

Hence inverse of one-one onto function is unique.

EXAMPLES

Example 3.10. Show that

$$(g\,o\,f)^{-1} = f^{-1}\,o\,g^{-1}$$

where $f : A \to B, g : B \to C$ and $f^{-1} : B \to A\; g^{-1} : C \to B$ are one-one onto functions.

Solution : We know that the cave position of two *o* one-onto functions is also one-one onto. Thus $(g\,o\,f)$ is one-one onto, so it is invertiable. Let $z \in C$ such that $g(y) = z$.

$\forall\, y \in B, y = g^{-1}(z)$ (as g is one-one onto so investible)

Also let $y = f(x) \Rightarrow x = f^{-1}(y)$ (as f is one-one onto so invertible)

Now $(g\,o\,f)(x) = g(f(x)) = g(y) = z$ ...(*i*)

$\Rightarrow$ $(g\,o\,f)^{-1}(z) = x$

now $(f^{-1}\,o\,g^{-1})(z) = f^{-1}(g^{-1}(z)) = f^{-1}(y) = x$

By (*i*) and (*ii*), we have

$$\boxed{(gof)^{-1} = f^{-1}og^{-1}} \qquad ...(ii)$$

Example 3.11. If $f : R \to R$ be a function defined as $f(x) = 3x + 7$, then find out $f^{-1}(y)$.

Solution : $f : R \to R$ defined by $f(x) = 3x = 7$ is one-one onto (check yourself) so it is invertible.

Let $y = f(x)$

$= 3x + 7$

$\Rightarrow$ $x = \dfrac{y-7}{3} \in R$

Since $f(x) = y \Rightarrow x = f^{-1}(y)$

So $\boxed{f^{-1}(y) = \dfrac{y-7}{3}}$ $\left(\text{as } x = \dfrac{y-7}{3}\right)$

Example 3.12. If A and B be two non-empty sets and $f : A \to B$ is one-one onto then for any subsets B_1 and B_2 of B

$$f^{-1}(B_1 \cap B_2) = f^{-1}(B_1) \cap f^{-1}(B_2)$$

Solution : Let x be an arbitrary element such that

$x \in f^{-1}(B_1 \cap B_2)$

$\Rightarrow f(x) \in B_1 \cap B_2$

$\Rightarrow f(x) \in B_1$ and $f(x) \in B_2$

$\Rightarrow x \in f^{-1}(B_1)$ and $x \in f^{-21}(B_2)$

(Since f is one-one two, so invertible)

$\Rightarrow x \in f^{-1}(B_1) \cap f^{-1}(B_2)$

So $f^{-1}(B_1 \cap B_2) \subseteq f^{-1}(B_1) \cap f^{-1}(B_2)$ ...(1)

Conversely, let

$x \in f^{-1}(B_1) \cap f^{-1}(B_2)$

$\Rightarrow x \in f^{-1}(B_1)$ and $x \in f^{-1}(B_2)$

$\Rightarrow f(x) \in B_1$ and $f(x) \in B_2$

$\Rightarrow f(x) \in (B_1 \cap B_2)$

$\Rightarrow x \in f^{-1}(B_1 \cap B_2)$

So $f^{-1}(B_1) \cap f^{-1}(B_2) \subseteq f^{-1}(B_1 \cap B_2)$ ...(2)

By (1) and (2), we conclude that

$$\boxed{f^{-1}(B_1 \cap B_2) = f^{-}(B_1) \cap f^{-1}(B_2)}$$

Example 3.13. If X and Y be two non-empty sets and $f: X \rightarrow Y$ is one-one onto function then for any sub-sets A and B of Y, prove that

(*a*) $A \subseteq B \Rightarrow f^{-1}(A) \subseteq f^{-1}(B)$

(*b*) $f^{-1}(A - B) = f^{-1}(A) - f^{-1}(B)$

(*c*) $f^{-1}(Y - A) = X - f^{-1}(A)$

Solution : (*a*) Given $A \subseteq B$

$\Rightarrow y \in A$ and $y \in B$

then $\exists$ same $x \in X$ such that $f(x) = y$

So if $y \in A \Rightarrow f(x) \in A \Rightarrow x \in f^{-1}(A)$

and $y \in B \Rightarrow f(x) \in B \Rightarrow x \in f^{-1}(B)$

$\Rightarrow$ $f^{-1}(A) \subseteq f^{-1}(B)$

So if $\boxed{A \subseteq B \Rightarrow f^{-1}(A) \subseteq f^{-1}(B)}$

(*b*) Let x be an arbitrary element such that

$x \in f^{-1}(A - B)$

$\Rightarrow f(x) \in (A - B)$ (f is one-one onto, so invertible)

$\Rightarrow f(x) \in A$ and $f(x) \notin B$

$\Rightarrow x \in f^{-1}(A)$ and $x \notin f^{-1}(B)$

$\Rightarrow x \in (f^{-1}(A) - f^{-1}(B))$

So $f^{-1}(A - B) \subseteq f^{-1}(A) - f^{-1}(B)$ ...(1)

Conversely, let

$x \in f^{-1}(A) - f^{-1}(B)$

$\Rightarrow\ x \in f^{-1}(A)$ and $x \notin f^{-1}(B)$

$\Rightarrow\ f(x) \in A$ and $f(x) \notin B$

$\Rightarrow\ f(x) \in (A - B)$

$\Rightarrow\ x \in f^{-1}(A - B)$

So $f^{-1}(A) \cap f^{-1}(B) \subseteq f^{-1}(A - B)$ ...(2)

By (1) and (2), we conclude that

$$\boxed{f^{-1}(A - B) = f^{-1}(A) - f^{-1}(B)}$$

(c) Let x be any arbitrary element such that

$x \in f^{-1}(Y - A)$

$\Rightarrow\ x \in f^{-1}(Y)$ and $x \notin f^{-1}(A)$

Since f is onto so as f^{-1} thus $f^{-1}(Y) = X$ (*i.e.*, range set of f^{-1})

So $x \in f^{-1}(Y)$ and $x \notin f^{-1}(A)$

$\Rightarrow\ x \in X$ and $x \notin f^{-1}(A)$

$\Rightarrow\ x \in (X - f^{-1}(A))$

So $f^{-1}(Y - A) \subseteq X - f^{-1}(A)$ ...(1)

Similarly, we can show that

$$X - f^{-1}(A) \subseteq f^{-1}(Y - A) \qquad ...(2)$$

By (1) and (2), we conclude that

$$\boxed{f^{-1}(Y - A) = X - f^{-1}(A)}$$

Example 3.14. If $f : X \to Y$ be one-one onto mapping then

$$f^{-1}{}_0 f = I_x \text{ and } f_0 f^{-1} = I_y$$

where I_x and I_y are identity mapping of the set X and Y respectively.

Solution : Let x be an arbitrary element of set X, then

$$f(x) = y \quad \text{for} \quad y \in Y$$

$$\Rightarrow \quad f^{-1}(y) = x$$

so
$$\begin{aligned}(f^{-1}{}_0 f)\,x &= f^{-1}(f(x)) \\ &= f^{-1}(y) \\ &= x \\ &= I_x(x)\end{aligned}$$

so
$$\boxed{f^{-1}{}_0 f = I_x}$$

now for any arbitrary $y \in Y$. There is associated a unique $x \in X$ such that

$$f(x) = y \quad \text{or} \quad x = f^{-1}(y)$$

so
$$(f_0 f^{-1})y = f(f^{-1}(y))$$

$$= f(x)$$
$$= y$$
$$= I_y(y)$$

so $$f_0 f^{-1} = I_y$$

3.11 OPERATIONS OF FUNCTIONS

As we know there are operations like addition, substraction, multiplication and division for numbers. These operations can be performed over functions too.

3.11.1 Addition of two Functions

Let $f(x)$ and $g(x)$ be two functions, then their addition will be a function $(f+g)$ (x) is defined as

$$(f+g)(x) = f(x)+g(x)$$

3.11.2 Substraction of Two Functions

Let $f(x)$ and $g(x)$ be two functions, then their substraction is defined as

$$(f-g)(x) = f(x)-g(x)$$

3.11.3 Multiplication of Two Functions

If $f(x)$ and $g(x)$ are two functions, then their multiplication denoted as $(f \cdot g)(x)$ is defined as :

$$(f \times g)(x) \text{ or } (f \cdot g)(x) = [f(x)] \cdot [g(x)]$$

3.11.4 Division of Two Functions

If $f(x)$ and $g(x)$ are two functions, then their division, denoted as $(f/g)(x)$ is defined as :

$$(f/g)(x) = \frac{f(x)}{g(x)}$$

Example 1. *Given $f(x) = 3x + 2$ and $g(x) = 4 - 5x$ find $(f+g)(x)$, $(f-g)(x)$, $(f \cdot g)(x)$ and $(f/g)(x)$.*

Solution : (*i*)
$$(f+g)(x) = f(x)+g(x)$$
$$= (3x+2)(4-5x)$$
$$= (6-2x)$$

(*ii*)
$$(f-g)(x) = f(x)-g(x)$$
$$= (3x+2)-(4-5x)$$
$$= 8x-2$$

(*iii*)
$$(f \cdot g)(x) = f(x) \cdot g(x)$$
$$= (3x+2) \cdot (4-5x)$$
$$= -15x^2+2x+8$$

(*iv*)
$$(f/g)(x) = \frac{f(x)}{g(x)}$$

$$= \frac{3x+2}{4-5x}$$

or $$= \frac{(3x+2)(4+5x)}{(16-25x^2)} = \frac{15x^2+22x+8}{(16-25x^2)}$$ **Ans.**

Example 2. *Given $f(x) = 2x$, $g(x) = x + 4$ and $h(x) = 5 - x^3$ find $(f+g)(2)$, $(h-g)(2)$, $(f \cdot g)(2)$ and $\left(\frac{h}{g}\right)(2)$*

Solution : We know that for any two functions $f(x)$ and $g(x)$

$$(f+g)(x) = f(x) + g(x)$$
$$(f-g)(x) = f(x) - g(x)$$
$$(f \cdot g)(x) = f(x) \cdot g(x)$$
$$(f/g)(x) = f(x)/g(x)$$

Here in each case, given that $x = 2$ so.

$$(f+g)(2) = f(2) + g(2)$$
$$= (2.4) + (2+4) = 10$$
$$(h-g)(2) = h(2) - g(2)$$
$$= (5-(2)^3) - (2+4)$$
$$= -3 - 6 = -9$$
$$(f \cdot g)(2) = f(2) \cdot h(2)$$
$$= (2 \times 2) \cdot (5 - 2^3)$$
$$= -12$$

and $$\left(\frac{h}{g}\right)(2) = \frac{h(2)}{g(2)} = \frac{(5-2^3)}{(2+4)} = -\frac{1}{2}$$ **Ans.**

Example 3. *Given $f(x) = 3x^2 - x + 4$, find the simplified form of the following expression and evaluate at $h = 0$.*

$$\frac{f(x+h) - f(x)}{h}$$

Solution : Given $$f(x) = 3x^2 - x + 4$$
$$= 3(x^2 + 2xh + h^2) - (x+h) + 4$$
$$= 3x^2 + 6xh + 3h^2 - x - h + 4 \quad ...(1)$$

and then $$f(x+h) - f(x) = 6xh + 3h^2 - h \quad ...(2)$$

Then $$\frac{f(x+h) - f(x)}{h} = \frac{6xh + 3h^2 - h}{h}$$
$$= (6x + 3h - 1) \quad ...(3)$$

evaluate (3) at $h = 0$, thus

$$\left[\frac{f(x+h)-f(x)}{h}\right]_{h=0} = 6x - 1 \textbf{ Ans.}$$

3.12 GROWTH OF FUNCTIONS

One of the important criteria in evaluating algorithms is the time it takes to complete a job. To have a meaningful comparison of algorithms, the estimate of computation time must be independent of the programming language, compiler, and computer used; must reflect on the size of the problem being solved; and must not depend on specific instances of the problem being solved. The quantities often used for the estimate are the worst case execution time, and average execution time of an algorithm, and they are represented by the number of some key operations executed to perform the required computation. Since we are mostly interested in the behaviour of algorithms for large values of n, lower terms can be ignored compared with the highest term. The concept that is used to address these issues is something called **big-oh.**

3.12.1 Big-Oh

The following example gives the idea of one function growing more rapidly than another. We will use this example to introduce the concept the big-Oh.

Example : *$f(n) = 100\, n^2$, $g(n) = n^4$, the following table and figure show that $g(n)$ grows faster than $f(n)$ when $n > 10$. We say f is big-Oh of g.*

n	$f(n)$	$g(n)$
10	10,000	10,000
50	250,000	6,250,000
100	1,000,000	100,000,000
150	2,250,000	506,250,000

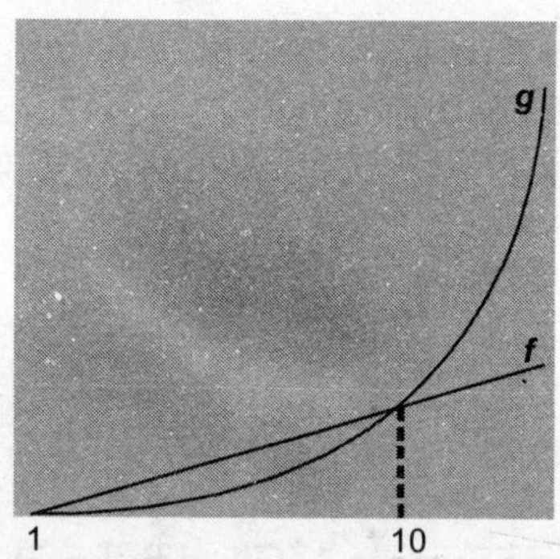

Definition (big-oh) : Let f and g be functions from the set of integers (or the set of real numbers) to the set of real numbers. Then $f(x)$ is said to be $O(g(x))$, which is read as $f(x)$ is big-oh of $g(x)$, if and only if there are constants C and n_0 such that

$$|f(x)| \leq C|g(x)|$$

whenever $x > n_0$.

Note that big-oh is a binary relation on a set of functions.

For example, $5x + 10$ is big-oh of x^2, because $5x + 10 < 5x^2 = 15x^2$ for $x > 1$. Hence for $C = 15$ and $n_0 = 1$, $|5x + 10| \leq C|x^2|$.

Similarly it can be seen that $3x^2 + 2x + 4 < 9x^2$ for $x > 1$. Hence $3x^2 + 2x + 4$ is $O(x^2)$. In general, we have the following theorem:

Theorem 1: $a_n x^n + ... + a_1 x + a_0$ is $O(x^n)$ for any real numbers $a_n, ..., a_0$ and any non negative number n.

Note: Let $f(x) = 3x^2 + 2x + 4$, $g(x) = x^2$, from the above illustration, we have that $f(x)$ is $O(g(x))$. Also, since $x^2 < 3x^2 + 2x + 4$, we can also get $g(x)$ is $O(f(x))$. In this case, we say these two functions are of the *same order*.

3.12.2 Growth of Combination of Functions

Big-oh has some useful properties. Some of them are listed as theorems here.

Let use start with the defintion of max function.

Definition(max function) : Let $f_1(x)$ and $f_2(x)$ be functions from a set A to a set of real numbers B. Then max $(f_1(x), f_2(x))$ is the function from A to B that takes as its value at each point x the larger of $f_1(x)$ and $f_2(x)$.

Theorem 2: If $f_1(x)$ is $O(g_1(x))$, and $f_2(x)$ is $O(g_2(x))$, then $(f_1 + f_2)(x)$ is $O(\max(g_1(x), g_2(x)))$.

From this theorem it follows that if $f_1(x)$ and $f_2(x)$ are $O(g(x))$, then $(f_1 + f_2)(x)$ is $O(g(x))$, and $(f_1 + f_2)(x)$ is $O(\max(f_1(x), f_2(x)))$.

Theorem 3: If $f_1(x)$ is $O(g_1(x))$, and $f_2(x)$ is $O(g_2(x))$, then $(f_1 * f_2)(x)$ is $O(g_1(x) * g_2(x))$.

3.12.3 Big-Omega and Big-Theta

Bigh-oh concerns with the "less than or equal to" relation between functions for large values of the variable. it is also possible to consider the "greater than or equal to" relation and "equal to" relation in a similar way. Big-Omega is for the former and big-theta is for the latter.

Defintion (big-omega): Let f and g be functions from the set of integers (or the set of real numbers) to the set of real numbers. Then $f(x)$ is said to be $\Omega(g(x))$, which is read as $f(x)$ is big-omega of $g(x)$, if there are constants C and n_0 such that

$$|f(x)| \geq C\,|g(x)|$$

whenever $$x > n_0.$$

Definition (big-theta): Let f and g be functions from the set of integers (or the set of real numbers) to the set of real numbers. Then $f(x)$ is said to be $\theta(g(x))$, which is read as $f(x)$ is Big-theta of $g(x)$, if $f(x)$ is $O(g(x))$, and $\Omega(g(x))$. We also say that $f(x)$ *is of order* $g(x)$.

For example, $3x^2 - 3x - 5$ is $\Omega(x^2)$, because $3x^2 - 3x - 5 \geq x^2$ for integers $x > 2$ $(C = 1, n_0 = 2)$.

Hence by Theorem 1 it is $\Theta(x^2)$.

In general, we have the following theorem:

Theorem 4: $a_n x^n + ... + a_1 x + a_0$ is $\Theta(x^n)$ for any real numbers $a_n, ..., a_0$ and any nonnegative number n.

3.12.4 Little-Oh and Little-Omega

If $f(x)$ is $O(g(x))$, but not $\Theta(g(x))$, then $f(x)$ is said to be $o(g(x))$, and it is read as $f(x)$ is **little-oh** of $g(x)$. Similarly for **little-omega**.

For example x is $o(x^2)$, x^2 is $o(2^x)$, 2^x is $o(x!)$, etc.

3.12.5 Calculation of Big-oh Relation Using Limit

Let $f(x)$ and $g(x)$ be functions from a set of real numbers to a set of real numbers. Then

1. If $\lim_{x\to\infty} f(x)/g(x) = 0$, then $f(x)$ is $o(g(x))$. Note that if $f(x)$ is $o(g(x))$, then $f(x)$ is $O(g(x))$.
2. If $\lim_{x\to\infty} f(x)/g(x) = \infty$, then $g(x)$ is $o(f(x))$.
3. If $0 < \lim_{x\to\infty} f(x)/g(x) < \infty$, then $f(x)$ is $\theta(g(x))$.
4. If $\lim_{x\to\infty} f(x)/g(x) < \infty$, then $f(x)$ is $O(g(x))$.

For example,

$$\lim_{x\to\infty} (4x^3 + 3x^2 + 5)/(x^4 - 3x^2 - 5x - 4)$$

$$= \lim_{x\to\infty} (4/x + 3/x^2 + 5/x^4)/(1 - 3/x - 5/x^3 - 4/x^4) = 0.$$

Hence

$$(4x^3 + 3x^2 + 5) \text{ is } o(4x^4 - 3x^3 - 5x - 4),$$

Let us see why these rules hold. Here we give a proof for 4. Others can be proven similarly.

Proof: Suppose $\lim_{x\to\infty} f(x)/g(x) = C_1 < \infty$.

By the definition of limit this means that

$\forall \varepsilon > 0, \exists\, n_0$ such that $|f(x)/g(x) - C_1| < \varepsilon$ whenever $x > n_0$.

Hence $-\varepsilon < f(x)/g(x) - C_1 < \varepsilon$

Hence $-\varepsilon + C_1 < f(x)/g(x) < \varepsilon + C_1$

In particular $f(x)/g(x) < \varepsilon + C_1$

Hence $f(x) < (\varepsilon + C_1)\, g(x)$

Let $C = \varepsilon + C_1$, then $f(x) < Cg(x)$ whenever $x > n_0$.

Since we are interested in non-negative functions f and g, this means that $|f(x)| \le C\,|g(x)|$

Hence $f(x) = O(g(x))$.

3.12.6 L Hospital (L Hôpital) s Rule

$\lim_{x\to\infty} f(x)/g(x)$ is not always easy to calculate. *For example,* take $\lim_{x\to\infty} x^2/3^x$. Since both x^2 and 3^x go to as x goes to ∞ and there is no apparent factor common to both, the calculation of the limit is not immediate. One tool we may be able to use in such cases is L'Hospital's Rule, which is given as a theorem below.

Theorem 5 (L Hospital)

If $\lim_{x\to\infty} f(x) = \infty$ and $\lim_{x\to\infty} g(x) = \infty$, and $f(x)$ and $g(x)$ have the first derivatives, $f'(x)$ and $g'(x)$, respectively, then $\lim_{x\to\infty} f(x) = f'(x)/g'(x)$.

This also holds when $\lim_{x\to\infty} f(x) = 0$ and $\lim_{x\to\infty} g(x) = 0$, intead of $\lim_{x\to\infty} f(x) = \infty$ and $\lim_{x\to\infty} g(x) = \infty$.

For example, $\lim_{x\to\infty} x/e^x = \lim_{x\to\infty} 1/e^x = 0$, because $(e^x) = e^x$, where e is the base for the natural logarithm.

Similary $\lim_{x\to\infty} \ln x/x = (1/x)/1 = \lim_{x\to\infty} 1/x = 0$.

Note that this rule can be applied repeatedly as long as the conditions are satisfied.

So, *for example,* $\lim_{x\to\infty} x^2/x^x = \lim_{x\to\infty} 2x/e^x = \lim_{x\to\infty} 2/x^x = 0$.

EXERCISES

1. Which f is not a function from R to R in the following equations, where R is the set of real numbers ? Explain why they are not a function.

   (*a*) $f(x) = \dfrac{1}{x}$

   (*b*) $f(x) = y$ such that $y^2 = x$

   (*c*) $f(x) = x^2 - 1$

2. Find the domain and range of the following functions.

   (*a*) The function that assigns to each bit string (of various lengths) the number of zeros in it.

   (*b*) The function that assigns the number of bits left over when a bit string (of various lengths) is split into bytes (which are blocks of 8 bits).

3. Determine whether each of the following functions from Z to Z is one-to-one, where Z is the set of integers.

   (*a*) $f(n) = n + 2$

   (*b*) $f(n) = n^2 + n + 1$

   (*c*) $f(n) = n^3 - 1$

4. Determine whether each of the following functions from Z to Z is onto.

   (a) $f(n) = n + 2$

   (*b*) $f(n) = n^2 + n + 1$

   (*c*) $f(n) = n^3 - 1$

5. Determine whether each of the following function is a bijection from R to R.

   (*a*) $f(x) = 2x + 3$

   (*b*) $f(x) = x^2 + 2$

6. Let $A_i = \{1, 2, 3, ..., i\}$ for $i = 1, 2, 3, ...$ Find $\bigcup_{i=1}^{n} A_i$

7. Let $A_i = \{i, i+1, i+2, ...\}$ for $i = 1, 2, 3, ...$. Find $\bigcap_{i=1}^{n} A_i$

8. Each of the following formulas defines a function from R to R. Find the range of each function
   (*i*) $f(x) = x^3$ (*ii*) $f(x) = x^2 + 1$

9. Classify the following as mappings of $R \to R$
   (*a*) $f(x) = \log x$ (*b*) $f(x) = +\sqrt{x}$
   (*c*) $f(x) = \tan x$; $(x \varepsilon R)$

10. Let $f: R \to R$ be given by $f(x) = x^2$, and $g: R \to R$ be given by $g(x) = x + 3$. Calculate $(fog)(2)$ and $(gof)(2)$. Hence, show that $fog \neq gof$.

11. If $f: R \to R$ be defined as $f(x) = |x|$, then prove that $fof = f$.

ANSWERS

1. (*a*) Not a function. $f(0)$ is not defined.
   (*b*) Not a function. f is not single valued. For example $f(1) = 1$ and -1.
   (*c*) It is a function.
2. (*a*) Domain : the set of bit strings. Range : the set of natural numbers.
   (*b*) Domain : the set of bit strings. Range : $\{0, 1, 2, 3, 4, 5, 6, 7\}$.
3. (*a*) and (*b*) are one-to-one.
   (*a*) Suppose that $f(n_1) = f(n_2)$
   Then $n_1 + 2 = n_2 + 2$. Hence $n_1 = n_2$.
   Hence f in this case is one-to-one.
   (*b*) *For example* $f(1) = 3 = f(-2)$. Hence f is not one-to-one.
   (*c*) Suppose that $f(n_1) = f(n_2)$.
   Then $n_1^3 - 1 = n_2^3 - 1$.
   Hence $n_1^3 = n_2^3$.

   Since $n_1 n_2 > 0$, if $n_1^3 = n_2^3$, then $n_1 = n_2$ follows.
   Hence f in this case is one-to-one.
4. (*a*) Yes.
   (*b*) No. *For example* nothing in Z is mapped to 0 by f.
   (*c*) No. *For example* nothing in Z is mapped to 1 by f.
5. (*a*) Yes. If $f(x_1) = f(x_1)$, then $2x_1 + 3 = 2x_2 + 3$. Hence $x_1 = x_2$. Hence f is one-to-one. For any y in R, $x = (y - 3)/2$ is in R and it is mapped to y by f. Hence f is onto. Hence f is a bijection.
   (*b*) This f is neither one-to-one nor onto. Hence it is not a bijection.
6. $\{1, 2, 3, ..., n\}$
7. A_n.
8. (*i*) $\forall x \in R, 3\sqrt{x} \in / R$ and thus

$$f\left(3\sqrt{x}\right) = \left(3\sqrt{x}\right)^3 = x$$

   so range of $f = R$.
   (*ii*) range of $f = [1, \infty)$
9. (*a*) not a function (*b*) not a function (*c*) not a function

Chapter

4 ALGEBRAIC STRUCTURES

4.1. ALGEBRAIC STRUCTURES

Now at this stage, we are well aware that Binary operation is a relation between two elements of the set.

If Z is a set of all integers then : operation '+' is a Binary operation on Z, since if $a, b \in Z$ then $a + b \in Z$. Similarly '–' is also binary operation on Z. Division is not binary operation on Z. In general we can say that if S is a set and $a, b \in S$ and $a * b \in S$ then we can say that '*' is binary operation on set S.

"A set G with binary operation '*' is said to be algebraic structure <G, *>".

For examples :

<N, +> is algebraic structure.

<N, *> is algebraic structure

<Z, –> is algebraic structure

<Z, /> is not an algebraic structure.

<N, –> is not an algebraic structure.

• An algebraic structure <G, *> is said to be associative, if the associative law holds in G w.r.t. the binary relation *, that is if $a, b, c \in G \Rightarrow (a * b) * c = a * (b * c)$.

For example, <N, +>, <N, ·>, <Q, +>, <Q, ·> <R, +>, <R, ·> are associative algebraic structures; while <Q, –>, <R, ÷> are not associative.

• An algebraic structure <G, *> is said to have an identity element w.r.t. *a* binary operation *, if for all $a \in G$, $\exists$ an element $e \in G$ such that $a * e = e * a$. An algebraic structure may or may not have an identity element with respect to binary operation *. *For example,* algebraic structure <N, +> does not have identity element as $0 \notin N$, while <N, ·> has the identity element $1 \in N$.

• Group, ring, field etc. are examples of various algebraic structures we come across in algebra.

4.1.1. Semigroup

An algebraic structure <G, *> is a semigroup if the following two axioms are satisfied.

G1 : Closure Property holds, that is $a, b \in G \Rightarrow a * b \in G$.

G2 : Associative law holds, that is $a, b, c \in G \Rightarrow$

$$(a * b) * c = a * (b * c).$$

Example 4.1. <Z, –> algebraic structure is not a semigroup.

Solution. Let $5, 3, 7 \in Z$

But $(5 - 3) - 7 \neq 5 - (3 - 7)$, so it is not following $G2$. Clearly <Z, –> is not a semigroup.

Example 4.2. Algebraic structure <Z⁺, ⊛> (where ⊛ is defined as : $a, b, \in Z^+$, $a \circledast b = b^a$) is not a semigroup.

Solution. Let $x = a, y = b, z = c$ and $a, b, t \in Z^+$ if <Z⁺, ⊛> is semigroup then it must follow $G2$.

$$(a \circledast b) \circledast c = a \circledast (b \circledast c)$$

$$(b^a) \circledast c = a \circledast (c^b)$$

$$c^{b^a} \neq (c^b)^a = c^{ba}$$

Clearly $G2$ is not followed then <Z+, > is not a semigroup.

Example 4.3. Algebric structure <Q, ÷> is not a semigroup (where Q stands for rational numbers).

Solution. Let $x = \frac{a}{b}, y = \frac{c}{d}, z = \frac{e}{f}$

Let say <Q, ÷> follows $G2$ then

$$(x \div y) \div z = x \div (y \div z)$$

$$\Rightarrow \quad \left(\frac{a}{b} \div \frac{c}{d}\right) \div \frac{e}{f} = \frac{a}{b} \div \left(\frac{c}{d} \div \frac{c}{f}\right)$$

$$\Rightarrow \quad \left(\frac{a}{b} \frac{d}{c}\right) \div \frac{e}{f} = \frac{a}{b} \div \left(\frac{cf}{de}\right)$$

$$\Rightarrow \quad \left(\frac{adf}{bce}\right) \neq \frac{ade}{bcf}$$

So our $G2$ is not satisfied, <Q, ÷> is not a semigroup.

4.1.2. Monoids

A semigroup <G, *> is monoid it contains an identity element or it follows $G1$, $G2$ and $G3$ existence of right and left identity element,

$e \in G, \exists$ such that $\quad a * e = a \quad$ (right identity), $\forall a \in G$

$\qquad e * a = a \quad$ (left identity) $\forall a \in G$

where e is identity element of G.

Example 4.4. <Z, +> is a monoid, since 0 is the identity element of '+'.

Example 4.5. <N, +> is not monoid since there is no identity element in *N* of '+', else it is a algebraic structure and semigroup.

Example 4.6. Prove that <S, ⊕> is a semigroup and monoid.

where $S = \{0, 1, 2\}$ **and**

$$\oplus = \begin{cases} a+b & \text{if } a+b \le 2 \\ a+b-3 & \text{if } a+b > 2 \end{cases}$$

Solution. Let us see first operation table

⊕	0	1	2
0	0	1	2
1	1	2	0
2	2	0	1

Observations :

1. *G*1 is followed (closure property satisfied) so <*S*, ⊕> is algebraic structure.
2. $(1 \oplus 2) \oplus 0 = 1 \oplus (2 \oplus 0)$

   $0 = 0$

   So *G*2 is followed (associative law holds)

   So <*S*, ⊕> is semigroup.
3. $0, 1, 2 \in S$ and 0 is identity element. So <*S*, ⊕> is monoid also.

Example 4.7. Prove that <S, *> is algebraic structure and semigroup but not a monoid.

where * is defined on $S = \{i, j, k\}$ **as :**

$$i * i = i \qquad i * j = i \qquad i * k = i$$
$$j * j = j \qquad j * k = j \qquad j * i = j$$
$$k * k = k \qquad k * i = k \qquad k * j = k$$

Solution. $S = \{i, j, k\}$ and by the operation table :

*	*i*	*j*	*k*
i	*i*	*i*	*i*
j	*j*	*j*	*j*
k	*k*	*k*	*k*

Observations :

1. Closure property holds so <*s*, *> is algebraic structure.
2. $(i * j) * k = i * (j * k)$

$$(i * k) = (i * j)$$
$$i = i$$

So G2 holds, <S, *> is semigroup.

3. *Here i, j, k are identity elements inspite of that <S, *> is not a monoid since more than one identity element exists.*

$$i * j = i \qquad (j \text{ is identity element})$$
$$j * k = j \qquad (k \text{ is identity element})$$
$$k * i = k \qquad (i \text{ is identity element})$$

4.1.3. Sub-Semigroups

Let <G, *> is a semigroup and $A \subseteq G$ such that <A, *> is also semigroup then <A, *> is sub semigroup of <G, *>.

For example <Z, +> is a semigroup and A = set of all event integers,

<A, +> is also semigroup and $A \subseteq Z$ so <A, +> is sub semigroup of <Z, +>.

$$B = \text{Set of all odd integers}$$

<B, +> is not semigroup but $B \subseteq Z$

So <B, +> is not sub semigroup of <Z, +>.

4.1.4. Addition and Multiplication Modulo *m*

We now define two types of composition known as *addition modulo m* and *multiplication modulo m*, where *m* is a fixed positive integer.

(*a*) If *a*, *b* are two integers, and *r* is the least non-negative remainder obtained by dividing the sum $a + b$ by *m*, then we define the composition *addition modulo m* ($+_m$) by the relation :

$$a + b \pmod{m} = r, \text{ where } 0 \leq r < m.$$

e.g., $13 + 7 \pmod 6 = 2$, since $13 + 7 = 6 \times 3 + 2$.

(*b*) If *r* be the least non-negative remainder when the product $a \times b$ is divided by *m*, then we define the composition *multiplication modulo m* ($\times_m$) by the relation :

$$a \times b \pmod{m} = r, \text{ where } 0 \leq r \leq m, \text{ and } a, b \in I$$

e.g., $8 \times 7 \pmod 5 = 1$, as $8 \times 7 = 5 \times 11 + 1$.

If S = {0, 1, 2, 3, 4, 5}, then under addition modulo 6 and under multiplication modulo 6, the composition tables are as given below :

+	0	1	2	3	4	5
0	0	1	2	3	4	5
1	1	2	3	4	5	0
2	2	3	4	5	0	1
3	3	4	5	0	1	2
4	4	5	0	1	2	3
5	5	0	1	2	3	4

$\times_6$	0	1	2	3	4	5
0	0	0	0	0	0	0
1	0	1	2	3	4	5
2	0	2	4	0	2	4
3	0	3	0	3	0	3
4	0	4	2	0	4	2
5	0	5	4	3	2	1

4.1.5. Groups

A group is a non-empty set G under a binary operation * (or o) if the following four axioms are satisfied.

G1. Closure property : $\forall\, a, b \in G, a * b \in G$

G2. Associative law : $\forall\, a, b, c \in G, (a * b) * c = a * (b * c)$

G3. Existence of right and left identity element : There exists an element $e \in G, \exists$ such that

$$a * e = a \text{ (right identity)}, \forall\, a \in G.$$
$$e * a = a \text{ (left identity)}, \forall\, a \in G.$$

G4. Existence of right and left inverse element : For every element $a \in G, \exists$ an element $a' \in G$ such that

$$a * a' = a \text{ (right inverse)}, \forall\, a \in G.$$
$$a' * a = a \text{ (left inverse)}, \forall\, a \in G.$$

where e is identity element of G.

Existence of identity element in G3 and existence of inverse of an element in G4 occur as left identity in G3 and left inverse in G4 or, right identity in G3 and right inverse in G4, but not in any manner. If in an algebraic structure $\langle G, * \rangle$, axioms G1 and G2 hold, e be a left identity and each element a $\in G$ may possess a right invese, then G may fail to be a group under the binary operation *.

Similarly, G having right identity and left invese for each of its elements may fail to be a group.

The group G with the binary operation * is sometimes denoted by $\langle G, * \rangle$. In particular the group $\langle G, + \rangle$ is called an *additive group*, the binary operation being addition; and the group $\langle G, \cdot \rangle$ is called a *multiplicative group*, the binary operation being multiplication.

4.1.6. Abelian Group

A group $\langle G, * \rangle$ is said to be abelian or commutative, if commutative law with respect to the operation * hold in G *i.e.*, if $a * b = b * a$, for all $a, b \in G$.

In particular, an *additive abelian group is called a module.*

A group which is not abelian group is called a non-abelian group.

4.1.7. Order of a Group

A natural characteristic of a group $\langle G, * \rangle$ is the number of elements it contains. We call this the order of the group. If the group $\langle G, * \rangle$ be such that the set G contains n elements, $n \in N$, then it is called a ***finite group of order*** n. If G is an infinite set, then the group $\langle G, * \rangle$ will be called an ***infinite group***.

Example 4.8. The set of integer I is group with respect to the operations of addition. For

(*i*) closure property holds in I, the sum of any two integers being an integer ;

(*ii*) addition is associative since $(x + y) + z = x + (y + z)$ for any three integers x, y, z.

(*iii*) the integer 0 is the identity element since $a + 0 = 0 + a = 0$, a being an integer.

(*iv*) the inverse of any element $a \in I$, is its negative *i.e.* $-a \in I$, since $a + (-a) = (-a) + a = 0$.

Hence, $\langle I, +\rangle$ is a group.

Moreover, as $a + b = b + a$, $a, b \in I$, it is an abelian group (or module).

Example 4.9. If Q_0 be the set of all non-zero rational numbers, then the algebraic structure $\langle Q_0, \cdot\rangle$ forms a multiplicative abelian group. For we have

G1. $a, b \in Q_0 \Rightarrow a \cdot b \in Q_0$.

G2. $a \cdot (b \cdot c) = (a \cdot b) \cdot c \; \forall \; a, b, c \in Q_0$.

G3. $I \in Q_0$ is the *identity element*, $a \cdot 1 = 1 \cdot a = a$ for $a \in Q_0$.

G4. For all $a \in Q_0$, $\frac{1}{a}$ or $a^{-1} \in Q_0$ such that $a \cdot \frac{1}{a} = \frac{1}{a} \cdot a = 1$.

$\frac{1}{a}$ or a^{-1} *is called the inverse of a.*

Moreover, for all a, b $\in Q_0$, $a \cdot b = b \cdot a$ holds good. Hence $\langle Q_0, \cdot\rangle$ is a multiplication abelian group.

Example 4.10. Let $G = (1)^{1/4} = \{1, -1, +i, -i\}$ and binary operation * is defined in following operation table :

*	1	–1	+*i*	–*i*
1	1	–1	*i*	–*i*
–1	–1	+1	–*i*	+*i*
+*i*	*i*	–*i*	–1	+1
–*i*	–*i*	+*i*	+1	–1

No let us see observation about <G, *>

1. G1 (closure property satisfied) over * so <G, *> is **algebraic structure.**
2. G2 (Associative low holds) is satisfied over *, that is

$$\begin{bmatrix} 1 * (i) * -1) & = & (1 * i) * (i) \\ 1 * 1 & = & 1 * (-i) \\ 1 & = & 1 \end{bmatrix}$$

   So <G, *> is **semigroup.**

3. G_3 is satisfied.

   From operation table it is clear that 1 is identity element so <G, *> is **monoid.**

4. G_4 is satisfied

   Inverse element exists so <G, *> is **Group**, as $1 * 1 = 1$, $(-1) * (-1) = 1$, $i * (-i) = 1$

5. Commutative law holds (by changing row and columns, we get same) so <G, *> is **Abelian group.**

Example 4.11. Let $G = (1)^{1/3} = \{1, w, w^2\}$ $(w^3 = 1)$ and operation table is as follows :

*	1	w	w^2
1	1	w	w^2
w	w	w^2	1
w^2	w^2	1	w

We can analyze in similar fashion as we did *for example* 4.10 and can say that <G, *> is Abelian group.

Example 4.12. Let $G = \{0, 1, 2, 3, 4, 5\}$ is a set, prove that G is an abelian group under operation $+_6$ (Additive modulo 6).

Solution. $G = \{0, 1, 2, 3, 4, 5\}$

Let us design operation table as follows :

$+_6$	0	1	2	3	4	5
0	0	1	2	3	4	5
1	1	2	3	4	5	0
2	2	3	4	5	0	1
3	3	4	5	0	1	2
4	4	5	0	1	2	3
5	5	0	1	2	3	4

Observation according to above operation table :

1. $G1$ holds so $(G, +_6)$ is **algebraic structure.**
2. $G2$ holds (associative law holds) so $(G, +_6)$ is **semigroup.**
3. $G3$ holds (0 is the identity element) so $(G, +_6)$ is **monoid.**
   as $0^{-1} = 0, 1^{-1} = 5, 2^{-1} = 4, 3^{-1} = 3, 4^{-1} = 2$ and $5^{-1} = 1$
4. $G4$ holds so $(G, +_6)$ is **group.**
5. There is no change after changing rows and columns (commulative holds) so $(G, +_6)$ is **abelian group.**

Example 4.13. Prove that $G = \{1, 2, 3, 4, 5, 6\}$ is an abelian group under operation $*_7$ (multiplicative modulo 7).

Solution. Given $G = \{1, 2, 3, 4, 5, 6, 7\}$, let us design operation table as follows :

$*_7$	1	2	3	4	5	6
1	[1]	2	3	4	5	6
2	2	4	6	[1]	3	5
3	3	6	2	5	[1]	4
4	4	[1]	5	2	6	3
5	5	3	[1]	6	4	2
6	6	5	4	3	2	[1]

Let us analyze the operation table by the following observations :

1. $G1$ holds (closure property satisfied) so $(G, *_7)$ is algebraic structure.

2. $G2$ holds (associative law) satisfied as :

$$(1 *_7 2) *_7 3 = 1 *_7 (2 *_7 3)$$
$$(2 *_7 3) = 1 *_7 6$$
$$6 = 6$$

So $(G, *_7)$ is semigroup.

3. $G3$ holds (1 is identity element) so $(G, *_7)$ is monoid.
4. $G4$ holds so $(G, *_7)$ is group.
5. G is commutative over $*_7$. So $(G, *_7)$ is an abelian group.

Theorem 4.1. In a group $\langle G, * \rangle$, a left identity element is also a right identity element, and a left inverse of an element is also its right inverse.

Proof. Let e be a left identity in group $\langle G, * \rangle$. Then for all $a \in G, e * a = a \in G$ ...(*i*)

Also let $a' \in G$ be the left inverse of an element $a \in G$.

Also let $a'' \in G$ be the left inverse of a'.

So $$a' * a = e \quad \text{...(ii)}$$

and $$a'' * a' = e \quad \text{...(iii) [by G4]}$$

Then $(a' * a) * a' = e * a' = a'$, by (*i*)

or $$a' * (a * a') = a' \quad \text{...(by G2)}$$

Again $$a'' * (a' * (a * a')) = a'' * a' = e, \text{ by } (iii)$$

or $$(a'' * a) * (a * a') = e \quad \text{(by G2)}$$

or $$e * (a * a') = e \quad \text{[by (iii)]}$$

So $$a * a' = e \quad \text{...(iv)} \quad (\because e * e = e)$$

Hence a' is also the right inverse of the element $a \in G$.

Now
$$a * e = a * (a' * a)$$
$$= (a * a') * a \quad \text{(by G2)}$$
$$= e * a \quad \text{[by (iv)]}$$
$$= a \quad \text{[by (i)]}$$

Hence e is also the right identity element in G. Similarly it can be proved that in a group $\langle G, * \rangle$, a right identity is also a left identity, and right inverse of an element in G is also its left inverse.

Thus an identity element, and an inverse of an element in a group are both-sided.

Theorem 4.2. The identity element and the inverse of an element in a group $\langle G, * \rangle$ are both unique.

Proof. (1) If possible, let e and e' be two identity elements in $\langle G, * \rangle$. We have seen that identity element are both-sided.

Then $e * e' = e'$, (*e* being a left identity)

Also $e * e' = e$, (*e* being a right identity)

Hence it follows that $e' = e$, which reveals the uniqueness of identity element in a group.

(*ii*) If possible, let a' and a'' be two inverses of an element a in $\langle G, *\rangle$. Here also we note that a' and a'' are both-sided. Then we have

$$a' = a' * e = a' * (a * a'') = (a' * a) * a'' = e * a'' = a''.$$

Hence the result.

Theorem 4.3. For any three elements $a, b, c \in \langle G, *\rangle$.

(*i*)if $a * b = a * c$, then $b = c$ (left cancellation law)

and (*ii*)if $a * c = b * c$, then $a = b$ (right cancellation law).

Proof. Let e be the identity element, and a^{-1}, b^{-1}, c^{-1} be the inverse elements of a, b, c respectively in $\langle G, *\rangle$.

We note that $e, a^{-1}, b^{-1}, c^{-1}$ are all unique.

Now when $a * b = a * c$, we have

$$a^{-1} * (a * b) = a^{-1} * (a * c)$$

or $$(a^{-1} * a) * b = (a^{-1} * a) * c \quad \text{[by G2]}$$

or $$e * b = e * c \quad \text{[by G4]}$$

or $$b = c \quad \text{[by G3]}$$

Again if $a * c = b * c$, then we have

$$(a * c) * c^{-1} = (b * c) * c^{-1}$$

or $$a * (c * c^{-1}) = b * (c * c^{-1}) \quad \text{[by G2]}$$

or $$a * e = b * e, \quad \text{[by G4]}$$

So $$a = b \quad \text{[by G3]}$$

Hence the theorem.

Theorem 4.4. For any two elements $a, b \in \langle G, *\rangle$, (*i*) $(a^{-1})^{-1} = a$ and (*ii*) $(a * b)^{-1} = b^{-1} * a^{-1}$.

Proof. Let e be the identity element in $\langle G, *\rangle$.

Then $a * a^{-1} = a^{-1} * a = e$ and $(a^{-1})^{-1} * a^{-1} = a^{-1} * (a^{-1})^{-1} = e$

As $a^{-1} * a = e$ and $a^{-1} * (a^{-1})^{-1} = e$ for $a, a^{-1} \in G$, by the left cancellation law, it follows that $(a^{-1})^{-1} = a$, which prove (*i*).

Again $(a * b)^{-1} * (a * b) = e$, where $a * b, (a * b)^{-1} \in G$.

Now $$(b^{-1} * a^{-1}) * (a * b) = b^{-1} * \{(a^{-1} * a) * b\}$$
$$= b^{-1} * (e * b)$$
$$= b^{-1} * b = e.$$

So $(a * b)^{-1} = b^{-1} * a^{-1}$, since the inverse of an element is unique. Hence follows the result (*ii*).

Theorem 4.5. In a group $\langle G, *\rangle$, the equations $a * x = b$ and $y * a = b$ have unique solutions $x = a^{-1} * b$ and $y = b * a^{-1}$, where $a, b \in G$.

Proof. We have $a * (a^{-1} * b) = (a * a^{-1}) * b$
$$= e * b$$
$$= b.$$

and $$(b * a^{-1}) * a = b * (a^{-1} * a) = b * e = b.$$

where e is the idenity element in G.

Thus as $a * x = b$ and $a * (a^{-1} * b) = b$, it follows that $x = a^{-1} * b$; and as $y * a = b$ and $(b * a^{-1}) * a = b$, it follows that $y = b * a^{-1}$. By virtue of the uniqueness of inverse and identity elements, it follows that the solutions are also unique.

4.1.8 Subgroup

A non-empty set H of a group G is called a subgroup of G if the composition in G is also a composition in H and for this composition H itself is a group.

In other, words, let $(G, *)$ be a group and $H \subseteq G$. Then $(H, *)$ is called a subgroup of $(G, *)$ if $(H, *)$ is itself a group.

Since every set is a subset of itself, so if G is a group then G itself is a subgroup of G. Also if e is the identity element of G, then the subset of G containing only one element *i.e.*, e is also a subgroup of G. These two are subgroups of any group and called as *trivial* or *improper* subgroups.

For example $(R, +)$ is a group. Set of complex numbers $C \subseteq R$ and $(C, +)$ also forms a group. Thus $(C, +)$ is a subgroup of the group $(R, +)$.

Theorem 4.6. The idenity element of a subgroup is the same as that of the group.

Proof. Let S be a subgroup G under $*$. Then by definition, S and G are both groups under the same operation $*$ and $S \subseteq G$. Let e and e' be the identity elements of G and S respectively. Let $a \in S$. Then $a \in G$. We have $a * e' = a$ and $a * e = a$, for the existence of identity element of S and G.

So $a * e = a * e'$ which gives $e = e'$ (by cancellation law)

Hence, the result.

Theorem 4.7. The inverse of an element of a subgroup of a group is the same as that of the same element belonging to the group itself.

Proof. Let S be a subgroup of G under $*$, and e be their identity element. Let $a \in S$. Then $a \in G$.

Let a' and a'' be the inverses of the element a of $\langle S, *\rangle$ and $\langle G, *\rangle$ respectively.

Then $a * a' = e$, where $a' \in S$,

and $a * a'' = e$, where $a'' \in G$.

So $a * a' = a * a''$; for $a, a', a'' \in G$.

$\Rightarrow$ $a' = a''$, (by left cancellation law).

Hence the theorem.

Theorem 4.8. If $\langle G, *\rangle$ be a group and $S \subseteq G$, then S will be a subgroup of G if and only if (*i*) $a, b \in S \Rightarrow a * b \in S$ (*ii*) $a \in S \Rightarrow a^{-1} \in S$.

Proof. If S is a subgroup of G under $*$, then S itself is a group under $*$. Hence it is evident by the axioms $G1$ and $G4$ that $a, b \in S \Rightarrow a \in S \in a^{-1} \in S$.

Conversely : Let us suppose that $S \subseteq G$ and conditions (i) and (ii) hold. In order to establish that S is a subgroup of G, all that is needed is to verify that $e \in S$, where e is the idenity element of G and that the associative law holds for the elements of S.

Since the associative law holds in G and $S \subseteq G$, the law holds all the more for elements of S (the property being hereditary).

Also by (ii), $a \in S \Rightarrow a^{-1} \in S$ and then by (i) $e = a * a^{-1} \in S$. This completes the proof.

Theorem 4.9. A subset S of a group G under a binary operation $*$ is a subgroup of G, if and only if, $a \in S$ and $b \in S \Rightarrow a * b^{-1} \in S$.

Proof. If S is a subgroup of G under $*$, then the elements $a, b \in S$ implies that $b^{-1} \in S$ and hence $a * b^{-1} \in S$.

Conversely, let $\quad a, b \in S \Rightarrow a * b^{-1} \in S$.

Then for any $b \in S$, $b * b^{-1} = e \in S$. Thus identity exists and $e \in S$, $b \in S \Rightarrow e * b^{-1} = b^{-1} \in S$.

Hence, inverse exists.

Now to show clousure property let a, $b \in S$. Then as shown above $b \in S \Rightarrow b^{-1} \in S$. Therefore, applying the given condition, we have

$$a \in S,\ b^{-1} \in S \Rightarrow a * (b^{-1})^{-1} = a * \mathrm{b} \in S$$

$\therefore$ S is closed with respect to the composition. Also we know that the elements of S are from the G and the composition in is associative so it is also associative in S. Hence S is a subgroup of G under the operation $*$.

Example 4.15. Prove that the intersection of any two subgroups of a group is a subgroup of the group.

Let G be any group and H_1, H_2, be its two subgroups under a binary operation $*$. The intersection $H_1 \cap H_2$ is non-empty, since at least the identity element is common to both.

Let $a, b \in H_1 \cap H_2$. Then $a, b \in H_1$ and H_2.

Again as $a, b \in H_1$ and H_1 is a subgroup of G, therefore, $a^{-1} * b \in H_1$. For the same reason $a^{-1} * b \in H_2$.

$$\therefore \qquad a^{-1} * b \in H_1 \cap H_2.$$

Thus $a, b \in H_1 \cap H_2 \Rightarrow a^{-1} * b \in H_1 \cap H_2$ and $(H_1 \cap H_2) \subseteq G$.

Hence it follows that $H_1 \cap H_1$ is a subgroup of G.

Note: Union of two subgroups is not necessarily a sub-group.

For example, let $(G, +)$ be a group of integers.

Then $S_1 = \{\ldots, -6, -4, -2, 0, 2, 4, 6, \ldots\}$

and $S_2 = \{\ldots, -9, -6, -3, 0, 3, 6, 9, \ldots\}$

are both subgroups of C.

Now $S_1 \cup S_2 = \{..., -9, -6, -4, -3, -2, 0, 2, 3, 4, 6, 9, ...\}$

is not a subgroup of *g*.

Clearly $(S_1 \cup S_2)$ is not closed with respect to addition as $2 \in S_1 \cup S_2, 3 \in S_1 \cup S_2$ but $2 + 3 = 5 \notin S_1 \cup S_2$.

4.1.9. Homomorphism of Group

Let <G, *> and <G′, o> be two groups then a mapping $f: G \to G'$ is called a homomorphism if for each $a, b \in G$, we have

$$f(a * b) = f(a) \circ f(b)$$

4.1.10. Isomorphism of Group

A well defined mapping $f: G \to G'$ on groups <G, *> and <G′, o> is said to be isomorphism if

(*i*) *f* is one-one;

(*ii*) *f* is onto and;

(*iii*) *f* is homomorphism.

We write $G \cong G'$ and call *G* and *G′* isomorphic groups or *G* is isomorphic to *G′*.

Example 4.14. Let $P = \{0, 1, 2, 3\}$ and $Q = \{1, 3, 7, 9\}$ be two sets. $<P, +_4>$ and $<Q, *_{10}>$ are algebic structure.

A mapping $f: P \to Q$ is such that, *f* is defined as follows :

$$f(0) = 1, f(1) = 3, f(2) = 9, f(3) = 7.$$

Decide the homomorphism between two given algebraic structure.

Solution. $P = \{0, 1, 2, 3\}$ and $Q = \{1, 3, 7, 9\}$ are given sets, let us design operation tables of *P* and *Q* on $+_4$ and $*_{10}$ respectively.

$+_4$	0	1	2	3
0	0	1	2	3
1	1	2	3	0
2	2	3	0	1
3	3	0	1	2

$*_{10}$	1	3	7	9
1	1	3	7	9
3	3	9	1	7
7	7	1	9	3
9	9	7	3	1

It can be easily seen from the both operation tables that $<P, +_4>$ and $<Q, *_{10}>$ are groups.

Now let us established homomorphism :

$a = 2, b = 3$

$$f(2 +_4 3) = f(1) = 3$$

$$f(2) *_{10} f(3) = 9 *_{10} 7 = 3$$

So, $$f(2 +_4 3) = f(2) *_{10} f(3)$$

So given mapping $f: P \to Q$ is homomorphism.

Now let us see some theorems based on the concepts of group and semigroup.

4.1.11. Permutation Group

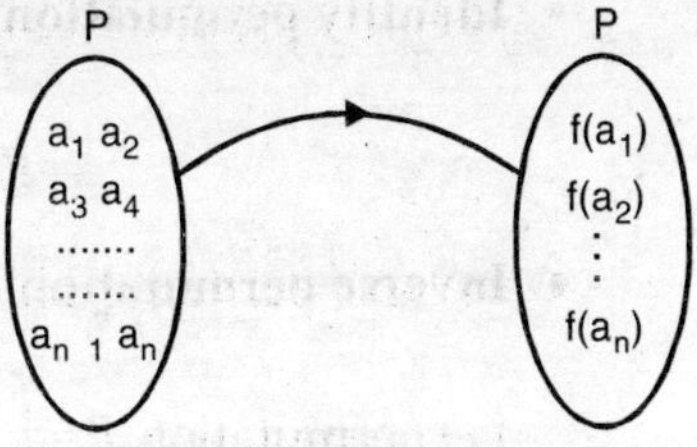

Fig. 4.1. *f* is one-one and onto mapping.

Let *P* be a finite set which has '*n*' distinct elements.

then $f: P \to P$, is known as permutation of *P* if

(*i*) *f* is one-one and;

(*ii*) *f* is onto.

Both set are same then $f(a_1), f(a_2) \dots f(a_n) \in$ domain set (*P*).

$$f = \begin{pmatrix} a_1 & a_2 & a_3 \dots\dots & a_n \\ f(a_1) & f(a_2) & f(a_3) & f(a_n) \end{pmatrix}$$

$$= \begin{pmatrix} a_1 & a_2 & a_3 & \dots\dots & a_n \\ b_1 & b_2 & b_3 & \dots\dots & b_n \end{pmatrix}$$

said to be permutation of *P*.

Example 4.16. Let $P = \{1, 2, 3\}$

$$P_1 = f = \begin{pmatrix} 1 & 2 & 3 \\ \downarrow & \downarrow & \downarrow \\ 3 & 1 & 3 \end{pmatrix} \leftarrow \text{Image}$$

$$P_2 = f = \begin{pmatrix} 1 & 2 & 3 \\ \downarrow & \downarrow & \downarrow \\ 3 & 1 & 2 \end{pmatrix} \leftarrow \text{Image}$$

Here $f: P \to P$ is one-one and onto and all the images should be belongs to *P* (same set).

- **Degree of permutation** is the distinct number of elements in the finite set.
- **Symmetric set of permutations** is the set of all possible permutations of *P* having n element.

  Let $P = \{a_1, a_2, \dots, a_n\}$

  then $n\,!$ permutations are possible.

Example 4.17. $P = \{1, 2, 3\}$

then degree of *P* is 3.

Total number of permutations $[P_n] = 3\,! = 6$

$$P_1 = \begin{pmatrix} 1 & 2 & 3 \\ 1 & 2 & 3 \end{pmatrix}, P_2 = \begin{pmatrix} 1 & 2 & 3 \\ 2 & 1 & 3 \end{pmatrix}, P_3 = \begin{pmatrix} 1 & 2 & 3 \\ 3 & 1 & 2 \end{pmatrix}$$

$$P_4 = \begin{pmatrix} 1 & 2 & 3 \\ 1 & 3 & 2 \end{pmatrix}, P_5 = \begin{pmatrix} 1 & 2 & 3 \\ 3 & 2 & 1 \end{pmatrix}, P_6 = \begin{pmatrix} 1 & 2 & 3 \\ 2 & 3 & 1 \end{pmatrix}$$

- **Identity permutation** is the permutation, which have the same image.

$$P_1 = \begin{pmatrix} 1 & 2 & 3 \\ 1 & 2 & 3 \end{pmatrix} \text{ is identity permutation.}$$

- **Inverse permutation** is defined as follows :

Let permutation $P = \begin{pmatrix} a_1 & a_2 \ldots a_n \\ b_1 & b_2 \ldots b_n \end{pmatrix}$

then inverse permutation of P is $P^{-1} = P = \begin{pmatrix} b_1 & b_2 \ldots b_n \\ a_1 & a_2 \ldots a_n \end{pmatrix}$

- **Equality of permutations** is decided by their images. Two permutations are said to be equal if their corresponding images are equal.

Let $f : P \to P$ and $g : P \to P$ be two permutation then they will be equal if

$$f(P) = g(P)$$

$$P_1 = \begin{pmatrix} 1 & 2 & 3 \\ 2 & 3 & 1 \end{pmatrix} \text{ under } f,$$

$$P_2 = \begin{pmatrix} 2 & 1 & 3 \\ 2 & 3 & 1 \end{pmatrix} \text{ under } g,$$

Here $\qquad f(a) = g(a)$

So P_1 and P_2 are equal.

- **Product of Permutation** is possible with same degree permutation.

Let P_1 and P_2 are two permutation of same degree.

$$P_1 = \begin{pmatrix} 1 & 2 & 3 & 4 & 5 \\ 3 & 4 & 5 & 1 & 2 \end{pmatrix} \text{ of degree 5.}$$

$$P_2 = \begin{pmatrix} 1 & 2 & 3 & 4 & 5 \\ 2 & 3 & 5 & 1 & 4 \end{pmatrix} \text{ of degree 5}$$

then
$$P_1P_2 = \begin{pmatrix} 1 & 2 & 3 & 4 & 5 \\ 3 & 4 & 5 & 1 & 2 \\ \downarrow & \downarrow & \downarrow & \downarrow & \downarrow \\ 5 & 1 & 4 & 2 & 3 \end{pmatrix} = \begin{pmatrix} 1 & 2 & 3 & 4 & 5 \\ 5 & 1 & 4 & 2 & 3 \end{pmatrix}$$

4.1.11.1. Group of Permutations

Set P_n is a set of all permutations, form a group under binary operation product of permutation.

$$P = (1\ 2\ 3)$$

Total permutation = 6

$$P_n = (P_1, P_2, P_3, P_4, P_5, P_6) \text{ form a group.}$$

Let us verified it by following example :

Let $P = (1, 2, 3)$

$P_6 = (P_1, P_2, P_3, P_4, P_5, P_6)$ forms a group under product operation.

$$P_1 = \begin{pmatrix} 1 & 2 & 3 \\ 1 & 2 & 3 \end{pmatrix}, P_2 = \begin{pmatrix} 1 & 2 & 3 \\ 2 & 1 & 3 \end{pmatrix}, P_3 = \begin{pmatrix} 1 & 2 & 3 \\ 3 & 1 & 2 \end{pmatrix}$$

$$P_4 = \begin{pmatrix} 1 & 2 & 3 \\ 1 & 3 & 2 \end{pmatrix}, P_5 = \begin{pmatrix} 1 & 2 & 3 \\ 2 & 3 & 1 \end{pmatrix}, P_6 = \begin{pmatrix} 1 & 2 & 3 \\ 3 & 2 & 1 \end{pmatrix}$$

$$P_1P_2 = \begin{pmatrix} 1 & 2 & 3 \\ 2 & 1 & 3 \end{pmatrix}, P_1P_5 = \begin{pmatrix} 1 & 2 & 3 \\ 2 & 3 & 1 \end{pmatrix}, P_3P_6 = \begin{pmatrix} 1 & 2 & 3 \\ 3 & 2 & 1 \end{pmatrix}$$

Observations :

1. **Closure:** G1 is satisfied (closure property exist) so $<P_6, *>$ is algebraic structure.
2. **Associativity :** $(P_1 * P_2) * P_3 = P_1 * (P_2 * P_3)$

$$\begin{pmatrix} 1 & 2 & 3 \\ 2 & 1 & 3 \end{pmatrix} * \begin{pmatrix} 1 & 2 & 3 \\ 3 & 1 & 2 \end{pmatrix} = P_1 * \begin{pmatrix} 1 & 2 & 3 \\ 3 & 1 & 2 \end{pmatrix}$$

$$\begin{pmatrix} 1 & 2 & 3 \\ 3 & 1 & 2 \end{pmatrix} = \begin{pmatrix} 1 & 2 & 3 \\ 3 & 1 & 2 \end{pmatrix}$$

So clearly $G2$ holds, $<P_6, *>$ is semigroup.

3. **Identity :** P_1 is identity permutation so $G3$ holds, $<P_6, *>$ is monoid.

4. **Inverse :** Inverse exist as $P_4P_4^{-1} = \begin{pmatrix} 1 & 2 & 3 \\ 1 & 3 & 2 \end{pmatrix}\begin{pmatrix} 1 & 3 & 2 \\ 1 & 2 & 3 \end{pmatrix}$

$$= \begin{pmatrix} 1 & 2 & 3 \\ 1 & 2 & 3 \end{pmatrix} = \text{Identity permutation}$$

G4 holds so $<P_6, *>$ is group.

4.1.11.2. Cyclic Permutations

Every permutation can be expressed as product of **disjoint cycles.**

For example :

$$P = \begin{pmatrix} 1 & 2 & 3 & 4 & 5 & 6 \\ 2 & 3 & 1 & 4 & 6 & 5 \end{pmatrix}$$

have three distinct cycles, (1, 2, 3), (4) and (5, 6).

$$P_1 = \begin{pmatrix} 1 & 2 & 3 & 4 & 5 & 6 \\ 2 & 3 & 4 & 5 & 6 & 1 \end{pmatrix} \text{ has cycle}$$

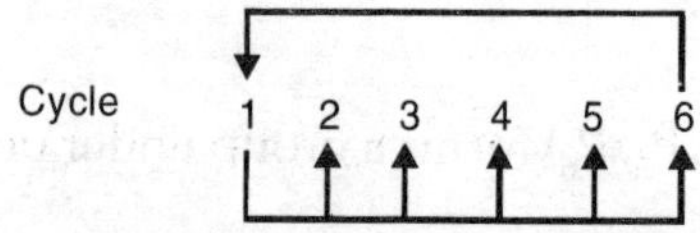

- **Length of cycle** is number of element in the cycle. Suppose a cycle is (3, 4, 5) then length of cycle is three.
- **Transposition :** A cycle of length two is called a transposition.

For example the cycle (14) is a transposition. It will represent a permutation in which the image of 1 is 4 the image of 4 is 1 and the remaining missing elements are unchanged.

If the transposition (2 5) is a permutation of degree 5 on 5 symbols then the permutation corresponding to the given transposition is

$$\begin{pmatrix} 1 & 2 & 3 & 4 & 5 \\ 1 & 5 & 3 & 4 & 2 \end{pmatrix}$$

Note :

(*i*) A cycle of length one means that the image of the elment involved is the element itself and the missing elements are left unchanged.

(*ii*) Every cycle of length one will represent the idensity permutation.

(*iii*) Two cycles are said to be disjoint if they have no symbol in common. For example (1 2 4) and (3 5 9) are disjoint cycles while (1 3 5) and (2 5 6 9) are not disjoint.

- **Even and odd permutations** are decided on the basis of number of transpositions. If number of transpositions are even then permutation is known as even permutation, otherwise it is odd permutation.
- **Some properties of permutations** are as follows :

1. A permutation cannot be both even and odd simultaneously.
2. Every permutation can be expressed as a product of disjoint cycles.

Let $P = \begin{pmatrix} 1 & 2 & 3 & 4 & 5 & 6 & 7 \\ 1 & 4 & 7 & 6 & 5 & 2 & 3 \end{pmatrix}$ be a permutation of degree 7 on set (1, 2, 3, ..., 7).

Then $p = (1)\,(5)\,(2\;4\;6)\,(3\;7)$

Since cycles of length one represent idenity permutation therefore we can omit them.

Thus $p = (2\;4\;6)\,(37)$

We can also write p as (3 7) (2 4 6) as product of disjoint cycles is commutative. That is if p_1 and p_2 are two disjoint cycles then $p_1p_2 = p_2p_1$.

Let $p_1 = (2\;4\;6)$ and $p_2 = (3\;7)$ be two permutations on 7 symbols then.

$$p_1p_2 = \begin{pmatrix} 1 & 2 & 3 & 4 & 5 & 6 & 7 \\ 1 & 4 & 3 & 6 & 5 & 2 & 7 \end{pmatrix} \begin{pmatrix} 1 & 2 & 3 & 4 & 5 & 6 & 7 \\ 1 & 2 & 7 & 4 & 5 & 6 & 3 \end{pmatrix}$$

$$= \begin{pmatrix} 1 & 2 & 3 & 4 & 5 & 6 & 7 \\ 1 & 4 & 7 & 6 & 5 & 2 & 3 \end{pmatrix}$$

and $$p_2p_1 = \begin{pmatrix} 1 & 2 & 3 & 4 & 5 & 6 & 7 \\ 1 & 2 & 7 & 4 & 5 & 6 & 3 \end{pmatrix}\begin{pmatrix} 1 & 2 & 3 & 4 & 5 & 6 & 7 \\ 1 & 4 & 3 & 6 & 5 & 2 & 7 \end{pmatrix}$$

$$= \begin{pmatrix} 1 & 2 & 3 & 4 & 5 & 6 & 7 \\ 1 & 4 & 7 & 6 & 5 & 2 & 3 \end{pmatrix}$$

3. Every cycle of 'n' can be expressed as product of $(n-1)$ transpositions.
4. Out of n ! permutations on n symbols, half of them will be even permutation and half of them will odd permutations.

4.2. COSETS

Consider of group G and H as a subgroup of G. Let $a \in G$. Then the set $Ha = \{ha : h \in H\}$ is called a right coset of H in G generated by element a. Similarly, the set $aH = \{ah : h \in H\}$ is called a left coset of H in G generated by a.

Notes: (*i*) Both the cosets Ha and aH are subsets of set G.

(*ii*) If $e \in G$ is the identify element then $H \cdot e = H = e \cdot H$. That is H itself is a right as well as left coset.

(*iii*) Since H is a subgroup of G, so $e \in H$ is also the identify element of H. Thus for any $a \in H$, $e \cdot a$ is an element of the right coset Ha of H in G, *i.e.*, $e \cdot a \in Ha$ or $a \in Ha$ or $a \in Ha$. Therefore Ha is a non-empty set. Similarly aH (left coset) is also non-empty set.

(*iv*) If group G is abelian then $\forall\ h \in H$ we have

$$a \cdot h = h \cdot a$$

Therefore the right coset Ha will be equal to the corresponding left coset aH. But if G is not abeliam, then we may have $aH = Ha$ or $aH \neq Ha$.

(*v*) If G is a group with additive composition then the right coset of H in G generated by a is defined as

$$H + a = \{h + a : h \in H\}$$

and similarly the left coset of H in G generated by a is

$$a + H = \{a + h : h \in H\}$$

4.2.1 Properties of Cosets

Theorem 4.10. Consider a group G and H is a subgroup of G. Let $a \in G$ then $a \in Ha$ and $a \in aH$.

Proof : Let $e \in G$ be the identify element. Then $e \in H$ (as H is a subgroup of G). Then for $a \in G$ and $e \in H$.

$a \cdot e \in a \cdot H$ (by the definition of coset)

$\Rightarrow\ a \in aH$

also, $a \in G$ and $e \in H$

$e \cdot a \in H \cdot a$

$\Rightarrow \; a \in Ha$

Thus, the cosets Ha and aH are non-empty.

Theorem 4.11. Let G be a group and H be its subgroup. Then for $h \in H, Hh = H = hH$.

Proof : Let $h \in H$ consider any arbitrary element $h^1 \in H$. Then $h^1 \cdot h \in Hh$, is any arbitrary element of Hh. But H is a subgroup of G, so by closure property $h^1 \cdot h \in H$.

$\Rightarrow$ every element of Hh is also an element of H

$\Rightarrow Hh \subseteq H$ ...(1)

Again, for $h' \in H$

$$\begin{aligned} h' &= h' \cdot e && \text{(by definition of identity)} \\ &= h' \cdot (h^{-1} h) && \text{(by definition of inverse)} \\ &= (h' \cdot h^{-1}) \cdot h && (h' \in H, h' \in H \Rightarrow h' h^{-1} \in H \text{ by closure property}) \\ &\in Hh \end{aligned}$$

$\Rightarrow$ every element of H is also an element of Hh

$\Rightarrow \; H \subseteq Hh$ ...(2)

By (1) and (2) $Hh = H$. Similarly, we can show that $hH = H$.

Theorem 4.12. Let G be a group and H be a subgroup of G. Then for any $a, b \in G$.

(*i*) $Ha = Hb$ if and only if $ab^{-1} \in H$

(*ii*) $aH = bH$ if and only if $a^{-1} b \in H$

Proof : Let $e\, Ha = Hb$

then $a \in Ha \Rightarrow a \in Hb$

$\Rightarrow \; ab^{-1} \in Hb\, b^{-1}$

$\Rightarrow \; ab^{-1} \in He$ (as $bb^{-1} = e \in G$)

$\Rightarrow \; ab^{-1} \in H$

Again, let $ab^{-1} \in H$

$\Rightarrow \; Ha\, b^{-1} = H$ ($h \in H \Rightarrow Hh = H$)

$\Rightarrow \; Ha\, b^{-1} b = Hb$

$\Rightarrow \; Ha\, e = Hb$ (as $b \cdot b^{-1} = e$)

$\Rightarrow \; Ha = Hb$

This $Ha = Hb$ if and only if $ab^{-1} \in H$. Similarly, we can show that $aH = bH$ iff $a^{-1} b \in H$.

Theorem 4.13. Any two left (right) cosets of a subgroup are either disjoint or identical.

Proof : Let G be a group and H be a subgroup of G.

Again let Ha and Hb are two right cosets of H in G, such that $Ha \cap Hb \neq \phi$ *i.e.*, they are not disjoint.

$\Rightarrow$ at least one element say $y \in Ha \cap Hb$

$\Rightarrow$ $y \in Ha$ and $y \in Hb$

Let $y = h_1 a$ and $y = h_2 b$ for some $h_1, h_2 \in H$

$\Rightarrow$ $h_1 a = h_2 b$

$\Rightarrow$ $h_1^{-1} h_1 a = h_1^{-1} h_2 b$

$\Rightarrow$ $ea = h_1^{-1} h_2 b$

$\Rightarrow$ $a = h_1^{-1} h_2 b$ ...(*i*)

Since H is a subgroup of G, so if $h_1, h_2 \in H$

$\Rightarrow h_1^{-1} h_2 \in h$

So for any $h \in H$, we have

$$ha = h(h_1^{-1} h_2 b)$$
$$= (h\, h_1^{-1} h_2)\, b \qquad (\because hh_1^{-1}, h_2 \in H)$$

$\Rightarrow$ $ha \in Hb$

So $Ha \subseteq Hb$ (as $ha \in Ha$)

Similarly, we can show that $Hb \subseteq Ha$. So by the two, $Ha = Hb$. So if $Ka \cap Hb \neq \phi$, then $Ha = Hb$. A similar argument can be given in case of two left cosets.

4.2.2 Lagrange s Theorem

The order of each subgroup of a group is a divisior of the order of the group.

Proof : Let G be a finite group of order n. Consider a subgroup. H of group G such that $O(H) = m$.

Consider $h_1, h_2, ..., h_m \in H$, the finite elements of subgroup H.

Let an element $a \in G$. Then Ha is a right coset of H in G and

$$Ha = \{h_1 a, h_2 a, ..., h_m a\}$$

has m-distinct elements as $h_i a = h_j a \Rightarrow h_i = h_j$. Thus each right coset of H in G has m distinct elements.

Any two right cosets of H in G are disjoint. Since G is a finite group, so the number of right cosets of H in G will also be finite. Let it be k. Then the union of these k – right cosets of H in G is equal to G.

That is, if $Ha_1, Ha_2, ..., Ha_k$ are the right cosets of H in G, then

$$G = Ha_1 \cup Ha_2 \cup ... \cup Ha_k$$

Thus, the number of elements in G

= number of elements in Ha_1 + number of elements in Ha_2 + ... +. Number of elements in

(Since two distinct right cosets Ha_k are mutually disjoint).

$$\Rightarrow \qquad o(G) = o(Ha_1) + o(Ha_2) + ... + o(Ha_k)$$

$$\Rightarrow \qquad n = m + m + ... + m\ (k - \text{times})$$

$$\Rightarrow \qquad n = n \cdot k \quad \Rightarrow k = \frac{n}{m}$$

or $$k = \frac{o(G)}{o(H)}$$

Thus order of each subgroup of a finite group G is a divisor of the order of the group.

4.2.3 Normal Subgroup

Definition 1. A subgroup H of a group G is said to be normal if for any $x \in G$ $x^{-1} Hx \subseteq H$, i.e., $x^{-1} hx \in H$, for $h \in H$.

Definition 2. A subgroup H of a group G is said to be normal if the right coset and the left coset of H in G concides (or equal). i.e., $Ha = aH$, for every $a \in G$.

Note : Every subgroup of an abelian group is normal as commutative property holds good in abelian group. A proon of this is given below.

Theorem 4.14. Every subgroup of an abelian group is normal.

Proof : Let a be an abelian group and H be a subgroup of G. Then for any $a \in G$ and $h \in H$.

$$ha = ah \qquad (G \text{ is abelian})$$

$$\Rightarrow \quad a^{-1} ha = a^{-1} \cdot ah$$

$$= e \cdot h \qquad (\text{as } x^{-1} x = e)$$

$$\Rightarrow \quad a^{-1} ha = h$$

Since $\quad h \in H \Rightarrow a^{-1} ha \in H$

Thus H is a normal subgroup of G.

Theorem 4.15. The intersection of two normal subgroups of a group is also a normal subgroup, *i.e.*, if H_1 and H_2 are two normal subgroups of G then $H_1 \cap H_2$ is also a normal subgroup of G.

Proof : Given that H_1 and H are two normal subgroups of G. Then we know that $H_1 \cap H_2$ is a subgroup of G. Now to prove $H_1 \cap H_2$ is a normal subgroup of G we need to show that for any $a \in G$ and $h \in H_1 \cap H_2$, $a^{-1} ha \in H_1 \cap H_2$.

Now, $a \in G$ and $h \in H_1 \cap H_2$

$\Rightarrow \quad h \in H_1$ and $h \in H_2$

Since H_1 and H_2 both are normal subgroups of G.

$$a \in G, h \in H_1 \Rightarrow a^{-1} ha \in H_1$$

and $\quad a \in G, h \in H_2 \Rightarrow a^{-1} ha \in H_2$

$$\Rightarrow \quad a \in G, h \in H_1 \cap H_2, a^{-1} ha \in H_1 \cap H_2$$

Thus $H_1 \cap H_2$ is a normal subgroup of G.

Theorem 4.16. A subgroup H of G is normal if and only if every right coset of H is a left coset of H *i.e.*, for $a \in G$, $aH = Ha$.

Proof : Let G be a group and H be a subgroup of G. Then

$$a^{-1} Ha = H, \forall a \in G$$

$\Rightarrow \quad a(a^{-1} Ha) = aH$

$\Rightarrow \quad (aa^{-1}) Ha = aH$ (by associativity)

$\Rightarrow \quad Ha = aH$ (as $aa^{-1} = e$)

That is, every left coset of H is G is a right coset of H in G. Conversely, let every right coset of H in G is a left coset of H in G.

i.e., $Ha = aH, \ \forall \ a \in G$

$\Rightarrow \ a^{-1} Ha = a^{-1}(aH)$

$\Rightarrow \ a^{-1} Ha = H$ (as $a^{-1} a = e$)

$\Rightarrow$ H is a normal sub group of G

Hence, H is a normal subgroup of a iff $Ha = aH$.

Theorem 4.17. A subgroup H of a group G is a normal subgroup if and only if $a^{-1} Ha = H$, for every $a \in G$.

Proof : Let G be a group and H be a normal subgroup of G.

Now $a \in G$ and $h \in H \Rightarrow a^{-1} ha \in H$

$\Rightarrow \quad a^{-1} Ha \subseteq H$ ...(1)

again let

$a^{-1} Ha \subset H, \ \forall \ a \in G$

as $\quad a \in G \Rightarrow a^{-1} \in G$ and $h \in H$

So by definition of normal subgroup

$(a^{-1})^{-1} ha^{-1} \in H$

$\Rightarrow \ (a^{-1})^{-1} Ha^{-1} \subset H$

$\Rightarrow \ aHa^{-1} \subset H$

$\Rightarrow \ a^{-1}(aHa^{-1})\, a \subset a^{-1} Ha$

$\Rightarrow \ (a^{-1} a)\, H\, (a^{-1} a) \subset a^{-1} Ha$

$\Rightarrow \ eHe \subset a^{-1} Ha$ (as $aa^{-1} = e = a^{-1} a$)

$\Rightarrow \ H \subset a^{-1} Ha$ ...(2)

By (1) and (2), we get

$$\boxed{H = a^{-1}Ha}$$

Conversely, let $H = a^{-1} Ha \ \forall \ a \in G$ and H being subgroup of G. Then to show H is normal we need to show that for every $a \in G$ and $h \in H$.

$a^{-1} ha \in H$

given that

$a^{-1} Ha = H, \ \forall \ a \in G$

$\Rightarrow \ a^{-1} Ha \subset H$

$\Rightarrow \ a^{-1} ha \in H$, for every $h \in H$ (by definition of subsets)

Thus H is a normal subgroup. Hence, the subgroup H of G is normal iff $a^{-1} Ha = H$.

Theorem 4.18. A subgroup H of group G is normal iff the product of any two right (left) cosets of H in G is again a right (left) coset of H in G.

Proof : Let G be a group and let H be a normal subgroup of G. Again let Ha_1 and Ha_2 are two right cosets of H in G, $\forall\, a_1, a_2 \in G$. How if $h_1, h_2 \in H$, then $h_1 a_1 \in Ha_1$ and $h_2 a_2 \in H a_2$.

Then $h_1 a_1 h_2 a_2 \in Ha_1\, Ha_2$

Also $h_1 a_1 h_2 a_2 = h_1(a_1 h_2) a_2$

We know that for same

$h_2 \in H, a_1 H = Ha_1$

$\Rightarrow\ a_1 h_2 = h_2 a_1$

So $h_1(a_1 h_2) a_2 \ = \ h_1(h_2 a_1) a_2$

$= \ h_1 h_2 (a_1 a_2) \in H a_1 a_2$

So $Ha_1\, Ha_2 \subset H a_1 a_2$ ...(1)

Now for any

$h \in H, h(a_1 a_2) \in H a_1 a_2,\ \forall\, a_1 a_2 \in G$

$h(a_1 a_2) \ = \ (ha_1) a_2$

$= \ (ha_1)(e a_2) \in Ha\, Hb$ (as $e \in H$)

$\Rightarrow\ Ha_1 a_2 \subset Ha_1 Ha_2$ ...(*ii*)

By equations (1) and (2), we have

$$Ha_1 a_2 = Ha_1\, Ha_2$$

Thus H is normal if product of two right cosets of H in G is again a right coset of H in G. Conversely, let the product of any two right cosets of H in G is again a right weight of H in G. Let $x \in G \Rightarrow x^{-1} \in G$ (as G is a group). So let Hx and Hx^{-1} be the two right cosets of H in G. Then $Hx\, Hx^{-1}$ is also a right coset of H in G.

How $x \in Hx$ and $x^{-1} \in Hx^{-1}$

$\Rightarrow\quad x \cdot x^{-1} \in Hx\, Hx^{-1},\ \forall\ x \in G$

$\Rightarrow\quad e \in Hx\, Hx^{-1}$

But H itself is a right coset of H in G containing the identity element e and we know that any two right cosets of H are either disjoint or identical.

$\Rightarrow\ H = Hx\, Hx^{-1},\ \forall\, x \in G$ (as $e \in H$ and $e \in Hx\, Hx^{-1} \Rightarrow H \cap Hx\, Hx^{-1} \neq \phi$)

So $H_1 x\, h_2 x^{-1} \in H,\ \forall\ x \in G$ and $h_1, h_1 \in H$

$\Rightarrow\ h_1^{-1}(h_1 x\, h_2 x^{-1}) \in h_1^{-1} H = H$ (as $h_1 \in H, h_1^{-1} \in H \Rightarrow h_1^{-1} H = H$)

$\Rightarrow\ xh_2 x^{-1} \in H,\ \forall\ x \in G$ (as $h_1^{-1} h_1 = e$) and $h_2 \in H$

$\Rightarrow H$ is a normal subgroup of G

EXAMPLES

Example 4.18. Let H be a finite subgroup of G. Then show that H and any coset Ha have same number of elements.

Solution : Let $H = \{h_1, h_2 \ldots h_k\}$, has k distinct elements. Then $Ha = \{h_1 a, h_2 a, \ldots, h_k a\}$ k-distinct elements so $h_i \neq h_j$. Thus $h_i a \neq h_j a$. So Ha also has k-distinct elements in it.

Example 4.19. Consider a group G. Let H be a subgroup of G and H be a normal subgroup of G. Prove that HM is a subgroup of G.

Solution : Since H and M both are subgroups of $G \Rightarrow e \in H$ and $e \in M$.

$\Rightarrow \; e = e \cdot e \in HM$

$\Rightarrow \; HM$ is not empty

Let $x, y \in HM$. Then $x = h_1 m_1$ and $y = h_2 m_2$ where $h_1, h_2 \in H$ and $m_1, m_2 \in M$.

So $\qquad xy = h_1 m_1 h_2 m_2 = h_1 h_2 (h_2^{-1} m_1 h_2) m_2$

Since M is normal subgroup of $G \Rightarrow h_2^{-1} m_1 h_2 \in M$ and since H and M are subgroup, $h_1 h_2 \in H$ and $(h_2^{-1}, m_1 h_2) m_2 \in M$.

So $h_1 h_2 (h_2^{-1} m_1 h_2) m_2 \in HM$ (by closure property)

$\Rightarrow \; xy \in HM$

We also have that $x^{-1} = (h_1 m_1)^{-1} = m_1^{-1} h_1^{-1} \neq h_1^{-1} (h_1 m_1^{-1} h_1^{-1})$.

Since M is normal $\Rightarrow h_1 m_1^{-1} h_1^{-1} \in M$ and also $h_1^{-1} \in H$.

$\Rightarrow \; h_1^{-1} (h_1 m_1^{-1} h_1^{-1}) \in HM$

or $\; x^{-1} \in HM$

Hence HM is closed under inverses. Consequently HM is a subgroup.

Example 4.20. Let H be a subset of G such that $H = \{a \in G : ag = ga \;\forall\; g \in G\}$. Show that H is normal subgroup of G.

Solution : First of all we need to show that H is a subgroup of G and then a normal subgroup. Clearly set H is non-empty as

$$e \in G \Rightarrow e \cdot g = g \cdot e, \;\forall\; g \in G$$

Let $\qquad a_1, a_2 \in H$

$$\Rightarrow \qquad a_1 g = g\, a_1 \text{ and } a_2 g = g\, a_2, \forall\; g \in G$$

How
$$\begin{aligned} a_1 a_2 g &= a_1 (a_2 g) \\ &= a_1 (g\, a_2) \\ &= (a_1 g)\, a_2 \\ &= g\, (a_1 a_2) \end{aligned}$$

$\Rightarrow a_1 a_2 \in H.$

Hence H is closed for composition. Again $a_2 \in H \Rightarrow a_2 g = g\, a_2 \;\; \forall\; g \in G$

$$\begin{aligned} \Rightarrow \qquad a_2^{-1}(a_2 g) &= a_2^{-1}(g\, a_2) \\ \Rightarrow \qquad (a_2^{-1} a_2)\, g &= (a_2^{-1} g)\, a_2 \\ \Rightarrow \qquad g &= (a_2^{-1} g)\, a_2 \qquad (\text{as } a_2^{-1} a_2 = e) \\ \Rightarrow \qquad g\, a_2^{-1} &= (a_2^{-1} g)\, a_2 a_2^{-1} \\ \Rightarrow \qquad g\, a_2^{-1} &= a_2^{-1} g \\ \Rightarrow \qquad a_2^{-1} &\in H \end{aligned}$$

Hence, for every element $a_2 \in H, a_2^{-1} \in H$.

How $\quad a_1 \in H, a_2 \in H \Rightarrow a_1 \in H, a_2^{-1} \in H$

then $(a_1 a_2^{-1}) g = a_1(a_2^{-1} g)$
$= a_1(g a_2^{-1})$
$= (a_1 g) a_2^{-1}$ (associativity)
$= (g a_1) a_2^{-1} = g a_1 a_2^{-1}$

$\Rightarrow \quad a_1 a_2^{-1} \in H$

So H is a subgroup of G. How co-claim that H is normal. Let $a \in G$ and $h \in H$.

Then $a^{-1} Ha = a^{-1} ha = a^{-1}(ha) = a^{-1}(ah)$
$= (a^{-1} a) h = eh = h \in H$

Thus $a^{-1} ha \in H$, $\forall x \in G$ and $h \in H$. Hence H is a normal subgroup of G.

4.2.4 Quotient Group

Let a be a group and H be a normal subgroup of G. Let Ha is a coset of G, $\forall a \in G$. The set of all cosets of H in G is denoted by G/H. Then this set G/H is a group under the binary composition defined as, for any $Ha, Hb \in G/H$.

$$Ha\,Hb = H\,ab \;\forall\, a, b \in G$$

and is called quotient group or factor group.

Note : If $(G, +)$ is a group and $(H, +)$ is a normal subgroup of G then G/H is a factor group or quotient group under the binary composition.

$$(H + a) + (H + b) = H + (a + b)$$

where $H + a, H + b \in G/H$ and $a, b \in G$.

4.2.5 Cyclic Groups

A group G with binary operator '·' (multiplication) is said to be cyclic if there exists an element $a \in G$ such that for any element $x \in G$ is of the form a^n for some integer n.

The element a is called generator of the cyclic group and denoted as

$$G = (a)$$

i.e., group G is generated by element a. For example i is the generator for the multiplicative group $a = \{1, -1, i, -i\}$ as $(i)^1 = i$, $(i)^2 = -1$; $(i)^3 = -i$ and $(i)^4 = 1$. The another generator of this group is $-i$ also.

Example 4.21. Find out-the generator(s) for the multiplicative group $a = \{1, w, w^2\}$.

Solution : There are two generators of this group, w and w^2, as

$$(w)^1 = w,\ (w)^2 = w^2 \text{ and } (w)^3 = 1$$

and $(w^2)^1 = w^2\ (w^2)^2 = w^4 = w^3 \cdot w = 1 \cdot w = w$

and $(w^2)^3 = w^6 = (w^3)^2 = (1)^2 = 1$

4.2.6 Properties of Cyclic Group

Theorem 4.19. Prove that every cyclic group is an abelian group.

Proof : Let $G = (a)$ be a cyclic group generated by a. Let x_1 and x_2 be two elements of group G. Then there eixsts two integers n_1 and n_2 such that $x_1 = a^{n_1}$ and $x_2 = a^{n_2}$.

now $\quad x_1 \cdot x_2 = a^{n_1} \cdot a^{n_2} = a^{n_1+n_2} = a^{n_2+n_1} = a^{n_2} \cdot a^{n_1} = x_2 \cdot x_1$

Thus we have $x_1 \cdot x = x_2 \cdot x_1$, $\forall\, x_1, x_2 \in G$. Hence $G = (a)$ is an abelian group.

Theorem 4.20.s Every subgroup of cyclic group is also cyclic.

Proof : Let H be a subgroup of a cyclic group $G = (a)$. Let m be the least positive integer such that $a^m \in H$. Now we shall show that H is a cyclic group generated by a^m.

Clearly $H \subseteq G$, and G is generated by a then any element of H is of the form a^n, $n \in I$.

Then by division algorithm, there exist integers q and r such that

$$n = mq + r, \qquad 0 \le r < m$$

so $\quad a^n = a^{(mq+r)} = a^{mq} \cdot a^r$

or $\quad a^r = (a^m)^{-q} \cdot a^n$

Clearly $a^r \in H$ as $a^n \in H$ (assumption) and a^m and all it's integral powers belongs to H.

Thus $\quad a^r \in H, \qquad 0 \le r < m$

But, since m is the least positive integer such that $a^m \in H$, so if $a^r \in H, 0 \le r < m$ then $r = 0$. Thus $n = m \cdot q$

and $\quad a^n = a^{m \cdot q}$

$\Rightarrow$ every element $a^n \in H$ is of the form $(a^m)^q$

Hence $H = (a^m)$ is a cyclic group generated by a^m.

Corollary : If $G = (a)$ is a cyclic group of order n, generated by a, then every subgroup of G is generated by an element a^m where m is a divisor of n.

Proof : Since $G = (a)$ of order n, *i.e.*, $a^n = e$. Then we know that any subgroup H of G is also cyclic group generated by a^m. Also $e \in H \Rightarrow a^n = e = (a^m)^r$, $r \in I$.

$\Rightarrow \quad n = m \cdot r$

$\Rightarrow \quad r = \dfrac{n}{m}$

Thus m divides n.

Theorem 4.21. If $G = (a)$, is a cyclic group generated by a, then G is also generated by a^{-1}.

Proof : Let $G = (a)$ cyclic group generated by a. Let $a^r \in G$, $r \in I$. Also we can write $a^r = (a^{-1})^{-r}$.

Since $-r \in I$, thus each element of G is generated by a^{-1}. Hence a^{-1} is also a generator of G>

Theorem 4.22. Every infinite cyclic group has two and only two generaters.

Proof : Let $G = (a)$ be an infinite cyclic group generated by a, then a^{-1} is also an generator of G. If $a = a^{-1}$, *i.e.*, the two generators are same.

Then $\quad a \cdot a^{-1} = a^2 = e$

$\Rightarrow \quad o(a) = 2$

$\Rightarrow \quad o(G) = 2$

$\Rightarrow$ G is a finite cyclic group

But G is infinite cyclic group, so $a \neq a^{-1}$. Now we need to show that a and a^{-1} are the only generators of G. So let a^m be a generator of group G, where $m \neq \pm 1$.

Now since $a \in G$ ans a^m is a generator of G, then there exists an integer n such that

$$a = (a^m)^n = a^{mn}$$

or

$$a \cdot a^{-1} = a^{mn} \cdot a^{-1}$$

$$e = a^{mn-1}$$

$\Rightarrow$ $$o(a) = mn - 1$$

$\Rightarrow$ $o(a)$ = is finite

$\Rightarrow$ G is finite which is a contradiction. hence a^m cannot be the generator of a unless

$$m = 1 \text{ or } m = -1$$

Hence G has exactly two generators namely a and a^{-1}.

Example 4.22. Show that <$G = \{i^4, i^2, i^0, i^3\}, *$> is cyclic group.

Solution. Given group is <$G = \{i^4, i^2, i^0, i^3\}, *$> and here $-i$ is another generator. So given group is cyclic group.

Example 4.23. Show that $G = \{1, w, w^2\}$ is cyclic group.

Solution. Given group is $G = \{1, w, w^2\}$

Here two generators exists, w and w^2

$$1 = (w)^3$$

$$w = (w)^4$$

$$w^2 = (w)^2$$

and $$w \in G$$

Same thing for $w^2 \in G$.

Example 4.24. Show that group <$G = \{1, 2, 3, 4, 5, 6\}, *_7$> is cyclic group.

Solution. Given group is <$G = \{1, 2, 3, 4, 5, 6\}, *_7$>. Let us find the operation table as follows :

$*_7$	1	2	3	4	5	6
1	1	2	3	4	5	6
2	2	4	6	1	3	5
3	3	6	2	5	1	4
4	4	1	5	2	6	3
5	5	3	1	6	4	2
6	6	5	4	3	2	1

$$3^2 = 3 *_7 3 = 2$$

$$3^3 = 3 *_7 (3 *_7 3) = 6$$

$$3^4 = 2^2 *_7 3^2 = 4$$

$$3^5 = 3^2 *_7 3^3 = 2 *_7 6 = 5$$

$$3^6 = 3^3 *_7 3^3 = 1$$

$$3^7 = 3^3 *_7 3^4 = 3$$

Here 3 is a generator and $3 \in G$. So given group is cyclic group.

Note : Every cyclic group have at least two generator, generator and inverse of it. In above example inverse of three is 5.

If inverse of 3 is 5 then 5 is also a generator as follows :

$$\left.\begin{aligned} 5^6 &= 1 \\ 5^4 &= 2 \\ 5^5 &= 3 \\ 5^2 &= 4 \\ 5 &= 5 \\ 5^3 &= 6 \end{aligned}\right\} \text{Here operation is } *_7$$

Example 4.25. Show that $<Z^+, +>$ is cyclic group.

Solution. Here $<Z^+, +>$ is group and 1 is generator, $1 \in Z^+$

$$0 = 1^0 \qquad \text{(1 is added 0 times)}$$

$$1 = 1^1$$

$$2 = 1^2 = 1 + 1$$

$$3 = 1^3 = 1 + 1 + 1$$

$$4 = 1^4 = 1 + 1 + 1 + 1 + ...$$

Example 4.26. Show that $<Z^-, +>$ is cyclic group.

Solution. Given group is $<Z^-, +>$.

$$(1^{-1}) = \text{Inverse of 1 under the addition}$$

$$(1^{-2}) = (1^2)^{-1} = [1 + 1]^{-1} = -2$$

$$(1^{-3}) = (1^3)^{-1} = [1 + 1 + 1]^{-1} = -3$$

So, $\quad Z = [1]^-$ generator $= [-1]$

So, $<Z, +>$ is cyclic group.

- Non-generator element of a cyclic group forms a subgroups (cyclic subgroups).

  For example $<G = \{1, 2, 3, 4, 5, 6\}, *_7\}$ is a cyclic group.

  $G = \{3\}, G = \{5\}$ so 1, 2, 4, 6 are not generators. But they may generate subset of G.

  $2^1 = 2, 2^2 = 4, 2^3 = 1$ identity element so $0(2) = 3$

  $H_1 = \{1, 2, 4\} \subseteq G$

  and H_1 is also cyclic or cyclic subgroup.

$4^1 = 4, 4^2 = 2, 4^3 = 1$ (identity element) so $0(4) = 3$

$H_2 = \{1, 2, 4\}$ is subset of G.

$H_2 = \{4\}$ so it is also cyclic.

$6^1 = 6, 6^2 = 1$

$H_3 = \{1, 6\} \quad 0(6) = 2$

$H_3 = \{6\}$ is generator

So $H_3 = \{1, 6\}$ is cyclic.

Here 2 and 4 are inverse

3 and 5 are inverse

So inverse generates some groups (may be subset).

EXERCISES

1. Show that the following are groups.
   (a) the set C of all complex numbers w.r.t. addition of complex numbers,
   (b) the set of all rational numbers w.r.t. addition.
2. Show that $(ab)^{-1} = b^{-1} a^{-1}, \forall a, b \in G$.
3. If $a, b \in G$, then the equation $ax = b$ and $ya = b$ have unique solutions in G.
4. Show that the set $G = \{x + y\sqrt{2} \mid x, y \in Q\}$ is *a* group w.r.t. addition.
5. Show that the set $G = \{1, w, w^2\}$ where w is an imaginary cube root of unity is a group w.r.t. multiplication.
6. Prove that the set of all nth root of unity, forms a finite abelian group of order n w.r.t. multiplication.
7. Distinguish between an abelian and a non-abelian group. Give an example of each.
8. Show that the set of matrice $\begin{bmatrix} \cos\alpha & \sin\alpha & 0 \\ -\sin\alpha & \cos\alpha & 0 \\ 0 & 0 & 0 \end{bmatrix}$ where a is a real number, forms a group under matrix multiplication.
9. Prove that the set $G = \{0, 1, 2, 3, 4, 5\}$ is a finite abelian group of order 6 w.r.t. addition modulo 6.
10. Prove that if $a^2 = a, a \in G$. Then $a = e$.
11. Prove that if G is an abelian group, then for all $a, b \in G$, and all integers n : $(ab)^n = a^n b^n$.
12. Show that every cyclic group is an abelian group.
13. Show that every group of prime order is cyclic.
14. Construct the tables for addition and multiplication residue modulo 6 for the set $S = \{0, 1, 2, 3, 4, 5\}$ and hence verify that S is an additive group but not a multiplicative one.

Chapter

5 RINGS AND FIELDS

5.1. INTRODUCTION

In the previous chapter, we have seen that group is an algebraic structure with only one binary operation defined in it. Here, we shall study algebraic structures equipped with two binary operations. To a certain extent the two operations are mainly addition and multiplication denoted by the ordinary symbols '+' and ' · ' respectively. The simplest algebraic structure $<R, +, \cdot>$ is a ring.

5.2. RING

A non-empty set R together with two binary operations '+' and ' · ' defined in it, is said to be a ring, if the following axioms are satisfied.

➤ **Axioms due to addition (+)**

*R*1. **Closure property :** $a, b \in R \Rightarrow a + b \in R$.

*R*2. **Associative law :** $a + (b + c) = (a + b) + c$, for $a, b, c \in R$.

*R*3. **Existence of identity element :** $0 \in R$ such that

$a + 0 = 0 + a$. for any $a \in R$.

'0' being the additive identity element in R.

*R*4. **Existence of inverse of an element :** For any $a \in R$,

$a + (-a) = (-a) + a = 0, -a \in R$,

0, being the additive identity element in R. '$-a$' being the additive inverse element in R.

*R*5. **Commutative law :** $a + b = b + a$, for $a, b \in R$.

➤ **Axioms due to multiplication (·)**

*R*6. **Closure law :** $a, b \in R \Rightarrow a \cdot b \in R$.

*R*7. **Associative law :** $a \cdot (b \cdot c) = (a \cdot b) \cdot c$, for $a, b, c \in R$.

➤ **Axioms due to addition (+) and Multiplication (·)**

*R*8. **Distributive laws :**

Left distributive law : $a \cdot (b + c) = a \cdot b + a . c$, for $a, b, c \in R$.

Right distributive law : $(a + b) \cdot c = a \cdot c + b \cdot c$, for $a, b, c \in R$.

The axioms $R1$ to $R5$ merely state that R is an abelian or commutative group under the operation of addition (+), *i.e.*, R is additively an abelian group. The axioms $R6$ and $R7$ state that R is a semigroup under the operation of multiplication ($\cdot$), *i.e.*, R is multiplicatively a semigroup. The axiom $R8$ serves to inter relate the two operations in R.

Thus, a set R, equipped with two binary opeations *viz.*, addition and multiplication, is said to form a ring, if

(*a*) R is additively an abelian group.

(*b*) It is multiplicatively a semi-group and

(*c*) Distributive laws hold in it. It is written as $<R, +, \cdot>$.

The additive identity element 0, called the zero of the ring R, is unique and so is the additive inverse of an element $a \in R$, denoted by $(-a)$ and called the negative (or minus) a. Since, the equation $a + x = b$ has a unique solution in any additive abelian group, it is observed that subtraction is possible and is unique in a ring R. If the multiplication in a ring R is such that $a \cdot b = b \cdot a$, for $a, b \in R$, then R is called a ***commutative ring***.

It may very well happen or even not happen that there is an element e in the ring R such that $a \cdot e = e \cdot a = a$ for every $a \in R$. If there is such element e in R, then the ring R is described as a ***ring with unity*** *element e*. As under an operation identity element in a set when exists is unique, it follows that the unit element of a ring when exists is also unique. Without any confusion the unity element may well *be* denoted by the number 1.

Example 5.1. If R *be* the set of integers, '+' is the usual addition and ' $\cdot$ ' is the usual multiplication of integers, then we can see that R is a commutative ring with unity element 1.

(*d*) If R has unity element 1, then for any $a \in R$,

$$\begin{aligned} a + (-1) \cdot a &= 1 \cdot a + (-1) \cdot a = [1 + (-1)] \cdot a \\ &= 0 \cdot a = 0 \end{aligned}$$

whence we get $(-1) \cdot a = -a$.

(*e*) In particular $a = -1$ in (*d*), we have

$$(-1) \cdot (-1) = 1.$$

Note. By virtue of this theorem, the computation with negatives and 0 is the same as we always have in the past and conveniently we can write

$$a + (-b) \text{ as } a - b$$

We can prove that

$$a \cdot (b - c) = a \cdot b - a \cdot c \text{ and } (b - c) \cdot a = b \cdot a - c \cdot a$$

For
$$\begin{aligned} a \cdot (b - c) &= a \cdot (b + (-c)) = a \cdot b + a \cdot (-c) \\ &= a \cdot b + (-a \cdot c) \\ &= a \cdot b - a \cdot c \end{aligned}$$

and using left distributive law we get

$$(b - c) \cdot a = b \cdot c - c \cdot a.$$

5.3 CLASSES OF RINGS

Although rings are a direct generalization of the integers, certain arithmetical facts to which we are familiar in the ring of integers need not hold in general.

Example 5.2. If R is the set of even integers, then under the usual operations of '+' and ' · ', R is a commutative ring but has no unity element.

Theorem 5.1. If R is a ring, then for all $a, b \in R$.

(a) $a \cdot 0 = 0 \cdot a = 0$

(b) $a \cdot (-b) = (-a) \cdot b = -(a \cdot b)$

(c) $(-a) \cdot (-b) = a \cdot b$.

If, in addition, R has unity element 1, then

(d) $(-1) \cdot a = -a$, and

(e) $(-1) \cdot (-1) = 1$.

(a) If $a \in R$, then $a \cdot 0 = a(0 + 0) = a \cdot 0 + a \cdot 0$ (using the right distributive law) and since R is additively a group, the equation $a \cdot 0 = a \cdot 0 + a \cdot 0$

$\Rightarrow$ $$a \cdot 0 = 0$$

Similarly, by using left distributive law, we get $0 \cdot a = 0$.

(b) For $a, b \in R$, we have $a \cdot b + a \cdot (-b) = a \cdot (b + (-b)) = a \cdot 0 = 0$ (by the right distributive law and result (i).

$\therefore$ $$a \cdot (-b) = -(a \cdot b)$$

Similarly, $(-a) \cdot b = -(a \cdot b)$.

(c) We have,
$$(-a) \cdot (-b) = -[(a \cdot b)] \quad \text{[by (ii)]}$$
$$= -[-(a \cdot b)] \quad \text{[by (ii)]}$$
$$= (a \cdot b).$$

For instance, $a \cdot b = 0$ with neither a nor b being zero, where $a, b \in R$.

5.3.1 Zero Division

If R is a ring, then a ($\neq 0$) $\in R$ is said to be a zero-divisor, if there exists, $a, b \in R$, $b \neq 0$ such that $a \cdot b = 0$.

In that case, b is also called **zero-divisor**. It may be noted that 0 is the additive identity element of R.

5.3.2 Ring with Divisors of Zero

A ring $< R, +, \cdot >$ is said to be a ring with zero-divisors, if for some pair of elements, $a, b \in R$; $a \neq 0$, $b \neq 0$, we have $a, b = 0$.

5.3.3 Ring Without Divisions of Zero

If a ring $<R, +, \cdot>$ be such that for some pair of elements $a, b \in R$, $a \cdot b = 0 \Rightarrow$ either $a = 0$ or $b = 0$, then it is called a ring without divisors of zero (or zero divisors).

5.3.4 Integral Domain

A commutative ring having unity element and no zero-divisors is called an integral domain.

In other words, an algebraic structure $<R, +, \cdot>$ is an integral domain, if (i) it is a commutative ring, (ii) it has unity element and (iii) is without any zero divisor.

For example, the ring of integers is an integral domain. But, the ring of even integers with zero element does not contain the unity element and hence, it is not an integral domain, although it has no divisor of zero.

It is possible to express the condition of a ring R being with or without zero-divisors by the help of cancellation laws. We say that cancellation laws hold in a ring R, if

$$a \neq 0, a \cdot b = a \cdot c \Rightarrow b = c \quad \text{(left cancellation law)}$$

and

$$a \neq 0, b \cdot a = c \cdot a \Rightarrow b = c \quad \text{(right cancellation law)}$$

where $a, b, c \in R$.

Theorem 5.2. A ring R is without any divisor of zero, if and only if, the cancellation laws hold in R.

First, we suppose that R is without any divisor of zero. Let $a, b, c \in$, such that $a \cdot b = a \cdot c, a \neq 0$.

Then, $a \cdot b = a \cdot c \Rightarrow a \cdot b - a \cdot c = 0$

$\Rightarrow \quad a \cdot (b - c) = 0 \quad (a \neq 0)$

$\Rightarrow \quad b - c = 0$

$\Rightarrow \quad b = c$

Similarly, we can show that

$b \cdot c = c \cdot a \Rightarrow (b - c) \cdot a = 0$

$\Rightarrow \quad b - c = 0 \quad$ (since $(a \neq 0)$)

$\Rightarrow \quad b = c.$

$\therefore$ The cancellation laws hold in R.

We now suppose that cancellation laws hold in R. If possible, let $a, b = 0, a \neq 0, b \neq 0$.

We have, $a \cdot b = a \cdot 0 \Rightarrow b = 0$, by the left cancellation law, which is a contradiction. Hence, the result follows.

5.3.5 Division Ring or Skew-field

A ring with at least two elements is called a division ring or skew field, if it is a ring with unity element and is such that every non-zero element is invertible.

Thus, a skew-field is additionally a group, and the set of all non-zero elements of it forms a multiplicative non-commutative group.

5.3.6 Field

A commutative division ring is called a field. By virtue of its importance, we shall discuss the field-properties in a separate article to follow.

Theorem 5.3. A commutative ring R is an integral domain, if and only if, for any non-zero element $a \in R$, $a \cdot u = a \cdot v \Rightarrow u = v$ where $u, v \in R$.

Let R *be* a commutative ring and $a \cdot u = a \cdot v$

$\Rightarrow \quad u = v$ where $a\ (\neq 0) \in R$ and $u, v \in R$.

Then, for $a \neq 0, a \cdot b = a \cdot 0 \Rightarrow b = 0$ (taking $u = b$ and $v = 0$). Hence, R has no zero-divisor and consequently R is an integral domain.

Conversely, let R *be* an integral domain, and let $a(\neq 0) \in R, u, v \in R$. Now, $a \cdot u = a \cdot v \Rightarrow a \cdot (u - v) = 0$. Since, R is an integral domain and $a \neq 0$, a cannot *be* a divisor of zero.

$\therefore \quad u - v = 0 \quad i.e., \; u = v$

Hence, $a \cdot u = a \cdot v \Rightarrow u = v$, where $a \neq 0$.

Theorem 5.4. A division ring does not contain any zero-divisor.

Let R be a division ring, and let $a, b, \in R$ such that $a \cdot b = 0$ and $a \neq 0$. Since in a division ring every non-zero element is invertible, a^{-1} exists and $a^{-1} \in R$ where $a \cdot a^{-1} = a^{-1} \cdot a = 1$.

$\therefore \quad b = 1 \cdot b = (a^{-1} \cdot a) \cdot b = a^{-1}(a \cdot b) = a^{-1} \cdot 0 = 0$

Thus, for $a \neq 0$, $a \cdot b = 0 \Rightarrow b = 0$.

Hence, R has no divisor of zero.

Theorem 5.5. A ring R is a division ring if and only if, the equations $a \cdot x = b$ and $y \cdot a = b$ have unique solutions in R where $a\ (\neq 0)$, $b \in R$.

Let R be a division ring. As the non-zero elements of R are invertible, R is multiplicatively a group. Therefore, the equations $a \cdot x = b$ and $y \cdot a = b$ have unique solutions in R, provided $b \neq 0$. If $y = 0$, since R has no zero-divisors.

Conversely, let R *be* a ring and the equations $ax = b$ and $y \cdot a = b$ have unique solutions in R, where $a(\neq 0)$, $b \in R$.

Then, if $b = a$, we have $a \cdot x = a$ and $y \cdot a = a$. Now, $a \cdot x^2 = a \cdot (x \cdot x) = (a \cdot x) \cdot x = a \cdot x$. Hence, $x^2 = x$, *i.e.*, $x - 1$ ($\because x \neq 0$). Likewise, we can prove that $y = 1$. Hence, R has the unity element 1. Further $a \cdot x = 1$ and $y \cdot a = 1$ implies that $a^{-1} \in R$, where $a\ (\neq 0) \in R$. Hence R is a division ring.

5.4. FIELD AND ITS AXIOMS

A set F (having at least two elements) equipped with two binary operations *viz.* addition (+) and multiplication (·), is said to form a field, if the following axioms are satisfied.

➤ **Properties due to addition (+)**

F1. Closure law : For any two elements $a, b \in F$,

$\Rightarrow \quad a + b \in R.$

F2. Associative law : For any three elements, $a, b\ c \in F$

$$a + (b + c) = (a + b) + c.$$

F3. Existence of additive elements : The element $0 \in F$ such that for all $a \in F$, $a + 0 = 0 + a = a$, 0 being the additive identity element of F.

F4. Existence of additive inverse of an element : For all $0 \in F$, the element $-a \in F$ such that $a + (-a) = (-a) + a = 0$. $-a$ is called the additive inverse of a in F.

F5. Commutative law : For all $a, b, \in F$, $a + b = b + a$.

➤ **Properties due to multiplication (·)**

F6. Closure law : $a, b \in F \Rightarrow a \cdot b \in F$.

F7. Associative law : $a, b, c \in F \Rightarrow a \cdot (b \cdot c) = (a \cdot b) \cdot c$.

F8. Existence of multiplicative identity : The element $1 \in F$, such that $a \cdot 1 = 1 \cdot a = a$ for all $a \in F$ is called the multiplicative identity element.

F9. Existence of multiplicative inverse of an element : For all $a \in F, a \neq 0$ $\Rightarrow$ $a^{-1} \in F$ such that

$$a \cdot a^{-1} = a^{-1} \cdot a = 1$$

Thus, a^{-1} is called the inverse of a and *vice-versa.*

F10. Commutative law : For all $a, b \in F$,

$$a \cdot b = b \cdot a.$$

➤ **Properties due to addition (+) and multiplication (·)**

F11. Distributive laws : For any three elements $a, b, c \in F$

$$a \cdot (b + c) = a \cdot b + a \cdot c, \quad \text{(left distributive law)}$$

and $$(a + b) \cdot c = a \cdot c + b \cdot c, \quad \text{(right distributive law)}$$

In other words, the algebraic structure $<F, +, \cdot>$ is called a field, if

(*i*) F is an additive abelian group.

(*ii*) F' is a multiplicative abelian group, where $F' = F - \{0\}$ and

(*iii*) Distributive laws hold in F.

Recalling to the definition of an integral domain, we can easily see that an integral domain F is a field, if every non-zero element $a \in F$ have a multiplicative inverse a^{-1} such that $a \cdot a^{-1} = a^{-1} . a = 1$. One can also say that a ring is a field, if it forms abelian groups under both addition and multiplication with distributive laws satisfied in it.

As all non-zero elements in the ring of integers do not possess multiplicative inverse, the ring of integers is not a field.

Theorem 5.6. A field is necessarily an integral domain.

Let F *be* a field. We are to prove that F has no divisor of zero, *i.e.*, for $a, b \in F$, $a . b = 0 \Rightarrow a = 0$ or $b = 0$.

If $a \neq 0$, when a^{-1} exists and we have $a^{-1}, (a \cdot b) = (a^{-1} \cdot a) \cdot b = 1 \cdot b = b$. Since, $a \cdot b = 0$, therefore, $b = a^{-1} \cdot 0 = 0$.

Similarly, we can show that $a = 0$, if $b \neq 0$.

Hence, the result follows.

Note. Converse of this theorem is not true. *e.g.*, the ring of all integers is an integral domain but not a field. However, the following result holds.

Theorem 5.7. A finite integral domain is a field.

Let F *be* a finite integral domain with n distinct elements. F being an integral domain, multiplication is commutative in F and it admits of no zero-divisors. We shall not show that every non-zero element of F possesses a multiplicative inverse.

Let a $(\neq 0) \in F$. Then, F being an integral domain, $a \cdot x = a \cdot y \Rightarrow x = y$ for $x, y \in F$.

But, $x \neq y$, since F contains n distinct elements. Therefore, $a \cdot x \neq a \cdot y, a \neq 0$.

Hence, there exists an element $x(\neq 0) \in F$ such that $a \cdot x = 1$, for every a $(\neq 0) \in F$. Hence, the result follows.

5.5. SUB-RINGS AND SUB-FIELDS

If the algebraic structure $<R, +, \cdot>$ a ring and $S \subseteq R$ such that $<S, +, \cdot>$ is also a ring, then S is called a sub-ring of R.

For example, the set of integers is a sub-ring of the set of rational numbers.

If an algebraic structure $<F, +, \cdot>$ *be* a field and $S \subseteq F$ such that $<S, +, \cdot>$ is either a ring or a field, then S is called a sub-ring or sub-field (as the case may *be*) of the field F.

For example, the ring of integers is sub-ring of the field of real numbers and the field of rational numbers is a sub-field of the field of real numbers.

EXAMPLES

Example 5.3. Let Q *be* the set of rational numbers, then show that $<Q, +, \cdot>$ is a ring. It is an integral domain ? Is it a field ?

Solution. (*a*) Under addition (+) :

(*i*) The sum of two rational numbers is a rational number.

$\therefore$ Q is closed under addition. (Closure law)

(*ii*) For any three rational number, $a, b \in Q$.

$a + (b + c) = (a + b) + c$. (Associative law)

(*iii*) The rational number 0 is such that $a + 0 = 0 + a = a$ for all $a \in Q$. The number '0' is the additive identity element in Q. (Existence of additive identity element).

(*iv*) For every rational number a, there is a rational number $-a$ such that $a + (-a) = 0$. The number $-a$ is called the additive inverse (negative) of a. Since $0 + 0 = 0$, therefore, 0 is the negative of itself. (Existence of additive inverse elements).

(*v*) For any two rational numbers a, b,

$$a + b = b + a. \quad \text{(Commutative law).}$$

Thus, $(Q, +)$ is an abelian group.

(*b*) Under Multiplication $(\cdot)$:

(*vi*) Q is closed with respect to multiplication, since

$$a, b \in Q \Rightarrow a \cdot b \in Q. \quad \text{(Closure law)}$$

(*vii*) Associative law holds in Q under multiplication, since $a, b, c \in Q \Rightarrow a \cdot (b \cdot c) = (a \cdot b) \cdot c$. (Associative law)

(*c*) Under addition and multiplication :

(*viii*) For any three rational numbers a, b, c;

$$a \cdot (b + c) = a \cdot b + a \cdot c$$

and

$$(a + b) \cdot c = a \cdot c + b \cdot c. \quad \text{(Distributive laws)}$$

Thus, $<Q, +, \cdot>$ is a ring.

(*ix*) Since for any two rational numbers of a and b, we have $a \cdot b = b \cdot a$, therefore, multiplication is *commutative* in Q.

Thus, $<Q, +, \cdot>$ is a commutative ring.

(*x*) The rational number 1 such that $a \cdot 1 = 1 \cdot a = a$ for any $a \in Q$. The element 1 is the multiplicative identity element in Q. (Existence of multiplicative identity element)

Thus, $<Q, +, \cdot>$ is a commutative ring with unity element.

(*xi*) Now, for any two rational numbers, a, b,

$$a \cdot b = 0 \Rightarrow \text{ either } a = 0 \text{ or } b = 0.$$

$\therefore$ $<Q, +, \cdot>$ is a commutative ring without any zero-divisor.

Hence, $<Q, +, \cdot>$ is also an integral domain.

(*xii*) If a *be* a non-zero rational number, then $a^{-1}\left(\text{or } \frac{1}{a}\right)$ is also a rational number.

$$\therefore \qquad a \cdot a^{-1} = a^{-1} \cdot a = 1$$

$\therefore$ a^{-1} is the multiplicative inverse of a ($\neq 0$). (Existence of multiplicative inverse)

By virtue of the conditions (*i*) to (*x*) and (*xii*), it is concluded that the algebraic structure $<Q, +, \cdot>$ is also a field.

Example 5.4. If *C be* set of complex numbers of the form *a* + *ib*, where *a*, *b* are real numbers, then show that $<c, +, \cdot>$ is a field.

Solution. (*a*) We use the rotation of ordered pair of real numbers and unit $a + ib = (a, b)$ where $a, b \in IR$.

(*a*) Under addition (+) :

(*i*) C is a closed under addition, since,

$(a, b) + (c, d) = (a + b, c + d) \in C$, where

$(a, b) \in C$ and $(c, d) \in C$, (Since, $a, b, c, d \in IR \Rightarrow a + b, b + d \in IR$). (Closure law)

(*ii*) Now, $a, b, c, d, e, f \in IR \Rightarrow a + c, b + d, a + c + e, b + d + f \in IR$

As $(a, b) + \{(c, d) + (e, f)\} = (a, b) + (c + e, d + f) = (a + c + e, b + d + f)$.

and $\{(a, b) + (c, d)\} + (e, f) = \{(a, c, b, d)\} + (e, f) = (a + c + e, b + d + f)$.

$\therefore$ Associative law under addition holds in C.

(*iii*) As $0 \in IR$ and $0 = 0 + i0 = (0, 0)$, therefore, $(0, 0) \in C$ such that $(a, b) + (0, 0)$

$$= (a + 0, b + 0) = (a, b), \text{ where } (a, b) \in C \text{ and } a, b \in IR.$$

$\therefore$ $(0, 0)$ is the additive identity element in C.

(*iv*) If $(a, b) \in C$, then $(-a, -b) \in C$ and $(a, b) + (-a, -b) = (a - a, b - b) = (0, 0)$. So, $(-a, -b)$ is the additive inverse of (a, b) in C.

(*v*) Also, $(a, b) \in C$ and $(c, d) \in C$

$\Rightarrow$ $(a, b) + (c, d) = (a + c, b + d) = (c + a, d + b) = (c, d) + (a, b)$.

$\therefore$ Addition is commutative in C.

[**Note** that $<C, +, \cdot>$ is an infinte abelian group].

(*b*) Under multiplication (·) :

(*vi*) If $a, b, c, d \in IR$, then $(a, b) \in C$ and $(c, d) \in C$.

Now, $(a, b), (c, d) = (ac - bd, bc + ad)$, where $ac - bd, bc + ad \in IR$.

$\therefore$ $(a, b) \cdot (c, d) \in C$, *i.e.*, C is closed under multiplication.

(*vii*) For all $(a, b), (c, d), (e, f) \in C$, we have

$$\begin{aligned}[(a, b) \cdot (c, d)] \cdot (e, f) &= (ac - bd, bc + ad) \cdot (e, f) \\ &= [(ac - bd)\, e - (bc + ad) f, (bc + ad)\, e + (ac - bd) f] \\ &= [a(ce - df) - b(de + cf), b(ce - df) + a(de + cf)] \\ &= (a, b)\,(ce - df, de + cf) \\ &= (a, b)\,\{(c, d) \cdot (e, f)\}.\end{aligned}$$

$\therefore$ Multiplication is associative in C.

(*viii*) $(1, 0) \in C$ and is such that $(a, b) \cdot (1, 0) = (a, b) = (1, 0) \cdot (a, b)$ for $(a, b) \in C$.

$\therefore$ $(1, 0)$ is the multiplicative identity element in C.

(*ix*) Let $(0, 0) \neq (a, b) \in C$. Then there exists $(c, d) \in C$ such that $(a, b) \cdot (c, d) = (1, 0) = (c, d) \cdot (a, b)$.

where $c = \frac{a}{a^2 + b^2}$ and $b = -\frac{b}{a^2 + b^2}$.

Further as $(a, b) \neq (0, 0)$, at least one of a and b is non-zero so that $a^2 + b^2 \neq 0$. So, c and d are defined and at least one of c and d is non-zero.

$\therefore$ (c, d), *i.e.*, $\left(\frac{a}{a^2 + b^2}, -\frac{b}{a^2 + b^2}\right)$ is the inverse of $(a, b) \neq (0, 0)$.

(*x*) For $(a, b), (c, d) \in C$, we have

$$\begin{aligned}(a, b) \cdot (c, d) &= (ac - bd, bc + ad)\\ &= (ca - bd, cb + da)\\ &= (c, d) \cdot (a, b)\end{aligned}$$

$\therefore$ Multiplication is commutative in C.

(*c*) Under addition and Multiplication :

(*xi*) Further if $(a, b), (c, d), (e, f) \in C$, then

$$\begin{aligned}(a, b) \cdot \{(c + d) + (e + f)\} &= (a, b) \cdot (c + e, d + f)\\ &= \{a(c + e) - b\,(d + f), b(c + e) + a(d + f)\}\\ &= \{(ac - bd) + ae - bf), (bc + ad) + (be + af)\}\\ &= (ac - bd, bc + ad) + (ae - bf, be + af)\\ &= (a, b) \cdot (c, d) + (a, b) \cdot (e, f)\end{aligned}$$

$\therefore$ Left distributive law holds in C.

Similarly, the right distributive law also holds in C, and is the same as the left distributive law, since multiplication is commutative in C.

Hence, $<C, +, \cdot>$ is *a* field.

Example 5.5. Prove that the set $S = \{0, 1, 2, 3\}$ forms *a* ring under addition and multiplication modulo 4, but not a field.

Solution. We construct the tables below for the composition or addition modulo 4 and multiplication modulo 4.

+	0	1	2	3
0	0	1	2	3
1	1	2	3	0
2	2	3	0	1
3	3	0	1	2

·	0	1	2	3
0	0	0	0	0
1	0	1	2	3
2	0	2	0	2
3	0	3	2	1

From the table, we observe

(*i*) the set S is closed under the operations of addition and multiplication modulo 4,

(*ii*) the operations are also associative and commutative,

(*iii*) '0' is the additive identity and '1' is the multiplicative identity, (*iv*) each element of S has an additive inverse and

(*v*) the distributive law holds in S.

Hence S is a commutative ring under addition and multiplication modulo 4.

Further the ring contains unity elements, but not every non-zero element has an inverse, since we observe that $2.2 = 0$. So, the ring has divisors of zero. Thus, S is a commutative ring but not a division ring, and so it is not a field.

EXERCISES

1. Examine whether the following sets from rings with respect to usual operation of addition and multiplication; state also wherever possible the nature of the ring.
   (*i*) the set of all natural numbers;
   (*ii*) the set of all integers;
   (*iii*) the set of all positive rational numbers;
   (*iv*) the set of all rational numbers;
   (*v*) the set of all even integers;
   (*vi*) the set of all real numbers;
   (*vii*) the set of all complex numbers of the form $x + iy$, where x and y are real numbers.
   (*viii*) the set of all real numbers of the form $x + y\sqrt{2}$, where x, y are integers.
2. (*a*) Examine which of the sets in (*i*)—(*viii*) forms a field.
   (*b*) Define and illustrate with examples :
   (*i*) Commutative ring,
   (*ii*) Ring with unity,
   (*iii*) Ring with a zero divisor,
   (*iv*) Integral domain
   (*v*) Division ring,
   (*vi*) Field,
   (*vii*) Sub-ring of a field.
3. Show that the set of all even integers forms a commutative ring, but not a field.
4. Prove that the set {0, 1} forms a field under addition and multiplication modulo 2.
5. Prove that in a ring R, each of the equations $a + x = b$ and $y + a = b$ have unique solutions.
6. If $S = \{0\}$, then show that $<S, +, \cdot>$ is ring. (this is called a null ring or a trivial ring).
7. Show that the set of all elements of the form $a + bw$, where a and b are rational numbers and w is an imaginary cube root of unity, is a sub-field of all complex numbers.
8. Show that the set of all those integers which are some multiples of a given integer is a ring and is a sub-ring of a ring of integers.

■■■

Chapter

6 POSETS, HASSE DIAGRAMS AND LATTICES

6.1. PARTIALLY ORDERED SET (OR POSET)

A relation R on a set A is called a partial order or partial ordering if it is

(*a*) **Reflexive :** For any $a \in A, aRa$, *i.e.*, $(a, a) \in R$.

(*b*) **Anti symmetric :** If aRb, bRa then $a = b$.

(*c*) **Transitive :** If aRb, bRc then aRc.

A set A together with a partial ordering R is called a partially ordered set (or poset) and it is denoted by (A, R).

For the realtion R, we use a relation $\leq$ ("less than or equal") called the usual order.

Note. The symbol "$\leq$" is different from the usual less than or equal to symbol.

$a \leq b$ is read as "a precedes b".

$a < b$ is read as "a strictly precedes b".

$a \geq b$ is read as "a succeeds b".

$a > b$ is read as "a strictly succeeds b".

Example 6.1. The set of integers Z with the usual ordering "$\leq$" is a poset.

Since (*a*) $a \leq a$ for every integer a,

So "$\leq$" is reflexive.

(*b*) If $a \leq b$ and $b \leq a$ where $a, b \in Z$ then $a = b$.

So "$\leq$ is antisymmetric".

(*c*) If $a \leq b$, $b \leq c$ where $a, b, c \in Z$ then $a \leq c$

So "$\leq$: is a transitive.

Hence $(Z, \leq)$ is a poset.

6.2. COMPARABILITY

If $(A, \leq)$ is a poset, the elements a and b of A are said to be comparable if $a \leq b$ or $b \leq a$. Also, if neither $a \leq b$ nor $b \leq a$, then a and b are called non-comparable.

Example 6.2. The relation of divisibility, *i.e.*, aRb if $a \mid b$, is *a* partial order on Z+. The integers 2 and 4 are comparable while 2 and 5 are non-comparable.

6.3. TOTALLY ORDERED (OR LINEARLY ORDERED) SET

An ordered set A is said to be linearly or totally ordered, if every pair of elements in A are comparable. We also say that A is a chain or Toset.

Example 6.3. The positive integers Z+ with the usual order '≤' is linearly ordered.

The set N of natural numbers under divisibility and the set of collection of sets with 'contained in' $\subseteq$ relation are not linearly ordered *i.e.*, these are not chains.

6.4. WELL ORDERED SET

A poset $(A, \leq)$ is called well ordered set if it is totally ordered and every non-empty subset of A has a least element.

Example 6.4. The set of all natural numbers ordered by the relation '<' is well ordered but the set Q of all rational numbers is not well-ordered by the relation '<'.

6.5. PRODUCT SETS AND ORDER

Let A and B are ordered sets. Let R be the relation on the product set $A \times B$ defined by $(a, b)\, R\, (a', b')$ whenever $a \leq a'$ and $b \leq b'$. Then R is a partial order on $A \times B$ and this ordering of $A \times B$ is called the product order.

6.6. LEXICOGRAPHIC ORDERING

If $(A, \leq)$ is a poset, then we say $a < b$ if $a \leq b$ but $a \neq b$. Suppose now that $(A, \leq)$ and $(B, \leq)$ are posets. Then the partial order on $A \times B$ denoted by '<', is defined as :

$(a, b) < (a', b')$ if $a < a'$ or if $a = a'$, and $b \leq b'$.

This ordering is called "lexicographic" or "dictionary" order.

If $(A, \leq)$ and $(B, \leq)$ are linearly ordered sets, then the lexicographic order < on $A \times B$ is also a linear order.

Example 6.5. (*i*) $(3, 4) < (3, 5)$ is lexicographic ordering constructed from the usual relation ≤ on the set of real numbers.

(*ii*) $(c, x) < (a, t)$ is lexicographic ordering constructed from the usual relation ≤ on the set of English alphabets $A = \{a, b, c, ..., x, y, z\}$.

Note. The digraph of a partial order has no cycle of length greater than 1.

6.7. HASSE DIAGRAMS

Let A be a finite set. The diagraph of a partial order on A has only cycles of length 1. Since a partial order is reflexive, every vertex in the diagraph of the partial order is contained in a cycle of length 1. To simplify the diagraph, we shall delete all such cycles from the diagraph. Consider the following diagraph.

This digraph can be simplified to the following digraph :

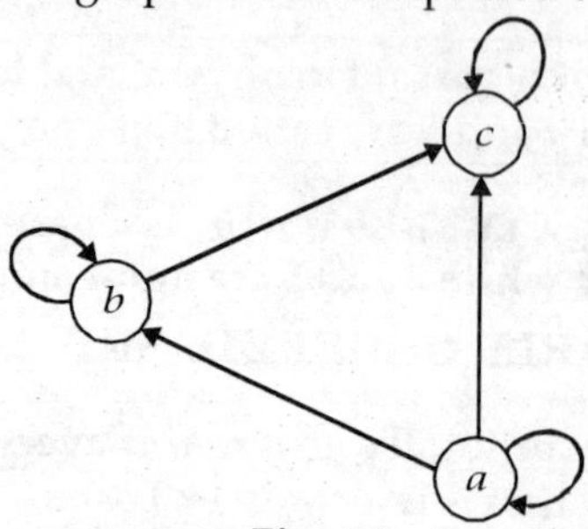

Fig. 6.1.

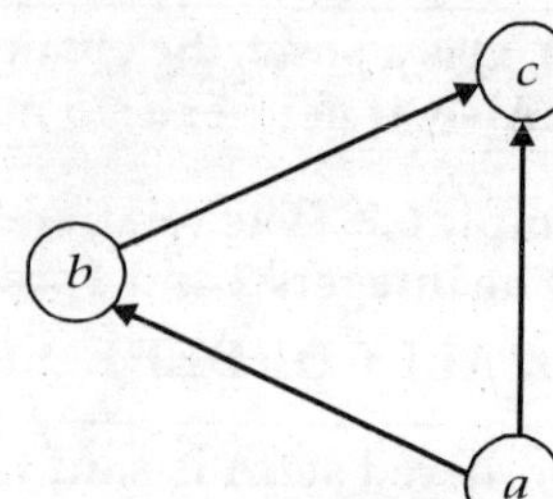

Fig. 6.2.

We shall now also eliminate all edges that are implied by the transitive property.

Thus if $a \le b$ and $b \le c \Rightarrow a \le c$ and we delete the edge from a to c. Hence **Fig. 6.2** reduces to the following digraph.

We also agree to draw the diagraph of a partial order with all edges pointing upward, so that arrows may be deleted from the edges. Finally, we replace the circles representing the vertices by dots. Thus the diagram shown in **Fig. 6.4.** gives the final form of the digraph shown in **Fig. 6.1.** The resulting diagram of a partial order, much simpler than its diagraph, is called the *Hasse diagram* of the partial order of the poset.

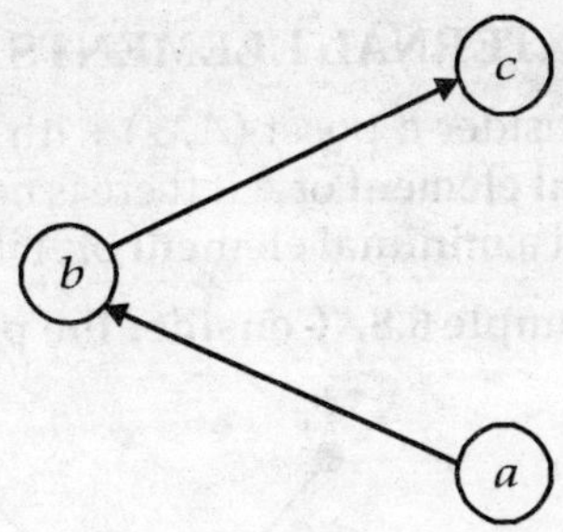

Fig. 6.3.

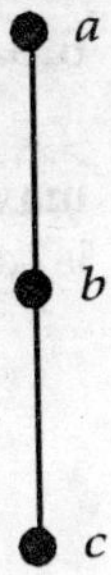

Fig. 6.4.

Example 6.6. Let $S = \{1, 2, 3, 4, 12\}$. Consider the partial order of divisibility on S. *i.e.* $a, b \in S$ iff $a \mid b$. Then, the Hasse diagram of the poset $(S, \le)$ is given below :

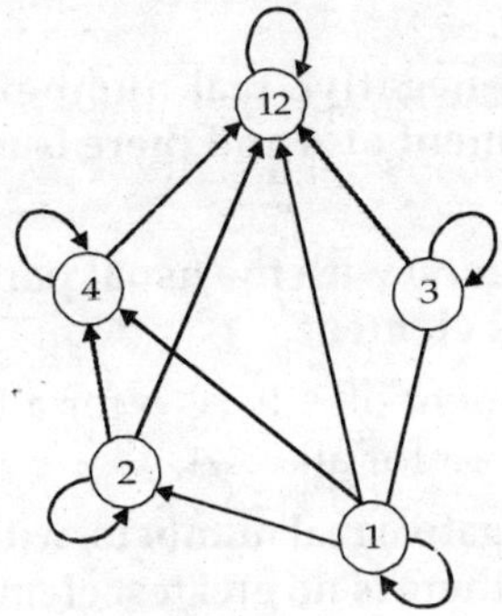

Fig. 6.5. Diagraph of S

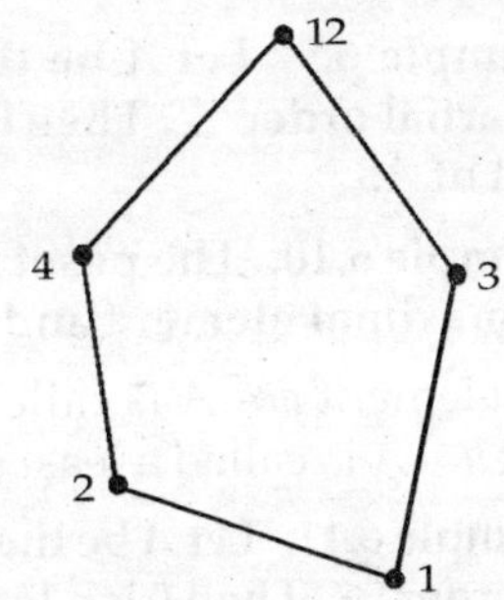

Fig. 6.6. Hasse diagram of S.

Example 6.7. Let $S = \{a, b, c\}$ and $A = P(S)$, is a poset with the partial order '$\subseteq$' (set inclusion) and $A = [\phi, \{a\}, \{b\}, \{c\}, \{a, b\}, \{a, c\}, \{b, c\}, \{a, b, c\}]$. The Hasse diagram of A is given below :

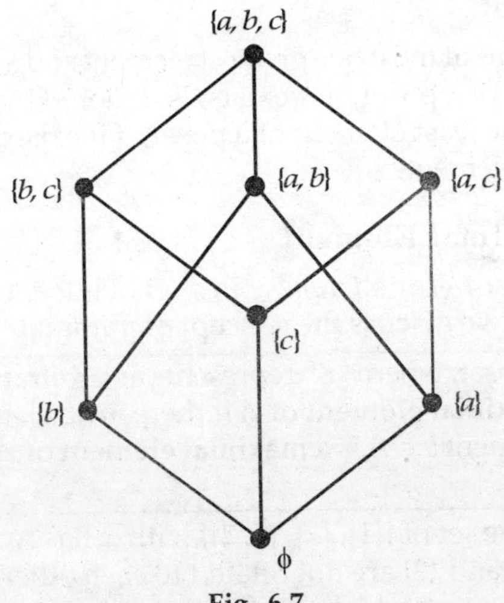

Fig. 6.7.

6.8. EXTERNAL ELEMENTS OF PARTIALLY ORDERED SETS

Consider a poset $(A, \leq)$ with partial order '$\leq$'. An element $a \in A$ is called a **maximal element** of A if there is no element c in A such that $a < c$. An element $b \in A$ is called a **minimal element** of A if there is no element c in A such that $c < b$.

Example 6.8. Consider the poset *A* whose Hasse diagram is as follows :

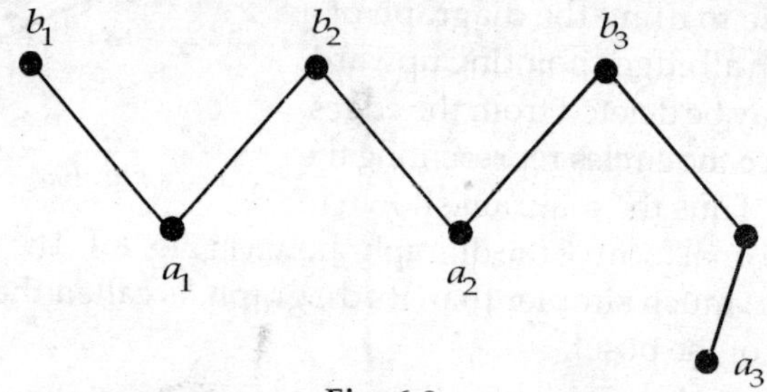

Fig. 6.8.

The elements b_1, b_2, b_3 are maximal elements of A and the elements a_1, a_2, and a_3 are minimal elements. Since there is no line between a_2 and a_3, we can conclude neither $a_3 \leq a_2$ nor $a_2 \leq a_3$.

Example 6.9. Let *A* be the poset of non-negative real numbers with the usual partial order '$\leq$'. Then 0 is minimal element of *A* and there is no maximal element of *A*.

Example 6.10. The poset *Z* (set of all integers) with the usual partial order $\leq$ has no maximal element and has no minimal element.

An element $a \in A$ is called a greatest element of A if $x \leq a$ for all $x \in A$. An element $a \in A$ is called a least element of A if $a \leq x$ for all $x \leq A$.

Example 6.11. Let *A* be the poset of non-negative real numbers, with the usual partial order '$\leq$'. Then 0 is a least element and there is no greatest element.

Example 6.12. Let $S = \{a, b, c\}$, consider the poset $A = P(S)$ with the partial order '$\leq$' (containment). Then the empty set is atleast element of *A* and the set *S* is a greatest element of *A*.

Example 6.13. The poset *Z* with the usual partial order has neither a least nor a greatest element.

Result. *A* poset has at most one greatest element and at most one least element. The greatest element of a poset, if it exists, is denoted by 1 and is called the unit element. Similarly, the least element of a poset, if it exists, is denoted by *O* and is often called the zero element.

6.8.1 Minimal Maximal Element

"The concept of *least/greatest number* in a set of integers can be generalized for a general poset. Here we discuss the concept of minimal/maximal elements."

Let $<A, \leq>$ be a poset, where "$\leq$" represents an arbitrary partial order. Then an element $b \in A$ is a **minimal element** of A if there is no element $a \in A$ that satisfies $a \leq b$. Similarly an element $b \in A$ is a **maximal element** of A if there is no element $a \in A$ that satisfies $b \leq a$.

Example 6.14. The set of $\{\{1\}, \{2\}, \{1, 2\}\}$ with $\subseteq$ has two minimal elements $\{1\}$ and $\{2\}$. Note that $\{1\}$, and $\{2\}$ are not related to each other in $\subseteq$. Hence we cannot say which is "smaller than" which, that is, they are not comparable.

6.8.2 Least/Greatest Element

Let $<A, \leq>$ be a poset. Then an element $b \in A$ is the **least element** of A if for every element $a \in A, b \leq a$. Similarly an element $b \in$ A is called greater if for every $a \in A, a \leq b$.

Note that the least greatest element of a poset is unique if one exists because of the antisymmetry of $\leq$.

Example 6.15. The poset of the set of natural numbers with the less-than-or-equal-to relation has the least element 0.

Example 6.16. The poset of the powerset of $\{1, 2\}$ with $\subseteq$ has the least element $\emptyset$

6.8.3 Upper and Lower Bounds

Consider a Poset A and a Subset B of A. An Element $a \in A$ is Called an Upper Bound of B if $b \leq a$ for all $b \in B$. An Element $a \in A$ is Called a Lower Bound of B if $a \leq b$ for all $b \in B$.

Example 6.17. Consider the poset $A = \{a, b, c, d, e, f, g, h\}$ whose Hasse-diagram is shown as follows :

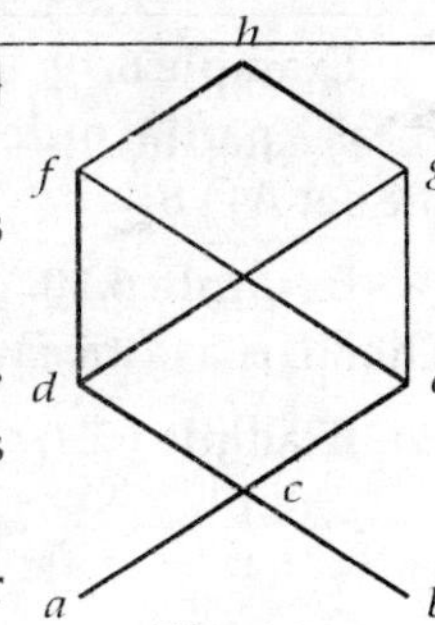

Fig. 6.9.

Find all upper and lower bounds of the following subsets of A :

(a) $B_1 = \{a, b\}$ **(b)** $B_2 = \{c, d, e\}$ **(c)** $B_3 = \{f, g\}$.

Solution. (a) B_1 has no lower bounds. Its upper bounds are c, d, e, f, g and h.

(b) The upper bounds of B_2 are f, g and h, and its lower bounds are c, a and b.

(c) The upper bound of B_3 is h and its lower bounds are a, b, c, d and e.

Thus a subset B of a poset A may or may not have upper or lower bounds (in A). Also an upper or lower bound of B may or may not belong to B itself.

6.8.4 Least Upper Bound and Greatest Lower Bound

Let A be a poset and B a subset of A. An element a $\in A$ is called a least upper bound ($L \cup B$) of B if a is an upper bound of B and $a \leq a_1$ whenever a_1 is an upper bound of B. Thus $a = L \cup B(B)$ if $b \leq a$ for all $b \in B$, and if whenever $a_1 \in A$ is also an upper bound of B, then $a \leq a_1$.

Similarly, an element $a \in A$ is called a Greatest Lower Bound (GLB) of B if a is a lower bound of B and $a_1 \leq a$, whenever a_1 is a lower bound of B. Thus $a =$ GLB (B) if $a \leq b$ for all $b \in B$ and if whenever $a_1 \in A$ is also a lower bound of B, then $a_1 \leq a$.

Example 6.18. Let A be the poset and $A = \{a, b, c, d, e, f, g, h\}$ whose Hasse diagram is shown before also *i.e.*

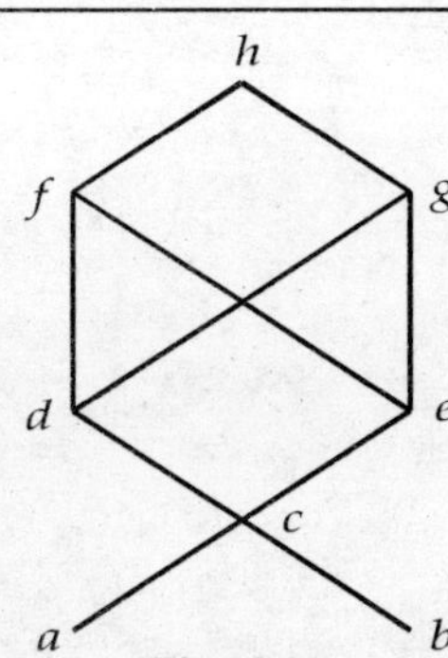

Fig. 6.10.

Find all upper bounds and all greatest lower bounds of (a) $B_1 = \{a, b\}$; (b) $B_2 = \{c, d, e\}$; (c) $B_3 = \{f, g\}$.

Solution. (a) Since B_1 has no lower bounds, it has no greatest lower bounds. And, least upper bound, LUB $(B_1) = c$.

(b) Greatest Lower Bound GLB $(B_2) = c$ and since f and g are non-comparable, B_2 has no least upper bound.

(c) Since d, e are non-comparable, B_3 has no greatest

lower bound and upper bound is only h so it is least upper bound of B_3.

Note. A subset B of A *i.e.*, poset with usual order '$\leq$' has at most one LUB and at most one GLB.

6.9. LATTICES

A lattice is a poset $(L, \leq)$ in which every subset $\{a, b\}$ consisting of two elements has a least upper bound and a greatest lower bound.

The Least Upper Bound (LUB) of $\{a, b\}$ is denoted by $a \vee b$ and is known as join of a and b.

The Greatest Lower Bound (GLB) of $\{a, b\}$ is denoted by $a \wedge b$ and is known as meet of a and b. Thus, a lattice is a mathematical structure with two binary operations, join and meet.

Example 6.19. **Let the set $A = \{1, 2, 3\}$, then $L = P(A)$ *i.e.* power set of A under '$\subseteq$' as a partial order relation is a lattice where $A \vee B$ is the set $A \cup B$ and $A \wedge B$ is the set $A \cap B$.**

Example 6.20. **Consider the poset $(Z^+, \leq)$ where $a \leq b$ iff a/b for $a, b \in Z^+$. Then L is a lattice with $a \vee b$ = LCM (a, b) and $a \wedge b$ = GCD (a, b).**

Example 6.21. **The set $L = \{1, 2, 3, 5, 6, 12\}$ under divisibility forms a lattice.**

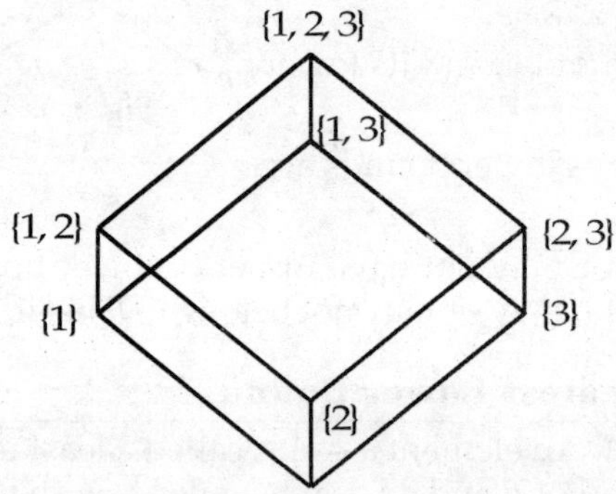

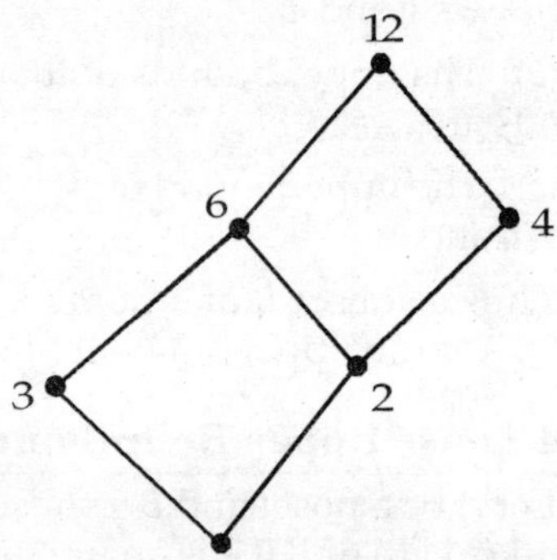

Fig. 6.11. Hasse diagram of Example 6.16. Fig. 6.12. Hasse diagram of Example 6.17.

Example 6.22. **The poset $\{2, 3, 4, 6\}$ under divisibility is not a lattice as $2 \wedge 3$ and $4 \vee 6$ does not exist.**

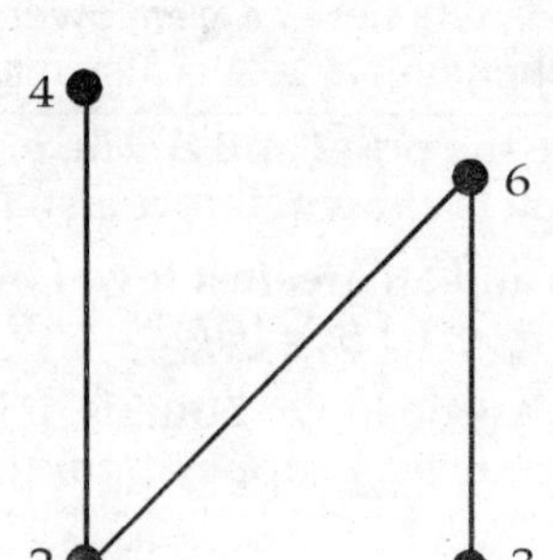

Fig. 6.13. Hasse diagram of Example 6.19.

6.9.1. Sublattice

Let $(L, \leq\}$ be a lattice. *A* non-empty subset *S* of *L* is called a sublattice of *L* if $a \vee b \in S$ and $a \wedge b \in S$ whenever $a, b \in S$.

***i.e.*, *S* is a sublattice of *L* iff *S* is closed under the operation $\wedge$ and $\vee$ of *L*.**

Example 6.23. Consider the positive integers *Z+* under the operation of divisibility. Then the set D_n of all division of $n > 1$ is a sublattice of *Z+*.

Example 6.24. Consider the lattice $(L, \leq)$ as shown in the figure :

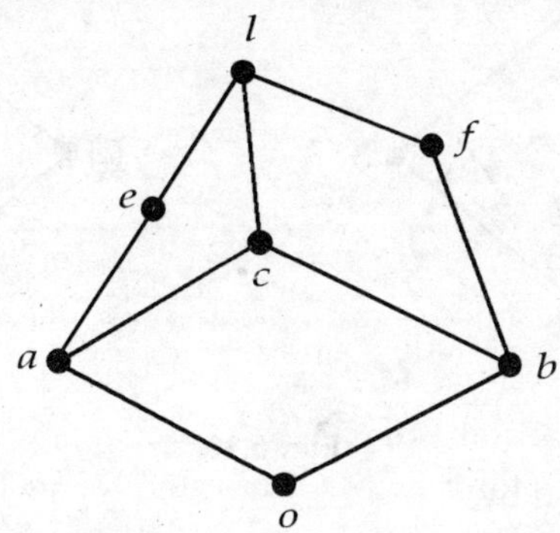

Fig. 6.14.

Consider the following subsets of this lattice :

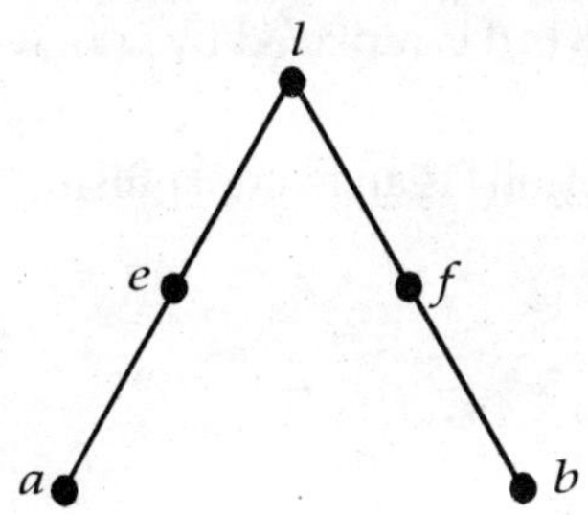

S_1

Fig. 6.15.

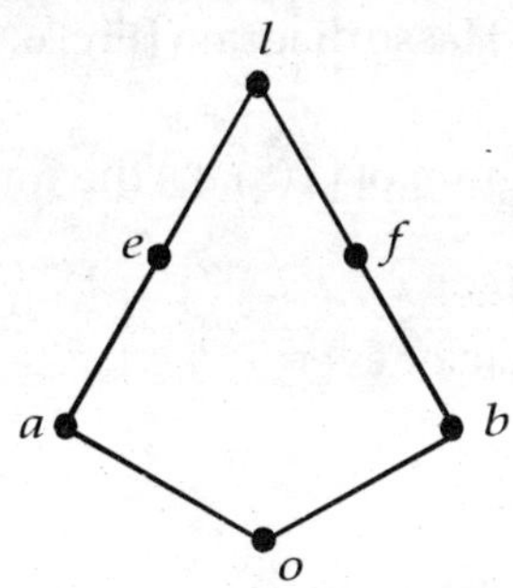

S_2

Fig. 6.16.

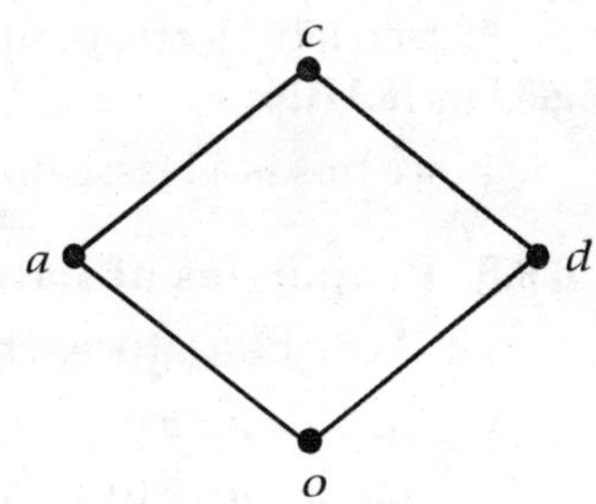

S_3

Fig. 6.17.

Then subset S_1 **Fig. 6.15** is not a sublattice of *L* since $a \wedge b \notin S_1$ and $a \vee b \notin S_1$.

The subset S_2 *i.e.*, **Fig. 6.16** is not sublattice of *L* since $a \vee b \notin S_2$.

The subset S_3 *i.e.*, **Fig. 6.17** is a sublattice of *L*.

Example 6.25. Consider the lattice *L* in the following figure :

Consider the subsets $L_1 = \{a, b, c, g\}$

$L_2 = \{a, b, e, g\}$, $L_3 = \{b, d, f, g\}$, $L_4 = \{a, d, f, g\}$

Then L_1 is not a sublattice since $b \vee c = d \notin L_1$, L_4 is also not a sublattice since $d \wedge f = b \notin L_4$. But L_2 and L_3 are sublattices.

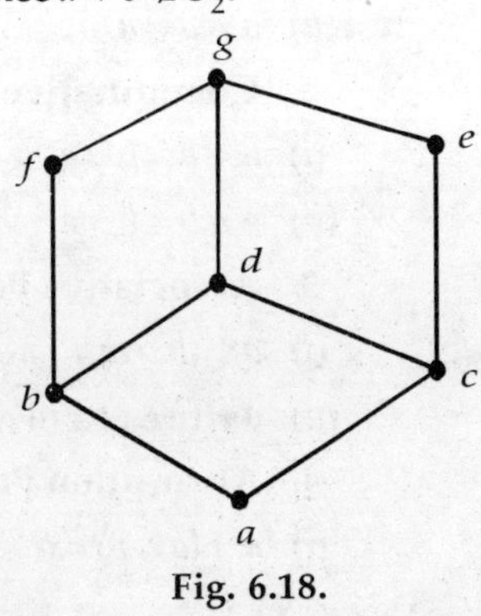

Fig. 6.18.

6.9.2. Isomorphic Lattices

Two lattices L_1 and L_2 are said to be isomorphic if there is a bijective function $f: L_1 \to L_2$ such that

$$f(a \wedge b) = f(a) \wedge f(b) \text{ and } f(a \vee b) = f(a) \vee f(b), \text{ for any elements } a, b \text{ in } L.$$

Example 6.26. Let L_1 be the lattice D_6 (divisors of 6) and let L_2 be the lattice $P(S)$ under the relation of containment, where $S = \{a, b\}$.

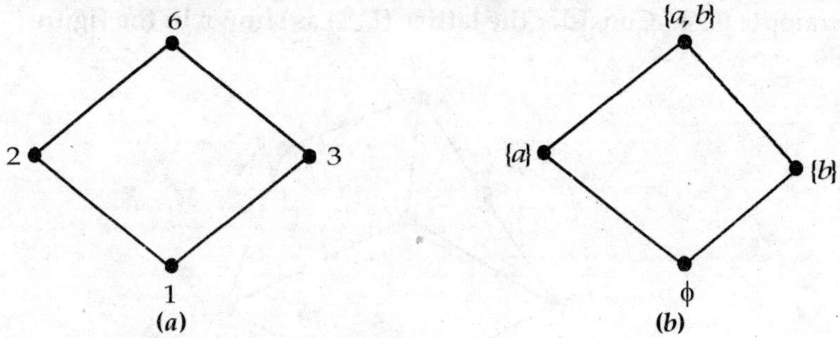

Fig. 6.19.

Let $f: L_1 \to L_2$ defined by

$$f(1) = \phi, f(2) = \{a\}, f(3) = \{b\}, f(6) = \{a, b\}$$

then clearly f is a one-one function.

If each label $a \in L_1$ of the Hasse diagram **[Fig. 6.18 (*a*)]** is replaced by $f(a)$, we get **Fig. [6.17(*b*)]**.

Since this is a Hasse diagram of $[P(S), \subseteq)$ the function f is an isomorphism.

6.9.3. Properties of Lattices

(*a*) Let L be a lattice. Then for every $a, b \in L$

(*i*) $a \vee b = b$ iff $a \le b$

(*ii*) $a \wedge b = a$ iff $a \le b$

(*iii*) $a \wedge b = a$ iff $a \vee b = b$.

(*b*) Let L be a lattice. Then

1. Idempotent Property:

(*i*) $a \vee a = a$

(*ii*) $a \wedge a = a$

2. Commutative Property:

(*i*) $a \vee b = b \vee a$

(*ii*) $a \wedge b = b \wedge a$

3. Associative Property:

(*i*) $a \vee (b \vee c) = (a \vee b) \vee c$

(*ii*) $a \wedge (b \wedge c) = (a \wedge b) \wedge c$

4. Absorption Property:

(*i*) $a \vee (a \wedge b) = a$

(ii) $a \wedge (a \vee b) = a$.

(c) Let L be a lattice. Then, for every a, b and $c \in L$.

1. If $a \leq b$, then

(i) $a \vee c \leq b \vee c$

(ii) $a \wedge c \leq b \wedge c$

2. $a \leq c$ and $b \leq c$ iff $a \vee b \leq c$

3. $c \leq a$ and $c \leq b$ iff $c \leq a \wedge b$

4. if $a \leq b$ and $c \leq d$, then

(i) $a \vee c \leq b \vee d$

(ii) $a \wedge c \leq b \wedge d$.

5. $a \wedge b \leq a$, $b \leq a \vee b$.

(d) If $0, u \in L$, then

(i) $0 \wedge a = 0, 0 \vee a = a$

(ii) $u \wedge a = a, u \vee a = u$.

6.9.4. Special Types of Lattices

➤ **Bounded Lattices**

"A lattice L is said to be bounded if it has a greatest element I and a least element 0."

In such lattices, we have

$$a \vee I = I, \quad a \wedge I = I$$
$$a \vee 0 = a, \quad a \wedge 0 = 0$$

for any $a \in L$, and $0 \leq a \leq I$.

Example 6.27. (a) The lattice $P(S)$ of all subsets of a set S, is bounded. It's greatest element is S and its least element is ϕ.

(b) The lattice Z^+ under the partial order of divisibility, is not a bounded lattice since it has a least element, *i.e.*, 1 but no greatest element.

(c) The lattice Z (set of all integers) under the partial order '$\leq$' is not bounded since it has neither a greatest nor a least element.

➤ **Distributive Lattices**

A lattice L is said to be distributive if for any elements a, b and c of L, the following distributive properties, are satisfied.

(i) $a \wedge (b \vee c) = (a \wedge b) \vee (a \wedge c)$.

(ii) $a \vee (b \wedge c) = (a \vee b) \wedge (a \vee c)$.

If L is not distributive, then L is non-distributive.

Example 6.28. (a) The power set P of a set S is a lattice under the operations $\cap$ and $\cup$. This lattice $P(S)$ is a distributive lattice, since

$$a \cap (b \cup c) = (a \cap b) \cup (a \cap c)$$

and

$$a \cup (b \cap c) = (a \cup b) \cap (a \cup c)$$

for any sets a, b, c in P.

(*b*) The lattice represented by the following figure is a distributive lattice.

Since, if we consider the triplet $\{a, b, d\}$, then

$$a \wedge (b \vee d) = a \wedge I = a$$

and $$(a \wedge b) \vee (a \wedge d) = a \vee a = a$$

Also, $$a \vee (b \wedge d) = a \vee a = a$$

and $$(a \vee b) \wedge (a \vee d) = b \wedge d = a$$

Hence $$a \wedge (b \vee d) = (a \wedge d) \vee (a \wedge d)$$

and $$(a \vee (b \wedge d) = (a \vee b) \wedge (a \vee d)$$

Thus, distributive property holds for $\{a, b, d\}$.

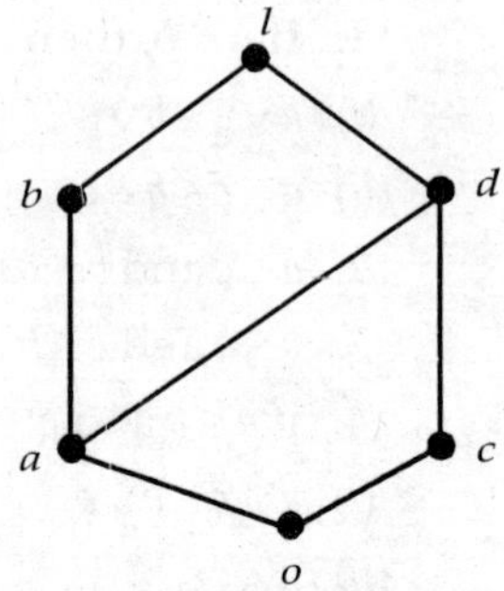

Fig 6.20.

(*c*) Consider the lattices represented by the following figures :

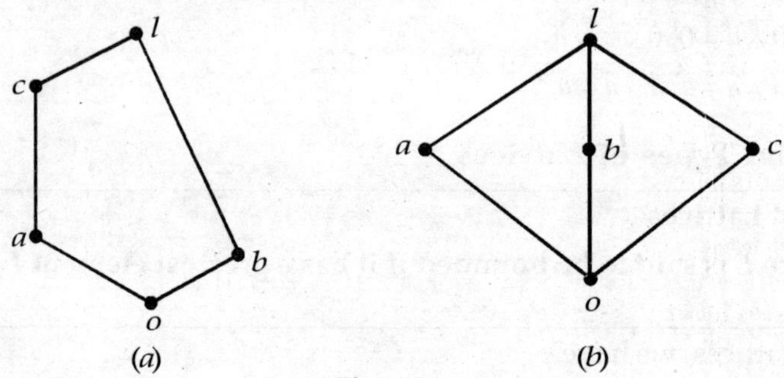

Fig. 6.21.

These lattices are non-distributive lattices. Since for lattice in **Fig. 6.21(*a*)** we have

$$c \wedge (a \vee b) = c \vee I = c$$

and $$(c \wedge a) \vee (c \wedge b) = a \vee o = a$$

For lattice in Fig. 6.21(*b*), we have

$$a \wedge (b \vee c) = a \wedge I = a$$

and $$(a \wedge b) \vee (a \wedge c) = o \vee o = o$$

> **➤ Complement in lattice**
>
> **Let L be a bounded lattice with greatest element I and least element O, and let $a \in L$. An element $a' \in L$ is called a complement of a if**
>
> $$a \vee a' = I \quad \text{and} \quad a \wedge a' = O$$

Note. Complements need not exist and need not be unique. If a is a complement of a', then a' is a complement of a *i.e.* the relation is symmetric.

Example 6.29. (*i*) Consider a lattice given by following **Fig. 6.22**

Since $c \vee b = I$ and $c \wedge b = 0$, the element b is a complement of c.

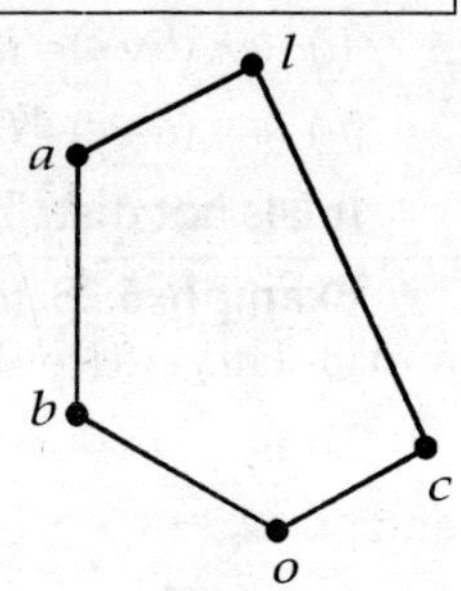

Fig. 6.22.

Also, since $a \vee c = I, a \wedge c = 0$, the element a is the complement of c.

(*ii*) Consider a lattice represented by following **Fig. 6.23.**

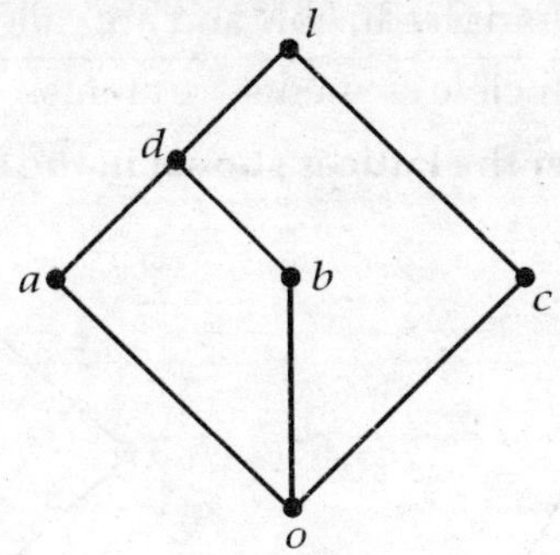

Fig. 6.23.

The complement of the element a is c since, $a \vee c = I, a \wedge c = 0$.

b is not a complement of 'a' since $a \vee b = d \neq 1$.

> **➤ Complemented lattice**
>
> **A lattice L is called *complemented* if it is bounded and if every element in L has a complement.**

Example 6.30. (*i*) The lattice $P(S)$ of all subsets of a non-empty set S is complemented and each $a \in P(S)$ has a complement so that $a \vee b = S$ and $a \wedge b = \phi$.

(*ii*) $D_{30} = \{1, 2, 3, 4, 5, 10, 15, 30\}$ *i.e.*, set of all divisors of 30 is a complemented lattice since every element in D_{30} has a complement.

> **➤ Join-Irreducible Element**
>
> Let L be a lattice with lower bound (least element) 0. An element a ($\neq 0$) in L is said to be *join irreducible* if $a = a_1 \vee a_2$ implies $a = a_1$ or $a = a_2$.

Also a is a join-irreducible iff it has a unique immediate predecessor.

Consider the following figures :

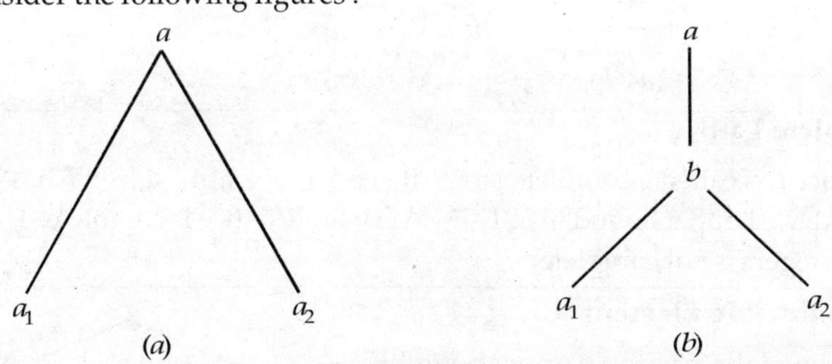

Fig. 6.24.

In **Fig. 6.24(*a*)**, $a = a_1 \vee a_2$, thus a is not join-irreducible.

In **Fig. 6.24(*b*)**, $a = a \vee a_1$ and $a = a \vee a_2$

$$a_1 \vee a_2 = b \text{ and not } a.$$

Thus, a is join irreducible.

➤ ATOM

Let L be a lattice with least element (lower bound) 0. Those elements which immediately succeed 0 are join irreducible and are called atoms.

Note. Not all join-irreducible elements are atoms.

Example 6.31. Consider the lattices shown in the following figures :

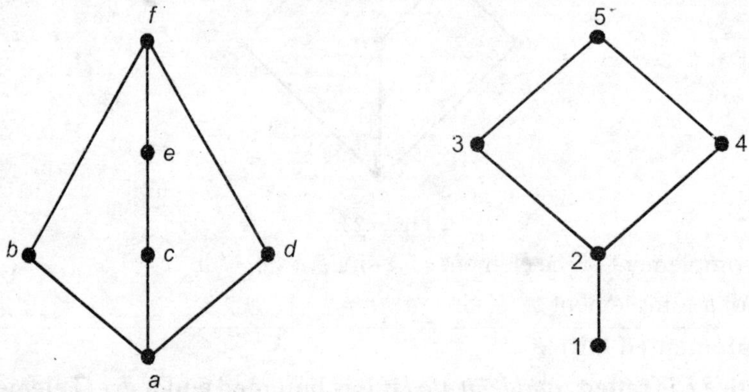

(*a*) (*b*)

Fig. 6.25.

In **Fig. 6.25(*a*),** a is the least element (*i.e.*, the 0 element). The elements which immediately succeed a are b, c, and d, hence they are the atoms.

In **Fig. 6.25(*b*)** 1 is the least element and there is one atom *i.e.* element '2' which immediately succeeds 1.

➤ Duality

The dual of any statement in *a* lattice $(L, \wedge, \vee)$ is defined to be the statement obtained by interchanging $\wedge$ and $\vee$ by $\vee$ and $\wedge$ respectively.

Example 6.32. Dual of the statement

$$(a \wedge b) \vee c = (b \vee c) \wedge (c \vee a)$$

is

$$(a \vee b) \wedge c = (b \wedge c) \vee (c \wedge a)$$

➤ Complete Lattice

A lattice L is called a complete lattice if every non empty subset L has greatest bound and least upper bound in L. Thus, every finite lattice is complete. The lattice $(Z, \leq)$ of integers is not complete.

Meet-Irreducible Element

Let L be a lattice with an upper bound I. An element a in the lattice L is said to be meet-irreducible if $a = a_1 \wedge a_2$ implies $a = a_1$ or $a = a_2$ and similarly, $a \neq I$ is meet-irreducible iff a has only one immediate successor.

Example 6.33. The meet-irreducible elements of the above lattice are a, b, c, d and e.

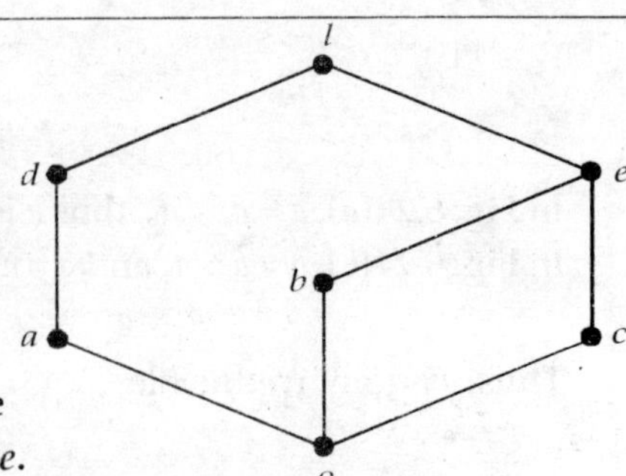

Fig. 6.26.

EXAMPLES

Example 6.34. Let P be any collection of sets. Is the relation of set inclusion '$\subseteq$' a partial order on P?

Solution. Yes, it is a partial order since set inclusion is reflexive, antisymmetric and transitive. For any sets A_1, A_2, A_3 in P, we have

(*i*) $A_1 \subseteq A_1$

(*ii*) If $A_1 \subseteq A_2$ and $A_2 \subseteq A_1$ then $A_1 = A_2$

(*iii*) If $A_1 \subseteq A_2, A_2 \subseteq A_3$ then $A_1 \subseteq A_3$.

Example 6.35. Consider the set Z of integers with the relation divisibility. Is this relation a partial ordering of Z?

Solution. No, the relation divisibility is not a partial ordering of Z since if 3/ –3 and –3/3 but $3 \neq -3$ hence, relation is not antisymmetric.

Thus, it is not a partial ordering of Z.

Example 6.36. Let $N = \{1, 2, 3, ...\}$ with the partial order divisibility. State whether each of the following subsets of N are totally ordered.

(*a*) {2, 6, 24} **(*b*) {3, 5, 15}.**

Solution. (*a*) Since 2 divides 6, which divides 24, thus, the set is totally ordered.

(*b*) Since 3 and 5 are not comparable, $3 \div 5$ and $5 \div 3$, thus the set is not totally ordered.

Example 6.37. Let $S = \{a, b, c\}$ and $P(S)$ is the power set of S. Draw the Hasse diagram of the poset $P(S)$ with the partial order '$\subseteq$' (containment).

Solution. $P(S) = \{\phi, \{a\}, \{b\}, \{c\}, \{a, b\}, \{b, c\}, \{a, c\}, \{a, b, c\}\}$

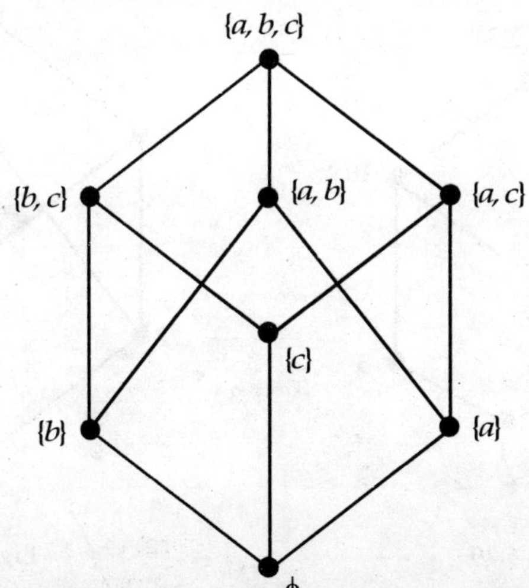

Fig. 6.27. Hasse diagram of *P*(*S*).

Example 6.38. Determine all maximal and minimal elements of the poset :

(*a*) $S = \{x \mid x$ is a real number and $0 < x \leq 1\}$ with the usual partial order $\leq$.

(*b*) $S = \{2, 3, 4, 6, 8, 24, 48\}$ with the partial order of divisibility.

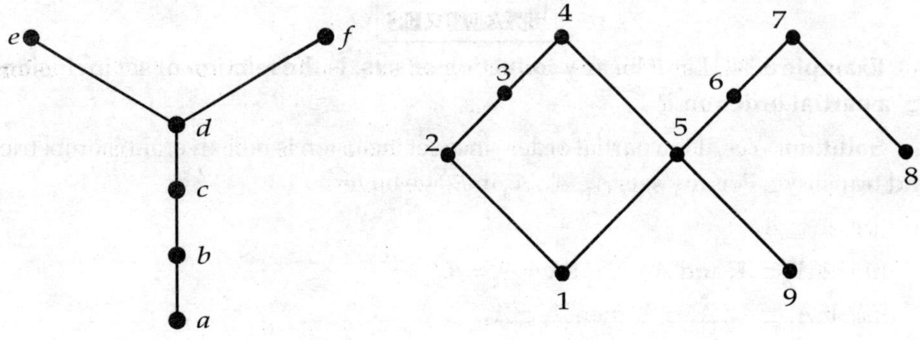

(a) (b)

Fig. 6.28.

Solution. (a) Maximal = 1, Minimal = none,

(b) Maximal = 48, Minimal = {2, 3}

(c) Maximal = {e, f}, Minimal = {a}

(d) Maximal = {4, 7}, Minimal = {1, 9}

Example 6.39. Let *n* be a positive integer and let D_n be the set of all positive divisors of *n*. Are D_{20} and D_{30} lattices ?

Solution. D_n is a lattice under the relation of divisibility since

$$a \vee b = \text{LCM}(a, b) \text{ and } a \wedge b = \text{GCD}(a, b)$$

if $a \leq b$, and $a, b \in Z^+$

Thus, D_{20} and D_{30} are lattices. Their Hasse diagram is shown below :

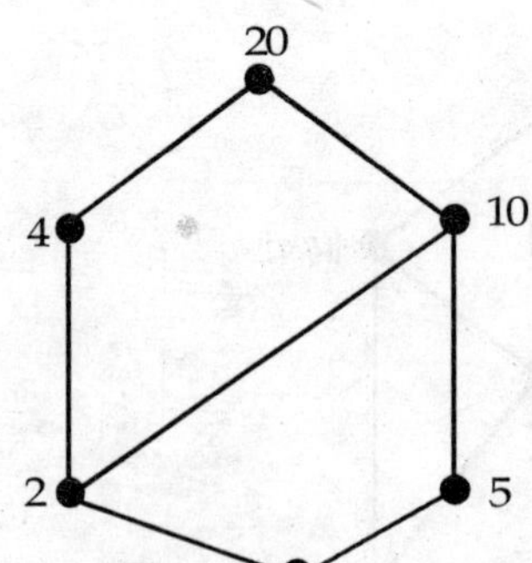

Fig. 6.29.

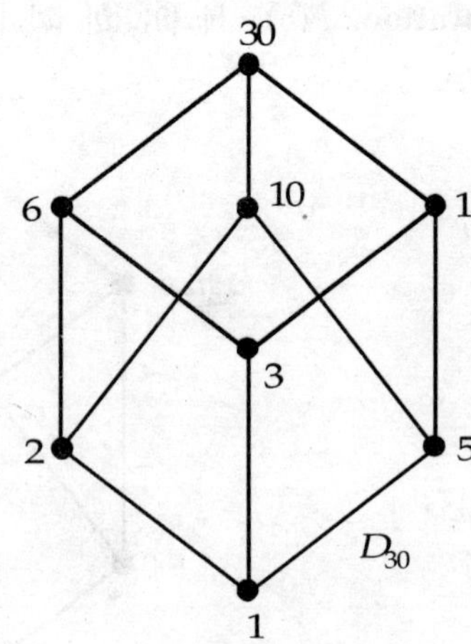

Fig. 6.30.

Example 6.40. Are the two lattices shown by Figs. 6.30 and 6.31 below isomorphic ?

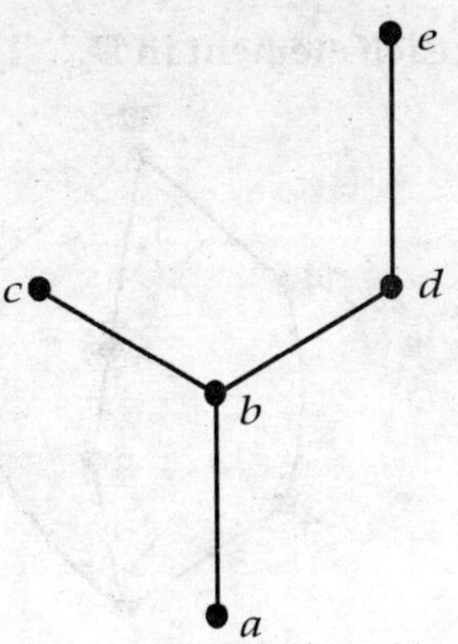

Fig. 6.31.

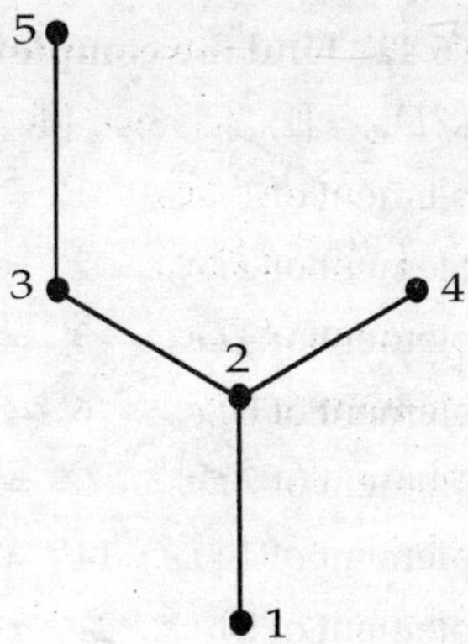

Fig. 6.32.

Solution. Yes, these are isomorphic lattices. Consider the mapping

$$f = \{(a, 1), (b, 2), (c, 4), (d, 3), (e, 5)\}$$

Now consider $f(c \wedge d) = f(b) = 2$

and $f(c) \wedge f(d) = 4 \wedge 3 = 2$

So $f(c \wedge d) = f(c) \wedge f(d)$

Also, consider $f(c \vee d) = f(e) = 5$

and $f(c) \vee f(d) = 4 \vee 3 = 5$

So $f(c \vee d) = f(c) \vee f(d)$

Hence, f is isomorphic.

Example 6.41. Let L be a bounded distributive lattice. Then prove if a complement exists, it is unique.

Solution. Let a_1, a_2 be complements of the element $a \in L$. Then

$$a \vee a_1 = I, a \vee a_2 = I$$

$$a \wedge a_1 = 0, a \wedge a_2 = 0$$

Using the distributive laws, we have

$$\begin{aligned} a_1 &= a_1 \vee 0 = a_1 \vee (a \wedge a_2) \\ &= (a_1 \vee a) \wedge (a_1 \vee a_2) \\ &= (a \vee a_1) \wedge (a_1 \vee a_2) \\ &= I \wedge (a_1 \vee a_2) = a_1 \vee a_2. \end{aligned}$$

Also,

$$\begin{aligned} a_2 &= a_2 \vee 0 \\ &= a_2 \vee (a \wedge (a_1) \\ &= (a_2 \vee a) \wedge (a_2 \vee a_1) \\ &= (a \vee a_2) \wedge (a_1 \vee a_2) \\ &= I \wedge (a_1 \vee a_2) \\ &= a_1 \vee a_2. \end{aligned}$$

Hence, $a_1 = a_2$.

Thus, complement is unique for a bounded distributive lattice.

Example 6.42. Find the complement of each element in D_{42}.

Solution. $D_{42} = \{1, 2, 3, 6, 7, 14, 21, 42\}$

The complement of 1 *i.e.,* $1' = 42$

The complement of 2 *i.e.,* $2' = 21$

The complement of 3 *i.e.,* $3' = 14$

The complement of 6 *i.e.,* $6' = 7$

The complement of 7 *i.e.,* $7' = 6$

The complement of 14 *i.e.,* $14' = 3$

The complement of 21 *i.e.,* $21' = 2$

The complement of 42 *i.e.,* $42' = 1.$

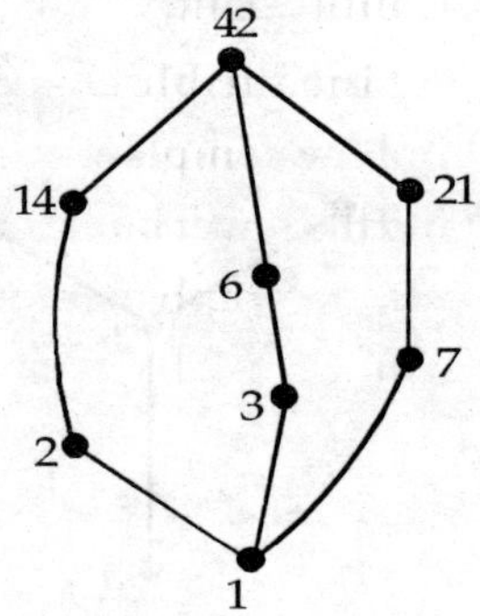

Fig. 6.33. Hasse diagram of D_{42}.

Example 6.43. Find the atoms for the two lattices given below :

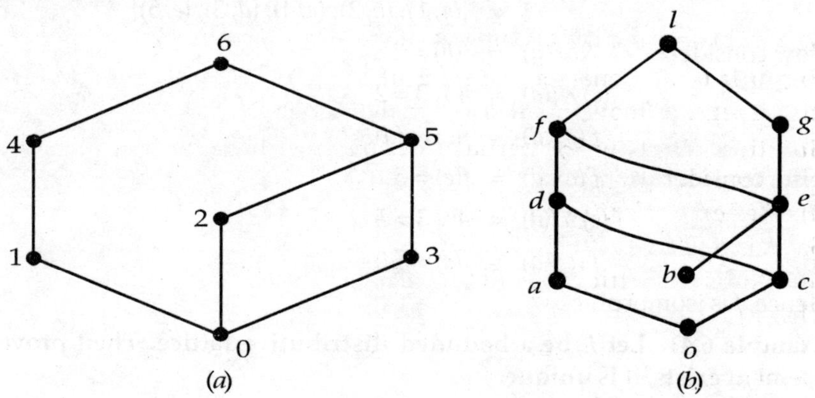

Fig. 6.34.

Solution. (*a*) The atoms are 1, 2, 3.

(*b*) The atoms are *a*, *b*, *c*.

Since these elements, immediately succeed '0'.

Example 6.44. Let L be a bounded distributive lattice. Then complements are unique if they exist. Using it, show that the following lattice is not distributive.

Solution. The element 3 has two complements 2 and 5, hence it is not distributive.

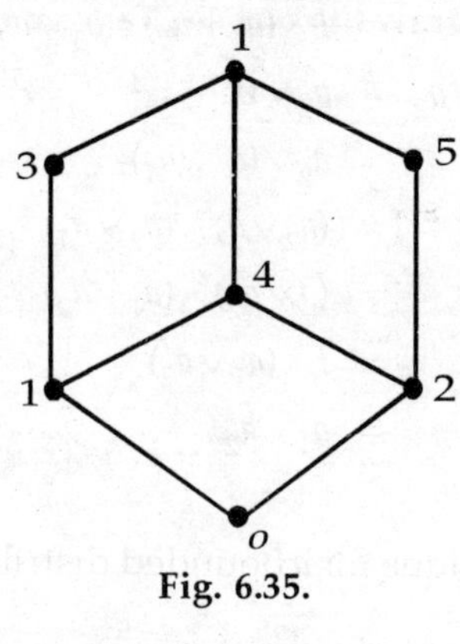

Fig. 6.35.

Example 6.45. Consider the lattice D_{30} = {1, 2, 3, 4, 5, 6, 10, 15, 30} under the order divisibility. Find

(*a*) Join irreducible elements and atoms

(*b*) Find the complements of 2 and 10, if exist.

(*c*) Find the lower bound and upper bound of D_{30}.

(*d*) Is D_{30} a complemented lattice.

Solution. Hasse diagram of D_{30} is given below :

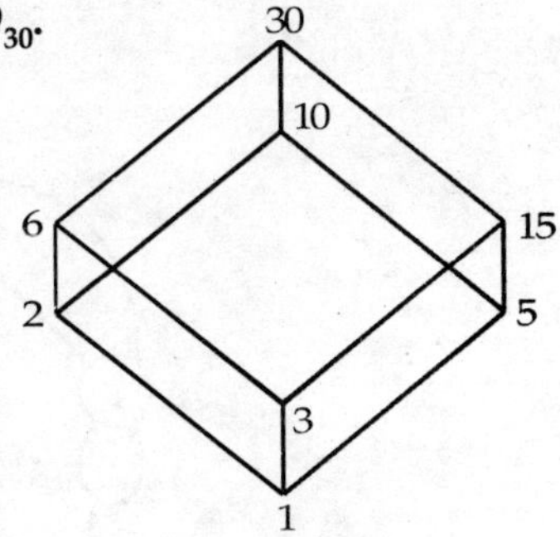

Fig. 6.36.

(*a*) The elements 2, 3 and 5 are join irreducible.

(*b*) There are no complements of 2 and 10 in D_{30}.

(*c*) The lower bound is 1 and upper bound is 30.

(*d*) No, D_{30} is not a complemented lattice since every element does not have complement.

Example 6.46. Consider the set *Z* of integers. Let *aRb* if $a = b^r$ for some positive integer *r*. Show that *R* is a partial order on *Z*.

Solution. To show *R* is partial order on Z, we have to show *R* is reflexive, antisymmetric and transitive.

(*i*) **Reflexive :** $aRa \Rightarrow a = a'$

So *R* is reflexive.

(*ii*) **Antisymmetric :** Let *aRb* and *bRa* then $a = b^r, b = a^s$

Then, $b = (b^r)^s = b^{rs}$

There are three posibilities : $rs = 1, b = 1, b = -1$.

If $rs = 1 \Rightarrow r = 1$ and $s = 1 \Rightarrow a = b$.

If $b = 1$ then $a\ 1^r = 1 = b \Rightarrow a = b$

If $b = 1$ then $a = (-1)^1$ (Since $a \neq 1$), $\Rightarrow a = b$.

In all cases, $a = b$, thus *R* is antisymmetric.

(*iii*) Let $aRb, bRc \Rightarrow a = b^r, b = c^s$

$\Rightarrow\ a = (c^s)^r = c^{rs}$

$\Rightarrow\ aRc$

Hence, *R* is transitive.

Therefore, *R* is a partial order on Z.

Example 6.47. Let *A* = {1, 2, 3, 4, 5, 6, 8, 9, 12, 18, 24}. Consider the partial order of divisibility on *A*. Draw the Hasse diagram of the poset $(A, \leq)$.

Solution.

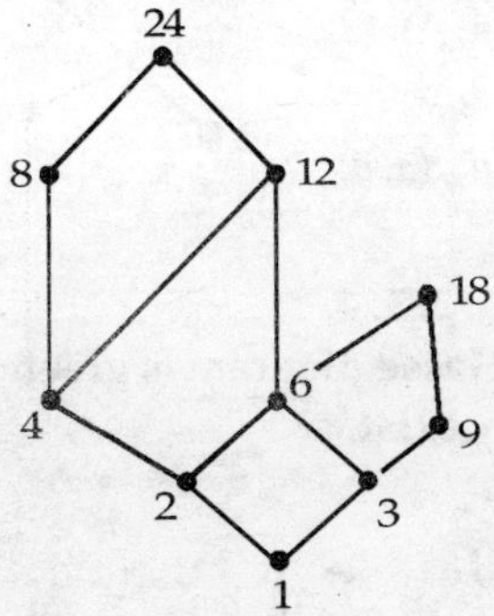

Fig. 6.37.

Example 6.48. Determine the Hasse diagram of the partial order having the given diagraph :

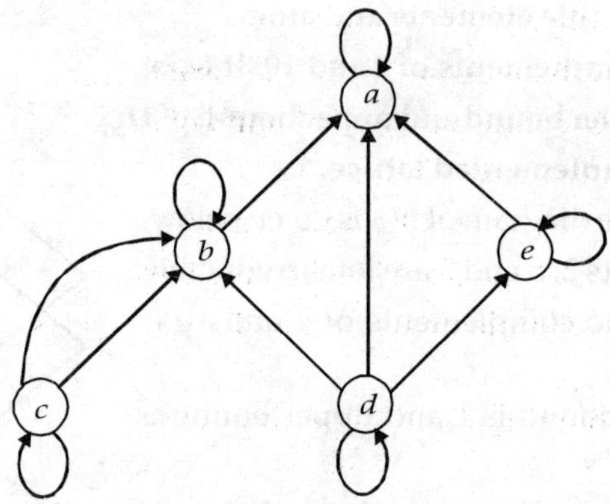

Fig. 6.38.

Solution. Removing the edges showing reflexivity and transitivity we have the following Hasse diagram :

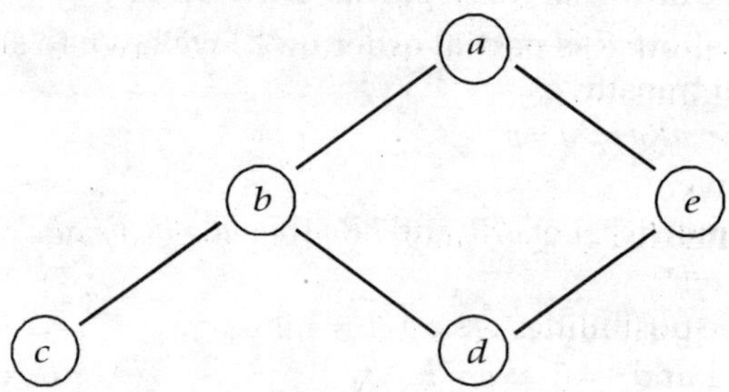

Fig. 6.39.

Example 6.49. Determine the ordered pairs in the relation from the Hasse diagram on the set S :

(a) **S = {1, 2, 3, 4}**

(b) **S = {a, b, c, d}**

Solution. Ordered pairs in the relation on set S are :

(a) {(1, 1), (2, 2), (3, 3), (4, 4), (2, 3), (2, 4), (1, 3), (3, 4), (1, 4)}

(b) {(a, a), (b, b), (c, c), (d, d), (a, b), (a, c), (a, d), (b, c), (b, d), (c, d)}.

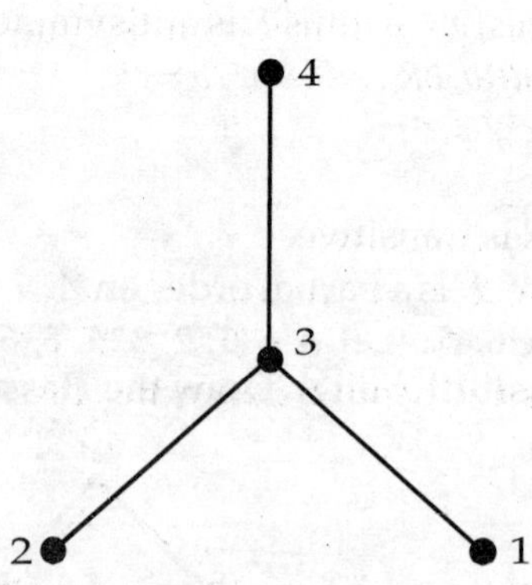

Fig. 6.40.

Fig. 6.41.

Example 6.50. Let $A = \{1, 2, 3, 4, 5, ...11\}$ be the poset whose Hasse diagram is given below. Find the LUB and the GLB of $A_1 = \{6, 7, 10\}$, if they exist.

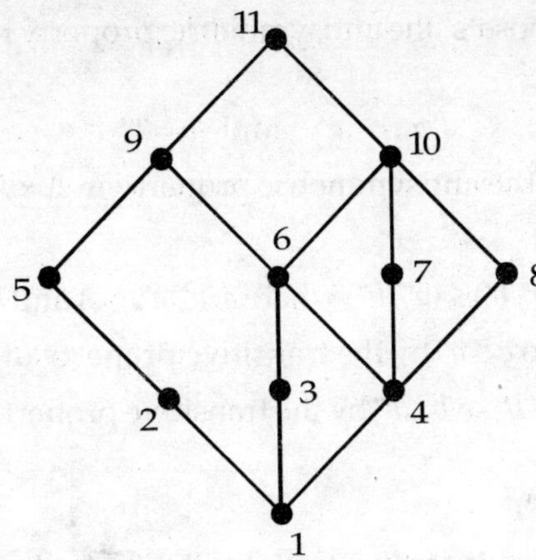

Fig. 6.42.

Solution. LUB $(A_1) = 10$, GLB $(A_1) = 4$.

Example 6.51. Find *(i)* all upper bounds, *(ii)* all lower bounds, *(iii)* the least upper bound and *(iv)* the greatest lower bound of B.

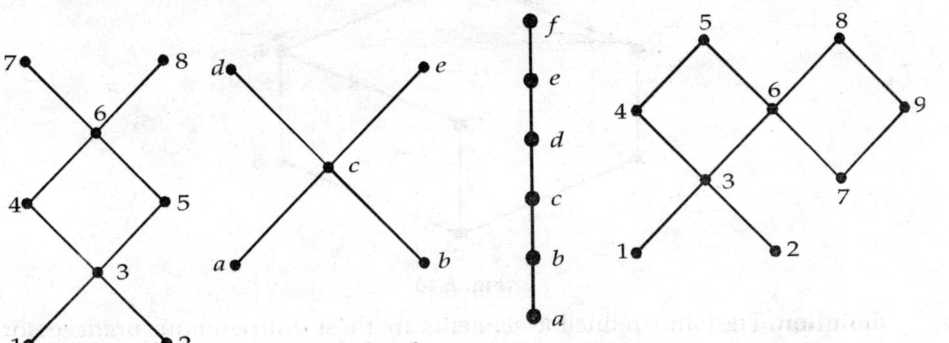

(a) $B = \{3, 4, 5\}$ (b) $B = \{a, b, c, d\}$ (c) $B = \{b, c, d\}$ (d) $B = \{3, 4, 6\}$

Fig. 6.43.

Solution.

(a)	(i) 6, 7, 8	(ii) 1, 2, 3	(iii) 6	(iv) 3
(b)	(i) d, e	(ii) a, b	(iii) None	(iv) None
(c)	(i) d, e, f	(ii) b, a	(iii) d	(iv) b
(d)	(i) 4, 5, 6	(ii) 1, 2, 3	(iii) None	(iv) 3

Example 6.52. If $(A, \leq)$ and $(B, \leq)$ are posets, then prove that $(A \times B, \leq)$ is a poset with partial order '$\leq$' defined by

$$(a, b) \leq (a', b') \text{ if } a \leq a' \text{ in } A \text{ and } b \leq b' \text{ in } B.$$

Solution. If $(a, b) \in A \times B$, then $(a, b) \leq (a, b)$ since $a \leq a$ in A and $b \leq b$ in B, so '$\leq$' satisfies the reflexive property in $A \times B$.

Next, let $(a, b) \leq (a', b')$ and $(a', b') \leq (a, b)$ where $a, a' \in A$ and $b, b' \in B$.

Then $\qquad a \leq a'$ and $a' \leq a$ in A

and $\qquad b \leq b'$ and $b' \leq b$ in B

Since A and B are posets, the antisymmetric property of the partial-order in A and B implies that

$$a = a' \quad \text{and} \quad b = b'$$

Hence '$\leq$' satisfies the antisymmetric property in $A \times B$.

Now, suppose that

$(a, b) \leq (a', b')$ and $(a', b') \leq (a'', b'')$ where $a, a', a'' \in A$ and $b, b', b'' \in B$.

Then $a \leq a', a' \leq a''$ so $a \leq a''$ by the transitive property of the partial order in A.

Similarly, $b \leq b', b' \leq b''$ so $b \leq b''$ by the transitive property of the partial order in B.

Hence, $(a, b) \leq (a'', b'')$.

Thus, transitivity holds for the partial order in $A \times B$.

Hence, $A \times B$ is a poset.

Example 6.53. Find the join-irreducible elements of the following lattice :

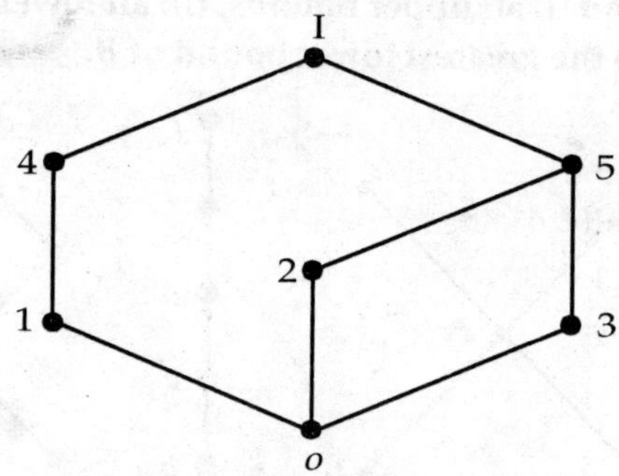

Fig. 6.44

Solution. The join-irreducible elements are those with a unique predecessor. Hence the elements 1, 2, 3 and 4 are join-irreducible.

Example 6.54. Is the following lattice a distributive lattice ?

Solution. The element 'a' is a complement of d and *vice versa.* Other elements do not have complements. Thus all complements which exist are unique.

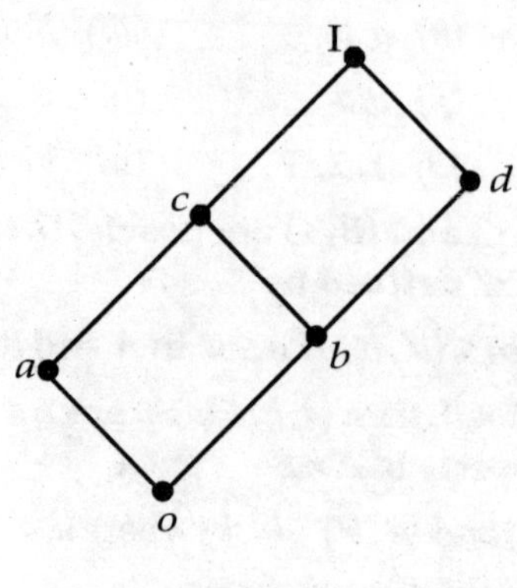

Fig. 6.45.

Example 6.55. Consider the lattice $D_{12} = \{1, 2, 3, 4, 6, 12\}$ under the order divisibility. Find

(a) The lower bound and upper bound of D_{12}.

(b) The complement (s) of 4 and 6.

(c) Is D_{12} a complemented lattice ?

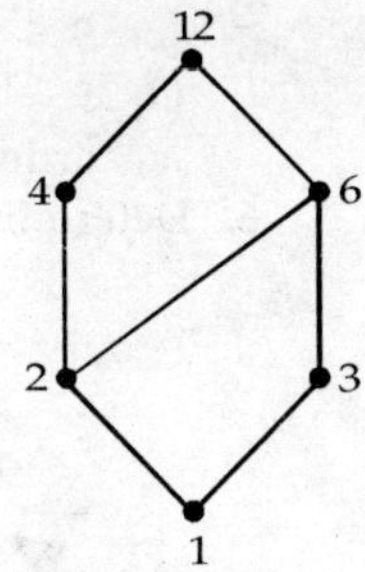

Fig. 6.46. Hasse diagram of D_{12}.

Solution. (a) The lower bound is 1 and upper bound is 12.

(b) The complement of 4 is 3, 6 has no complement.

(c) D_{12} is not a complemented lattice since 6 has no complement.

EXERCISES

1. Determine whether the relation R is a partial order on the set S
   (a) $S = Z$ and aRb iff $a = 2b$
   (b) $S = R$ (set of real nos.), and aRb iff $a \leq b$.
2. Determine the Hasse diagram of the relation R where $S = \{1, 2, 3, 4\}$ and $R = \{(1, 1), (1, 2), (2, 2), (2, 4), (1, 3), (3, 3), (3, 4), (1, 4), (4, 4)\}$.
3. Consider the partial order of divisibility on the set S. Determine which posets are totally ordered :
   (a) $S = \{1, 2, 3, 5, 6, 10, 15, 30\}$ (b) $S = \{2, 4, 8, 16, 32\}$
   (c) $S = \{3, 6, 12, 36, 72\}$ (d) $S = \{1, 2, 3, 4, 5, 6, 10, 12, 15, 30, 60\}$.
4. Determine the Hasse diagram of the partial order having the given diagraph :

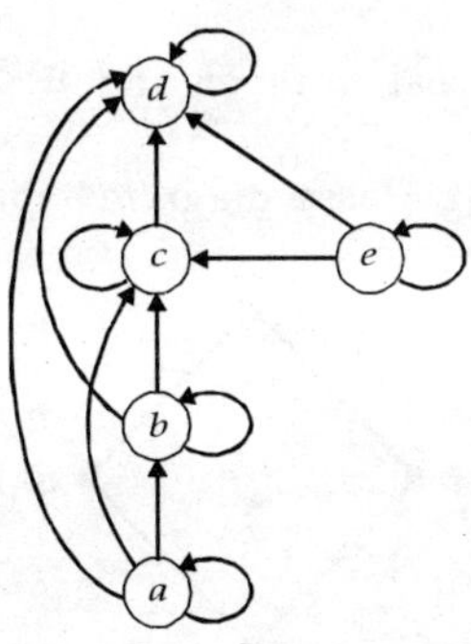

Fig. 6.47.

5. Determine all maximal and minimal elements of the poset :

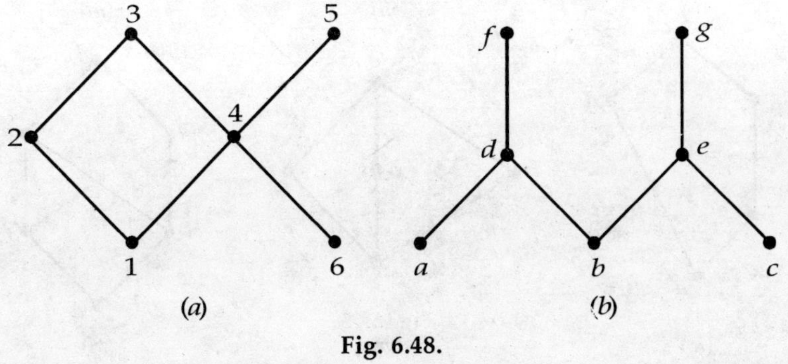

Fig. 6.48.

(c) $S = R$ with usual partial order $\leq$.

(d) $S = \{x \mid x$ is a real number and $0 \leq x < 1\}$ with the usual partial order $\leq$.

6. Determine the greatest and least elements, if they exist, of the poset

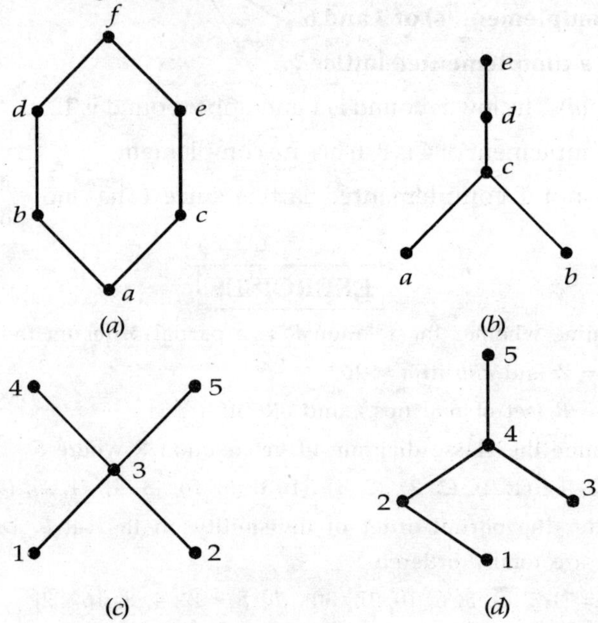

Fig. 6.49.

(e) $A = \{x \mid x$ is a real number and $0 < x < 1\}$ with the usual partial order $\leq$.

(f) $A = \{x \mid x$ is a real number and $0 \leq x \leq 1\}$ with the usual partial order $\leq$.

7. Which of the following Hasse diagram represent lattices ?

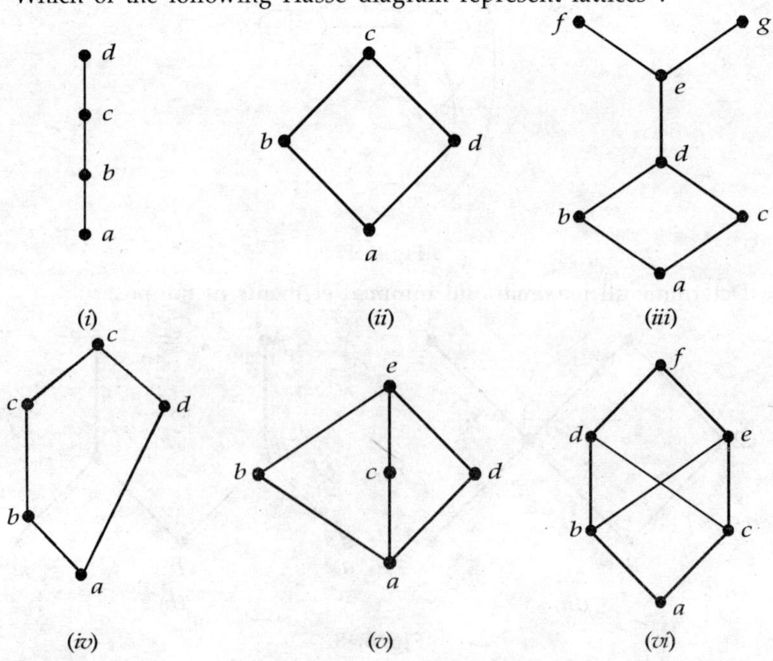

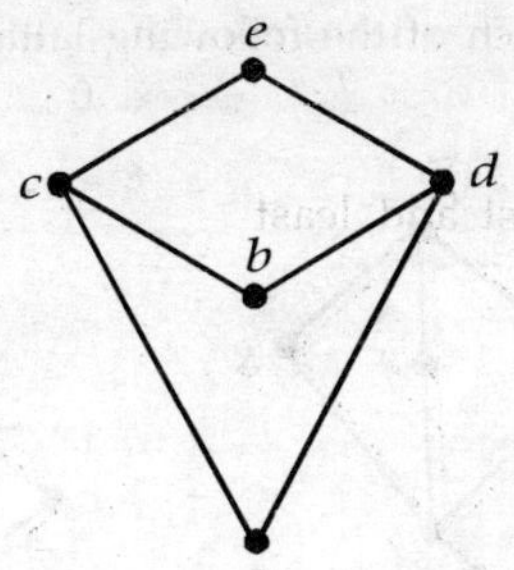

(vii)

Fig. 6.50.

8. If $(L_1, \leq)$ and $(L_2, \leq)$ are lattices, then prove $(L, \leq)$ is a lattice, where $L = L_1 \times L_2$ and the partial order '$\leq$' of L is the product partial order.

9. Are the following lattices isomorphic ?

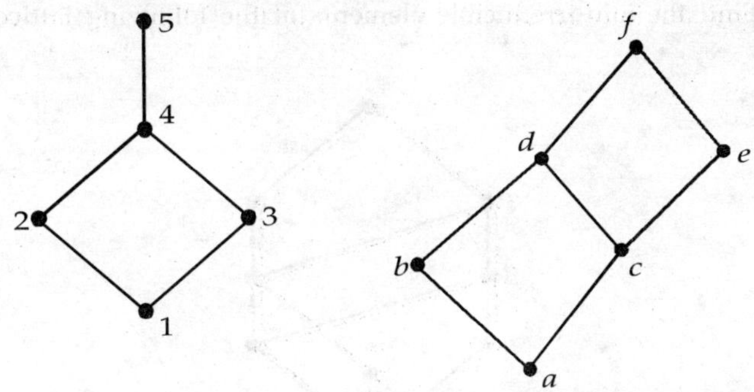

Fig. 6.51.

10. Determine whether the following Hasse diagram represents a lattice :

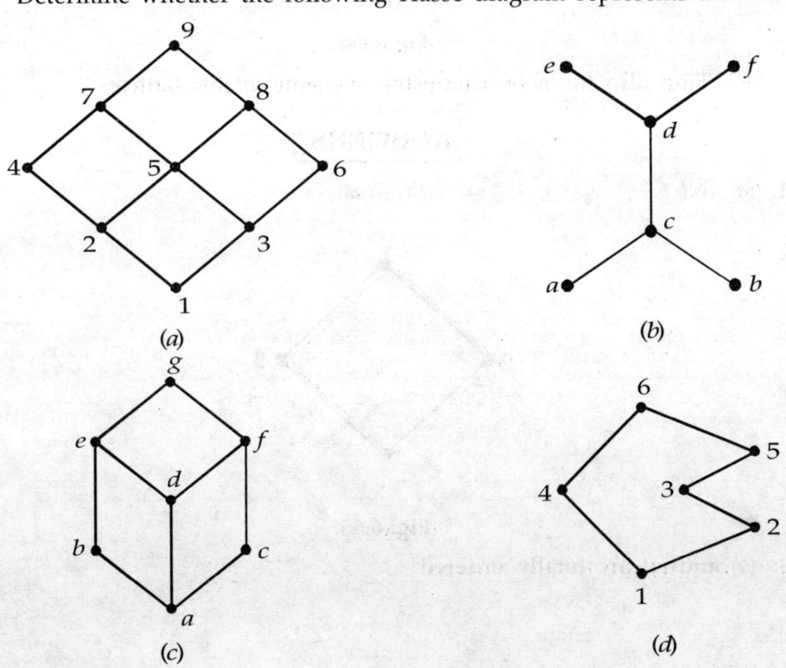

Fig. 6.52.

11. Determine whether each of the following lattice is distributive, complemented, both or neither.

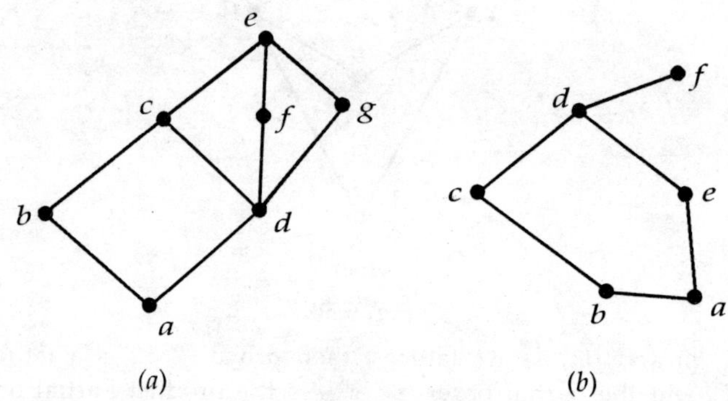

Fig. 6.53.

12. (*i*) Find the join-irreducible elements of the following lattice :

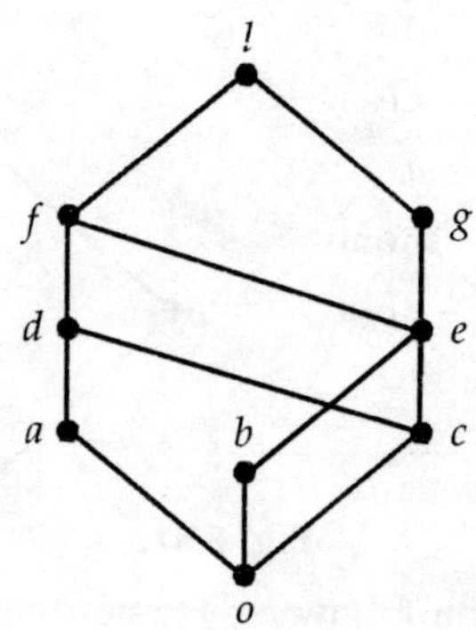

Fig. 6.54.

(*ii*) Find also the meet-irreducible elements of this lattice.

ANSWERS

1. (*a*) No (*b*) Yes

2.

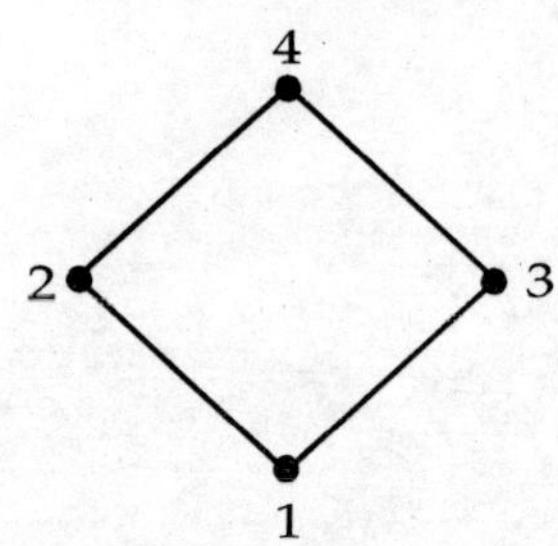

Fig. 6.55.

3. (*b*) and (*c*) are totally ordered

4.

Fig. 6.56.

5. Maximal	Minimal
(*a*) 2, 3, 5	1, 6
(*b*) *f*, *g*	*a*, *b*, *c*
(*c*) none	none
(*d*) none	0

6. Greatest	Least
(*a*) *f*	*a*
(*b*) *e*	none
(*c*) none	none
(*d*) 5	1
(*e*) none	none
(*f*) 1	0

7. (*i*), (*ii*), (*iv*) and (*v*) represent lattices.

9. No.

10. (*a*) and (*c*) are lattices.

11. (*a*) Neither (*b*) Neither

12. (*i*) *a*, *b*, *c* and *g* (*ii*) *a*, *b*, *d*, *f* and *g*.

Chapter

7 BOOLEAN ALGEBRA

7.1. INTRODUCTION

The circuits in computers and other electronic devices have inputs, each of which is either a 0 or a 1, and produce outputs that are also 0's and 1's. Circuits can be constructed using any basic element that has two different states. Such elements include switches that can be in either 'on' or 'off' position etc. In 1938 Claude Shamon showed how the basic rules of logic, first given by George Boole (1815-1864) in 1854 in his "An Investigation of the Laws of Thought," could be used to design circuits. These rules form the basis for Boolean Algebra. Here we give some basic properties of Boolean algebra and its applications.

Boolean algebra provides the operations and the rules for working with the set {0, 1}. The three operations in Boolean algebra $\vee$, $\wedge$ and $'$ are commonly used. Here $'$ denotes the complement of an element, and defined by $0' = 1$ and $1' = 0$. Also the operator $\vee$ or 'OR' has the following values :

$$1 \vee 1 = 1, 1 \vee 0 = 1, 0 \vee 1 = 1, \ 0 \vee 0 = 0.$$

The operator $\wedge$ or 'AND' has the following values :

$$1 \wedge 1 \ = \ 1, 1 \wedge 0 = 0, 0 \wedge 1 = 0, 0 \wedge 0 = 0.$$

Example 7.1. Find the value of $(1 \wedge 0) \vee (0 \vee 1)'$.

Solution. By definition $(1 \wedge 0) \vee (0 \vee 1)' = (1 \wedge 0) \vee (0 \wedge 1)$

$$= \ 0 \vee 0 = 0.$$ **Ans.**

7.2. BOOLEAN EXPRESSIONS AND BOOLEAN FUNCTIONS

Let $B = \{0, 1\}$. The variable x is called a *Boolean variable* if it assume values only from B. A function from B^n, the set $\{(x_1, x_2, ..., x_n) \ ; x_i \in B, 1 \le i \le n\}$, to B is called a *Boolean function of degree n*. The values of a Boolean function are often displayed in tables.

Boolean functions can be represented using expressions made up from the variables and Boolean operations. The *Boolean expressions* in the variable $x_1, x_2, ..., x_n$ are defined as follows $0, 1, x_1, x_2, ..., x^n$ are Boolean expressions.

If E_1 and E_2 are Boolean expressions, then E'_1, $E \wedge E_2$ and $E_1 \vee E_2$ are all Boolean expressions. Thus each Boolean expression represents a Boolean function.

The values of this function are obtained by substituting 0 and 1 for the variables in the expression.

Example 7.2. Find the values of the Boolean function represented by $f(x, y, z) = (x \wedge y) \vee z'$.

Solution. The values of the function $f(x, y, z) = (x \wedge y) \vee z'$ are displayed by the table :

x	y	z	$x \wedge y$	z'	$(x \wedge y) \vee z'$
1	1	1	1	0	1
1	1	0	1	1	1
1	0	1	0	0	0
1	0	0	0	1	1
0	1	1	0	0	0
0	1	0	0	1	1
0	0	1	0	0	0
0	0	0	0	1	1

$$\begin{bmatrix} \textbf{Note} & 1 \vee 1 = 1, \; 1 \vee 0 = 1, \; 0 \vee 1 = 1, \; 0 \vee 0 = 0 \\ \textbf{and} & 1 \wedge 1 = 1, \; 1 \wedge 0 = 0, \; 0 \wedge 1 = 0, \; 0 \wedge 0 = 0 \end{bmatrix}$$

The Boolean functions F and G of n variables are *equal* or *equivalence* iff $F(b_1, b_2, ..., b_n) = G(b_1, b_2, ..., b_n)$ whenever $b_1, b_2, ..., b_n \in B$.

The complement of Boolean function F is the function f', where

$$f'(x_1, x_2, ..., x_n) = [f(x_1, x_2, ..., x_n)]'.$$

Let F and G be Boolean function of degree n. The *Boolean sum* $F \vee G$ and *Boolean product* $F \wedge G$ are defined by

$$(F \vee G)(x_1, x_2, ..., x_n) = F(x_1, x_2, ..., x_n) \vee G(x_1, x_2, ..., x_n)$$

$$(F \wedge G)(x_1, x_2, ..., x_n) = F(x_1, x_2, ..., x_n) \wedge (x_1, x_2, ..., x_n).$$

Example 7.3. How many different Boolean functions of degree n are there ?

Solution. There are 2^n different n-tuples of 0's and 1's. Since a Boolean function is an assignment of 0 or 1 to each of these 2^n different n tuples, thus by product rule of counting there are 2^{2^n} different Boolean functions.

7.3. DUALITY

> **The *dual* of a Boolean expression is obtained by interchanging Boolean sums and Boolean products and interchanging 0s and1s.**

Example 7.4. Find the dual of $x \wedge (y \vee 0)$.

Solution. The dual of $x \wedge (y \vee 0)$ is $x \vee (y \wedge 1)$.

➤ **Principle of Duality**

The dual of a Boolean function F represented by a Boolean expression is the function represented by the dual of this expression. This dual functions, denoted by F^d, does not depend upon the particular Boolean expression used to represent F. An identity between function represented by Boolean expressions remains valid when the duals of both sides of the identity are taken. This result is called the duality principle.

7.4. BOOLEAN ALGEBRA (DEFINITION)

By a *Boolean algebra*, we mean a system consisting of a non-empty set B together with two binary operations $\wedge$ and $\vee$ and a unary operation, 'satisfying the following properties for all $x, y, z \in B$:

1.	$\left.\begin{array}{l} x \vee 0 = x \\ x \wedge 1 = x \end{array}\right\}$	Identity laws
2.	$\left.\begin{array}{l} x \vee x' = 1 \\ x \wedge x' = 0 \end{array}\right\}$	Compliment laws
3.	$\left.\begin{array}{l} (x \vee y) \vee z = x \vee (y \vee z) \\ (x \wedge y) \wedge z = x \wedge (y \wedge z) \end{array}\right\}$	Associative laws
4.	$\left.\begin{array}{l} x \vee y = y \vee x \\ x \wedge y = y \wedge x \end{array}\right\}$	Commutative law
5.	$\left.\begin{array}{l} x \vee (y \wedge z) = (x \vee y) \wedge (x \vee z) \\ x \wedge (y \vee z) = (x \wedge y) \vee (x \wedge z) \end{array}\right\}$	Distributive law

7.5. BOOLEAN ALGEBRA HOMOMORPHISM

If $(A, +, \cdot, ', 0, 1)$ and $(B, \wedge, \vee, -, 0', 1')$ are two Boolean algebras, a function $h : A \to B$ is called a *Boolean algebra homomorphism* if h preserves the two binary operations and the unary operations in the following sense $\forall\ a, b \in A$:

(*a*) $h(a + b) = h(a) \vee h(b)$

(*b*) $h(a \cdot b) = h(a) \wedge h(b)$

(*c*) $h(a') = h'(a)$.

A Boolean homomorphism $h : A \to B$ is a Boolean isomporphism if h is one-to-one onto B.

➤ **Note.** For a lattice L to be a Boolean algebra, it must have two properties. **First**, it must be *complemented*. For a lattice to be complemented it must have a least element 0 and a greatest element 1 and for every element x of the lattice, there must exist an element x' such that $x \vee x' = 1$ and $x \wedge x' = 0$. **Second,** it must be *distributive*. This means that for every x, y and z in L, $x \vee (y \wedge z) = (x \vee y) \wedge (x \vee z)$ and $x \wedge (y \vee z) = (x \wedge y) \vee (x \wedge z)$. Showing that a complemented - distributive lattice is a Boolean algebra.

7.5.1 Identities of Boolean Algebra

There are many identities in Boolean algebra. These are useful in simplifying the design circuits. The most important of these are :

Table 7.1. Boolean Identities.

Identity	*Name*
$(x')' = x$	Law of the double complement
$x \vee x = x$	
$x \wedge x = x$	Idempotent laws
$x \vee 0 = x$	
$x \wedge 1 = x$	Identity laws
$x \vee 1 = 1$	
$x \wedge 0 = 0$	Dominance laws
$x \vee y = y \vee x$	
$x \wedge y = y \wedge x$	Commutative laws
$x \vee (y \vee z) = (x \vee y) \vee z$	
$x \wedge (y \wedge z) = (x \wedge y) \wedge z$	Associative laws
$x \vee (y \wedge z) = (x \vee y) \wedge (x \vee z)$	
$x \wedge (y \vee z) = (x \wedge y) \vee (x \wedge z)$	Distributive law
$(x \wedge y) = x' \vee y'$, $(x \vee y)' = x' \wedge y'$	De-Morgan's laws

Example 7.5. Prove the *absorption law* $x \wedge (x \vee y) = x$, using the identities of Boolean algebra.

Solution. To prove $x \wedge (x \vee y)$

$$= (x \vee 0) \wedge (x \vee y) \quad \text{[Identity law for Boolean sum]}$$
$$= x \vee (0 \wedge y) \quad \text{[Distributive law]}$$
$$= x \vee (y \wedge 0) \quad \text{[Commutative law]}$$
$$= x \vee 0 \quad \text{[Dominance law]}$$
$$= x.$$

Hence the proof.

Note. $x \wedge (x \vee y) = x$ is called an *absorption law* since absorbing $x \vee y$ into x leaves x unchanged.

7.5.2 Some Theorems on Boolean Algebra

Theroem 1. (*i*) Additive identity is unique.

(*ii*) Multiplicative identity is unique.

(*iii*) For each $a \in B$ its complement a' is unique.

Proof. (*i*) Let 0 and 0_1 be two additive identities $\in B$, then,

$$0_1 + 0 = 0_1 = 0 + 0_1 \quad [\because 0 \text{ is additive identity}]$$

and

$$0 + 0_1 = 0 = 0_1 + 0 \quad [\because 0_1 \text{ is additive identity}]$$

But we have,

$$0_1 + 0 = 0 + 0_1 \quad \text{(By cummutativity)}$$

$$\Rightarrow \quad 0_1 = 0$$

Therefore additive identity is unique in B.

Similarly (*ii*) may be proved.

(*iii*) Let b and c be two complements of $a \in B$, then

$$a \cdot b = 0 \quad \& \quad a \cdot c = 0$$

$$a + b = 1 \quad \& \quad a + c = 1$$

$$\begin{aligned} b &= b + 0 && [\because 0 \text{ is additive identity in } B] \\ &= b + a \cdot c && [\because c \text{ is complement of } a] \\ &= (b + a) \cdot (b + c) && \text{(distributive law)} \\ &= (a + b) \cdot (b + c) && \text{(by commutativity)} \\ &= 1 \cdot (b + c) && [\because a' = b \text{ and } a + a' = 1] \\ &= b + c. && \text{(1 is mutiplicative identity)} \end{aligned}$$

Similarly,

$$\begin{aligned} c &= c + 0 \\ &= c + a \cdot b && [\because b \text{ is complement of } a] \\ &= (c + a) \cdot (c + b) \\ &= (a + c) \cdot (b + c) \\ &= 1 \cdot (b + c) && [\because c = a' \text{ and } a + a' = 1] \\ &= b + c \end{aligned}$$

Comparing, the above two results we find that,

$$b = c.$$

Therefore the two complements of a are equal.

Hence the complement of a is unique. so, every element of B has unique complement in it.

Theorem 2. Indempotent laws.

(*i*) $a + a = a$

(*ii*) $a \cdot a = a, \forall \ a \in B$.

Proof. (*i*)

$$\begin{aligned} a + a &= (a + a) \cdot 1 && [\because \text{I is identity w.r.t. '–'}] \\ &= (a + a) \cdot (a + a') && [\because a + a' = 1] \\ &= a + a \cdot a' && \text{(by distributive law)} \\ &= a + 0 && [\because a \cdot a' = 0] \\ &= a \cdot && \text{(0 is identity w.r.t. '+')} \end{aligned}$$

Hence, $\quad a + a = a \cdot$

(*ii*) $\quad a \cdot a = a \cdot a + 0 \quad$ (0 being additive identity)

$$
\begin{aligned}
&= a \cdot a + a \cdot a' && [\because a \cdot a' = 0] \\
&= a \cdot (a + a') && \text{(by distributivity)} \\
&= a \cdot 1 && [\because a + a' = 1] \\
&= a && [\because 1 \text{ is multiplicative identity}]
\end{aligned}
$$

Hence, $a \cdot a = a \ \forall \ a \in B.$

Theorem 3. Involution law

$\forall \ a \in B, (a')' = a.$

Proof. Let a be any arbitrary element of B, then

$$
\begin{aligned}
(a')' &= 1 \cdot (a')' \\
&= (a + a') \cdot (a')' && [\because a + a' = 1] \\
&= a \cdot (a')' + a' \cdot (a')' && \text{(by distributivity)} \\
&= a \cdot (a')' + 0 && [\because (a')', \text{is complement of } a'] \\
&= 0 + a \cdot (a')' && \text{(by commutativity)} \\
&= a \cdot a' + a \cdot (a')' && [\because a \cdot a' = 0] \\
&= a \cdot [a' + (a')'] && \text{(by distributivity)} \\
&= a \cdot 1 && [\because a' + (a')' = 1] \\
&= a && \text{(1 is multiplicative identity)}
\end{aligned}
$$

Therefore we have,

$$(a')' = a$$

Theorem 4. **(*i*)** $\ a + 1 = 1, \forall \ a \in B,$

(*ii*) $\ a \cdot 0 = 0, \forall \ a \in B,$

Proof. Let a be any arbitrary element of B, then,

(*i*)

$$
\begin{aligned}
a + 1 &= 1 \cdot (a + 1) && [\because 1 \text{ is multiplicative idenity}] \\
&= (a + a') \cdot (a + 1) && [\because a + a' = 1] \\
&= a + (a' \cdot 1) && \text{(by distributivity)} \\
&= a + a' && [\because a' \cdot 1 = a'] \\
&= 1. && [\because a + a' = 1]
\end{aligned}
$$

Hence $a + 1 = 1.$

(*ii*)

$$
\begin{aligned}
a \cdot 0 &= 0 + a \cdot 0 && \text{(by 0 being additive identity)} \\
a \cdot 0 &= a \cdot a' + a \cdot 0 && [\because a \cdot a' = 0] \\
&= a \cdot (a' + 0) && \text{(by distributivity)} \\
&= a \cdot a' && [\because a' + 0 = a'] \\
&= a. && [\because a \cdot a' = 0]
\end{aligned}
$$

Hence $a \cdot 0 = 0.$

Theorem 5. Absorption law :

(i) $a + (a \cdot b) = a$,

(ii) $a \cdot (a + b) = a, \ \forall \ a, b \in B$.

Proof. Let a, b be two arbitrary elements of B then,

(i)
$$
\begin{aligned}
a + a \cdot b &= a \cdot 1 + a \cdot b && \text{(by 1 being multiplicative identity)}\\
&= a \cdot (1 + b) && \text{(by distributivity)}\\
&= a \cdot (b + 1) && \text{(by commutativity)}\\
&= a \cdot 1 && [\because b + 1 = 1]\\
&= a.
\end{aligned}
$$

Hence $a + a \cdot b = a, \ \forall \ a, b \in B$.

(ii)
$$
\begin{aligned}
a \cdot (a + b) &= (a + 0) \cdot (a + b) && [\because 0 \text{ is additive identity}]\\
&= a + (0 \cdot b) && \text{(by distributivity)}\\
&= a + (b \cdot 0) && \text{(by cummutativity)}\\
&= a + 0 && [\because b \cdot 0 = 0, \ \forall \ b \in B]\\
&= a
\end{aligned}
$$

Hence $a \cdot (a + b) = a$.

Theorem 6. (i) $0' = 1$,

(ii) $1' = 0$.

Proof. (i)
$$
\begin{aligned}
0' &= 0 + 0' && [\because 0 \text{ is additive identity}]\\
&= 1. && [\because a + a' = 1]
\end{aligned}
$$

Hence $0' = 1$.

(ii)
$$
\begin{aligned}
1' &= 1.1' && [\because 1 \text{ is multiplicative identity}]\\
&= 0. && [\because a \cdot a' = 0]
\end{aligned}
$$

Hence $1' = 0$.

Theorem 7. De-Morgan's law : Let $[B, +, \cdot, ']$ be a Boolean algebra and a, b be any two elements of B. Then,

(i) $(a + b)' = a' \cdot b'$

(ii) $(a \cdot b)' = a' + b', \ \forall \ a, b \in B$.

Proof. (i) States that $(a' \cdot b')$ is the complement of $a + b$.

Therefore, we have to prove that

$$(a + b) + a' \cdot b' = 1$$

and
$$(a + b) \cdot (a' \cdot b') = 0.$$

where 1 is multiplicative and 0 is additive identity in B.

Now, $(a + b) + a' \cdot b'$

$$
\begin{aligned}
&= [(a + b) + a'] \cdot [(a + b) + b'] && \text{(by distributivity)}\\
&= [a + (b + a')] \cdot [a + (b + b')] && \text{(by associativity)}\\
&= [a + (a' + b)] \cdot [a + (b + b')] && \text{(by commutativity)}\\
&= [(a + a') + b] \cdot [a + (b + b')] && \text{(by associativity)}\\
&= (1 + b) \cdot (a + 1) && [\because a + a' = 1, b + b' = 1]
\end{aligned}
$$

$$= 1.1 \qquad [\because a+1=1, b+1=1]$$

$$= 1.$$

Therefore, $(a+b)+a'\cdot b' = 1.$ ...(*i*)

Also, $(a+b)\cdot(a'\cdot b') = a\cdot(a'\cdot b')+b\cdot(a'\cdot b')$ (by distributivity)

$$= (a\cdot a')\cdot b'+(b\cdot b')\cdot a'$$

(by cummutativity and associativity)

$$= 0\cdot b'+0\cdot a' \qquad [\because a\cdot a'=0=b\cdot b']$$

$$= b'\cdot 0+a'\cdot 0 \qquad \text{[by cummutativity)}$$

$$= 0+0 \qquad [\because a\cdot 0=0 \text{ etc.}]$$

$$= 0 \qquad [\because 0 \text{ is additive identity}]$$

Hence $(a+b)\cdot(a'\cdot b') = 0.$ ...(*ii*)

By (*i*) and (*ii*) we conclude that $(a'\cdot b')$ is the complement of $(a+b)$ consequently,

$$(a+b)' = a'\cdot b'$$

(*ii*) In this part we have to prove that $a'+b'$ is the complement of $(a\cdot b)$. Therefore we have to prove

$$(a\cdot b)+(a'+b') = 1$$

and $(a\cdot b)\cdot(a'+b') = 0$

Now $(a\cdot b)+(a'+b') = [a+(a'+b')]\cdot[b+(a'+b')]$ [by distributivity)

$$= [(a+a')+b']\cdot[(b+b')+a']$$

(by associativity and cummutativity)

$$= (1+b')\cdot(1+a') \qquad [\because a+a'=1 \text{ and } b+b'=1]$$

$$= (b'+1)\cdot(a'+1) \qquad \text{(by commutativity)}$$

$$= 1.1 \qquad [\because a+1=1]$$

$$= 1.$$

Therefore, $(a\cdot b)+(a'+b') = 1$ ...(*i*)

and $(a\cdot b)\cdot(a'+b') = (a\cdot b)\cdot a'+(a\cdot b)\cdot b'$ (by distributivity)

$$= a\cdot(b\cdot a')+a\cdot(b\cdot b') \qquad \text{(by associativity)}$$

$$= a\cdot(a'\cdot b)+a\cdot(b\cdot b') \qquad \text{(by commutativity)}$$

$$= (a\cdot a')\cdot b+a(b\cdot b') \qquad \text{(by associativity)}$$

$$= 0\cdot b+a\cdot 0 \qquad [\because a\cdot a'=0=b\cdot b']$$

$$= b\cdot 0+a\cdot 0 \qquad \text{(by commutativity)}$$

$$= 0+0 \qquad [\because a\cdot 0=0,\ \forall a\in B]$$

$$= 0. \qquad \text{(0 being additive identity)}$$

Thus, $(a\cdot b)\cdot(a'+b') = 0$ ...(*ii*)

By (*i*) and (*ii*) we conclude that $a'+b'$ is the complement of a, b, consequently

Note : Operator v and ^ are replaced by '+' and '·' respectively.

ILLUSTRATIVE EXAMPLES

Example 7.6. Let $[B, +, \cdot, ']$ be a Boolean algebra and $a, b \in B$, then show that:

(i) $a + (a + b) = a + b$,

(ii) $a \cdot (a \cdot b) = a \cdot b$,

(iii) $a \cdot b = a \Rightarrow a \cdot b' = 0$,

(iv) $a \cdot b' = b \Rightarrow a + b = b$,

(v) $a + b = b \Rightarrow a \cdot b = a$,

(vi) $a + b = a + b$,

(vii) $a + a \cdot b = a' + b$.

Solution : (*i*) $a + (a + b)$

$= (a + a) + b$ (by associativity)

$= a + b.$ $[\because a + a = a]$

Hence, $a + (a + b) = a + b.$ **Proved.**

(*ii*) $a \cdot (a \cdot b) = (a \cdot a) \cdot b$ (by associativity)

$= a \cdot b$ (by idempotent law)

$\Rightarrow$ $a \cdot (a \cdot b) = a \cdot b$ **Proved**

(*iii*) $a \cdot b' = (a \cdot b) b'$ $[\because a \cdot b = a;$ given$]$

$= a \cdot (b \cdot b')$ [by associativity]

$= a \cdot 0$ $[\because b \cdot b' = 0]$

$= 0.$ $[\because a \cdot 0 = 0]$

Hence, $a \cdot b = a \Rightarrow a \cdot b' = 0.$ **Proved.**

(*iv*) $a + b = (a + b) \cdot 1$ $[\because$ 1 is multiplicative identity$]$

$= (a + b) \cdot (b + b')$ $[\because b + b' = 1]$

$= (b + a) \cdot (b + b')$ (by commutativity)

$= b + (a \cdot b')$ (by distributivity)

$= b + 0$ (given that $a \cdot b' = 0$)

$= b.$ $[\because$ 0 is additive identity$]$

Therefore, $a \cdot b' = 0 \Rightarrow a + b = b.$ **Proved.**

(*v*) $a \cdot b = a \cdot (a + b)$ $[\because a + b = b,$ given$]$

$= a \cdot a + a \cdot b$ (by distributivity)

$= a + a \cdot b$ (by idempotent law *i.e.*, $a \cdot a = a$)

$= a.$ (by absorption law $a + a \cdot b = a$)

Therefore, $a + b = b \Rightarrow a \cdot b = a.$ **Proved.**

(*vi*) $a + b = (a + b) \cdot 1$ $[\because a + a' = 1]$

$= a + (b \cdot a')$ (by distributivity)

$= a + a' \cdot b$ (by commutativity)

Hence, we have

$$a + a' \cdot b = a + b.$$ **Proved**

$$
\begin{aligned}
(vii) \qquad a' + a \cdot b &= (a' + a) \cdot (a' + b) && \text{(by distributivity)} \\
&= (a + a') \cdot (a' + b) && \text{(by commutativity)} \\
&= 1 \cdot (a' + b) && [\because a + a' = 1] \\
&= a' + b && [\because 1 \text{ is multiplicative identity}]
\end{aligned}
$$

Hence, $a' + a \cdot b = a' + b.$ **Proved**

Example 7.7. In a Boolean algebra $[B, +, \cdot, ']$. Prove that:

***(i)* $(a + b)' + (a + b')' = a'.$**

***(ii)* $(a + b) \cdot (a' + c) = a' \cdot b + a \cdot c.$**

Solution: (*i*) By De-Morgan's law,

$$
\begin{aligned}
(a \cdot b)' &= a' + b', \text{ we have} \\
(a + b)' + (a + b')' &= [(a + b) \cdot (a + b')]' \\
&= (a + b \cdot b')' && \text{(by distributivity)} \\
&= (a + 0)' && [\because b \cdot b' = 0] \\
&= (a)' && [\because a + 0 = a] \\
&= a^2.
\end{aligned}
$$

Hence, we get,

$$(a + b)' + (a + b')' = a'. \qquad \textbf{Proved.}$$

$$
\begin{aligned}
(ii) \text{ L.H.S. } (a + b) \cdot (a' + c) &= a \cdot (a' + c) + b \cdot (a' + c) && \text{(by distributivity)} \\
&= (a \cdot a' + a \cdot c)' + (b \cdot a' + b \cdot c) && \text{(by distributivity)} \\
&= (0 + a \cdot c) + b \cdot a' + b \cdot c && [\because a \cdot a' = 0] \\
&= (a \cdot c + 0) + b \cdot a' + b \cdot c \\
&= a \cdot c + b \cdot a' + b \cdot c \\
&= ac + ba' + bc \cdot 1 \\
&= ac + ba' + bc\,(a + a') \\
&= ac + ba' + bc \cdot a + bc \cdot a' \\
&= ac + bc \cdot a + a'b + a'bc \\
&= ac\,(1 + b) + a'\,b\,(1 + c) \\
&= ac \cdot 1 + a'\,b \cdot 1 \\
&= a \cdot c + a' \cdot b.
\end{aligned}
$$

Thus, we get

$$(a + b) \cdot (a' + c) = a \cdot c + a' \cdot b. \qquad \textbf{Proved.}$$

Example 7.8. Simplify the following:

(*a*) (*i*) $(a + b) \cdot a' \cdot b'$,

(*ii*) $(a + b + c)'$,

(*iii*) $a \cdot b + b \cdot c + b \cdot c'$

(*iv*) $a \cdot b \cdot c + a' + c'$.

(*v*) $a \cdot b + [(a + b') \cdot b]'$

(*vi*) $[a + (a' \cdot b')] \cdot [a' + (a \cdot b)]$

(*b*) Show that:

$$a \cdot b + b \cdot c + c \cdot a = (a + b) \cdot (b + c) \cdot (c + a)$$

Solution: (*a*) (*i*) $\qquad (a + b)\,a' \cdot b' = (a + b) \cdot (a' \cdot b')$

$= (a+b) \cdot (a+b)'$

[by De-Morgan's law $(a+b)' = (a' \cdot b')$]

$= 0.$ $[\because a \cdot a' = 0]$

Thus, $(a+b) \cdot a' \cdot b' = 0.$ **Ans.**

(*ii*) $(a+b+c)' = [(a+b)+c]'$

$= (a+b)' \cdot c'$ (by De-Morgan's law)

$= (a' \cdot b') \cdot c'$ (" ")

$= a' \cdot b' \cdot c' \cdot$

Hence, $(a+b+c)' = a' \cdot b' \cdot c'.$ **Ans.**

Note : Proceeding in the above manner, we can generalise the De-Morgan's theorem as

$$(a+b+c+ \ldots +k)' = a' \cdot b' \cdot c' \ldots\ldots k', \quad \forall\, a, b, c \ldots\ldots k \in B.$$

(*iii*) $a \cdot b + b \cdot c + b \cdot c' = b \cdot a + b \cdot c + b \cdot c'$ (by commutativity)

$= b \cdot (a + c) + b \cdot c'$ (by distributivity)

$= b \cdot (a + c + c')$ (" ")

$= b\,[a + (c + c')]$ (by associativity)

$= b \cdot [a + 1]$ $[\because c + c' = 1]$

$= b \cdot 1$ $[\because a + 1 = 1]$

$= b.$ [$\because$ 1 is multiplicative identity]

Thus, $a \cdot b + b \cdot c + b \cdot c' = b.$ **Ans.**

(*iv*) $a \cdot b \cdot c + a' + b' + c' = a \cdot b \cdot c + (a \cdot b \cdot c)'$ (by De-Morgan's law)

$= 1.$ $[\because a + a' = 1]$ **Ans.**

(*v*) $a \cdot b + [(a + b') \cdot b]' = a \cdot b + [b \cdot (a + b')]'$ (by commutativity)

$= a \cdot b + [b \cdot a + b \cdot b']'$ [by distributivity]

$= a \cdot b + [a \cdot b + 0]'$ [$\because b \cdot a = a \cdot b$ and $b \cdot b' = 0$]

$= a \cdot b + [a \cdot b]'$

$= 1.$ $[\because a + a' = 1]$

(*vi*) $[a + (a'\,b')]\,[a' + (a \cdot b)] = [(a + a')\,(a + b')]\,[(a' + a) \cdot (a' + b)]$ (by distributivity)

$= [1 \cdot (a + b')]\,[1 \cdot (a' + b)]$ $[\because a + a' = 1]$

$= (a + b')\,(a' + b)$ $[\because 1 \cdot a = a]$

$= a \cdot a' + a \cdot b + b' \cdot a' + b' \cdot b$ (by distributivity)

$= 0 + a \cdot b + b' \cdot a' + 0$ $[\because a \cdot a' = 0]$

$= a' \cdot b' + a \cdot b$ (by commutativity)

(*b*) R.H.S. $= (a + b) \cdot (b + c) \cdot (c + a)$

$= (a + b) \cdot [(b + c) \cdot (c + a)]$ (by associativity)

$= (a + b) \cdot [(c + b)\,(c + a)]$ (by commutativity)

$= (a + b)\,(c + b \cdot a)$ [by distributivity]

$= a \cdot c + b \cdot c + a\,(b \cdot a) + b \cdot (b \cdot a)$ (by distributivity)

$= a \cdot c + b \cdot c + (a \cdot a) \cdot b + (b \cdot b) \cdot a$

(by commutativity and associativity)

$= a \cdot c + b \cdot c + a \cdot b + b \cdot a$ $[\because a \cdot a = a$ etc.$]$

$= a \cdot c + b \cdot c + (a \cdot b + a \cdot b)$ (by commutativity)

$= a \cdot c + b \cdot c + a \cdot b$ $[\because a + a = a]$

$= a \cdot b + b \cdot c + c \cdot a$ (by commutativity)

$=$ L.H.S. **Proved.**

Example 7.9. Show that for the Boolean algebra $[B, +, \cdot, ']$:

(i) $p \cdot q \cdot r + p \cdot q' \cdot r + p' \cdot q \cdot r = p \cdot q + q \cdot r + r \cdot p, \ \forall\, p, q, r \in B,$

(ii) $a \cdot b + a \cdot b' + a' \cdot b + a' \cdot b' = 1, \ \forall\, a, b \in B.$

Solution : (*i*) L.H.S. $= p \cdot q \cdot r + p \cdot q \cdot r' + p \cdot q' \cdot r + p' \cdot q \cdot r$

$= p \cdot q\,(r + r') + p \cdot q' \cdot r + p' \cdot q \cdot r$ (by distributivity)

$= p \cdot q \cdot 1 + p \cdot q' \cdot r + p' \cdot q \cdot r$ $[\because r + r' = 1]$

$= p \cdot q + p \cdot q' \cdot r + p' \cdot q \cdot r$ [$\because$ 1 is multiplicative identity]

$= p \cdot [q + q' \cdot r] + p' \cdot q \cdot r$ [by distributivity]

$= p'\,(q + q') \cdot (q + r)] + p' \cdot q \cdot r$ (" ")

$= p[1\,(q + r)] + p' \cdot q \cdot r$ $[\because q + q' = 1]$

$= p \cdot (q + r) + p' \cdot q \cdot r$ [$\because$ 1 is multiplicative identity]

$= p \cdot q + p \cdot r + p' \cdot q \cdot r$

$= p \cdot q + (p + p' \cdot q) \cdot r$ (by distributivity and associativity)

$= p \cdot q + [(p + p') \cdot (p \cdot q) \cdot r$ (by distributivity)

$= p \cdot q + [1 \cdot (p + q)] \cdot r$ $[\because p + p' = 1]$

$= (p \cdot q) + (p + q) \cdot r$

$= p \cdot q + p \cdot r + q \cdot r$ (by distributivity)

$= p \cdot q + q \cdot r + r \cdot p$ (by commutativity)

$=$ R.H.S. **Proved**

(*ii*) L.H.S. $= a \cdot b + a \cdot b' + a' \cdot b + a' \cdot b'$

$= a \cdot (b + b') + a' \cdot (b + b')$ (by distributivity)

$= a \cdot 1 + a' \cdot 1$ $[\because b + b' = 1]$

$= a + a'$ [$\because$ 1 is multiplicative identity]

$= 1$ $[\because a + a' = 1]$

$=$ R.H.S. **Proved.**

Example 7.10. Show that :

(i) $a \cdot b + a' \cdot b' = (a' + b) \cdot (a + b')$

(ii) $a' + a \cdot b = a' + b$

Solution : (*i*) $a \cdot b + a' \cdot b' = (a \cdot b + a') \cdot (a \cdot b + b')$ (by distributive law)

$= \{(a + a') \cdot (b + a')\} \cdot \{(a + b') \cdot (b + b')\}$

(by distributive law)

$$= \{1 \cdot (a' + b)\} \cdot \{(a + b') \cdot 1\}$$

[$\because a + a' = 1$ and by commutativity]

$$= (a' + b) \cdot (a + b'). \qquad [\because a \cdot 1 = a\cdot] \textbf{ Proved.}$$

(*ii*) $$a' + a \cdot b = (a' + a) \cdot (a' + b)$$
$$= 1 \cdot (a' + b)$$
$$= a' + b$$

Example 7.11. Prove that if $\{B, +, \cdot, '\}$ be a Boolean algebra, then

$$\mathbf{(a + b) + c = a + (b + c), \ \forall\ a, b, c \in B.}$$

Solution: Let $(a + b) + c = x$

and $a + (b + c) = y$

Then $ax = a \cdot [(a + b) + c]$

$= a \cdot (a + b) + a \cdot c$ (by distributive law)

$= a + a \cdot c$ (by absorption law $a + a \cdot b = a$) ...(*i*)

and $a, y = a \cdot \{a + (b + c)\}$

$= a$ [by absorption law $a \cdot (a + b) = a$] ...(*ii*)

Comparing (*i*) and (*ii*), we get

$ax = a, y$ ...(X)

Also, $a' x = a' \cdot [(a + b) + c]$

$a' x = a' \cdot (a + b) + a' \cdot c$ (by distributive law)

$= (a' \cdot a + a' \cdot b) + a' \cdot c$ (" ")

$= (0 + a' \cdot b) + a' \cdot c$ [$\because a' \cdot a = 0$]

$= a' \cdot b + a' \cdot c$ ...(*iii*)

and $a' \cdot y = a' \cdot [a + (b + c)]$

$= a' \cdot a + a' \cdot (b + c)$ (by distributive law)

$= 0 + (a' \cdot b + a' \cdot c)$

(by distributive law and $a' \cdot a = 0$)

$= a' \cdot b + a' \cdot c$ [$\because a + 0 = 0$] ...(*iv*)

Comparing (*iii*) and (*iv*), we have

$a' \cdot x = a' \cdot y$ ...(*Y*)

Again, by (*X*) and (*Y*), we have,

$ax + a' x = a \cdot y + a' \cdot y$

$\Rightarrow (a + a') x = (a + a') \cdot y$ (by distributive law)

$\Rightarrow 1 \cdot x = 1 \cdot y$ [$\because a + a' = 1$]

$\Rightarrow x = y$ [$\because 1, x = x$]

$\therefore (a + b) + c = a + (b + c).$ **Proved.**

Example 7.12. If $a, b, c \in B$, then prove that

$$\mathbf{ab + \{c\,(a' + b')\} = a \cdot b + c.}$$

Solution: L.H.S. $= a \cdot b + c \cdot (a' + b')$

$= a \cdot b + c \cdot (ab)'$ (by De-Morgan law)

$$= (a \cdot b + c) \cdot (a \cdot b + (a \cdot b)') \quad \text{(by distributive law)}$$
$$= (a \cdot b + c) \cdot 1 \quad \text{(by complement law)}$$
$$= ab + c \quad \text{(by identity law)}$$
$$= \text{R.H.S.}$$

Example 7.13. Prove that

$$[a + a' b]\,[a' + a \cdot b] = b.$$

Solution: L.H.S. $= [a + a' \cdot b] \cdot [a' + a \cdot b]$

$$= (a + a') \cdot (a + b) \cdot (a' + a) \cdot (a' + b) \quad \text{(by distributive law)}$$
$$= 1 \cdot (a + b)\, 1 \cdot (a' + b) \quad \text{(by compliment law)}$$
$$= (b + a) \cdot (b + a') \quad \text{(by commutative law)}$$
$$= b + a \cdot a' \quad \text{(by distributive law)}$$
$$= b + 0$$
$$= b = \text{R.H.s.}$$

Example 7.14. In Boolean algebra $[B, +, \cdot, ']$ prove that

(i) $\{x'\,(x + y)\}' + \{y \cdot (y + x')\}' + \{y\,(y' + x)\}' = 1$

(ii) $(xy' + yz) \cdot (xz + yz') = xz$

(iii) $(x + yz) \cdot (y' + x) \cdot (y' + z') = x(y' + z')$

Solution: *(i)* $\{x' \cdot (x + y)\}' + \{y \cdot (y + x')\} + \{y\,(y' + x)\}'$

$$= (x')' + (x + y)' + y' + (y + x')' + y' + (y' + x)'$$
$$= x + (x' \cdot y') + y' + (y'\,(x') + y' + (y')'\,x'$$
$$= x + x'y' + y' + y'x + y' + yx'$$
$$= x + y'\,x + x'y' + y' + y' + yx'$$
$$= x(1 + y') + y'\,(1 + x') + y' + yx'$$
$$= x \cdot 1 + y' \cdot 1 + y' + yx'$$
$$= x + y' + (y' + y)\,(y' + x')$$
$$= x + y' + 1 \cdot (y' + x')$$
$$= x + y' + y' + x'$$
$$= x + x' + y' + y'$$
$$= 1.$$

(ii) $(xy' + yz) \cdot (xz + yz')$

$$= (xy' + y)\,(xy' + z)\,(xz + y)\,(xz + z')$$
$$= (x + y)\,(y' + y)\,(x + z)\,(y' + z)\,(x + y)\,(z + y)\,(x + z') \cdot (z + z')$$
$$= (x + y) \cdot 1\,(x + z)\,(y' + z)\,(x + y)\,(z + y)\,(x + z') \cdot 1$$
$$= (x + y)\,(x + y)\,(x + z)\,(x + z')\,(z + y)\,(z + y')$$
$$= (x + y)\,(x + z \cdot z')\,(z + y \cdot y')$$
$$= (x + y)\,(x + 0)\,(z + 0)$$
$$= (x + y) \cdot xz$$
$$= x \cdot x \cdot z + x \cdot y \cdot z$$
$$= x \cdot z + x \cdot y \cdot z$$

$= x \cdot z + y \cdot x' \cdot z$

$= x \cdot z\,(1 + y)$

$= x \cdot z \cdot 1$

$= x \cdot z.$

(*iii*) $(x + yz)\,(y' + x)\,(y' + z')$

$= (x + y)\,(x + z)\,(x + y')\,(y' + z')$

$= (x + y)\,(x + y')\,(x + z)\,(y' + z')$

$= (x + y \cdot y')\,(x + z)\,(y' + z')$

$= x \cdot (x + z)\,(y' + z')$ $\quad [\because y \cdot y' = 0, x + 0 = x]$

$= (x \cdot x + x \cdot z)\,(y' + z')$

$= (x + x \cdot z)\,(y' + z')$

$= x(1 + z)\,(y' + z')$

$= x \cdot 1\,(y' + z')$ $\quad [\because 1 + z = 1]$

$= x \cdot (y' + z').$

Example 7.15. In Boolean algebra $[B, +, \cdot, ']$ prove that

(i) $\boldsymbol{a + b = 0 \leftrightarrow a = 0, b = 0}$

(ii) $\boldsymbol{ab' = 0 \leftrightarrow a' + b = 1}$

(iii) $\boldsymbol{a + b = a + c}$ ***and*** $\boldsymbol{a \cdot b = a \cdot c \Rightarrow b = c.}$

Solution : (*i*) Let $a = 0$ and $b = 0$, then $a + b = 0 + 0 = 0$.

Conversely. If $a + b = 0$, then

$$a = a + 0$$

$= a + b \cdot b'$ (by complement law)

$= (a + b)\,(a + b')$ (by distributive law)

$= 0\,(a + b')$ $\quad [\because a + b = 0, \because 0 \cdot a = 0\ \forall\ a \in B]$

$= 0$

$$b = b + 0$$

$= b + a \cdot a'$

$= (b + a)\,(b + a')$

$= (a + b)\,(a' + b)$

$= 0\,(a' + b)$

$= 0$

(*ii*) $\quad ab' = 0$

$\leftrightarrow \quad (ab')' = 0'$

$\leftrightarrow \quad a' + (b')' = 1$ (since $0' = 1$)

$\leftrightarrow \quad a' + b = 1$

(*iii*) Let $\quad a + b = a + c$ and $a \cdot b = a \cdot c$, then

$b = b + ba$ (by absorption law)

$= b \cdot b + b \cdot a$

$= b \cdot (b + a)$

$$\begin{aligned}
&= b \cdot (a+b) \\
&= b \cdot (a+c) && \text{(given } a+b=a+c\text{)} \\
&= b \cdot a + b \cdot c \\
&= a \cdot b + b \cdot c \\
&= a \cdot c + b \cdot c && \text{(given } a \cdot b = a \cdot c\text{)} \\
&= (a+b)\, c \\
&= (a+c) \cdot c && \text{(given } a+b=a+c\text{)} \\
&= a \cdot c + c \cdot c \\
&= a \cdot c + c \\
&= c + c \cdot a \\
&= c \cdot (1+a) && (\because 1+a=1, c \cdot 1 = c) \\
&= c.
\end{aligned}$$

7.5.3 Theorems on Partial Ordered Relations in Boolean Algebra

Theorem 1. In Boolean algebra $[B, +, \cdot, ']$ the relation "$\leq$" is a partial ordered relation.

Proof. Since for $a, b \in B$,

$$a \leq b \leftrightarrow a \cdot b' = 0. \qquad \text{(definition)}$$

Hence, if $a \leq b$, then we have

$$a, b = a \text{ and } a + b = b. \qquad ...(i)$$

Now,

1. Reflexivity : $\because\ \forall\ a \in B$, we have

$$a + a = a \qquad \text{(by idempotent law)}$$

and $\quad a \cdot a = a$

Hence $\quad a \leq a$

Therefore, '$\leq$' is reflexive

2. Anti-symmetric. Since, by definition of $a \leq b$, $\forall\ a, b \in B$ we have,

If $\quad a \leq b, a + \text{I}b = b$

and $\quad b \leq a, b + a = a$

Therefore, $\quad a + b = b + a \qquad$ (by commutativity)

$\leftrightarrow \quad b = a$

Therefore the relation '$\leq$' is anti-symmetric, because

$$a \leq b \text{ and } b \leq a \leftrightarrow a = b$$

i.e., $\quad aRb$ and $bRa \leftrightarrow a = \text{I}b$

3. Transitive. Let a, b, c be any three elements of B, then

$$a \leq b, b \leq c \leftrightarrow a + b = b, b + c = c$$

Therefore,

$$\begin{aligned}
a + c &= a + (b + c) \\
&= (a + b) + c && \text{(by associativity)} \\
&= b + c && [\because a + b = b] \\
&= c && [\because b + c = c]
\end{aligned}$$

Thus, $a + c = c \Rightarrow a \leq c$

So, $a \leq b, b \leq c \Rightarrow a \leq c$

Therefore the relation '≤' is transitive. Since, the relation '≤' is reflexive, anti-symmetric and transitive, hence it is partial ordered relation in the Boolean algebra B.

Thus theorem is proved.

Theorem 2. For any element a of a Boolean algebra *B*, show that

(*i*) $a \leq 1$ (*ii*) $0 \leq a, \ \forall \ a \in B$.

Proof. (*i*) Since we know that for any element $a \in B$,

$$a \cdot 0 = 0$$

Hence $a \cdot 1' = 0,$ $[\because 1' = 0]$

Thus, $a \cdot 1' = 0 \Rightarrow a \leq 1.$ $[\because a \leq b \Rightarrow a \cdot b' = 0]$

(*ii*) $a \in B \Rightarrow a' \in B$

Since, $a' \cdot 0 = 0$

$\Rightarrow$ $0 \cdot a' = 0$ (by commutativity)

$\Rightarrow$ $0 \leq a$

Thus in above theorem we find for all $a \in B$,

$$0 \leq a \leq 1$$

Theorem 3. If *a, b* ∈ *B* then,

$$\boldsymbol{a \leq b \Rightarrow b' \leq a'}$$

In other words this theorem gives.

If *a* + *b* = *b* then *b*′ + *a*′ = *a*′.

Proof.
$$\begin{aligned} b' + a' &= (b \cdot a)' && \text{(by De-Morgan's law)} \\ &= (a \cdot b)' && \text{(by commutativity)} \\ &= [a \cdot (a + b)]' && [\because b = a + b] \\ &= (a)' && \text{[by absorption law]} \\ &= a' \end{aligned}$$

Hence, $b' + a' = a'$

$\Rightarrow$ $b' \leq a'$ **Proved.**

Theorem 4. In a Boolean algebra [*B*, +, ·, ′] prove that :

(*i*) $\boldsymbol{a + b = \text{lub}\{a, b\}}$

***i.e.,* *a* + *b* is equal to the least upper bound of *a* and *b*.**

(*ii*) $\boldsymbol{a \cdot b = \text{glb}\{a, b\}}$

***i.e.,* *a* · *b* is equal to the greatest lower bound of *a* and *b*.**

Proof. (*i*) Since, for all $a, b \in B$, we have

$$\begin{aligned} a + (a + b) &= (a + a) + b && \text{[by associativity]} \\ &= a + b && [\because a + a = a] \end{aligned}$$

Thus, $a + (a + b) = a + b \Rightarrow a \leq (a + b)$ ...(*i*)

Similarly,

$$b + (a + b) = (b + a) + a \qquad \text{(by associativity)}$$

$$= (a+b)+b \qquad \text{(by commutativity)}$$
$$= a+(b+b) \qquad \text{(by associativity)}$$
$$= a+b \qquad [\because (b+b)=b]$$

Thus, $\quad b+(a+b) = a+b$

$\Rightarrow \quad b \leq a+b \qquad$ ...(*ii*)

By (1) and (2) we conclude that $(a + b)$ is the uper bound of a and b.

Let c be any other upper bound of a and b then,

$$a \leq c \text{ and } b \leq c$$

$\Rightarrow \quad a+c = c \text{ and } b+c=c$

Now $\quad (a+b)+c = a+(b+c) \qquad$ (by associativity)

$$= a+c \qquad [\because b+c=c]$$
$$= c \qquad [\because a+c=c]$$

Hence $\quad a+b \leq c$

Thus, we conclude that $a + b$ is the **least upper bound** of a and b *i.e.*,

$$a+b = \text{lub}\{a, b\}. \qquad \textbf{Proved}$$

(*ii*) Similarly, we have,

$$a+a,b = a \text{ and } b+a\cdot b=b \qquad \text{(by absorption law)}$$

$\Rightarrow \quad a\cdot b+a = a \text{ and } a\cdot b+b=b \qquad$ (by commutative law)

Therefore, $\quad a\cdot b \leq a \text{ and } a\cdot b \leq b.$

Hence, $a\,b$ is the **lower bound** of a and b.

Let c be some other lower bound of a and b, then,

$$c \leq a, c \leq b \Rightarrow c+a=a, c+b=b$$

Now, $\quad c+a\cdot b = (c+a)\cdot(c+b) \qquad$ (by distributivity)

$$= ab. \qquad [\because c+a=a, c+b=b]$$

Hence, $\quad c \leq ab.$

Therefore, $a \cdot b$ is the greatest lower bound of a and b, *i.e.*,

$$a\cdot b = \text{glb}\{a, b\}. \qquad \textbf{Proved.}$$

ILLUSTRATIVE EXAMPLES

Example 7.16. In a Boolean algebra $[B, +, \cdot, ']$ prove that :

***(i)* $a+b \leq c \leftrightarrow a \leq c, b \leq c$.**

***(ii)* $a \leq b, a \leq c \Rightarrow a \leq bc$.**

Solution : (*i*) Since, we know that,

$$a \leq c \Rightarrow a+c=c, a\cdot c=a$$

and $\quad b \leq c \Rightarrow b+c=c, b\cdot c=b$

then we are to prove that,

$$a+b \leq c$$

i.e., $\quad (a+b)+c = c$

Now, $\quad (a+b)+c = a+(b+c) \qquad$ (by associativity)

$$= a + c \qquad [\because b \le c \Rightarrow b + c = c]$$

$$= c \qquad [\because a \le c \Rightarrow a + c = c]$$

Thus, $(a + b) + c = c$

$\Rightarrow$ $a + b \le c$ ...(*i*)

Again, if $(a + b) \le c$, then to prove that

$$a \le c, b \le c$$

Since, $(a + b) \le c$

$\Rightarrow$ $(a + b) \cdot c = a + b$ $[\because a \le b \Rightarrow a \cdot b = a]$

$\Rightarrow$ $a \cdot c + b \cdot c = a + b$ (by distributivity)

$\Rightarrow$ $a \cdot c = a$ and $b \cdot c = b$

$\Rightarrow$ $a \le c$ and $b \le c$. ...(*ii*)

Combining (*i*) and (*ii*) we have,

$$(a + b) \le c \leftrightarrow a \le c, b \le c.$$ **Proved.**

(*ii*) $a \le b, a \le c \leftrightarrow a \cdot b' = 0$ and $ac' = 0$ ...(*i*)

Now, $a \cdot (b \cdot c)' = a \cdot (b' + c')$ (by De-Morgan's law)

$= a \cdot b' + a \cdot c'$ (by distributive)

$= 0 + 0$ [by equation (*i*)]

$= 0$

$\Rightarrow$ $a \le b \cdot c$ **Proved.**

Example 7.17. If $a, b \in B$, where $[B, +, \cdot, ']$ is a Boolean algebra, then prove that,

$$b \le a' \leftrightarrow a \cdot b = 0$$

Solution: $b \le a'$

$\Rightarrow$ $b + a' = a'$

$\Rightarrow$ $b = a'$. $[\because a' + a' + a'$ idempotent law$]$

Therefore, $a \cdot a' = 0$

$\Rightarrow$ $a \cdot b = 0$ $[\because b = a']$

Thus, $b \le a' \Rightarrow a \cdot b = 0$ ...(*i*)

Again, $a \cdot b = 0$

$\Rightarrow$ $b = a'$ $[\because a \cdot a' = 0]$

Therefore, $b + a' = a' + a'$, $[\because b = a']$

$= a'$ (Idempotent law)

Thus, $b + a' = a' \Rightarrow b \le a'$. ...(*ii*)

Combining the results (*i*) and (*ii*) we get,

$$b \le a' \leftrightarrow a \cdot b = 0.$$ **Proved.**

Example 7.18. If $a = b$, then prove that:

$$a \cdot b' + a' \cdot b = 0.$$

Solution: $a \cdot b' + a' \cdot b = b \cdot b' + a' \cdot a$ $[\because a = b]$

$= b \cdot b' + a \cdot a'$ (by commutative law)

$= 0 + 0$ $(\because a \cdot a' = 0 \;\forall\, a \in B]$

$$= 0. \quad \text{(as 0 is additive identity)}$$

So, $a \cdot b' + a' \cdot b = 0$ if $a = b$. **Proved.**

Example 7.19. *Prove that in any Boolean algebra B, the dual of $a \leq b$ is $b \leq a$.*

Solution : '$\leq$' is called ordered relation which is defined by,

$$a \leq b \leftrightarrow a \cdot b' = 0, \ \forall\, a, b \in B.$$

Also, the dual of $a \cdot b' = 0$ is $a + b' = 1$

Hence,
$$\begin{aligned} a + b' = 1 &\leftrightarrow (a + b')' = 1' \\ &\leftrightarrow a' \cdot (b')' = 0 && [\because 1' = 0] \\ &\leftrightarrow a' \cdot b = 0 && [\because (b')' = b] \\ &\leftrightarrow b \cdot a' = 0 && \text{(by commutative law)} \\ &\leftrightarrow b \leq a && \text{(by definition of '}\leq\text{')} \end{aligned}$$

Hence, dual of $a \leq b$ is $b \leq a$. **Proved.**

Theorem 7.1. In a Boolean algebra, the following result holds :

(*i*) $(x')' = x$

(*ii*) $(x \wedge y)' = x' \vee y'$

(*iii*) $(x \vee y)' = x' \wedge y'$ **De-Morgan's law**

(*iv*) $x \leq y \Leftrightarrow x' \geq y'$

(*v*) $x \leq y \Leftrightarrow x \wedge y' = 0 \Leftrightarrow x' \vee y = 1.$

Proof. (*i*) To prove $(x')' = x$

$$\begin{aligned} & x \wedge x' = 0, \ x \vee x' = 1 \\ \Rightarrow \quad & x' \wedge x'' = 0, \ x' \vee x'' = 1 \\ \Rightarrow \quad & x \wedge x' = x'' \wedge x', \ x \vee x' = x'' \vee x' \\ \Rightarrow \quad & (x')' = x. \end{aligned}$$

(*ii*)
$$\begin{aligned} (x \wedge y) \wedge (x' \vee y') &= \{(x \wedge y) \wedge x'] \vee [(x \wedge y) \wedge y'] \\ &= [(x \wedge x') \wedge y] \vee [x \wedge (y \wedge y')] \\ &= [0 \wedge y] \vee [x \wedge 0] \\ &= 0 \vee 0 = 0. \end{aligned}$$

Also
$$\begin{aligned} (x \wedge y) \vee (x' \vee y') &= (x' \vee y') \vee (x \wedge y) \\ &= [(x' \vee y') \vee x] \wedge [x' \vee y') \vee y] \\ &= [(x' \vee x) \vee y'] \wedge [x' \vee (y' \vee y)] \\ &= [1 \vee y'] \wedge [x' \vee 1] \\ &= 1 \wedge 1 \\ &= 1. \end{aligned}$$

Hence $(x \vee y)' = x' \wedge y'$ **Proved.**

(*iii*) As (*ii*) part.

(*iv*)
$$\begin{aligned} x \leq y &\Rightarrow x = x \wedge y \\ &\Rightarrow x' = (x \wedge y)' = x' \vee y' \end{aligned}$$

$$\Rightarrow y' \leq x'.$$

$$y' \leq x' \Rightarrow y'' \geq x'' \Rightarrow y \geq x$$

Thus $\quad x \leq y \Rightarrow x' \geq y'.$

(*v*) $\quad x \leq y \Rightarrow x \wedge y' \leq y \wedge y'$

$$\Rightarrow 0 \leq x \wedge y' \leq 0$$

$$\Rightarrow x \wedge y' = 0.$$

Let $x \wedge y' = 0$. Then $x = x \wedge 1$

$$= x \wedge (y \wedge y')$$

$$= (x \wedge y) \vee (x \wedge y')$$

$$= (x \wedge y) \vee 0$$

$$= x \wedge y$$

$\Rightarrow \quad x \leq x \wedge y.$

Thus $x \leq y \Leftrightarrow x \wedge y' = 0.$

Similarly the next part can be proved.

Theorem 7.2. A Boolean algebra is self dual.

Proof. Let B be a Boolean algebra.

Define a map $f : B \to B$ such that $f(x) = x'$.

Then f is well defined, for each $x \in B$, x' exists and is unique.

Now for $x, y \in B$,

$$f(x) = f(y) \Rightarrow x' = y'$$

$$\Rightarrow \quad (x')' = (y')'$$

$\Rightarrow \quad x = y$ showing that f is one-one

Now we show f is onto.

For any $y \in B$, y' is its pre-image under f showing that f is onto. Also for $x, y \in B$, $f(x) = x'$ and $f(y) = y'$.

Then $\quad f(x \wedge y) = (x \wedge y)' = x' \vee y' = f(x) \vee f(y)$

and $\quad f(x \vee y) = (x \vee y)' = x' \wedge y' = f(x) \wedge f(y)$

showing that f is a dual-homomorphism.

Thus f is a dual-isomorphism and hence B is self-dual.

Theorem 7.3. Let B be a bounded distributive lattice and $x \in B$, then B is one-one homomorphism into $[0, a] \times [a, 1]$.

Proof. Define $f : B \to [0, a] \times [a, 1]$ such that

$$f(x) = (x \wedge a, x \vee a)$$

Here f is well-defined since $x \wedge a \in [0, a]$ and $x \vee a \in [a, 1]$.

Let $\quad x = y \Rightarrow x \wedge a = y \wedge a \,;\, x \vee a = y \vee a$

$$\Rightarrow (x \wedge a, x \vee a) = (y \wedge a, y \vee a)$$

$$\Rightarrow f(x) = f(y)$$

i.e., f is well-defined.

Now we show f is one-one.

If $f(x) = f(y)$. Then $(x \wedge a, x \wedge a) = (y \wedge a, y \vee a)$

$\Rightarrow \quad x \wedge a = y \wedge a, x \vee a = y \vee a$

$\Rightarrow \quad x = y$

i.e., f is one-one.

So
$$f(x \wedge y) = ((x \wedge y) \wedge a, (x \wedge y) \vee a)$$
$$= ((x \wedge a) \wedge (y \wedge a), (x \vee a) \wedge (y \vee a))$$

and
$$f(x) \wedge f(y) = (x \wedge a, x \vee a) \wedge (y \wedge a, y \vee a)$$
$$= ((x \wedge a) \wedge (y \wedge a), (x \vee a) \wedge (y \vee a))$$

$$\Rightarrow \quad f(x \wedge y) = f(x) \wedge f(y).$$

Similarly, $\quad f(x \vee y) = f(x) \vee f(y)$

Hence, f is one-one homomorphism.

Note. The above theorem can be re-written as "if B be a Bounded distributive lattice and $a \in B$, the B can be imbedded into $[0, a] \times [a, 1]$.

Theorem 7.4. If B is a Boolean algebera and $a \in B$, then B is isomorphic to $[0, a] \times [a, 1]$.

Proof. Define $f: B \to [0, a] \times [a, 1]$ such that

$$f(x) = (x \wedge a, x \vee a)$$

Clearly f is well defined.

7.6. ATOM (DEFINITION)

A non-zero element a in a Boolean algebra $[B, \wedge, \vee, 1]$ is called an *atom* if for every $x \in B$, $x \wedge a = a$ or $x \wedge a = 0$.

The condition that $x \wedge a = a$ tells us that x is a successor of a *i.e.*, $a \leq x$, as shown in **Fig. 7.1**(*a*) :

The condition $x \wedge a = 0$ is true only when x and a are 'not connected'. This occurs when x is another atom or if x is a successor or atoms different from a, as shown in **Fig. 7.1** (*b*).

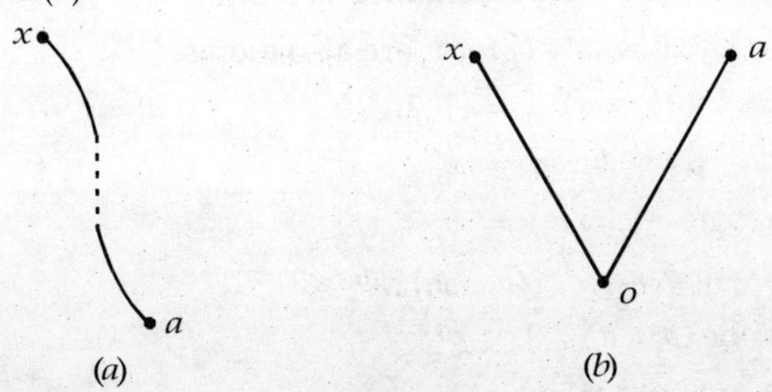

Fig. 7.1.

So by previous theorem, f is one-one homomorphism.

Now, we show f is onto.

Let $(y, z) \in [0, a] \times [a, 1]$. Then

$$0 \le y \le a, a \le z \le 1. \quad ...(i)$$

Take $x = y \vee (z \wedge a')$. Then

$$\begin{aligned} f(x) &= f(y \vee (z \wedge a')) \\ &= (((y \vee (z \wedge a')) \wedge a, y \vee (z \wedge a') \vee a) \\ &= ((y \wedge a) \vee ((z - a') \wedge a), y \vee ((z \vee a) \wedge (a' \vee a)) \\ &= (y \vee (z \wedge 0), y \vee (z \vee 1)) \\ &= (y, z) \quad \text{[by (}i\text{)]} \end{aligned}$$

$\Rightarrow$ f is onto.

Hence f is an isomorphism.

Theorem 7.5. Let B be a finite Boolean algebra (with more than one element). Let $x \in B$ be a non-zero element and suppose $a_1, a_2, ..., a_r$ are the atoms in B such that $a_i \le x$, then $x = a_1 \vee a_2 \vee a_3 \vee ... \vee a_r$ and this representation is unique.

Proof. Since $a_1 \le x, a_2 \le x, a_3 \le x, ..., a_r \le x$.

We get $a_1 \vee a_2 \vee a_3 \vee ... \vee a_r \le x \vee x \vee x \vee ... \vee x = x$

Put $a_1 \vee a_2 \vee a_3 \vee ... \vee a_r = y$.

Now we claim $x \wedge y' = 0$.

Suppose $x \wedge y' \ne 0$, *i.e.*, $x \wedge y' > 0$.

Then $\exists$ an atom a such that $0 < a \le x \wedge y'$.

$\Rightarrow \; a \le x, y'$.

$a \le x$ means $a = a_i$ for some i.

Thus $a = a_i \le a_1 \vee a_2 \vee ... \vee a_r = y$

So $a \le y', a \le y \Rightarrow a \le y \wedge y' = 0$, which is not possible for $a > 0$.

Thus $x \wedge y' = 0$.

$\Rightarrow \; x \le y = 0$.

Hence $x = y = a_1 \vee a_2 \vee a_3 \vee ... \vee a_r$.

Now we show that the representation is unique.

Suppose $x = b_1 \vee b_2 \vee ... \vee b_k$ for b_j are also atoms.

Then $\quad b_j \le x \; \forall \; j = 1, 2, ... k.$

$\Rightarrow \quad b_j \wedge x = b_j$

$\Rightarrow \quad b_j \wedge (a_1 \vee a_2 \vee .... \vee a_r) = b_j$

$\Rightarrow \; (b_j \wedge a_1) \vee (bj \wedge a_2) \vee ... (b_j \wedge a_r) = b_j > 0$

$\Rightarrow$ at least one $(b_j \wedge a_i) \ne 0$.

But b_j and a_i are both atoms.

Thus $b_j = a_i$, each b_j is some a_i *i.e.*, representation is unique.

Theorem 7.6. Let B be a finite Boolean algebra and let S = set of all atoms of B then B is isomorphic to $P(S)$, (power of S).

Proof. Define a map $f: B \to P(s)$ such that $f(0) = \phi, f(x) = \{a_1, a_2, ..., a_r\}$, where a_i are atoms such that $a_i \leq x\ (x \neq 0)$.

Then f is well defined, also $x = a_1 \vee a_2 \vee a_3 \vee .... \vee a_r$.

Now $f(x) = f(y)$

$\Rightarrow \{a_1, a_2, ..., a_r\} = \{b_1, b_2, ..., b_s\}\ a_i \leq x, b_j \leq y$

$\Rightarrow r = s$ and $a_i = b_j$

$\Rightarrow r = s$ and $a_i = b_j$ (without loss of generality)

$\Rightarrow x = a_1 \vee a_2 \vee a_3 \vee ... a_r = b_1 \vee b_2 \vee b_3 \vee ... \vee b_s = y$

$\Rightarrow f$ is one-one.

For any $\{a_1, a_2, ..., a_r\} \in P(S)$. Let $x = a_1 \vee a_2 \vee ... \vee a_r$. Then x is the required pre-image showing f is onto.

Now consider, $f(x \wedge y)$ for $x, y \in B$.

Let $f(x \wedge y) = \{c_1, c_2, ..., c_k\}$ where $c_i \leq x \wedge y$.

Then $c_i \leq x, c_i \leq y$,

and $x \wedge y = a \vee c_2 \vee c_3 \vee .... \vee c_t$.

If $f(x) \wedge f(y) = \{a_1, a_2, ..., a_r\} \cap \{b_1, b_2, ... b_s\}$.

Then $a_i \leq x, b_j \leq y, x = a_1 \vee a_2 \vee, ... \vee a_r, y = b_1 \vee b_2 \vee ... \vee b_s$

Suppose $\{a_1, a_2, ..., a_r\} \cap \{b_1, b_2, ... b_s\} = \{d_1, d_2, ..., d_k\}$

Then $d_i \leq x, d_i \leq y$ and so, $d_i \leq x \wedge y$.

i.e., each d_i is some c_j.

Again as each $c_i \leq x$, it is some a_i,

and $c_i \leq y$, it is some b_j.

Showing that $f(x \wedge y) = f(x) \wedge f(y)$.

Similarly, $f(x \vee y) = f(x) \vee f(y)$.

Hence, f is an isomorphism to $P(S)$.

7.7. DISJUNCTION NORMAL FORM (DNF) (SUM OF PRODUCTS)

A Boolean function (expression) is said to be *disjunction normal form* (DNF) in n variables $x_1, x_2, ..., x_n$ if it can be written as join of terms of the type $f_1(x_1) \wedge f_2(x_2) \wedge ... \wedge f_n(x_n)$, where $f_i(x_i) = x_i$ or x_i's for all $i = 1, 2, ...n$ and no two terms are same.

In this case, terms of the type

$f_1(x_1) \wedge f_2(x_2) \wedge ... \wedge f_n(x_n)$ are called **minterms or minimal polynomials.**

Note. A normal form is also called a *Canonical form.*

A minterm has the value 1 for one and only one combination of values of its variables.

The sum of minterms that represents the function is called the *sum-of-products expansion* or the *disjunctive normal form* of the Boolean function.

➤ Complete Disjunctive Normal Form

If a *CDNF* form in a variable contains all the 2^n minterms, then it is called the complete *DN* form in variables.

If number of variables is 2, say x, y then there are 2^n $(n = 2)$ possible minterms, namely

$$x \wedge y, x' \wedge y, x \wedge y', x' \wedge y'.$$

If number of variables is three, then number of minterms is $2^3 = 8$.

7.8. CONJUNCTIVE NORMAL FORM (CNF) (PRODUCT OF SUMS)

A Boolean function f is said to be *conjunctive normal form* (CN form) in n variables $x_1, x_2, x_3, ..., x_n$ if f is meet of terms of the type

$$f_1(x_1) \vee f_2(x_2) \vee f_3(x_3) \vee ... \vee f_n(x_n)$$

where $f_i(x_i) = x_i$ or x'_i, $\forall\, i = 1, 2, ..., n$, and no two terms are same.

Also 0 and 1 are said to be in *CN* form.

Example 7.7. Put the function $f[(x \wedge y')' \vee z'] \wedge (x' \vee z')'$ in the *DN*-form.

Solution. We have,

$$\begin{aligned} f &= [(x' \vee y'') \vee z'] \wedge [x' \wedge z''] \\ &= [x' \vee y \vee z'] \wedge [z' \wedge x] \\ &= [x' \wedge z' \wedge x) \vee (y \wedge z' \wedge x) \vee (z' \wedge z' \wedge x) \\ &= 0 \vee (x \wedge y \wedge z') \vee (x \wedge z') \\ &= (x \wedge y \wedge z') \vee [(x \wedge z') \wedge (y \vee y')] \\ &= (x \wedge y \wedge z') \vee [(x \wedge z' \wedge y) \vee (x \wedge z' \wedge y')] \end{aligned}$$

or

$$f = (x \wedge y \wedge z') \vee (x \wedge y' \wedge z')$$

Theorem 7.7. Every Boolean function can be put in disjunction normal form.

Proof. We establish the result by taking the following steps :

(*i*) If primes occur outside brackets, then open the brackets by using De-Morgan's laws *i.e.*, $(x \wedge y)' = x' \vee y'$, $(x \vee y)' = x' \wedge y'$.

(*ii*) Open all brackets by using distributivity and simplify using any of the identities like idempotency, absorption etc.

(*iii*) If any of the terms does not contain a certain variable x_i (or x'_i), then take meet of that term with $x_i \vee x'_i$. Do this with each such term.

Now open brackets and drop all terms of the type $x \wedge x' = 0$. Again if the terms occur more than once, these can be omitted because of idempotency. Thus the resulting expression is in *DN* form.

Hence we conclude that every function in a Boolean algebra is equal to a function in *DN* form.

Theorem 7.8. Every Boolean function can be expressed in the *DN* form. This representation is unique (in a given number of variables).

Proof. For part first, see previous theorem.

We now show that this representation is unique.

Suppose $f = A_1 \vee A_2 \vee A_3 \vee .... \vee A_n$.

$f = B_1 \vee B_2 \vee B_3 \vee ... \vee B_m$

where $A_1, A_2, ..., A_n$ and $B_1, B_2, ..., B_m$ are minterms, be two representations of a function f in DN form. Clearly $A_1, A_2, ..., A_n, B_1, B_2, ..., B_n$ are distinct by definition.

In general if X and Y be two distinct minterms, then $X \wedge Y = 0$ as X would always contain atleast one x_i such that Y contains x_i's.

Now $f = A_1 \vee A_2 \vee ... \vee A_n = B_1 \vee B_2 \vee ... \vee B_m$

$\Rightarrow A_i \leq B_1 \vee B_2 \vee ... B_m$; $i = 1, 2, 3, ...n$

$\Rightarrow \quad A_i = A_i \wedge (B_1 \vee B_2 \vee ... \vee B_m)$

$\Rightarrow A_i = (A_i \wedge B_i) \vee (A_i \wedge B_2) \vee ... \vee (A_i \wedge B_m) \; \forall \, i = 1, 2, ... n.$

Now if A_i does not equal any $B_1, B_2, ..., B_m$, then RHS is zero which implies $A_i = 0$, which is not true.

Thus A_i equals some B_j (it can not be equal to two or more B_j's as B_j's are all distinct).

Similarly, each B_j is equal to some A_i.

Hence, the two representations of f are same.

Let us see some examples now :

Example 7.20. If A, B, C are lattices such that $B \cong C$, then $A \times B \cong A \times C$.

Solution. Let $f : B \rightarrow C$ be the given isomorphism.

Define $g : A \times B \rightarrow A \times C$ such that $g((a \times b)) = (a, f(b))$

$$\begin{aligned}
\text{Then} \qquad & g((a,b) = g((c,d)) \\
\Leftrightarrow \qquad & (a, f(b)) = (c, f(d)) \\
\Leftrightarrow \qquad & a = c, f(b) = f(d) \\
\Leftrightarrow \qquad & a = c, \quad b = d \qquad \text{[as } f \text{ is well-defined 1–1 map]} \\
\Leftrightarrow \qquad & (a,b) = (c,d).
\end{aligned}$$

Again for any $(x, y) \in A \times C$, as $y \in C, f : B \rightarrow C$ is onto, $\exists b \in B$ such that $f(b) = y$.

Now $g((x, b)) = (x, f(b) = (x, y)$ and thus g is onto.

$$\begin{aligned}
\text{Also} \qquad g((a,b) \wedge (c,d) &= g(a \wedge c, b \wedge d) = (a \wedge c, f(b \wedge d)) \\
&= (a \wedge c, f(b) \wedge (f(d)) \\
&= (a, f(b)) \wedge (c, f(d)) \\
&= g((a,b) \wedge g((c,d))
\end{aligned}$$

Similarly, $\quad g((a,b) \vee (c,d) = g((c,b) \vee g(c,d))$

Thus g is an isomorphism.

Example 7.21. Put the function

$f = [(x' \wedge y) \vee (x \wedge y \wedge z') \vee (x \wedge y' \wedge z) \vee (x' \wedge y' \wedge z' \wedge t) \vee t']'$ in the *DN* form.

Solution. For simplicity we write + for $\vee$ and . for $\wedge$.

$$\begin{aligned} \text{Thus } f &= [(x'\,y) + (x\,y\,z') + (x\,y'\,z) + (x'\,y'\,z'\,t) + t'] \\ &= (x'\,y)' \cdot (x\,y\,z')' \cdot (x\,y'\,z)' \cdot (x'\,y'\,z'\,t)' \cdot t \\ &= (x + y') \cdot (x' + y' + z) \cdot (x' + y + z') \cdot (x + y + z + t') \cdot t \\ &= (xx' + xy' + xz + y'x' + y'\,y' + y'z) \cdot (x' + y + z') \cdot (xt + yt + zt + tt') \\ &= (xy' + xz' + y'x' + y' + y'z)\,(x'yt + x'zt + yxt + yt + yzt + z'xt + z'xt + z'yt) \\ &= xy'z't + xyzt + y'x'zt + y'ztx' + y'z'xt + y'zx't \\ &= xy'z't + xyzt + x'y'zt. \end{aligned}$$

Example 7.22. Write the function $x \vee y'$ in *DNF* in three variables x, y, z.

Solution.

$$\begin{aligned} x + y' &= [x(y + y') \cdot (z + z')] + [y' \cdot (x + x') \cdot (z + z')] \\ &= [(xy + xy') \cdot (z + z')] + [(y'x + y'x') \cdot (z + z')] \\ &= (xyz) + (xyz') + (xy'z) + (xy'z') + (xy'z) \\ &\quad + (xy'z') + (x'y'z) + (x'y'z') \end{aligned}$$

or

$$\begin{aligned} x + y' &= xyz + xyz' + xy'z' + xy'z + xy'z' + x'y'z' \\ &\quad + x'y'z + x'y'z'. \end{aligned}$$

Example 7.22(a). Find the Boolean expression for the function f given by

$$f(x, y, z) = \begin{cases} \mathbf{1} \text{ when } & x = z = 1,\ y = 0 \\ \mathbf{0} \text{ otherwise } & x = 1,\ y = z = 0 \end{cases}.$$

Solution. The function is specified by the minterms

$$(x \wedge y' \wedge z) \text{ and } (x \wedge y' \wedge z')$$

i.e., the function in the DN form is $(x \wedge y' \wedge z) \vee (x \wedge y' \wedge z')$.

Example 7.23. Find the Boolean expression that defines the function f given by

$$\begin{aligned} f(0, 0, 0) &= 0, f(1, 0, 0) = 1, f(0, 1, 0) = 1, f(1, 0, 1) = 1. \\ f(0, 0, 1) &= 0, f(1, 1, 0) = 0, f(0, 1, 1) = 0, f(1, 1, 1) = 1. \end{aligned}$$

Solution. Consider those values of $f(x, y, z)$ which are equal to 1.

The minterms corresponding to

$f(0, 1, 0), f(1, 0, 0), f(1, 0, 1)$ and $f(1, 1, 1)$ will be

$$(x' \wedge y \wedge z'), (x \wedge y' \wedge z'), (x \wedge y' \wedge z), (x \wedge y \wedge z).$$

Hence the function in DN-form is

$f(x, y, z) = (x' \wedge y \wedge z') \vee (x \wedge y' \wedge z') \vee (x \wedge y' \wedge z) \vee (x \wedge y \wedge z)$

Simplifying we have

$$\begin{aligned} f(x, y, z) &= (x'yz') + (xy'z') + (xy'z) + (xyz) \\ &= (x'yz') + x[y' \cdot (z' + z)] + (xyz) \\ &= (x'yz' + x[y' \cdot (z' + z) + yz] \end{aligned}$$

or

$$f = (x'yz') + x[y' + yz]$$

$$= (x'yz') + x[(y' + y) \cdot (y' + z)]$$
$$= (x'yz') + x \cdot (y' + z)$$
$$= (x'yz') + [(xy') + (xz)]. \textbf{ Ans.}$$

Example 7.24. Show that $f(x, y) = [x \wedge f(1, y)] \vee [x' \wedge f(0, y)]$.

Solution. Since any function f (in two variables) in complete DN form is

$$f(x, y) = (x \wedge y) \vee (x \wedge y') \vee (x' \wedge y) \vee (x' \wedge y')$$
$$= [x \wedge (y \vee y')] \vee [x' \wedge (y \vee y')] \quad ...(i)$$

Put $x = 1$, $x' = 0$, we get

$$f(1, y) = [1 \wedge (y \vee y')] \vee [0 \wedge (y \wedge y']$$
$$= y \vee y'.$$

Thus (i) gives

$$f(x, y) = [x \wedge f(1, y)] \vee [x' \wedge f(0, y)] \quad \textbf{Proved.}$$

Example 7.25. Put the Boolean function $f = [(x \wedge y') \vee z'] \wedge [x' \wedge z]'$ in the CN form.

Solution. $f(x, y) = [(x \wedge y')' \vee z'] \wedge [x' \wedge z]'$

or $f(x, y) = [(xy')' + z'] \cdot (x'z)'$

$$= [(x' + y) + z'] \cdot (x + z')$$
$$= (x' + y + z') \cdot [(xz') + (y \cdot y')]$$
$$= (x' + y + z') \cdot [(xz' + y) \cdot (xz' + y')]$$
$$= (x' + y + z') \cdot [(x + y) \cdot (z' + y) \cdot (x + y')(z' + y')]$$
$$= (x' + y + z') \cdot [\{(x + y) + (z \cdot z')\} \cdot \{(z' + y) + (x \cdot x')\} \cdot \{(x + y') + (zz')\}] \cdot \{(z' + y') + (xx')\}]$$
$$= (x' + y + z') \cdot (x + y + z) \cdot (x + y + z') \cdot (z' + y + x) \cdot (z' + y + x') \cdot (x + y' + z) + (x + y' + z') + (z' + y' + x) \cdot (z' + y' + x')$$
$$= (x + y + z) \cdot (x' + y + z') \cdot (x + y + z') \cdot (x + y' + z) \cdot (x + y' + z')(x' + y' + z')$$

Example 7.26. Put the function $x \wedge (y \vee z)$ in the CN form.

Solution. $x \wedge (x \vee z) = [x \vee (y \wedge y')] \wedge [(y \vee z) \vee (x \wedge x')]$

$$= (x \vee y) \wedge (x \vee y') \wedge (y \vee z \vee x) \wedge (y \vee z \vee x')$$
$$= (x \vee y) \vee (z \wedge z') \wedge (x \vee y') \vee (z \wedge z') \wedge (x \vee y \vee z) \wedge (x' \vee y \vee z)$$
$$= (x \vee z \vee z) \wedge (x \vee y \vee z') \wedge (x \vee y' \vee z) \wedge (x \vee y' \vee z') \wedge (x \vee y \vee z) \wedge (x' \vee y \vee z)$$
$$= (x \vee y \vee z) \wedge (x \vee y \vee z') \wedge (x \vee y' \vee z) \wedge (x \vee y' \vee z') \wedge (x' \vee y \vee z).$$

Example 7.27. Find the DN form of the function whose CN form is

$$f = (x + y + z) \cdot (x + y + z') \cdot (x + y' + z) \cdot (x + y' + z') \cdot (x' + y + z).$$

Solution. Since $f = (f')'$. Therefore,

$$f = [\{(x + y + z) \cdot (x + y + z') \cdot (x + y' + z) \cdot (x + y' + z') \cdot (x' + y + z)\}']'$$

$$= [(x+y+z)' + (x+y+z')' \cdot (x+y'+z)' + (x+y'+z')' + (x'+y+z)']'$$
$$= [(x'y'z') + (x'y'z) + (x'yz') + (x'yz) + (xy'z')]'$$
$$= (xyz) + (xy'z) + (xyz').$$

Note. By similar steps we can find the CN form of a function from its DN-form.

Example 7.28. Find the CN form of the function $f = [x \wedge (y' \vee z)] \vee z'$ and then find its DN form from it.

Solution.
$$\begin{aligned} f &= (x \wedge (y' \vee z)) \vee z' \\ &= (x \vee z') \wedge ((y' \vee z) \vee z') \\ &= x \wedge z' \\ &= (x \vee z') \vee (y \wedge y') \\ &= (x \vee y \vee z') \wedge (x \vee y' \vee z'). \end{aligned}$$

Now
$$\begin{aligned} f &= (f')' = [\{(x \vee y \vee z') \wedge (x \vee y' \vee z')\}']' \\ &= [(x \vee y \vee z')' \wedge (x \vee y' \vee z')']' \\ &= [(x' \wedge y' \wedge z) \vee (x' \wedge y \vee z)]' \\ &= (x \wedge y \wedge z) \vee (x' \wedge y \wedge z') \vee (x \wedge y \wedge z') \vee (x \wedge y' \wedge z) \\ &\quad \vee (x \wedge y' \wedge z') \vee (x' \wedge y' \wedge z'). \end{aligned}$$

Example 7.29. Find out the disjunctive and conjunctive normal form of the polynomial.

$$f(x, y, z) = [x + (x' + y')'] \cdot [x' + (y' \cdot z')']$$

Solution : $[x + (x' + y)'] \cdot [x + (y' \cdot z')']$

$$= [x + \{(x')' \cdot y'\}] \cdot [x + \{(y')' + (z')'\}]$$

(by De-Morgan's law)

$$= (x + x \cdot y') \cdot (x + y + z) \qquad [\because (x')' = x]$$
$$= x \cdot (x + y + z) \quad [\because a + a \cdot b = a] \text{ (absorption law)}$$
$$= x \cdot (x + y) + xz \qquad \text{(by distributivity)}$$
$$= x + xz \qquad [\because a \cdot (a + b) = a] \text{ (absorption law)}$$
$$= x.$$
$$= x \cdot 1 \cdot 1$$
$$= x \cdot (y + y') \cdot (z + z')$$
$$= (x \cdot y) + (x \cdot y') \cdot (z + z') \qquad \text{(by distributivity)}$$
$$= x \cdot y \cdot z + x \cdot y \cdot z' + x \cdot y' \cdot z + x \cdot y' \cdot z'$$

which is disjunctive normal form of the given function

Again, since we have

$$\begin{aligned} f(x, y, z) &= x \\ &= x + 0 + 0 \\ f(x, y, z) &= x + y \cdot y' + z \cdot z' \qquad [\because y \cdot y' = 0 = z \cdot z'] \\ &= (x + y) \cdot (x + y') + z \cdot z' \\ &= \{(x + y) + z \cdot z'\} \cdot \{x + y') + z \cdot z'\} \text{(by distribution)} \end{aligned}$$

$$= (x + y + z) \cdot (x + y + z') \cdot (x + y' + z) \cdot (x + y' + z')$$

(by distributivity)

which is conjunctive normal form of the given function.

Example 7.30. Write the conjunctive normal form of the function,

$$(x \cdot y' + x \cdot z)' + x'$$

Solution : Let,

$$
\begin{aligned}
f(x, y, z) &= (x \cdot y' + x \cdot z)' + x' \\
&= (x \cdot y')' \cdot (x \cdot z)' + x' && \text{(by De-Morgan's law)} \\
&= (x' + y) \cdot (x' + z') + x' && \text{(by De-Morgan's law)} \\
&= (x' + y + x') \cdot (x' + z' + x') && \text{(by distribution)} \\
&= \{(x' + x') + y\} \cdot \{(x' + x') + z'\} \\
& && \text{(by commutativity and associativity)} \\
&= (x' + y) \cdot (x' + z') && [\because x' + x' = x'] \\
&= (x' + y + 0) \cdot (x' + z' + 0) \\
&= (x' + y + z \cdot z') \cdot (x' + z' + y \cdot y') && [\because z \cdot z' = 0 = y \cdot y'] \\
&= \{(x' + y) + z \cdot z'\} \cdot \{(x' + z') + y \cdot y'\} \\
&= (x' + y + z) \cdot (x' + y + z') \cdot (x' + z' + y) \cdot (x' + z' + y') \\
& && \text{(by distributivity)} \\
&= (x' + y + z) \cdot (x' + y + z')\,(x' + y' + z')
\end{aligned}
$$

which is the conjunctive normal form of the given function.

Example 7.31. Find out the conjunctive normal form of the following functions in three variables *a*, *b* and *c* :

***(a)* *a*,** ***(b)* $(a + b) \cdot (a' + b')$**

Solution : *(a)*

$$
\begin{aligned}
f(a, b, c) &= a \\
&= a + 0 && [\because a + 0 = a] \\
&= a + b \cdot b' && [\because b \cdot b' = 0] \\
&= (a + b) \cdot (a + b') && \text{(by distributivity)} \\
&= (a + b + 0) \cdot (a + b' + 0) \\
&= \{(a + b) + c \cdot c'\} \cdot \{(a + b') + c \cdot c\} \\
&= (a + b + c) \cdot (a + b + c') \cdot (a + b' + c) \cdot (a + b' + c) \\
& && \text{(by distributivity)}
\end{aligned}
$$

which is conjunctive normal form of a Boolean function of single variable *a* in three variables.

(b) Let the given function be denoted by $f(a, b, c)$ then

$$
\begin{aligned}
f(a, b, c) &= (a + b) \cdot (a' + b') \\
&= (a + b + 0)\,(a' + b' + 0) \\
&= (\overline{a + b} + c \cdot c') \cdot (\overline{a' + b'} + c \cdot c') \\
&= (a + b + c) \cdot (a + b + c') \cdot (a' + b' + c) \cdot (a' + b' + c')
\end{aligned}
$$

which is conjunctive normal form of the function.

Example 7.32. Write down the function $f(x, y)$ given in conjunctive normal form to its disjunctive normal form, where

$$f(x, y) = (x + y') \cdot (x' + y) \cdot (x' + y')$$

Solution : $(x + y') \cdot (x' + y) \cdot (x' + y')$

$$= (x + y') \cdot (x' + y \cdot y') \qquad \text{(using distributive law)}$$
$$= (x + y') \cdot (x' + 0) \qquad [\because y \cdot y' = 0]$$
$$= (x + y') \cdot x'$$
$$= xx' + y' \cdot x'$$
$$= 0 + x' \cdot y'$$
$$= x' \cdot y'$$

which is disjunctive normal form of the given function.

Example 7.33. Change the disjunctive normal form $x \cdot y' + x' \cdot y' + x' \cdot y$ into conjunctive normal form.

Solution :
$$x \cdot y' + x' \cdot y' + x' \cdot y = (x + x') \cdot y' + x' \cdot y$$
$$= 1 \cdot y' + x' \cdot y$$
$$= y' + x' \cdot y$$
$$= (y' + x') \cdot (y' + y)$$
$$= (x' + y') \cdot 1 = (x' + y')$$

Example 7.34. Write down the disjunctive normal form in minimum possible variables of the following functions :

(*i*) $x + x' \cdot y$,

(*ii*) $x \cdot y' + x \cdot z + x \cdot y$,

(*iii*) $(u + v + w) \cdot (u\,v + u'\,w)'$

Solution : (*a*)
$$x + x' \cdot y = x \cdot 1 + x' \cdot y$$
$$= x \cdot (y + y') + x' \cdot y$$
$$= x \cdot y + x \cdot y' + x' \cdot y$$

which is the required disjunctive normal form in minimum possible number of variable.

(*b*) $x \cdot y' + x \cdot z + x \cdot y = x \cdot y' \cdot 1 + x \cdot z \cdot 1 + x \cdot y \cdot 1$

$$= x \cdot y' \cdot (z + z') + x \cdot z \cdot (y + y') + x \cdot y \cdot (z + z)$$
$$= x \cdot y' \cdot z + x \cdot y' \cdot z' + x \cdot z \cdot y + x \cdot z \cdot y' + x \cdot y \cdot z + x \cdot y \cdot z$$
$$= x \cdot y' \cdot z + x \cdot y' \cdot z' + x \cdot y \cdot z + x \cdot y' \cdot z + x \cdot y \cdot z + x \cdot y \cdot z' \qquad \text{(by commutativity)}$$
$$= \{(x \cdot y' \cdot z) + (x \cdot y' \cdot z)\} + \{(x \cdot y \cdot z) + (x \cdot y \cdot z)\} + \{x \cdot y' \cdot z' + x \cdot y \cdot z'\}$$
$$= x \cdot y' \cdot z + x \cdot y \cdot z + x \cdot y' \cdot z' + x \cdot y \cdot z' \qquad [\because a + a = a]$$
$$= x \cdot y \cdot z + x \cdot y' \cdot z + x \cdot y \cdot z' + x \cdot y' \cdot z'$$

which is the required disjunctive normal form

(*c*) The given function is,

$(u + v + w) \cdot (u\,v + u'\,w')$

$$= (u+v+w)\cdot\{(u\,v)'\cdot(u'\cdot w)'\}$$
$$= (u+v+w)\cdot\{(u'+v')\cdot(u+w')\}$$
$$= (u+v+w)\cdot\{u'\cdot u+u'\cdot w'+v'\cdot u+v'\cdot w'\}$$
$$= (u+v+w)\cdot(0+u'\cdot w'+v'\cdot u+v'\cdot w')$$
$$= (u+v+w)\cdot(u'\cdot w'+v'\cdot u+v'\cdot w')$$
$$= u\cdot u'\cdot w'+u\cdot v'\cdot u+u\cdot v'\cdot w'+v\cdot u'\cdot w'+v\cdot v'\cdot u+v\cdot v'\cdot w'$$
$$+w\cdot u'\cdot w'+w\cdot v'\cdot u+w\cdot v'\cdot w'$$
$$= 0+u\cdot v'+u\cdot v'\cdot w'+u'\cdot v\cdot w'+0+0+0+u\cdot v'\cdot w+0$$

(by associativity commutativity and $u\cdot u'=0$ etc.)

$$= u\cdot v'+u\cdot v'\cdot w'+u'\cdot v\cdot w'+u\cdot v'\cdot w$$
$$= u\cdot v'\cdot 1+u\cdot v'\cdot w'+u'\cdot v\cdot w'+u'\cdot v\cdot w$$
$$= u\cdot v'\,(w+w')+u\cdot v'\cdot w'+u'\cdot v\cdot w'+u\cdot v'\cdot w$$
$$= u\cdot v'\cdot w+u\cdot v'\cdot w'+u\cdot v'\cdot w'+u'\cdot v\cdot w'+u\cdot v'\cdot w$$
$$= \{(u\cdot v'\cdot w)+(u\cdot v'\cdot w)\}+\{(u\cdot v'\cdot w')+(u\cdot v'\cdot w')\}+u'\cdot v\cdot w'$$
$$= u\cdot v'\cdot w+u\cdot v'\cdot w'+u'\cdot v\cdot w' \qquad [\because a+a=a]$$

which is the required disjunctive normal form in minimum variables.

Example 7.35. Convert the function,

$$f(x,y,z) = (x+y)\cdot(x+z')+(y+z')$$

into disjunctive normal form.

Solution : $f(x,y,z) \equiv (x+y)\cdot(x+z')+(y+z')$

$$\equiv (x+y\cdot z')+y+z' \qquad \text{(by distributive law)}$$
$$= x+(y\cdot z'+y)+z' \qquad \text{(by associative law)}$$
$$= x+y+z' \qquad \text{(by absorption law)}$$
$$= x\cdot 1+y\cdot 1+z'\cdot 1$$
$$= x\cdot(y+y')+y\cdot(x+x')+z'\cdot(x+x') \qquad [\because y+y'=1]$$
$$= x\cdot y+x\cdot y'+x\cdot y+x'\cdot y+z'\cdot x+z'\cdot x'$$

(by distributive and commutative law)

$$= x\cdot y+x\cdot y'+x'\cdot y+z'\cdot x+z'\cdot x' \qquad [\because x\cdot y+x\cdot y=x\cdot y]$$
$$= x\cdot y\cdot 1+x\cdot y'\cdot 1+x'\cdot y\cdot 1+z'\cdot x\cdot 1+z'\cdot x'\cdot 1$$
$$= x\cdot y\cdot(z+z')+x\cdot y'\cdot(z+z')+x'\cdot y\cdot(z+z')$$
$$+z'\cdot x\cdot(y+y')+z'\cdot x'\cdot(y+y')$$

(Since first term is missing z hence $1=z+z+z'$) etc.

$$= x\cdot y\cdot z+x\cdot y\cdot z'+x\cdot y'\cdot z+x\cdot y'\cdot z'+x'\cdot y\cdot z+x'\cdot y\cdot z'$$
$$+z'\cdot x\cdot y+z'\cdot x\cdot y'+z'\cdot x'\cdot y+z'\cdot x'\cdot y'$$

(by distributive law)

$$= x\cdot y\cdot z+(x\cdot y\cdot z'+x\cdot y\cdot z')+x\cdot y'\cdot z+(x\cdot y'\cdot z'+x\cdot y'\cdot z')$$
$$+x'\cdot y\cdot z+(x'\cdot y\cdot z'+x'\cdot y\cdot z')+x'\cdot y'\cdot z'$$

(by commutative and associative laws)

$$= x\cdot y\cdot z+x\cdot y\cdot z'+x\cdot y'\cdot z+x\cdot y'\cdot z'+x'\cdot y\cdot z+x'\cdot y\cdot z'+x'\cdot y'\cdot z'$$

$$= x \cdot y \cdot z + x \cdot y' \cdot z + x' \cdot y \cdot z + x \cdot y \cdot z' + x \cdot y' \cdot z' + x' \cdot y \cdot z' + x' \cdot y' \cdot z'$$

which is the required disjunctive normal form of the given function.

Example 7.36. Convert the function

$$f(x, y, z) = (x \cdot y' + xz') + x'$$

into disjunctive normal form.

Solution : $f(x, y, z) = (xy' + xz')' + x'$

$= (xy')' \cdot (xz')' + x'$ (by De-Morgan law)

$= [x' + (y')'] [x' + (z')'] + x'$ (by De-Morgan law)

$= [x' + y] [x' + z] + x'$

$= x' + y \cdot z + x'$ (by distributive law)

$= x' + x' + yz$ (by associative law)

$= x' + yz$

$= x' \cdot 1 + 1 \cdot yz$

$= x' \cdot (y + y') + (x + x') yz$

$= x' \cdot y + x' \cdot y' + x \cdot y \cdot z + x' \cdot y \cdot z$

$= x' \cdot y \cdot 1 + x' \cdot y' \cdot 1 + x \cdot y \cdot z + x' \cdot y \cdot z$

$= x' \cdot y (z + z') + x' \cdot y' \cdot (z + z') + x \cdot y \cdot z + x' \cdot y \cdot z$

$= x' \cdot y \cdot z + x' \cdot y \cdot z' + x' \cdot y' \cdot z + x' \cdot y' \cdot z + x' \cdot y' \cdot z'$

Example 7.37. Convert the function

$$f(x, y, z) = (x + y + z) (xy + x' z)'$$

into conjuntive and disjunctive normal form. Also write the complement of conjunctive normal form.

Solution : (*i*) Conjuntive normal form.

$f(x, y, z) = (x + y + z) (x \cdot y + x' \cdot z)'$

$= (x + y + z) [(x \cdot y)' \cdot (x' \cdot z)']$

$= (x + y + z) (x' + y') \cdot ((x')' + z')$

$= (x + y + z) (x' + y') (x + z')$

$= (x + y + z) \cdot (x' + y' + 0) \cdot (x + z' + 0)$

$= (x + y + z) (x' + y' + z \cdot z') (x + z' + y \cdot y')$

$= (x + y + z) (x' + y' + z) (x' + y' + z')$
$\cdot (x + y + z') \cdot (x + y' + z')$

(*ii*) Disjunctive normal form

$f(x, y, z) = (x + y + z) (x \cdot y + x' \cdot z)'$

$= (x + y + z) \cdot [(x \cdot y)' \cdot (x' \cdot z)']$

$= (x + y + z) (x' + y') (x + z')$

$= (x + y + z) (x' \cdot x + x' \cdot z' + x \cdot y' + y' \cdot z')$

$= (x + y + z) (x' \cdot z' + x \cdot y' + y' \cdot z')$

$= x \cdot x' \cdot y' + x \cdot x \cdot y' + x \cdot y' \cdot z' + x' \cdot y \cdot z'$
$+ x \cdot y \cdot y' + y \cdot y' \cdot z' + x' \cdot z \cdot z' + x \cdot y' \cdot z + y' \cdot z \cdot z'$

$$= 0 \cdot z' + x \cdot y' + x \cdot y' \cdot z' + x' \cdot y \cdot z' + x \cdot 0 + 0 \cdot z' + x' \cdot 0 + x \cdot y' \cdot z + y' \cdot 0$$
$$= x \cdot y' + x \cdot y' \cdot z' + x' \cdot y \cdot z' + x \cdot y' \cdot z$$
$$= x \cdot y' \cdot 1 + x \cdot y' \cdot z' + x' \cdot y \cdot z' + x \cdot y' \cdot z$$
$$= x \cdot y' \cdot (z + z') + x \cdot y' \cdot z' + x' \cdot y \cdot z' + x \cdot y' \cdot z$$
$$= x \cdot y' \cdot z + x \cdot y' \cdot z' + x \cdot y' \cdot z' + x' \cdot y \cdot z' + x \cdot y' \cdot z$$
$$= x \cdot y' \cdot z + x \cdot y' \cdot z' + x' \cdot y \cdot z'$$

Complete conjunctive normal form

$$(x + y + z)\,(x + y + z')\,(x + y' + z)\,(x' + y + z)\,(x + y' + z')\,(x' + y + z')\,(x' + y' + z)\,(x' + y' + z')$$

Hence complement conjuctive normal form f is

$$(x + y + z')\,(x' + y + z)\,(x + y' + z)\,(x + y' + z')$$

Example 7.38. Convert the function

$$f(x, y) = (xy' + xy)' + x'$$

into conjunction normal form.

Solution:

$$f(x, y) = (xy' + xy)' + x'$$
$$= [x\,(y' + y)]' + x'$$
$$= [x \cdot 1]' + x'$$
$$= x' + x'$$
$$= x'$$
$$= x' + 0$$
$$= x' + y \cdot y'$$
$$= (x' + y) \cdot (x' + y')$$

7.9 BOOLE S EXPANSION THEOREM

Theorem. Any function $f(a, b, c, ..., n)$ can be expanded as given below.

$$f(a, b, c, ..., n) = f(1, 1, 1 ..., 1) \cdot a \cdot b \cdot c ... n + f(0, 1, 1, ..., 1) \cdot a' \cdot b \cdot c ..., n + f(1, 0, 1, ... 1)\, a \cdot b' \cdot c ... n + ... + ... + f(0, 0, 0, ..., 0)\, a', b', c' ..., n'$$

where 0, 1 are the values given to the variables $a, b, c, ... n$.

Proof. Since, by theorem 1 we know that every function can be expressed in additive normal form, hence, let us have,

$$f(a, b, c, ..., n) = \alpha_1 \cdot a \cdot b \cdot c ... n + \alpha_2 \cdot a' \cdot b' \cdot c ..., n + ... + \alpha_m\, a' \cdot b' \cdot c' ... n \qquad ...(i)$$

where α_1 and α_2 etc. are either 0 or 1. Also, by the theorem 3, we know that only one term of the additive normal form of the function will assume the value 1 for any arbitrary allotment of values 0 and 1 to the variables. consequently for that particular allotment, let $f(a, b, c, ... n)$ of L.H.S. of (1) have the value 1 and R.H.S. of (1) becomes α.

Thus by the definition of equality of two functions we have

$$1 = \alpha \cdot 1$$

$\Rightarrow \qquad \alpha = 1$

with the similar arguments on arbitrary assignment of values 0 or 1 to the variables, we will be able to evaluate all α_1 or α_2 *i.e.*, co-efficients of R.H.S. of (1).

Putting these values at their respective places in (1) we will get the Boole's expansion of the function.

Sunch type of expansions of $f(a, b, c, ..., k)$ is know as Boole expansions.

ILLUSTRATIVE EXAMPLES

Example 7.39. Construct the function f and g given by the following truth table :

a	b	c	x	f	g	y
1	1	1	$a \cdot b \cdot c$	1	0	$a' \cdot b' \cdot c'$
1	1	0	$a \cdot b \cdot c'$	0	0	$a' \cdot b' \cdot c$
1	0	1	$a \cdot b' \cdot c$	1	0	$a' \cdot b \cdot c'$
0	1	1	$a' \cdot b \cdot c$	0	1	$a \cdot b' \cdot c'$
1	0	0	$a \cdot b' \cdot c'$	0	0	$a' \cdot b \cdot c$
0	1	0	$a' \cdot b \cdot c'$	0	1	$a \cdot b' \cdot c$
0	0	1	$a' \cdot b' \cdot c$	1	0	$a \cdot b \cdot c'$
0	0	0	$a' \cdot b' \cdot c'$	0	0	$a \cdot b \cdot c$

Solution : Then according to Boole's expansion theorem use only those rows, to construct the function, which have the value 1 under f or g. Thus we have

$$f = 1 \cdot (a \cdot b \cdot c) + 1 \cdot (a \cdot b' \cdot c) + 1 \cdot (a' \cdot b' \cdot c)$$
$$= a \cdot b \cdot c + a \cdot b' \cdot c + a' \cdot b' \cdot c$$

Simplifying this we get

$$f = a \cdot c\,(b + b') + a' \cdot b' \cdot c$$
$$= a \cdot c \cdot 1 + a' \cdot b' \cdot c \qquad [\because b + b' = 1]$$
$$= a \cdot c + a' \cdot b' \cdot c$$
$$= (a + a' \cdot b') \cdot c$$
$$= c \cdot (a + a' \cdot b')$$
$$= c \cdot \{(a + a') \cdot (a + b')\}$$
$$= c \cdot \{1 \cdot (a + b')\}$$
$$= c \cdot a + c \cdot b'$$
$$= a \cdot c + b' \cdot c$$

Similarly for the function g, we get

$$g = 1(a \cdot b' \cdot c') + 1\,(a \cdot b' \cdot c)$$
$$= a \cdot b' \cdot c' + a \cdot b' \cdot c$$
$$= a \cdot b' \cdot 1$$
$$= a \cdot b'.$$

Example 7.40. Construct the functions from the table given below :

x	y	z	f	T
1	1	1	0	$x' \cdot y' \cdot z'$
1	1	0	1	$x \cdot y \cdot z'$
1	0	1	1	$x \cdot y \cdot z$
0	1	1	0	$x' \cdot y \cdot z$
0	1	0	0	$x' \cdot y \cdot z'$
0	0	1	1	$x' \cdot y' \cdot z$
0	0	0	0	$x \cdot y \cdot z$

where T stands for the term of the function

Solution : Using Boole's expansion theorem

$$
\begin{aligned}
f &= 0 \cdot (x' \cdot y' \cdot z') + 1\,(x' \cdot y' \cdot z') + 1\,(x' \cdot y \cdot z) + 0\,(x \cdot y' \cdot z') \\
&\qquad + 0\,(x, y' \cdot z) + 1 \cdot (x \cdot y \cdot z') + 0\,(x \cdot y \cdot z) \\
&= x' \cdot y' \cdot z + x' \cdot y \cdot z' + x' \cdot y \cdot z + x \cdot y \cdot z' \\
&= x' \cdot y' \cdot z + x' \cdot y\,(z' + z) + x \cdot y \cdot z' \\
&= x' \cdot y' \cdot z + x' \cdot y \cdot 1 + x \cdot y \cdot z' && [\because z' + z = 1] \\
&= x' \cdot y' \cdot z + x'y + x \cdot y \cdot z' \\
&= x' \cdot y' \cdot z + x' \cdot y + x \cdot z' \cdot y && \text{(by commutativity)} \\
&= x' \cdot y' \cdot z + (x' + x \cdot z') \cdot y && \text{(by distributivity)} \\
&= x' \cdot y' \cdot z + y \cdot (x' + x \cdot z') \\
&= x' \cdot y' \cdot z + y \cdot (x' + x) \cdot (x' + z') \\
&= x' \cdot y' \cdot z + y \cdot 1 \cdot (x' \cdot z') && [\because x + x' = 1] \\
&= x' \cdot y' \cdot z + y \cdot (x' + z') && [\because y \cdot 1 = 1] \\
&= x' \cdot y' \cdot z + y \cdot x' + y \cdot z' \\
&= x' \cdot y' \cdot z + x' \cdot y + y \cdot z' \\
&= x' \cdot (y' \cdot z + y) + y \cdot z' \\
&= x' \cdot (y' + y) \cdot (z + y) + y \cdot z' \\
&= x' \cdot 1 \cdot (z + y) + y \cdot z' \\
&= x' \cdot (y + z) + y \cdot z' \\
&= x' \cdot y + x' \cdot z + y \cdot z'
\end{aligned}
$$

7.10 CIRCUITS

One of the major applications of Boolean algebra is to the electrical network consisting of switches that involves two state of devices. The simplest example of such a device being an ordinary on-off switch. The switch can assume two states 'closed' or 'open' (ON or OFF). In the first case the current flows and in the second case the current does not flow.

The basic elements of the circuits are called gates (arrangement of switches). Each type of gates implements a Boolean operation. In this section we define three types of gates. Using these gates and applying the properties of Boolean algebra we design circuits in the simplest form. The circuits that we will study give output that depends only on the input and not on the current state of the circuit.

There are two basic ways in which switches are generally interconnected. These are **'in series'** and **'in parallel'**. The switches are arranged in series is nothing but using **AND gate** and the switches are arrenged in parallel is **OR gate**. They are reperented by the **Fig. 7.2**(*a*) and (*b*) and there is another gate that we will study is invertal that is related to complement of element (switches) is given in **Fig. 7.2**(*c*).

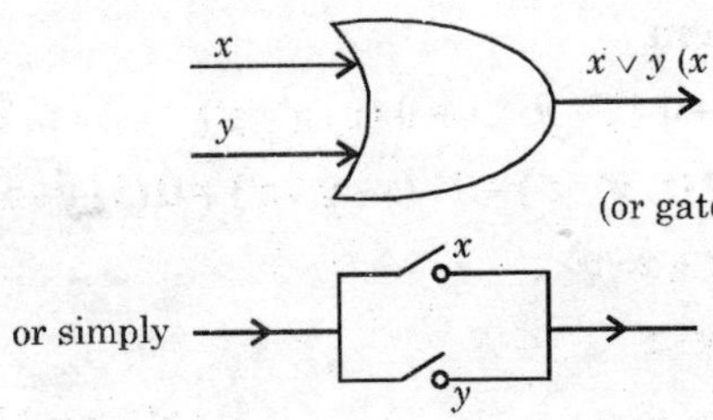

Fig. 7.2 (*a*)

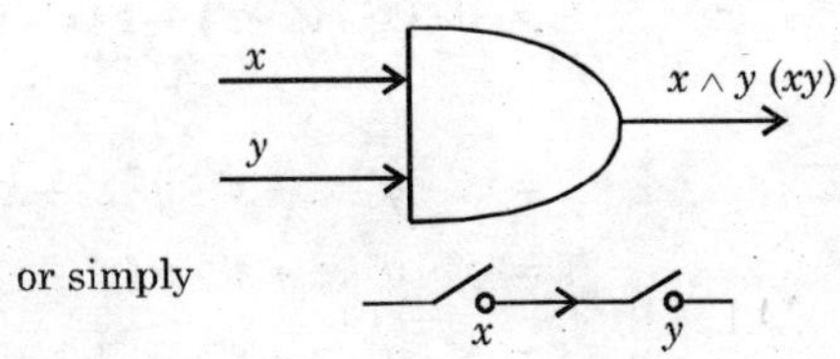

Fig. 7.2(*b*)

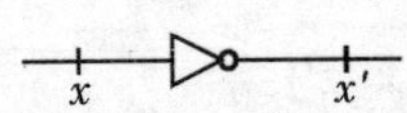

Fig. 7.2.(c)

The input to the circuit (gates) is shown on the left side of the entering element and output is shown on the right side, leaving the element. Closed state of the circuit means current is flowing and open state means current is not flowing. Two switches a and b are said to be connected 'in series' or by AND gate, if the current can pass only when both are in the closed state and current does not flow if any one or both are open see Fig 7.4(*b*).

Two switches a and b are said to be connected 'in parallel' or by 'OR gate' if current flows when any one or both are closed and current does not pass when both are open. see Fig 7.4(*a*)

If a switch x is open then we denote x' for closed switch and is reprsented by the third gate that is **inverter.**

As we have done in chapter 2, the compound statements can be constructed by using simple statement. Similarly using the same logic we can make some compound circuits. We will do this with the help of Boolean algebra.

Let 0 denotes open circuit (current does not pass)

and 1 denotes closed circuit (current passes)

therefore we can consider the Boolean algebra as circuit system $\{\{0, 1\}, \vee, \wedge, '\}$.

7.11. REPRESENTATION OF CIRCUITS

Now we consider how a boolean function can represent a circuit and how a circuit can be represented by a Boolean function. We use the basic as given in the Section 7.4.

Example 7.41. Draw the circuit represented by the boolean function

$$f(x, y, z) = (x \vee y) \wedge z$$

Solution. $x \vee y$ [OR gate] are represented by the parallel circuits

$\therefore$ $x \vee y$ is given as

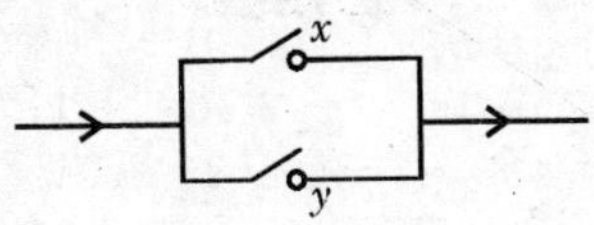

Fig. 7.3(*a*)

Now $(x \vee y) \wedge z$ is given by (AND gate) or switches in series therefore the required circuit is

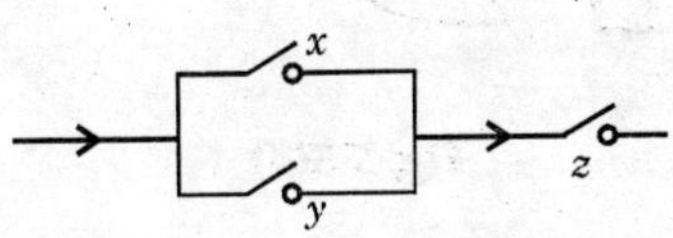

Fig. 7.3(*b*)

ALITER

This circuit can be given in terms of gates.

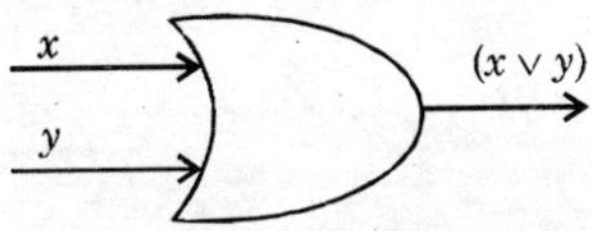

$x \vee y$ (OR gate)

Fig. 7.4(*a*)

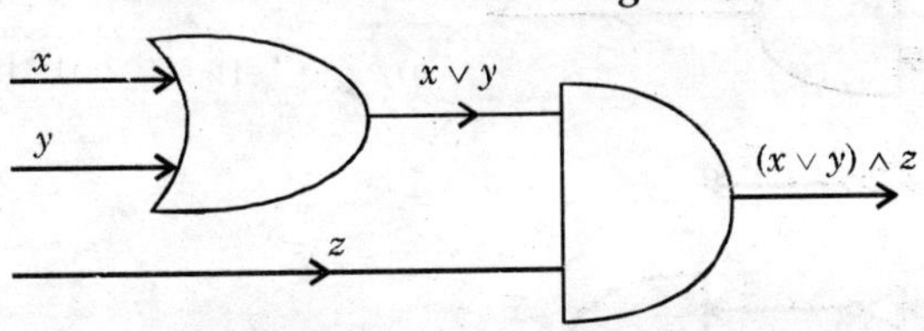

[By apply AND gate]

Fig. 7.4(*b*)

Example 7.42. Now readers are required to draw the circuit represented by the Boolean function $f = x \vee (y \wedge z)$ and give the condition that when the current will flow.

Solution : By applying the rule of precedence

$(y \wedge z)$

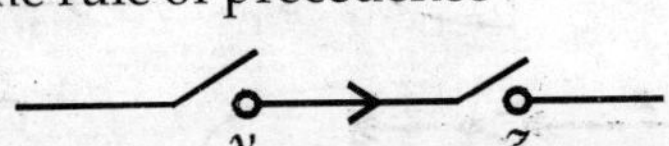

Fig. 7.5(*a*)

$x \vee (y \wedge z)$

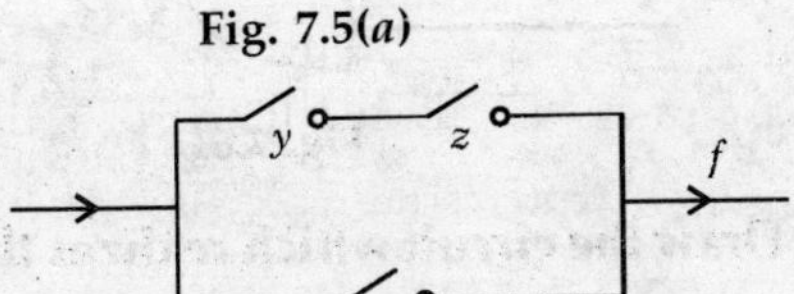

Fig. 7.5(*b*)

ALITER

The required circuit can be given in terms of gates

$(y \wedge z)$

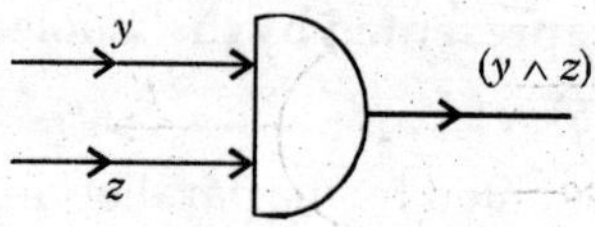

(AND gate)

Fig. 7.5(*c*)

$x \vee (y \wedge z)$

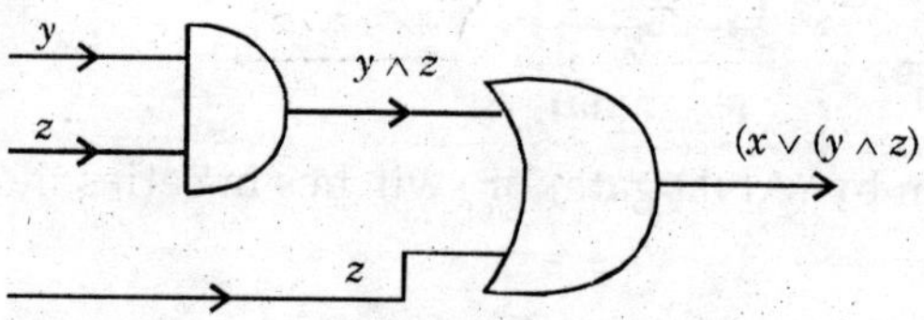

(OR GATE)

Fig. 7.5(*d*)

Example 7.43. Draw the circuit represented by the following Boolean function.

$$f : xy + \bar{x}\, y$$

Solution: $f(x, y, z) = xy + \bar{x}\, y$

$$= (x \wedge y) \vee (\bar{x} \wedge y)$$

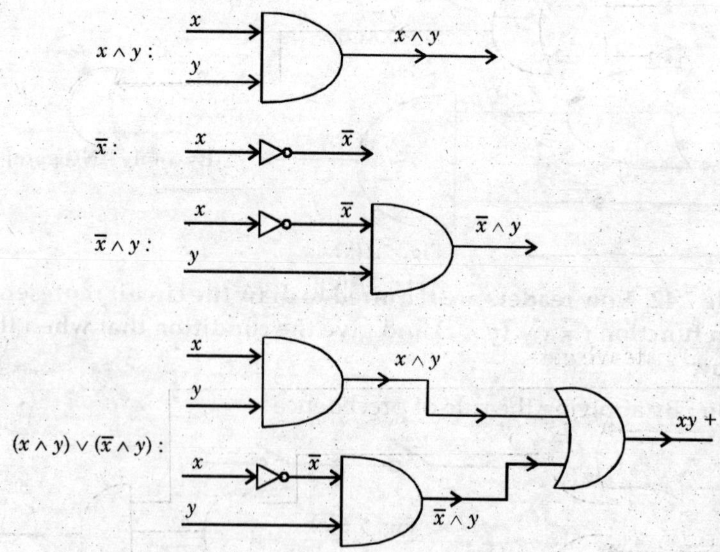

Fig. 7.6.

Example 7.44. Draw the circuit which realizes the function

$$f(x, y, z, t) = x \wedge [(y \vee t') \vee (z \vee (x \vee t \vee z'))] \wedge y$$

Solution : It is to be noticed that while drawing a circuit one should use rule of precedence

$y \vee t'$

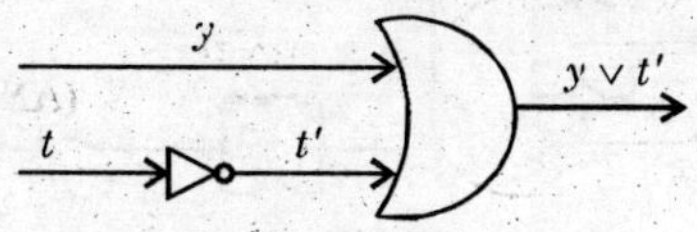

Fig. 7.7(*a*)

$x \vee t \vee z'$

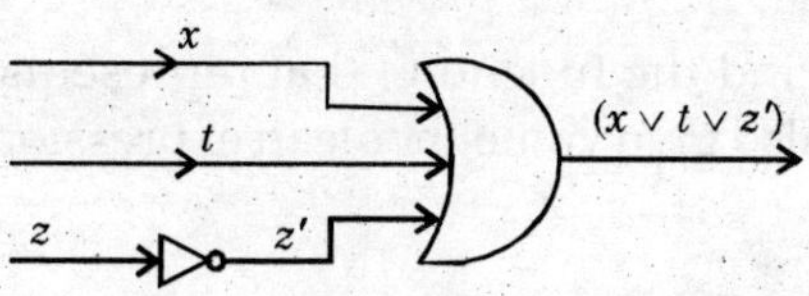

Fig. 7.7(*b*)

$z \wedge (x \vee t \vee z')$

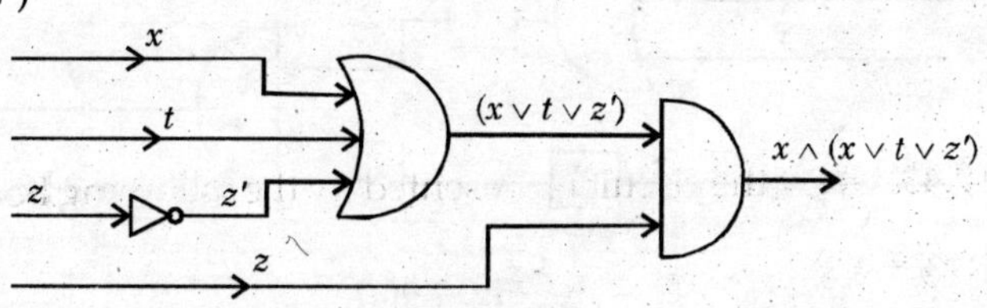

Fig. 7.7(*c*)

By (*a*) and (*c*)

$$g = (y \vee t') \vee (z \wedge (x \vee t \vee z'))$$

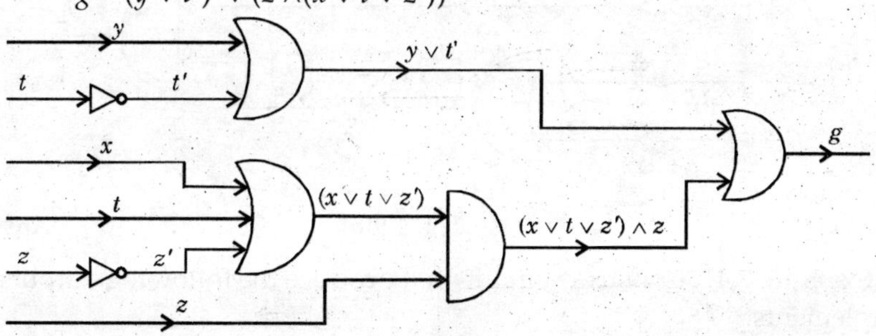

Fig. 7.7(*d*)

$f : x \wedge g \wedge y$

Using AND gate we get

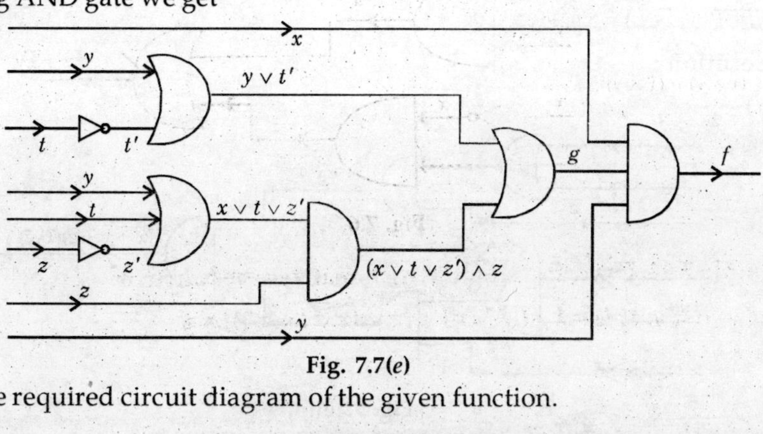

Fig. 7.7(*e*)

is the required circuit diagram of the given function.

Example 7.45. Find the functon f that represents the circuit.

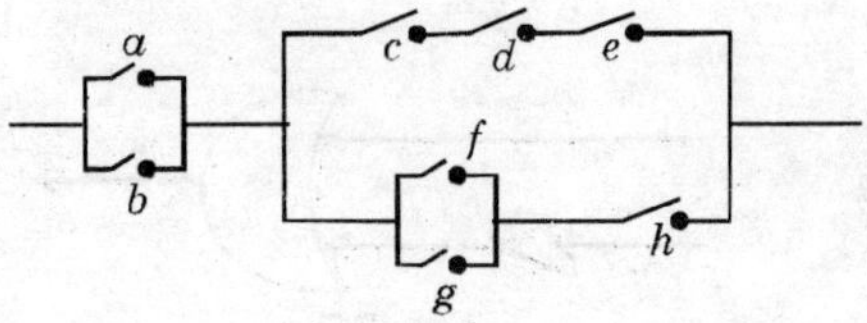

Fig. 7.8.

Example 7.46. Find the function f that represents the circuits given by the **Fig. 7.9** (*a*), (*b*) and also give value [boolean expression] in the blocks marked as [?].

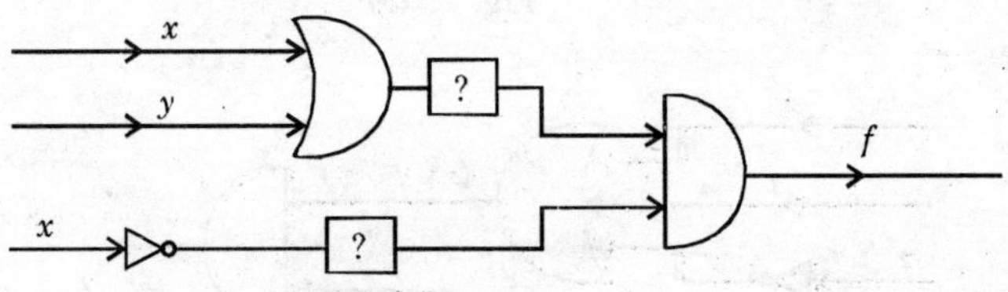

Fig. 7.9(*a*) **Ans.** $(x + y)\,\overline{x}$

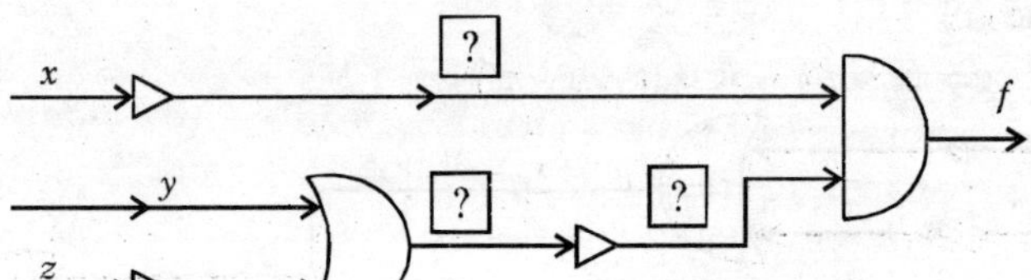

Fig. 7.9(*b*) **Ans.** $\overline{x}\,(\overline{y} + \overline{z})$

Example 7.47. Construct circuitsthat produce the following outputs and their complements.

(*i*) $P(x, y, z) = (x + y + z)\,(\overline{x}\,\overline{y}\,\overline{z})$

(*ii*) $P(a, b, c, d, e, f, g, h) = (a \vee b) \wedge ((c \wedge d \wedge e) \vee ((f \vee g) \wedge h))$

(*iii*) $P(x, y, z) = xy + xz + yz$

Solution :

(*i*)

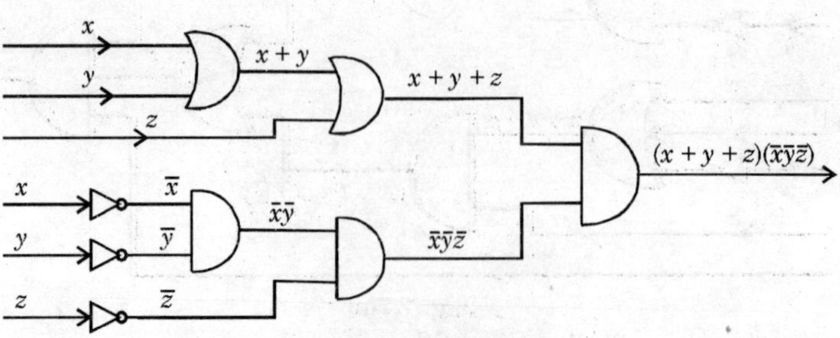

Fig. 7.10(*a*).

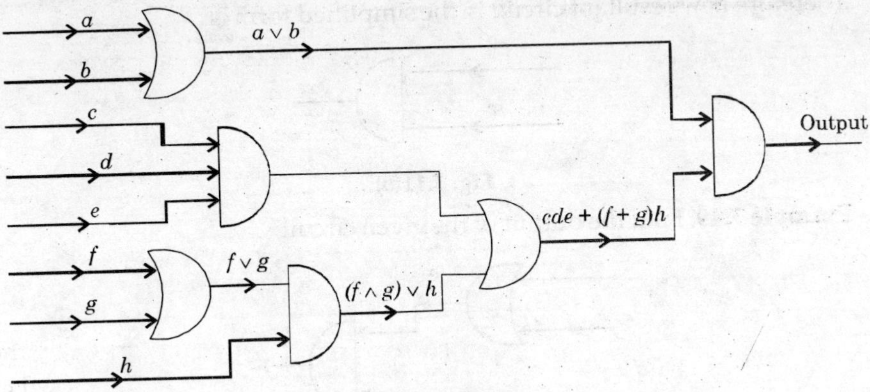

Fig. 7.10(b).

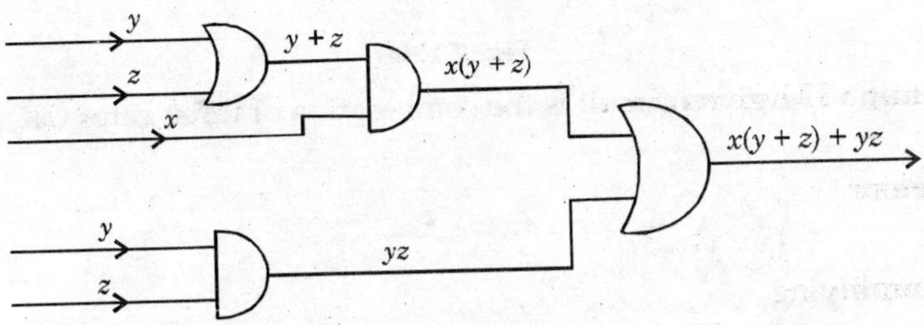

Fig. 7.10(c).

7.12. SIMPLIFICATION OF CIRCUITS

The efficiency of a circuit depends on the number and arrangement of its gates. Simlification of a circuit would normally mean the least complicated circuit with minimum cost and best result, *i.e.*, economical circuit. The simplification of circuits would mean lesser number of switches, which we achieve by using different properties of Boolean algebra.

Example 7.48. Simplify the circuit given by the function :

$$f(x, y, z) = x\,y\,z + x\,\overline{y}\,z$$

The circuit diagram of given function is

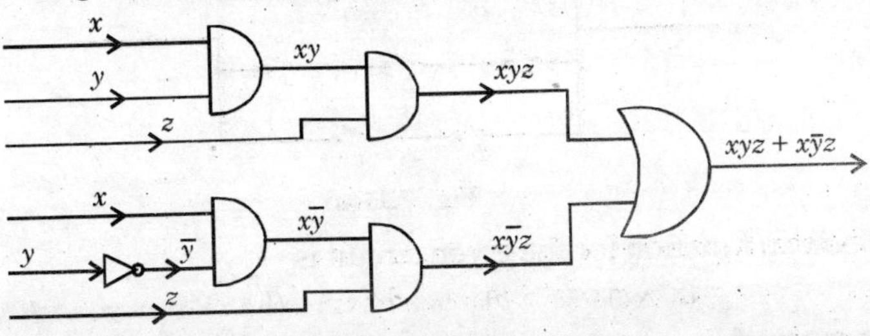

Fig. 7.11(a).

As in the above function switches x and z are repeated

Now simplify

$$\begin{aligned} f(x, y, z) &= x(y + \overline{y})z \\ &= x(1)z \\ &= xz \end{aligned}$$

Therefore new resultant circuit in the simplified form is

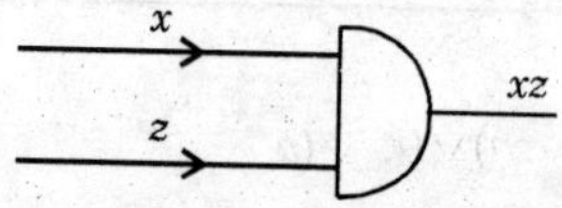

Fig. 7.11(*b*).

Example 7.49. Find the output of the given circuit.

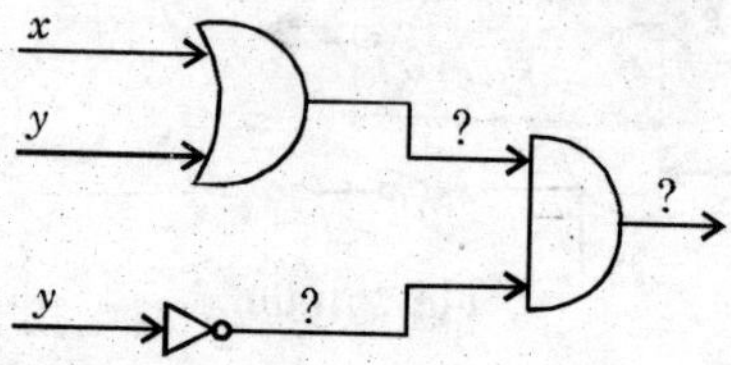

Fig. 7.12(*a*)

Solution : The given circuit is the combination of three gates OR, AND and invertor.

Therefore

$$(x \vee y) \wedge \bar{y}$$

On simplifying

$$(x \wedge \bar{y}) \vee (y \wedge \bar{y})$$

$$(x \wedge \bar{y}) \vee 0 = x \wedge \bar{y}$$

Therefore the simplified circuit is

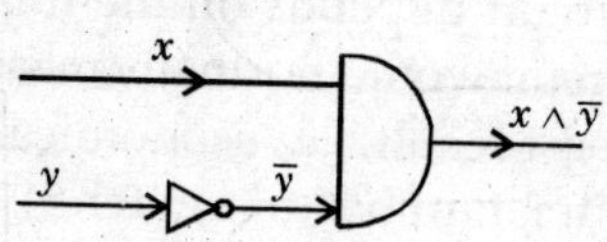

Fig. 7.12(*b*)

Example 7.50. Simplify the circuit

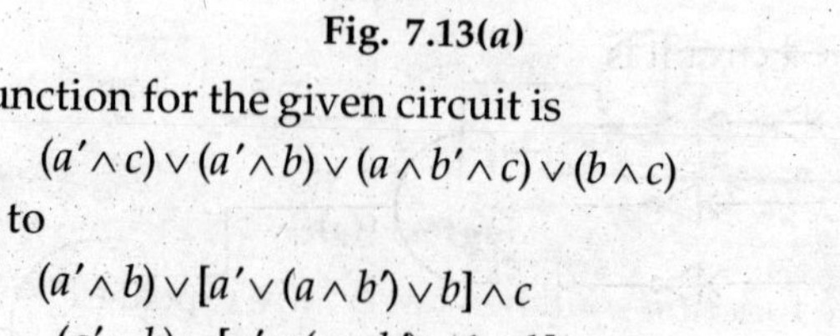

Fig. 7.13(*a*)

The Boolean function for the given circuit is

$$(a' \wedge c) \vee (a' \wedge b) \vee (a \wedge b' \wedge c) \vee (b \wedge c)$$

which is equal to

$$\begin{aligned}
& (a' \wedge b) \vee [a' \vee (a \wedge b') \vee b] \wedge c \\
= \; & (a' \wedge b) \vee [a' \vee (a \wedge b') \vee 1 \wedge b] \wedge c \\
= \; & (a' \wedge b) \vee [a' \vee (a \wedge b') \vee (a \vee a') \wedge b] \wedge c \\
= \; & (a' \wedge b) \vee [a' \vee (a \wedge b') \vee (a \wedge b) \vee (a' \wedge b)] \wedge c \\
= \; & (a' \wedge b) \vee [a' \vee (a \wedge (b' \vee b)) \vee (a' \wedge b)] \wedge c
\end{aligned}$$

$$= (a' \wedge b) \vee [a' \vee (a \wedge 1) \vee (a' \wedge b)] \wedge c$$
$$= (a' \wedge b) \vee [a' \vee a \vee (a' \wedge b)] \wedge c$$
$$= (a' \wedge b) \vee ((1 \vee (a' \wedge b)) \wedge c)$$
$$= (a' \wedge b) \vee ((1 \wedge c) \therefore \vee (a' \wedge b) \wedge c)$$
$$= ((a' \wedge b) \vee c) \vee ((a' \wedge b) \wedge c)$$
$$= (a' \wedge b) \vee c$$

$\therefore$ Simplified circuit

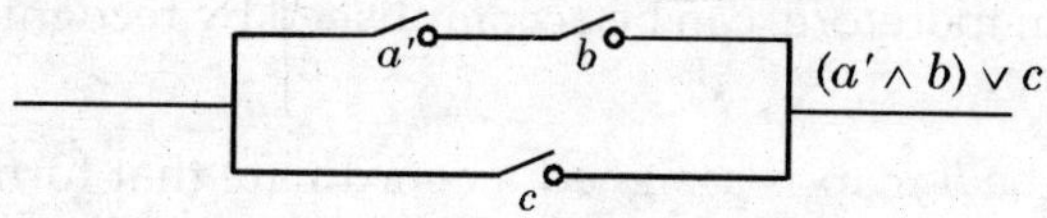

Fig. 7.13(*b*)

Readers can make the circuit diagram by the use of gates.

Example 7.51. Give output of the gives circuit and also simplify if

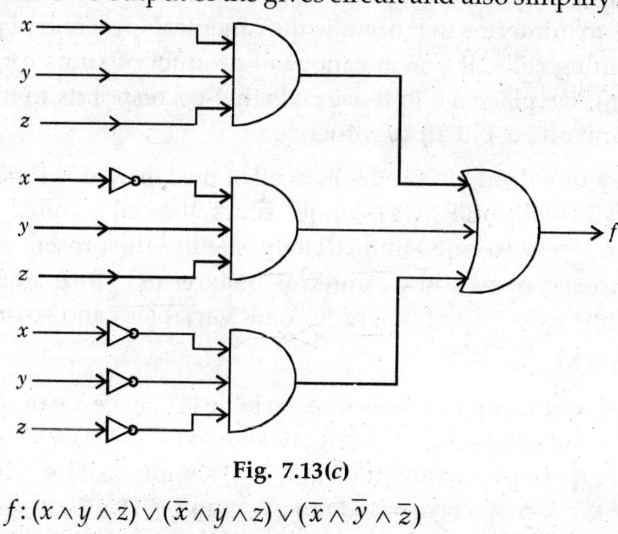

Fig. 7.13(*c*)

$$f : (x \wedge y \wedge z) \vee (\overline{x} \wedge y \wedge z) \vee (\overline{x} \wedge \overline{y} \wedge \overline{z})$$
$$[(x \vee \overline{x}) \wedge (y \wedge z)] \vee (\overline{x} \wedge \overline{y} \wedge \overline{z})$$
$$[1 \wedge (y \wedge z)] \vee (\overline{x} \wedge \overline{y} \wedge \overline{z}]$$

$(y \wedge z) \vee (\overline{x} \wedge \overline{y} \wedge \overline{z})$ is the simplified form

$\therefore$ The simplified circuit is

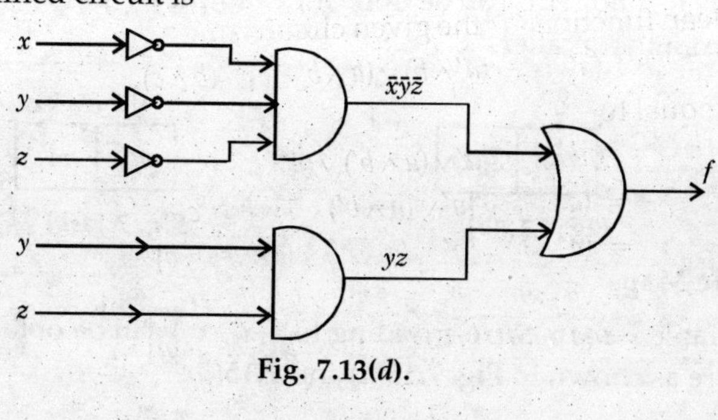

Fig. 7.13(*d*).

7.13. KARNAUGH MAP

A karnaugh map (or *k*-map for short) is a graphical method for representing the truth table of a Boolean function.

If n is the number of variables, then *k*-map has 2^n cells one for each of the 2^n minterms (maxterms) and, therefore, one for each now of the truth table. Moreover, these cells one arranged such that the minterm (maxterms) are displayed in a geometric pattern so that logically adjacent terms are visibly apparent.

Minimization, therefore, can be accomplished by recognizing basic pattern on the *k*-map.

Each cell of the *k*-map is assigned a coordinate that forms an n-bit binary number corresponding to a minterm (maxterms) of n literals. Two cells are adjacent if their corresponding binary combinations differ by only one bit. If a function F is expressed in canonical Sum-of-Products (SOP) form, then we complete the *k*-map of F by placing a 1 in those cells whose co-ordinates corresponds to minterms that occur is the canonical form and placing a 0 in each of the remaining cells. If F is in canonical product of sums form (conjunctive normal form), we place a 0 in those cells that corresponds to maxterms of the canonical form and a 1 in all the others.

If the functional value 1 occurs in a cell of the *k*-map of a function F, then the cell in called a 1-cells of F ; if a 0 value occurs, the cell is called a o-cell. Hence, two adjacent 1-cells (o-cells) immediately identify the presence of a redundant variable in the corresponding minterms (maxterms) ; four adjacent 1-cells (o-cells) imply the existence of two redundant variables ; and so on.

One-Variable Maps

Consider a function F of only one variable (x). Since x can take on only two values, the *k*-map needs only two cells, as shown in 7.14 (a). The region where x = 0 is the left half of the map and the value $F(0)$ should be entered in that cell. The right half of the *K*-map corresponds to $x = 1$, and $F(1)$ is centered in that cell. In the **Fig 7.14**(b) the locations of minterms are indicated so that a 1 is entered in the left cell if the minterms m_0 appears in the canonical form of F and a 1 is entered in the right cell if the minterm m_1 appears.

Higher order *K*-maps can be obtained by successively doubling the size of the 1- variable *K*-map. This can be done for example by simply "folding out" the larger map from the smaller.

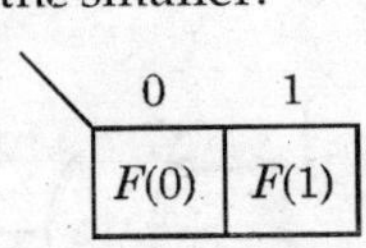

Fig. 7.14(a)

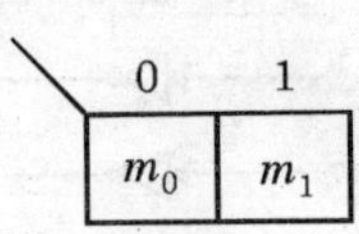

Fig. 7.14(b)

Two-variable Maps

A 2- variable k-map corresponding to $F(x_1, x_2)$ can be obtained by the fold-out procedure as shown in **Fig. 7.15**(a) and **7.15**(b).

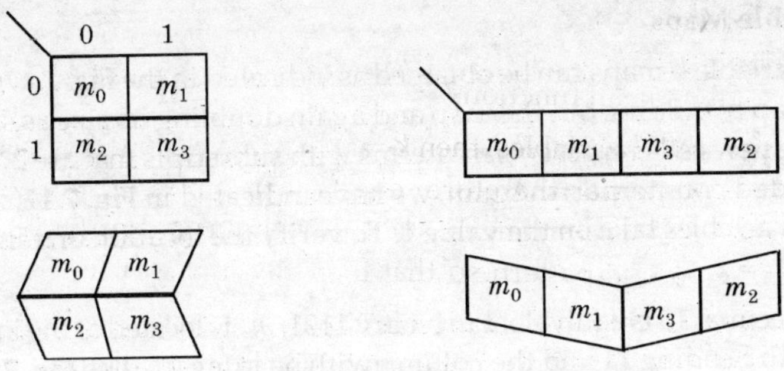

(a) (b)

Fig. 7.15.

The cells correspond to the minterms $m_o = x_1'x_2'$, $m_1 = x_1'x_2$, $m_2 = x_1x_2'$, $m_3 = x_1x_2$.

This particular arrangement forces minterms differing in only one variable to be adjacent. **Figure 7.15**(*a*), and 7.15(*b*) depict general form of 2 variable *k*-maps and **figure 7.15**(*b*) shows the *k*-map for $F(x_1, x_2) = \Sigma$ (2, 3).

The representation of $F(x_1, x_2)$ in Fig. 7.16(*a*) and (*b*) is useful as each conveys the information that certain minterms are logically adjacent to others.

It should be noted that in the 2-variable *k*-map ; Every minterm is adjacent to the two other minterms. Similarly in the 3-variable *k*-map every minterms will adjacent to 3 other minterms and in 4-variable *k*-map every minterms will adjacent to 4 other minterms. Therefore in general in *n*-variable *k*-map minterm is adjacent to *n* other minterms.

Three-Variable Map

The 3-variable *k*-map of **Fig. 7.16**(*a*) is produced by a fold out from the 2-variable map as given by the **Fig. 7.16**(*b*).

In the **Fig 7.16**(*b*) the cell assigned to $m_5 = x_1 x_2' x_3$ corresponds to row 1 and column 01 because binary 101 is equal to decimal 5. In other words, m_5 appears at the intersection of the row marked x_1 and column belonging to $x_2' x_3$.

The **Fig 7.16**(*c*) shows the *k*-map for

$F = \Sigma(0, 3, 5, 7)$

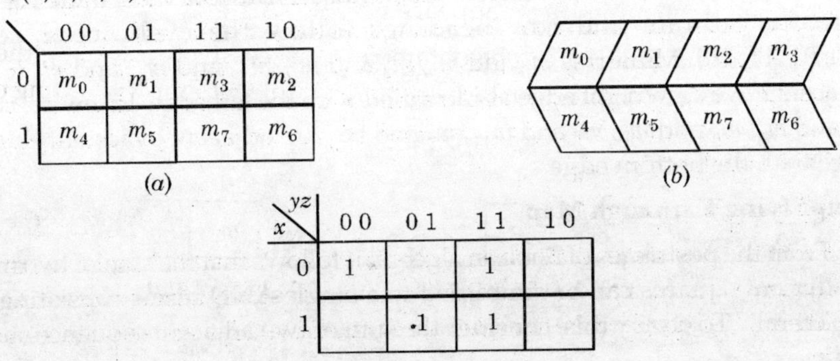

(a) (b)

F = Σ(0, 3, 5, 7)

(c)

Fig. 7.16.

Four Variable Maps

Four variable k-maps can be obtained as indicated in the **Fig. 7.17**(a) from the 3 variable map by folding out the map and again doubling its size as shown in **Fig. 7.17**(b). The new cells represents minterms with subscripts that are 2^3 greater than their unfolded counterparts. Again, we have indicated in **Fig. 7.17**(c) the regions where the variables take on the value 1. To verify the location of minterms, let us consider $m_{13} = x_1 x_2 x'_3 x_4$.

Since decimal 13 is equivalent to binary 1101, m_3 is located at the intersection of the row with heading 11 and the column with heading 01. The **Fig. 7.17**(c) shows the k-map for $F = \Sigma(1, 6, 7, 9, 12, 15)$.

wx \ yz	$y'z'$	$y'z$	yz	yz'
$w'x'$	m_0	m_1	m_3	m_2
$w'x$	m_4	m_5	m_7	m_6
wx	m_{12}	m_{13}	m_{15}	m_{14}
wx'	m_8	m_9	m_{11}	m_{10}

(a)

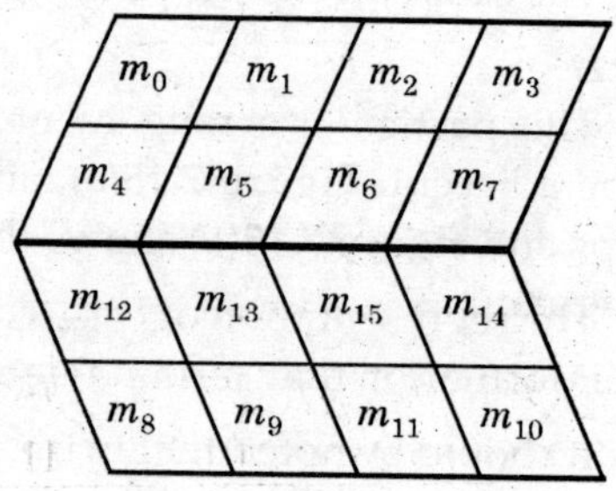

(b)

wx \ yz	$y'z'$	$y'z$	yz	yz'
$w'x'$		1		
$w'x$			1	1
wx	1		1	
wx'		1		

(c)

Fig. 7.17(a)

Every minterm is the 4-variable k-map is adjacent to four other minterms because both the column and row headings follow the cyclic code sequence {0 0, 01, 11, 10}. Minterms m_o and m_2, m_4 and m_6, m_{12} and m_{14} and m_8 & m_{10} and adjacent if we view right edge as identified with the left edge. Likewise minterms m_o and m_8, m_1 and m_0, m_3 and m_{11}, m_2 and m_{10} are adjacent by identifying the top edge with the bottom edge.

Simplifying Karnaugh Map

From the postulates of Boolean algebra, it follows that the sum of two minterms in adjacent squares can be simplified to a single AND terms consisting of only two literals. To clarify this, consider the sum of two adjacent sequence such as m_5 and m_7.

$$m_5 + m_7 = xy'z + xyz = xz(y' + y)$$

$$= xz$$

Here the two squares differ by the variable y, which can be removed when the sum of the two minterms in formed. Thus any two minterms in adjacent squares that are O Red together will cause a removal of the different variable. The following example explains the procedure for minimizing a Boolean function with a map.

Karnaugh Map

Each square represents one minterm. Since any Boolean function can be expressed as a sum of minterms, it follows that a Boolean function is recognized graphically in map from area enclosed by those squares whose minterms are included in the function.

Two Variable Map

x \ y	y'	y
x'	$x'y'$	$x'y$
x	xy'	xy

Four minterms for two variables [$2^2 = 4$]

Three Variable Map

Eight minterms for three variables [$2^3 = 8$]

x \ yz	$y'z'$ 00	$y'z$ 01	yz 11	yz' 10
0	$x'y'z'$	$x'y'z$	$x'yz$	$x'yz'$
1	$xy'z'$	$xy'z$	xyz	xyz'

Four Variable Map

wx \ yz	00	01	11	10
00	$w'x'y'z'$	$w'x'y'z'$	$w'x'yz$	$w'xyz'$
01	$w'xy'z'$	$w'xy'z$	$w'xyz$	$w'xyz'$
11	$wxy'z'$	$wxy'z$	$wxyz$	$wxyz'$
10	$wx'y'z'$	$wx'y'z$	$wx'yz$	$wx'yz'$

Four Binary Variables – 2^4 – minterms.

Example 7.52. Simplify the boolean f^n

(*i*) $F = x'yz + x'yz' + xy'z' + xy'z$

x \ yz	$y'z'$ 00	$y'z$ 01	yz 11	yz' 10
x' 0	m_0	m_1	1 m_2	1 m_3
x 1	1 m_4	1 m_5	m_6	m_7

$$f(x, y, z) = x'yz + x'yz' + xy'z' + xy'z$$
$$= x'y(z + z') + xy'(z' + z) = x'y + xy'$$

(*ii*) $F = x'yz + xy'z' + xyz + xyz'$

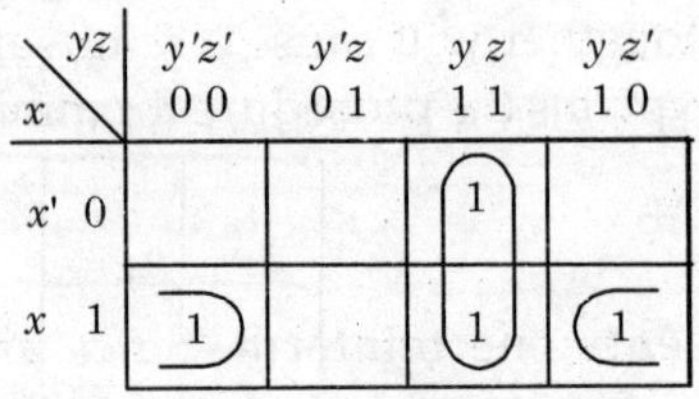

$$\begin{aligned} f(x, y, z) &= x'yz + xyz + xy'z' + xyz' \\ &= (x' + x)\,yz + x(y' + y)\,z' \\ &= yz + xz' \end{aligned}$$

(*iii*) $F = A'C + A'B + AB'C + BC$

$= A'C(B + B') + A'B(C + C') + AB'C + BC(A + A')$

As there is only a pair of 2 so multiply *e.g.*, $A'C$, *i.e.*,

$$\begin{aligned} A'C &= (A'C) \,.\, 1 = (A'C)\,(1 + 0) \\ &= (A'C)\,(B + B') \\ &= A'CB + A'CB' \end{aligned}$$

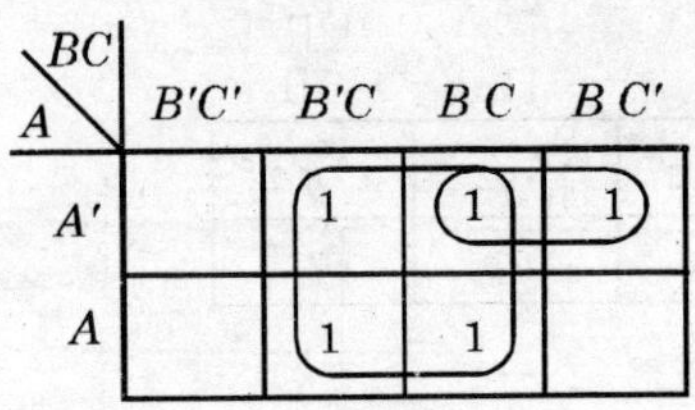

Applying the prucedure same as above we wiil get

$$F = C + A'B$$

(*iv*) $F(x, y, z) = \Sigma(0, 2, 4, 5\ 6)$

x \ yz	y'z'	y'z	yz	yz'
x'	x'y'z'	x'y'z	x'yz	x'yz'
x	xy'z'	xy'z	xyz	xyz'

Now in the form of Binary Notation

000	001	011	010
= 0	= 1	= 3	= 2
100	101	111	110
= 4	= 5	= 7	= 6

$$\begin{aligned} e.g.\ 001 &= 0 \times 2^2 + 0 \times 2' + 1 \times 2^0 = 1 \\ 011 &= 0 \times 2^2 + 1 \times 2' + 1 \times 2^0 = 3 \end{aligned}$$

So, (1) reduces to

x \ yz	00	01	11	10
0	0	1	3	2
1	4	5	7	6

Now $$F(x, y, z) = \Sigma(0, 2, 4, 5, 6)$$

x \ yz	00	01	11	10
0	1			1
1	1	1		1

Applying the procedure same as above.

So $$F = z' + xy'$$

Similarly, for the Four variable map

wx \ yz	$y'z'$	$y'z$	yz	yz'
$w'x'$	0	1	3	2
$w'x$	4	5	7	6
wx	12	13	15	14
wx'	8	9	11	10

Example 7.53. Simplify the Boolean expression

(*i*) $F(w, x, y, z) = \Sigma(0, 1, 2, 4, 5, 6, 8, 9, 12, 13, 14)$

$wxyz \quad y'z' \quad y'z \quad yz \quad yz'$

wx \ yz	$y'z'$	$y'z$	yz	yz'
$w'x'$	1	1		1
$w'x$	1	1		1
wx	1	1		1
wx'	1	1		

$$f = y' + w'z' + xz'$$

(*ii*) $F = A'B'C' + B'CD' + A'BCD' + AB'C'$

$$= A'B'C'(D+D') + B'CD'(A+A') + A'BCD' + AB'C'(D+D')$$

AB \ CD	C'D'	C'D	CD	CD'
A'B'	1	1		1
A'B				1
AB				
AB'	1	1		1

Ans. $F = B'C' + A'CD' + B'D'$

Example 7.54. Simplify the Boolean function

$$f(w,x,y,z) = \Sigma(1,3,7,11,15)$$

and don't care conditions

$$f(w,x,y,z) = \Sigma(0,2,5)$$

wx \ yz	y'z'	y'z	y z	y z'
w'x'		1	1	
w'x			1	
w x			1	
w x'			1	

Ans. $f = yz + w'x'z$

$$= (y + w'x')z$$

Example 7.55. Obtain the simplified expressions in sum of product for the following Boolean functions.

$$F(x,y,z) = \Sigma(2,3,6,7)$$

x \ yz	0 0	0 1	1 1	1 0
0	0	0	1	1
1	0	0	1	1

Ans. $f(x,y,z) = y$

Example 7.56. Show the sum-of-product for the Karnaugh map given below:

(*a*)

	$\bar{z}\,\bar{w}$	$\bar{z}\,w$	$z\,w$	$z\,\bar{w}$
$\bar{x}\,\bar{y}$	0	0	0	0
$\bar{x}\,y$	0	0	0	1
$x\,y$	1	1	1	1
$x\,\bar{y}$	1	1	1	1

Ans. $f(x, y, z, w) = x + y\,\overline{zw}$

(*b*)

	$\bar{z}\,\bar{w}$	$\bar{z}\,w$	$z\,w$	$z\,\bar{w}$
$\bar{x}\,\bar{y}$	1	1	1	1
$\bar{x}\,y$	1	1	1	0
$x\,y$	1	0	0	0
$x\,\bar{y}$	1	0	0	0

Ans. $f(x, y, z, w) = \bar{z}\,\bar{w} + \bar{x}\,\bar{y} + \bar{x}\,w$

SOLVED PROBLEMS

Problem 7.1. Write the Boolean expression that represents the combinatorial circuit, write the logic table and write the output of each gate symbolically.

Solution:

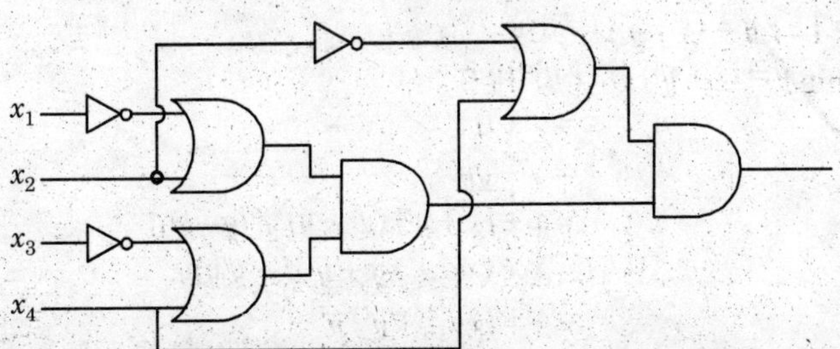

Fig. 7.18

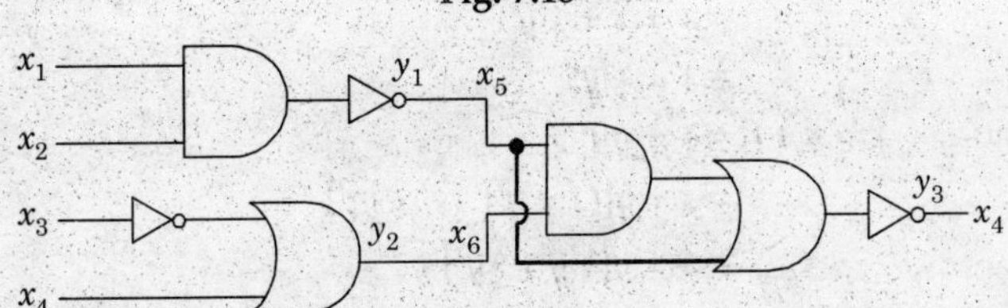

Fig. 7.19

Problem 7.2. Show that the combinatorial circuits to following are equivalent.

Solution:

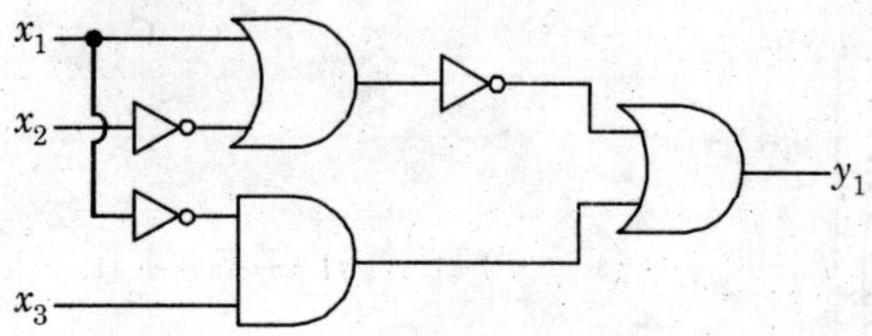

Fig. 7.20

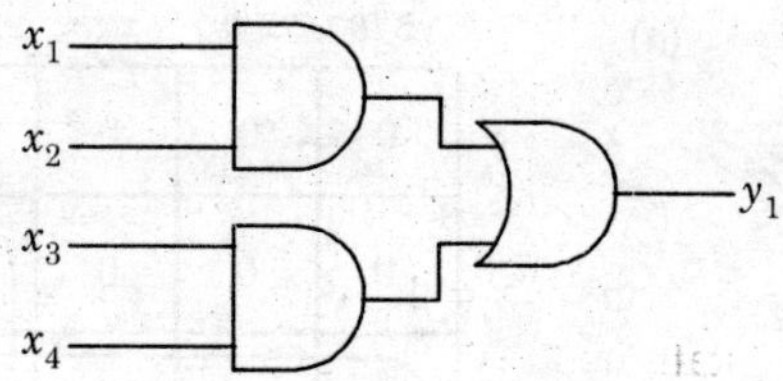

Fig. 7.21

Problem 7.3. Write the dual of each statement and also prove it.

(*a*) If $x + y = x + z$ and $x' + y = x' + z$ then $y = z$

(*b*) $x = 0$ if and only if $y = xy' + x'y$ for all y.

Solution: (*a*) *Dual* : if $xy = xz$ and $x'y = x'z$ then $y = z$

Proof: Assume $xy = xz$ and $x'y = x'z$

then
$$\begin{aligned} y &= y \,.\, 1 = y(x + x') = yx + yx' = xy + x'y \\ &= xz + x'z \\ &= zx + zx' \\ &= z(x + x') \\ &= z \,.\, 1 \\ &= z \end{aligned}$$

(*b*) *Dual* : $x = 1$ if and only if $y = (x + y')\,(x' + y)\ \forall y$

To prove this, we must recall the definition of q if and only if p or $p \leftrightarrow q$.

This is logically equivalent to $(p \to q) \wedge (q \to p)$

So we must prove that

$x = 1 \to y = (x + y')\,(x' + y)$ and then prove that $y = (x + y')\,(x' + y) \to x = 1$

Assume $x = 1$. Then
$$\begin{aligned} (x + y')\,(x' + y) &= (1 + y')\,(1' + y) \\ &= (1 + y')\,(0 + y) \\ &= 1(y) = y. \end{aligned}$$

So $x = 1 \to y = (x + y')\,(x' + y)$

Assume $y = (x + y')\,(x' + y)$ then
$$\begin{aligned} x &= x + 0 \\ &= x + yy' \\ &= x + (x + y')\,(x' + y)\,y' \text{ (given)} \\ &= x + (xx' + xy + y'x' + y'y)y' \\ &= x + (xy + x'y')\,y' \\ &= x + (xyy' + x'y'y') \\ &= x + x + x'y' \\ &= x + x'y' \end{aligned}$$

In addition,
$$\begin{aligned} x &= x + 0 = x + yy' \\ &= x + y[(x + y')'\,(x' + y)'] \\ &= x + y[x'y + xy'] \\ &= x + x'y \end{aligned}$$

So we have $x = x + x'y'$ and $x = x + x'y$

Since $x = x + x = x + x'y' + x + x'y$

$= x + x'y' + x'y$

$= x + x' (y' + y)$

$= x + x' = 1$

Since $x = 1 \rightarrow (x + y') (x' + y) = y$ and $y = (x + y') (x' + y) \rightarrow x = 1$ thus the statement.

Problem 7.4. Find the disjunctive normal form of each functions and draw the combinatorial circuit corresponding to the disjunctive normal form.

(*a*)

x	y	z	$f(x, y, z)$
1	1	1	0
1	1	0	0
1	0	1	0
1	0	0	1
0	1	1	1
0	1	0	1
0	0	1	1
0	0	0	0

(*b*)

w	x	y	z	$f(w, x, y, z)$
1	1	1	1	1
1	1	1	0	0
1	1	0	1	1
1	1	0	0	0
1	0	1	1	0
1	0	1	0	0
1	0	0	1	0
1	0	0	0	1
0	1	1	1	1
0	1	1	0	0
0	1	0	1	0
0	1	0	0	0
0	0	1	1	1
0	0	1	0	0
0	0	0	1	0
0	0	0	0	0

Solution: (*a*)

(*b*)

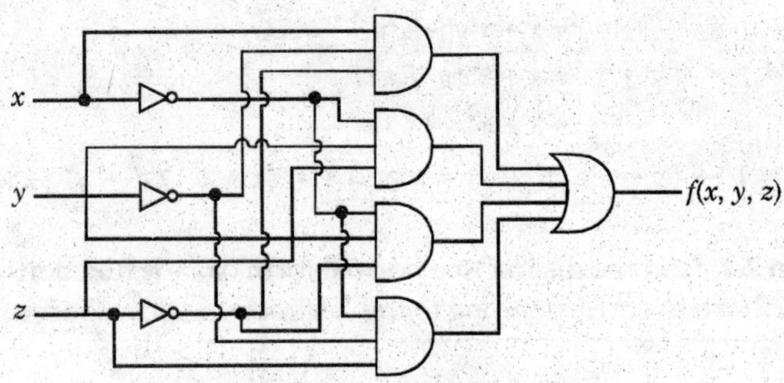

Fig. 7.22.

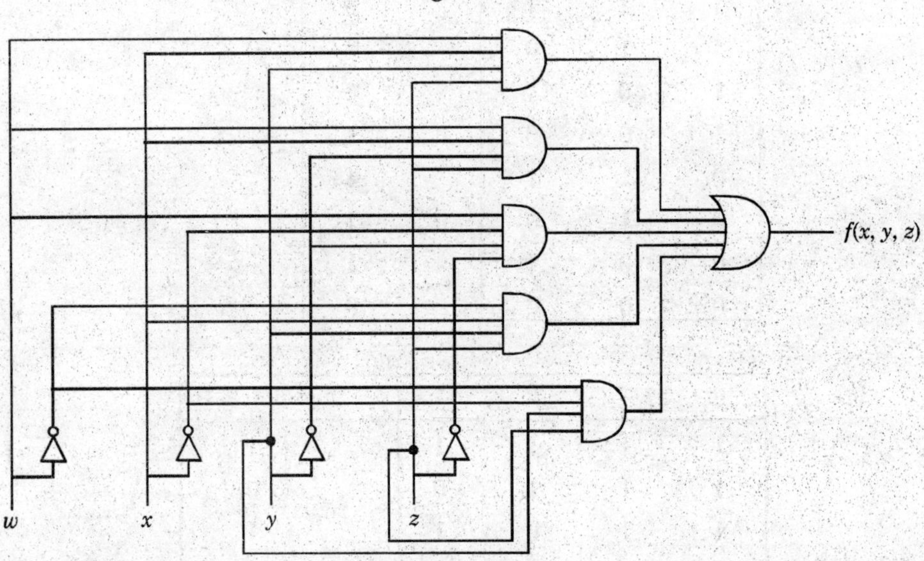

Fig. 7.23.

Problem 7.5. Write $x \downarrow y$ in terms of $\uparrow$

Solution:

$$\begin{aligned} x \downarrow y &= \overline{x \vee y} \\ &= \bar{x}\ \bar{y} \\ &= \overline{\bar{x} \uparrow \bar{y}} \\ &= (x \uparrow y) \uparrow (x \uparrow y) \\ &= (x \uparrow x) \uparrow (y \uparrow y) \uparrow (x \uparrow x) \uparrow (y \uparrow y) \end{aligned}$$

So $x \downarrow y = (x \uparrow x) \uparrow (y \uparrow y) \uparrow (x \uparrow x) \uparrow (y \uparrow y)$

EXERCISES

1. Show that in a Boolean algebra :
   (*i*) $x \vee (x' \wedge y) = x \vee y,\ x \wedge (x' \vee y) = x \wedge y.$
   (*ii*) $x = 0 \Leftrightarrow y = (x \wedge y' \vee (x' \wedge y),\quad \forall\quad y.$
   (*iii*) $x = y \Leftrightarrow (x \wedge y') \vee (x' \wedge y) = 0.$

2. If x, y are atoms in the finite Boolean algebra such that $x \wedge y \neq 0$, then show $x = y$.
3. Prove that order of a Boolean algebra cannot be a prime (> 2).
4. Show that in a Boolean algebra :
   (i) $(x' \vee y') \vee (x \vee y \wedge z') = (y \wedge z') \vee (x' \vee y')$
   (ii) $(x' \vee z) \vee (x' \wedge y) \vee (x \vee y' \wedge z) \vee (y \wedge z) = (x' \wedge y) \vee z.$
5. Prove the generalised De-Morgan's laws, in a Boolean algebra :
   $(x_1 \vee x_2 \vee x_3 \vee \ldots \vee x_n)' = x'_1 \wedge x'_2 \wedge x'_3 \wedge \ldots \wedge x'_n.$
   and $(x_1 \wedge x_2 \wedge x_3 \wedge \ldots \wedge x_n)' = x'_1 \vee x'_2 \vee x'_3 \vee \ldots \vee x'_n.$
6. Express the following Boolean functions in the DN (CN) form in minimum number of variables :
   (i) $x \wedge (y \vee z)$
   (ii) $(x \wedge y \wedge z) \vee [(x \vee y) \wedge (x \vee z)]$
   (iii) $(x'_1 \wedge x_2)\ (x_1 \wedge x'_2 \wedge x'_3) \vee (x_2 \wedge x_3)$
   (iv) $[(x' \wedge y) \vee (x \wedge z')] \wedge [(x \vee (y \wedge z)]'$
   (v) $(x \wedge y) \vee (x \wedge z) \vee (y' \wedge z)$
7. Express (i) $x' \wedge y$ (ii) $(x' \wedge z) \vee (x \wedge z')$ in the DN (CN) form in three variables.
8. Let $A = \{0, 1\}$ and F = set of all functions from $A^n \to A$. Define for $f, g \in F$ such that

$$(f \vee g)\ (x) = f(x) \vee g(x),\ (f \wedge g)\ (x) = f(x) \wedge g(x).$$
$$f'(x) = (f(x)' \cdot O(x) = 0.\ I(x) = 1\ \ \forall\ x \in A$$

   Show that $(F, \wedge, \vee, ')$ forms a Boolean algebra.
9. Draw the circuit represented by the function
   (i) $(a' \wedge b' \wedge c) \vee (a' \wedge b \wedge c)$ (ii) $(a \wedge (a \vee b) \vee (b \wedge (a' \vee b))$
   (iii) $(a \wedge b') \vee (a' \wedge b)$.
10. Find the Boolean functions that represent the following circuits :

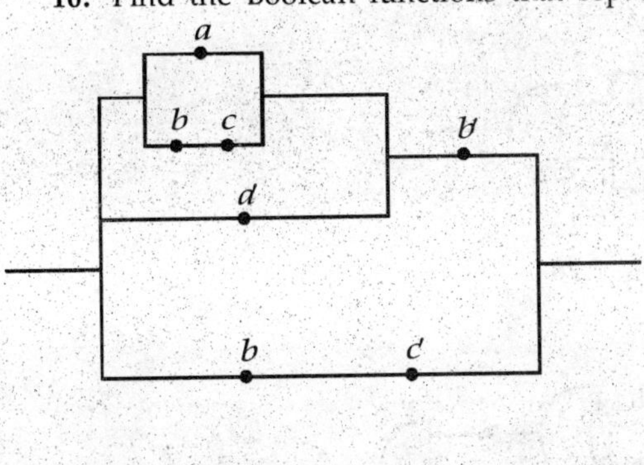

Fig. 7.24 (a).

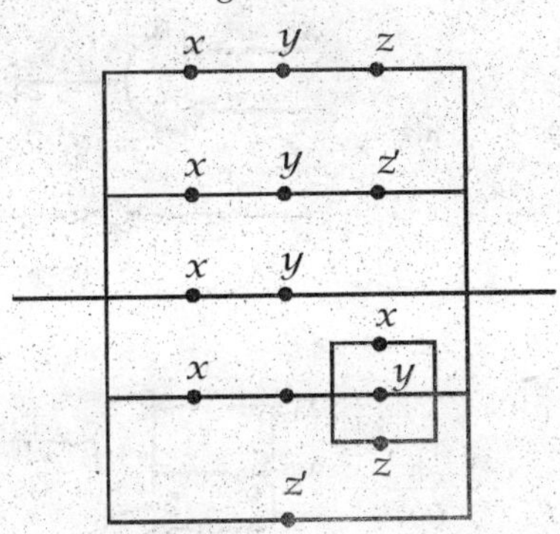

Fig. 7.24 (b).

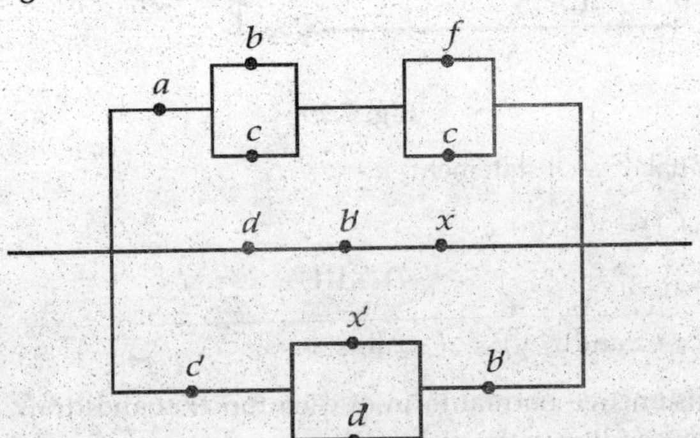

Fig. 7.24(c).

11. Simplify the circuit represented by

$$(a \wedge c' \wedge b') \vee (a \wedge b' \wedge d) \vee (a \wedge c \wedge d').$$

12. Find the Boolean expression for the logic circuit given the figures below and simplify them if possible.

(*a*)

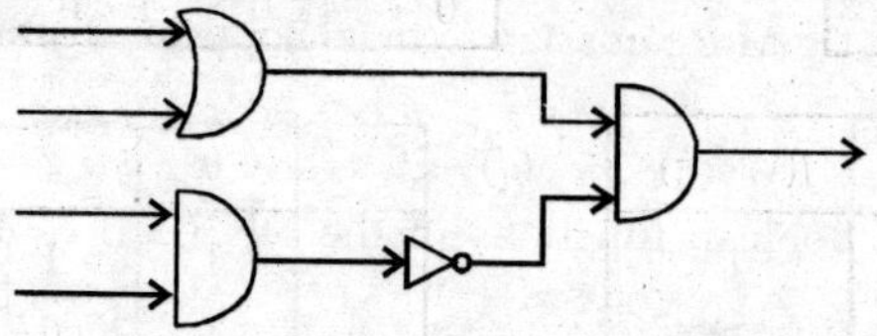

Fig. 7.25

(*b*)

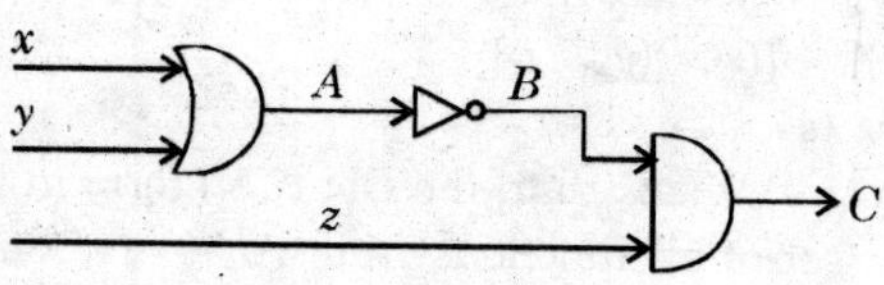

Fig. 7.26

(*c*)

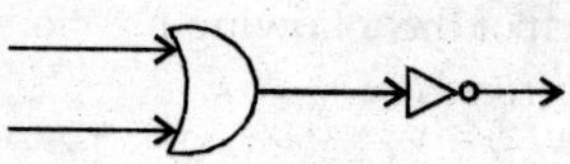

Fig. 7.27

(*d*)

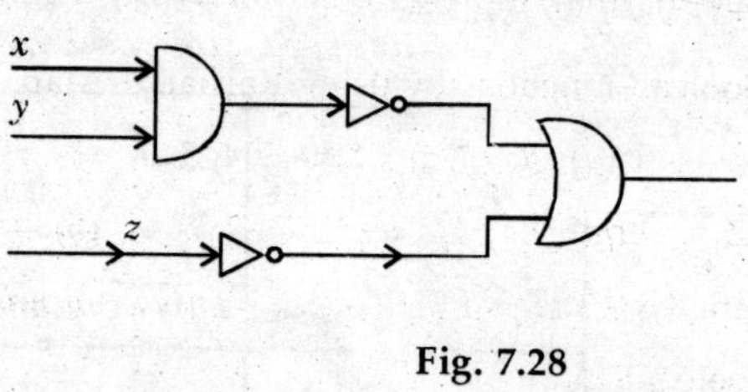

Fig. 7.28

(*e*)

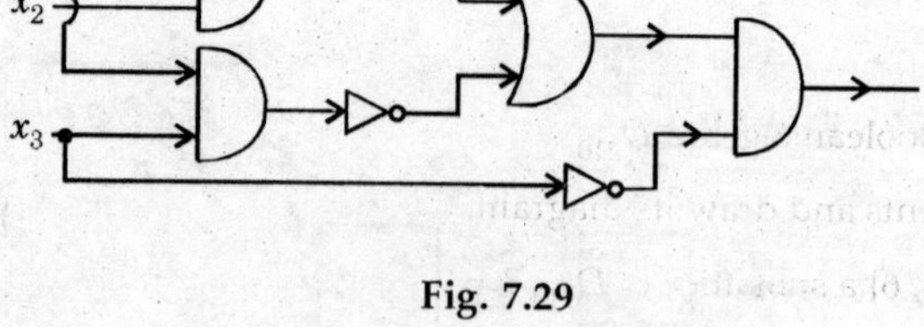

Fig. 7.29

13. Write the dual of each statement

(*a*) $(x + y)(x + 1) = x + xy + y$

(*b*) $(x' + y)' = xy$

(*c*) If $x + y = x + z$ and $x' + y = x' + z$ then $y = z$

14. Find the disjunctive normal form of each function and draw the combinatorial circuit corresponding to the disjunctive normal form and give the simplified form also.

(a)

x	y	$f(x, y)$
1	1	1
1	0	0
0	1	1
0	0	1

(b)

x	y	$f(x, y)$
1	1	0
1	0	1
0	1	0
0	0	1

(c)

x	y	z	$f(x, y, z)$
1	1	1	1
1	1	0	1
1	0	1	0
1	0	0	1
0	1	1	0
0	1	0	0
0	0	1	1
0	0	0	1

(d)

x	y	z	$f(x, y, z)$
1	1	1	1
1	1	0	1
1	0	1	1
1	0	0	0
0	1	1	0
0	1	0	1
0	0	1	1
0	0	0	1

15. Find the Disjunctive Normal form for each of the following functions.

(a) $f(x, y) = \left(x \vee y\right) \wedge \left(\bar{x} \vee \bar{y}\right)$ (b) $f(x, y, z) = x \vee y \wedge x$

(c) $f(x, y, z) = \left(\bar{x}\, y + \overline{xz}\right)\left(\overline{x \vee yz}\right)$ (d) $f(x, y, z) = x + \left(\bar{y} + (x\bar{y} + (x\bar{y} + x\bar{z})\right)$

16. Simplify the following Boolean functions by using Karnaugh Map.

(i) $f(x, y, z) = \Sigma\,(3, 4, 6, 7)$ (ii) $f(x, y, z) = \Sigma\,(0, 2, 4, 5, 6)$

(iii) $X'Z + X'Y + XY'Z + YZ$ (iv) $f(x, y, z, w) = \Sigma(1, 3, 7, 11, 15)$

17. Define the Boolean algebra. Give two examples and justify your answer also.

18. Write the dual of each Boolean equation

1. $(a \cdot 1) \cdot (0 + a') = 0$
2. $a + ab' = a + b$
3. $a(a' + b) = ab$
4. $(a + 1)\,(a + 0) = a$
5. $a(a' + b) = ab$

19. Consider the Boolean algebra D_{210}

(a) List its elements and draw its diagram.

(b) Is $X = \{1, 2, 3, 6\}$ a sublattice of D_{210} ?

(c) Is $X = \{1, 2, 3, 6\}$ a subalgebra of D_{210} ..... ?

(d) Find the number of subalgebras of D_{210}

20. Consider the following lattices :

(a) D_{20} (b) D_{36}

(c) D_{100} (d) D_{130}

Which of them are Boolean algebras and what are their atoms.

21. Express in the principle disjunctive normal form.

(The following Boolean function)

(*i*) $f(a, b, c) = (a' + b)' + a'c$ (*ii*) $f(x, y, z) = y(x + yz)'$

(*iii*) $f(x, y, z) = x(xy' + x'y + y'z)$ (*iv*) $f(x, y, z) = (x + y'z)(y + z')$

(*v*) $f(x, y, z) = (x'y)'(x' + xyz')$ (*vi*) $f(x, y) = (xy')'(x + y)'$

22. Use Karnaugh map to obtain the following functions in the simplified form as sum of product

(*a*) $x'z' + y'z' + yz' + xyz$

(*b*) $xy + x'y'z' + x'yz'$

(*c*) $F(x, y, z, t) = \Sigma(7, 13, 14, 15)$

(*d*) $F(x, y, z, t) = \Sigma(2, 3, 12, 13, 14, 15)$

23. Using Karnaugh map find a minimal sum of product expression of the following function

$F(A, B, C, D) = \Sigma(0, 1, 4, 5, 9, 11)$ ***(End-term Exam., GGSIPU, 2001)***

ANSWERS

6. (*i*) $(x \wedge y \wedge z) \vee (x \wedge y \wedge z') \vee (x \wedge y' \wedge z)$

(*ii*) $(x \vee y \vee z) \wedge (x \vee y' \vee z) \wedge (x \vee y \vee z')$ in CN-form

(*iii*) $(x'_1 \wedge x_2 \wedge x_3) \vee (x'_1 \wedge x_2 \wedge x'_3) \vee (x_1 \wedge x'_2 \wedge x'_3) \vee (x_1 \wedge x_2 \wedge x_3)$ in DN-form.

(*iv*) $(x' \wedge y \wedge z') \vee (x' \wedge y' \wedge z) \vee (x' \wedge y' \wedge z')$

(*v*) $(x \wedge y \wedge z) \vee (x \wedge y \wedge z') \vee (x \wedge y' \wedge z) \vee (x' \wedge y' \wedge z)$ in DN form.

7. (*i*) $(x \wedge y \wedge z) \vee (x' \wedge y \wedge z')$

(*ii*) $(x' \wedge y \wedge z) \vee (x' \wedge y' \wedge z) \vee (x \wedge y \wedge z') \vee (x \wedge y' \wedge z')$.

14. (*a*) $y + x'y'$

(*b*) y'

(*c*) $xyz + y'$

(*d*) $xy + x'y' + xy'z + x'yz'$

16. (*i*) z'

(*ii*) $z' + xy'$

(*iii*) $z + x'y$

(*iv*) $z' + wy'$

x	y	z	$f = x \wedge y \wedge z$
0	1	1	0
0	1	1	0
1	1	0	0
1	1	0	0
1	1	0	0
1	1	0	0
1	0	1	0
1	0	1	0

21. (*a*)

x	y	z	xz	x'	$x'y$	$x_2 + x'y$
1	1	1	1	0	0	1
1	1	0	0	0	0	0
1	0	1	1	0	0	1
1	0	0	0	0	0	0
0	1	1	0	1	1	1
0	1	0	0	1	1	1
0	0	1	0	1	0	0
0	0	0	0	1	0	0

(*b*) Same as (*a*).

Chapter

8 GRAPHS

8.1 INTRODUCTION

Graph theory is an applied branch of mathematics. Due to simplicity it has a wide range of applications in operations research, genetics, physical, biological and social sciences, engineering, computer science and in many other areas. The large portion of Graph theory have been motivated by recreational mathematics and the study of games. Generally graphs can be used in two situations. Firstly since graph is a very simple, convenient and natural way of representating the relationship between objects. In many problems such a pictorial representation may be all that is needed. Example of such applications are chemical molecule

C_2H_6 can be represented as

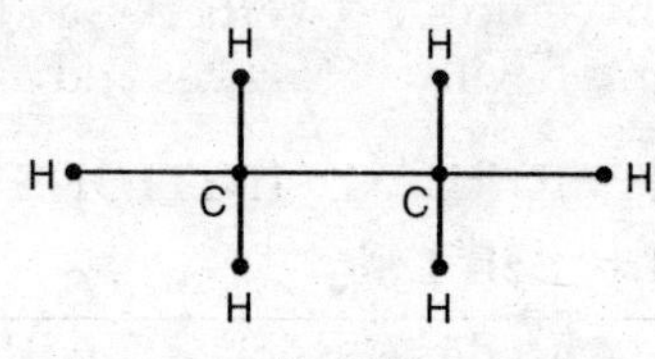

Secondly, we take the graph as mathematical model, solve the appropriate graph-theoretic problem, and then interpret the solution in terms of the original problem. Königsberg Bridge problem is one of the best example of solving puzzels using graphs. The origin of graph theory can be traced back to Euler's work on this problem (1736), which subsequently led to the concept of an Eulerian graph. Euler was the first person, who presented the very first paper in graph theory.

8.2 KÖNIGSBERG BRIDGE PROBLEM

Königsberg (now renamed Kalinigrad and in West Soviet Russia), the capital of East Prussia, was located on the banks of Pregil River. There were two islands, formed by the river, (which were) connected with each other and to the banks of the river with seven bridges shown in **Fig. 8.1.**

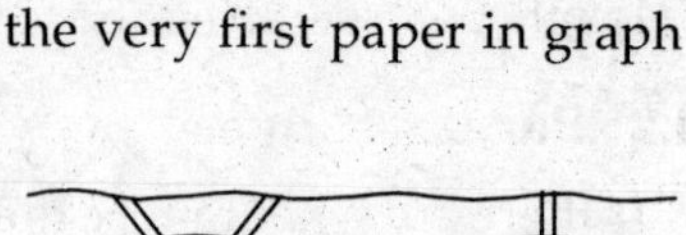

Fig. 8.1 Königsberg bridge problem

The problem was to traverse each of the seven bridges *exactly* once, starting from any of the four land areas and returning back to the starting point.

Leonard Euler (Swiss Mathematician) solved the problem by means of a graph, shown in **Fig. 8.2,** in which land areas represented by vertices and bridges by edges.

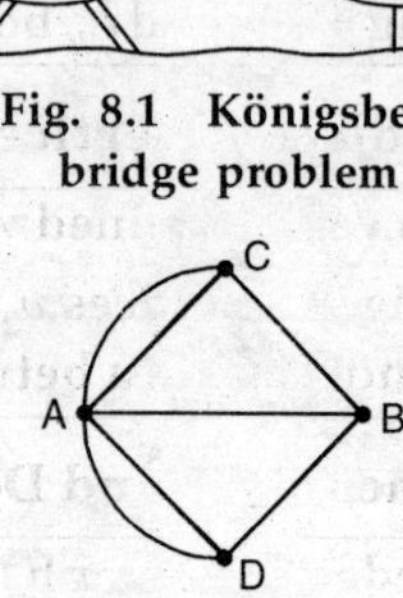

Fig. 8.2 Graphical Representation of Königsberg Bridge Problem

Euler proved that there exist no solution for this problem. The problem is similar with the problem of drawing a figure on paper without lifting the pen and without retracing a line. More applications of graphs will be discussed in the next chapters.

8.3 GRAPH

Informally a graph is a pictorial representation consisting of points called vertices and lines, called edges; each edge joins exactly two vertices.

Definition of Graph

A graph $G = (V, E)$ consist of a finite set denoted by V or $V(G)$ and a collection E or $E(G)$ of unordered pairs (u, v) of distinct elements from V. Each element of V is called a vertex or node or point and each element of E is called an edge or a line or a link.

The cardinality of V, *i.e.*, the number of vertices, is called the **order** of graph G and denoted by $|V|$. The cardinality of E, *i.e.*, the number of edges, is called the **size** of the graph and denoted by $|E|$. Let $V(G) = \{v_1, v_2, \ldots v_n\}$ and $E(G) = \{e_1, e_2, \ldots, e_m\}$, be the set of vertices and edges of a graph G. Each edge $e_k \in E(G)$ is identified with an unordered pair (v_i, v_j) of vertices. The vertices v_i and v_j are called the end vertices of e_k. Fig. 8.3 represents a graph with six vertices and ten edges.

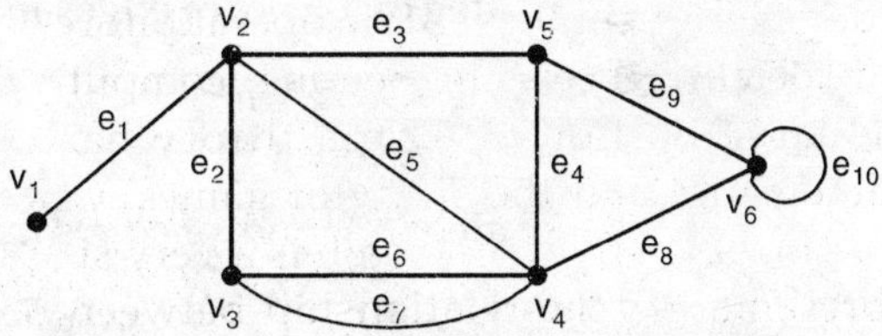

Fig. 8.3

8.4 MORE DEFINITIONS

8.4.1 Self Loop

An edge associated with vertex pair (v_i, v_i) is called a self-loop. In Fig. 8.3 edge e_{10} is a self loop starting and ending at vertex v_6.

In other words, an edge is called a self-loop if its terminal vertices are same.

8.4.2 Parallel Edges

If there are two or more then two edges in a graph G are associated with a given vertex pair, say, (v_i, v_j), then the edges are referred as parallel edges.

In Fig. 8.3 e_6 and e_7, both associated with vertex pair (v_3, v_4) are parallel edges.

8.4.3 Adjacent Vertices

Two vertices joined with an edge are called adjacent vertices.

In Fig. 8.3 vertices v_2 and v_4 are adjacent vertices. But v_1 and v_5 are not as there is not edge in a between v_1 and v_5.

8.4.4 Incidence and Degree

An edge e of a graph G is said to be incident with a vertex v and vice versa if v is the end vertex of e. Any two non-parallel edges say e_i and e_j are said to be adjacent if e_i and e_j are incident with a common vertex v. *For example*, e_1 and e_2 in Fig. 8.3 are adjacent.

The **degree** of any vertex v of G is the number of edges incident with vertex v. Each self-loop is counted twice. Degree of a vertex is always a positive number and is denoted as **deg (v)**. The **minimum** and **maximum degree** of vertices in $V(G)$ are denoted by $\delta(G)$ and $\Delta(G)$, respectively.

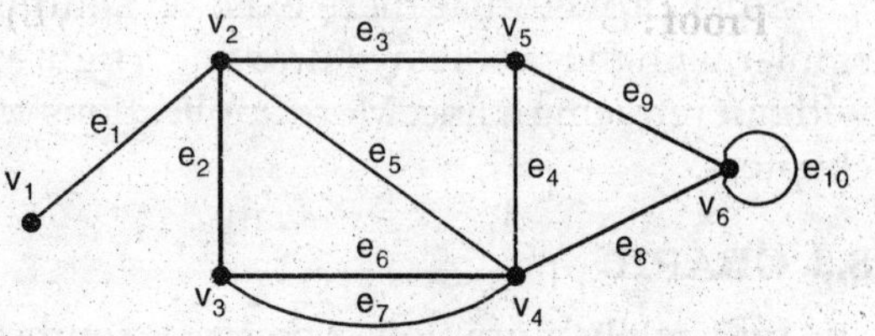

Fig. 8.4 Graph with 6 vertices to edges

In **Fig. 8.4.**

$$\deg(v_1) = \text{no. of edges incident with vertex } v_1$$
$$= 1 \text{ (only } e_1 \text{ is incident with } v_1)$$
$$\deg(v_2) = \text{no. of edges incident with } v_2 \text{ i.e.,}$$
$$= (e_1, e_2, e_3, e_5)$$
$$= 4$$
$$\deg(v_6) = \text{no. of edges incident with } v_6 \text{ i.e.,}$$
$$= (e_8, e_9, e_{10}, e_{10}); (e_{10} \text{ is a self-loop})$$
$$= 4$$

Similarly, $\deg(v_3) = 3 \quad \deg(v_4) = 5 \quad \deg(v_5) = 3$

Clearly, $\delta(G) = 1$ and $\Delta(G) = 5$

8.4.5 Isolated Vertex

A vertex which is not incident with any edge is called an isolated vertex. In other words a vertex with degree zero is called an isolated vertex. In Fig. 8.5, v_5 is the isolated vertex.

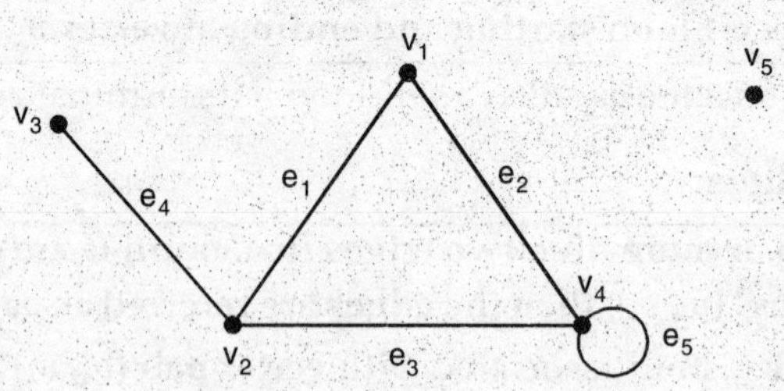

Fig. 8.5

8.4.6 Pendant Vertex

A vertex with degree one is called a pendant vertex or an end vertex. In Fig. 8.5, v_3 is the pendant vertex.

Note: Two adjacent edges are said to be in **series** if their common vertex is of degree two.

In **Fig. 8.5** e_1 and e_2 are in series, *i.e.*, degree of the common vertex v_1 is two.

Theorem 8.1. Prove that, the sum of degrees of all vertices in G is twice the number of edges in G.

Proof: Let us consider a graph $G = (V, E)$ with n vertices $v_1, v_2, ..., v_n$ and e edges. Then we have to show that

$$\deg(v_1) + \deg(v_2) + ... + \deg(v_n) = 2e$$

Since each edge is associated with two vertices and hence contributes two degree in graph G. So if there are e edges in a graph G then total degree of G will be equal to $2e$, and is equal to the sum of degrees of all vertices in G.

i.e.

$$\deg(v_1) + \deg(v_2) + ... + \deg(v_n) = \text{Total degree of } G = 2e$$

or $$\sum_{i=1}^{n} 2\deg(v_i) = 2e$$

Hence the theorem.

Theorem 8.2. Prove that, the number of vertices having odd degree in a graph G is always even.

Proof: Let us consider a graph $G = (V, E)$ with n vertices $(v_1, v_2, ..., v_n)$ and e edges. Out of these n vertices, let us separate the vertices of even degree and vertices of odd degree. So, the total degree of graph G can be calculated by taking the sum of degree of all even vertices and degrees of all odd vertices.

i.e., $$\text{Degree of } G = \sum_{\text{even}} \deg(v_j) + \sum_{\text{odd}} \deg(v_k) \quad ...(8.1)$$

and $$\text{Degree of } G = \sum_{i=1}^{n} \deg(v_i) = 2e \quad ...(8.2)$$

So by equations (8.1) and (8.2)

$$\sum_{\text{even}} \deg(v_j) + \sum_{\text{odd}} \deg(v_k) = 2e \quad ...(8.3)$$

Since the left side of equation (8.3) is even and the first part of right side is also even, (being the sum of even degrees or numbers). So the second part of right side *i.e.,* $\sum_{\text{odd}} \deg(v_k)$ must also be an even number.

i.e., $$\sum_{\text{odd}} \deg(v_k) = \text{even number}$$

Since each $\deg(v_k)$ is odd, so the total number of terms in the sum must be even to make the overall sum an even number. Hence the theorem.

Example 8.1. A graph G has 21 edges, 3 vertices of degree 4 and other vertices are of degree 3. Find the number of vertices in G.

Solution: Let there be n vertices in G. Out of these n vertices 3 are of degree 4 and $(n-3)$ vertices are of degree 3.

So, degree of $G = 3 \times 4 + 3 \times (n-3)$ ...(8.4)

But from theorem 8.1, we know that

degree of $G = 2 \times e = 2 \times 21$ ...(8.5)

Hence, $12 + 3(n-3) = 42$ (by eqns. (8.4) and (8.5))

$$(n-3) = \frac{42-12}{3} = 10$$

$$n = 13$$

i.e., the total number of vertices in G are 13.

Example 8.2. A graph G has 8 edges. Find the number of vertices, if the degree of each vertex is 2.

Solution: Let there be n vertices in graph G.

So, degree of $G = \sum_{i=1}^{n} \deg(v_i)$ degree of each vertex in G is 2

So degree of $G = 2 \,.\, n = 2 \,.\, e$ (by theorem 8.1)

$\Rightarrow$ $2 \,.\, n = 2 \,.\, 8 = 16$

$\Rightarrow$ $n = 8$

i.e., the total number of vertices in G are 8.

8.5 TYPES OF GRAPH

8.5.1 Simple Graph

A graph without self-loops and parallel edges is called a simple graph.

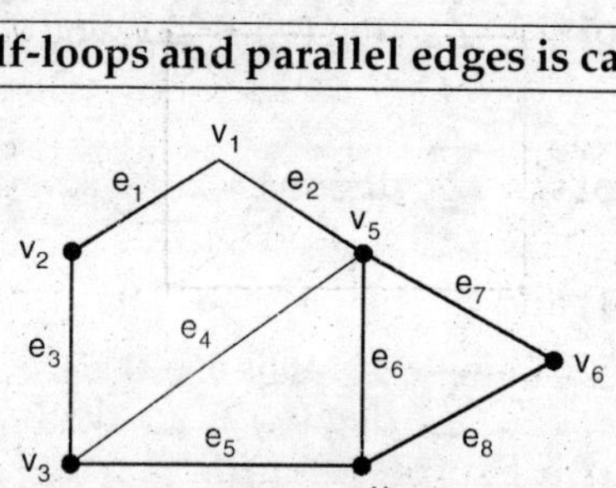

Fig. 8.6. A simple graph

8.5.2 Finite and Infinite Graph

A graph having finite number of vertices and finite number of edges is called a finite graph, otherwise it is an infinite graph.

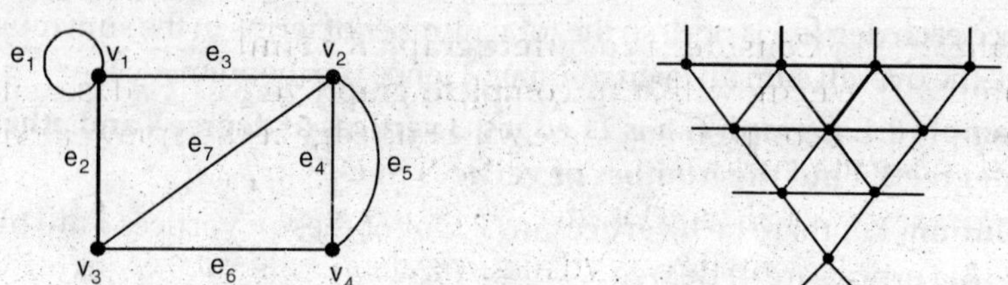

(*a*) Finite graph (*b*) Portion of infinite graph.

Fig. 8.7

8.5.3 Trivial Graph

A finite graph with one vertex and no edges is called a trivial graph.

8.5.4 Null Graph

A graph of order *n* and size zero is called a null graph. In other words a graph $G = (V, E)$ is called a null graph if $E(G) = \{\phi\}$ *i.e.*, empty.

Fig. 8.8. Null graph

8.5.5 Complete or Full Graph

If in a simple graph there exist an edge between each and every pair of vertices, then the graph is said to be a complete or full graph. A complete graph of *n* vertices is denoted by K_n.

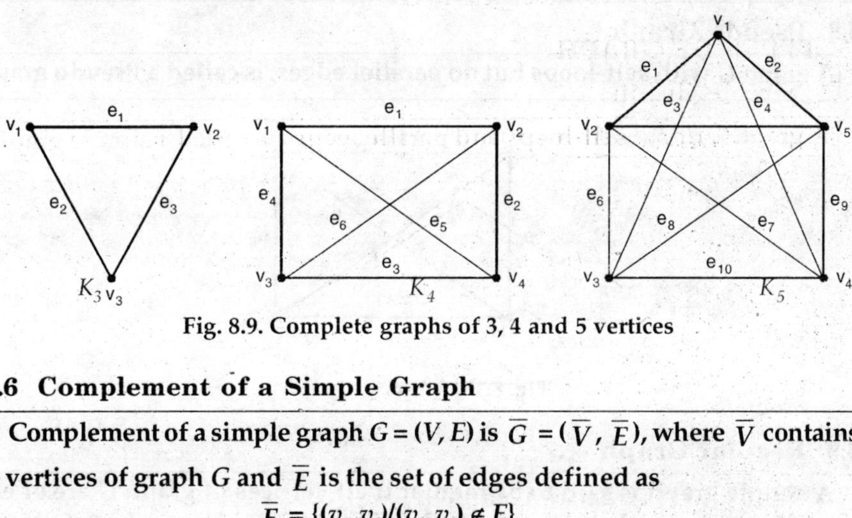

Fig. 8.9. Complete graphs of 3, 4 and 5 vertices

8.5.6 Complement of a Simple Graph

Complement of a simple graph $G = (V, E)$ is $\overline{G} = (\overline{V}, \overline{E})$, where $\overline{V}$ contains all the vertices of graph G and $\overline{E}$ is the set of edges defined as

$$\overline{E} = \{(v_i, v_j)/(v_i, v_j) \notin E\}$$

***i.e.*, if two vertices are not adjacent in G (there is no edge between them) they will be adjacent in $\overline{G}$.**

Example 8.3. Consider a complete graph K_3. Find $\overline{K_3}$.

Solution : We know that in complete graph vertex is ad2jacent to other vertices thus in it's complement, there will be no edge between any pair of vertices.

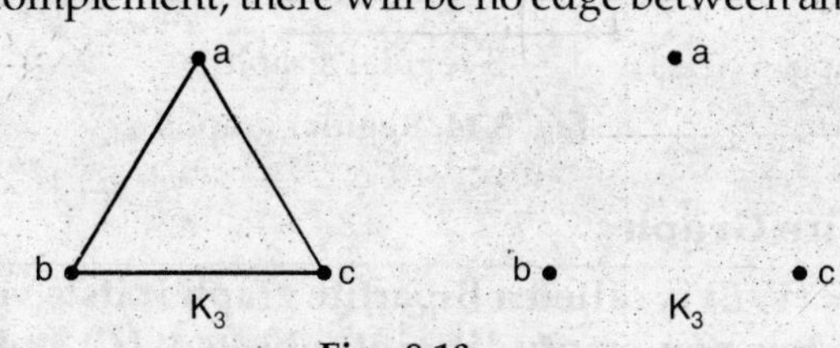

Fig. 8.10

Thus $\overline{K_3}$ will be a null graph.

Example 8.4. Find the complement $\overline{G}$ of the given simple graph G.

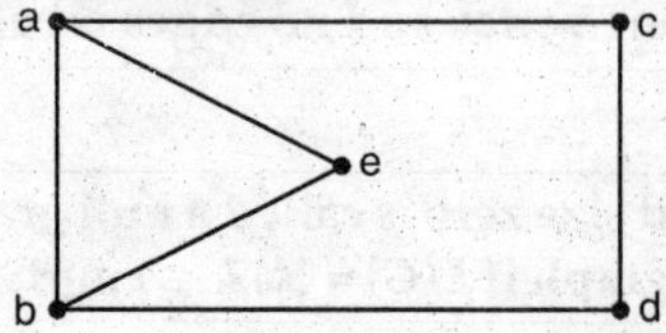

Fig. 8.11

Solution : Clearly, vertex a is not adjacent to d, vertex b is not adjacent to c and vertex e is not adjacent to c and d both. So $\overline{G} = (\overline{V}, \overline{E})$ will be

$$\overline{V} = \{a, b, c, d, e\}$$
$$E = \{(a, d), (b, c), (e, c), (e, d)\}$$

Thus,

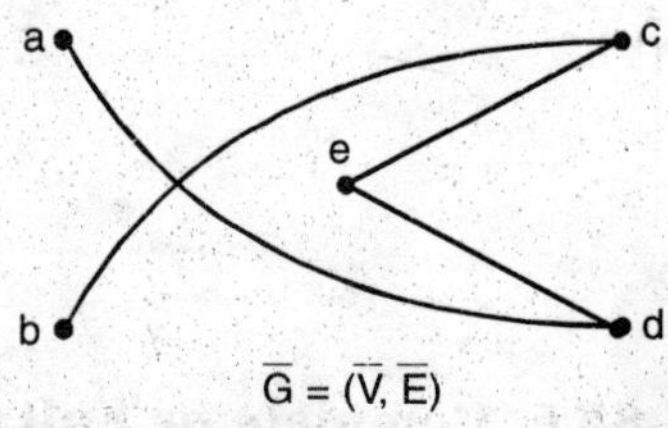

Fig. 8.12.

8.5.7 Multigraph

A graph G with some parallel edges but no loops, is called a multigraph. The best example of multi-graph is a road map.

8.5.8 Pseudo Graph

A graph G with self-loops but no parallel edges, is called a Pseudo graph.

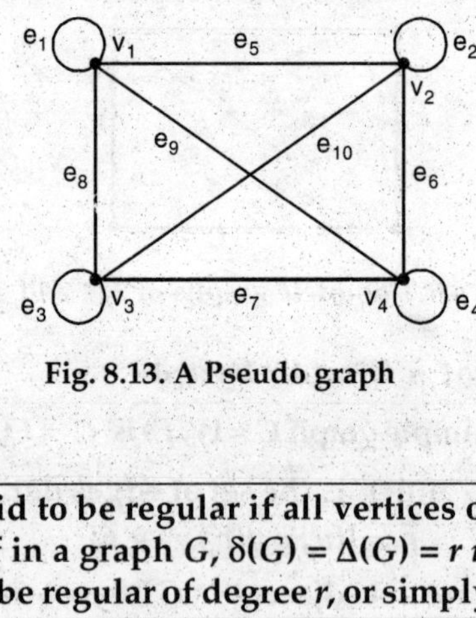

Fig. 8.13. A Pseudo graph

8.5.9 Regular Graph

A simple graph is said to be regular if all vertices of graph G are of equal degree. In other words if in a graph G, $\delta(G) = \Delta(G) = r$ *i.e.*, each vertex having degree r then G is said to be regular of degree r, or simply r-regular.

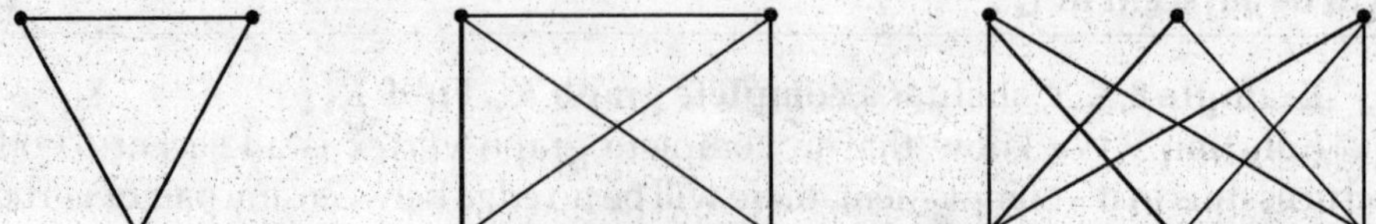

2-regular graph 3-regular graph 3-regular graph

Fig. 8.14. Regular graphs

8.5.10 Bipartite Graph

A graph $G = (V, E)$ is called a Bipartite graph if its's vertex set $V(G)$ can be partitioned into two non-empty disjoint subsets $V_1(G)$ and $V_2(G)$ in such a way

that each edge $e \in E(G)$ has it's one end point in $V_1(G)$ and other end point in $V_2(G)$. The partition $V = V_1 \cup V_2$ is called a bipartition of G.

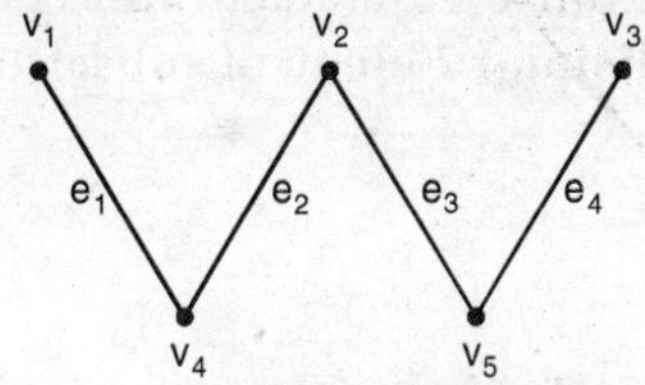

$$V_1(G) = [v_1, v_2, v_3\}$$
$$V_2(G) = \{v_4, v_5\}$$
$$V_1 \cap V_2 = \phi$$

Fig. 8.15. Bipartite graph

If each vertex of $V_1(G)$ is joined with each vertex $V_2(G)$, then the graph G is called **complete bipartite graph** and is denoted by $K_{m,n}$ when m and n are the number of vertices in $V_1(G)$ and $V_2(G)$ respectively

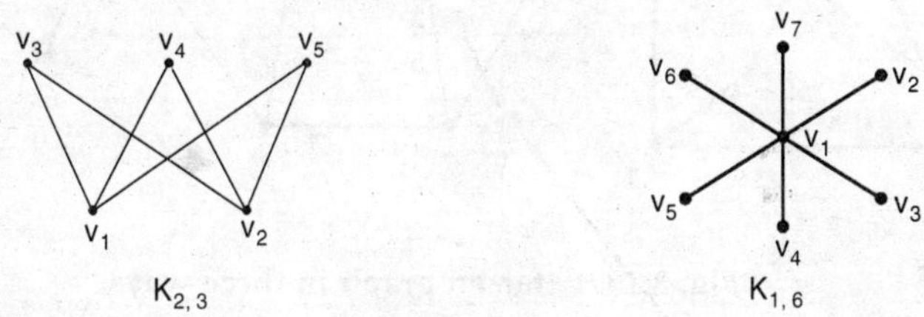

Fig 8.16.

$K_{1,n}$ is called a **Star** graph.

8.5.11 Labelled Graph

If the vertices and edges of a graph G are labelled with name or data then the graph is called a labelled graph. The difference between labelled and unlabelled graphs is more apparent when one tries to count them.

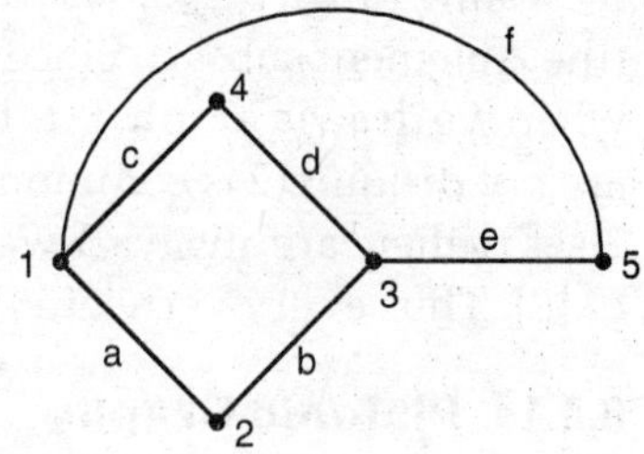

Fig. 8.17. Labelled graph.

8.5.12 Weighted Graph

When in a graph G, some additional information is given by assigning positive number called weight to the vertices and (or) edges, the graph is called weighted graph.

One example of such type of graph is the graph of railway tracks between different cities. The vertices represent cities and railway tracks presented by edges. The positive number assigned to each edge shows the length of a railway track between two cities and known as weight of the edge. The weighted graph is shown in **Fig. 8.18.**

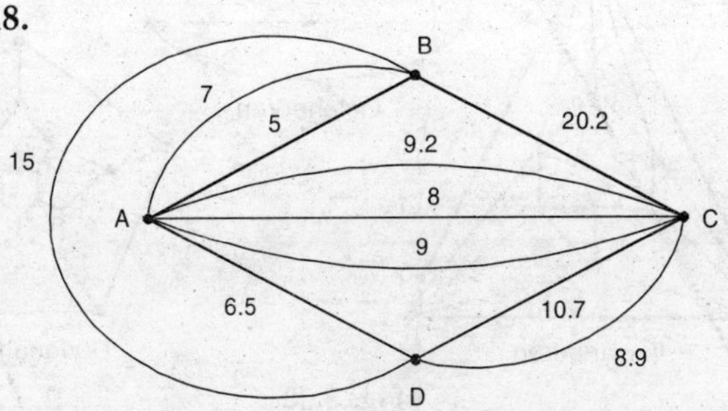

Fig. 8.18. Weighted graph representing the traffic density of different roads between different cities.

8.5.13 Peterson Graph

A simple graph *G* whose vertices are the **2-element subsets** of a 5-element set and whose edges are the pairs of **disjoint 2-element subsets** is called **Petersen Graph. (Fig. 8.19)**

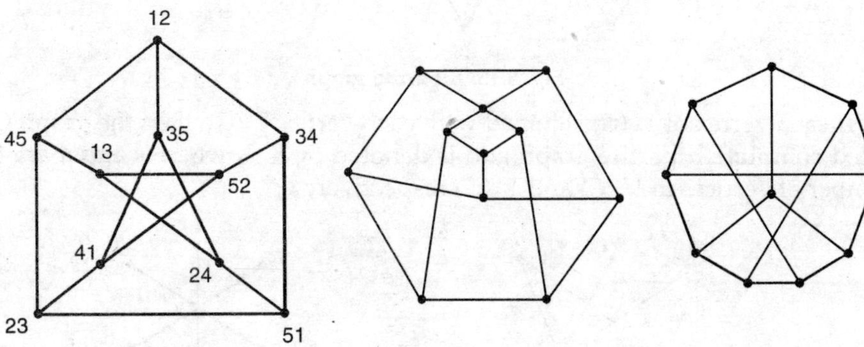

Fig. 8.19. Petersen graph in three ways.

Structure of the Peterson Graph

Using [5] = {1, 2, 3, 4, 5} as our 5-element set, write the pair (x, y) as xy or yx (the 2-element subset). Since 12 and 45 are disjoint so they are adjacent vertices when we draw a graph. But 12 and 24 are not adjacent in Petersen graph as they are not disjoint (2 is common in 12 and 24). One can easily observe that for each 2-set xy there are always 3-ways to pick a 2-set from the remaining three elements of [5]. Thus every vertex in Petersen graph has **degree 3**.

8.5.14 Platonic Graphs

In Fig. 8.20, the graphs of the edges of the five platonic solids are shown. These graphs are called Platonic graphs.

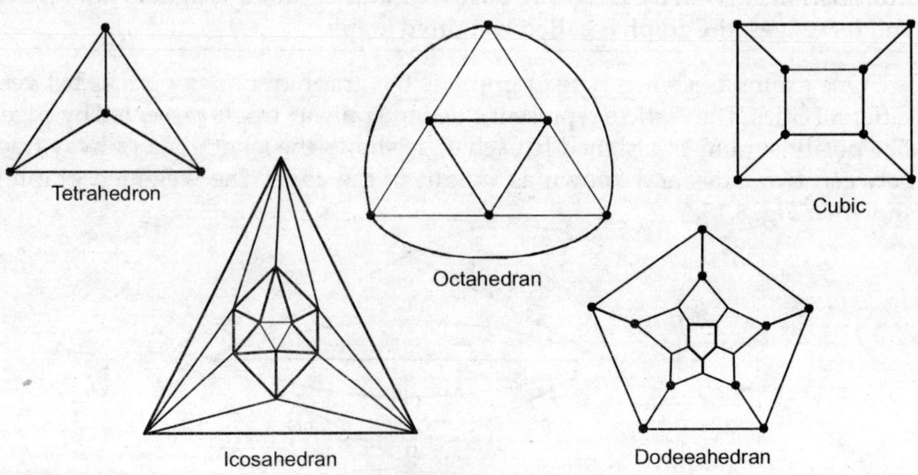

Fig. 8.20

Platonic graphs are formed by the vertices and edges of the five regular (platonic) solids. These solids are tetrahedron, octahedron cubic, dodeeahedron and icosahedron.

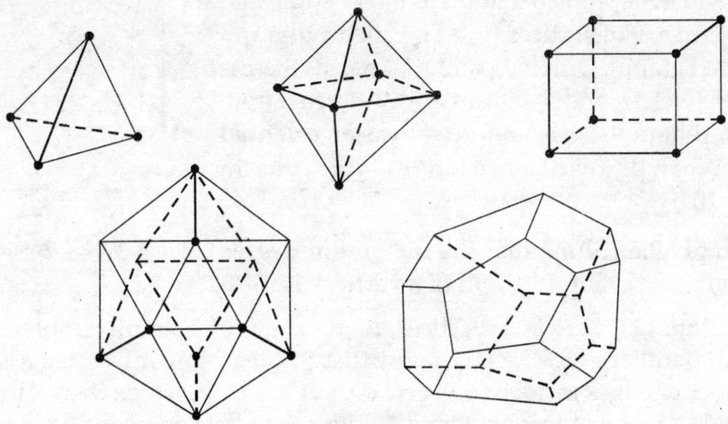

Fig. 8.21

8.5.15 Cycle Graph

A graph consist of n vertices ($n \geq 3$) $v_1, v_2, \ldots, v_n$ and edges (v_1, v_2), $(v_2, v_3) \ldots (v_{n-1}, v_n)$ and $(v_n - v_1)$ is called a cycle graph denoted by C_n.

In other words a 2-regular graph of n ($n \geq 3$) vertices is called a **cycle graph**.

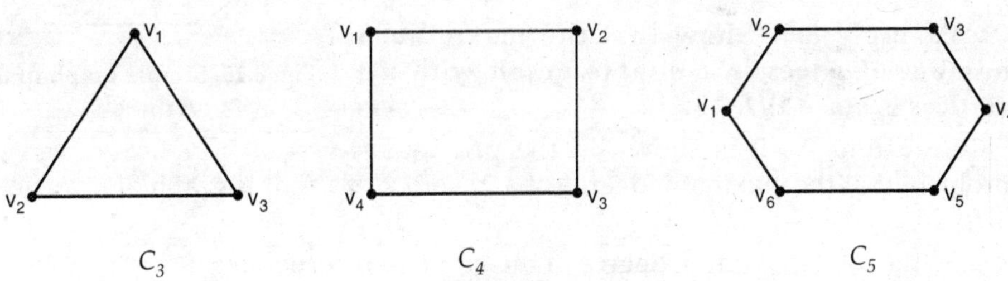

Fig. 8.22

It is clear that in cycle graph there will be equal number of vertices and edges.

8.5.16 Wheel Graph

A wheel graph W_n is obtained by introducing an additional vertex to the cycle graph C_{n-1}, ($n \geq 3$) and joining this new vertex to each of the ($n - 1$) vertices in C_{n-1} by new edges.

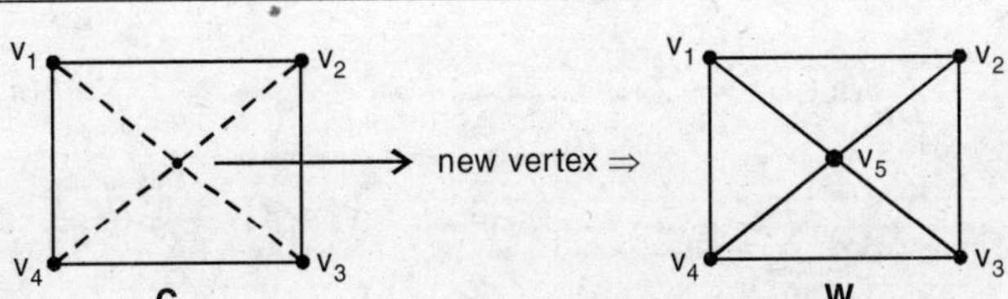

Fig. 8.23

8.5.17 Digraphs

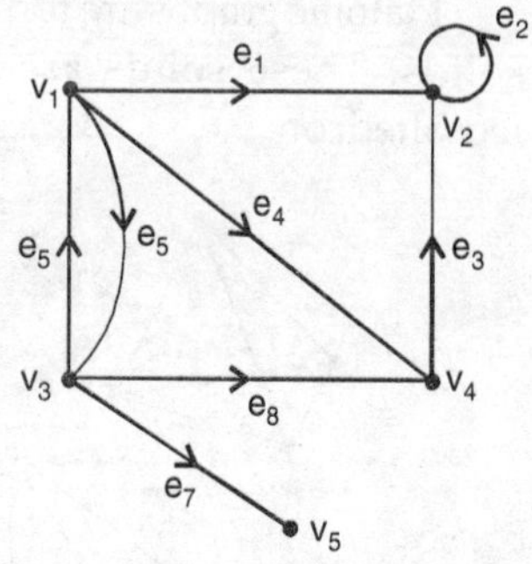

Fig. 8.24. Directed graph.

A graph G together with a set of vertices V, a set of edges E and a mapping δ that maps every edge onto some *ordered pair* of vertices (v_i, v_j) is called a **digraph** (or directed graph). Ordered pair (v_i, v_j) means there is an edge between v_i and v_j with an arrow directed from v_i to v_j. A diagraph is also referred to as an **oriented graph**. We will read more about digraphs in Chapter-10.

Example 8.5. Show that the maximum degree of any vertex in a simple graph of n vertices is $n - 1$.

Solution: Let there be n vertices, $v_1, v_2, \ldots, v_n$ of a simple graph G (no self-loops and parallel edges). As we know that the degree of any vertex v is equal to the number of edges incident on vertex v. Let $v_i \in G, i \leq n$ then there can be at the most $(n-1)$ edges incident on v_i (join v_i with rest of the vertices). So the maximum degree of $v_i = n - 1$ *i.e.*, the maximum degree of any vertex in a simple graph of n vertices is $n - 1$.

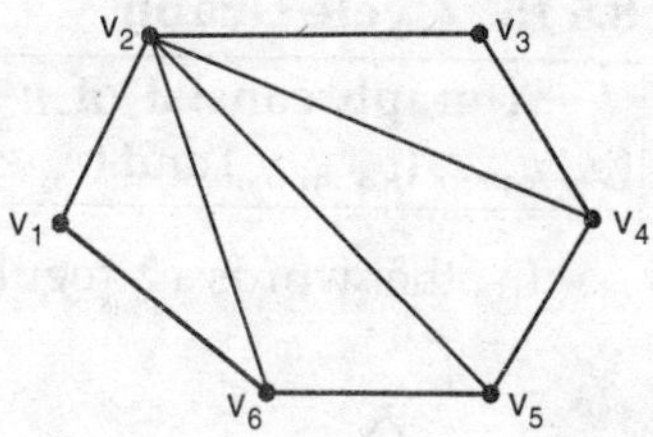

Fig. 8.25. Simple graph of 6 vertices

For example in graph G, given in Fig. 8.25.

v_2 is joined with each and every vertex of the graph and hence the maximum deg of $v_2 = 6 - 1 = 5$, whereas degree of $v_3 = 2$ which is not maximum as it is not joined with all the vertices of graph G.

Example 8.6. Show that the maximum number of edges in a simple graph with n vertices is $n(n - 1)/2$.

Solution: As it is shown in the previous problem that the maximum degree of any vertex in a simple graph of n vertices is $(n - 1)$.

So the total maximum degree of the graph of n vertices.

$$= n \, . \, (n-1) \qquad \ldots(1.4.1)$$

and hence the maximum number of edges $e = \dfrac{n(n-1)}{2}$. (by Theorem 1.1)

Example 8.7. Find the size of a k-regular graph.

Solution: Given that the graph is k-regular *i.e.*,

$$d(G) = \Delta(G) = k$$

In other words $d(v_i) = k$ for all $v_i \in V(G)$.

We know that

$$\sum_{i=1}^{n} \deg(v_i) = 2e \qquad (e \text{ –no. of edges})$$

So here, $$\delta(G) \sum_{i=1}^{n} k = 2 \, . \, e$$

$$\Rightarrow \qquad n \, . \, k = 2 \, . \, e \;\Rightarrow\; e = \frac{n(n-1)}{2}$$

Hence the size of G, *i.e.*, the number of edges in a k-regular graph is $\dfrac{n \, . \, k}{2}$.

8.6 SUBGRAPHS

A graph $g(v, e)$ is called a subgraph of $G = (V, E)$ if $v(g) \subseteq V(G)$ and $e(g) \subseteq E(G)$ and each edge of g has the same end vertices in g as in G.

For example, the graph in **Fig. 8.26**(*b*) is a **subgraph** of the graph in **Fig. 8.26**(*a*).

One can see easily that each edge of subgraph g having the same end vertices in g as in G. (edge 11, having f and g as end vertices in both the graphs).

Some Observations: From the definition of subgraph following observations can be made:

(*i*) Every graph is its own subgraph.

(*ii*) A single vertex in a graph G is a subgraph of G and a single edge together with its end vertices is also a subgraph of the graph G.

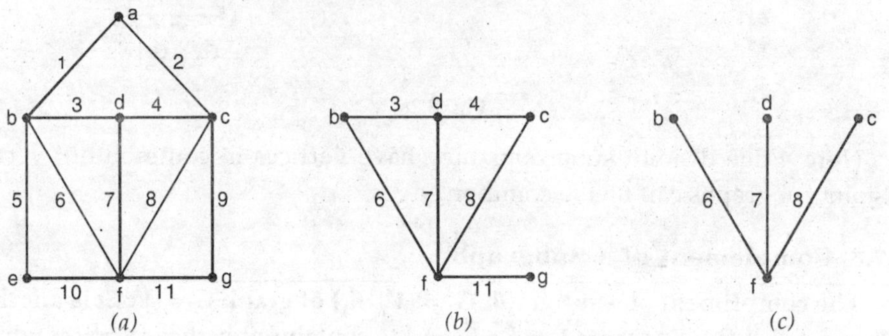

Fig. 8.26. Subgraph (g_1) of graph (g)

(*iii*) A subgraph g_1 of a subgraph g of G is a subgraph of G.

For example, **Fig. 8.26**(*c*) is a subgraph of the graph in **Fig. 8.26**(*b*), which is a subgraph of the graph shown in **Fig. 8.26**(*a*) and hence **Fig. 8.26**(*c*) is a subgraph of the graph shown in **Fig. 8.26**(*a*).

8.7 TYPES OF SUBGRAPH

There are two types of subgraphs.

8.7.1 Vertex Disjoint Subgraph

Any two subgraphs $G_1 = (V_1, E_1)$ and $G_2 = (V_2, E_2)$ of a graph $G_1 = (V, E)$ are said to be vertex disjoint if $V_1(G_1) \cap V_2(G_2) = \phi$, *i.e.*, there is no vertex common between G_1 and G_2.

The graphs shown in **Fig. 8.27** are the vertex disjoint subgraphs of the graph shown in **Fig. 8.26**(*a*).

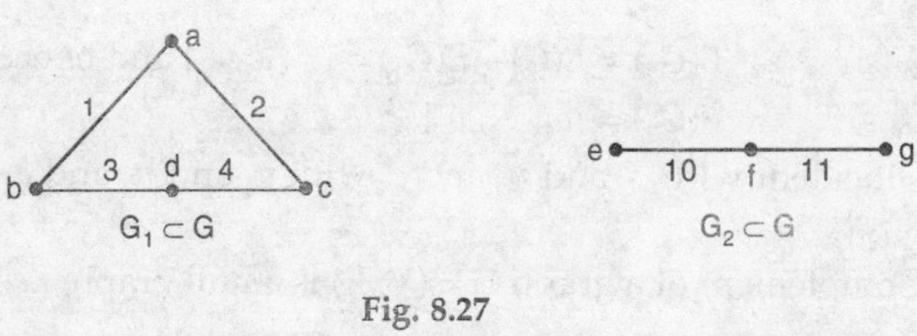

Fig. 8.27

8.7.2 Edge Disjoint Subgraphs

Any two subgraphs $G_1 = (V_1, E_1)$ and $G_2 = (V_2, E_2)$ of a graph $G = (V, E)$ are said to be edge disjoint if $E_1(G_1) \cap E_2(G_2) = \phi$, *i.e.*, G_1 and G_2 do not have nay edges in common between G_1 and G_2.

The graphs shown in **Fig. 8.28** are edge disjoint subgraphs of the graph shown in **Fig. 8.26.**

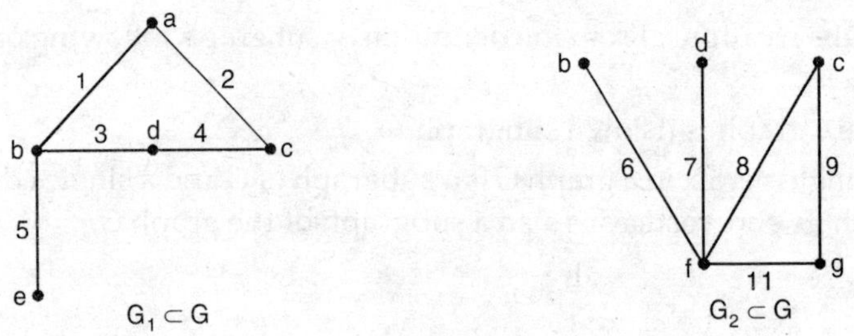

Fig. 8.28.

Note: Edge disjoint subgraphs may have vertices in common but vertex disjoint subgraphs can't have common edges.

8.7.3 Complement of a Subgraph

The compliment of a subgraph $G_1 = (V_1, E_1)$ of graph $G = (V, E)$ is another subgraph $G_2 = (V_2, E_2)$ where $E_2 = E - E_1$ and V_2 contains only those vertices which are incident with the edges in E_1.

For example, graph G_2 in **Fig. 8.29**(*c*) is the complement of the subgraph G_1 in **Fig. 8.29**(*b*). Graph *G is* shown in **Fig. 8.29**(*a*).

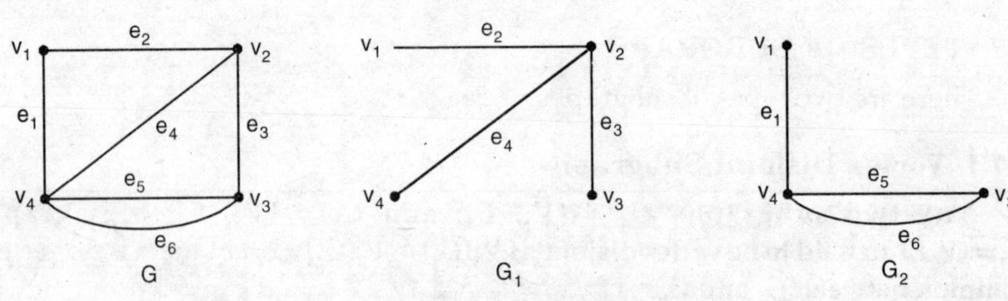

(*a*) (*b*) (*c*)

Fig. 8.29.

$$E(G_2) = E(G) - E(G_1) = \{e_1, e_5, e_6\} \text{ and hence}$$
$$V(G_2) = \{v_1, v_4, v_3\}$$

[$\because$ e_1 is incident with v_1 and v_4 and e_5 with v_4 and v_3 and e_6 is also incident with v_4 and v_3]

Note: Complement of a graph G = (V, E) is a null graph, *i.e.*, $G = (V, \{\phi\})$

8.7.4 Spanning Subgraph

A subgraph G_1 of a graph G is called **spanning subgraph** of G if G_1 contains all vertices of G. The graph G and its spanning subgraph G_1 is shown in **Fig. 8.30.**

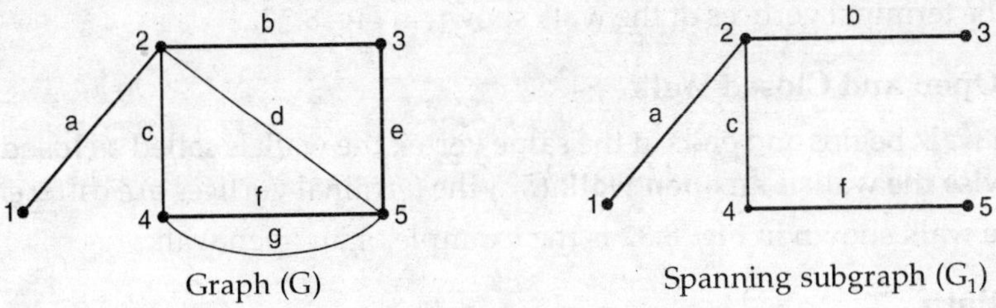

Fig. 8.30

8.7.5 Subgraph Introduced by a Set

Let $G = (V, E)$ be a graph and S be the set of vertices. Then the subgraph G_1 of G **induced** by set S is obtained by taking S as the set of vertices of graph G_1 and (*i.e.,* $V_1 = S$) the set of edges of G_1 contains all those edges of G that joins the vertices of set S in G.

Fig. 8.31 shows of subgraph G_1 of the graph G induced by the set $S = \{2, 3, 4, 6\}$.

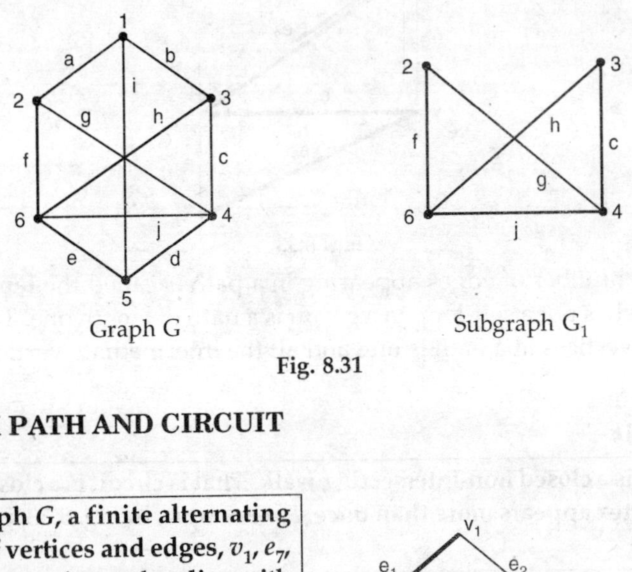

Fig. 8.31

8.8 WALK PATH AND CIRCUIT

8.8.1 Walk

In a graph G, a finite alternating sequence of vertices and edges, $v_1, e_7, v_2, e_2, \ldots e_m, v_k$ starting and ending with vertices such that each edge in the sequence is incident with the vertices following and preceding it, is called a walk. In a walk no edge appears more then once however a vertex may appear more than once.

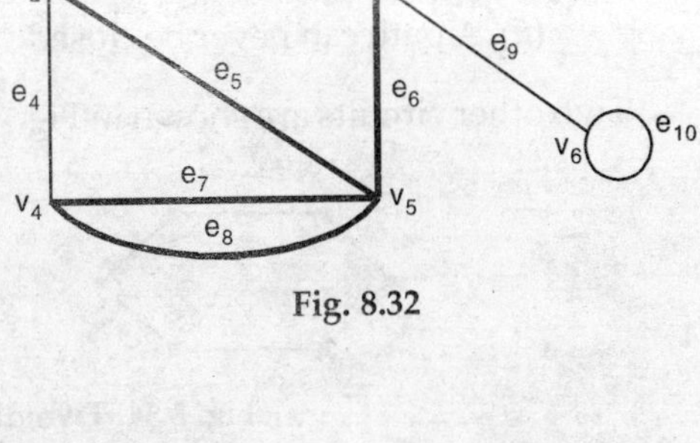

Fig. 8.32

For instance, the sequence $v_1, e_1, v_2, e_3, v_3, e_6, v_5, e_8, v_4, e_7, v_5, e_5, v_2$ is a walk shown with bold lines in **Fig. 8.32.**

There can be a number of such sequence in a given graph. Another sequence could be $v_4, e_7, v_5, e_6, v_3, e_9, v_6, e_{10}, v_6$.

Starting and ending vertices of a walk are called it's **terminal vertices.** v_1 and v_2 are the terminal vertices of the walk shown in **Fig. 8.32.**

8.8.2 Open and Closed Walk

If a walk begins and ends at the same vertex the walk is called a **closed walk.** Otherwise the walk is an **open walk** (*i.e.,* the terminal vertices are different).

The walk shown in **Fig. 8.32** is the example of an open walk.

8.8.3 Path

A path (or an elementary path) is an open walk in which no vertex appears mere than one. That is a path is an open, non-intersecting walk.

In **Fig. 8.33** sequence $v_1, e_3, v_4, e_6, v_3, e_7, v_5$ is a path but $v_1, e_3, v_4, e_4, v_2, e_5, v_2$ is not a path because in this sequence vertex v_2 is appearing twice Thus we cannot include a self-loop in a path.

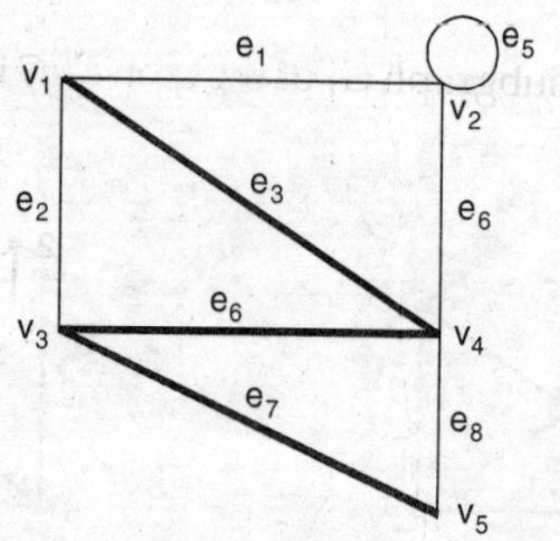

Fig. 8.33

The total number of edges appearing in a path is called the **length of path.** An edge which is not a self-loop in a graph is a path of length one. The **degree** of the terminal vertices in a path is **one** and all the intermediate vertices will have degree **two.**

8.8.4 Circuit

A circuit is a closed non-intersecting walk. That is circuit is a closed walk in which no vertex appears more than once. *For example,* in Fig. 8.33, v_2, e_5, v_2 is a circuit.

Notes: (*i*) Every self-loop is a circuit but not *vice-versa*

(*ii*) A path can never be closed

Two other circuits are shown in Fig. 8.34.

Fig. 8.34. Two different circuits

The degree of every vertex in the circuit is two. These degrees are contributed by the edges present in the circuit only, not by all the edges of the graph. (If circuit is a sub-graph of a graph).

We can summarize all the definitions of this section as shown in **Fig. 8.35**.

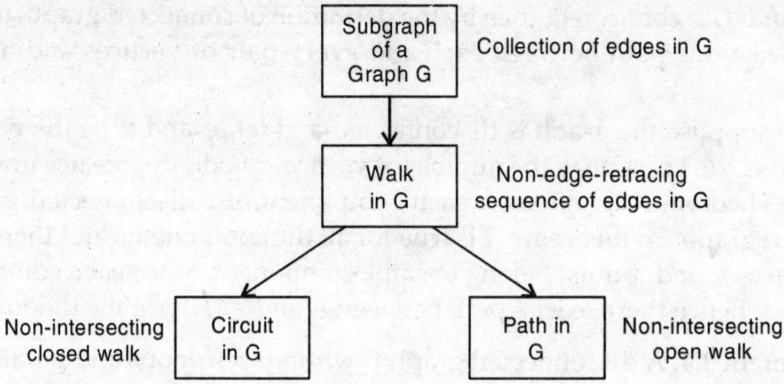

Fig. 8.35

8.9 CONNECTED GRAPHS, DISCONNECTED GRAPHS AND COMPONENTS

A graph G is said to be **connected** if for any pair of vertices (v_i, v_j) of G the two vertices are *reachables* from one another. Two vertices v_i and v_j are said to be **reachable** if there exist a path from vertex v_i to vertex v_j (or from v_j to v_i) in G. The minimum length of a path from v_i to v_j is called **geodesic**.

i.e., geodesic = min. (lengths of different paths between (v_i, v_j))

Definition: A graph G is said to be connected if there exist at least one path between each and every pair of vertices in G. Otherwise, the graph G is disconnected.

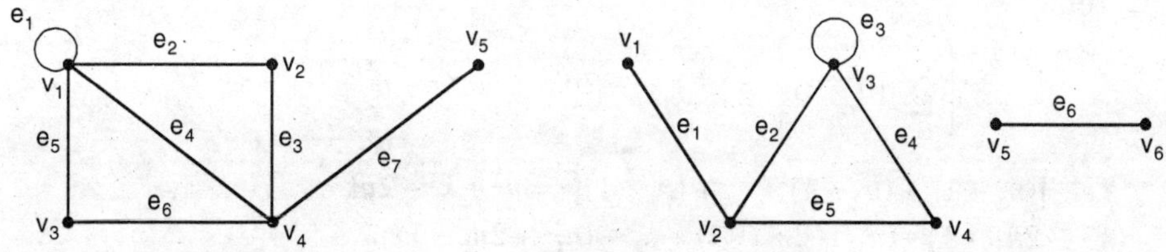

(*a*) Connected graph (*b*) Disconnected graph.

Fig. 8.36

The graph shown in **Fig. 8.36**(*a*) is connected as one can reach any vertex from any other vertex by travelling along the edges but in **Fig. 8.36**(*b*), the graph is disconnected because vertices v_5 and v_6 are not reachable by any other vertex of the graph.

One can easily see that the disconnected graph shown in **Fig. 8.36**(*b*) consists of two connected graphs. Each of these is a connected Subgraph of the graph G. So a disconnected graph consists of two or more connected graphs and each of the connected subgraph is called **component** of the disconnected graph.

The disconnected graph shown in **Fig. 8.36**(*b*) consists of two components.

A null graph of n vertices is a disconnected graph consists of n-components, each component contains one vertex and no edge.

Theorem 8.3. If a graph G (connected or disconnected) has exactly two vertices of odd degree, there must be a path joining the two vertices.

Proof: If G is connected, then by the definition of connected graph, there will exist at least one path between each and every pair of vertices and hence the theorem.

Now suppose the graph is disconnected and let v_i and v_j be the vertices of odd degree. We know that the number of vertices of odd degree in a graph must be even (Theorem 8.2) and since each component of a disconnected graph is a connected graph. So theorem 8.2 is true for all the components and therefore the two vertices v_i and v_j must belong to same component. Since each component is connected, hence there exist a path between v_i and v_j. Hence the theorem.

Theorem 8.4. A disconnected graph G without self-loops and parallel edges with n vertices and k components can have at most $\dfrac{(n-k)(n-k+1)}{2}$ edges.

Proof: Let $n_1, n_2, \ldots, n_k$ be the number of vertices in each of the k connected components of a disconnected graph G.

and $n_1 + n_2 + n_3 + \ldots + n_k = n,\ n_i \geq 1,\ i = 1$ to k

i.e., each component has at least one vertex. For proving the result, let us take one equality

$$\sum_{i=1}^{k} (n_i - 1) = n - k$$

$$\text{i.e.,} (n_1 - 1) + (n_2 - 1) + \ldots + (n_k - 1) = (n_1 + n_2 + \ldots + n_k) - (1 + 1 + \ldots + 1)\ k \text{ times}$$
$$= n - k$$

Squaring both sides,

$$\left(\sum_{i=1}^{k} (n_i - 1)\right)^2 = n^2 + k^2 - 2nk$$

$$\text{i.e.,}\ [(n_1 - 1) + (n_2 - 1) + \ldots + (n_k - 1)]^2 = n^2 + k^2 - 2nk$$

$$(n_1^2 - 2n_1 + 1) + (n_1^2 - 2n_2 + 1) + \ldots + (n_k^2 - 2n_k + 1)$$
$$+ \text{non-negative cross terms} = n^2 + k^2 - 2nk$$

or $$\sum_{i=1}^{n} (n_i^2 - 2n_i) + k + \text{non-negative cross terms} = n^2 + k^2 - 2nk$$

$$(\because n_i - 1 \geq 0,\ \forall\ i)$$

Therefore, $$\sum_{i=1}^{n} n_i^2 \leq n^2 + k^2 - 2nk + 2n - k = n^2 - (k-1)(2n - k) \quad \ldots (a)$$

Since we know that the maximum number of edges in a simple connected graph is given by $\dfrac{n(n-1)}{2}$, where n is the number of vertices in the graph. So here in

ith component the maximum number of edges are $\frac{n_i(n_i-1)}{2}$.

So the maximum no. of edges in G is

$$\frac{1}{2}\sum_{i=1}^{k} n_i(n_i-1) = \frac{1}{2}\sum_{i=1}^{k} n_i^2 - \frac{1}{2}\sum_{i=1}^{k} n_i$$

$$= \frac{1}{2}\sum_{i=1}^{n} n_i^2 - \frac{n}{2}$$

$$\leq (n^2-(k-1)(2n-k)) - \frac{n}{2} \qquad \text{(by eqn. (a))}$$

$$= \frac{1}{2}(n-k)(n-k+1)$$

Hence the theorem.

8.10 CUT-VERTEX AND BRIDGES

In a connected graph G a vertex v is called a cut-vertex if it's removal makes the graph G disconnected That is v is a cut-vertex of G if G-v is disconnected. Removal of a vertex v from a graph G also removes (or delete) all the edges incident on v.

An edge e in a connected graph G is called a bridge if the removal of e makes the graph G disconnected. In other words an edge $e \in G$, G is connected graph, called a bridge if $G-e$ is a disconnected graph.

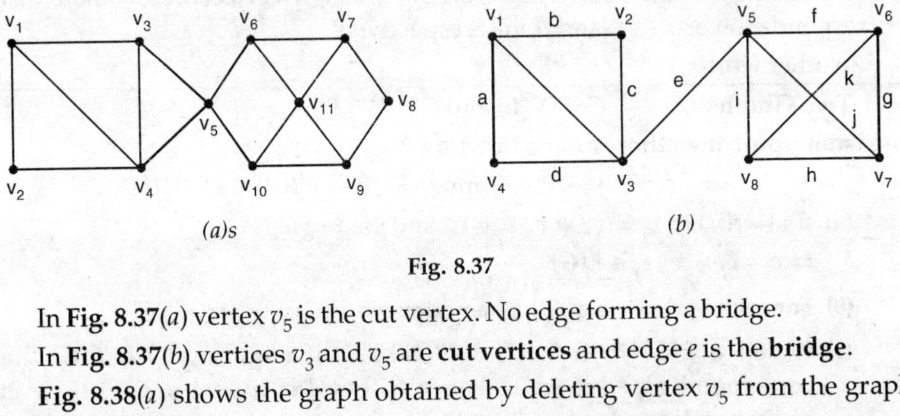

(a)s *(b)*

Fig. 8.37

In **Fig. 8.37**(*a*) vertex v_5 is the cut vertex. No edge forming a bridge.

In **Fig. 8.37**(*b*) vertices v_3 and v_5 are **cut vertices** and edge e is the **bridge**.

Fig. 8.38(*a*) shows the graph obtained by deleting vertex v_5 from the graph shown in **Fig. 8.37**(*a*).

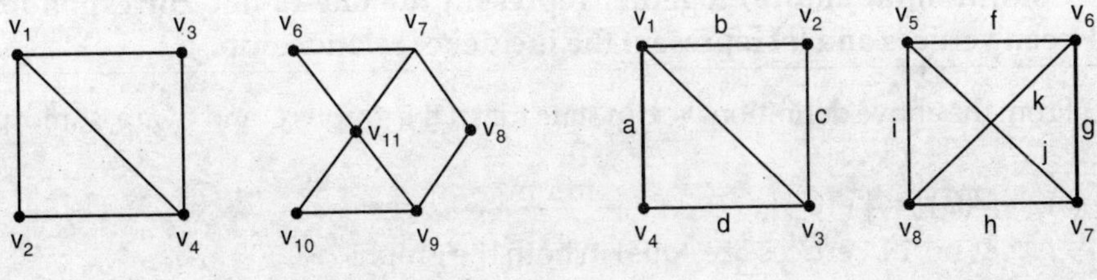

(a) $(G-v_5)$ *(b)* $(G-e)$

Fig. 8.38

8.11 DISTANCE AND DIAMETER

In a connected graph $G = (V, E)$ the distance between two vertices v_i and v_j is the length of the shortest path between v_i and v_j and diameter is the maximum distance between any two vertices.

Distance between v_i and v_j is denoted by $d(v_i, v_j)$ where as diameter of a graph is denoted as dia (G), and defined as

$$\text{dia}(G) = \max d(v_i, v_j),\ \forall\ v_i, v_j \in V$$

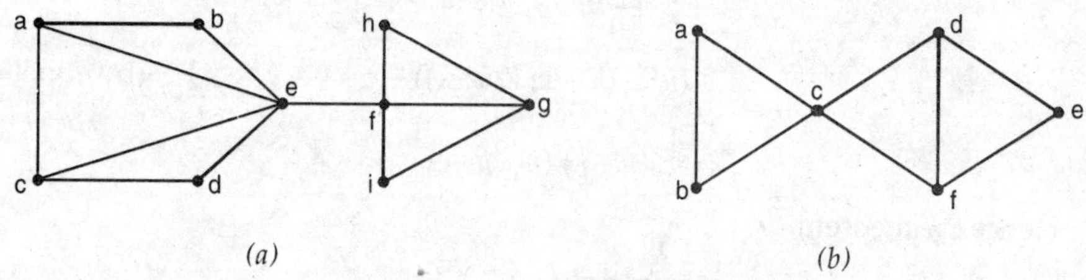

Fig. 8.39

The distance between vertices a and f in **Fig. 8.39**(a) is 2 and dia $(G) = 6$. In **Fig. 8.39**(b), the dia $(G) = 4$.

8.12 ISOMORPHISM

Two graphs $G = (V, E)$ and $G' = (V', E')$ are said to be **isomorphic** with each other if there is a **one-to-one** correspondence between there vertices and edges such that if e is an edge between vertices v_i and v_j in G then there exist an edge e', corresponding to edge e, between vertices v_i' and v_j' in G'. Vertices v_i' and v_j' in G' corresponds to vertices v_i and v_j in G respectively.

In other words,

Two Graphs $G = (V, E)$ and $G' = (V', E')$ are isomorphic then there exist a function.

$$f : V(G) \to V'(G') \text{ and } g : E(G) \to E'(G') \text{ such that}$$

(a) $f(v_i) = f(v_j) \Rightarrow v_i = v_j,\ \forall\ v_i, v_j \in G$, and $g(e_i) = g(e_j)$ $\Rightarrow e_i = e_j\ \forall\ e_i, e_j \in E(G)$

(b) **For each vertex $u \in v(G')$, $\exists$ a vertex $v \in V(G)$.**

(c) $(v_i, v_j) \in E(G) \Leftrightarrow (v_i', v_j') \in E'(G')$ **i.e., if edge** $(v_i, v_j) \in E(G)$ **then corresponding edge** $(v_i', v_j') \in E'(G')$ **and if** $(v_i, v_j) \notin E(G)$ **then the corresponding edge** $(v_i', v_j') \notin E(G')$.

Condition (a) and (b) together represent the one-to-one correspondence between vertices and (c) represent the incidence relationship.

From the above definition we can state that if the graphs G and G' are isomorphic then

(a) $|V(G)| = |V(G')|$
i.e., no. of vertices are equal in both the graphs.

(b) $|E(G)| = |E(G')|$, no. of edges are equal in both the graphs.

(c) Both the graphs will have an equal no. of vertices with a specific degree.

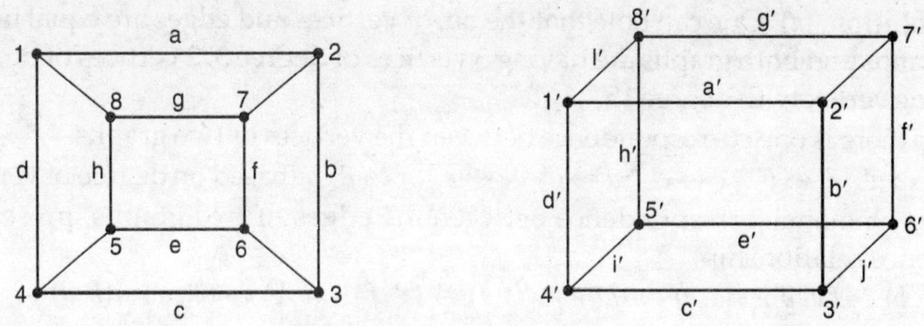

(*a*) Two isomorphic graphs.

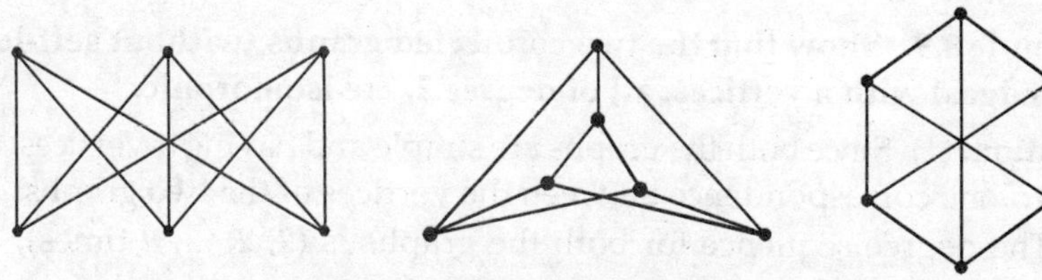

(*b*) Three isomorphic graphs.

Fig. 8.40

The graphs shown in **Fig. 8.40**(*a*) and (*b*) are isomorphic with each order.

From the above discussion one can see that isomorphic graphs are the same graphs except the labelling of vertices and edges, drawn differently. But it is not always easy to determine the isomorphism between given graphs.

The graphs shown in **Fig. 8.41** appears to be isomorphic and also they have same number of vertices and edges but still they are not **isomorphic**.

Fig. 8.41

Because, in **Fig. 8.41**(*a*) there are two pendant vertices 1 and 3 adjacent to 2 while in **Fig. 8.41**(*b*) there is only one pendant vertex *a* adjacent to *b*. Vertices 2 and *b* corresponds to each other because they have equal degree in both the graphs.

Criterion for checking whether the two graphs are isomorphic or not is still an unsolved problem in graph theory.

Example 8.8. Show that the following graphs are isomorphic to each other.

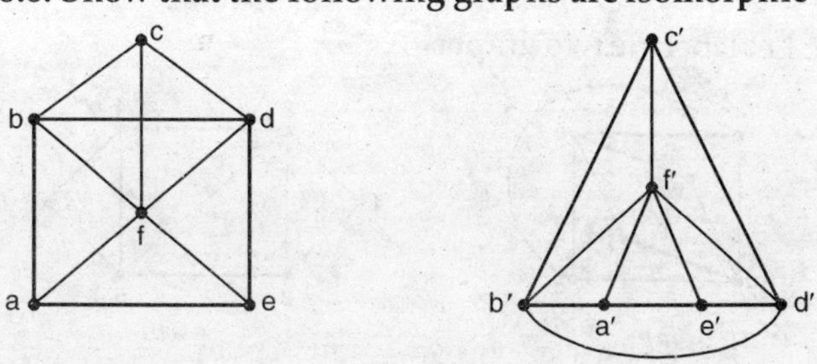

Fig. 8.42

Solution: (*a*) One can note that the no. of vertices and edges are equal in both the graphs and both graphs are having 3 vertices of degree 3, 2 vertices of degree 4 and one vertex with degree 15.

So there is one correspondence between the vertices of two graphs

$a \leftrightarrow a' \quad b \leftrightarrow b' \quad c \leftrightarrow c' \quad d \leftrightarrow d' \quad e \leftrightarrow e' \quad f \leftrightarrow f'$ (based on degree of vertices)

(*b*) One-one correspondence between the edges of two graphs, preserving incidence relationship,

$(a, b) \leftrightarrow (a', b') \quad (a, e) \leftrightarrow (a', e') \quad (a, f) \leftrightarrow (a', f') \quad (b, c) \leftrightarrow (b', c') \quad (b, d) \leftrightarrow (b', d')$ and so.

So the given graphs are isomorphic.

Example 8.9. Show that the two connected graphs (without self-loop and parallel edges) with *n* vertices, all of degree 2, are isomorphic.

Solution. (1) Since both the graphs are simple and having n-vertices, so there is a one-to-one correspondence between the vertices of the two graphs.

(2) The degree sequence for both the graphs is (2, 2, ..., n times), and the graph is connected.

By the above two points one can conclude that the two graphs are cycle graphs of n-vertices.

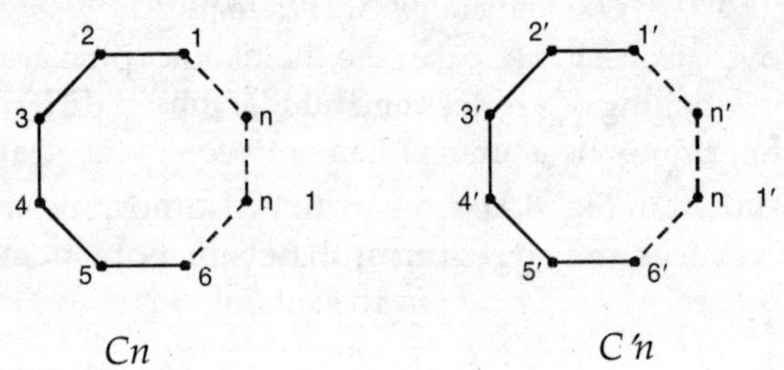

Fig. 8.43

Hence, the two graphs are isomorphic.

Example 8.10. Check whether the following graphs are isomorphic?

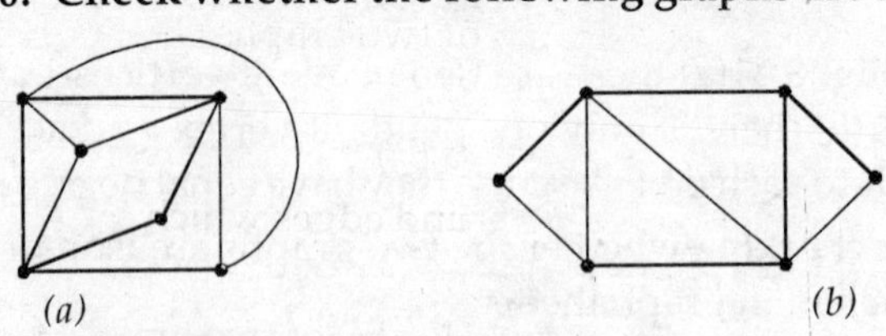

Fig. 8.44

Solution: Let label the two graphs

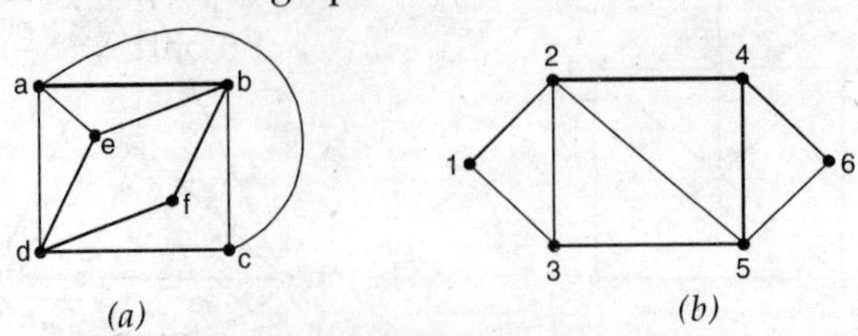

Fig. 8.45

Though the number of vertices in both the graphs are same but the given graphs are not isomorphic because

(*i*) graph (*a*) has 10 edges while graph (*b*) has 9

(*ii*) in graph (*a*) there are 3 vertices of degree 4 (*a*, *d* and *b*) and one vertex of deg. 2 (*f*) but in graph (*b*) there are only two vertices of degree 4 (2 and 5) and two vertices of degree 2 (1 and 6).

8.13 HOMEOMORPHIC GRAPHS

Two graphs G_1 and G_2 are said to be **homeomorphic** if one graph can be obtained from the other by merging the edges in series or by inserting a vertex of degree two in any edge of the graph i.e., creating edges in series.

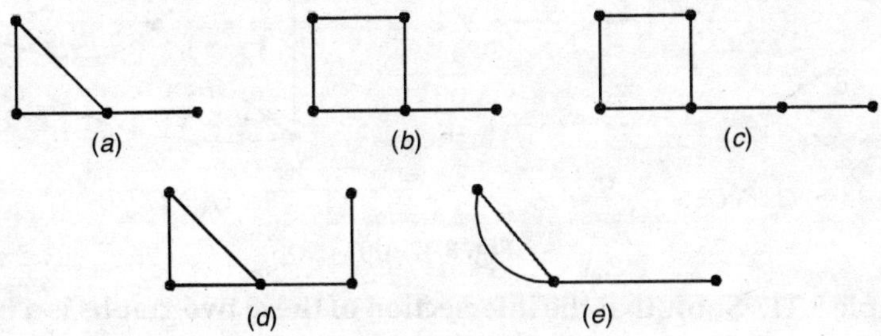

Fig. 8.46. Homeomorphic graphs

8.14 OPERATIONS ON GRAPHS

Since we are defining graphs in terms of sets of vertices and edges, so we can employ set-theoretic operations on the sets of vertices and edge to define operations between graphs.

(*i*) **Union:** Let $G_1 = (V_1, E_1)$ and $G_2 = (V_2, E_2)$ be two graphs then their union is another graphs $G_3 = (V_3, E_3)$ can be defined as

$$G_3 = G_1 \cup G_2 = (V_1 \cup V_2, E_1 \cup E_2)$$

(*ii*) **Intersection:** The intersection of two graphs $G_1 = (V_1, E_1)$, $G_2 = (V_2, E_2)$ is another graph $G_3 = (V_3, E_3)$ defined as

$$G_3 = (V_1 \cap V_2, E_1 \cap E_2)$$

i.e., G_3 contains only those vertices and edges which are common in both the graphs G_1 and G_2.

(*iii*) **Ring Sum:** The ring sum $G_3 = (V_3, E_3)$ of two graphs $G_1 = (V_1, E_1)$ and $G_2 = (V_2, E_2)$ is defined as

$$G_3 = G_1 \oplus G_2 = (V_1 \cup V_2, (E_1 \cup E_2) - (E_1 \cap E_2))$$

i.e., G_3 contains all the vertices of G_1 and G_2 and only those edges which are either in G_1 or in G_2 but not in both.

Note: For a graph G, $G \cup G = G$, $G \cap G = G$ and $G \oplus G =$ null graph.

Fig. 8.47 shows the two graphs. Their union is shown in **Fig. 8.47**(*a*), intersection in **Fig. 8.47**(*b*) and Ring sum in **Fig. 8.47**(*c*).

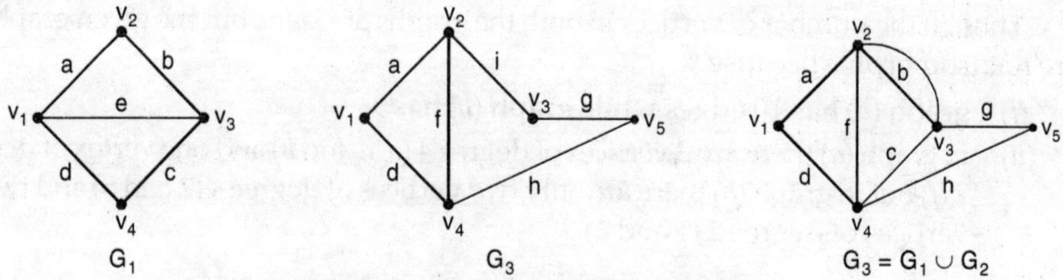

Fig. 8.47 (a) Union of G_1 and G_2

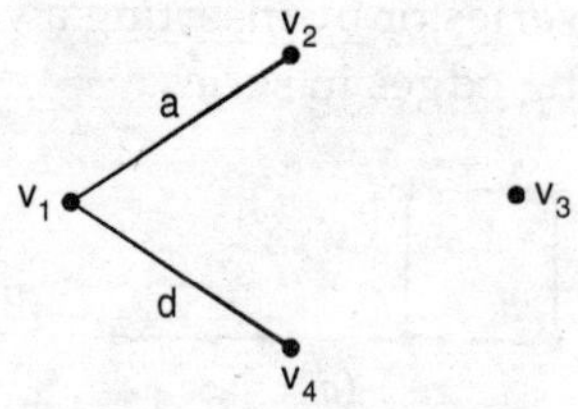

(b) $G_1 \cap G_2 = G_3$

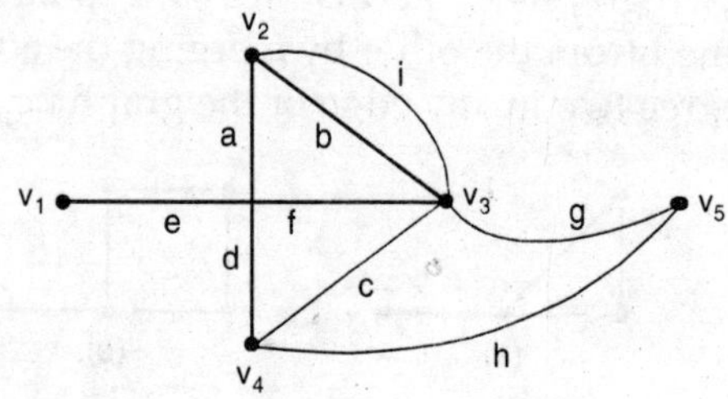

(c) $G_3 = G_1 \oplus G_2$

Fig. 8.47. ***(b)***

Example 8.11. Show that the intersection of these two graphs is a null graph.

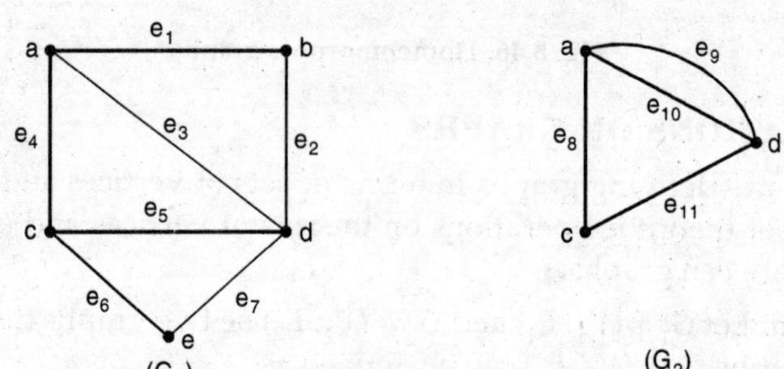

Fig. 8.48.

Solution: Let $G_3 = G_1 \cap G_2 = (V_3, E_3)$

then $V_3 = V_1 \cap V_2 = (a, b, c, d, e) \cap (a, c, d)$

$= (a, c, d)$

and $E_3 = E_1 \cap E_2 = (e_1, e_2, e_3, e_4, e_5, e_6, e_7) \cap (e_8, e_9, e_{10}, e_{11})$

$= \phi$ (*i.e.*, no edge in common)

i.e., G_1 and G_2 are edge disjoint graphs.

So $G_3 = ((a, c, d), \{\phi\})$

Graph of G_3 is shown in **Fig. 8.46.**

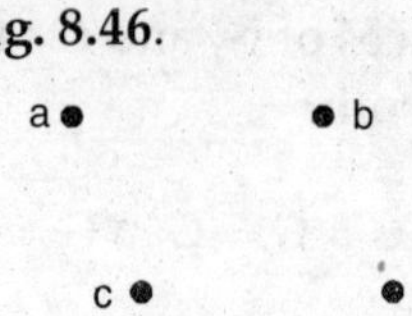

Fig. 8.49. Graph G_3, a null graph

Hence the intersection of 2 edge disjoint graphs is a null graph.

SUMMARY

In this chapter some basic concepts of graph theory have been introduced and some elementary results have been obtained. Also we have discussed subgraphs– a graph is part of another graph, walk path and circuit in graph. We have discussed the property of isomorphism in the graphs and also some operations on the graph.

Some special types of graphs such as peterson's graph and platonic graphs are also discussed.

EXERCISES

1. Define: (*i*) multigraph, (*ii*) wheel graph, (*iii*) Peterson's graph (*iv*) Bipertite graph (*v*) Tripertile graph.
2. Prove, if every vertex of a simple, graph G has degree 2, then G is a cycle.
3. Determine whether the peterson graph is bipertite?
4. Consider the graph G given below.

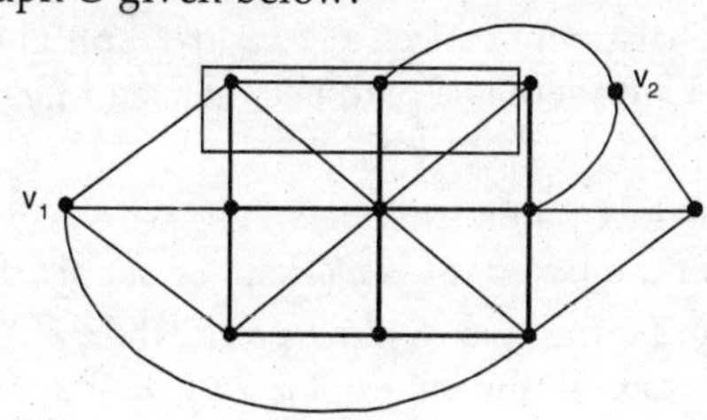

Graph G

Fig. 8.50.

(*i*) Find all the walks between v_1 and v_2.

(*ii*) Find all the paths between v_1 and v_2. Calculate their lengths also.

(*iii*) Count the number of cycles of length 3, 4 and 5 in the graph G.

5. Construct the wheel graphs W_7 and W_9. Also calculate the size of W_n.
6. Construct the following graphs:
   (*i*) K_7 (*ii*) $K_{3, 5}$ (*iii*) $K_{3, 3, 3}$.
7. Find out the value of n for which the following graphs are regular.
   (*i*) W_n (*ii*) $K_{m, n}$ (*iii*) K_n
8. Draw all the non-isomorphic graphs (simple) of 6-vertices, and 7-vertices.
9. Consider a connected graph G. Prive that G-e is connected if e is an edge in a circuit C in G.
10. Check whether the following graphs are connected. If not find out the number of components.

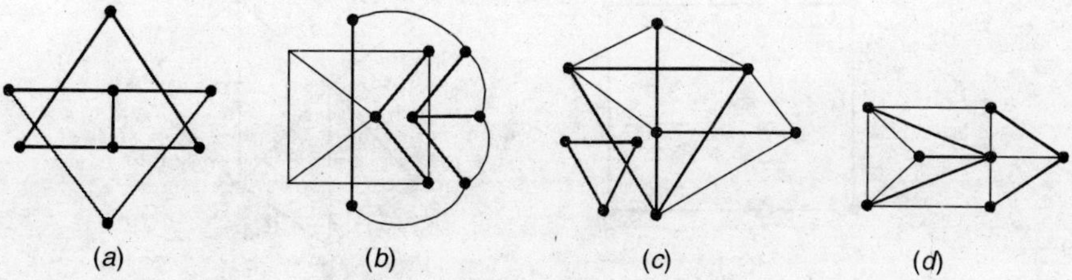

(*a*) (*b*) (*c*) (*d*)

Fig. 8.51.

11. Show that a graph G is self complementary if it has $4n$ of $4n+1$ vertices, n being a positive integer.

12. Consider a graph G with 12 edges. If G has 6 vertices of degree 3 and the other vertices have degree less than 3, then calculate the minimum number of vertices G can have.

13. Draw any two self-complementary graphs.

14. Prove that in a graph G if $\delta(G) = \frac{n-1}{2}$, where n is the number of vertices in G, then G is connected.

15. Consider a graph G (non-directed) or order 13 such that each vertex has degree 7 or 8. Prove that there are at least 8 vertices of degree 7 or 7 vertices of degree 8.

16. Calculate the number of edges in the following graph G if G has

(*i*) 16 vertices, each of degree 2.

(*ii*) 3 vertices of degree 4, 2 vertices of degree 3 and other 4 vertices of degreee one.

17. Draw all the edge-disjoint subgraphs of (*i*) K_3 (*ii*) $K_{2,3}$ (*iii*) W_5.

18. Consider the three distinct vertices x, y, z in graph G.Prove that there exist a path between z and x if there is a path between x and y and also a path between y and z.

19. If there are two distinct paths exist between two vertices v_i and v_j, then show that the ring sum of the two path is a circuit or set of circuits in graph G.

[Hint: Let p_1 and p_2 be the two distinct paths, then $p_1 \oplus p_2$ will contain all the vertices of p_1 and p_2 and all the edges of p_1 and p_2. If p_1 and p_2 are vertex disjoint (except v_i and v_j) then $p_1 \oplus p_2$ is a circuit. If v_k is common between p_1 and p_2 then $p_1 \oplus p_2$ contains two circuits one is between v_i and v_k another between v_k and v_j**]**.

20. Let v be a cut-vertex of a simple connected graph G (i.e., removal of v from G, disconnectes G). Prove that $\overline{G} - v$ is connected, where $\overline{G}$ is the complement of G.

21. Show that the there exist no cycle of odd length in bipertitle graphs.

22. Determine whether the statements below are true or false:

(*i*) Every disconnected graph has an isolated vertex.

(*ii*) A graph is connected iff some vertex is connected to all other vertices.

(*iii*) Every u, v-walk (i.e., walk between vertices u and v) contains a u, v-path.

(*iv*) Removal of cut-edge in a graph G increases the number of components of G.

(*v*) Every pendent vertex is a cut-vertex.

23. Count the cycles of length n in K_n, and also count the cycles of length $2n$ in $K_{n,n}$.

24. For the two graphs G_1 and G_2 given in Fig. 8.52 obtain.

(*i*) $G_1 \cup G_2$ (*ii*) $G_1 \cap G_2$ and (*iii*) $G_1 \oplus G_2$.

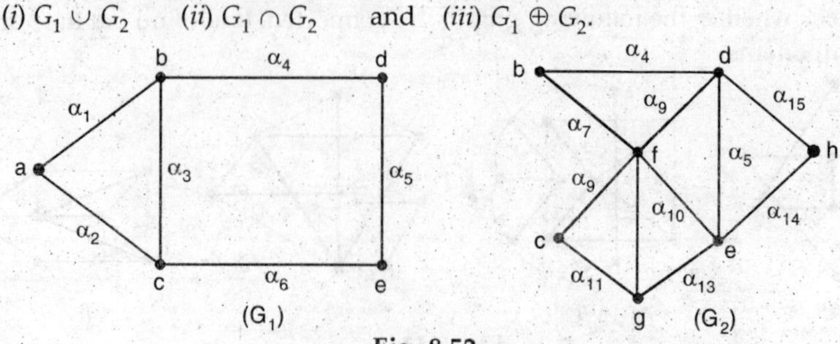

Fig. 8.52

25. Draw the complement of the graph G given in **Fig. 8.53.**

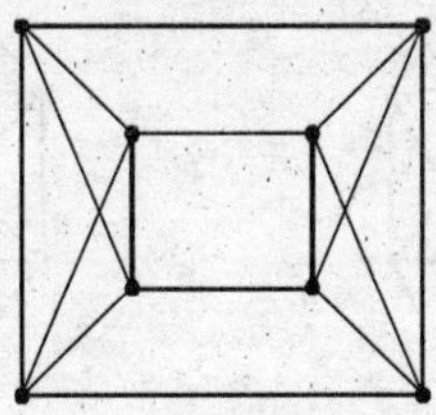

Fig. 8.53. Graph G

26. Determine which pairs of graphs below are isomorphic.

(*a*)

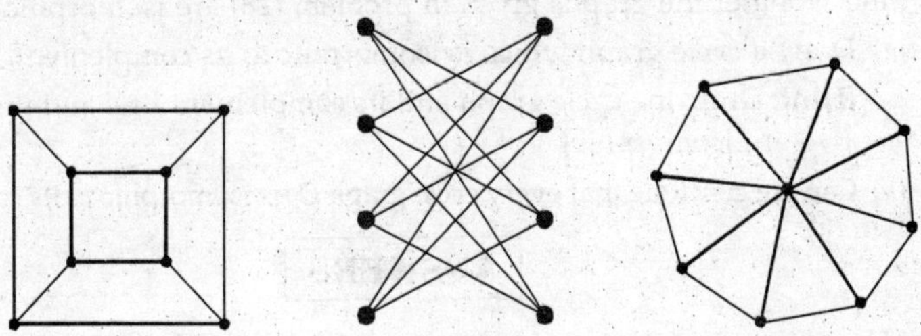

(*i*) (*ii*) (*iii*)

Fig. 8.54.

(*b*)

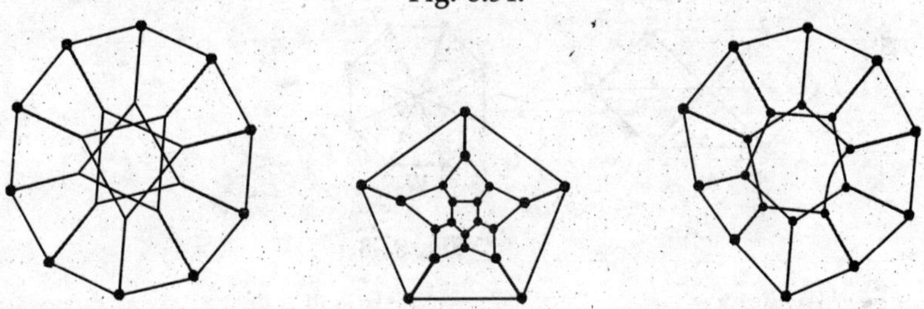

(*i*) (*ii*) (*iii*)

Fig. 8.55.

27. Show that the graphs below are all isomorphic. Also prove that they all are drawings of peterson graph.

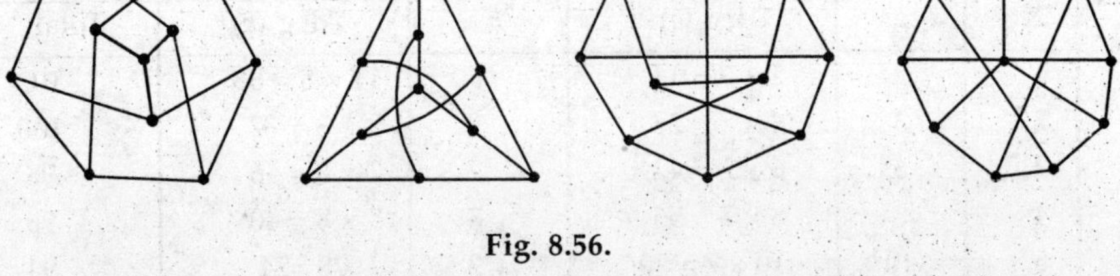

Fig. 8.56.

{**Hint:** Use definition of peterson graph. Also if two vertices are nonadjacent in peterson graph, they have exactly one common neighbour].

28. Determine whether the graphs below are bipertite.

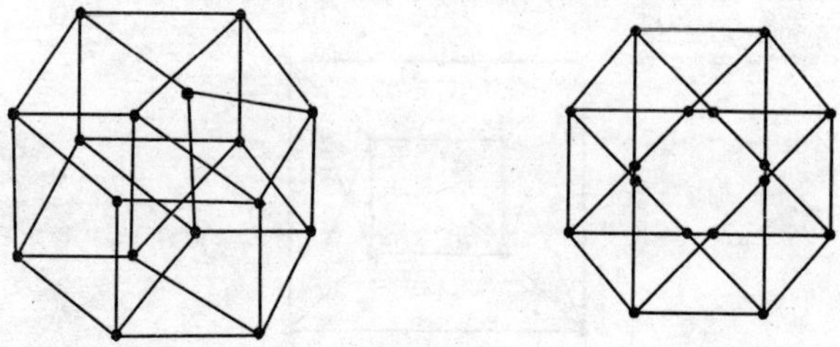

(*a*) (*b*)

Fig. 8.57.

29. Find whether the graphs given in problem (28) are isomorphioc?

30. (*a*) Draw a cycle graph which is isomorphic to its complement.

[**Hint:** Draw the cycle graph and its complement first and then show that the two are isomorphic]

(*b*) Can we conclude that every cycle graph C_n is isomorphic to its complement $\overline{C}_n$?

ANSWERS

1. No, peterson graph is not bipertite as it has a cycle of odd length, i.e. 5 and there is no circuit of odd length in bipertite graph.

5.

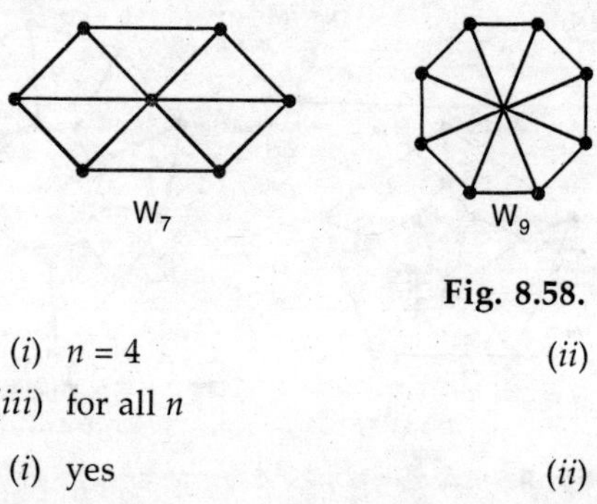

W_7 W_9

Fig. 8.58.

7. (*i*) $n = 4$ (*ii*) $m = n$

(*iii*) for all n

10. (*i*) yes (*ii*) No, two components

(*iii*) No, three components (*iv*) Yes

12. $n \geq 9$

15. Order = $|V| = 13$. *a* vertices of degree 7 and *b* vertices of degree 8. So a is always an even number.

No.	a	Σdeg (*a*)	*b*	Σdeg (*b*)	Total
1.	2	2 × 7 = 14	11	11 × 8 = 88	102
2.	3	4 × 7 = 28	9	9 × 8 = 72	100
3.	6	6 × 7 = 42	7	7 × 8 = 56	98
4.	8	8 × 7 = 56	5	5 × 8 = 40	96
5.	10	10 × 7 = 70	3	3 × 8 =24	94
6	12	12 × 7 = 84	1	1 × 8 = 8	92

So from (1), (2) and (3) we have at least 7 vertices of degree 8 and from (4), (5) and (6) we have at least 8 vertices of degree 7.

16. (*i*) $e = 16$

(*ii*) $e = 10$

17. *K*

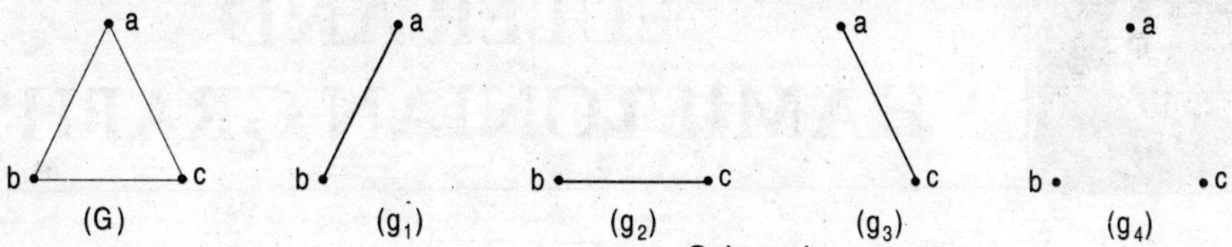

Subgraphs

22. (*i*) false (*ii*) True (*iii*) True (*iv*) True (*v*) False

26. (*a*) yes (*b*) No.

■■■

Chapter

9 EULER AND HAMILTONIAN GRAPHS

9.1 INTRODUCTION

In this chapter we are going to discuss the two very important graphs, Euler and Hamiltonian, and their properties. As mentioned in chapter-8 that Euler is the father of graph theory. He solved the very famous Königsberg bridge problem in 1736 in his very first paper of graph theory. In the same paper he introduced and solved a more general problem: In what type of graph G is it possible to trace a closed walk containing all the edges of graph G exactly once and named those graphs as Euler graphs.

The word **Hamiltonian** is derived from the Irish mathematician **Sir William Rowan Hamilton**. He invented a game consisting of a solid regular dodecahedron in 1857 and the concept is used in solving the very famous **travelling salesman problem**.

9.2 EULER GRAPH

In a graph G, if a closed walk contains all the edges of the graph G, then the walk is called an Euler line or Euler cycle.

A graph G is called as Euler graph if it contains an Euler line in G.

We know by the definition of walk, that the walk traces each edge exactly once and a walk is always connected. So in an Euler line each edge appears exactly once and the Euler graph is always connected. So it is assumed that an Euler graph do not have isolated vertices.

9.2.1 Euler Path

An open walk in a graph G is called an Euler path if it contains all the edges of graph G. Given is a very important theorem, which will help us in finding whether or not a given graph is Euler.

Theorem 9.1. A given connected graph G contains an Euler line (*i.e.*, graph G is Euler graph) if all the vertices of G are of even degree.

Proof: Suppose G is a connected graph with n vertices $(v_1, v_2, ..., v_n)$ and m-edges $(e_1, e_2, ..., e_m)$ also suppose that G is an Euler graph. Therefore G must contain an Euler line (which is a closed walk). Let's assume that $v_1 e_1 v_2 e_2 v_3 e_3 e_4 ... e_i v_i ... e_{m-1} v_m e_m v_1$ is the Euler line in the given graph.

It can be seen easily that every time the line traces a new vertex v it goes through two new edges incident on v. *For example* if walk taken above, meets a new vertex say v_2, it goes through two edges e_1 and e_2 which are incident on v_2 with e_1 we entered v_2 and with e_2 we exited. This is true with the terminal vertex also as we "exited" and "entered" the same vertex at the starting and end of the walk respectively. So every time we trace a vertex it's degree increases by 2. Thus if *G* is an Euler graph, the degree of every vertex is even.

Conversely, let us assume that all the vertices of *G* are of even degree. To prove that *G* is an Euler graph, we need to construct an Euler line in *G*. Start with an arbitrary vertex *v* and tracing the edges of *G* such that no edge appears more than once. Since *G* contains all the vertices of even degree, we can exit from the same vertex we enter. Which means that the walk cannot stop at any vertex but only *v*.

Now if the walk *w*, say starting and ending at *v*, traced all the edges of the graph *G*, walk is an Euler line and the graph is an Euler graph. But if the walk *w* does not contains all the edges of *G*, then remove all the edges which are in walk *w* from the graph *G*, we are left with a graph *G′* which is a subgraph of *G*. The degree of each vertex in G′ is even because graph G and walk *w* have all their vertices of even degree. *For example,* consider the graph shown in **Fig. 9.1**(*a*).

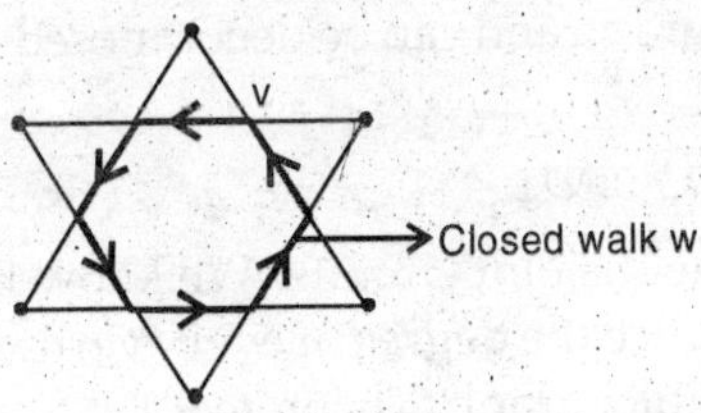

Fig. 9.1 (*a*) Graph G

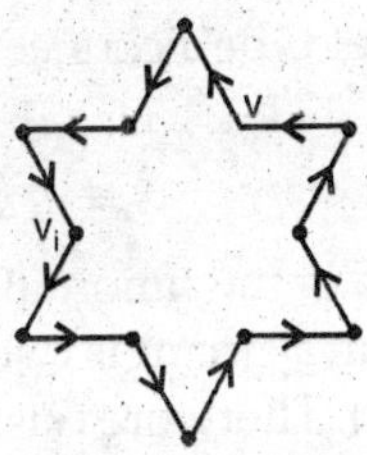

Fig. 9.1 (*b*) G′ = G – w

One can see that all vertices of *G* are of even degree and *G* is connected. One of the closed walk *w* starting and ending at an arbitrary vertex *v* shown by bold lines. Walk *w* does not contains all the edges of *G*. Subgraph *G′* obtained by removing walk *w* from graph *G* is shown in **Fig. 9.1**(*b*).

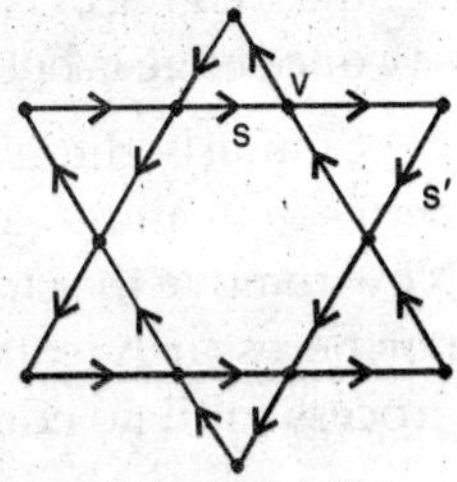

Fig. 9.1 (*c*)

Again all the vertices in *G′* are also of even degree. Now starting again with a vertex v_i which is common to both walk *w* and subgraph *G′*, we construct a closed walk *w′* which terminate at vertex v_i. This walk *w′* can be combined with the walk *w* to from a new walk which begins and ends at vertex *v* and contains more edges then *w*.

The walk *w′* is shown in **Fig. 9.1**(*b*) with bold lines starting and ending at v_i. The combined walk (*i.e.*, *w* and *w′*) is shown in **Fig. 9.1**(*c*), which starts and ends at vertex *v*.

This process is repeated until we obtain a closed walk (*i.e.*, Euler line) that traces each edge of graph *G* exactly once. Hence the graph *G* is an Euler graph.

The graph shown in **Fig. 9.1**(*c*) is an Euler graph as it contains an Euler line which starts and ends at vertex *v* and traces all the edges exactly once.

Example 9.1. Which of the following graphs are Eulerian?

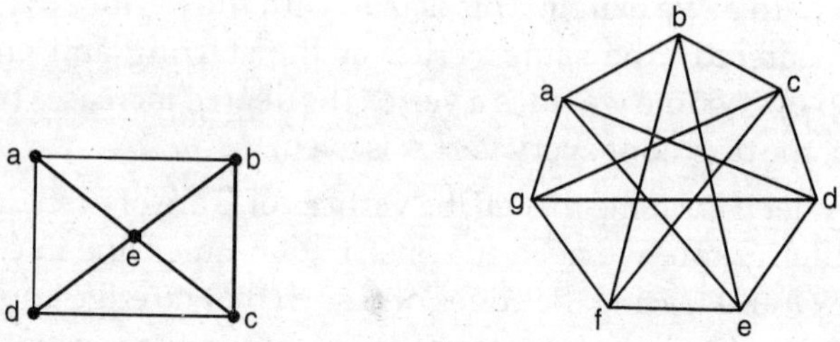

(a) (b)

Fig. 9.2

Solution. In **Fig. 9.2** (*a*) 4 vertices are of odd degree and one vertex is of 4 degree. By the Theorem 9.1, If a graph *G* is Eulerian then all the vertices of *G* are of even degree. So the graph shown in **Fig. 9.2**(*a*) is not *Eulerian* while the graph shown in **Fig. 9.2**(*b*) has all vertices of even degree (*i.e.*, 4), Hence **Eulerian.**

Theorem 9.2. A connected graph *G* is an Euler graph if it can be decomposed into edge disjoint circuits.

Proof: Let *G* be a connected graph and can be decomposed into edge disjoint circuits $c_1, c_2, ..., c_n$.

then $$G = c_1 \cup c_2 \cup ... \cup c_n$$

That is, *G* is the union of edge disjoint circuits. We know that the degree of every vertex in a circuit is two. Hence the degree of every vertex in *G* is a multiple of 2 (*i.e.*, even). Therefore *G* is an Euler graph (Theorem 9.1).

Conversely, let us suppose that *G* is an Euler graph. Now we need to show that it can be decomposed into edge disjoint circuits. Consider an arbitrary vertex v_1 in *G*. Since degree of v_1 is even, there must be at least two edges incident on v_1. Let us say once edge is between v_1 and v_2.

Similarly there are at least two edges incident on v_2 and one of the edge is between v_2 and v_3. In this way we are tracing a circuit starting and ending at v_1. Now remove the circuit from the graph *G*, we are left with a subgraph of *G* (need not necessarily connected), whose vertices are of even degree. Continue this process until no edge is left. Hence the theorem.

Example 9.2. Show that the graph shown in Fig. 9.3 is Eulerian.

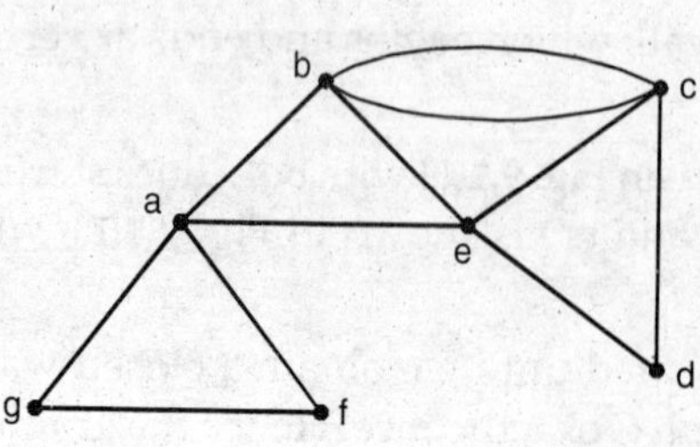

Fig. 9.3

Solution: As stated in Theorem 9.2, A connected graph *G* is Eulerian if it can be decomposed into edge disjoint circuits.

The graph shown in **Fig. 9.3** is connected. Start the tracing of a circuit from vertex a. The very first traced circuit circuit is (a, b, e, a). After removing the circuit from G, we are let with a graph G' which is shown in **Fig. 9.4**(a).

The other circuit will be (a, g, f, a), (b, c, b), (c, e, d, c). Hence the graph G is Eulerian.

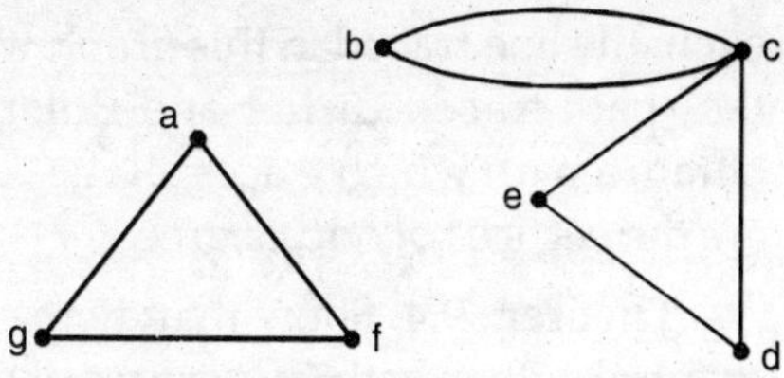

Fig. 9.4 G′ (disconnected graph)

9.3 UNICURSAL LINE (OPEN EULER LINE) AND GRAPH

In a graph G, an open walk which includes all the edges of the graph G, called a unicursal line or an open Euler line and the graph G called a unicursal graph.

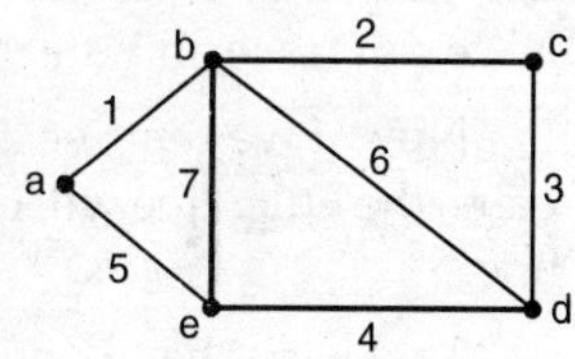

Fig. 9.5 A unicursal graph

That is the graph G containing a unicursal line called a unicursal graph. The graph in **Fig. 9.5** is a unicursal graph.

One of the unicursal line in the graph G shown in **Fig. 9.5** is $e5\,a1\,b7\,e4\,d6\,b2\,c3\,d$.

If we add an edge between the terminal vertices of the **unicursal line**. We shall get an **Euler line** and the graph becomes an Euler graph. Next theorem generalized the above observations.

Theorem 9.3. A connected graph G contains an unicursal line (or open Euler line or Eulerian path) if it has two vertices of odd degree.

Proof: Let us assume that the connected graph G contains a unicursal line, say $v_1\, e_1\, v_2 \ldots v_i e_i v_2 \ldots v_n$. We know that if we add one edge between its terminal vertices, we shall get an Euler line and the graph will become an Euler graph in which the degree of all the vertices will be even.

i.e., $\qquad \deg(v_1) = \text{even} = \deg(v_n)$

v_1 and v_n are the terminal vertices of unicursal line. But this is not the case with given graph.

So $\qquad \deg(v_1) = \deg(v_n) = \text{even} - 1 = \text{odd}$

Hence there are two vertices of odd degree.

Conversely, let G be a connected graph having two vertices of odd degree. Now we have to show that there exist a unicursal line in G. Let us start the tracing of unicursal line starting at one of the two odd degree vertices and going through the edges such that no edge appears more than once in the unicursal line. Whenever in this tracing the line meets a new vertex v of even degree it goes through two new edges, one through which we entered v another through which we exited. Continue this process until we obtain a unicursal line ending at the second vertex of odd degree. If the traced path does not contains all the edges of the graph, we remove the traced edges from the graph (obviously all the vertices of the remaining graph will be of even degree) and continue the same process in the remaining graph. As the unicursal line or path is connected so it must touch the remaining graph at least at one vertex. Since in the remaining graph the degree of all the vertices is even, the line traced in this graph will be an Euler line and by joining the two lines (*i.e.*, path traced earlier and Euler line traced in the remaining graph) we will obtain a path which traces all the edges of the graph and whose terminal vertices are the vertices of odd degree.

Theorem 9.4. Show that there exist m edge disjoint subgraphs of a connected graph G with exactly $2n$ vertices of odd degree, such that the m subgraphs together contain all the edges of the graph G and each subgraph is a unicursal graph.

Proof: Let G be a graph (connected) with $2m$ odd vertices say $v_1, v_2, ..., v_m, v_1', v_2', ..., v_m'$. If we add an edge between each vertex pair of the type (v_i, v_i') then in the new graph say G' (formed by new edges) the degree of each vertex will be even. That is we can trace an Euler line in G'.

Now if we remove all those m edges, which we just added between m vertex pairs, the Euler line will divide into m open walks and each of the walk is a unicursal line.

That is, we have divided the graph G into m subgraphs and each subgraph is a unicursal graph.

Hence the theorem.

Example 9.3. Show that the graph in Fig. 9.6 can be divided into 2 edge disjoint unicrusal graphs. Draw the graph also.

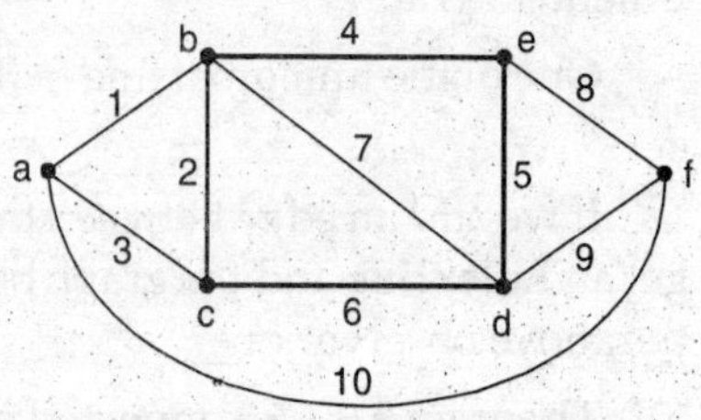

Fig. 9.6

Solution: As there are 4 vertices *i.e.*, (2×2) of odd degree so by theorem 9.4 there exist 2 unicrusal edge disjoint subgraphs of the given graph.

To find out 2 subgraphs (unicursal graphs) let us add two new edges between the vertex pair (c, e) and (a, f) as a, c, e and f are of odd vertices. The new graph G' obtained is shown in Fig. 9.7(a).

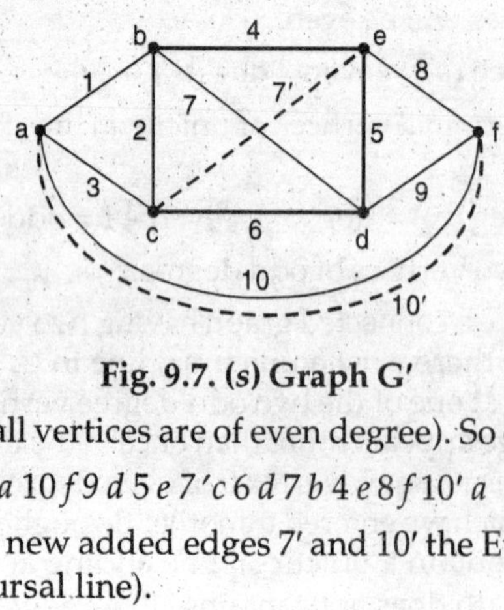

Fig. 9.7. (s) Graph G′

G' is an Euler graph (all vertices are of even degree). So one of the Euler line is

$$a\,1\,b\,2\,c\,3\,a\,10\,f\,9\,d\,5\,e\,7'\,c\,6\,d\,7\,b\,4\,e\,8\,f\,10'\,a$$

Now remove the two new added edges 7′ and 10′ the Euler line will split into two open walks (or unicursal line).

(1) $a1\ b2c\ 3a\ 10f\ 9d\ 5e$

(2) $c6\ d\ 7b\ 4e\ 8f$.

The graphs shown in **Fig. 9.7**(*b*) and **Fig. 9.7**(*c*) are the two subgraphs of the graph in **Fig. 9.6** with respect to unicrusal line (1) and (2) respectively.

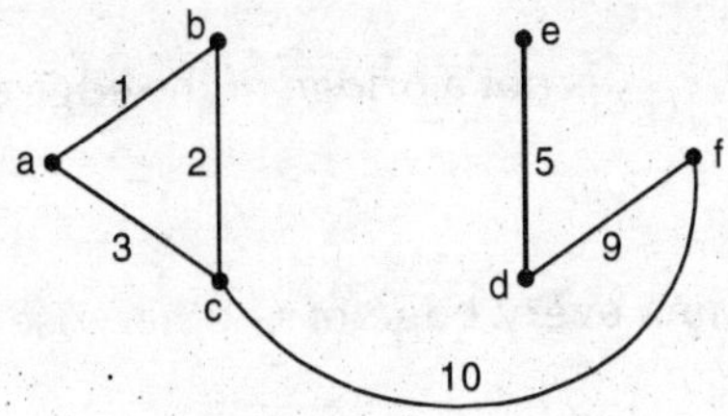

Fig. 9.7(b)

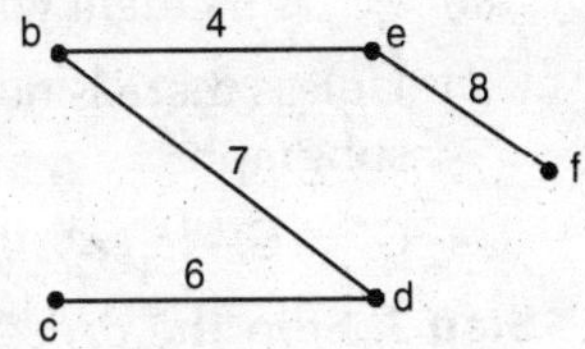

Fig. 9.7(c)

9.4 ARBITRARILY TRACEABLE GRAPHS

A graph G is called arbitrarily traceable from vertex v if we can construct an Euler line starting and ending at v by going through any edge incident on v. Which has not been previously traversed.

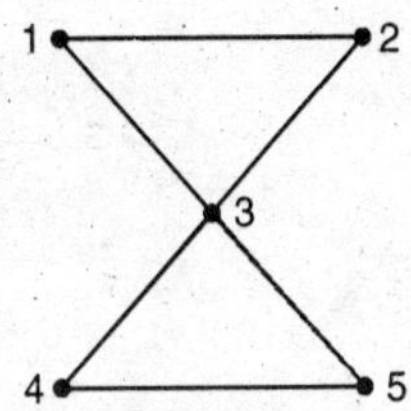

Fig. 9.8

For example the graph shown in **Fig. 9.8** is arbitrarily traceable from vertex 3.

But if we start with vertex 1, and trace paths 1, 2, 3. Now at vertex 3 we have three choices going to 1, 4 or 5. Now if we took the first choice (*i.e.*, going to vertex 1), we are left with a circuit 1, 2, 3, 1 not an Euler line. So this graph is not arbitrarily traceable from vertex 1.

Graph shown in **Fig. 9.9**(*a*) is arbitrarily traceable graph from all vertices and the graph in **Fig. 9.9**(*b*) is not arbitrarily traceable from any vertex.

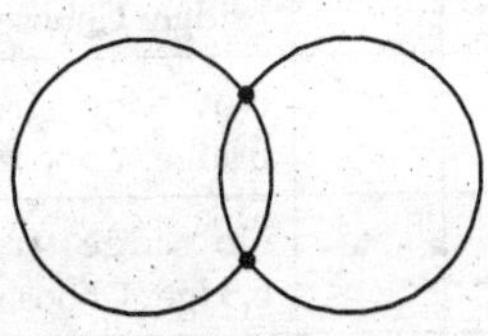

(*a*) Arbitrarily traceable graph from all vertices

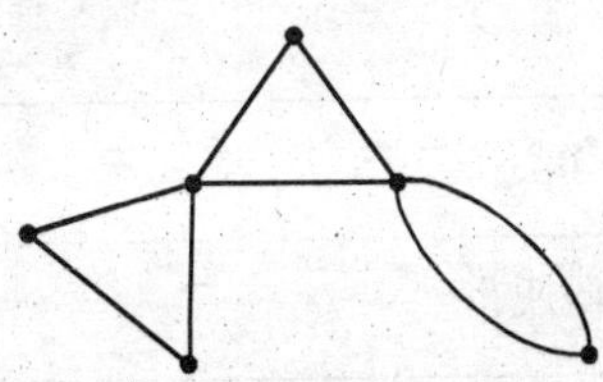

(*b*) Not arbitrarily traceable.

Fig. 9.9

One can draw the observation by seeing the graphs in **Fig. 9.9**(*a*) and **Fig. 9.9**(*b*) that a graph G is **arbitrarily traceable** from a vertex v if every circuit in G contains v.

9.5 FLEURY S ALGORITHM FOR CONSTRUCTING AN EULER LINE IN GRAPH G

This algorithm is used to construct an Euler line in an Euler graph. Algorithm's approach is to construct a trail that grows to the desired Euler line. The algorithm says :

Step 1: Choose an arbitrary vertex v_0 in the graph and set $\lambda_0 = v_0$.

Step 2: If the trail $\lambda_i = v_0, e_1, v_1, e_2, v_2, ..., e_i, v_i$ has been chosen, chose an edge e_{i+1} different from the traced edges $e_1, e_2, ..., e_i$ such that

(*a*) e_{i+1} is incident with v_i and

(*b*) Unless there is no other choice, e_{i+1} is not a bridge of the edge deleted subgraph

$$G' = G - \{e_1, e_2, ..., e_i\}$$

Step 3: Stop the process, if λ_i contains every edge of G otherwise repeat step 2.

Example 9.4. Using Fleury's algorithm construct an Euler line of the graph shown in Fig. 9.10.

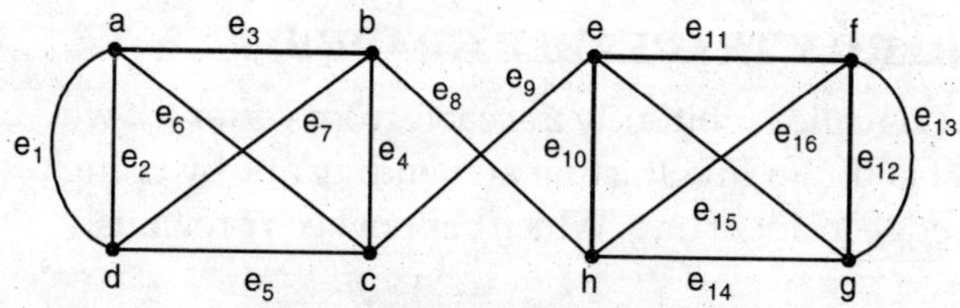

Fig. 9.10. Euler graph

Solution: Let us chose vertex a as the starting vertex *i.e.*, $v_0 = a$ and set $\lambda_0 = a$. The Euler line is obtained by repeating step 2. The results are shown in the Table 2.1 given in **Fig. 9.10**(*a*).

Table 2.1

λ : *Current path*	G_i	*Next Edge*	*Reasoning*
$\lambda_0 : a$	$G_0 = G$	e_1	No edge from a is a bridge. Choose any one.
$\lambda_1 : a, e_1, d$	$G_1 = G - e_1$	e_2	No edge from d is a bridge. Choose any one.
$\lambda_2 : a, e_1, d, e_2, a$	$G_2 = G_1 - e_2$	e_3	No edge from a is a bridge. Choose any one.
$\lambda_3 : a, e_1, d, e_2, a, e_3, b$	$G_3 = G_2 - e_3$	e_4	No edge from b is a bridge. Choose any one.
$\lambda_4 : a, e_1, d, e_2, a, e_3, b, e_4, c$	$G_4 = G_3 - e_4$	e_5	e_6 is a bridge. So choose e_5.
$\lambda_5 : a, e_1, d, e_2, a, e_3, b, e_4, c, e_5, d$	$G_5 = G_4 - e_5$	e_7	Only one choice from d.
$\lambda_6 : a, e_1, d, e_2, a, e_3, b, e_4, c, e_5, d, e_7, b$	$G_6 = G_5 - e_7$	e_8	Only one choice from b.
$\lambda_7 : a, e_1, d, e_2, a, e_3, b, e_4, c, e_5, d, e_7, b, e_8, h$	$G_7 = G_6 - e_8$	e_{10}	No edge from h is a bridge. Choose any one.
$\lambda_8 : a, e_1, d, e_2, a, e_3, b, e_4, c, e_5, d, e_7, b, e_8, h, e_{10}, e$	$G_8 = G_7 - e_{10}$	e_{11}	e_a is a bridge from e. Choose from e_{11} and e_{16}.
$\lambda_9 : a, e_1, d, e_2, a, e_3, b, e_4, c, e_5, d, e_7, b, e_8, h, e_{10}, e, e_{11}, f$	$G_9 = G_8 - e_{11}$	e_{12}	No edge from f is a bridge. Choose any one.

λ_{10} : $a, e_1, d, e_2, a, e_3, b, e_4, c, e_5, d, e_7\ b, e_8, h, e_{10}, e, e_{11}, f, e_{12}, g$	$G_{10} = G_9 - e_{12}$	e_{13}	No edge from g is a bridge. Choose any one.
λ_{11} : $a, e_1, d, e_2, a, e_3, b, e_4, c, e_5, d, e_7, b, e_8, h, e_{10}, e, e_{11}, f, e_{12}, g, e_{13}, f$	$G_{11} = G_{10} - e_{13}$	e_{15}	Only one choice left from f.
λ_{12} : $a, e_1, d, e_2, a, e_3, b, e_4, c, e_5, d, e_7, b, e_8, h, e_{10}, e, e_{11}, f, e_{12}, g, e_{13}, f, e_{15}, h$	$G_{12} = G_{11} - e_{15}$	e_{14}	Only one choice left from h.
λ_{13} : $a, e_1, d, e_2, a, e_3, b, e_4, c, e_5, d, e_7, b, e_8, h, e_{10}, e, e_{11}, f, e_{12}, g, e_{13}, f, e_{15}, h, e_{14}, g$	$G_{13} = G_{12} - e_{14}$	e_{16}	Only one choice left from g.
λ_{14} : $a, e_1, d, e_2, a, e_3, b, e_4, c, e_5, d, e_7, b, e_8, h, e_{10}, e, e_{11}, f, e_{12}, g, e_{13}, f, e_{15}, h, e_{14}, g, e_{16}, e$	$G_{14} = G_{13} - e_{16}$	e_9	Only one choice left from vertex e.
λ_{15} : $a, e_1, d, e_2, a, e_3, b, e_4, c, e_5, d, e_7, b, e_8, h, e_{10}, e, e_{11}, f, e_{12}, g, e_{13}, f, e_{15}, h, e_{14}, g, e_{16}, e, e_9, c$	$G_{15} = G_{14} - e_9$	e_9, e_7	Only one choice left from vertex c.
λ_{16} : $a, e_1, d, e_2, a, e_3, b, e_4, c, e_5, d, e_7, b, e_8, h, e_{10}, e, e_{11}, f, e_{12}, g, e_{13}, f, e_{15}, h, e_{14}, g, e_{16}, e, e_9, c, e_7, a$			

λ_{16} is the desired Euler line. **Fig. 9.11** shows the Euler line dotted constructed using Fleury's algorithm in the given graph.

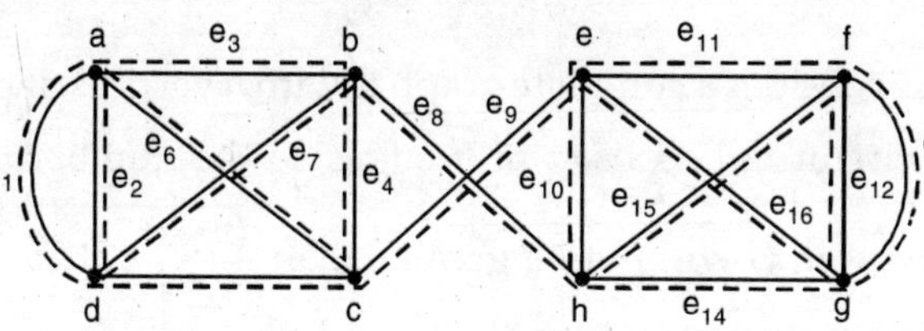

Fig. 9.11

9.6 HAMILTONIAN GRAPHS

As mentioned earilier that the Hamiltonian graphs are named after Sir William Hamilton who introudced a puzzle using a dodecahedron called the Icosran. In the puzzle each 20 vertices of dodecahedron was labelled by the name of some city and the objective of the puzzle was to construct a tour of all the cities going through edges such that no city appears more them one in the tour. That, one need to construct a closed walk such that each vertex of the graph is traced exactly once. This closed walk is called Hamiltonian circuit and the graph containing a Hamiltonian circuit is called a Hamiltonian graph.

9.6.1 Definition

In a graph G, a circuit that contains each vertex of G exactly once except the terminal vertices that appears twice, is known as Hamiltonian circuit.

A Hamiltonian graph *G* is the graph which contains a Hamiltonian circuit.

If we delete (or remove) one edge from the Hamiltonian circuit, we are left with a path, called Hamiltonian path.

The graphs in **Fig. 9.12**(*a*) and (*b*) contains Hamiltonian circuits.

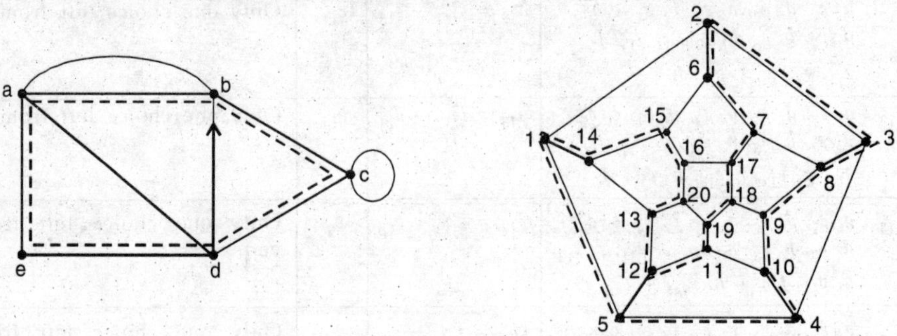

(*a*) (*b*)

Fig. 9.12 Two graphs containing Hamiltonian circuit

We are interested in finding the number of edge-disjoint Hamiltonian circuits in a graph but there is no such formula or technique exist for this task.

However, the number of edge-disjoint Hamiltonian circuits in a complete graph of odd vertices is given by Theorem 9.5.

Theorem 9.5. There are $\frac{(n-1)}{2}$ edge disjoint Hamiltonian circuits in a complete graph of n odd vertices, where $n \geq 3$.

Proof: Let us consider a complete graph G with n vertices and $\frac{n(n-1)}{2}$ edges. A Hamiltonian circuit in G consist of n edges so the number of edge-disjoint Hamiltonian circuit in G will not be greater than $\frac{(n-1)}{2}$. It can be shown that there are $\frac{(n-1)}{2}$ edge disjoint circuits in a graph of n odd vertices.

Fig. 9.13 shows the subgraph of complete graph of n vertices, which is a Hamiltonian circuit.

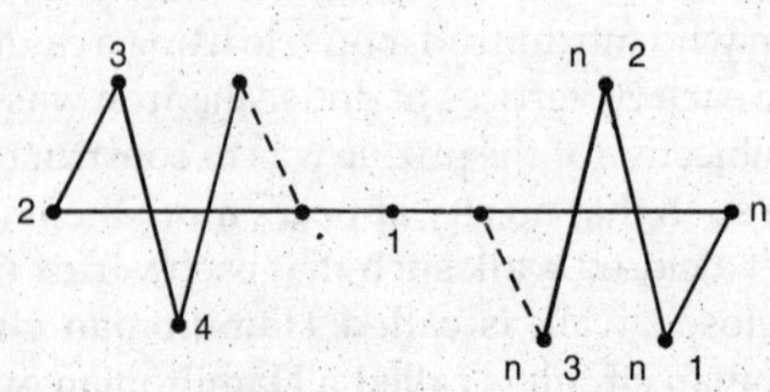

Fig. 9.13

If we keep the vertices fixed on a circle and rotate the polygonal pattern clockwise by

$$\frac{360}{(n-1)}, \frac{2.360}{(n-1)}, \ldots, \frac{(n-3)}{2} \cdot \frac{360}{(n-1)} \text{ degrees.}$$

It can be seen that in every rotation we are getting a new Hamiltonian circuit without any edge common to previous ones.

Thus, we have $\frac{(n-3)}{2}$ all edge-disjoint Hamiltonian circuits and also edge-disjoint among themselves. Hence the theorem.

Theorem 9.6. Prove that a graph G with n-vertices always has a Hamiltonian path if the sum of the degrees of every pair of vertices v_i, v_j in G satisfies the condition.

$$d(v_i) + d(v_j) \geq n - 1$$

Proof: To prove the result, we have to show that the graph is connected. Let us do this by the method of contradiction. Assume that the graph G can be partitioned into two components say G_1 and G_2. Let us assume also that vertex v_1 in G_1 and vertex v_2 is in G_2. The total number of vertices in G_1 and G_2 and n_1 and n_2 respectively such that $n_1 + n_2 = n$. The degree of v_1 and v_2 be utmost $n_1 - 1$ and $n_2 - 1$ respectively. Hence

$$d(v_1) + d(v_2) = n_1 - 1 + n_2 - 1 = n - 2 < n - 1$$

But the given condition is $d(v_1) + d(v_2) \geq n - 1$. Hence contradiction, so the graph must be connected. Let us construct a Hamiltonian path starting with a path of single edge.

Consider a path of length $m - 1$ where $m < n$ in graph G, passing through the vertices $v_1, v_2, ..., v_m$. We can extend the length of the path (or can construct a path of length m) by tracing a new vertex adjacent to either v_1 or v_m.

If the traced path is Hamiltonian than we are done otherwise, one can observe that there exist a vertex b in graph G which is adjacent to a vertex 'a' which is in the path P traced by extending the end vertices through edges at its fullest.

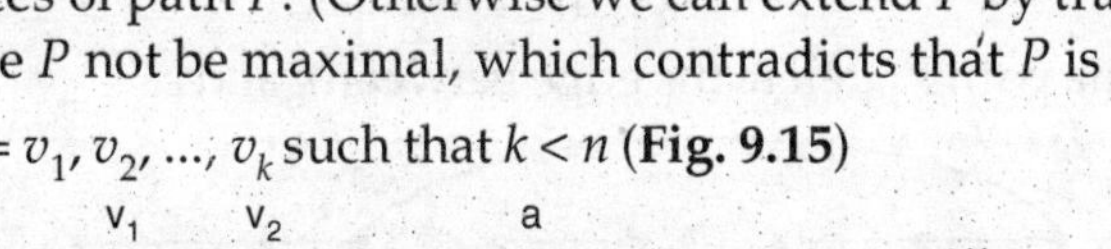

Fig. 9.14

As we know that P is maximal now so there is no vertex in G which is adjacent to the end vertices of path P. (Otherwise we can extend P by tracing that vertex and hence can be P not be maximal, which contradicts that P is maximal).

Again let $P = v_1, v_2, ..., v_k$ such that $k < n$ (**Fig. 9.15**)

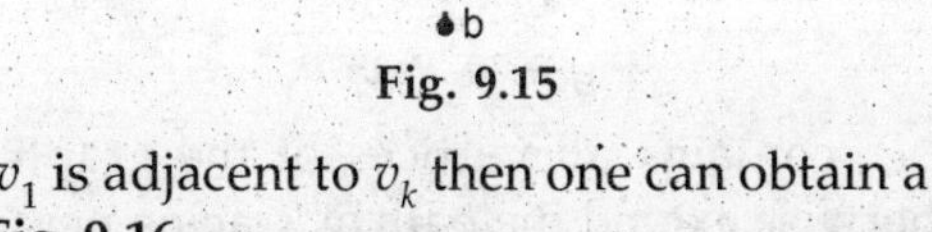

Fig. 9.15

Case I: Now if v_1 is adjacent to v_k then one can obtain a circuit by joining v_1 and v_k as shown in **Fig. 9.16**.

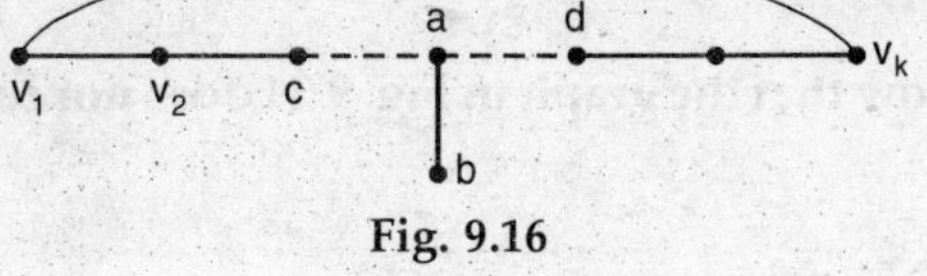

Fig. 9.16

So to obtain a path delete the edge (a, c) or (a, d) from the path P and join the vertices a and b. So we have obtained a new path which contains one more vertex then path P let us call the new path as P_1 (**Fig. 9.17**).

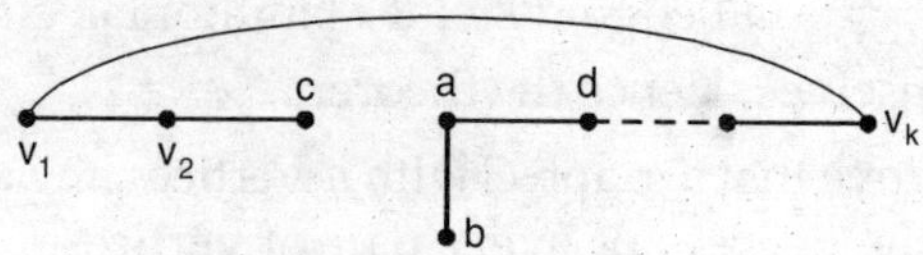

Fig. 9.17. Path P_1

Extend the path by tracing new vertices adjacent to the terminal vertices (*i.e.*, c and b) to obtain a maximal path P_1. If the extended path contains all vertices of the graph G the P_1 is the Hamiltonian path otherwise repeat the process till we get a Hamiltonian path.

Case II: Since P is maximal so no vertices of G which are not in the path P are adjacent to the end vertices of path P which means that the end vertices are adjacent to only those vertices which are in P. We have already discussed the case when v_1 and v_k are adjacent.

Now let us assume that v_1 and v_k are not adjacent then there exist integer i, $1 < i < k$ such that the edges (v_1, v_{i+1}) and (v_i, v_k) are in graph G (**Fig. 9.18**)

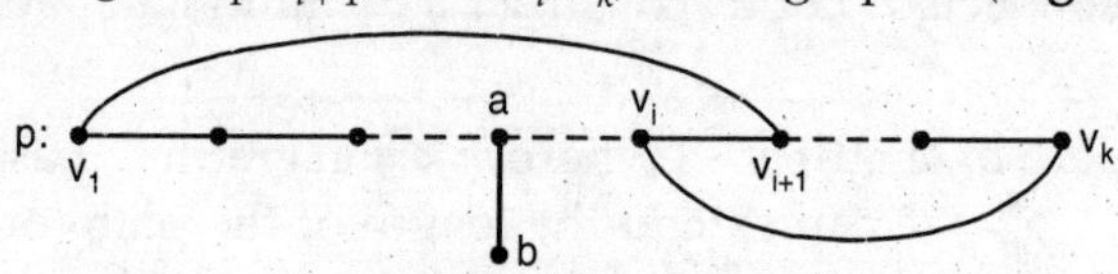

Fig. 9.18

If this is not true then v_1 is not adjacent to any vertex v_{i+1} in P which is adjacent to v_k. If degree of v_k is m then degree of v_1 is $k - m - 1$ (at most) then $\deg(v_1) + \deg(v_k) = k - m - 1 + m = k - 1 < n - 1$, a contradiction.

Now we can construct a circuit by deleting the edge (v_i, v_{i+1}).

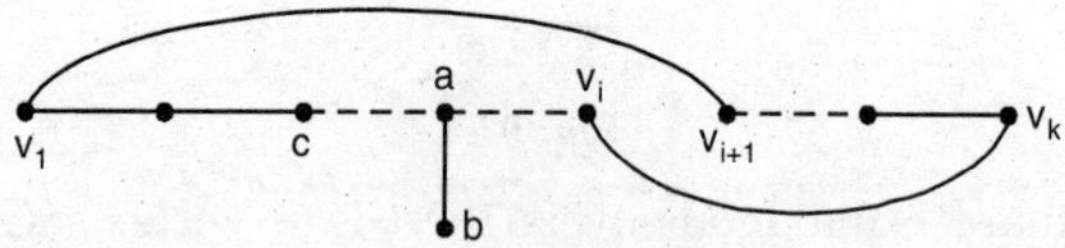

Fig. 9.19. Circuit

To construct the path P_1 delete the edge between a and c.

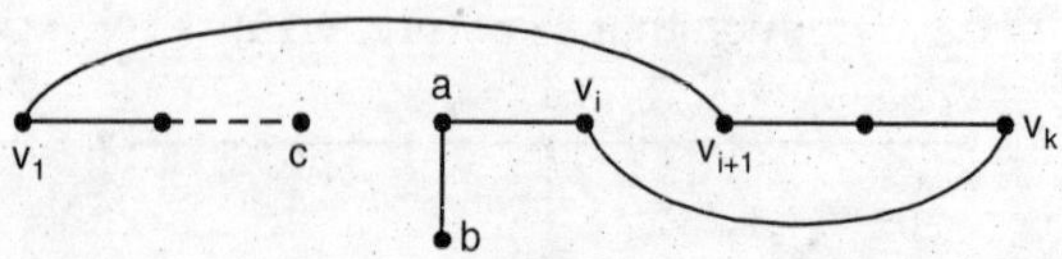

Fig. 9.20. Path P_1

Now if this path P_1 contains all vertices of the graph G, then P_1 is the Hamiltonian path. Otherwise extend the path by tracing new vertices which are adjacent to the end vertices of P_1 and repeat the process till a Hamiltonian path is obtained.

Example 9.5. Show that the graph in Fig. 9.21 does not contain a Hamiltonian circuit.

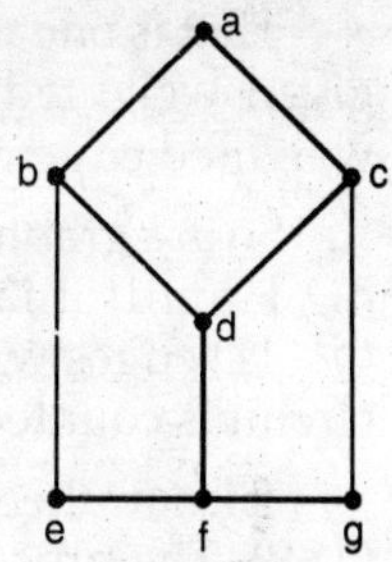

Fig. 9.21

Solution: From the **Fig. 9.21** it is clear that there are vertices of degree 2 and 3 only. We can observe that there exist edges between vertices of degree 2 and vertices of degree 3. As we know that a Hamiltonian circuit traces the vertices of the graph exactly once except the terminal vertices. That is the tracing will contain the vertices of degree 2 and vertices of degree 3 alternatively. Therefore we want to establish a one-to-one correspondence between the vertices and for this we need to have an equal number of vertices of degree 2 and degree 3. But in the given graph.

The no. of vertices having degree 2 = 3

and the no. of vertices having degree 3 = 4.

Clearly 3 ≠ 4 so we can not establish a one-to-one relationship between the vertices and hence the graph in **Fig. 9.21** does not contain a Hamiltonian circuit.

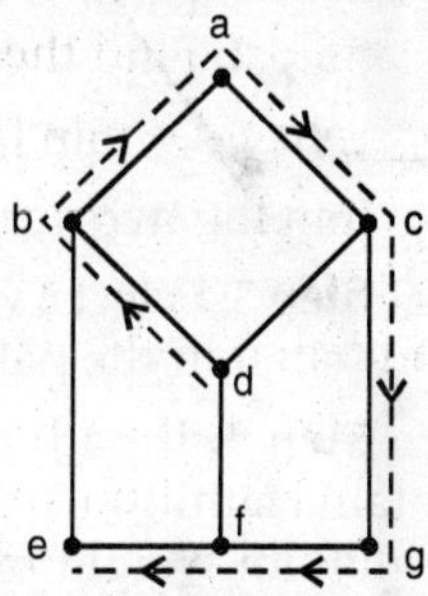

Fig. 9.22. Hamiltonian path

Moreover by the help of the above reasoning, a Hamiltonian path exist in the graph *G* if the difference between the vertices of degree 2 and the vertices of degree 3 is one (the terminal vertices in Hamiltonian path are different).

So in the given graph we have 4 vertices of degree 3 and 3 vertices of degree 2, the difference is 4 – 3 = 1.

Hence we can construct a Hamiltonian path in the given graph. This path is shown in **Fig. 9.22**.

Example 9.6. Show that the graph shown in Fig. 9.23 does not contain Hamiltonian path.

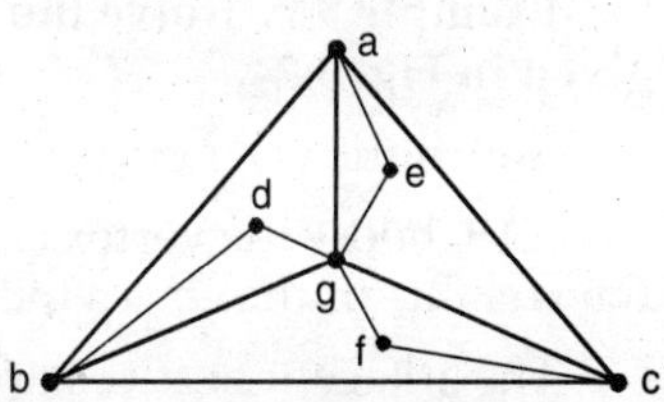

Fig. 9.23. Graph G

Solution: We know that, a graph *G* always has a Hamiltonian path if the sum of the degrees of every pair of vertices v_i and v_j in *G* satisfies the condition deg (v_i) + deg $(v_j) \geq n - 1$, where n is number of vertices in *G* (Theorem 9.6).

Here in the given graph let us take a vertex pair (d, e), then

$$\deg(d) = 2, \quad \deg(e) = 2$$

and hence $\deg(d) + \deg(e) = 4 < n - 1 = 7 - 1 = 6$.

So we are getting a contradiction. Hence the graph does not contain a Hamiltonian path.

9.7 TRAVELLING SALESMAN PROBLEM

The travelling-salesman problem is simply a combinational problem based upon the concept of Hamiltonian circuit. It states that "A salesman wants to visit some cities allotted to him cyclically. The distance betwen every pair of cities is known to him. His problem is to construct a tour that starts from his home town (city), passing through each city exactly once and return back to the starting point by covering the shortest distance."

That is one wants to construct a Hamiltonian circuit of minimum weight in a given weighted graph, in which cities are represented by vertices and each weighted edge represents the distance between the two cities.

So in a graph G of n-vertices there are $(n-1)!$ edges and hence the total number of Hamiltonian circuits in a complete graph of n vertices and $(n-1)!$ edges will be $(n-1)!/2$ (as out of total $(n-1)!$ choices each Hamiltonian circuit is counted twice, hence the total number *i.e.* $(n-1)!$ is divided by 2).

Till now there is no efficient (proper) algorithm for Travelling-Salesman Problem (TSP). The procedure we are going to discuss is known as "Nearest neighbour method" which produces the result close to the desired one.

9.7.1 Nearest Neighbour Methods

In the graphical representation of TSP vertices corresponds to the cities and weights of the edges denotes the distance between the cities.

Step 1: Choose an arbitrary vertex x as starting vertex and find all the vertices adjacent to vertex v, say $v_1, v_2, ..., v_n$.

Step 2: Find the vertex v_k that is closest to vertex v

$w(v, v_k) = \min \{w(v, v_1), w(v, v_2), ..., w(v, v_i)\}; 1 \le k \le i$

Join the two vertices v and v_k to form a path of length one say p.

Step 3: Select a vertex which is not selected earilier from the last traced vertex and join it in the path p to extend the path.

Step 4: If path p contains all the vertices of G. Then we are done. Now to obtain Hamiltonian circuit join the terminal vertices of the path p by the edge of minimum weight existing between the two vertices. Otherwise repeat steps 2 and 3.

Example 9.7. Solve the travelling-salesman problem given in Fig. 9.24.

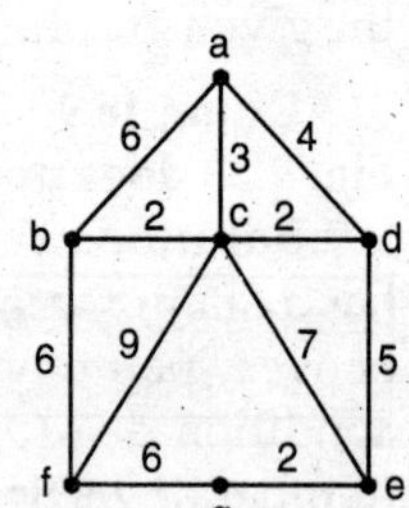

Fig. 9.24. A TSP

Solution: (1) Lets start with vertex a.

(2) Choose the vertex closest to vertex a, by calculating the weights of the edges incident from a.

The adjacent vertices are b, c and d.

$$w(a, b) = 6, w(a, c) = 3, \quad w(a, d) = 4$$

Min. weight = 3 *i.e.*, between a and c. So next vertex is c.

Join a and c to obtain a path of length one. (**Fig. 9.25**(*a*))

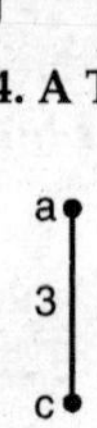

Fig. 9.25 (*a*)

From vertex c, two vertices are adjacent, that is there are two weighted edges, (c, b) and (c, d).

As we can see the weights are equal for both the edges so we can choose one of the edges arbitrary. Lets join the vertex d from c to obtain a path of length 2 (**Fig. 9.25**(*b*)) or choose vertex b from c

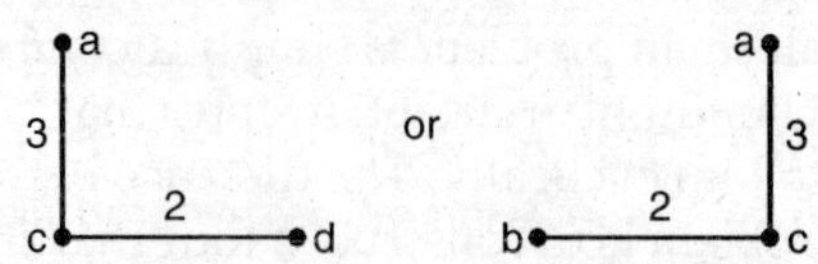

Fig. 9.25(*b*)

Now from vertex d there is a single choice to trace the vertex e through edge (d, e) (**Fig. 9.25**(c)).

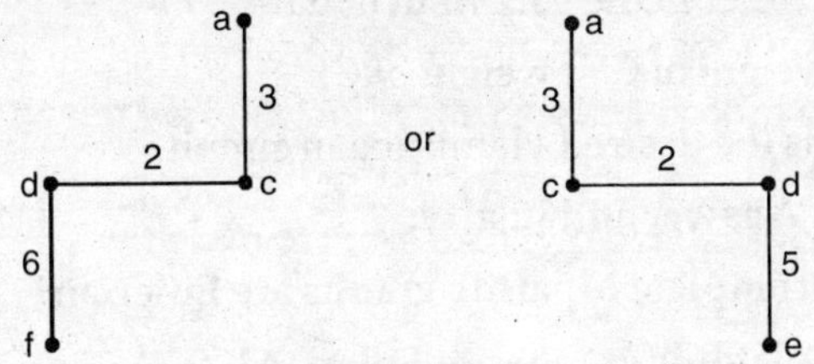

Fig. 9.25(c)

From e again only one edge *i.e.*, (e, g) and from g only one edge (g, f) (**Fig. 9.25**(d))

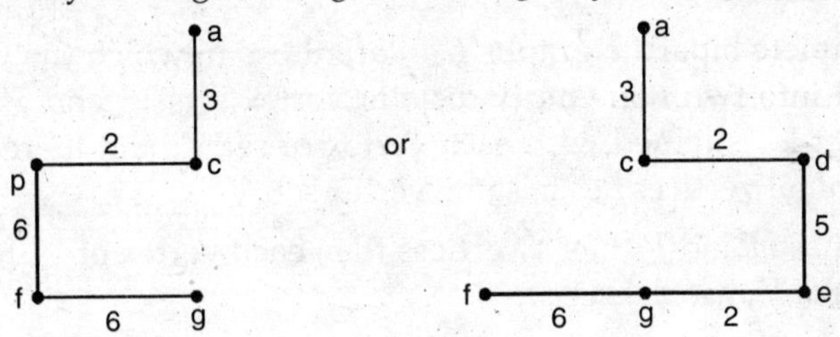

Fig. 9.25 (*d*)

The vertices adjacent to f are b and c minimum weight is with edge (f, b) so the next vertex which is included in the path is b. (also we can not trace c from f because in that case we will not have Hamiltonian circuit). The path is shown in **Fig. 9.25** (e).

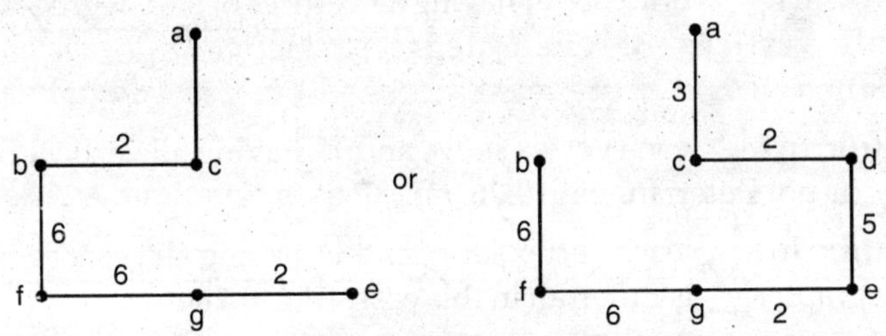

Fig. 9.25 (*e*)

It is clear that the path shown in **Fig. 9.25**(e) is the Hamiltonian path as it includes all the vertices of G. To obtain Hamiltonian circuit join the two terminal vertices of the Hamiltonian path and we are done.

The Hamiltonian circuit constructed is shown in **Fig. 9.25**(f).

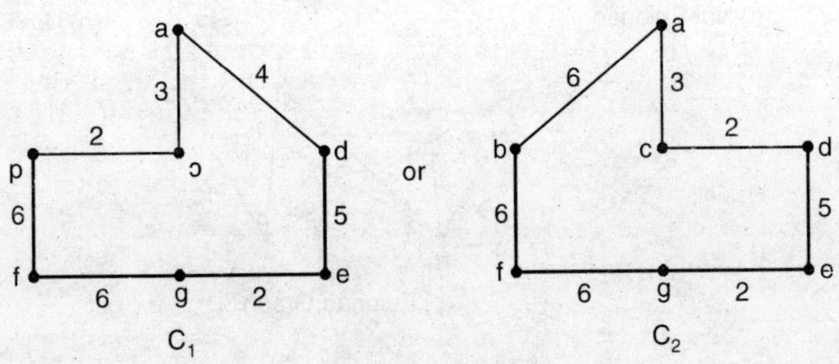

Fig. 9.25 (*f*)

Total weight of the circuit C_1 is $3 + 2 + 6 + 6 + 2 + 5 + 4 = 28$ total weight of the circuit C_2 is $3 + 2 + 5 + 2 + 6 + 6 + 6 = 30$. It can be observed that there is no other Hamiltonian circuit C' in the graph such that

$$\text{weight of } C' < \text{weight of } C_1$$

So circuit C_1 is the desired Hamiltonian circuit.

Example 9.8. Answer and justify: *(MCA, U.P.T.U., 2005-06)*

(*i*) Which complete bipartite graphs are Eulerian?

(*ii*) Which platonic graphs are Eulerian?

Solution: We know that a graph is Eulerian if all the vertices of G are of even degree (Theorem 9.1).

(*i*) Complete bipartite graphs $K_{m,n}$ are those in which the vertex set can be partitioned into two non-empty disjoint vertex sets V_1 and V_2 such that each vertex of set V_1 is joined with each vertex of set V_2 and there is no edge in G whose terminal vertices lies in same vertex set.

Let V_1 has m and V_2 has n vertices then each vertex of V_1 has degree n and each vertex of V_2 has degree m.

Now there are 3 possibilities:

(*i*) m is even no., n is odd no. (or *vice versa*).

(*ii*) both m and n are odd numbers.

(*iii*) both m and n are even numbers.

Case (*i*): In $K_{m,n}$, vertex set V_1 having all vertices of odd degree and vertex set V_2 having all vertices of even degrees. Hence $K_{m,n}$ is not Eulerian. **(Fig. 9.26** (*a*))

Case (*ii*): In $K_{m,n}$, each vertex set V_1 and V_2 having all vertices of odd degrees. Hence $K_{m,n}$ is not Eulerian. **(Fig. 9.26** (*b*))

Case (*iii*): In $K_{m,n}$, each vertex set V_1 and V_2 having all vertices of even degrees. Hence the graph $K_{m,n}$ is Eulerian in this case. **(Fig. 9.26**(*c*))

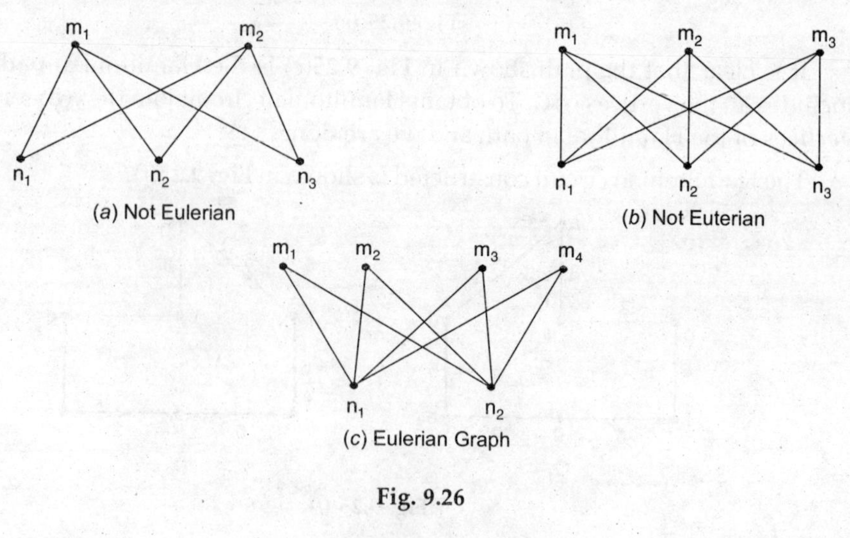

(*a*) Not Eulerian

(*b*) Not Euterian

(*c*) Eulerian Graph

Fig. 9.26

Hence a complete bipartite graph $K_{m,n}$ is Eulerian if m and n both are even numbers (*i.e.*, $|V_1| = |V_2|$ = even number)

(*ii*) **Platonic Graphs:** The graphs formed by the vertices and edges of the five regular platonic solids are called platonic graphs.

The five regular platonic solids are Tetrahedron octahedron, cubic, Icosahedron and Dodecahedron. The platonic graphs formed by these solids are shown in **Fig. 9.27.**

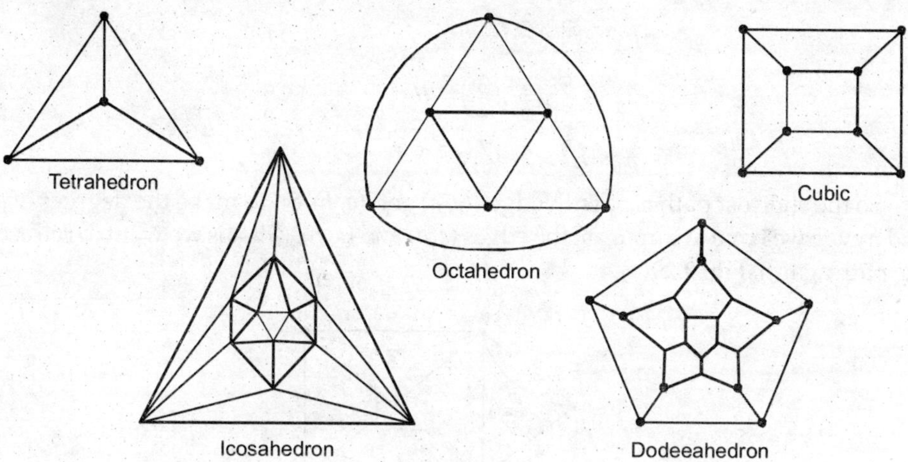

Fig. 9.27

It is clear that only octahedron is an Eulerian graph because all vertices in octahedron are of even degree.

9.8 CHINESE POSTMAN PROBLEM

Chinese postman problem is to find out an Euler line (cycle) in a connected weighted graph. If the degree of all the vertices is even then an Euler line can be constructed and weight is calculated by taking the sum of weights of the edges of the graph. Now when there are exactly two vertices of odd degree in the graph, the problem of constructing an Euler line gets more difficult.

To solve this problem let us find the two odd vertices as we cannot join then together. So we are interested in finding out the different paths (weighted) between them say $p_1, p_2, ..., p_k$. Calculate the weight of each path and select the one with minimum weight.

That is, we are retracing the shortest path between the two vertices to construct an Euler line. By duplicating the edges existing in the shortest path with the same weights.

Now degree of each vertex is even and we can easily construct the Euler line. Weight of the Euler line = Total weight of all the edge + weight of the shortest path between the two vertices

Example 9.9. Solve the Chinese problem given in Fig. 9.28.

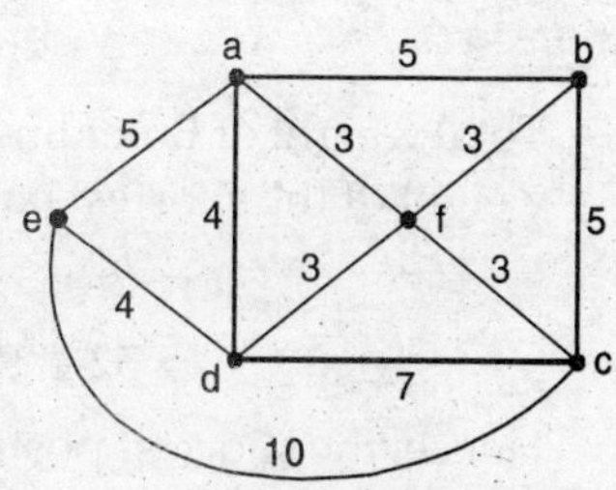

Fig. 9.28

Solution: Vertices *b* and *c* are having odd degree. So we can not trace the Euler line directly. Since the graph is too small, one can find out the weighed paths between the two vertices *b* and *c*.

Some of the paths are:

	Total weight
$p_1: e \overset{5}{-} a \overset{5}{-} b$	10
$p_2: e \overset{5}{-} a \overset{3}{-} f \overset{3}{-} b$	11
$p_3: e \overset{5}{-} a \overset{4}{-} d \overset{3}{-} f \overset{3}{-} b$	15
$p_4: e \overset{5}{-} a \overset{4}{-} d \overset{7}{-} c \overset{5}{-} b$	21

So the shortest path is p_1 with edges (*e*, *a*) and (*a*, *b*). So to make the degrees of *e* and *b* even we have to duplicate the edges (*e*, *a*) and (*a*, *b*). (That is we want to retrace the path again). **Fig. 9.29**(*a*).

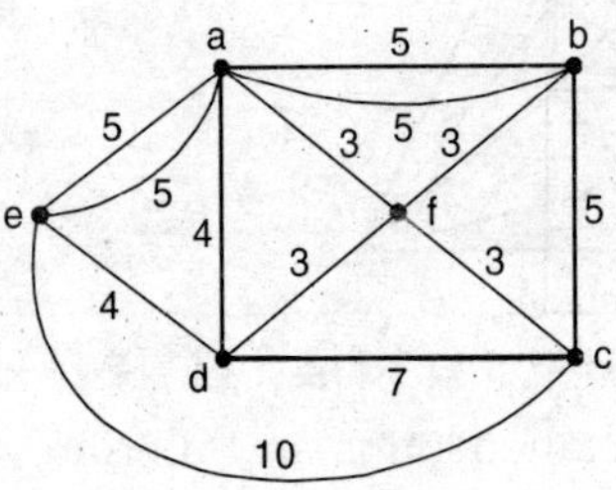

Fig. 9.29(*a*)

Now from **Fig. 9.29**(*a*) one can see that now all the vertices are of even degree. So we can construct the Euler line (**Fig. 9.29**(*b*)).

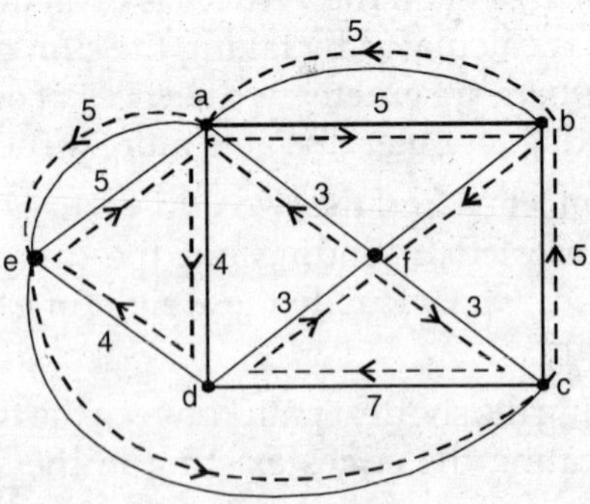

Fig. 9.29(*b*)

Total weight of Euler line = Sum of the weights of the edges of the given graph *G* + weight of the shortest path.

$$w = (5 + 3 + 3 + 3 + 3 + 5 + 5 + 4 + 4 + 7 + 10) + (10)$$

$$= 52 + 10 = 62$$

Hence, the shortest postman tour has weight of 62 units.

SUMMARY

In this chapter we have discussed the two very important graphs called Euler graphs and Hamiltonian graphs with special properties. Many physical problems can be represented by graphs and solved by observing the relevant properties of the corresponding graphs.

EXERCISES

1. Check whether the following graphs are eulerion. If yes find the Euler line.

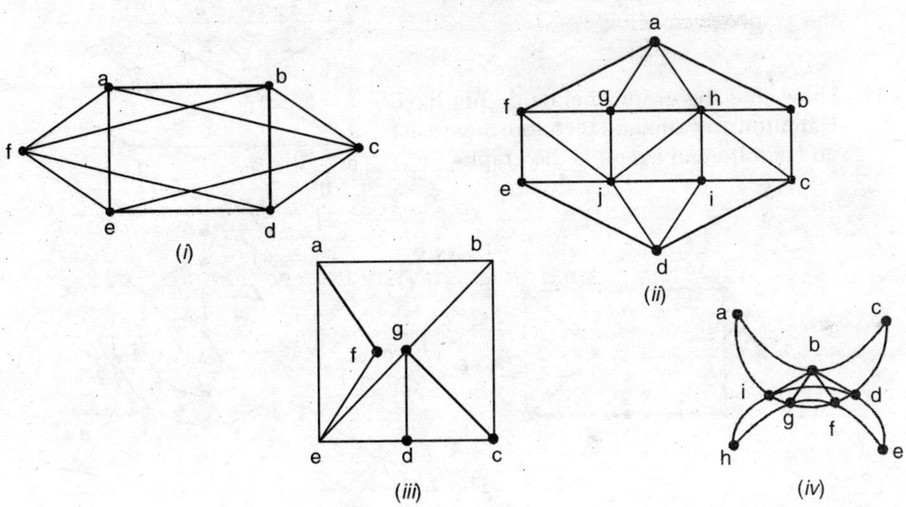

Fig. 9.30

2. Show that any Hamiltonian cycle in the graph shown below containing edge (v_1, v_2) must also contain the edge (v_5, v_8).

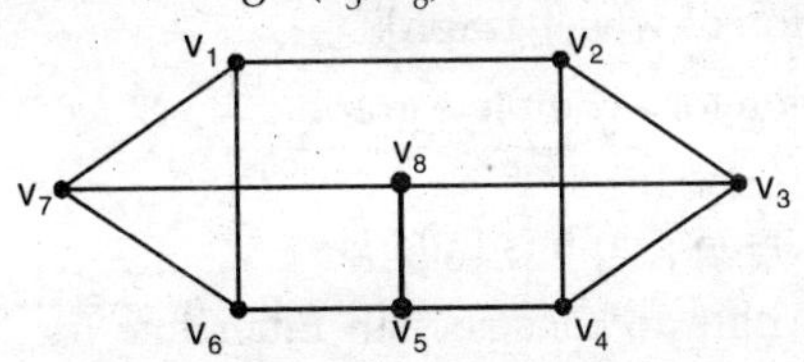

Fig. 9.31

[**Hint:** Construct a Hamiltonian circuit starting from vertex v_1, tracing edge (v_1, v_2). From v_2 we have two choices, trace (v_2, v_3) then (v_3, v_4), (v_4, v_5). Now from v_5 we have to trace (v_5, v_8) to trace v_8 then (v_8, v_7) (v_7, v_6) and (v_6, v_1). Similarly the other circuit can be traced starting from vertex v_1.

3. Check whether the graphs $K_{3,3}$, $K_{2,4}$ and $K_{2,3}$ has (*a*) Hamiltonian circuit (*b*) Hamiltonian path.
4. Give the sufficient conditions for a complete bipartite graph $K_{m,n}$. If $K_{m,n}$ contains (*a*) Hamiltonian circuit, (*b*) Euler circuit, and (*c*) Hamiltonian path.
5. Draw a graph containing both Euler line and Hamiltonian circuit.
6. Draw a graph containing Hamiltonian circuit but not Eulerian.
7. Prove or disprove :

   (*a*) Every Euler bipartite graph has an even number of edges.

(b) Every simple Euler graph with an even number of vertices has an even number of edges.

(c) If G is an Euler graph with edges c, d that share a common vertex, then G has an Euler circuit in which c, d appear consecutively.

8. Give an example of a graph that has an Euler circuit and a Hamiltonian circuit but the two circuits are not same.

9. How many Hamiltonian circuits exist in a complete graph of n-vertices, $n \geq 3$.

10. Show that in a graph no path can be longer than a Hamiltonian path (if it exist).

11. Solve the travelling salesman problem for the graph given below :

12. Show that the graphs below do not have Hamiltonian paths and therefore there exist no Hamiltonian circuit is the graphs.

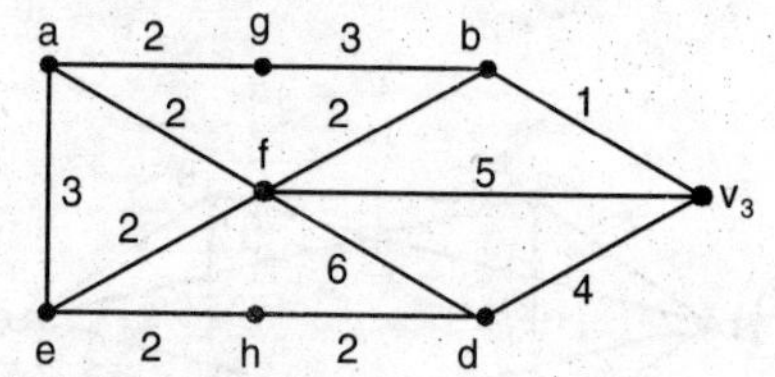

Fig. 9.32

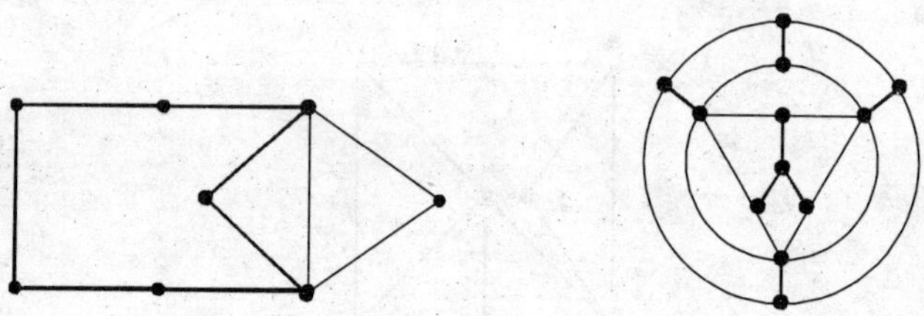

(i) (ii)

Fig. 9.33

13. Construct the minimal Hamiltonian circuit in the graph given in **Fig. 9.34**:

14. Show that W_n cannot have an Eulerian line.

[Hint: Use definition of wheel graph].

15. What is the condition for a complete graph of n vertices K_n to be Eulerian ?

16. Show that the Peterson graph is Eulerian.

17. Use Fleury's algorithm to construct an Euler line for the graphs given below:

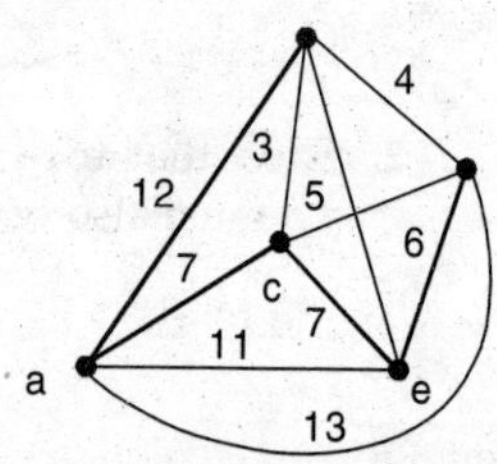

Fig. 9.34

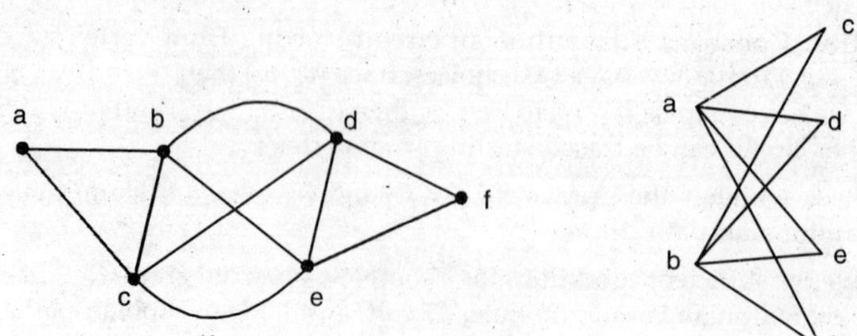

(i) (ii)

Fig. 9.35

18. Solve the Chinese postman problem for the following graph:

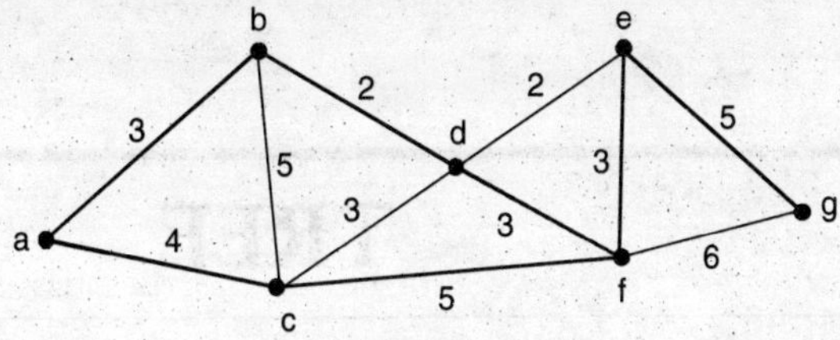

Fig. 9.36

ANSWERS

1. (*i*) Yes ; *a, b, c, d, b, f, d, e, c, a, e, f, a* (*ii*) No (*iii*) No (*iv*) Yes.

3. (*a*) $K_{3,3}$ (*b*) $K_{3,3}$, (*c*) $K_{2,4}$.

4. (*a*) $m = n$ (*b*) m and n are even (*c*) $m = n - 1$.

5. C_n = a cycle graph.

6.

Fig. 9.37

8.

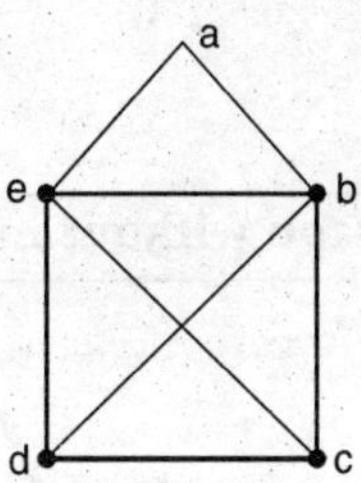

Fig. 9.38

Hamiltonian circuit: $a - b - c - d - e - a$

Euler line: *a, b, c, d, b, e, c, a, d, e, a.*

11. $a \underline{2} g \underline{3} b \underline{1} c \underline{4} d \underline{2} h \underline{2} e \underline{2} f \underline{2} a$; total weight = 18 units.

13. *a, c, b, d, e, a*; total weight = 31.

15. n should be odd.

18. *b* and *c* are the two vertices of odd degree. The length of the shortest path between *b* and *c* is *b, d, e* of weight 4. So repeat this path and solve the problem.

■■■

Chapter

10 TREE

10.1 INTRODUCTION

The role of tree in the study of graph theory is a very important concept. Trees are useful in describing any hierarchical structure.

In this chapter we will study the properties of tree and also study the a special kind of tree called binary tree and its application in graph theory and computer science.

10.2 TREES

A connected graph without any circuit is called a tree.

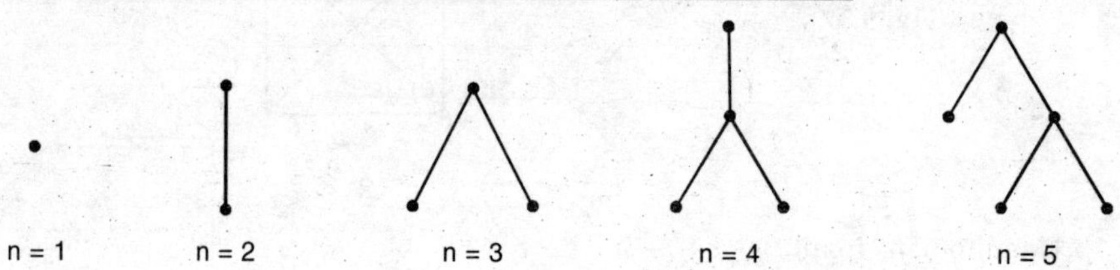

Fig. 10.1

Trees with one, two, three four and five vertices respectively are shown in **Fig. 10.1**. As we know that in a graph there must exist at least one vertex, so graph with one vertex is a tree.

From the definition of a tree it is clear that a tree is a simple graph (*i.e.*, without self loops and parallel edges).

Some Examples of Tree

(*i*) Family tree

(*ii*) Organizational set up

(*iii*) Decision tree, etc.

10.2.1 Properties of Tree

Theorem 10.1. Prove that there is one and only one path between every pair of vertices in a tree T.

Proof: Definition of tree states that tree is a connected circuitless graph. Since T is connected so there exist at least one path between every pair of vertices.

Let us consider that there are two paths between a vertex pair (v_i, v_j). (**Fig. 10.2**).

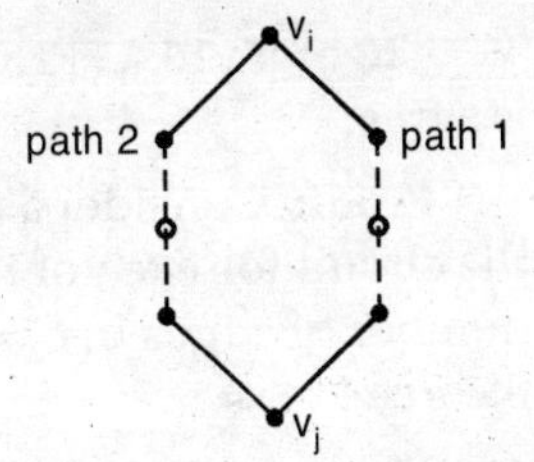

Fig. 10.2

The union of these two paths will contain a circuit in T and T can not be a tree. Hence there is one and only one path between every pair of vertices in a tree T.

Theorem 10.2. If in a graph G there is one and only one path between every pair of vertices, then G is a tree.

Proof: Since there exist a path between every pair of vertices, in graph G, so G is connected.

If there is a circuit in graph G with two or more vertices, then we can find out a vertex pair (v_i, v_j) (v_i, v_j are in circuit) such that there are two distinct paths between v_i and v_j. But in graph G there is one and only one path between every pair of vertices. Hence there is no circuit in G and therefore G is a tree.

Theorem 10.3. Prove that there are $n-1$ edges in a tree with n vertices.

Proof: (*a*) Let us assume that the number of edges in a tree of n vertices be less than $(n-1)$ (We know that there is no circuit in a tree) so this graph is disconnected. But by the definition of tree, tree is a connected circuitless graph. So the number of edges in a tree of n vertices cannot be less than $(n-1)$

Now let us assume that the number of edges in a tree of n vertices be more than $(n-1)$ say n. Then there will be at lest one circuit in the graph and hence will contradict the definition of tree.

Therefore, the number of edges in a tree of n vertices are $(n-1)$.

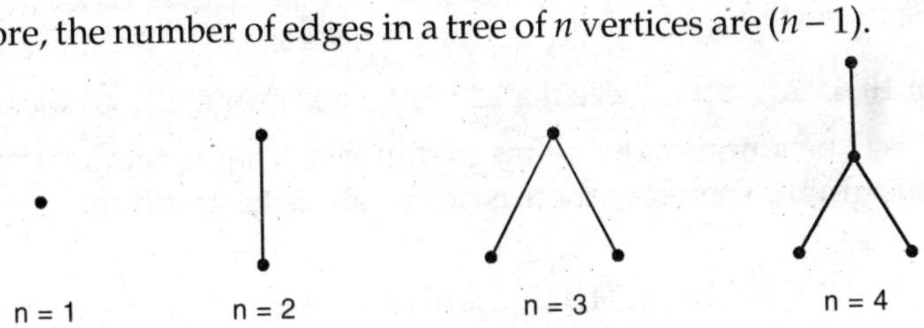

Fig. 10.3

(*b*) Let us prove this theorem by induction on the number of vertices. Graphs shown in **Fig. 10.3** are verifying the theorem.

Now let us assume that the theorem is valid or all the trees having $(n-1)$ vertices. That is the number of edges in a tree of $(n-1)$ vertices are $(n-2)$.

The graph shown in Fig. 10.4 is a tree T_{n-1} of $(n-1)$ vertices and $(n-2)$ edges.

Now a tree T_n with a vertices can be constructed by introducing a new vertex say v_i in T_{n-1}. Since T_n is connected, join v_i and any other vertex say v_j of T_{n-1} with an edge.

So total number of edges in T_n

= number of edges in T_{n-1} + new edge between v_i and v_j

$= n-2+1 = 1$

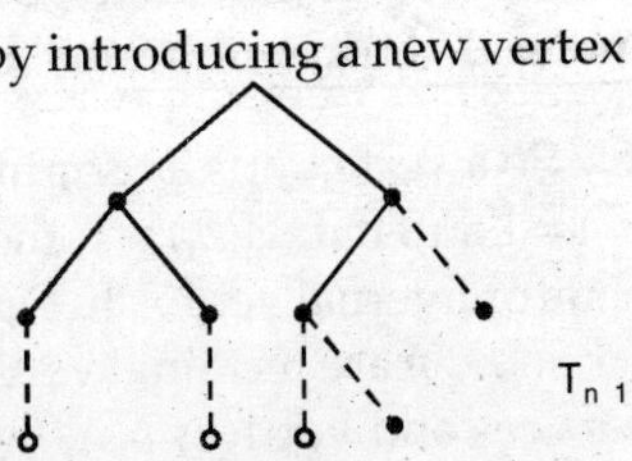

Fig. 10.4

Hence there are exactly $n-1$ edges in a tree with n vertices.

Theorem 10.4. Prove that any connected graph with n vertices and $n-1$ edges is a tree.

Proof: Consider a circuit of length k in G. There are k-vertices and k-edges on the circuit for each of the remaining $n-k$ vertices not on the circuit. So the total number of edges in $G = n - k + k = n > n - 1$ which gives a contraction. Hence the theorem.

Theorem 10.5. Prove that graph G with n vertices, $n-1$ edges and without circuits is connected.

Proof: Let us assume that graph G with n vertices $n-1$ edges and no circuit is disconnected. The there exist at least two components of G say G_1 and G_2. Each of the component is circuit less as G is circuitless. Now to make graph G connected we need to add an edge 'e' between two vertices v_i and v_j. Vertex. v_i is in component G_1 and v_j is in G_2.

Now the edges in $G = n - 1 + 1$ (*i.e.*, edge e) $= n$

So G is now a connected, circuitless graph with n vertices and n edges, which is not possible. As we have already proved in Theorem 10.3 that a connected, circuitless graph of n vertices has $n-1$ edges.

Hence, G must be connected.

10.2.2 Minimally Connected Graph

A graph G is said to be minimally connected if removal of one edge makes the graph disconnected.

Theorem 10.6. A graph is tree if and only if it is minimally connected.

Proof: Let G be a minimally connected graph. That is removal of one edge disconnects the graph. Therefore there is no circuit in the graph and hence graph G is a tree.

Conversely, let G be a tree. That is there exist one and only one path between every pair of vertices. We know that a path is always connected and removal of one edge disconnects the path, hence the graph G.

Therefore graph G (or tree) is minimally connected.

10.2.3 Pendant Vertex

A vertex of degree one in a tree T is called a pendant vertex or leaf or a terminal node and all the other vertices except from terminal vertices are called internal vertices or branch node.

So a vertex v_i is a **terminal vertex** if deg $(v_i) = 1$ and if deg $(v_j) > 1$, then v_i is a branch node or internal vertex. In **Fig. 10.2** vertices a, c, e, i, j, k, n are terminal vertices or pendant vertices and vertices $b, d, f\, g, h, l$ and m are internal vertices.

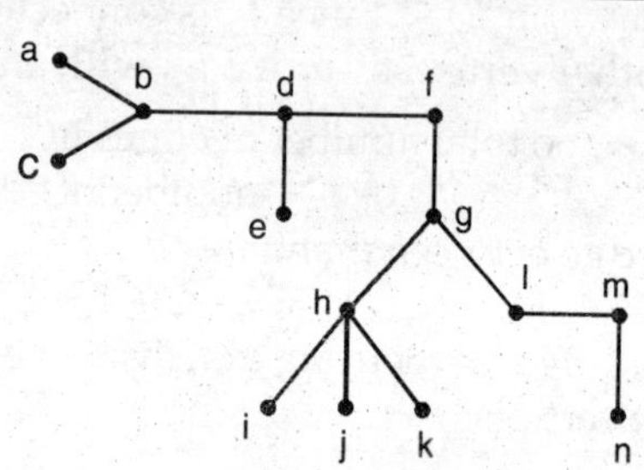

Fig. 10.5

Theorem 10.7. Prove that in a non-trivial tree T there are atleast two pendant vertices.

Proof: Let T be a non-trivial tree with n vertices, then total degree of T is given by the formula

$$\sum_{i=1}^{n} \deg(v_i) = 2 \,.\, e \qquad ...(1)$$

We know that there are $n-1$ edges in a tree of degree n, so by eq. (1)

$$\sum_{i=1}^{n} \deg(v_i) = 2 \,.\, (n-1) \qquad ...(2)$$

Now let us assume that there is only one vertex say v_1 of degree one in T; *i.e.*, only one pendant vertex.

$$\text{So} \qquad \sum_{i=1}^{n} \deg(v_i) = \deg(v_1) + \sum_{i=2}^{n} \deg(v_i)$$

$$= 1 + \sum_{i=2}^{n} \deg(v_i) \qquad ...(3)$$

Since T is a non-trivial tree so $\deg(v_i) \geq 2$ for all v_i ; $i \geq 2$ in T. So by eqs. (2) and (3).

$$2(n-1) = 1 + \sum_{i=2}^{n} \deg(v_i)$$

$$\sum_{i=2}^{n} \deg(v_i) = 2n - 2 - 1 = 2n - 3 \qquad ...(4)$$

$$\text{Since } \deg(v_i) \geq 2 \;\Rightarrow\; \sum_{i=2}^{n} \deg(v_i) \geq 2(n-1) = 2n-2$$

Hence $\qquad 2n - 3 \geq 2n - 2 \qquad$ (by eq. (4))

which is not possible.

Therefore in tree T there are at least two pendant vertices.

Example 10.1. Consider a tree T with 3 vertices of degree 2, 4 vertices of degree 3 and 3 vertices of degree 4. Calculate the number of pendant vertices in a tree be m.

Solution: Let the number of pendant vertices in a tree be m. Total number of vertices in a tree

$$T = m + 3 + 4 + 3$$

$$= (10 + m)$$

and hence the total number of edges

$$= (10 + m) - 1 = m + 9$$

Now the degree of T is given by the formula;

$$\sum_{i=1}^{n} \deg(v_i) = 2 \,.\, e \qquad ...(1)$$

Calculate the $\deg(v_i)$ given in Table 10.1.

Table 10.1

No. of vertices	Degree	Total degree
m	1	m . 1 = m
3	2	3.2 = 6
4	3	4.3 = 12
3	4	3.4 = 12
		$\sum_{i=1}^{n} \deg(v_i) = 30 + m$

$$\sum_{i=1}^{n} \deg(v_i) = 30 + m = 2 \,.\, e = 2 \,.\, (m+9)$$

$$\Rightarrow \qquad 30 + m = 2m + 18 = 12$$

Hence there are 12 pendant vertices in the given tree T.

Example 10.2. Draw all the non-isomorphic trees with 5 vertices.

Solution:

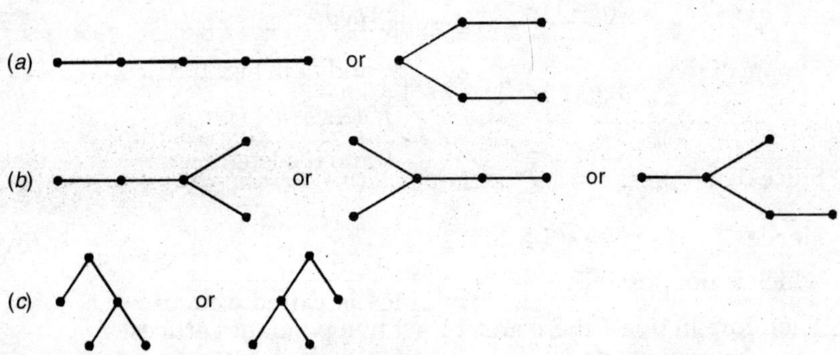

So there are only three non-isomorphic trees of 5 vertices.

Example 10.3. Show that there is tree with degree sequence (1, 1, 1, 1, 1, 1, 2, 3, 7).

Solution: The given degrees sequence is (1, 1, 1, 1, 1, 1, 2, 3, 7) so total number of vertices are 9. Hence edges are 9 – 1, *i.e.*, 8.

Degree of Tree $\quad T = 2\,.e = 2.8 = 16$

$$= \sum_{i=1}^{a} \deg(v_i)$$

$$= 1 + 1 + 1 + 1 + 1 + 1 + 2 + 3 + 7 = 18$$

Clearly $\qquad 16 \neq 18$

Hence no tree with given degree sequence.

Note: Degree sequence: Consider the vertices $(v_1, v_2, \ldots v_n)$ of graph G. Then the sequence $(d_1, d_2, d_3, \ldots d_n)$, where $d_i = \deg(v_i)$, is the degree sequence of graph G. Normally, we write the degree sequence of the graph in increasing order.

10.2.4 Distance in a Tree

In a connected graph G, the distance between two vertices v_i and v_j, denoted as $d(v_i, v_j)$ is defined as

$$d(v_i, v_j) = \text{length of the shortest path between } v_i \text{ and } v_j$$
$$= \text{number of edges in the shortest path between } v_i \text{ and } v_j$$

In a tree, there is one and only one path between every pair of vertices so it that case it is easier to calculate the length of the shortest path. But in a connected graph G which is not a tree to find out the distance between a vertex pair (v_i, v_j) one need to explore all the paths between v_i and v_j and then find out the shortest path.

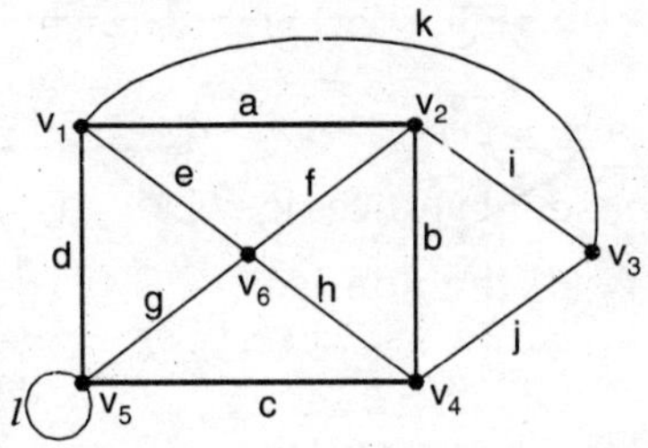

Fig. 10.7

Some of the paths between vertex v_1 and v_3 in Fig. 10.7 are

$$\lfloor a-i,\quad e-f-i, k, e-h-i, d-g-h-i \text{ etc.}\rfloor$$

There is only one shortest path *i.e.*, k and it's length is one. Hence the distance between vertex v_1 and v_3 is one.

10.2.5 Metric

A function $f(x, y)$ of two variables is called a metric if it satisfy certain conditions :

(*i*) Non-negativity: $f(x, y) \geq 0$ and $f(x, y) = 0$ if $x = y$.

(*ii*) Symmetry: $f(x, y) = f(y, x)$

(*iii*) Triangle Inequality: $f(x, y) \leq f(x, z) + f(z, y)$ for any z.

Obviously, the distance $d(v_i, v_j)$ between two vertices of *a* connected graph satisfies the first two conditions. The distance $d(v_i, v_j)$ is the shortest path between v_i and v_j and it cannot be greater than any other path between v_i and v_j which passes through a vertex v_k.

Hence $$d(v_i, v_j) \leq d(v_i, v_k) + d(v_k, v_j)$$

For example, the distance between v_1 and v_2 in Fig. 10.7 is one which is less than the length of the another between v_1 and v_2 which passes through vertex v_6 (*i.e.*, path $e-f$)

So clearly

$$d(v_1, v_2) \leq d(v_1, v_6) + d(v_6, v_2)$$
$$1 \leq 1 + 1$$
$$1 \leq 2$$

Hence $d(v_i, v_j)$ is a metric.

10.2.6 Eccentricity and Centres in a Tree

The eccentricity of a vertex v, denoted as $E(v)$, is the distance from v to the vertex farthest from v in graph G; that is

$$E(v) = \max_{v_i \in G} d(v, v_i)$$

A vertex having minimum eccentricity is called the centre of the graph G; that is,

Centre of G, $C(G) = \{v_i \mid v_i$ has minimum eccentricity$\}$

Example 10.4 Find out the centre (s) in the trees given in Fig. 10.8.

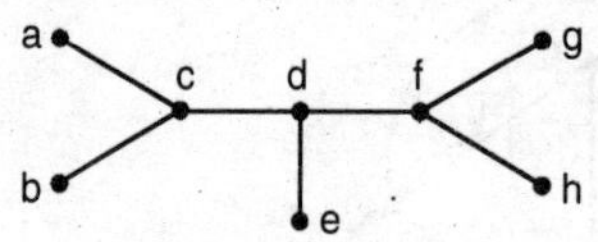

(*a*)

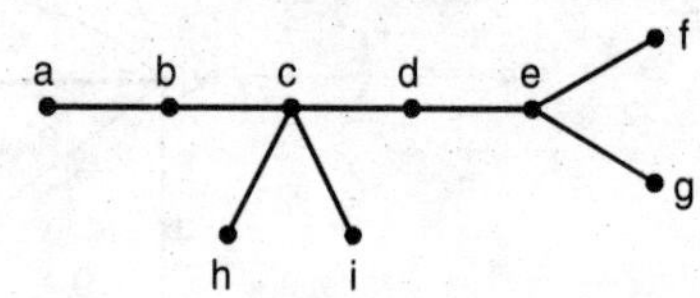

(*b*)

Fig. 10.8

Solution: The eccentricity of each vertex of the graph in **Fig. 10.8**(*a*) is shown in **Fig. 10.9**(*a*) next to the vertices.

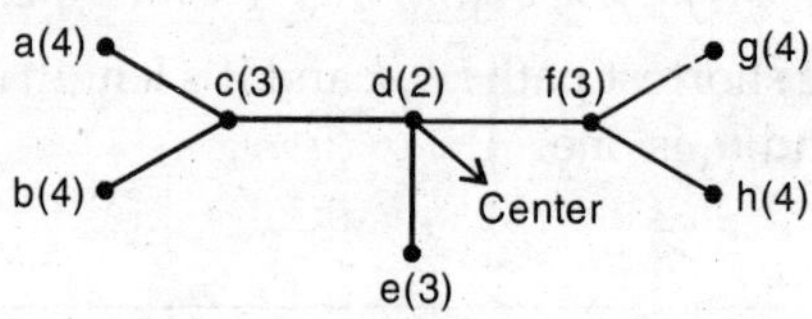

Fig. 10.9(a)

$$E(a) = \max_{v_i \in G} \{d(a, v_i)\}$$

$$= \max \{d(a, b), d(a, c), d(a, d), d(a, e), d(a, f), d(a, g), d(a, h)\}$$

$$= \max \{2, 1, 2, 3, 3, 4, 4\} = 4$$

Similarly calculated eccentricity of every vertex.

Vertex d is having the minimum eccentricity *i.e.*, 2 in the graph (or tree) and hence becomes the centre.

(*b*) Again eccentricity of each vertex of the graph in **Fig. 10.8**(*b*) is shown next to the vertex in the graph **Fig. 10.9**(*b*).

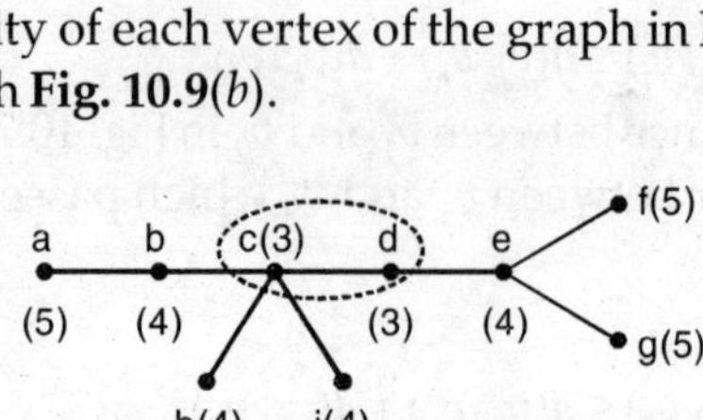

Fig. 10.9(b)

Here two vertices c and d are having minimum eccentricity 3.

Hence this tree has 2 centres.

Theorem 10.8. Show that every tree has either one or two centres.

Proof: A centre in a tree is a vertex with minimum eccentricity and eccentricity is the maximum distance of that vertex from any other vertex of the tree. So in a tree maximum distance between vertex v and v_i *i.e.*, $d(v, v_i)$ occurs only when v_i is a pendant vertex.

Now let us take a tree T with two or more than two pendant vertices and follow the process defined below:

(*i*) Remove all the pendant vertices from tree T to form a graph T' which is still a tree.

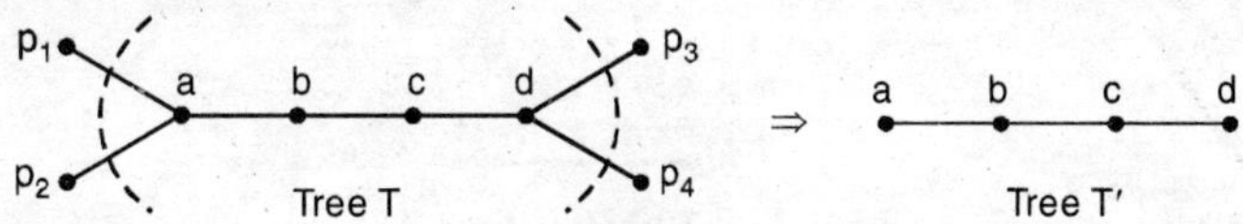

Fig. 10.10

p_1, p_2, p_3, p_4 are pendant vertices.

(*ii*) Calculate the eccentricity of each vertex in T and T'.

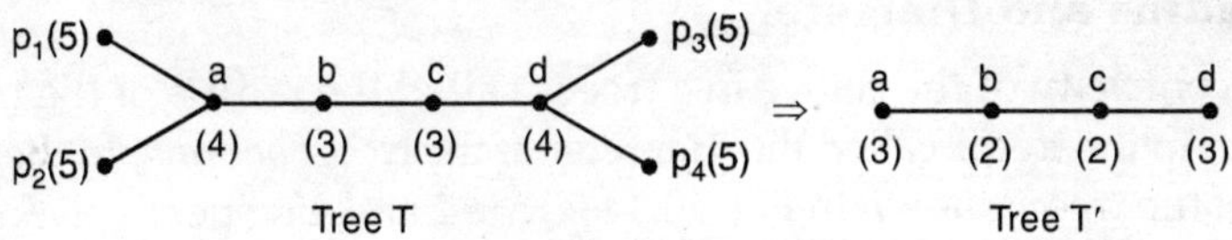

Fig. 10.11

It can be observe that the removal of pendant vertices tree T decreases the eccentricities of the remaining vertices in T' by one.

That is if an internal vertex in T has eccentricity say m, then its eccentricity in T' becomes $(m - 1)$.

So if a vertex has minimum eccentricity in T then after removal of pendant vertices that vertex is still having minimum eccentricity in T'. *For example*, in *Fig. 10.11*, vertices b and c are having minimum eccentricity (3) in T and minimum in T' (*i.e.*, 2) also.

This implies that b and c are the centres in T, as well as in T'

(*iii*) Repeat the process of removal of pendant vertices until there is left either a vertex which is the centre of tree T or an edge, whose terminal vertices are the two centres of T.

Hence the theorem.

Example 10.5. Use the procedure of Theorem 10.7 in finding out the centre(s) of the tree given in Fig. 10.12.

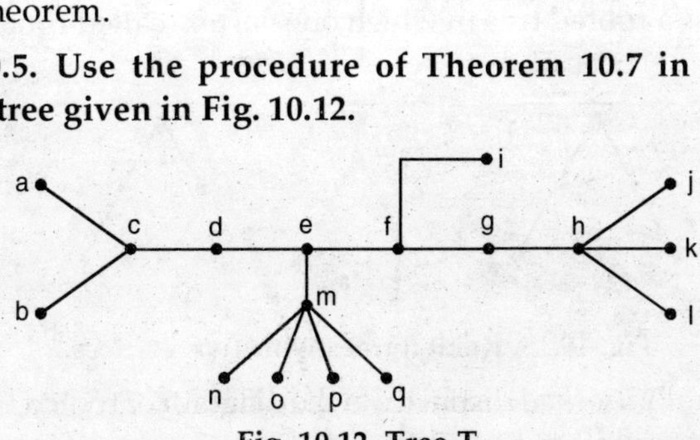

Fig. 10.12. Tree T

Solution: Remove all the pendant vertices ($a, b, i, j, k, l, n, o, p, q$) from tree T to obtain tree T' **Fig. 10.13**(a)

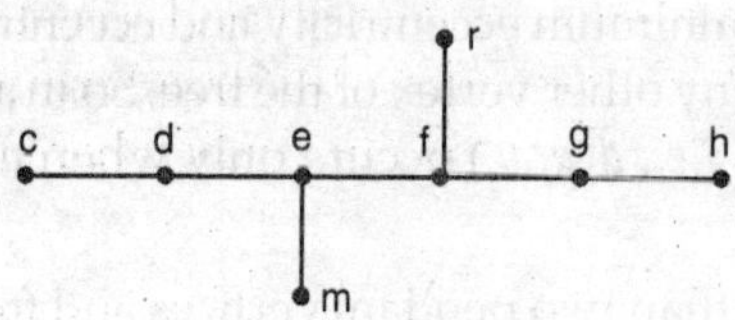

Fig. 10.13(a). Tree T′

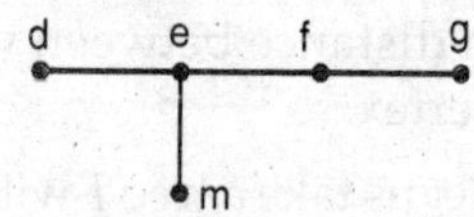

Fig. 10.13(b). Tree T″

Again remove the pendant vertices (c, m, r, h) from T' to obtain T''(**Fig. 10.13** (b)).

T'' obtained by removing all the pendant vertices of tree T is shown in **Fig. 10.13**(c).

Fig. 10.13(c). Tree T‴

Now, since we are left with an edge. Hence vertices e and f are the two centres of the tree T.

10.2.7 Radius and Diameter

The eccentricity of the centre in a tree is called the **radius** of the tree and the longest path in the tree is called the **diameter** of the tree. *For example*, the radius and diameter of the tree T shown in **Fig. 10.14** (a) are 2 and 4 respectively.

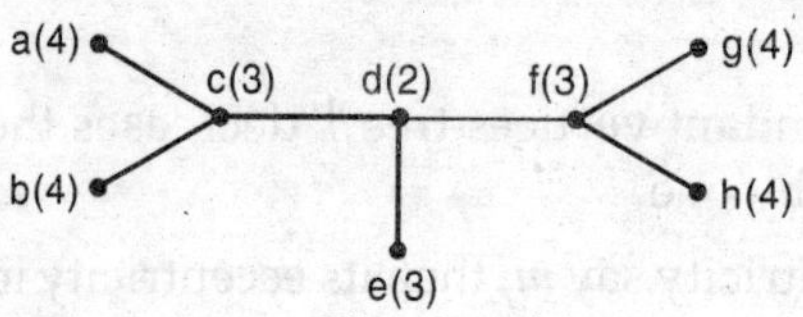

Fig. 10.14(a)

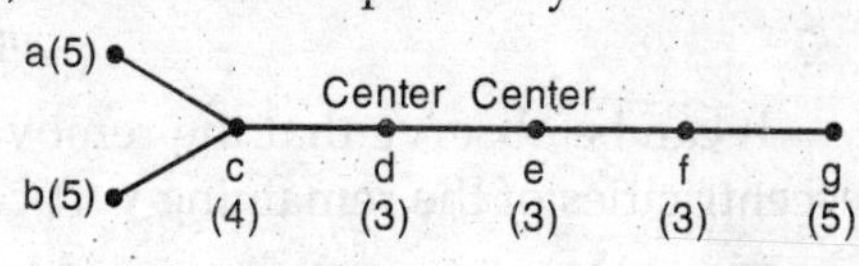

Fig. 10.14(b)

Diameter is also defined as the maximum eccentricity of a vertex in tree T.

As we know that diameter = 2. Radius, but it is not always true in the case of a tree, **Fig. 10.14**(b).

$$\text{radius} = \text{eccentricity of centre(s)} = 3$$

$$\text{diameter} = 5$$

and $$5 \neq 2.3 = 6$$

10.3 ROOTED TREE

A tree is called a **rooted tree** in which one vertex, called **root**, is distinguished from all the other vertices of the tree.

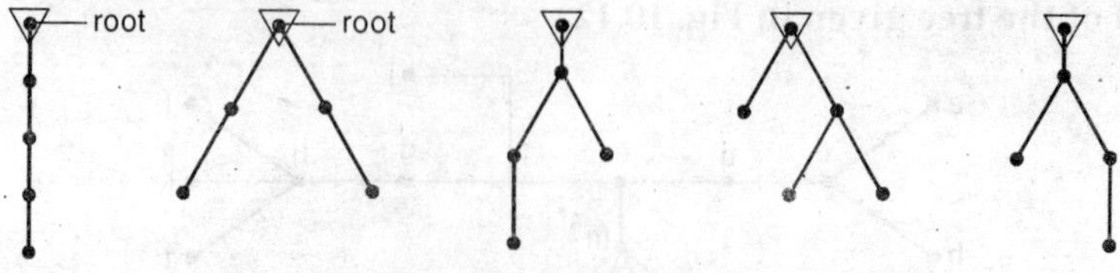

Fig. 10.15. Rooted trees with five vertices.

A root is generally worked distinctly in the diagram of rooted tree. The term tree generally means trees without any root.

10.4 BINARY TREE

A special class of rooted tree is binary tree (or binary rooted tree) in which every vertex has at most two children.

The two children of each vertex is distinguished as the left child and right child of the vertex. *For example* in **Fig. 10.16**(*a*) v_1 is the left child of v_r and v_2 is the right child.

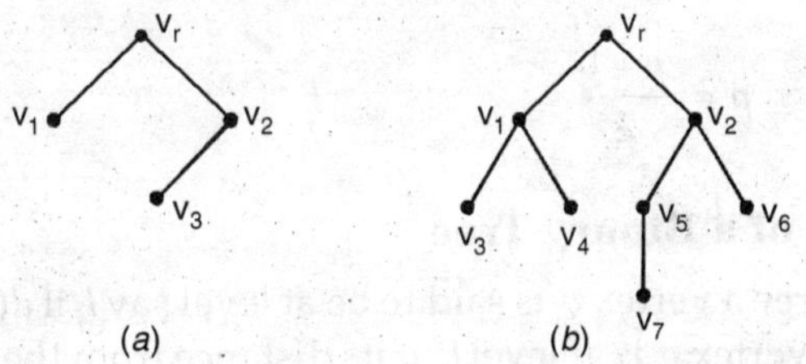

Fig. 10.16. Binary trees

10.4.1 Complete (or Full) Binary Tree

A tree is called a complete binary tree (or full binary tree) if there is exactly one vertex of degree two and each of the remaining vertices is of degree one or three.

Fig. 10.17 shows a full binary tree with 5 vertices.

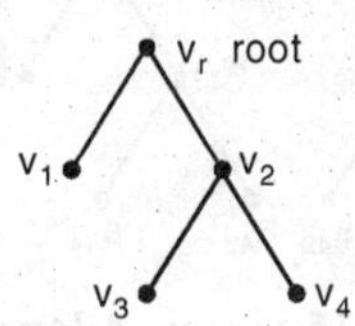

Fig. 10.17. Complete binary tree.

In a binary tree there is exactly **one vertex of degree** two and this property makes it different from the other vertices. Hence the vertex with **degree two** in a binary tree is called a **root** (because the root is always distinguished from all the other vertices). In **Fig. 10.17**, v_r is the root because degree of vertex v_r is two.

By the definition of binary tree one can observe that:

(*i*) The total number of vertices *n* in a tree is always odd.
Since there are total *n* vertices, one is of degree two and the other (*n*– 1) vertices are of degree one or three that is odd degree. We know that (from theorem 1.2) the number, of odd vertices in a graph is always even. Hence *n* is odd.

(*ii*) The total number of pendant vertices in a binary tree are $\frac{n+1}{2}$, where *n* is the total number of vertices.

Let p be the number of pendant vertices then number of vertices having degree three $= n - p - 1$. (*i.e.*, total vertices – pendant vertices – root)

We know that total number of edges in a tree are $n - 1$.

So total degree of the tree is given by

$$1 \cdot p + 3 \cdot (n - p - 1) + 2 = 2(n - 1) \quad \left(\text{by} \sum_{i=1}^{a} \deg(v_i) = 2e \right)$$

$$\Rightarrow \quad p = 2(n-1) - 3(n-p-1) - 2$$

$$\Rightarrow \quad p = 2n - 2 - 3n + 3p + 3 - 2$$

$$= -n - 1 + 3p$$

$$\Rightarrow \quad p = \frac{n+1}{2}$$

10.4.2 Height of a Binary Tree

In a binary tree a vertex v_i is said to be at level say l_i if $d(v_i, v_r) = l_i$, where v_r is the root. That is vertex v_i is at level l_i if its distance from the root is l_i.

Hence the level of root (v_r) in a binary tree is zero because $d(v_r, v_r) = 0$.

Fig. 10.18 shows a binary tree of 15 vertices

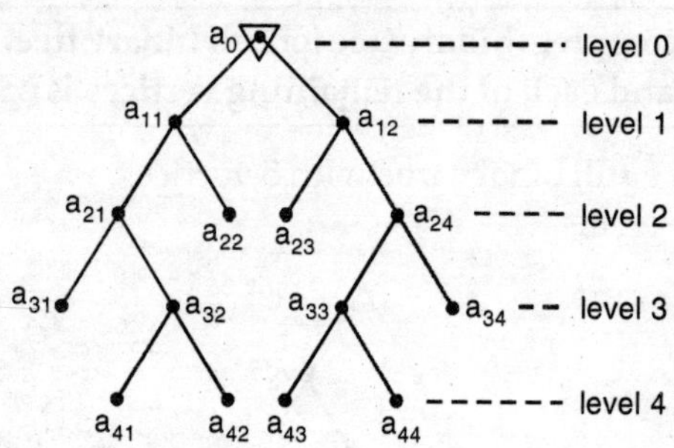

Fig. 10.18. A 15-vertex, 4-level binary tree

The maximum level, l_{max} of any vertex v in a binary tree is called the height of the tree.

That is height = l_{max} of any vertex. For example in **Fig. 10.18** vertices a_{41}, a_{42}, a_{43} and a_{44} are at maximum level *i.e.*, 4 and hence the height of this binary tree is 4.

Let us find out the maximum possible height and minimum possible height of a binary tree of n-vertices.

Suppose there are p-levels in a n-vertex binary tree then according to the definition of binary tree the maximum number of vertices possible in a p-level binary tree is:

$$n \leq 2^0 + 2^1 + 2^2 + \ldots + 2^p$$

There can be one vertex *i.e.*, not at level 0, at must two vertices at level 1, at most four vertices at level 2 and so on.

Level p is the minimum possible height *i.e.*, min l_{max} if there are exactly two children of every vertex in a binary tree. So here

$$n = 2^0 + 2^1 + \ldots + 2^p$$

$$n = 2^0 + 2^1 + 2^2 + \ldots + 2^p = \frac{(1-2^{p+1})}{(1-2)}$$
$$n = -1 + 2^{p+1}$$
$$2^{p+1} = (n+1) \Rightarrow p + 1 = \log_2 (n+1)$$
$$\Rightarrow \quad p = \lceil \log_2 (n+1) - 1 \rceil \quad i.e. \quad \min l_{max} = \lceil \log_2 (n+1) - 1 \rceil$$

where $\lceil n \rceil$ denotes the smallest integer greater than or equal to n.

On the other hand, to construct a binary tree with maximum possible height *i.e*, max l_{max}, we must have exactly two vertices at each level.

i.e., $\quad 2^0 + 2^1 + 2^1 + \ldots + 2\ (p)$ times $= n$

$1 + 2 + 2 + 2 + \ldots p$ times $= n$

$$p \cdot 2 = n - 1$$
$$\Rightarrow \quad p = \frac{n-1}{2}$$
i.e., $$\max l_{max} = \frac{n-1}{2}$$

Fig. 10.19, shows 13-vertex binary trees with min l_{max} and max l_{max} respectively.

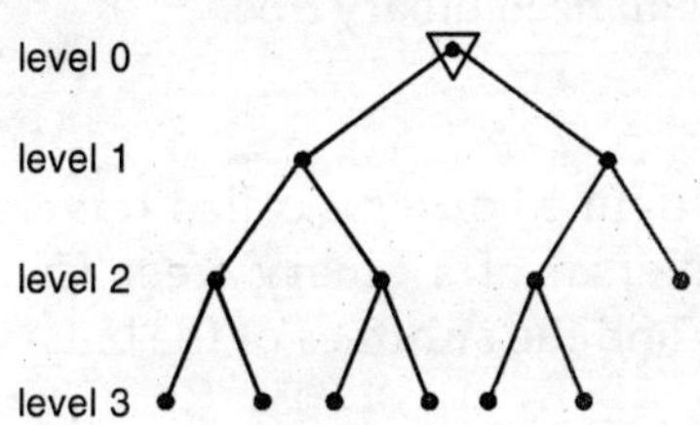

min lmax = ⌈log2(13 + 1) – 1⌉

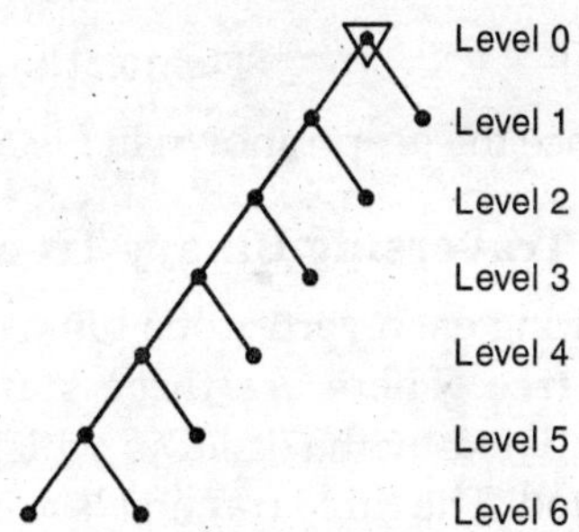

max lmax = $\frac{13-1}{2}$ **= 6**

Fig. 10.19

The term **path length** is calculated by taking the sum of the levels of all pendant vertices in a binary tree. The path lengths of the trees shown in **Fig. 10.19** are 20 *i.e.,* (2 + 3 + 3 + 3 + 3) and 21 *i.e.,* (1 + 2 + 3 + 4 + 5 + 6) respectively.

10.4.3 Height Balanced Binary Tree

A binary tree is called height balanced binary tree if the difference between the heights of the left subtree and right subtree of each vertex is at most one.

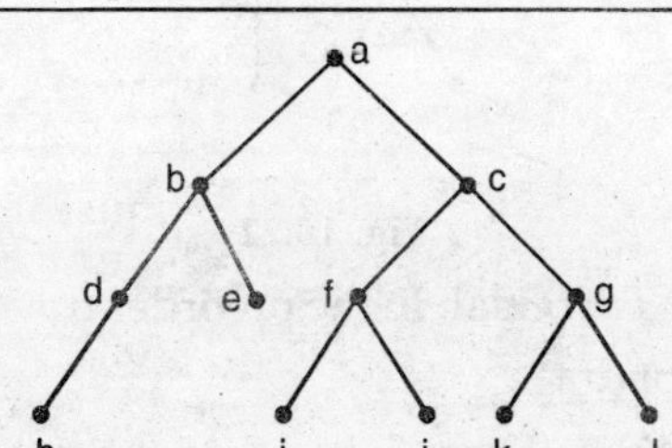

Fig. 10.20. A height balanced binary tree

The graph shown in **Fig. 10.20** is a height balanced binary tree.

The difference between the heights of the two subtrees of a is zero as the maximum level *i.e.*, l_{max} in both the subtrees is two (**Fig. 10.21**(*a*)).

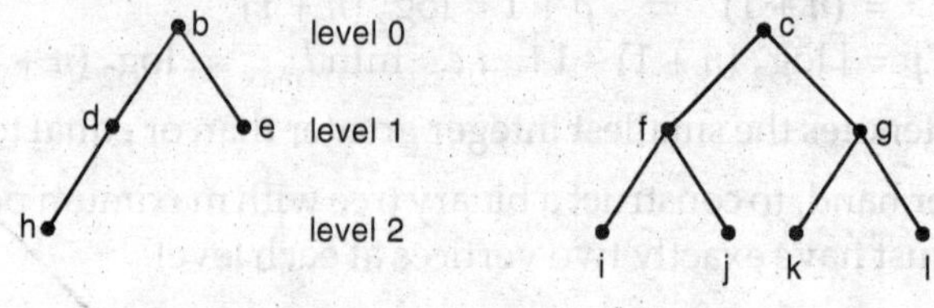

(*a*) (*b*)

Fig. 10.21. Left and right subtree of 'a'

Similarly the difference between the heights of the subtrees of vertex *c* is alsos zero and the difference between the heights of the subtrees of vertex *b* is one. (**Fig. 10.21**(*b*)).

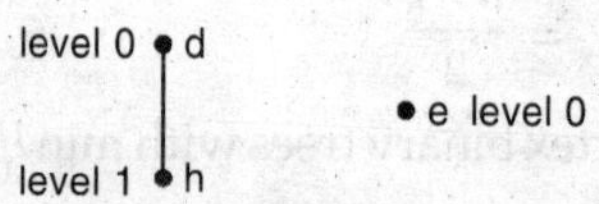

Fig. 10.21(b) Left and right subtree of 'b'

Hence the graph shown in Fig. 10.21 is a balanced binary tree.

10.4.4 Traversing Binary Trees

Visiting **each vertex** of a binary tree in a defined order is called **traversal of binary tree**. There are three standard traversal of a binary tree. They are differentiated by the order in which the **root** and the **subtrees** of the binary tree are visited. The three traversals are

(*i*) Preorder traversal (NLR) or node-left-right traversal.

(*ii*) In order traversal (LNR) or left-node-right traversal

(*iii*) Post order traversal (LRN) or left-right-node traversal

Example 10.6. For the binary tree shown in Fig. 10.22, process the (*i*) preorder (*ii*) in-order and (*iii*) post-order traversals.

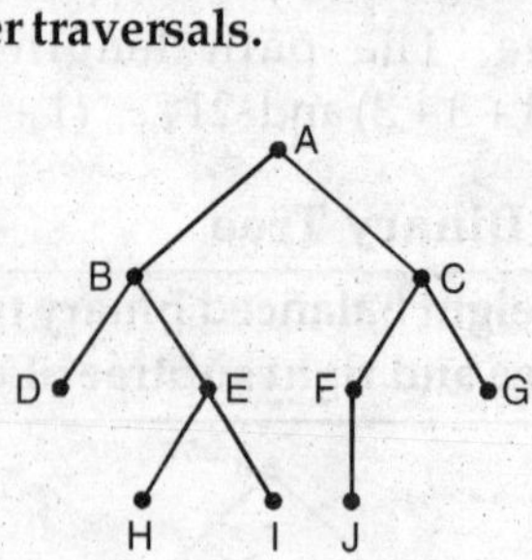

Fig. 10.22

Solution: (*i*) Preorder Traversal: In the preorder traversal of a binary tree *T*,

Step I: Process the root (v_r)

Step II: Then traverse the left tree T_L and then the right tree T_R in preorder.

That is process the root of the left tree first then the left subtree and right subtree of the left tree and then process the root of the right subtree T_R and then the left subtree and right subtree of T_R.

Step III: Stop, when each vertex is traversed in preorder.

For the given tree (**Fig. 10.22**) A is the root B and C are the left tree and right tree of A with root B and C respectively D and E respectively are the left and right subtrees of left tree B and similarly F and G are the left and right subtrees of the right tree C respectively.

The preorder traversal of tree T will process root A first, then traverse T_L and T_R. Preorder traversal of T_L visits the root B and then the left and right subtrees of B. That is now visit D (left subtree of left tree B) and then the preorder traversal of the right subtree E of the left tree B visits root E first and then H, and I. Similarly, the preorder traversal of T_R processes the root C and the root of the left subtree, *i.e.*, F then J and in the end vertex G which is the right subtree of the right tree T_R.

Hence, the preorder traversal of T is

$$A-B-D-E-H-I-C-F-J-G$$

(*ii*) In-order (LNR) Traversal:

Step I: (*a*) Find out the node or root of the tree and it's left and right subtrees.

(*b*) Process the left subtree T_L is in-order. That is process the left most child first node and then the right must child.

Step II: Now process the root or node (v_r).

Step III: Process the right subtree T_R in in-order.

Step IV: Stop when all the vertices are traversed.

Hence the in-order traversal for the graph given in Fig. 10.22 is

LNR Traversal:

$D-B-H-E-I-A-J-F-C-G.$

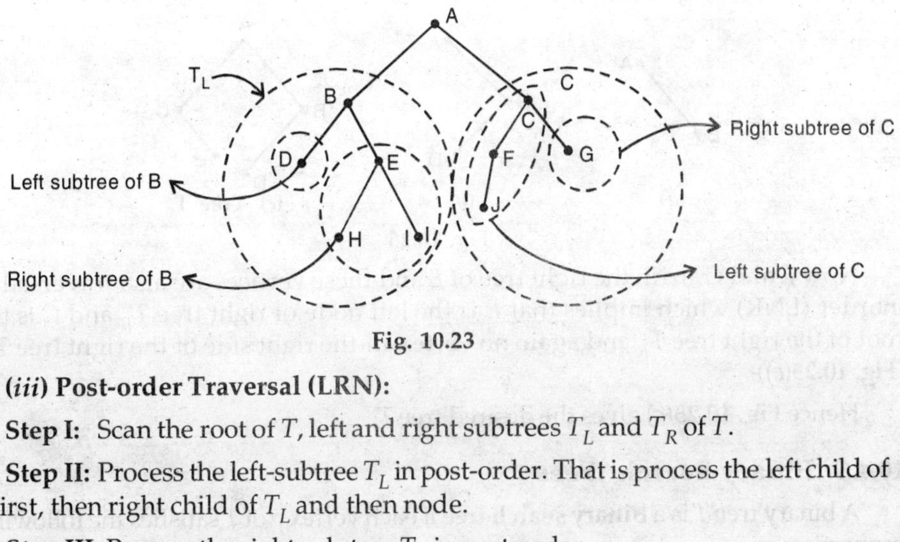

Fig. 10.23

(*iii*) Post-order Traversal (LRN):

Step I: Scan the root of T, left and right subtrees T_L and T_R of T.

Step II: Process the left-subtree T_L in post-order. That is process the left child of T_L first, then right child of T_L and then node.

Step III: Process the right sub-tree T_R in post-order.

Step IV: Process the root or node (v_r) in the end of the end of the traversal and stop.

Hence the post-order traversal thus obtained of the graph given in **Fig. 10.22** is

(LRN) Traversal: $D-H-I-E-B-J-F-G-C-A$

Example 10.7. List the order in which the vertices are processed using preorder, inorder and post order traversal for the tree in Fig. 10.24.

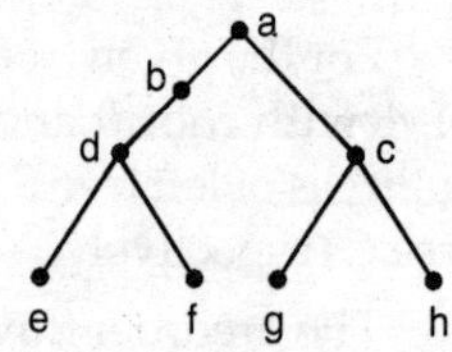

Fig. 10.24.

Solution: Preorder: *a b d e f c g h.*

Inorder: *e d f b a g c h.*

Post order: *e f d b g h c a.*

Example 10.8. Draw a binary tree *T* whose preorder and inorder traversals are *ABDCE* and *BDAEC* respectively.

Solution: Preorder (NLR) – *A B D C E*

⇒ *A* is the root.

Inorder (LNR) $\underbrace{BD}_{T_L}\ \underbrace{A}_{\text{root}}\ \underbrace{EC}_{T_R}$

⇒ *B-D* in the left tree and *E-C* in the right tree.

Lets start with preorder traversal, *A* is the root then comes *B* which means that *B* is in the left of *A*. (That is *B* is the root of left tree of *A*). **Fig. 10.25**(*a*).

Now inorder traversal is *B D A E C*

⇒ *B* is the root of T_L and *C* is the root of T_R

Since left tree of *A* is also traversed in order, so there should be a vertex prior to *B* in the inorder traversal. But as we can see that the traversal is starting from *B* itself implies that the left tree *B* does not have a left child in T_L and hence *D* is the right child of *B*. (**Fig. 10.25**(*b*)).

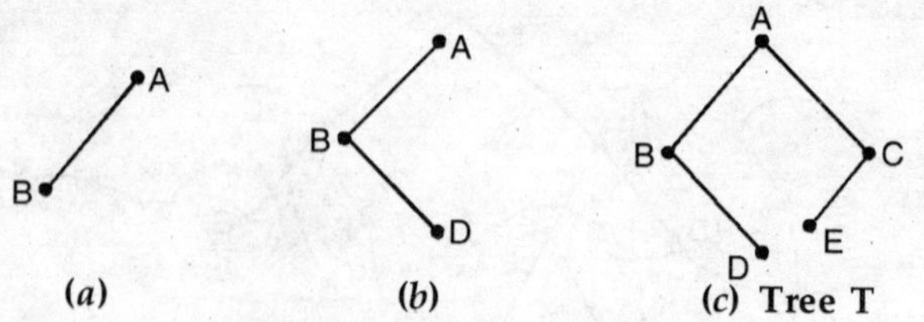

Fig. 10.25

Now *E* and *C* are in the right tree of *E* and these vertices are also traversed in inorder (LNR) which implies that *E* is the left node of right tree T_R and *C* is the root of the right tree T_R and again no vertex-in-the right side of the right tree T_R. (**Fig. 10.25**(*c*)).

Hence **Fig. 10.25**(*c*) gives the desired tree *T*.

10.4.5 Binary Search Trees

A binary tree *T* is a **binary search tree** if each vertex *v* of *T* satisfies the following properties:

(*i*) Each data item in the left subtree of v is less than the data item at vertex v.

(*ii*) Each data item in the right subtree of v is greater than the data item at v.

Hence the inorder traversal of binary searched tree gives the sorted list of the elements of T.

The inorder traversal of the binary searched tree given in **Fig. 10.26** is

$$10 - 15 - 20 - 21 - 22 - 40 - 50 - 55 - 60$$

which is the sorted list of the elements of the balanced binary tree.

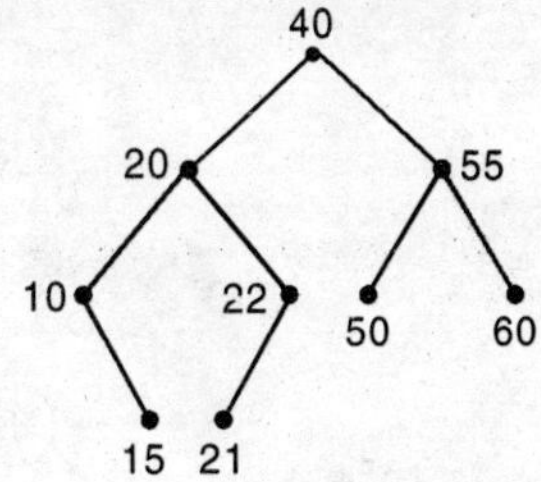

Fig. 10.26. A binary searched tree

Example 10.9. Draw the tree T if the following numbers are inserted into an empty binary search tree T: 52, 35, 46, 24, 79, 37, 62, 42.

Solution: Lets mark 52 as the root of the tree T

35, 24, 26, 47, 42 are less than 52 so all of them are in the left subtree of 52. Make a left child of 52 and label it 46 as 46 > 35, 46 > 24, 46 > 37, 46 > 42. **Fig. 10.27**(*a*) and so on.

Now the remaining number are 62 and 79 both are greater than 52 and in the right subtree of 52. The root of the right subtree is 62 and 79 is the label of the right child of 62. (**Fig. 10.27**(*b*)).

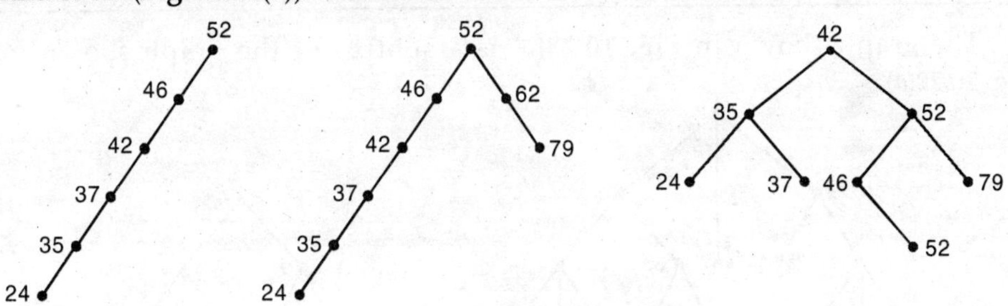

(*a*) (*b*) (*c*)

Fig. 10.27

Fig. 10.27(*b*) the desired binary search tree.

Another binary search tree is given in **Fig. 10.27**(*c*) rooted with a vertex labeled with 42.

In order traversal is 24 – 35 – 37 – 42 – 46 – 52 – 62 – 79.

Example 10.10. Represent the following arithmetic expression as a binary tree. **(U.P.T.U. 2002-03)**

$$(((a + b * c) * d - e/(f + g)) + h*i*j$$

Solution:

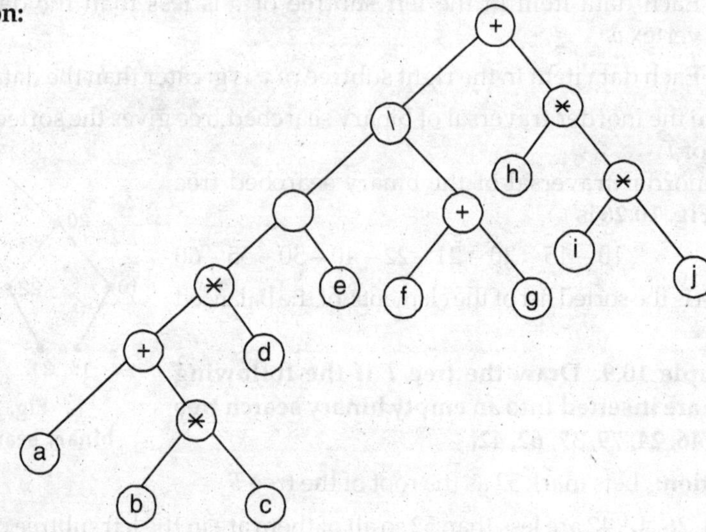

10.5 SUBTREE

A subtree $T' = (V', E')$ with v as the root is a subgraph of $T = (V, E)$, such that $v \in V'$ and contains all the descendants of v and E' contains edges emanating from v. A subtree that has a son of v as root is called the subtree of v.

The graph shown in **Fig. 10.28**(*b*) is a subtree of the graph shown in **Fig. 10.28**(*a*).

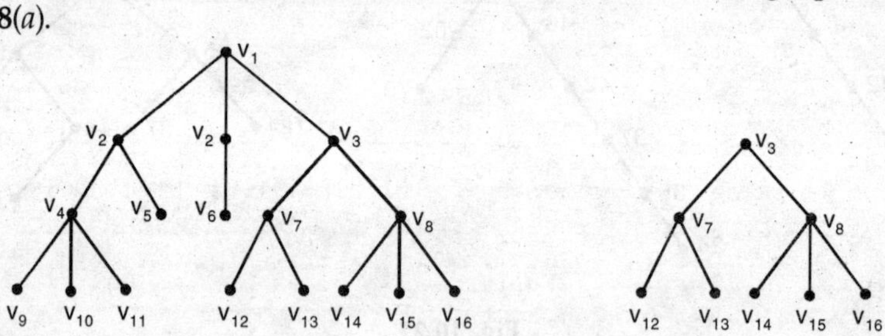

(*a*) Tree T, root is v1. (*b*) Subtree T′, root is v_3.

Fig. 10.28

Notes : In Fig. 10.28(*a*),

(*i*) v_{12} is the child of v_7 so is v_{13}.

(*ii*) v_7 is v_{12}'s and v_{13}'s parent *i.e.*, father.

(*iii*) v_1, v_3 and v_7 are v_{12}'s ancestors.

(*iv*) v_3, v_7 and v_{12} are descendants of v_1.

10.6 SPANNING TREE

A tree *T* which is a subgraph of a connected graph $G = (V, E)$ is called a spanning tree if *T* contains all the vertices of *G*.

Since *T* is a tree so it is free from circuits. Hence a circuitless connected subgraph *T* of a graph *G* is a **spanning tree** of *G* if *T* contains **all** the vertices of *G*.

For example, the subgraph T shown in **Fig. 10.29(*b*)** is a spanning tree of the graph G shown in Fig. 10.29(*a*).

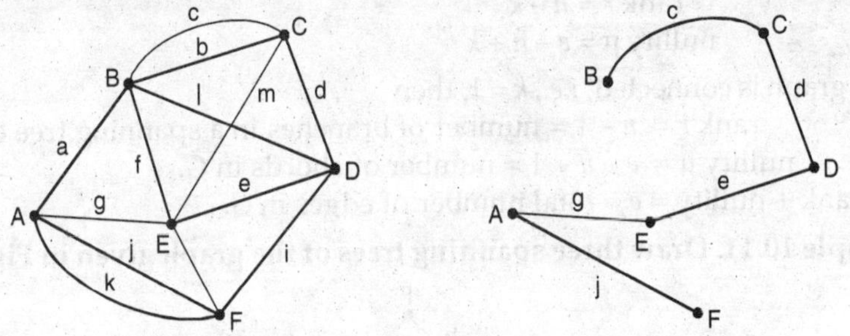

(*a*) Connected graph G (*b*) Spanning tree T.

Fig. 10.29

A given graph has numerous subgraphs, from e edges 2^e distinct combinations are possible and some of these subgraphs will be trees and some of these trees will be spanning trees. So for a given graph there are always more than one spanning trees.

Note: We cannot define spanning tree for a disconnected graph because tree is always connected. However each connected component of the disconnected graph will have a spanning trees.

10.6.1 Branch and Chord

An edge $e \in T$ (spanning tree) is called a branch of spanning tree T.

For example, edges c, d, e, g and j are the branches of the spanning tree T shown in **Fig. 10.29**(*b*).

An edge $e \notin T$ (spanning tree) is called a chord.

For example, for the given spanning tree T (**Fig. 10.29**(*b*)) of graph G (**Fig. 10.29**(*a*)) edges a, b, f, l, m, i and k are chords. Clearly set of chord and set of branches are two disjoint sets.

Hence we can say that $G = T \cup \overline{T}$

where T is the spanning tree of G and $\overline{T}$ is the complement of T in G.

For the given graph G and its spanning tree T, shown in **Fig. 10.29**(*a*) and **Fig. 10.29**(*b*) respectively.

Total number of edges in $G = 12 = e$

Total number of vertices in $G = 6 = n$

Total number of vertices in $T = 6 = n$

Total number of branches in $T = 5 = n - 1$

So total numbers of chords with respect to spanning tree T = total number of edges in G – Total number of branches in T

$$= 12 - 5 = 7$$

Hence in a connected graph G of n vertices and e edges, a spanning tree has $\boldsymbol{n - 1}$ edges or tree branches and $\boldsymbol{e - (n - 1)}$ chords.

10.6.2 Rank and Nullity

Let G be a graph with n vertices, e-edges and k-components (the three fundamental numbers) then the two other important numbers can be derived from these numbers. The two numbers are called **rank** and **nullity** respectively and defined as

$$\text{rank } r = n - k$$
$$\text{nullity } \mu = e - n + k$$

If the graph is connected; *i.e.*, $k = 1$, then

$$\text{rank } r = n - 1 = \text{number of branches in a spanning tree of } G.$$
$$\text{nullity } \mu = e - n + 1 = \text{number of chords in } G.$$

and rank + nullity = e = total number of edges in G.

Example 10.11. Draw three spanning trees of the graph given in Fig. 10.30.

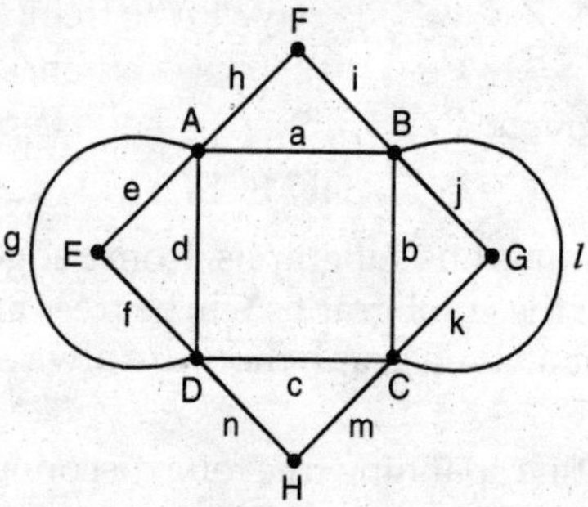

Fig. 10.30

Solution:

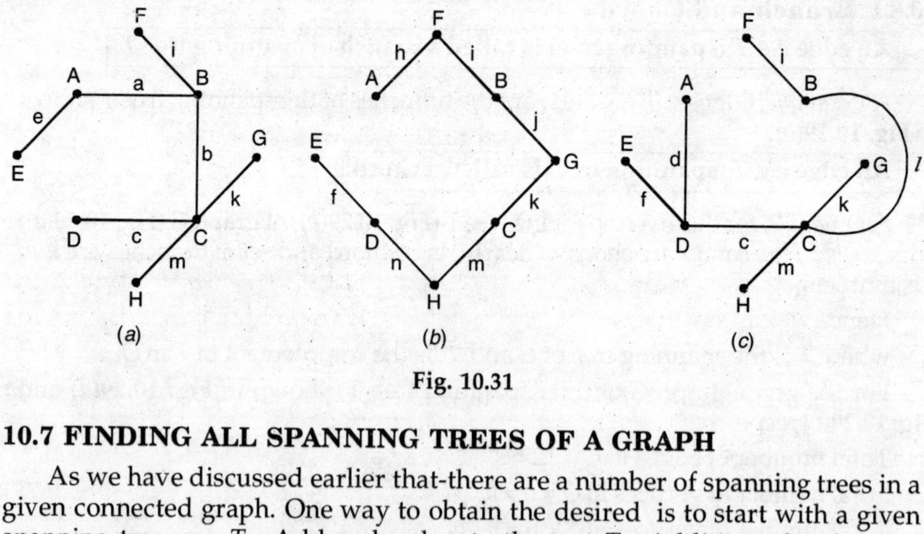

Fig. 10.31

10.7 FINDING ALL SPANNING TREES OF A GRAPH

As we have discussed earlier that-there are a number of spanning trees in a given connected graph. One way to obtain the desired is to start with a given spanning tree, say T_1. Add a chord c_1 to the tree T_1. Addition of c_1 forms a fundamental circuit. Removal of any branch, say b_1 from the fundamental circuit, just formed by chord c_1, will create a new spanning tree T_2 and so on.

This type of generation of one spanning tree from another through addition of a chord and detection of an appropriate branch, is called a **cyclic intechange** or **elementary tree transformation**.

Let us illustrate the procedure through an example. Consider the graph G shown in **Fig. 10.32**(a) and one of is spanning tree T shown in **Fig. 10.32**(b).

Chord set with respect of spanning tree T_1 is

$$C = (e, c, f, h)$$

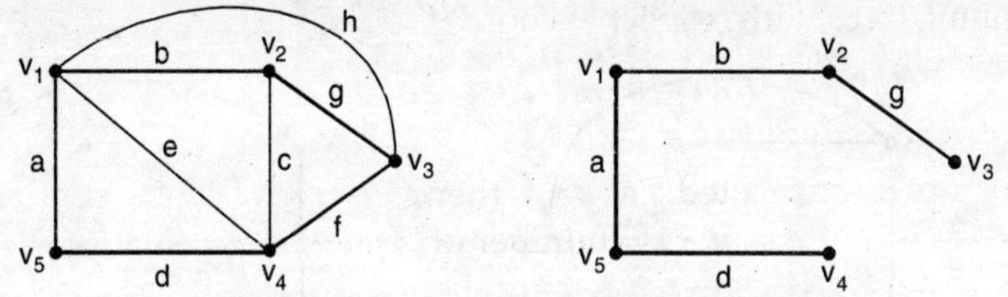

(a) Graph G (b) Spanning tree T_1,

Fig. 10.32

Now start with tree T_1, add chord, say e to the tree T_1. This forms a fundamental circuit (a, d, e) in **Fig. 10.33**(a). Removal of any branch, say a from the fundamental circuit (a, d, e) will create a new spanning tree T_2 (**Fig. 10.33**(a)).

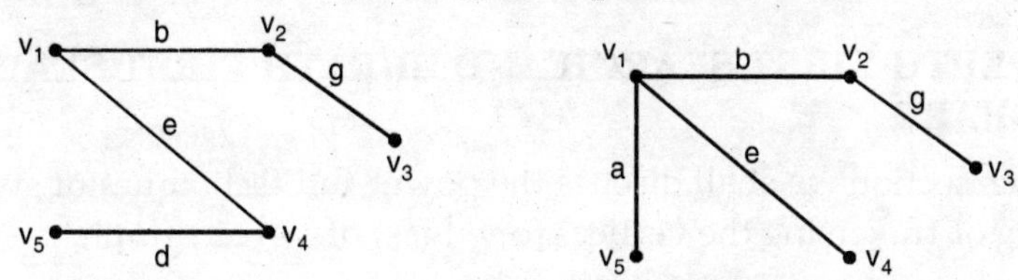

(a) Spanning tree T_2. (b) Spanning tree T_3.

Fig. 10.33

Instead of deleting branch a, if one deletes branch d from the fundamental circuit (a, d, e) then obtains another spanning tree T_3 (**Fig. 10.33**(b)).

After generating the two spanning trees, each with chord e in it, we can restart with tree T_1 and add a different chord (c, f, or h) and repeat the process of obtaining a new spanning tree. **Fig. 10.33**(c), (d) and **10.33**(e) shows the spanning trees formed by chord c, f and h respectively.

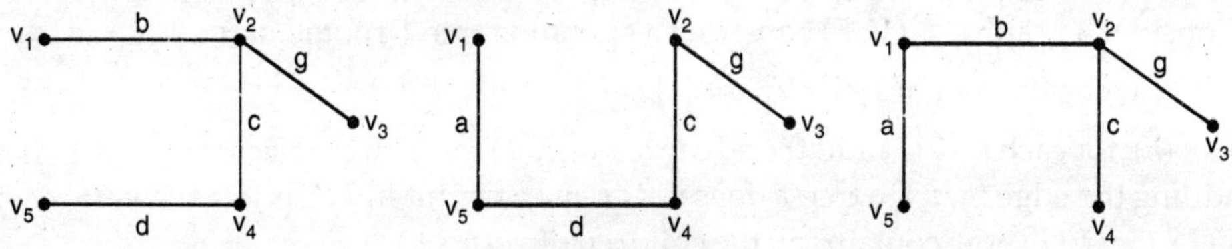

Fig. 10.33 (c)

Spanning trees with respect to chord 'c'

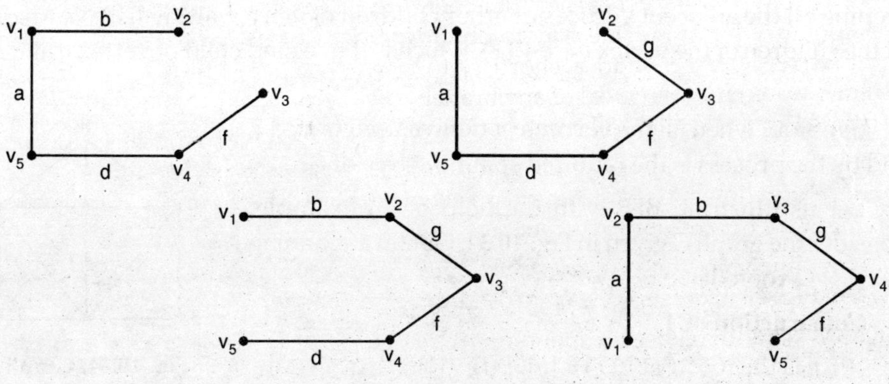

Fig. 10.33(d)

Spanning trees with respect to chord 'f'.

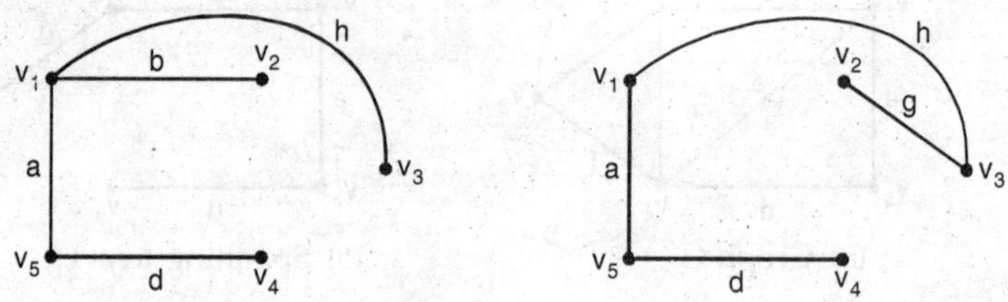

Fig. 10.33(e)

Spanning trees with respect to chord 'h'.

Thus we have obtained all the spanning trees for the given graph G.

10.8 DEPTH-FIRST SEARCH AND BREATH FIRST SEARCH IN A GRAPH

In this section we shall discuss the power full techniques of systematically searching or traversing the vertices (or edges) of a given graph.

In DFS (Depth-first-Search), we explore always from the most recently discovered vertex that has unexplored edges. This method is also called **bracktracing** whereas BFS (Breath-first-Search) explores from the oldest vertex. The basic difference between the two, *i.e.*, BFS and DFS, is that is DFS we maintain the list of vertices to be searched as Last In First Out (LIFO) "Stack" rather than queue. Both BFS and DFS generates a spanning tree T rooted at a vertex say v_o.

10.8.1 BFS

We are interested in finding out a spanning tree in a graph G using BFS. Consider a graph $G = (V, E)$ construct a spanning tree T rooted at $u \in V(G)$ as:

(*i*) x be the root of T. Set $V(T) = \{u\}$

(*ii*) for each $u \in V(T)$ add the edge $(u, v_i) \in E(T)$ for each v_i adjacent to u such that adding the edge (u, v_i) in tree T does not create a circuit in T. Update the vertex set $V(T)$. (Set $V(T)$ now contains all the children of vertex u).

(*iii*) Repeat step (*ii*) for each v_i in $V(T)$ one by one. That is, select the very first child say v_i, of u first and add all the adjacent vertices of the first child v_i in set $V(T)$, keeping all the adjacent vertices of other children of u in a waiting list. Now scan all the children of the vertex v_2, $\in V(T)$, which is the second child of vertex u and so on.

(*iv*) Stop, when all the vertices of G have been traced T find by the process is the required spanning tree of G.

Let us illustrate BFS with the help of an example. Consider the graph G given in **Fig. 10.34.** Obtain a spanning tree T of G rooted at u.

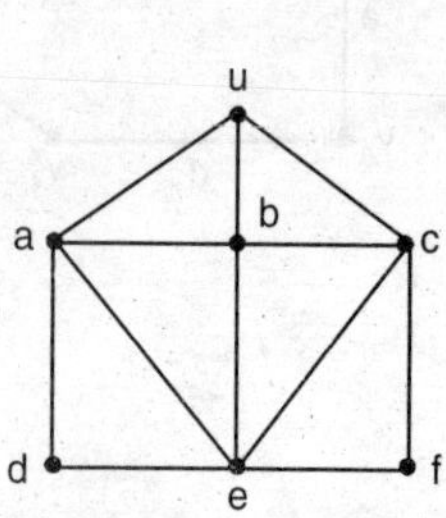

Fig. 10.34. Graph G

Construction of T

(*i*) u is the root. Set $V(T) = [u]$, $E(t) = \{\phi\}$

(*ii*) List of the vertices adjacent to u: $\{a, b, c\}$ none of the three is in $V(T)$. Traverse the edges (u, a), (a, b) and (u, c) and add then in $E(T)$ as three don't creat any circuit in T (**Fig. 10.35(*a*)**).

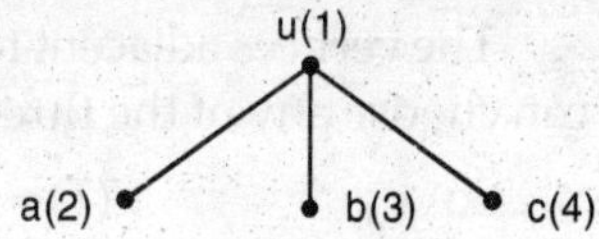

Fig. 10.35(a)

New $\quad V(T) = \{u\} \cup \{a, b, c\}$

$= \{u, a, b, c\}, E(T)$

$= \{(u, a), (a, b), (u, c)\}$

(*iii*) List of all vertices adjacent to $a = \{a, d, e\}$.

As $b \in V(T)$ and edge (a, b) creates a circuit in T so (a, b) cannot be included in T.

Vertices d and e have not been traversed before so $V(T) = \{u, a, b, c\} \cup \{b, d, e\}$ and new $E(T) = \{(u, a), (u, b), (u, c), (a, d), (a, e)\}$ (**Fig. 10.35(*b*)**).

Proceeding in the same manner, we have obtained a spanning tree T for G as:

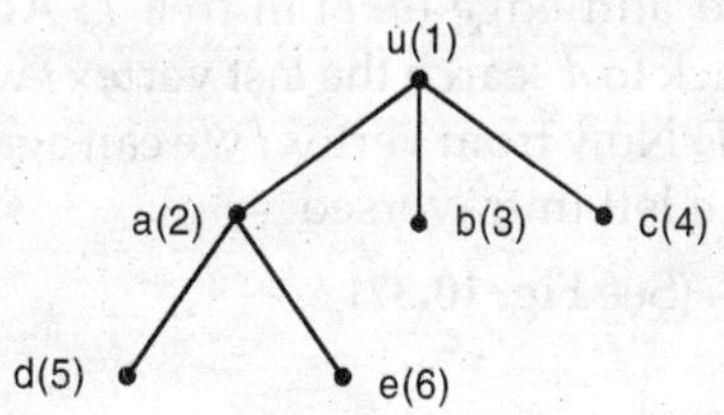

Fig. 10.35 (*b*)

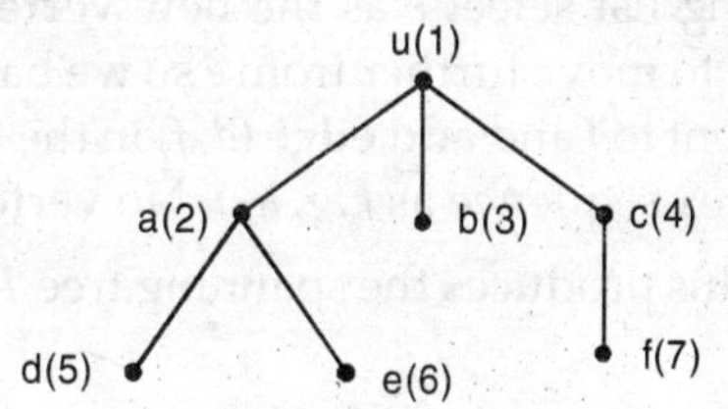

Fig. 10.35 (*c*)

10.8.2 DFS

Consider a graph $G = (V, E)$. Obtaining a spanning tree T in G rooted at vertex u as:

(*i*) Set $V(T) = \{u\}$, $E(T) = \{\phi\}$

(*ii*) for a vertex $u \in V(T)$, let L be all the set of all the vertices adjacent to u. Select one such vertex which is not traced before (keeping all other vertices of L in waiting list) say v_i. Add unexplored edge (u, v_i) in $E(T)$ such that adding (u, v_i) to T does not create cycle. If there is no such edge left go to step (*iv*)

Set $V(T) = \{u, v_i\}$ $E(T) = \{(u, v_i)\}$

(*iii*) Repeat step (*iii*) for latest searched v_i and update the sets $V(T)$ and $E(T)$ respectively.

(*iv*) Backtrack and terminated the process. If vertex $z \in V(T)$ is the parent of the latest searched v_i in tree T, repeat step (*ii*) for z (*i.e.*, construct a subtree T rooted at z using step 2). If there is no such vertex in $V(T)$ which is the parent of the new searched vertex v_i in T, terminate the process. T is thus obtained is the desired spanning tree T of G.

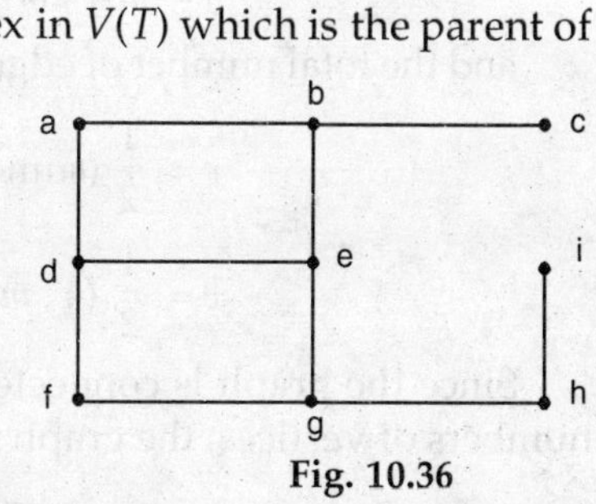

Fig. 10.36

Example. 10.12. Find a spanning tree T in the graph G given in Fig. 10.36 using DFS.

Solution: Let us arbitrarily start with vertex d on can choose any other vertex.

Set $V(T) = \{d\}$ $E(T) = \{\phi\}$

The vertices adjacent to d: $\{a, f, e\}$. Trace the vertex a and add edge (d, a) in T. (one can choose any of the three vertices).

So $\qquad V(T) = \{d, a\}$ and $E(T) = \{(d, a)\}$

So the new (or latest) searched vertex is a now trace one of the adjacent vertex of a by keeping vertices (f, e) in a waiting list and add the unexplored edge from a in tree T such that adding the edge does not create a circuit with previously selected edges in T. In this way we have obtained a sequence of vertices which are in T as:

$$S = \{d, a, b, c\}$$

We cannot go beyond C as there is no unexplored edge left in G which is not in T. So we backtrack to b. Adding edge (b, e) in T will create a circuit. So we back track to a. But, there is no edge left which is not in tree T from vertex a. So we back track to d now select any one of the two adjacent vertices of d that are in a waiting list select e as the new vertex and add edge (d, e) in tree T. Again no scope to move further from e so we backtrack to d search the last vertex f which is adjacent to f and add edge (d, f) in the tree T. Now from vertex f we can again find a vertex sequence as f, g, h, i. No vertex in a left untraversed.

This produces the spanning tree T in G. (See **Fig. 10.37**)

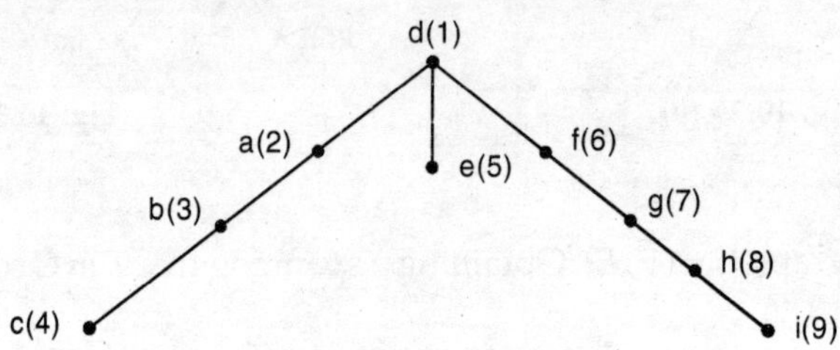

Fig. 10.37

10.9 ON COUNTING TREES

The problem of counting trees was discovered by Arthus Cayley in 1857, while he was trying to count the numbers of structured isomers of the saturated hydrocarbons. C_mH_{2m+2}. In fact Cayley was the first who discovered trees during this process. He represented the C_mH_{2m+2} molecule by using a connected graph in which a carbon atom was represented by a vertex of degree μ and a hydrogen atom by a vertex of degree one.

So the total number of vertices in such a graph is

$$n = m + 2m + 2 = 3m + 2$$

and the total number of edges

$$e = \frac{1}{2} \text{ (sum of the degrees)}$$

$$e = \frac{1}{2} (4 . m + 1 . (2m + 2)) = \frac{6m+2}{2} = 3m + 1$$

Since the graph is connected and the number of edges is one less than the numbers of vertices, the graph formed is a tree.

So the problem of finding the structural isomers of a given hydrocarbon becomes the problem of counting trees.

10.9.1 Labelled Graph

A graph in which each vertex is assigned a unique label or name is called a labelled graph.

Fig. 10.38 shows the three different labelled graphs.

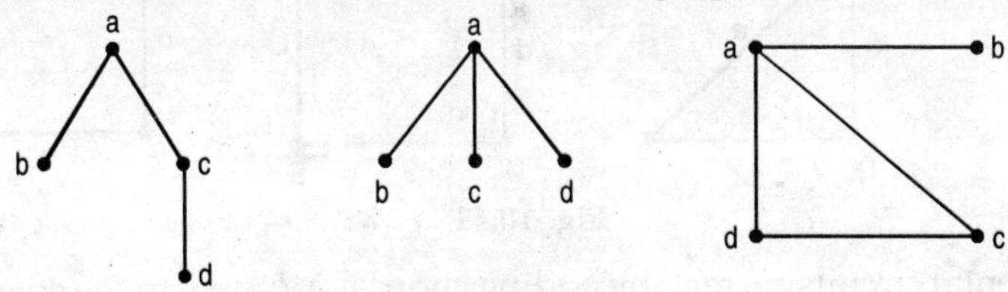

Fig. 10.38

When we are counting the number of different graphs, we are counting the labelled graphs.

Theorem 10.9. The number of labelled trees with n-vertices, when $n \geq 2$ is n^{n-2}.

Proof: The proof of this theorem is given in Chapter-9. This theorem was first stated and proved by Cayley and hence called *Cayley's theorem*.

The different trees for $n = 3$ vertices are shown in Fig. 10.39. The total number of different labelled trees with $n = 3$ vertices is $3^{3-2} = 3^1 = 3$ (according to Cayleys theorem).

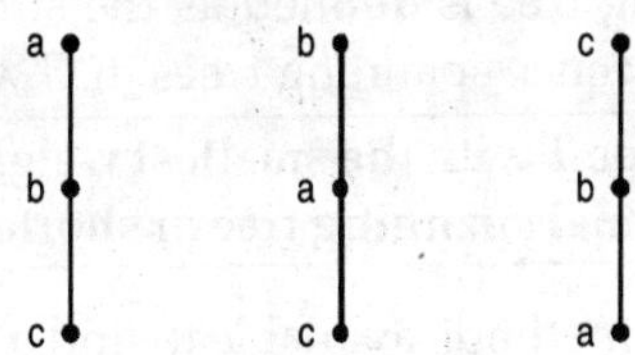

Fig. 10.39. All three of three labelled vertices

10.10 FUNDAMENTAL CIRCUIT

Let G be a connected graph and T be its spanning tree. A circuit formed by adding a chord to spanning tree T, is called a fundamental circuit.

Fig. 10.40(*b*) shows a spanning tree $T = \{(A, B, C, D, E), (a, c, d, g)\}$ of the graph G shown in **Fig. 10.40**(*a*).

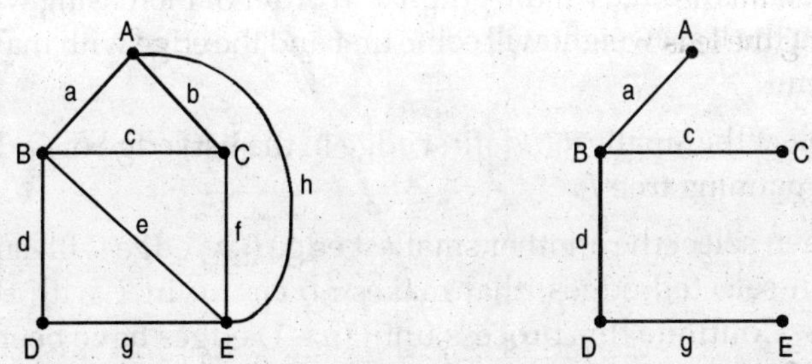

(*a*) Graph, G (*b*) Spanning Tree T.

Fig. 10.40

The chord set with respect to spanning tree T is

$$C = (b, e, f, h)$$

The fundamental circuits with respect to the chord set are given in **Fig. 10.41**.

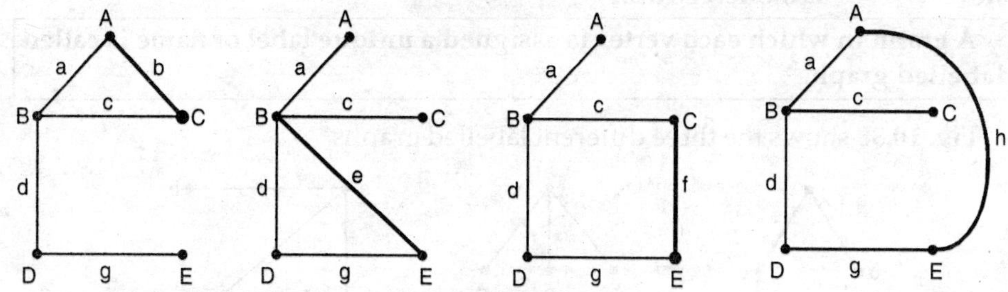

Fig. 10.41

Fundamental circuits formed by adding chord b, e, f and h to the spanning tree T **(Fig. 10.41)** respectively.

We have seen that the exist four fundamental circuits with respect to four chords.

So if G is a connected of e edges and n vertices and T is a spanning tree of G with $n-1$ **branches**, then there are exactly $(e - n + 1)$ **chords** in G with respect to T. Hence $(e - n + 1)$ **fundamental circuits**, with respect to T in G.

10.11 SPANNING TREES IN A WEIGHTED GRAPH

Let G be a connected weighted graph and T be the spanning tree of G. Then the weight of the spanning tree is defined as the sum of the weights of all the branches in T. So the difference spanning trees in G will have different weights.

Hence "A spanning tree T with the smallest weight, among all the spanning trees of G, is called a minimal spanning tree or shortest spanning tree."

There are a number of methods available to find a minimal spanning tree in a given graph. The two important algorithms to find out the minimal spanning tree in a given graph G are **Kruskal's** and **Prim's**.

10.11.1 Kruskal s Algorithm for a Shortest Spanning Tree

Let G be a graph with n-vertices and e-edges and w_i be the weight assigned to edge e_i. Then to find out a minimal spanning tree T with n vertices and $n-1$ edges, do the following (according to Kruskal's algorithm):

Step 1: List all the edges of the graph G in order of increasing weights. That is, the edge having the least weight will come first and the edge with maximum weight will be in the end.

Step 2: Select the smallest (*i.e.*, first edge in the list) edge of G. This is the first branch of the spanning tree T.

Step 3: Then select the another smallest edge (*i.e.*, edge with smallest weight among the non-selected edges) that makes no circuit in T with the previously selected edges. Continue this process until $(n-1)$ edges have been selected and these edges will constitute the desired minimal spanning tree.

Example 10.13. Find out the minimal spanning tree using Kruskal's algorithm in the graph G shown in Fig. 10.42.

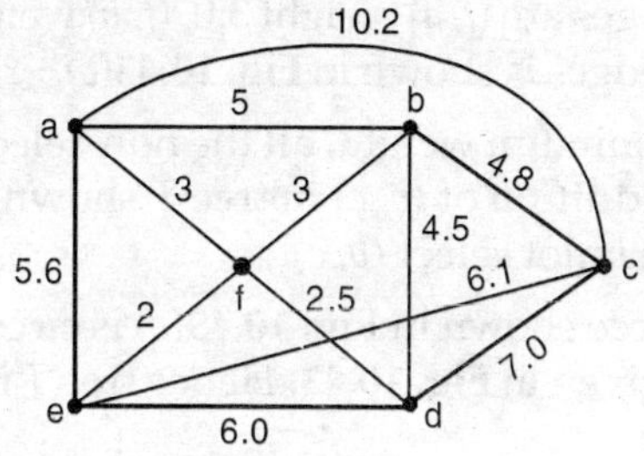

Fig. 10.42.

Solution: Step I: List all the edges of G in order of increasing weight.

Hence list l is:

Edge	Weight
(e, f)	2
(f, d)	2.5
(f, a)	3.0
(f, b)	3.0
(b, d)	4.5
(b, c)	4.8
(a, b)	5.0
(a, e)	5.6
(e, d)	6.0
(e, c)	6.91
(d, c)	7.0
(a, c)	10.2

Step 2: Select the edge with minimuml weight (*i.e.*, the first edge from the list). So the first selected edge is (e, f) with weight 2. Hence it is the first branch of spanning tree T (**Fig. 10.43**(*a*)).

Step 3: Select the next edge having minimal weight and add it in the spanning tree constructed so for such that the edge selected in step 3 does not create any circuit in spanning tree T with the previously selected edges.

So the next edge is (f, d) with weight (2.5) and one can easily see that it does not make any circuit in T (**Fig. 10.43**(*b*)).

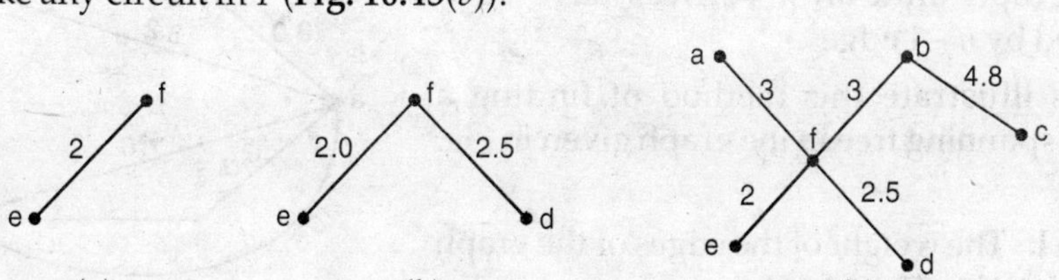

Fig. 10.43

Step 4: Go on selecting the edges with minimum weights such that no edge makes a circuit in T with the previously selected edges.

So the next selected edges are (f, a) weight 3.0, (f, b) weight 3.0, (e, c) weight 4.5. The tree formed by these edges is shown in **Fig. 10.43**(c).

The next edge with minimum weight, all the non-selected edges is edge (b, c) with weight 4.8. But the addition of (b, c) in tree T shown in Fig. 10.39(c) creates a circuit (b, f, e, c), so we cannot select (b, c).

One can see that the tree shown in **Fig. 10.43**(c) is circuitless and contains all the vertices of the graph given in **Fig. 10.43**. Hence tree T is the desired spanning tree with minimal weight.

The weight of the spanning tree obtained by Kruskal's algorithm is obtained by taking sum of the weights of the edges in tree T.

So the weight of tree $T = 2 + 3 + 3 + 2.5 + 4.8 = 15.3$

10.11.2 Prim s Algorithm

For a given weighted graph G with n-vertices, e-edges.

Step 1: Draw n-isolated vertices and label them $v_1, v_2, \ldots, v_n$. Write the given weights of the edges of G in an n by n table. As shown in **Fig. 10.44**.

$$\begin{array}{c} \\ v_1 \\ v_2 \\ v_3 \\ \vdots \\ v_n \end{array} \begin{array}{c} \begin{array}{ccccc} v_1 & v_2 & v_3 & \cdots & v_n \end{array} \\ \begin{bmatrix} - & w_{12} & w_{13} & - & w_{1n} \\ w_{21} & - & w_{23} & - & w_{2n} \\ w_{31} & - & - & - & w_{3n} \\ \vdots & \vdots & \vdots & \vdots & \vdots \\ w_{n1} & - & - & - & w_{n,n-1} \end{bmatrix} \end{array}$$

Fig. 10.44 Table containing weights of the given graph G

Step 2: Start from vertex v_1 and connect it to it's nearest neighbour by searching the vertex in row 1 which has the smallest entry. Let us assume that the new vertex is v_i.

Step 3: Consider v_1 and v_i as one subgraph and connect this subgraph to it's closest neighbour by searching the vertex (except v_1 and v_i) that has the smallest entry among all entries in row 1 and i row. Let the new vertex be v_j. Make sure that the vertex selected in the subgraph does not create any circuit.

Step 4: Now consider the tree formed by v_1, v_i and v_j as one subgraph and continue the process of finding the neighbour of the subgraph until all n vertices have been connected by $n - 1$ edges.

Let's illustrate this method of finding a minimal spanning tree in the graph given in Fig. 10.45

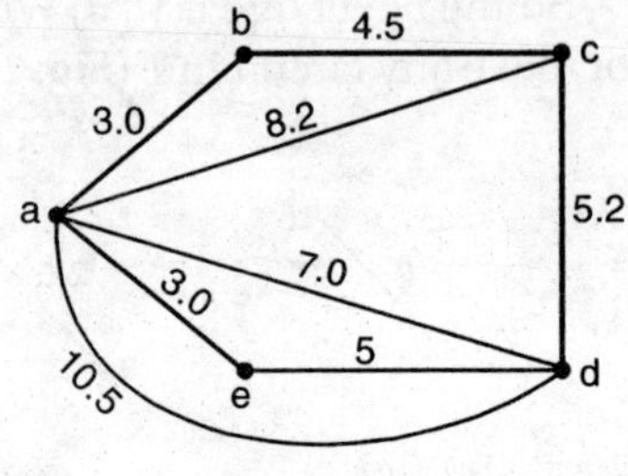

Fig. 10.45

Step I: The weight of the edges of the graph G is tabulated in **Fig. 10.46**(a).

$$\begin{array}{c} \\ a \\ b \\ c \\ d \\ e \end{array} \begin{array}{c} \begin{array}{ccccc} a & b & c & d & e \end{array} \\ \left[\begin{array}{ccccc} - & 3.0 & 8.2 & 7.0 & 3.0 \\ 3.0 & - & 4.5 & \infty & \infty \\ 8.2 & 4.5 & - & 5.2 & \infty \\ 3.0 & \infty & 5.2 & - & 5.0 \\ 3.0 & \infty & \infty & 5.0 & - \end{array}\right] \end{array}$$

Fig. 10.46(a)

Notes:

(*i*) The table containing the weight of the edges of the graph G is symmetric about the diagonal and the diagonal is empty.

(*ii*) If there is no edge between the two vertices then the corresponding entry in the table will be ¥, assuming that the weight of non-existing edges is very large. As there is no edge between vertices b and d, hence the corresponding entry in the table is . (**Fig. 10.46(*a*)**).

Step II: We start with vertex a and pick the smallest entry in row one, which is either (a, b) or (a, e). We can select any of them. Let us pick (a, b), (one can obtain a different spanning tree with minimum weight by selecting (a, e)).

Step III: Now we find the closest neighbour of the subgraph (a, b) by selecting the vertex which has smallest entry in the rows corresponding to vertices a and b.

It can be seen that the new vertex is c as it ha the smallest entry among all the entries in the rows corresponding to vertices a and b (*i.e.*, row 1 and row 2 in the table).

Now taking (a, b, c) as the new subgraph continue the process of finding the closest neighbour of the subgraph, such that no selected vertex creates a circuit in the subgraph and continue the process until all 5 vertices have been connected by 5-1 *i.e.*, 4 edges.

Subgraph	**Nearest Neighbour**
1. **(a, b, c)**	**d**
2. **(a, b, c, d)**	**e**
3. **(a, b, c, d, e)**	stop

We can see that the last subgraph (a, b, c, d, e) contains all the vertices of the G. Edges selected in this process are (a, b), (b, c), (c, d) and (d, e) and the graph (circuitless) obtained by this procedure is shown in **Fig. 10.46**(*b*)

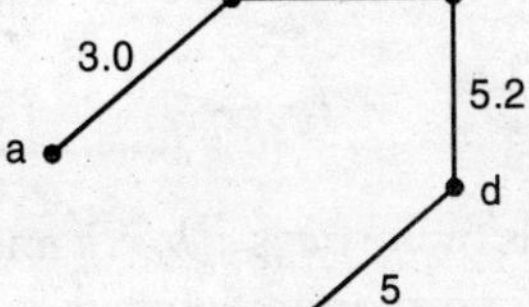

Fig. 10.46(b)

Hence the graph obtained is the minimal spanning tree with weight

$$3 + 4.5 + 5.2 + 5 = 17.7$$

10.12. DIJKSTRA S ALGORITHM

Dijkstra's Algorithm is one of the most efficient method of finding the shortest path between a specified vertex to another specified vertex in a given weighted graph (or digraph).

Algorithm: Label are the vertices of a given graph (or digraph) by permanent label and temporary label. Give a permanent label 0 to the starting vertex and a temporary label ∞ to all the other $(n - 1)$ vertices of the graph (or digraph) G.

The weight of an edge (x, y) is defined by $w(x, y)$ $w(x, y) \geq 0$ $w(x, y) = \infty$ if (x, y) is not our edge in G.

Itterations: (*i*) If a vertex v_j in G is not at a permanent label, give a new temporary lebel to v_i whose value is given by

min [old lebel of v_j, (old lebel of $v_i + d(v_i, v_j)$)]

v_i is the latest vertex with permanent label.

$d(v_i, v_j)$ is the distance between vertex v_i and v_j.

$d(v_i, v_j) = \infty$ if there is no edge between v_i and v_j.

(*ii*) Find out the smallest value among all the temporary lebels and make it permanent for the corresponding vertex. Ties can be broken arbitrarily.

(*iii*) Repeat step (*i*) and step (*ii*) until the destination vertex gets a permanent lebel.

Example 10.14. Find the shortest paths from vertex *u* to all the other vertices using dijkstra's algorithm in the given weighted graph.

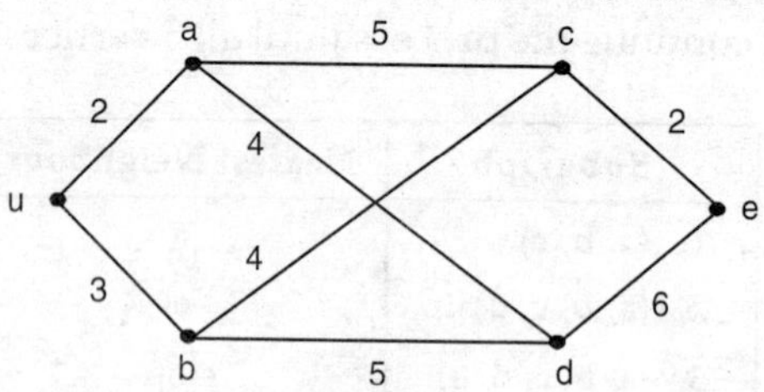

Fig. 10.47

Solution. Assign lebel 0 to vertex u and ∞ to all the other vertices of graph G.

u	a	b	c	d	e	
[0]	∞	∞	∞	∞	∞	Starting vertex u is lebeled 0.

[0] $\rightarrow$ permanent lebel

The new temporary lebels of vertices a, b, c, d and e are obtained by (calculated using the latest vertex having permanent lebel *i.e.*, u)

$$\text{min [old lebel of } v_j, \text{(old lebel of } u + d(u, j))]$$

where $\quad v_j \in (a, b, c, d, e)$

So new lebel (temporary) of vertex a is given by

$$\min [\text{old lebel of } a, (\text{old lebel of } u, d(u, a)]$$
$$= \min [\infty, (0 + 2) = 2$$

Similarly

new lebel of b = $\min [\infty, (0 + 3)] = 3$

new lebel of c = $\min [\infty, (0 + \infty)] = \infty$

(Since there is no edge between u and c so $d(u, c) = \infty$)

The permanent lebel of each vertex gives the shortest distance of that vertex from the source vertex (or starting vertex).

The following label shows the lebels of the vertices.

u	a	b	c	d	e	
$\boxed{0}$	∞	∞	∞	∞	∞	
$\boxed{0}$	2	3	∞	∞	∞	: All successors of u get lebelled
$\boxed{0}$	✓ $\boxed{2}$	3	∞	∞	∞	: smallest lebel becomes permanent
$\boxed{0}$	$\boxed{2}$	3	7	6	∞	: all successors of a get lebelled
$\boxed{0}$	$\boxed{2}$	✓ $\boxed{3}$	7	6	∞	: smallest lebel becomes permanent
$\boxed{0}$	$\boxed{2}$	$\boxed{3}$	7	✓ $\boxed{6}$	∞	: successors of b get lebelled
$\boxed{0}$	$\boxed{2}$	$\boxed{3}$	7	$\boxed{6}$	12	: smallest lebel becomes permanent
$\boxed{0}$	$\boxed{2}$	$\boxed{3}$	✓ $\boxed{7}$	$\boxed{6}$	12	: successors of d get lebelled
$\boxed{0}$	$\boxed{2}$	$\boxed{3}$	$\boxed{7}$	$\boxed{6}$	9	: smallest lebel permanent
$\boxed{0}$	$\boxed{2}$	$\boxed{3}$	$\boxed{7}$	$\boxed{6}$	$\boxed{9}$	: end

Illustration of the lebelling of vertices in row (6) of the table

New temporary lebel assigned to vertices c, d and e from the latest permanently lebelled vertex b as:

new temporary lebel of c = win [old lebel of c, (old lebel of $b + d(b, c)$)]

$$= \min [\infty, (3 + 4)] = 7$$

Similary

new temporary lebel of $d = \min [6, (0 + 5)] = 6$

and so on.

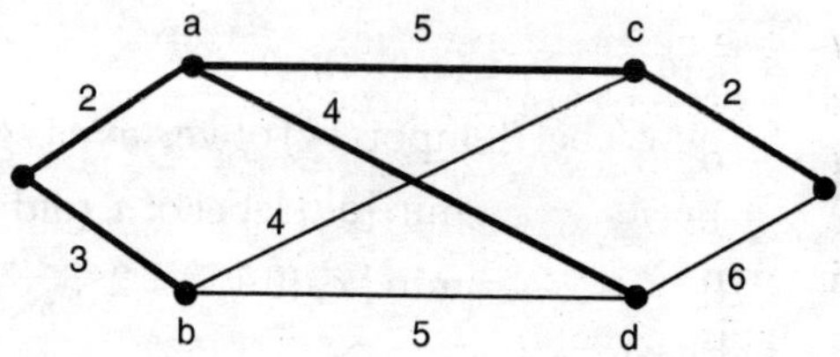

Fig. 10.48

Since all the vertices become permanent. So distance of the shortest path from u to all other vertices are given by the permanent lebel of each vertex. *For example,* permanent lebel of vertex e indicates that the length of the shortest path between u and e is 9. **Fig. 10.48** shows the shortest path from u to all other vertices in G.

Example 10.15. Obtain a shortest path between vertex s and t in the given weighted digraph using dijkstra's algorithm (Fig. 10.49)

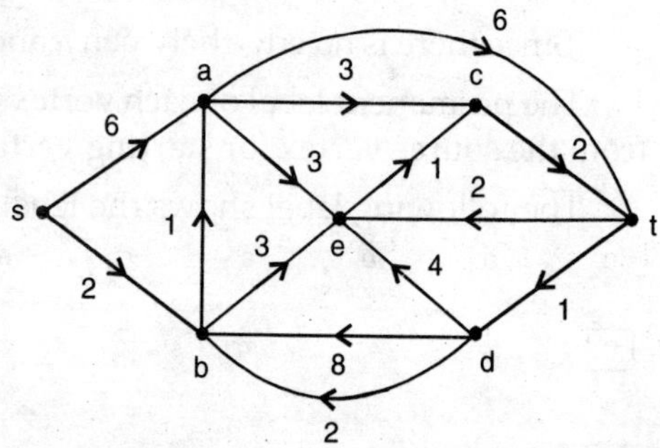

Fig. 10.49

Solution: Starting vertex s gets a permanent lebel 0 and a temporary lebel ∞ is assigned to all the other vertices.

The changing of temporary lebels in to permanent is given by following table.

s	a	b	c	d	e	t
0	∞	∞	∞	∞	∞	∞
0	6	2	∞	∞	∞	∞
0	6	✓ 2	∞	∞	∞	∞
0	3	2	∞	10	5	∞
0	✓ 3	2	∞	10	5	∞
0	3	2	6	10	5	9
0	3	2	6	10	✓ 5	9
0	3	2	6	10	5	9
0	3	2	✓ 6	10	5	9
0	3	2	6	10	5	8
0	3	2	6	10	5	✓ 8

As the destination vertex t gets a permanent lebel stop the process. The distance of the shortest path thus obtained between source vertex to destination vertex t is 8 and the path is given in **Fig. 10.49**(*a*).

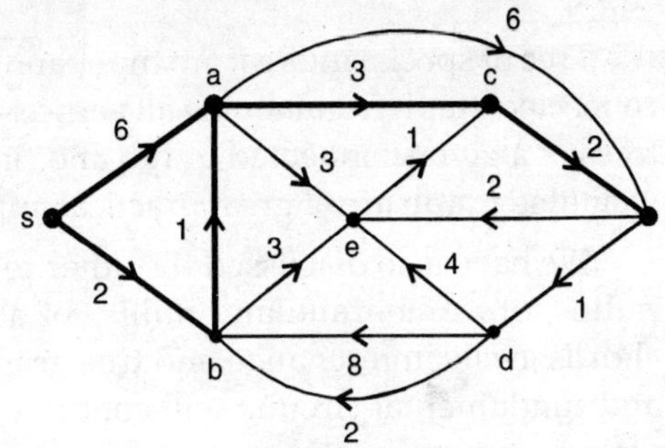

Fig. 10.49(a)

Note: To obtain the distance of the shortest paths between source vertex u and all the other vertices of the digraph continue the process of lebelling the vertices till all the vertices get a permanent lebel.

As there is only one vertex left in digraph a whose lebel is not permanent (see table), lets continue with the process of finding a new lebel, which is the permanent lebel of vertex d from vertex t which is the latest parmently lebelled vertex by calculating the new value for d as:

$$\min [\text{old lebel of } d, (\text{cold lebel of } t + d(t, d))]$$
$$= \min [10, (8 + 1)] = \min [10, 9] = 9$$

The following last row of the table gives the permanent lebels of all the vertices.

a	b	c	d	e	t
3	2	6	9	5	8

permanent lebel of each vertex represents the distance of the shortest directed path from source vertex s to the corresponding vertex.

The shortest paths from s to all other vertices are given in **Fig. 10.49**(*b*)

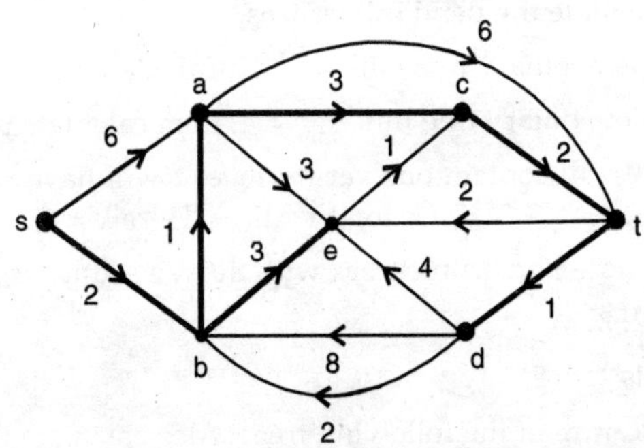

Fig. 10.49(b)

SUMMARY

This chapter introduced a particular type of connected graph called tree. Since tree has wide applications, it is the most important topic in graph theory. Different types of trees, such as weighted tree, labeled tree, rooted tree and binary tree, were discussed, together with their properties and applications.

Tree of special interest are the spanning trees that are the subgraphs of a given connected graph G containing all vertices of G. The problems of finding all spanning trees of a given connected graph and finding a shortest spanning tree in a given weighted graph are of great practical importance.

We have also discussed the other related concepts, such as centre, diameter, radius, of a tree, rank and nullity of a graph, fundamental circuits, branches, chords, cyclic interchange and tree graphs. The concept of trees, spanning trees and fundamental circuits will continue to appear from time to time in some of the proceeding chapters.

EXERCISES

1. Draw all the trees (*a*) labeled and (*b*) unlabeled vertices for $n = 2, 3, 4, 5$ and 6.
2. Draw all the unlabeled rooted trees of $n = 3, 4$ and 5 vertices.
3. Draw all different (*i.e.*, pair-wise disjoint) or non-isomorphic trees of $n = 3, 4, 5$ and 6 vertices.
4. Find the condition under which the diameter of a tree is equal to twice the radius.
5. Draw all complete binary trees with (*a*) 6 and (*b*) 8 pendant vertices.
6. Let T be an n-vertex tree having one vertex of each degree $i, 2 \le i \le k$; the remaining $n - k + 1$ vertices are leaves. Determine n in terms of k.

   [**Hint:** There is one vertex of degree 2, one of degree 3, one of degree 4 and so on, one of degree $k - 1$ and one vertex of degree k. Rest $n - (k - 1) = n - k + 1$ vertices of degree one. So $1.2 + 1.3 + 1.4 + ... + 1.(k - 1) + 1.k + (n - k + 1).1 = 2.e = 2.(n - 1)$].
7. Prove that G is a tree iff G is connected and every edge is a cut-edge.
8. Show that any tree is a bipartite graph.
9. A tree has two vertices of degree 2, three vertices of degree 3 and two vertices of degree 4. Calculate the pendant vertices.
10. Draw a complete binary tree with $n = 17$. Calculate max l_{max} and min l_{max}.
11. If for a complete binary tree min $l_{max} = 4$, then calculate the number of vertices.

    [**Hint:** Level 0 will contain one vertex, level 1 will have two vertices and so on. *i.e.*, $2^0 + 2^1 + 2^2 + ... + 2^k = n$ = total vertices. Here $k = 4$].
12. Draw all the non-isomorphic trees with degree sequences

    (*i*) (1, 1, 1, 1, 2, 2, 3, 3)

    (*ii*) (1, 1, 1, 1, 1, 1, 2, 3, 5).
13. Find out the centre of the following trees. Also calculate the radius of each tree.

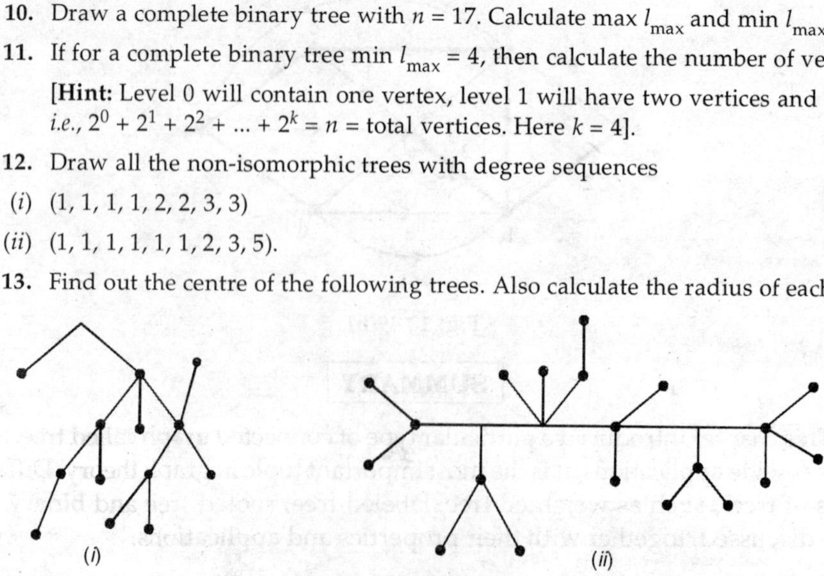

(*i*) (*ii*)

Fig. 10.50

14. Check whether the following pairs of trees are is amorphic.

(*a*)

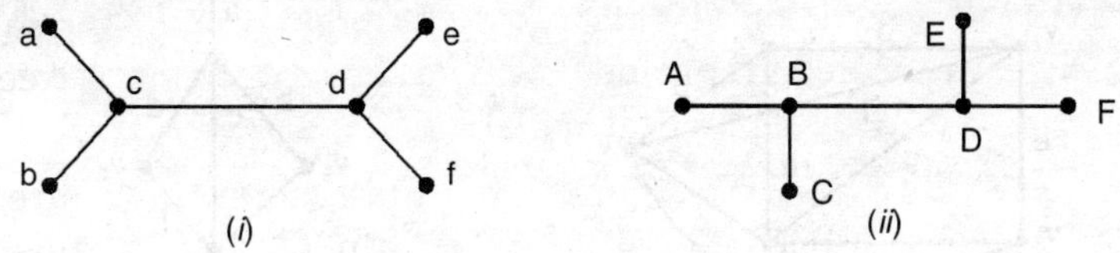

Fig. 10.51

(*b*)

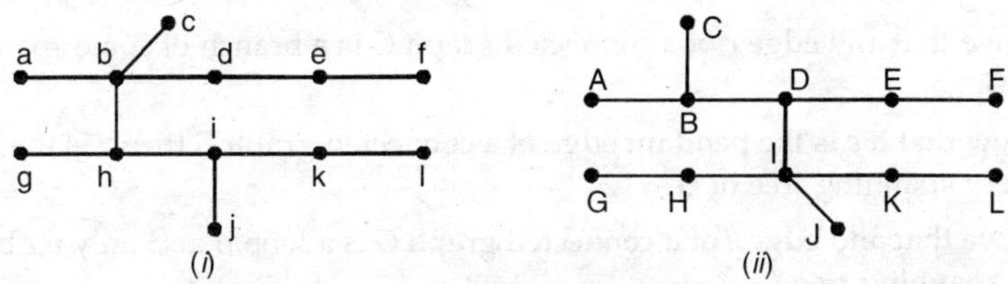

Fig. 10.52

15. List the order in which the vertices are processed using (*i*) pre-order (*ii*) in-order and (*iii*) post order. Traversal for the following:

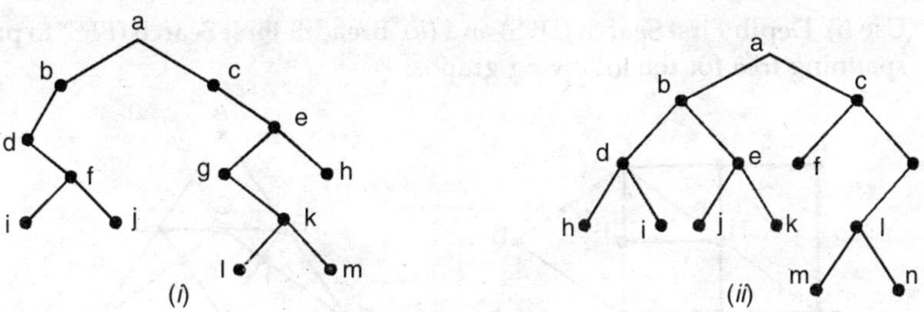

Fig. 10.53

16. If the pre-order, in-order and post order traversals of a tree are *A B D E G H C F*, *D B G E H A C F* and *D G H E B F C A* respectively, then draw the tree *T*.

17. Draw the final binary *T* if the following numbers are inserted into an empty binary search tree *T*.

Numbers: 52, 27, 17, 24, 42, 46, 32, 35, 62, 77, 92, 82.

18. Show that there is no tree with degree sequence

(*i*) (1, 1, 2, 2, 3, 3)

(*ii*) (1, 1, 1, 1, 1, 1, 2, 3, 7).

19. Show that a complete bipartite graph $K_{m,\,n}$ is a tree if $m = 1$.

20. Make a table with three columns labeled "Vertex", "Depth First Search" and "Breadth First Search". For each vertex in the given graph find how many steps it takes to find it using the two search methods.

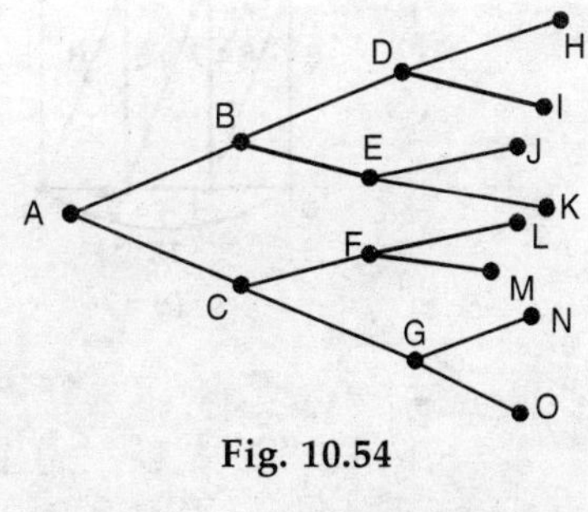

Fig. 10.54

21. Draw all the spanning trees of the following graphs:

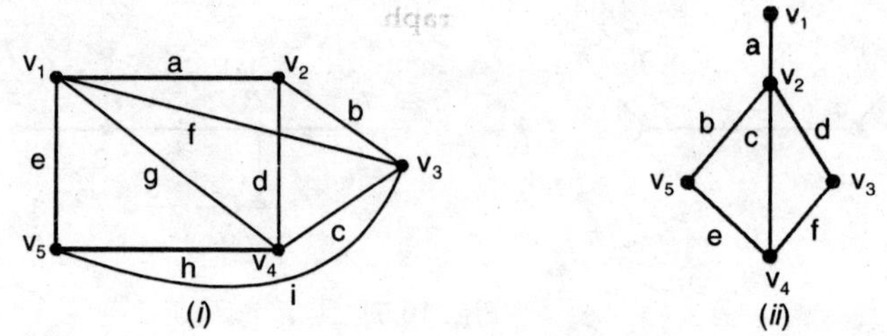

Fig. 10.55

22. Prove that any edge e of a connected graph G is a branch of some spanning tree of G.

23. Show that if e is the pendant edge of a connected graph G then e is the branch of every spanning tree of G.

24. Prove that and edge e of a connected graph G is a loop if and only if e belongs to no spanning tree.

25. Prove that any subgraph h of a connected graph G is contained in some spanning tree of G if h contains no circuit.

26. Show that a Hamiltonian path is a spanning tree.

27. Use (*i*) Depth First Search (DFS) and (*ii*) Breadth First Search (BFS) to produce a spanning tree for the following graphs:

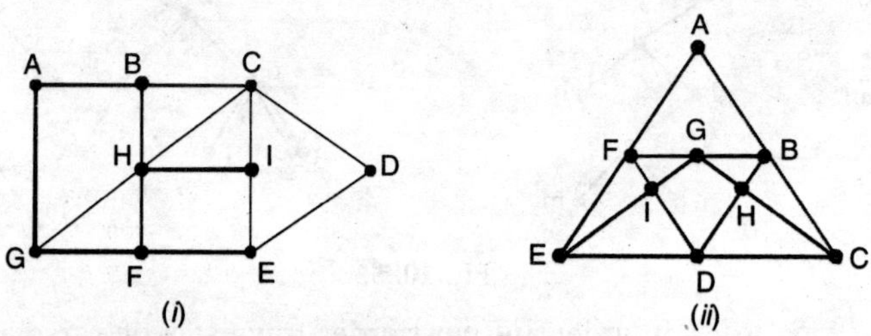

Fig. 10.56

28. Prove that in a tree every non-pendant vertex v is a cut-vertex.

29. Consider a connected weighted graph G such that e_m is the edge with the minimum weight among all the edges. Show that every minimal spanning tree of G contains e_m.

[Hint: Use definition of minimal spanning tree. Procedure of Kruskal algorithm]

30. Consider the following connected weighted graphs

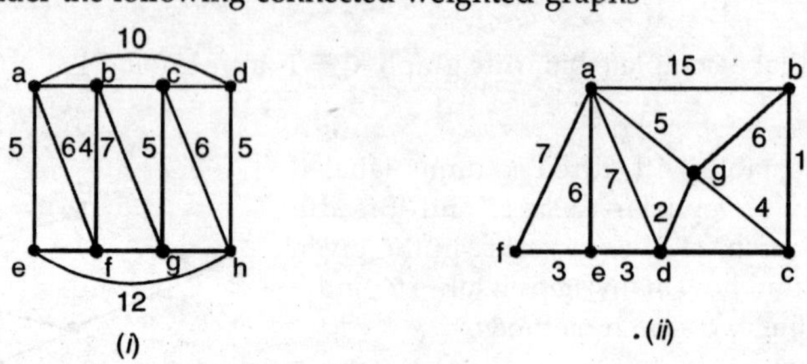

Fig. 10.57

Show that an edge e_m having the largest weight among all the edges does not belong to any shortest tree of the graph.

31. Use Kruskal algorithm to find out the minimal spanning tree T of the following graphs.

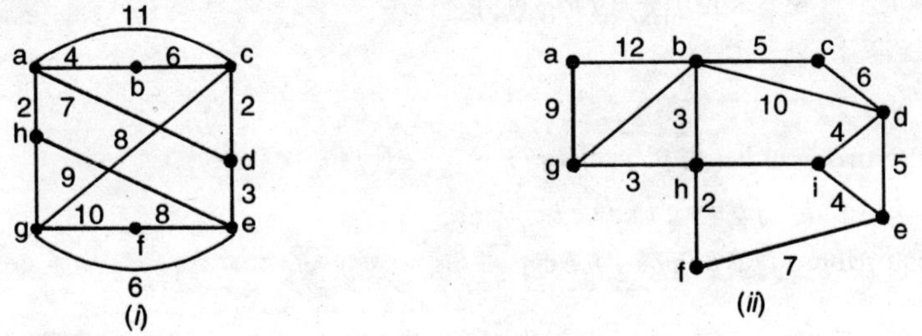

Fig. 10.57

32. Use Prim's algorithm to find out the minimal spanning tree of the following graphs:

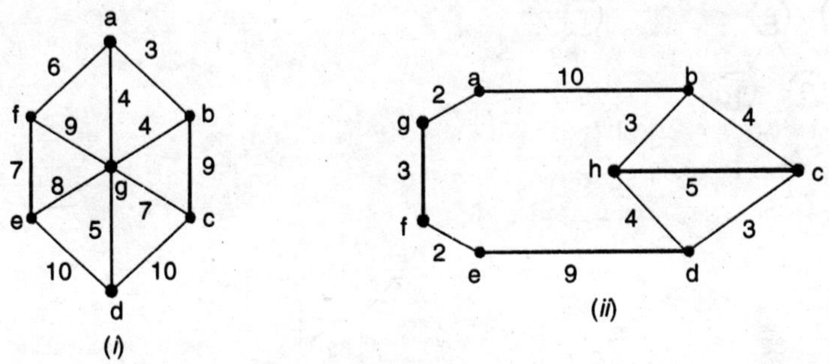

Fig. 10.58

33. Find out the numbers of spanning trees. For the following graph without listing them.

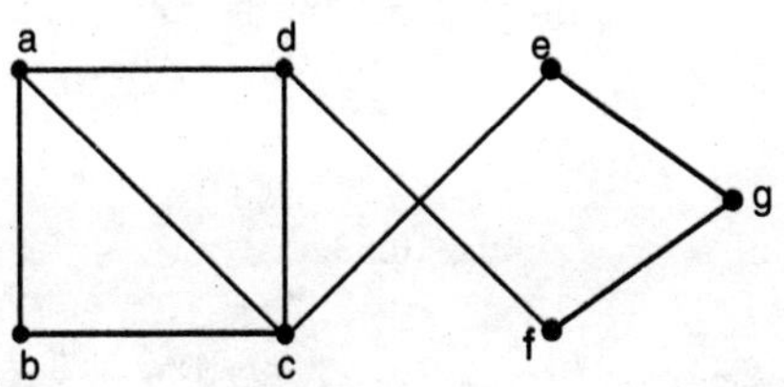

Fig. 10.59

34. Show that a path is its own spanning tree

ANSWERS

8. Consider a tree as given below:

it can be redrawn as

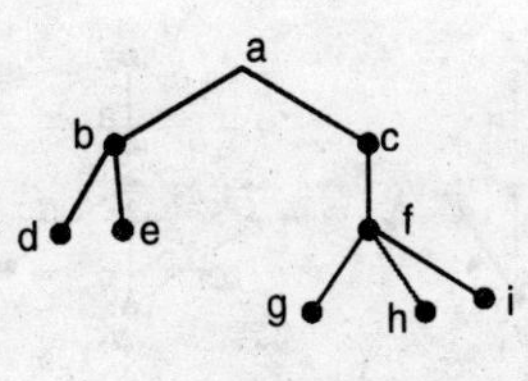

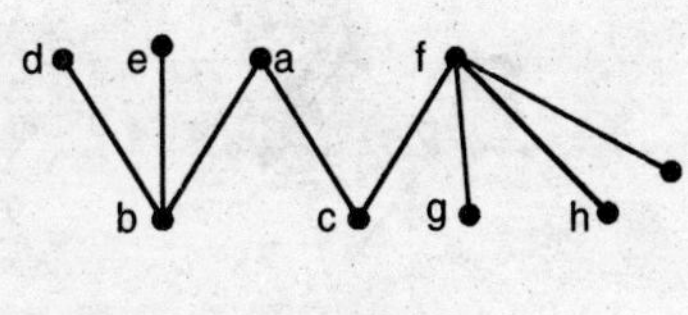

Fig. 10.60 **Fig. 10.61**

now we can partition the vertex set $V(T)$ as $V_1 = \{a, e, f, d\}$ and $V_2 = \{b, c, g, h, i\}$ and each edge e between the two sets has its one vertex in V_1 and another in V_2. Hence it is a bipartite graph.

9. 5

10. max $l_{max} = 8$; min $l_{max} = \lceil \log_2(18) - 1 \rceil$

14. (*a*) Yes

(*b*) No.

15. (*i*) Pre-order: *a b d f i j c e g k l m h*

In-order: *d i f j b a c g l k m e h.*

Postorder: *i j f d b l m k g h e c a*

16.

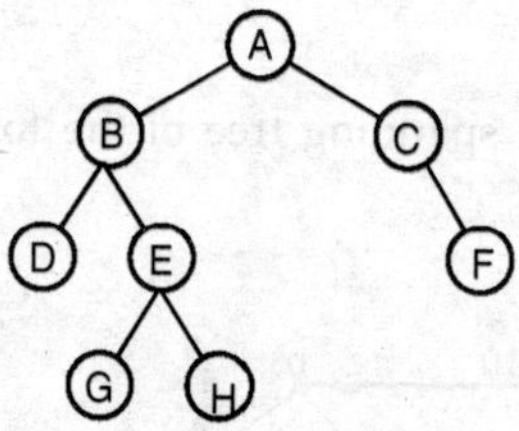

Fig 10.62

17.

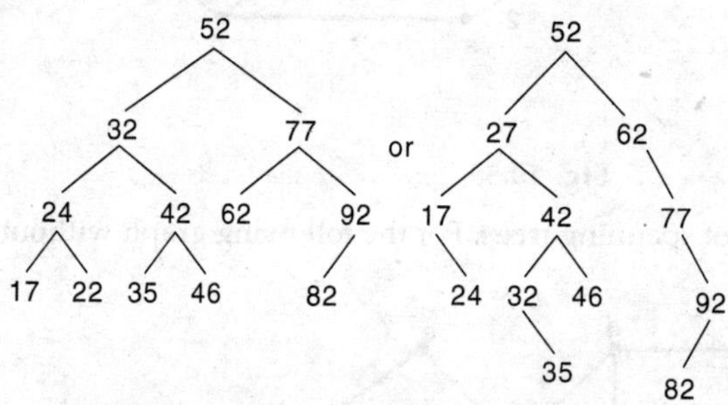

Fig 10.63

17. (*ii*)

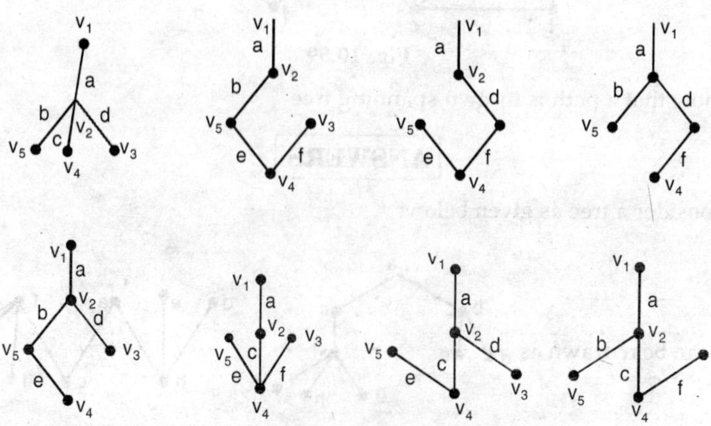

Fig 10.64

27. (*i*) DFS Spanning tree:

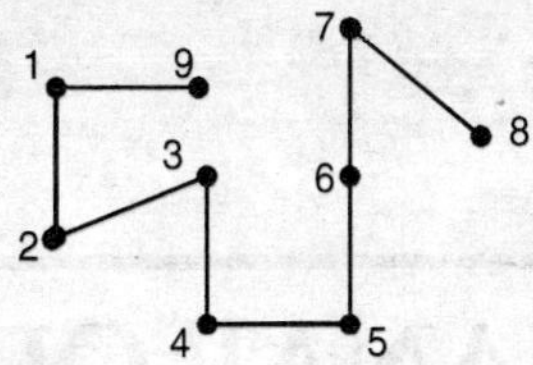

BFS Spanning tree:

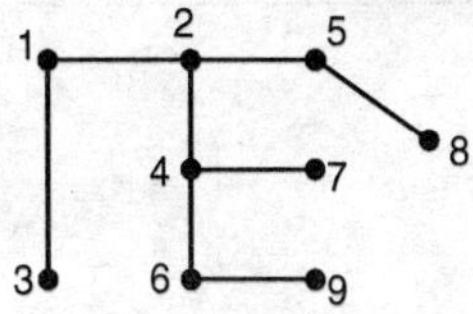

31. (*i*)

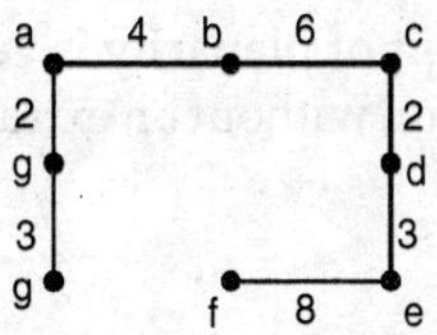

Total weight = 28

■■■

Chapter

11 PLANAR GRAPHS

11.1 INTRODUCTION

In this chapter we will study those graphs which can be drawn in a plane without cross over of the edges. The concept of planarity is very useful in many practical situations such as electric connection without an extra layer of insulation in the design of printed-circuit board.

11.2 PLANAR GRAPHS

A graph G is said to be planar if there exist some geometric representation of the graph G which can be drawn on a plane without crossover between its edges.

Otherwise the graph is non-planar. For instance, graph in **Fig. 11.1**(*a*) is planar and the graph in **Fig. 11.1**(*b*) is non-planar.

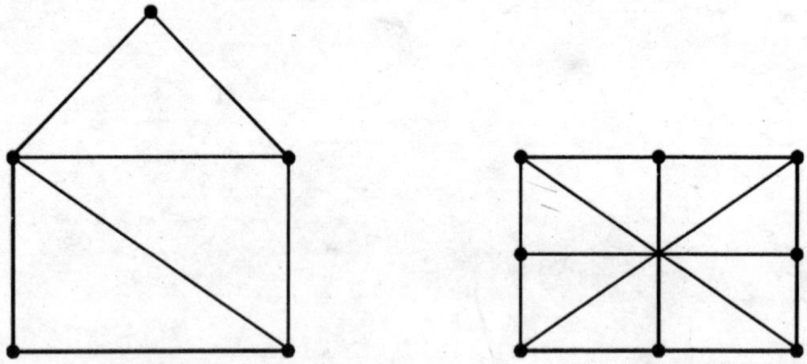

(*a*) Planar graph (*b*) Non-planar graph

Fig. 11.1

Let us draw a complete graph of 4-vertices. (**Fig. 11.2**(*a*))

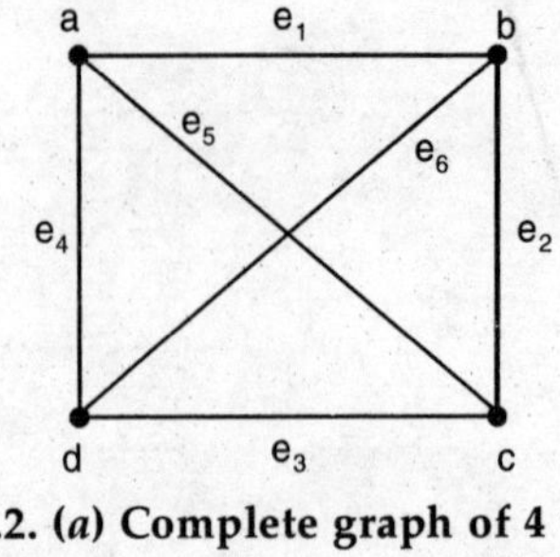

Fig. 11.2. (*a*) Complete graph of 4 vertices.

The graph seems to be non-planar. But a different embedding of the graph shows that it is a planar graph. (**Fig. 11.2** (*b*))

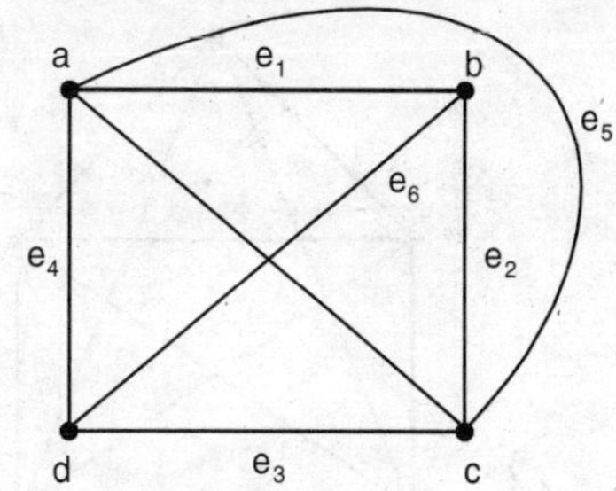

Fig. 11.2(*b*)

Thus, to declare that a graph *G* is non-planar, we have to show that **no possible geometric representation** of *G* can be embedded (drawing of a graph) on any surface such that no edges intersect in a plane.

In other words a graph *G* is planar if a graph *G*′, which is isomorphic to *G*, can be embedded in a plane. So by the above discussion we can conclude that the complete graph of 4-vertices is planar.

The **abstract** graph (or **combinatorial**) *G*′ with respect to geometrical graph *G* in **Fig. 11.2**(*a*) can be defined as

$$G' = (V, E, \phi)$$

where $V = \{a, b, c, d\}$

$$E = \{e_1, e_2, e_3, e_4\}$$

and the relationship between the two sets *V* and *E* is defined by the mapping.

$$\phi: E \to V$$

which consist of

$$\phi = \begin{cases} e_1 & \to & (a, b) \\ e_2 & \to & (b, c) \\ e_3 & \to & (c, d) \\ e_4 & \to & (a, d) \\ e_5 & \to & (a, c) \\ e_6 & \to & (b, d) \end{cases}$$

$e_1 \to (a, b)$ indicates that the edge e_1 from set *E* is mapped onto the ordered pair (a, b) of vertices from set *V*.

So the graph in **Fig. 11.2** (*a*) is just one representation of the graph *G*′ and not the graph *G*′ itself another possible representation (geometric) of *G*′ is shown in **Fig. 11.2(*b*).**

11.3 KURATOWSKI S TWO NON-PLANAR GRAPHS

It is not always possible to draw all the geometric representation on a plane to check whether the given graph is planar or non-planar.

Kuratowski (a mathematician) introduced the two non-planar graphs, called **Kuratowski's graphs**. The first Kuratowski's graph is **complete graph of 5-vertices (K_5)** and the second Kuratowski's graph is **regular connected graph of six vertices and nine edges** ($K_{3,3}$).

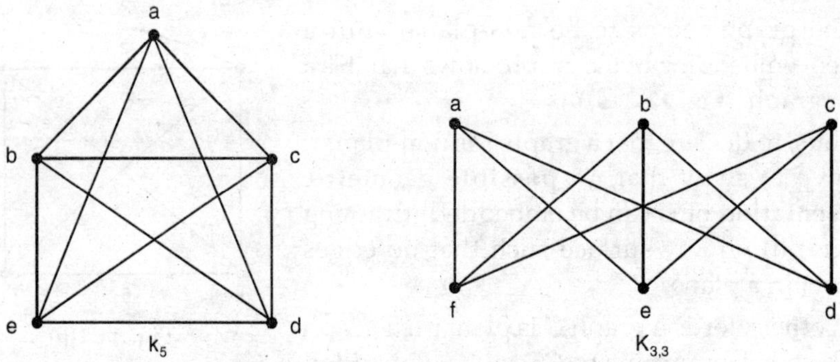

Fig. 11.3. Kuratowski's two graphs

Notes:

(*i*) Any graph isomorphic to any of the Kuratowski's graph is non-planar.

(*ii*) If a graph G contains any of the (or both) Kuratowski's graph then G is non-planar.

(*iii*) Both the graphs K_5 and $K_{3,3}$ are regular graphs.

(*iv*) K_5 (Kuratowski's first graph) is non-planar with the smallest number of vertices.

(*v*) $K_{3,3}$ (Kurotowski's second graph) is the non-planar graph with the smallest number of edges.

Example 11.1. Show that the complete graph of 5-vertices (K_5) is non-planar.

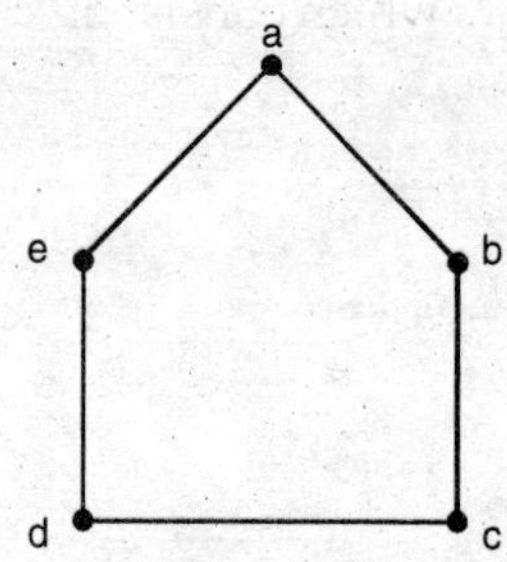

Fig. 11.4(*a*)

Solution: To show that K_5 is non-planar, we have to show that K_5 cannot be embedded in a plane.

Since K_5 is a simple connected graph which is complete that is every vertex is connected to every other vertex with an edge. So we must have a circuit of 5-vertices. See **Fig. 11.4**(*a*).

Since the graph is complete, vertex a must be connected with vertices *d* and *c* by means of edges. The edges may be drawn inside the pentagon without intersecting the previously drawn edges. See **Fig. 11.4**(*b*)

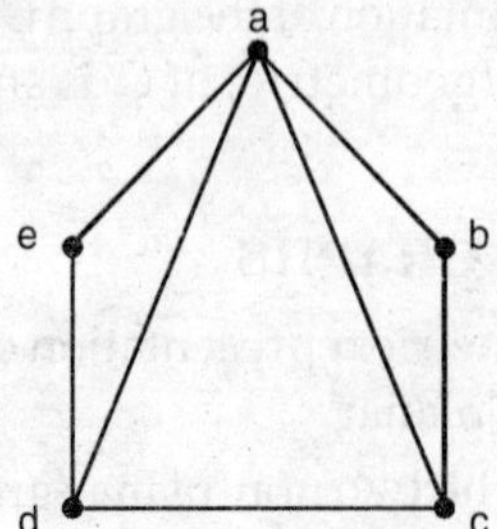

Fig. 11.4(*b*)

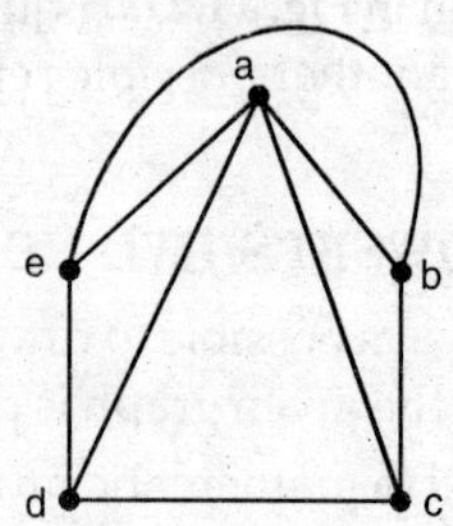

Fig. 11.4(*c*)

Now we have to draw an edge between vertices *e* and *b*. This edge cannot be drawn inside the pentagon without intersecting the previous edges, so we draw the edge outside the pentagon see **Fig. 11.4**(*c*).

Now to make the graph in **Fig. 11.4**(*c*) complete we have to draw edges between the unordered pairs (*e*, *c*) and (*b*, *d*).

Edge (*e*, *c*) cannot be drawn inside the pentagon without crossing the existing edges, so we connect vertex *e* and vertex *c* with an edge drawn outside the pentagon without intersecting the edge between *e* and *b*. See **Fig. 11.4**(*d*).

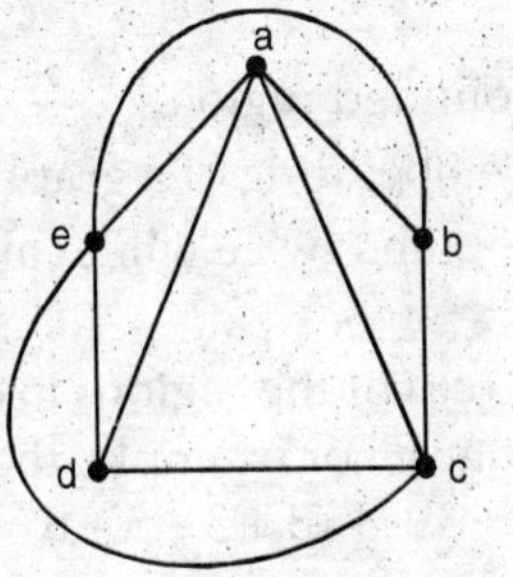

Fig. 11.4(d)

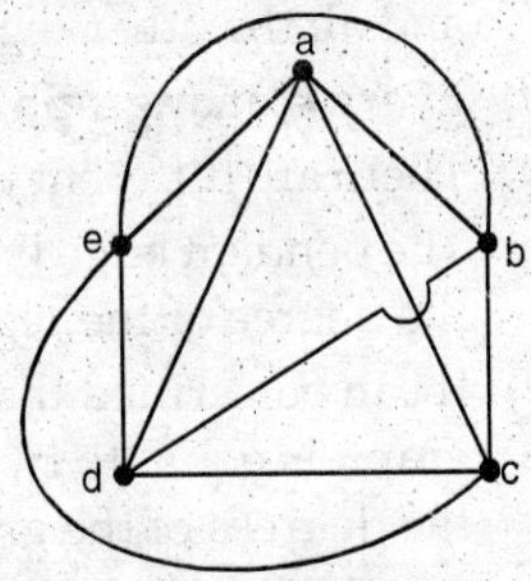

Fig. 11.4(e)

Now we are left with one edge that is edge between *b* and *d*. Now this edge cannot be drawn inside or outside the pentagon without any crossover. See **Fig. 11.4**(*e*).

Thus K_5 cannot be embedded in a plane and hence non-planar.

11.4 REGION AND ITS DEGREE

> **The planar representation of graph a divides the plane of paper into several *regions*, also called faces or meshes or windows.**
>
> **The degree of any region say *R*, denoted as deg (*R*), is the length of the closed walk which bounds the region *R*.**

For example, the graph in **Fig. 11.5** divides the plane of paper into seven regions $R_1, R_2, R_3, R_4, R_5, R_6$ and R_7.

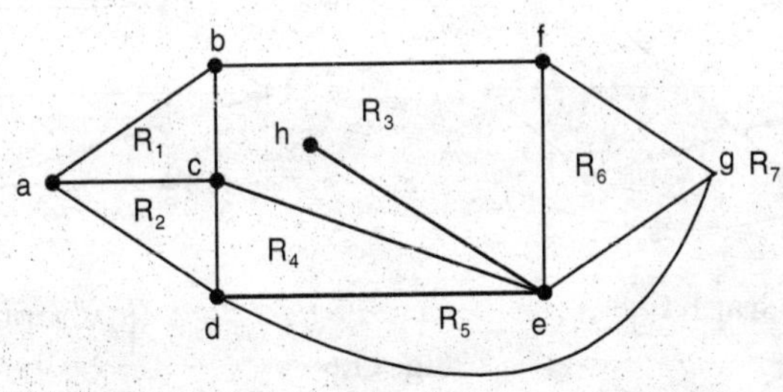

Fig. 11.5

Now

deg (R_1) =length of closed walk which bounds R_1 as one can see easily that R_1 is bounded by 3 edges (*a*, *b*) (*b*, *c*) and (*c*, *a*).

So, deg (R_1) = 3, deg (R_3) = 6

Similarly, deg (R_2) = 3, deg (R_4) = 3

deg (R_5) = 3, deg (R_6) = 5, deg (R_7) = 5

The closed walk for region R_3 is *b-f-e-h-c-b*, which means the edge (*e*, *h*) is occurring twice. The closed walk for region R_7 is *a-b-f-g-d-a*.

Notes:

(*i*) Region is not defined for a non-planar graph.

(*ii*) The outer region is also called an **infinite region** or **unbounded region**. Whereas the region lying inside the graph which is embedded in a plane is called **finite** or **bounded region**. In **Fig. 11.5** R_7 is unbounded region and all the other regions are bounded.

(*iii*) Every planar graph has exactly one unbounded region.

(*iv*) There are different regions for different embedding of a graph.

(*v*) The pendent edge lying inside a region counted twice while calculating the degree of the region.

(*vi*) It can be verified that the sum of the degrees of the regions in a planar graph is equal to twice the number of edges. For example, the sum of the degrees of the graph in **Fig. 11.5**, is $3 + 3 + 3 +3 + 3 + 5 + 6 = 26$.
and the total number of edges = 13.
So, $26 = 2 \times 13$.

Hence $\sum_{i=1}^{r} \deg(R_i) = 2.e$

11.5 KURATOWSKI S THEOREM

Theorem 11.1. A necessary and sufficient condition for a graph *G* to be planar is that *G* does not contain either of Kuratowski's two graphs or any graph homeomorphic to either of them as a subgraph of *G*.

Example. 11.2. Check whether the following graphs are planar or not. If yes, redraw such that no edges cross, if no, find a subgraph homeomorphic to either K_5 or $K_{3,3}$.

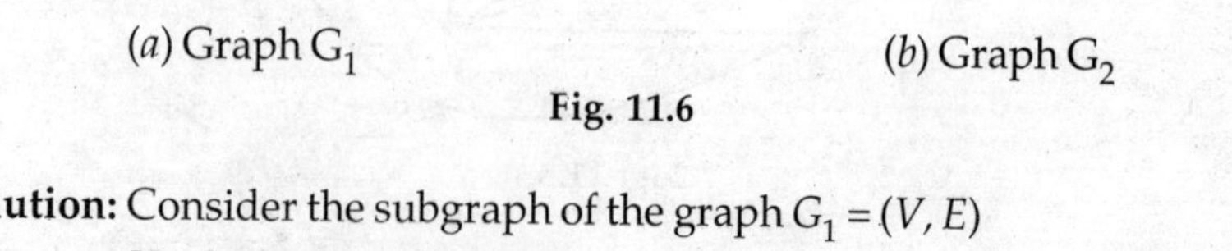

(*a*) Graph G_1

(*b*) Graph G_2

Fig. 11.6

Solution: Consider the subgraph of the graph $G_1 = (V, E)$

where $V = \{a, b, c, d, e, f\}$ and

$E = \{(a, b), (b, c), (c, d), (d, e), (e, f), (a, d), (b, e), (c, f)\}$

which is isomorphic to $K_{3,3}$. Hence G_1 is non-planar. Graph G_2 has 6-vertices and 11 edges and clearly it does not contain K_5 or $K_{3,3}$ as it's subgraph. Let's try to draw G_2 such that no edges cross. **Fig. 11.6**(*c*) the embedding of graph G_2. So it is planar.

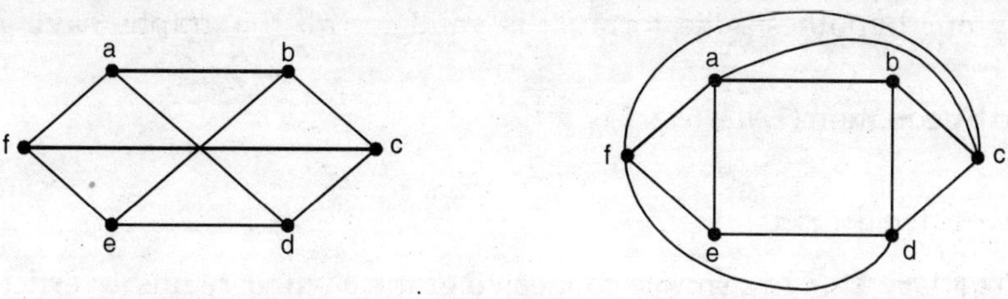

(c) Subgraph G_1' of graph G_1 (d) Graph G2

Fig. 11.6

11.6 EULER S FORMULA

We know that for a planar graph we may have different plane representations. Now the question is, whether the number of regions resulting from each embedding is same or not. The answer is the **Euler's formula** which gives the number of regions in any planar graph.

Theorem 11.2. (Euler's Formula): For any connected graph G with n-vertices and e-edges the total number of regions R is equal to $e - n + 2$.

i.e., $$R = e - n + 2 \quad \text{or} \quad n - e + r = 2$$

Proof: Let us prove the formula with mathematical induction. One can observe that the formula is true for a tree with n-vertices.

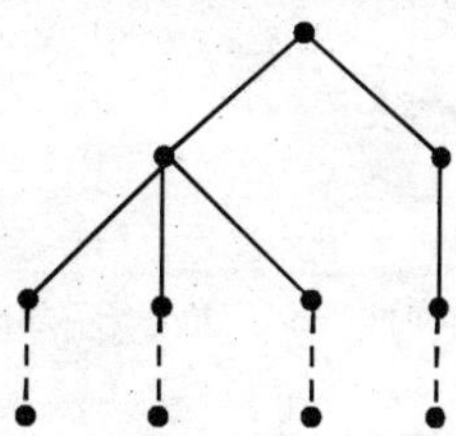

tree of n-vertices

Number of regions formed by the embedding of a tree with n-vertices and $n - 1$ edges is one.

i.e., $$R = 1$$

So $$e - n + 2 = (n - 1) - n + 2 = 1 = R$$

Hence the formula is valid for tree of n-vertices. That is, the formula is valid for $R = 1$.

Let us assume that the formula is valid for $R \geq 1$, say result is valid for $R = r$ regions.

i.e., $$r = e - n + 2 \quad \text{or} \quad n - e + r = 2$$

Let us show that the result holds good for $R = r + 1$ regions.

Suppose a connected graph G of n'-vertices and e'-edges has $r + 1$ regions. Then the removal of an edge common to the boundary of two regions reduces the number of edges by one. Also decreases the number of region by one. The number of vertices are same *i.e.,* n'.

So the formula becomes

$$n' - (e' - 1) + (r + 1 - 1) = n' - e' + 1 + r$$

$$= n' - (e' + 1) + r \qquad \text{...(1)}$$

By our hypothesis the formula is valid for all the graphs having region $R = r$.

So by equation (1) $n' - (e' + 1) + r$

$$= n - e + r = 2$$

Hence the theorem

Corollary 11.2. In a simple connected graph a with r-regins n-vertices and e-edges ($e > 2$) the following inequalities must hold:

(*i*) $e \geq \frac{3}{2} r$ (*ii*) $e \leq 3n - 6$

Proof: (*i*) Since $e > 2$, this implies that each region is bounded by at least 3 edges and each edge belongs to exactly two regions.

We know total degree of all the regions $= 2e$ ...(1)

Since each region is bounded by at least 3-edges.

So, degree of one region ≥ 3

Hence, degree of r regions $\geq 3 . r$ = Total degree of all the regions. ...(2)

$\Rightarrow$ By (1) and (2)

$$2e \geq 3r$$

$$\Rightarrow \quad e \geq \frac{3}{2} r$$

(*ii*) By Euler's formula, $r = 2 - n + e$

So inequality (*i*) becomes

$$e \geq \frac{3}{2} (2 - n + e) \quad \Rightarrow \quad 2e \geq 6 - 3n + 3e.$$

$$\Rightarrow \quad e \leq 3n - 6$$ **Proved.**

Example 11.3. Show that in a planar graph there is at least one vertex with degree less than or equal to 5.

Solution: Let G be a planar graph. Let us assume that all vertices of G have degree greater than or equal to 6.

i.e., $\deg (v_i) \geq 6$ for all $i \leq n$

So, deg (G) $\geq 6.n$, n is the number of vertices.

$$\Rightarrow \quad 2e \geq 6n \text{ or } n \leq \frac{e}{3} \quad ....(1)$$

also we know that $2e \geq 3.r$ for planar graphs.

$$\Rightarrow \quad r \leq \frac{2e}{3} \quad ...(2)$$

now by Euler's formula

$$e = r + n - 2$$

$$\leq \frac{2e}{3} + \frac{e}{3} - 2$$

$$\leq e - 2$$

or $$e \leq e - 2$$

which is a contradiction.

Hence our assumption is wrong. So there exist at least one vertex of degree less than or equal to 5 in a planar graph G. It also conclude that a planar graph G cannot be 6-connected.

11.7 DETECTION OF PLANARITY

To check whether the given graph G is planar or not is an important question and the answer is find out by drawing it. Which is obviously not a good technique. Lets try to find out some simple and efficient criteria. We take the following steps:

Step I: A disconnected graph is planar iff each of it's component is planar and a separable graph is planar iff each of its block is planar. Therefore for a given graph G, find out the set

$$G = \{G_1, G_2, ..., G_m\}$$

where each G_i is a component or non-separable block of G. Now we have to check each G_i for planarity.

Step II: Remove all self-loops as addition or removal of self-loops does not effect planarity.

Step III: Remove all parallel edges as addition or removal of parallel edge does not affect the planarity.

Step IV: Eliminate all the edges in series, as elimination of a vertex of degree two by merging two edges in series does not affect the planarity of the graph.

The connected non-separable graph G_i is reduced to a new graph S_i by repeated application of step-3 and step-4 on graph G_i. Theorem 11.3 gives the look of graph S_i.

Theorem 11.3. The graph S_i is

(*a*) A single edge, or

(*b*) A complete graph of four vertices, or

(*c*) A non-separable simple graph with $n \geq 5$ and $e \geq 7$.

If S_i is a single edge or S_i is a complete graph of 4-vertices then S_i is planar.

So, we need to investigate only simple, connected non-separable graphs of $n \geq 5$ and $e \geq 7$ with every vertex of degree 3 or more. To check planarity of these graphs.

(*i*) Check if $e \leq 3n - 6$ is satisfied or not.

If no then S_i is nonplanar, if yes then

(*ii*) Test S_i with *Kuratowski's theorem* (Theorem 11.1)

Example 11.4. Check whether the graph in Fig. 11.7 is planar or not.

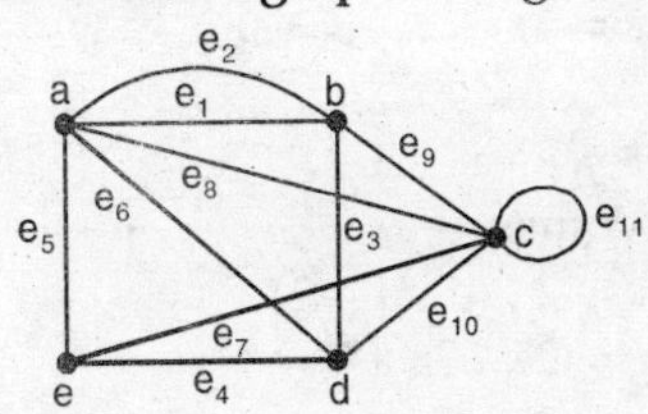

Fig. 11.7. Graph G

Solution: We will check the planarity of the graph by detection of planarity test:

The graph in **Fig. 11.7**(a) is obtained by applying step (2).

Graph G_2 is obtained by applying step (3) on graph G_1. See **Fig. 11.7(*b*).**

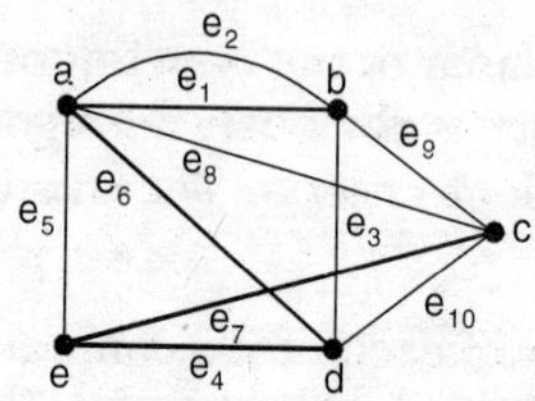

Fig. 11.7(a) Graph G1

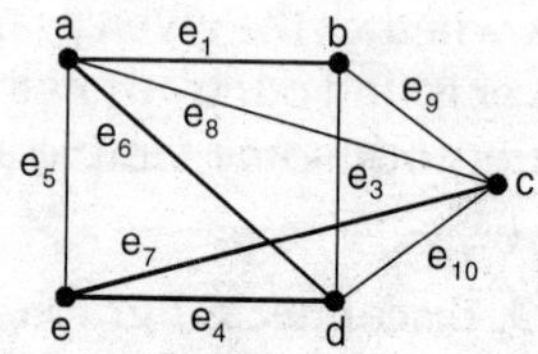

Fig. 11.7(*b*) Graph G2

Step IV cannot be applied on graph G_2 as there are no edges in series in G_2.

So G_2 is the reduced graph obtained by applying steps (2) and (3) on graph G.

As G_2 is not a single edge and G_2 is not also a complete graph of 4-vertices. So to check the planarity of G_2. Lets check the Euler's inequality.

$$e \leq 3n - 6$$

Number of edges in $G_2 = 9$

Number of vertices in $G_2 = 5$

and $$9 \leq 3.5 - 6$$

$$9 \leq 15 - 6 = 9$$

Inequality, satisfied so, G_2 is planar.

Hence the given graph G is also planar.

Example 11.5. Show that $K_{3,3}$ is non-planar.

Solution: $K_{3,3}$ is given in **Fig. 11.8**.

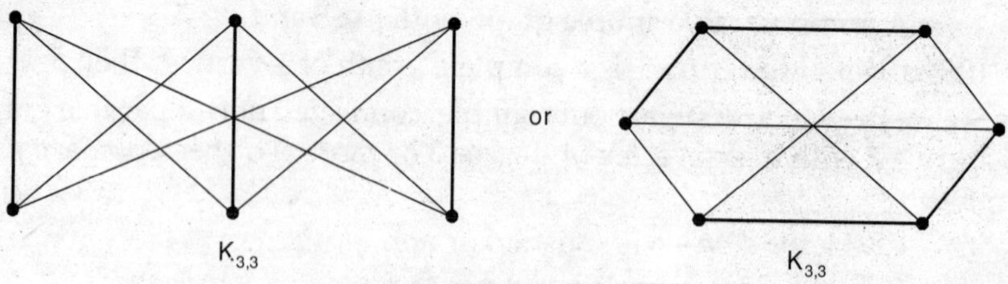

Fig. 11.8

Each region is bounded by 4 edges and each edge is associated with 2 regions.

So, $$2e = 4r \Rightarrow e = 2r. \quad ...(1)$$

By Euler's formula

$$r = e - n + 2.$$

then equation (1) becomes

$$e = 2 . (e - n + 2)$$

$$\Rightarrow \quad e = 2e - 2n + 4$$

$$\Rightarrow \quad e = 2n - 4 \quad ...(2)$$

How, number of edges in $K_{3,3} = 9$ and number of vertices are 6.

$\Rightarrow$ $e = 2.6 - 4$

$= 12 - 4 = 8$

$\Rightarrow$ $e = 8$ by equation (2)

which gives a contradiction.

Hence $K_{3,3}$ is non-planar.

Example 11.6. Show that $K_{2,2,3}$ is non-planar.

Solution:

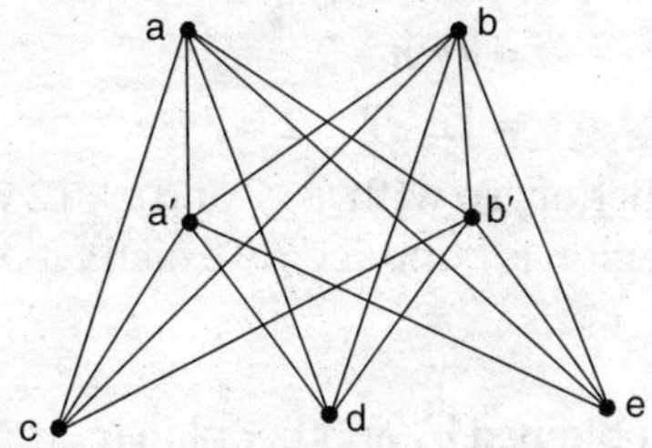

Fig. 11.9

Total number of vertices $n = 7$

Total degree of the graph $K_{2,2,3} = \deg(a) + \deg(b) + \deg(a') + \deg(b') + \deg(c) + \deg(d) + \deg(e)$

$$\deg(K_{2,2,3}) = 32$$

We know that $\sum_{i=1}^{n} \deg(v_i) = 2 \, . \, e$

So, number of edges $e = 16$.

One can observe that in $K_{2,2,3}$ each region is bounded by 3 edges.

So r-regions are bounded by $3r$-edges and each edge is associated with 2 regions.

$\Rightarrow$ $2 \, . \, e = 3 \, . \, r$

$\Rightarrow$ $r = \dfrac{2e}{3}$

By Euler's formula $r = 2 - n + e$

$\Rightarrow$ $\dfrac{2e}{3} = 2 - n + e$

or $2e = 6 - 3n + 3e$

or $e = 3n - 6$

But given that $e = 16$ and $n = 7$

So $16 = 3 \, . \, 7 - 6$

$\Rightarrow$ $16 = 21 - 6$

$\Rightarrow$ $16 = 15$

which is a contradiction.

So $K_{2,2,3}$ is non-planar.

Example 11.7. Can we draw a planar graph with $n = 7$ and $e = 12$ such that each region is bounded by exactly 3-edges ?

Solution : Given that $n = 7$ and $e = 12$.

Since, each region is bounded by exactly 3 edges and we know that each edge is associated with 2 regions.

So $$2 . e = 3 . r$$

$$\Rightarrow \quad r = \frac{2 . e}{3} = \frac{2 . 12}{3} = 8$$

By Euler's formula, for a planar graph.

$$r = e - n + 2$$
$$= 12 - 7 + 2 = 7$$

Which is a contradiction, so with $n = 7$ and $e = 12$ we cannot construct a planar graph in which each region is bounded by exactly 3-edges.

11.8 DUAL GRAPHS

A planar graph G^* obtained by another planar graph G is called the dual graphs of G. The dual graph is of two types: **Geometrical dual** and **combinational dual.**

11.8.1 Geometrical Dual

Consider a planar graph G in Fig. 11.10(a), with six regions R_1, R_2, R_3, R_4, R_5 and R_6.

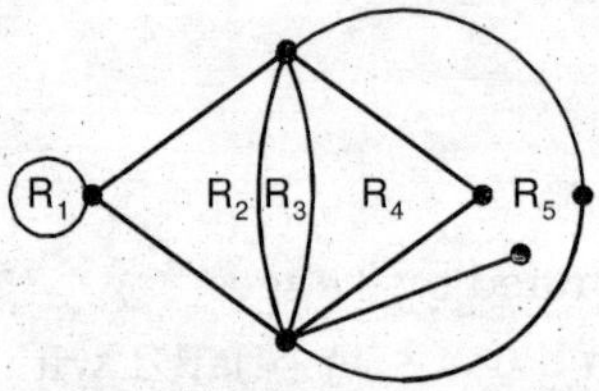

Fig. 11.10. (a) Planar Graph G

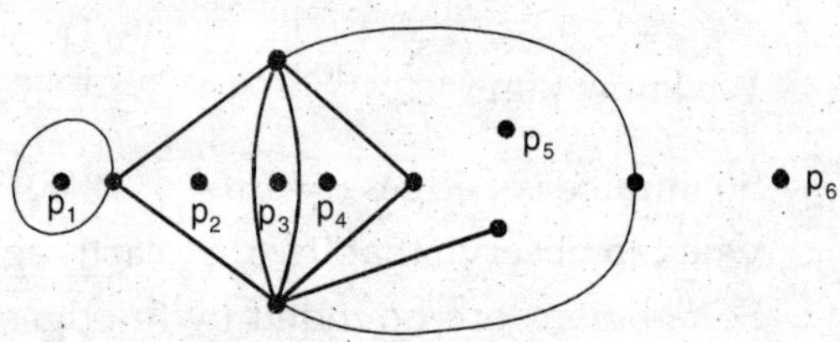

Fig. 11.10(b)

The geometrical dual G^* of G can be obtained by the following procedure :

Step I: With respect ot regions R_1, R_2, R_3, R_4, R_5 and R_6 place six points p_1, p_2, p_3, p_4, p_5 and p_6 in G respectively. See **Fig. 11.10(b)**.

Step II: Draw a line joining two points p_i and p_j that intersects the common edge between R_i and R_j exactly ones.

The two regions R_i and R_j are adjacent. If there is more than one common edge between R_i and R_j , draw line between p_i and p_j for each of the common edge.

Step III: For a pendent edge e, lying entirely in one region say R_k, draw a self-loop intersecting edge e exactly once at point p_k. p_k is the point placed in graph G with respect to region R_k.

The new graph G^* (in broken lines) in **Fig. 11.10(c)** is obtained by the procedure defined above is the geometrical dual of the graph G.

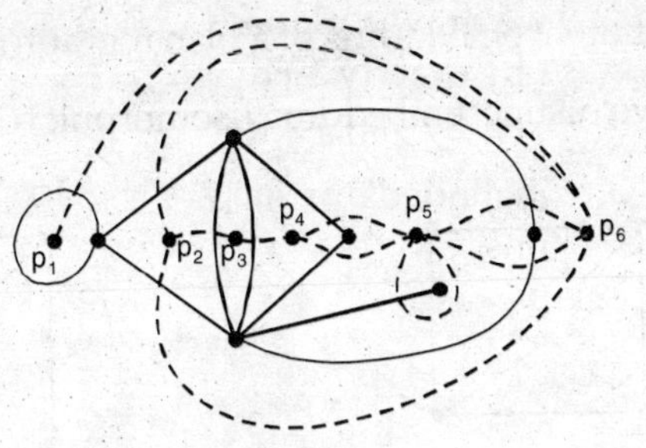

Fig. 11.10(*c*) Dual G* (in broken lines)

Fig. 11.10(*d*) shows the dual G^* of graph G.

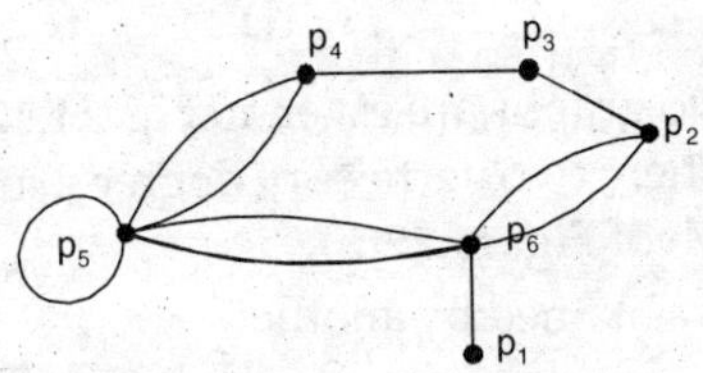

Fig. 11.10(*d*) Dual G*

Notes:

(*i*) An edge forming a self-loop in G^* corresponds to a pendent edge in G.

(*ii*) A pendent edge in G^* corresponds to a self-loop in G.

(*iii*) Parallel edges in G^* are produced by edges in series in G and vice-versa.

(*iv*) The degree of a region R_i in G is equal to the degree of vertex p_i in G^* and *vice-versa*.

(*v*) If e, n, R, r and μ denotes the number of edges, vertices, regions, rank and nullity of a connected planar graph G, then

$$e^* = e$$
$$n^* = R$$
$$R^* = n$$

where e^*, n^* and R^* are number of edges, vertices and regions in dual graph G^*.

$$\text{rank } r^* \text{ of } G^* = n^* - 1 \qquad (k = 1, \text{ as } G^* \text{ is connected})$$
$$= R - 1$$
$$= e - n + 2 - 1 = e - n + 1$$
$$= e - n + 1 \qquad \text{(by Euler's formula)}$$
$$= \mu$$

$$\Rightarrow \qquad r^* = \mu$$

$$\text{Nullity} \quad \mu^* \text{ of } G^* = e^* - n^* + 1$$
$$= e - R + 1$$
$$= e - (e - n + 2) + 1 \qquad \text{(By Euler's formula)}$$
$$= n - 1$$

$$\Rightarrow \qquad \mu^* = r$$

(*vi*) For a unique embedding of planar graph G the dual G^* wil be unique.

That is different embedding of a planar graph have different dual graphs.

Consider the two distinct embedding (isomorphic) of a graph. See **Figs. 11.11**(*a*) and **11.11**(*b*).

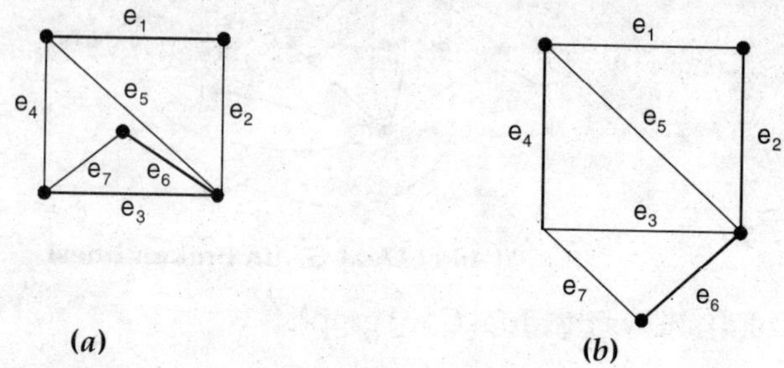

Fig. 11.11

The duals of the two graphs are given in **Fig. 11.12**(*a*) and (*b*). Clearly the duals are non-isomorphic. (There two vertices of degree 4 in **Fig. 11.10(*a*)** but no vertex of degree four in the graph of **Fig. 11.12(*b*)**.

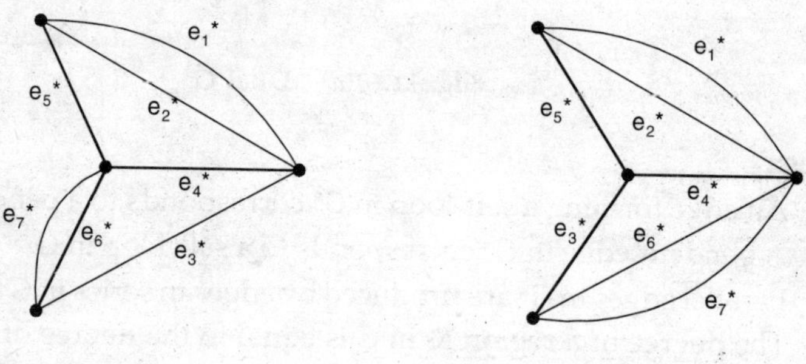

Fig. 11.12

Example 11.8. Show that the graphs shown below are isomorphic but their geometric duals are not isomorphic. **(U.P.T.U., 2002-03)**

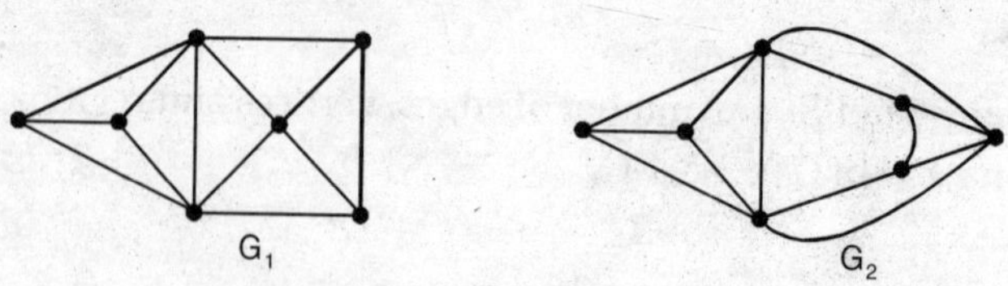

Fig. 11.13

Solution: Labeling of the vertices and edges of the given graphs is shown below:

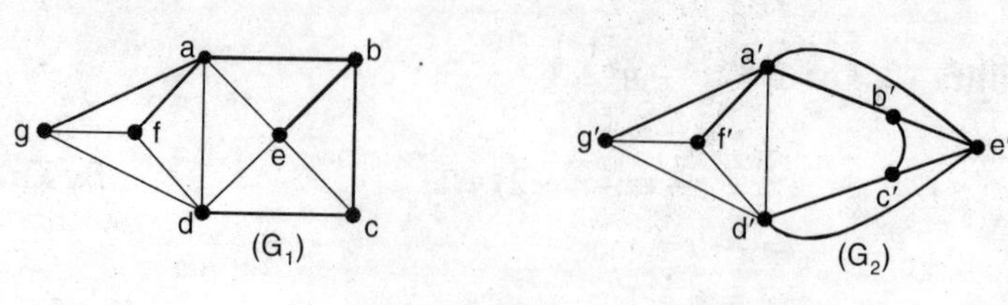

reconstruct the graph G_2 by taking the vertex e' inside the cycle $a'b'c'$ and d'.

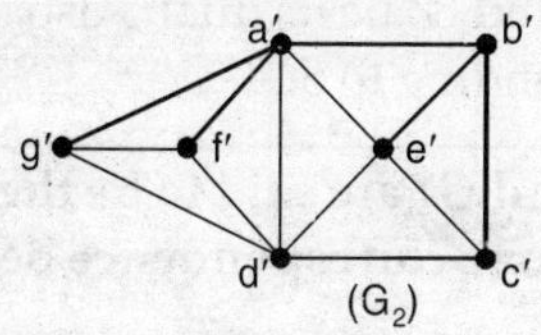

clearly G_2 is isomorphic to G_1.
Let us construct the dual of G_1.

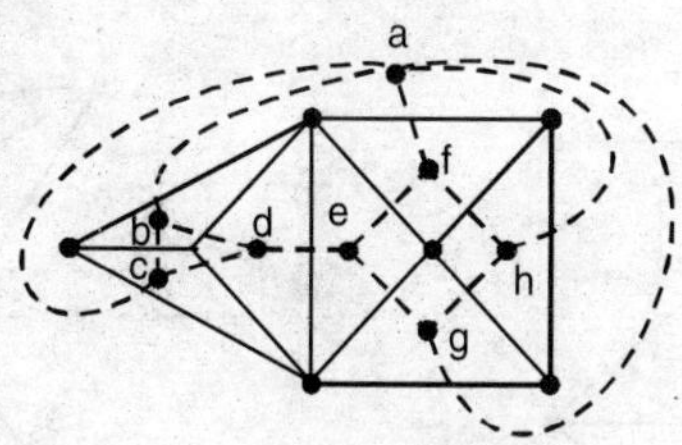

So dual of G_1, *i.e.*, G_1' is

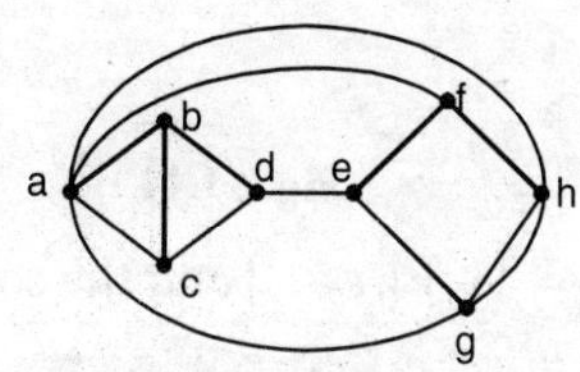

G_1'

Now dual of G_2 can be constructed as

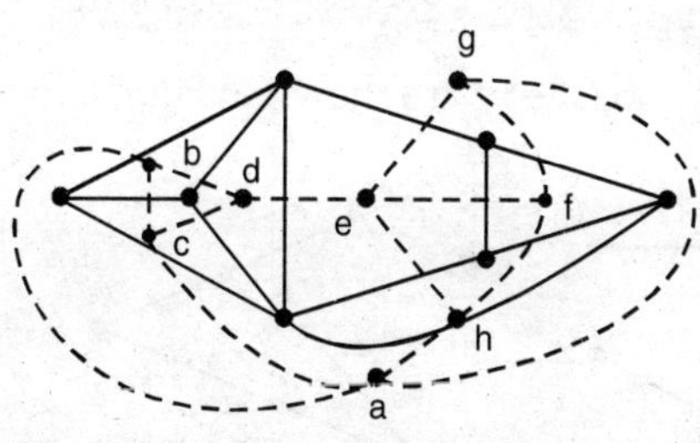

So G_2', i.e. dual of graph G_2 is

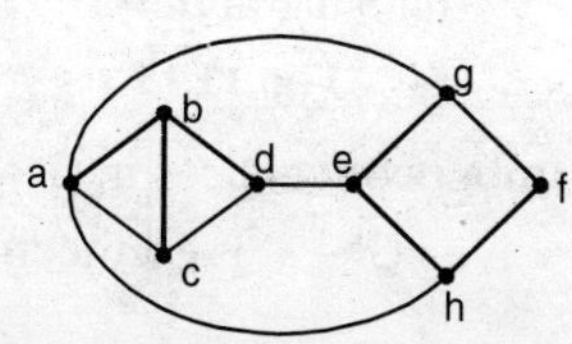

G_2'

Now one can see easily that in G_1' there is one vertex of degree 5 (*i.e.*, *a*) but in G_2' and G_2' are not isomorphic. Though the number of vertices in both the graphs are same.

11.8.2 Combinational Dual

Let G be a planar graph and G^* be its dual. Also let S be a subgraph of G and S^* be a subgraph of G^* corresponding to S.

Two planar graphs G and G^* are said to be the combinational dual of each other if there exist a one-to-one correspondence between the edges of G and G^* such that

$$\textbf{Rank of } (G^* - S^*) = \textbf{rank of } G^* - \textbf{nullity of } S.$$

For example, consider the two graphs G and G* shown in **Fig. 11.14**(*a*) and (*b*).

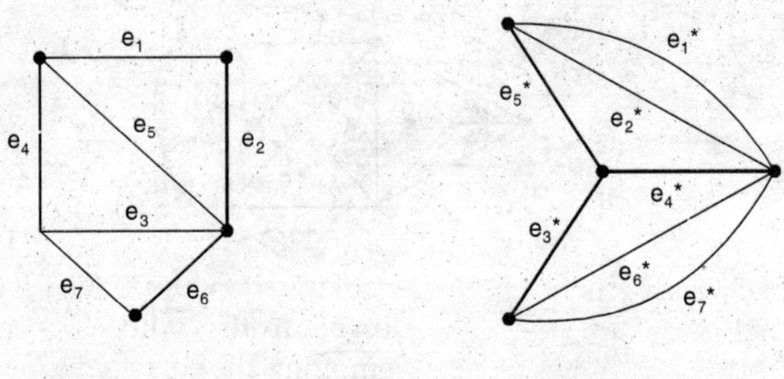

(*a*) Graph G **(*b*) Graph G***

Fig. 11.14

Consider subgraph $S = \{e_1, e_3, e_7, e_6\}$ and the corresponding subgraph

$$S^* = \{e_1^*, e_3^*, e_7^*, e_6^*\}$$

then $G^* - S^* = \{e_2^*, e_4^*, e_5^*\}$

Now rank of $G^* = n^* - 1$

$$= 4 - 1 = 3$$

The subgraphs S, S^* and $G^* - S^*$ are shown in Fig. 11.15(*a*) (*b*) and (*c*) respectively.

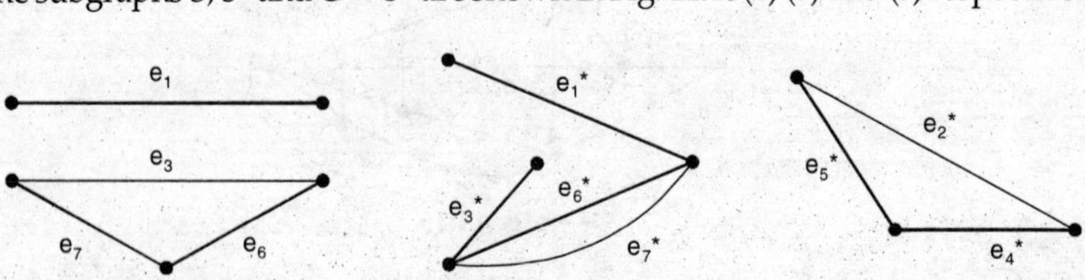

(*a*) Subgraph (*b*) Subgraph S* (*c*) Subgraph G* – S*

Fig. 11.15

Now rank of $G^* - S^*$ = numbers of vertices in $G^* - S^* - k$

($k = 1$, as $G^* - s^*$ is connected)

$\Rightarrow$ rank of $(G^* - S^*) = 3 - 1 = 2$

Nullity S = Number of edges in S – number of vertices in S + number of components in S.

$= 4 - 5 + 2$ ($k = 2$ S has two components)

$= 1$

$\Rightarrow$ rank of $(G^* - S^*)$ = rank of G^* – nullity of S

$= 3 - 1 = 2$

Here G and G^* are combinational dual of each other.

Theorem 11.4. A necessary and sufficient condition for two planar graphs G and G^* to be duals of each other is stated as: There is one-to-one correspondence between the edges in G and the edges in G^* such that a set of edges in G forms a circuit iff the corresponding set in G^* forms a cut-set.

Theorem 11.5. All duals of a planar graph G are 2-isomorphic. Every graph which is 2-isomorphic to a dual of G is also a dual of G.

11.9 DUAL OF SUBGRAPH

Let G be a planar graph and G^* be its dual. Consider edge e^* in G^* corresponding to edge e in graph G.

$G - e$ is a subgraph of G. The dual of $G - e$ can be obtained from the dual G^* of G in two ways:

Step (*i*): If e is lying on the boundary of two regions in G (*i.e.*, e is the common edge between two regions) then detection of e from G will merge the two regions. Then the dual $(G - e)^*$ is obtained by removing the corresponding edge e^* from G^* and fuse the two end vertices of e^* in $G^* - e^*$.

Step (*ii*): If e is not lying on the boundary of two regions. (*i.e.*, e is a pendent edge) then e^* is a self loop in G^*.

So $(G - e^*) = G^* - e^*$

Thus if G^* is the dual of G, then the dual of any subgraph of G can be obtained by successive application of steps (*i*) and (*ii*).

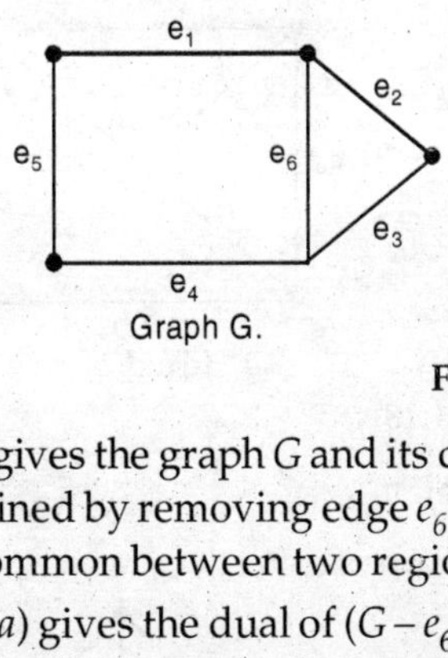

Graph G.

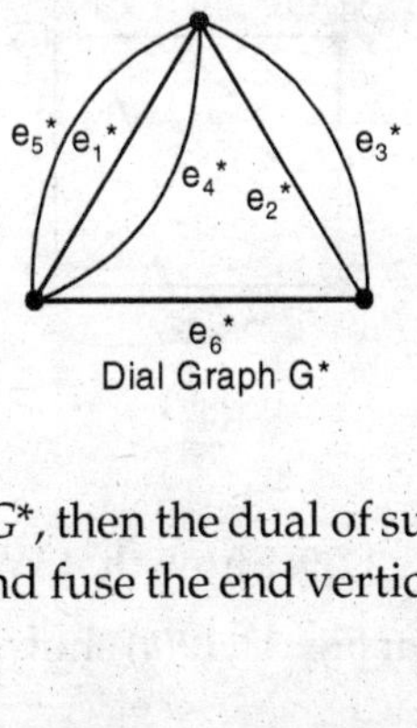

Dial Graph G*

Fig. 11.16

Fig. 11.16 gives the graph G and its dual graph G^*, then the dual of subgraph $G' = G - e_6$ is obtained by removing edge e_6^* from G^* and fuse the end vertices of e_6^* in $G^* - e^*$. (e_6 is common between two regions).

Fig. 11.16(*a*) gives the dual of $(G - e_6)$.

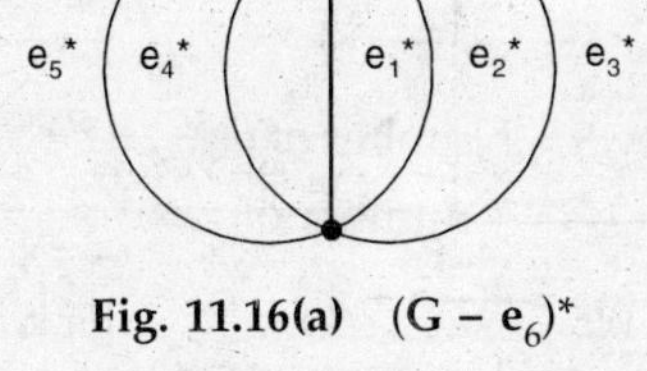

Fig. 11.16(a) $(G - e_6)^*$

11.10 DUAL OF A HOMEOMORPHIC GRAPH

Let G be a planar graph and G^* be it's dual. A graph G^H, homeomorphic to graph G can be obtained by (*i*) introducing a vertex of degree two in any edge of graph G, (*ii*) by merging two edges in series in G. In both the ways the dual of G^H is obtained by the following procedure.

Step (*i*): Let $e \in G$ is an edge. Introduction of any vertex of degree two in edge e, adds an edge parallel to e^* in G^*.

Step (*ii*): Merging two edges in series in G will eliminate one of the corresponding parallel edge in G^*.

Hence the dual of G^H can be obtained from G^* by the above method.

For example, **Fig. 11.16** shows the two graphs G and G^*. The graph in **Fig. 11.17**(*a*) is the homeomorphic graph G^H of G.

Fig. 11.17(*a*) Graph GH obtained by introducing **Fig. 11.17(b) Dual of GH vertex v of degree two in edge e5 of graph G**

The dual of G^H is given in **Fig. 11.17**(*b*).

11.11 SELF DUAL GRAPHS

A planar graph G is said to be **self-dual** if G is isomorphic to its own dual G^*.

For example, the complete graph of 4-vertices is a self-dual graph. See **Fig. 11.18.**

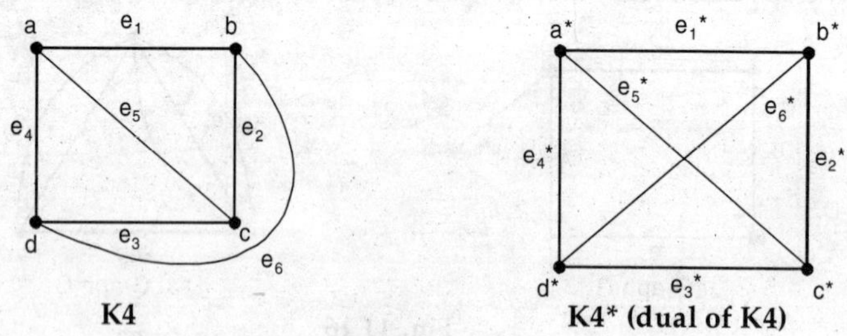

K4 **K4* (dual of K4)**

Fig. 11.18

Example 11.9. Show that W_5 is self-dual.

Solution: Fig. 11.19(*a*) shows a wheel graph of 5-vertices *i.e.*, W_5.

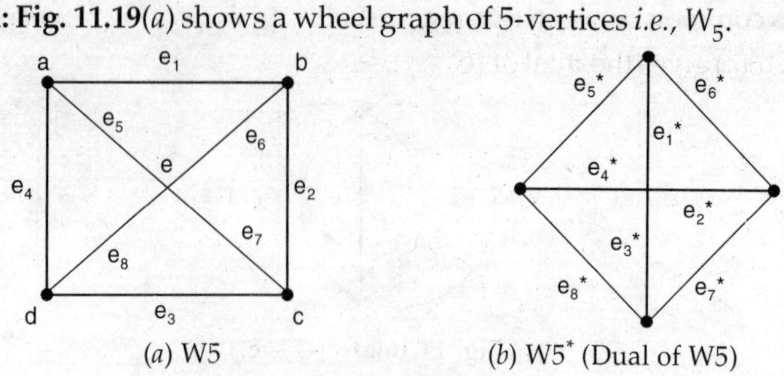

(*a*) W5 (*b*) W5* (Dual of W5)

Fig. 11.19

Dual of W_5 is given in **Fig. 11.19**(*b*) and it is again a wheel graph of 5-vertices. Hence W_5 is self dual.

11.12 THICKNESS AND CROSSING

Thickness of a graph *G* is the minimum number of planar subgraphs whose union is *G*.

Crossing number of graph *G* is the minimum number of intersection in drawing of graph *G* in a plane.

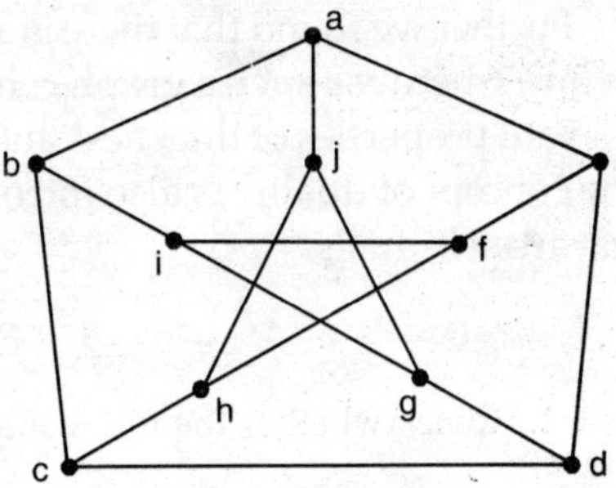

Fig. 11.20

For example, the thickness of G, is *one* and crossing number is *zero.*

Thickness of K_5 and $K_{3,3}$ is two crossing number of each Kuratowski's graph is one.

Example 11.10. Find the thickness and crossing number of Petersen's graph.

Solution : Consider a drawing of Petersen's graph

Redraw it to find the thickness of it.

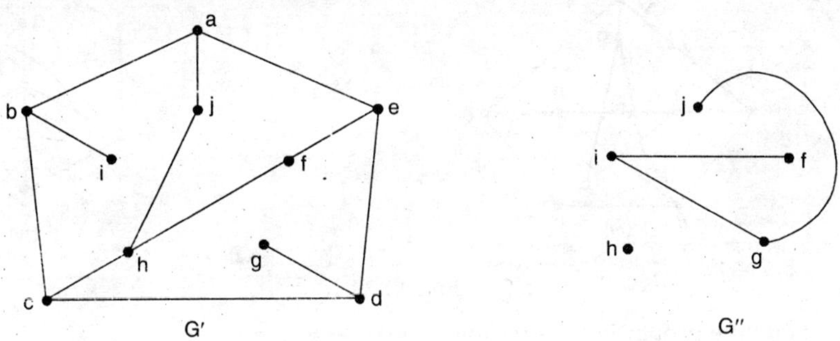

Fig. 11.21

As we can obtain two planar subgraphs *G*′ and *G*″ of *G* (Petersen's graph) whose union is *G*. Thus it's thickness is 2.

Again redraw it to find the crossing number.

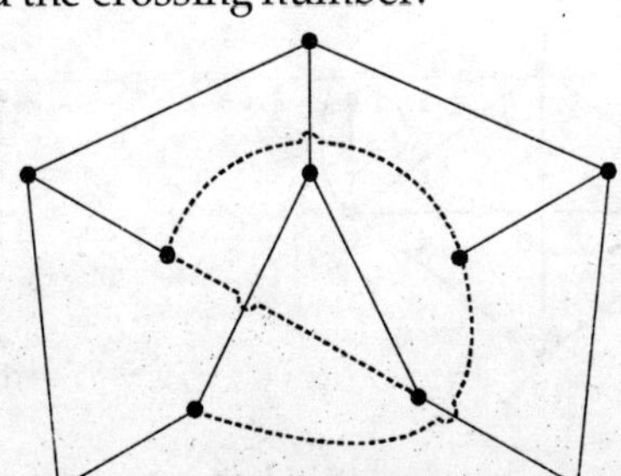

Fig. 11.22.

So, crossing number of Petersen's graph is = 3.

SUMMARY

In the chapter, we have discussed the concept of planarity which is very useful in many practical applications.

Further we found that there is a no need to investigate the graphs having edges $e > 3n - 6$ because such a graph is non-planar. For non-planar graphs, additional, relevant properties of thickness and crossing number were defined and discussed. The concept of duality is also introduced in planar graph and found some graphs that are self-duals.

EXERCISES

1. Check whether the following graphs are planar or not.

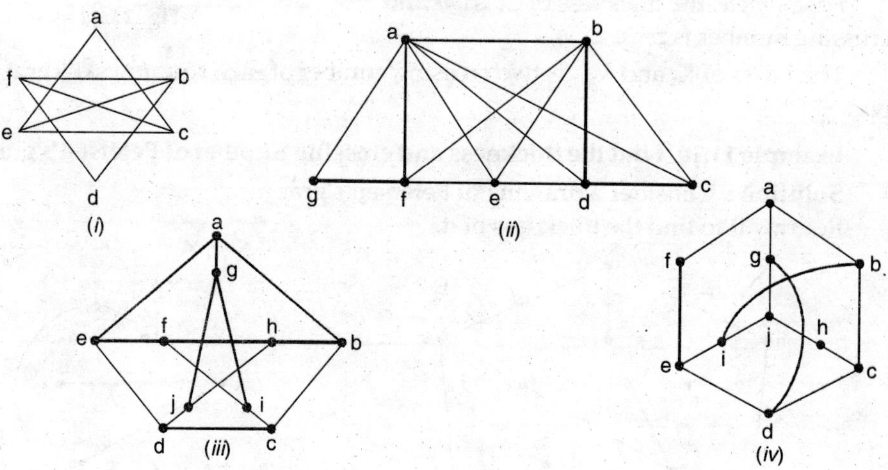

Fig. 11.23

2. State the properties of Kuratowski's graphs.

3. Show that, if every region of a simple planar graph with n-vertices and e-edges in a plane is bounded by k-edges then $e = k(n-2)/(k-2)$.

4. Find the degree of each region in the following graphs

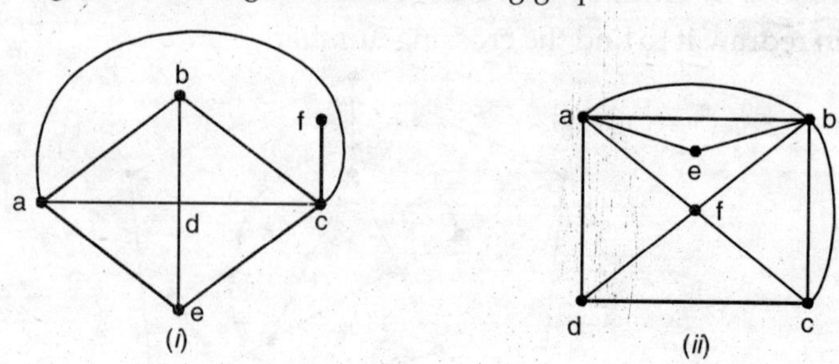

Fig. 11.24

5. Show that the following graphs are not planar by finding a subgraph homeomorphic to either of the Kuratowski's two graphs.

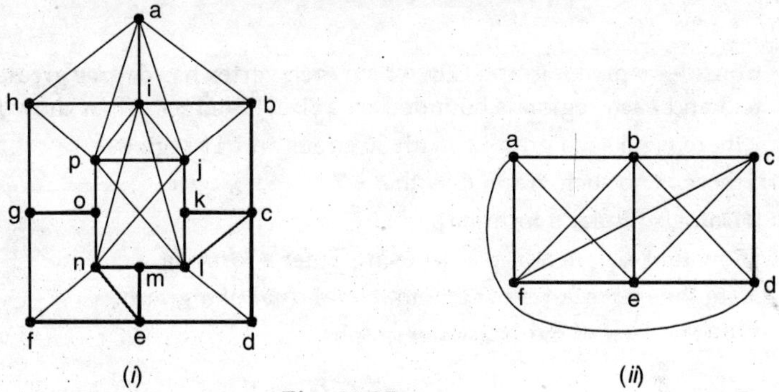

Fig. 11.25

6. Show that the Peterson's graph is non-planar.
   [Hint: Use Kuratowski's theorem]
7. Show that the complete tripartite graph $K_{1, 2, 3}$ is non-planar.
8. Count the number of vertices, edges and regions of each of the planar graphs shown below and verify the Euler's formula.

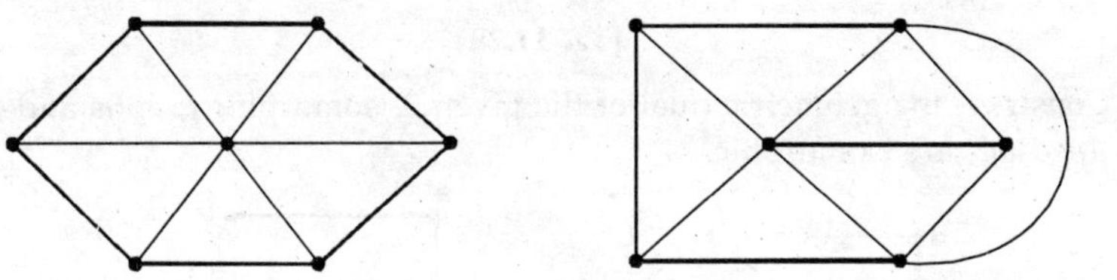

Fig. 11.26

9. Show that the wheel graph W_n is planar and hence calculate the number of edges, vertices and regions of W_n.
10. Prove that a planar graph of n vertices ($n \geq 4$) has a least 4-vertices with degree 5 or less. Also prove that there is no 6-connected planar graphs.
11. Find out the value of m and n for which the complete bipartite graph $K_{m, n}$ is planar.
12. Find the thickness and crossing number of each of the following graphs:

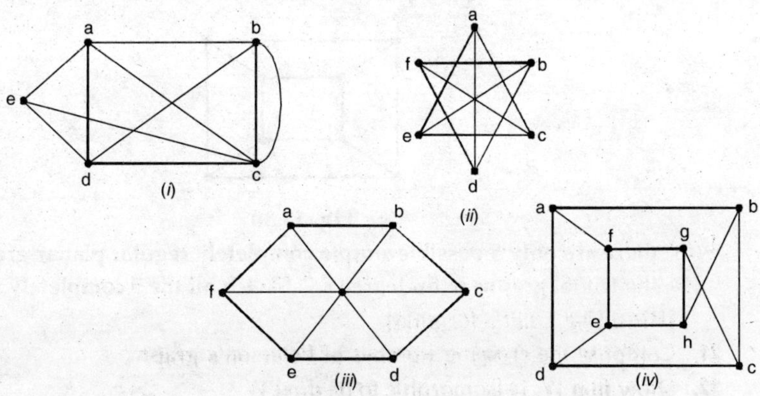

Fig. 11.27

[Hint: Check for planarity first by redrawing the graphs].

13. Draw a planar, connected graph such that
   (*i*) $e = 3n - 6$
   (*ii*) $e < 3n - 6$
14. Consider a planar graph G in which each vertex has degree greater than or equal to 3 and each region is bounded by at least 3-edges. Then show that
   (*i*) there is no such graph G with 30 edges and 11 regions.
   (*ii*) there is no such graph G with $e = 7$.
   [**Hint:** Use Euler's formula].
15. Show that $K_{3,3}$ is non-planar using Euler's formula.
16. State the importance of combinational dual of a graph.
17. Find the dual of the following graphs :

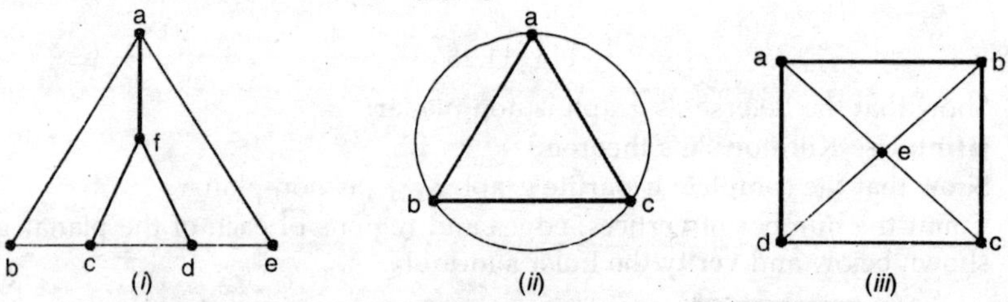

Fig. 11.28

18. Construct the geometric dual of the given 2-isomorphic graphs and show that the duals are isomorphic.

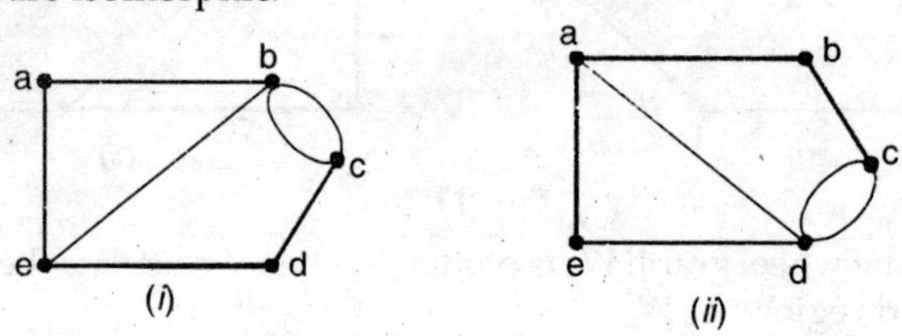

Fig. 11.29

19. Draw any two self-dual graphs.
20. Completely regular graphs are those planar graphs in which the degree of all the vertices are same and also every region is bounded by the same number of edges. Then show that :
   (*i*) the graph given is completely regular.

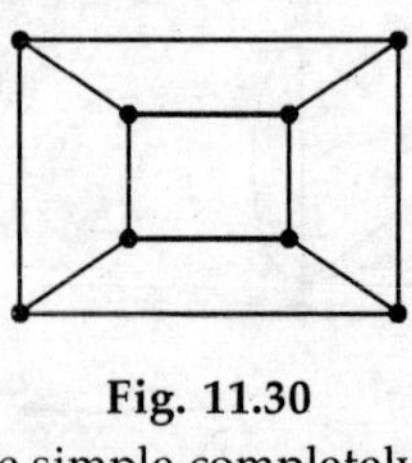

Fig. 11.30

   (*ii*) there are only 5 possible simple completely regular planar graphs, excluding the trivial graphs with degree £ 2. Sketch all the 5 completely regular graphs.
   [**Hint:** Use Euler's formula].
21. Compute the crossing number of Peterson's graph.
22. Show that W_n is isomorphic to its dual W_n^*.
23. Show that the edges forming a spanning tree in a planar graph G corresponds to the edges forming a set of chords in the dual G*. (*i.e.*, if there are 3-edges forming a spanning tree in G, then there are also 3-edges forming a chord set in G*)

[**Hint:** Draw a planar graph G and dual G^*. Consider an arbitrary spanning tree in G and verify the result].

24. Show that the two graphs are combinational dual of each other.

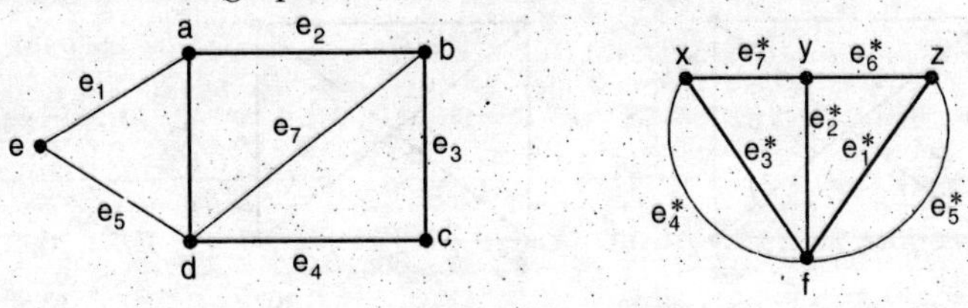

Fig. 11.31

25. Show that, dual of petersen's graph does not exist.

ANSWERS

1. (*i*) Yes (*ii*) No (*iii*) No (*iv*) No.

4. (*i*)

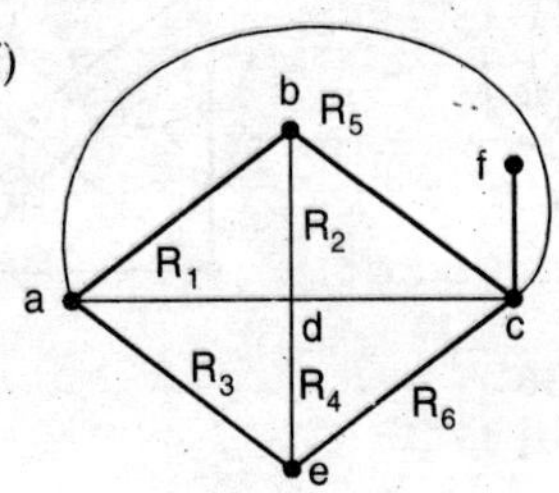

Fig. 11.32

$\deg(R_1) = 3 = \deg(R_2) = \deg(R_3) = \deg(R_4)$

$\deg(R_5) = 5$

$\deg(R_6) = 3$

5. (*i*) Subgraph homeomorphic to K_5 is

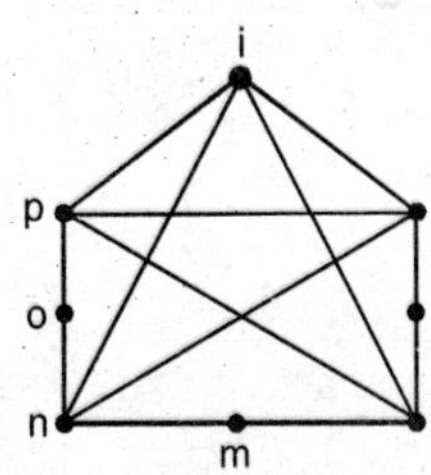

Fig. 11.33

(*ii*) Subgraph homeomorphic to $K_{3,3}$ is

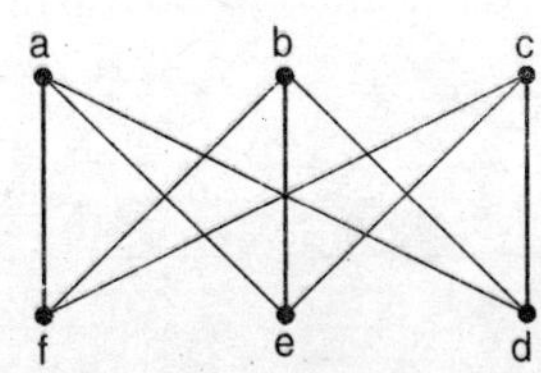

Fig. 11.34

9. Number of vertices in $W_n = n$ number of edges in $W_n = 2(n-1)$.
Thus $R = e - n + 2 = n$

12. (*i*) thickness = 1 crossing number = 0
(*ii*) thickness = 2 crossing number = 1

(*iii*) thickness = 2 crossing number = 1

(*iv*) thickness = 2 crossing number = 1

13. (*i*)

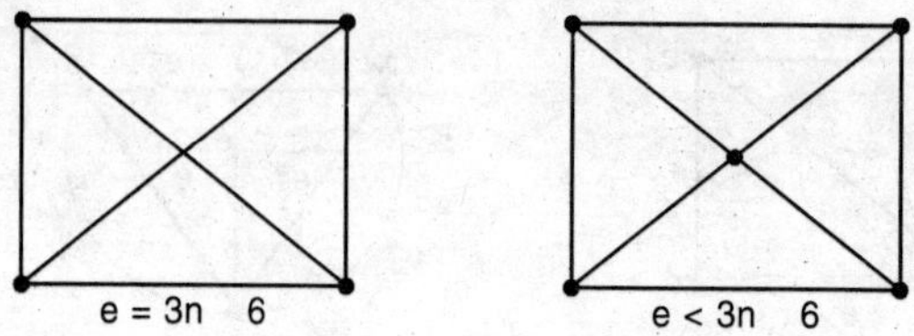

Fig. 11.35

17.

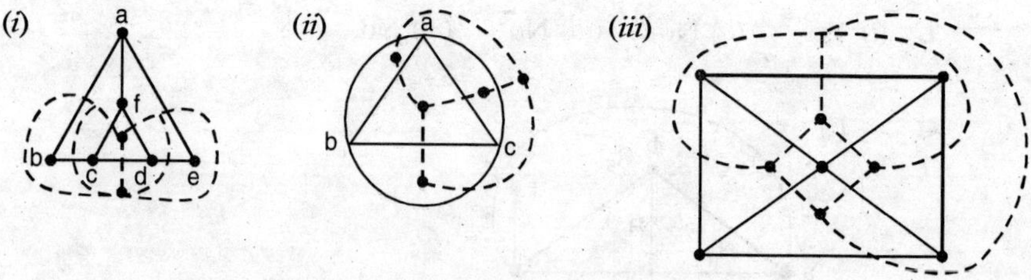

Fig. 11.36

Chapter

12 MATRIX REPRESENTATION OF GRAPHS

12.1 INTRODUCTION

Pictorial representation of a graph is a very convenient approach for a visual study of a graph but not very good for computer processing. Representation of graph as a matrix is a convenient way and a useful technique in computer processing.

In this chapter we shall consider the two very important and frequently used matrix representation of a graph. We shall also establish a correspondence between some graph-theoretic properties and matrix properties.

12.2 INCIDENCE MATRIX

Let G be a graph (without self-loops) with n-vertices, e-edges. Then an n by e matrix $A = [a_{ij}]$, n-rows correspond to the n-vertices and the e-columns correspond to the e-edges, is called incidence matrix of graph G.

The matrix element a_{ij} is defined as

$$a_{ij} = \begin{cases} 1 & \text{if } j\text{th edge } e_j \text{ is incident on } i\text{th vertex } v_i, \text{ and} \\ 0 & \text{otherwise.} \end{cases}$$

Incidence matrix A of graph G sometimes denoted as $A(G)$. Since $A(G)$ contains only two elements 0 and 1 so it is called *binary or (0, 1)-matrix* also.

A graph G and it's incidence matrix $A(G)$ is given in **Fig. 12.1** (*a*).

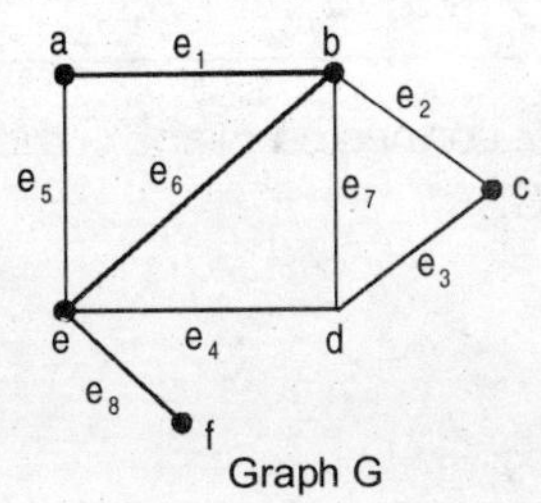

Graph G

$$A(G) = \begin{array}{c} \\ a \\ b \\ c \\ d \\ e \\ f \end{array} \begin{array}{c} e_1\ e_2\ e_3\ e_4\ e_5\ e_6\ e_7\ e_8 \\ \begin{bmatrix} 1 & 0 & 0 & 0 & 1 & 0 & 0 & 0 \\ 1 & 1 & 0 & 0 & 0 & 1 & 1 & 0 \\ 0 & 1 & 1 & 0 & 0 & 0 & 0 & 0 \\ 0 & 0 & 1 & 1 & 0 & 0 & 1 & 0 \\ 0 & 0 & 0 & 1 & 1 & 1 & 0 & 1 \\ 0 & 0 & 0 & 0 & 0 & 0 & 0 & 1 \end{bmatrix} \end{array}$$

Incidence matrix of G.

Fig. 12.1. (*a*)

12.2.1 Observation about Incidence Matrix *A* of Graph *G*

(*i*) Each column of *A* has exactly two 1's because each edge in *G* is incident on exactly two vertices.

(*ii*) The total number of 1's in a row gives the degree of the corresponding vertex in G.

(*iii*) A row with no 1's (*i.e.*, all 0's) represents an isolated vertex.

(*iv*) A row with exactly one '1' represents a pendant vertex.

(*v*) Two identical columns in a matrix represents parallel edges in graph G.

(*vi*) Permutation of any two rows or columns in an incidence matrix simply indicates the relabeling of the vertices and edges of the same graph.

(*vii*) If graph G is disconnected and has two connected components g_1 and g_2 then the incidence matrix of G can be written in a block diagonal form as

$$A(G) = \left[\begin{array}{c|c} A(g_1) & 0 \\ \hline 0 & A(g_2) \end{array}\right]$$

$A(g_1)$ and $A(g_2)$ are the incidence matrix of component g_1 and g_2 respectively.

For example, **Fig. 12.1** (*b*) shows a disconnected graph with two components and **Fig. 12.1**(*c*) shows it's incidence matrix in block diagonal form.

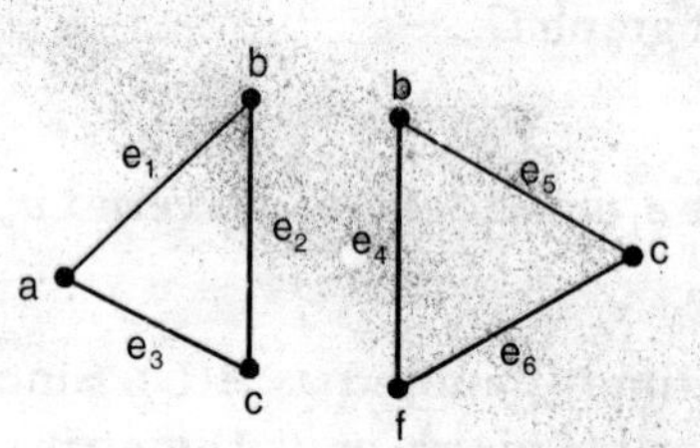

Fig. 12.1(b)

$$A(G) = \begin{array}{c} \\ a \\ b \\ c \\ d \\ e \\ f \end{array} \begin{array}{c} \begin{array}{cccccc} e_1 & e_2 & e_3 & e_4 & e_5 & e_6 \end{array} \\ \left[\begin{array}{ccc|ccc} 1 & 0 & 1 & 0 & 0 & 0 \\ 1 & 1 & 0 & 0 & 0 & 0 \\ 0 & 1 & 1 & 0 & 0 & 0 \\ \hline 0 & 0 & 0 & 1 & 1 & 0 \\ 0 & 0 & 0 & 0 & 1 & 1 \\ 0 & 0 & 0 & 1 & 0 & 1 \end{array}\right] \end{array}$$

Fig. 12.1(c)

12.2.2 Rank of the Incidence Matrix

The rank of a matrix is equal to the number of a linearly independent vectors in the matrix.

Theorem 12.1. If $A(G)$ is the incidence matrix of a connected graph G then rank of $A(G) = n-1$, where n is the number of vertices in G.

Proof: Consider

$$A(G) = \begin{bmatrix} A_1 \\ A_2 \\ \vdots \\ A_n \end{bmatrix}$$

where $A_1, A_2, ..., A_n$ are n-vertex in $GF(2)$ with respect to n-rows of incidence matrix. That is vector in the first row be called A_1, in the second row A_2, and so on. Since there are exactly two is in each column of A, the sum of all the vectors is 0

i.e., $A_1 + A_2 + ... + A_n = 0$ (modulo 2 sum of the corresponding entries)

Thus the n-vectors are not linearly independent.

Therefore rank of $A(G) < n$ *i.e.*, $A(G) \leq n - 1$.

Now let us consider the sum of any k-rows of these n-rows (or vectors) such that $k \leq n - 1$ since the graph g is connected so it can not be partitioned. In other words we cannot find a $k \times k$ submatrix of $A(G)$ for $k \leq n - 1$. Hence the addition of all vectors taken k at a time exhausts all possible linear combinations of $n - 1$ row vectors. That is there is a no linear combination of k-vectors of $A(G)$ equal to zero in $GF(2)$.

Thus rank of $A(G) \geq n - 1$.

Hence by the two results, rank of $A(G) = n - 1$.

12.2.3 Reduced Incidence Matrix

An $(n - 1)$ by e matrix A_R obtained by deleting one row from the incidence matrix A of a graph G is called reduced incidence matrix.

The vertex corresponding to the removed row in A_R is called **reference vertex**.

The rank of incidence matrix A_R is $n - 1$. Because there are $(n - 1)$ vectors in $A_R(G)$ and these $(n - 1)$ vectors are linearly independent.

Example 12.1. Show that, the reduced incidence matrix of a tree is non-singular.

Solution: A reduced incidence matrix A_R of a tree T is a $(n - 1) \times (n - 1)$ square matrix, obtained by detecting the nth row from the incident matrix.

Clearly, the $(n - 1)$ vectors in A_R are linearly independent.

So rank, $A_R = (n - 1)$ ...(*i*)

We know that a square matrix A is singular if

$$|A| = 0$$

i.e., if $|A| \neq 0$, A is non-singular and the inverse of A is possible.

Now if rank $(A) = m$, then there exist at least one minor of order m such that

$$|A'|_{m \times m} \neq 0$$

where A' is a minor (*i.e.*, determinant of m-rows and m-columns) of order m.

equation (*i*) says that rank $(A_R) = (n - 1)$

$\Rightarrow$ there is at least one minor of order $(n - 1)$ in A_R which is not zero

$\Rightarrow |A_R|_{(n-1) \times (n-1)} \neq 0$

$\Rightarrow A_R$ is non-singular.

12.3 CIRCUIT MATRIX

Consider a graph G with n-vertices and e edges. Let there be m-different circuits. Then a circuit matrix $B = [b_{ij}]$ of the graph G is a m by n matrix, where

$$B(G) = b_{ij} = \begin{cases} 1, & \text{if } i\text{th circuit includes } j\text{th edge} \\ 0, & \text{otherwise} \end{cases}$$

Consider the graph in **Fig. 12.2**(*a*). Graph has four circuits

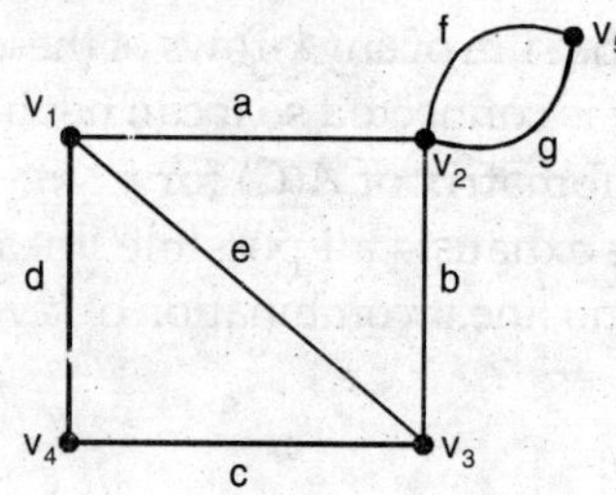

Fig. 12.2 (*a*) Graph G

$C_1 = (a, b, e)$, $C_2 = (d, e, c)$, $C_3 = (f, g)$ and $C_4 = (a, b, c, d)$. The corresponding circuit matrix $B(G)$ of the graph G is shown in **Fig. 12.2**(*b*)

$$B(G) = \begin{array}{c} \\ c_1 \\ c_2 \\ c_3 \\ c_4 \end{array} \begin{array}{c} \begin{matrix} a & b & c & d & e & f & g \end{matrix} \\ \begin{bmatrix} 1 & 1 & 0 & 0 & 1 & 0 & 0 \\ 0 & 0 & 1 & 1 & 1 & 0 & 0 \\ 0 & 0 & 0 & 0 & 0 & 1 & 1 \\ 1 & 1 & 1 & 1 & 0 & 0 & 0 \end{bmatrix} \end{array}$$

Fig. 12.2(*b*) Circuit matrix of graph *G*

12.3.1 Observations

(*i*) The number of 1's in each row is equal to the number of edges in the corresponding circuit.

(*ii*) Two graphs G_1 and G_2 will have same circuit matrix iff G_1 and G_2 are isomorphic.

(*iii*) All zero's in a column indicates that the corresponding edge of graph does not belongs to any circuit of graph G.

(*iv*) The row containing single 1 in circuit matrix indicates a self-loop in graph G.

(*v*) If graph G is disconnected (or separable) and consists of two components g_1 and g_2, then the circuit matrix $B(G)$ can be written in a block diagonal form as

$$B(G) = \left[\begin{array}{c|c} B(g_1) & 0 \\ \hline 0 & B(g_2) \end{array}\right]$$

where $B(g_1)$ and $B(g_2)$ are the circuit matrices of the components g_1 and g_2 respectively.

(*vi*) Permutation matrix of the circuit matrix corresponds to relabeling the rows and columns (*i.e.*, circuits and edges).

Theorem 12.2. Let A and B be the incidence matrix and circuit matrix respectively of a self-loop free graph G. Then show that every row of matrix B is orthogonal to every row of matrix A.

$$i.e., \quad A \cdot B^T = B \cdot A^T = 0 \qquad (\text{mod } 2).$$

Proof: To prove this result let us consider a circuit C and a vertex v in the graph G. Then exist two cases:

Case I: Let $v \notin C$, that is there is no edge incident on vertex v in circuit C.

Case II: Let $v \in C$, that is there are exactly two edges incident on vertex v in circuit C.

Now let us consider the ith row of matrix A and jth row of matrix B. Since the edges in both the matrices are arranged in same order, the non-zero entries in the corresponding positions occur only if the particular edge is incident on the ith vertex and is also in the jth circuit. This implies that a zero indicates that ith vertex not in the jth circuit and the dot product of the two row's is zero.

If the ith vertex is in jth circuit, then there will be exactly two is in the sum of the products of individual entries and $1 + 1 = 0$ (mod 2) (*i.e.*, the dot product of the two arbitrary rows).

Hence the theorem.

For example, consider the circuit matrix of the graph G (See **Fig. 12.2** (b)). The incidence matrix of the graph G in **Fig. 12.2**(a) is as follows:

$$A(G) = \begin{array}{c} \\ v_1 \\ v_2 \\ v_3 \\ v_4 \\ v_5 \end{array} \begin{array}{c} \begin{matrix} a & b & c & d & e & f & g \end{matrix} \\ \begin{bmatrix} 1 & 0 & 0 & 1 & 1 & 0 & 0 \\ 1 & 1 & 0 & 0 & 0 & 1 & 1 \\ 0 & 1 & 1 & 0 & 1 & 0 & 0 \\ 0 & 0 & 1 & 1 & 0 & 0 & 0 \\ 0 & 0 & 0 & 0 & 0 & 1 & 1 \end{bmatrix} \end{array} \quad \text{and} \quad B^T = \begin{array}{c} a \\ b \\ c \\ d \\ e \\ f \\ g \end{array} \begin{bmatrix} 1 & 0 & 0 & 1 \\ 1 & 0 & 0 & 1 \\ 0 & 1 & 0 & 1 \\ 0 & 1 & 0 & 1 \\ 1 & 1 & 0 & 0 \\ 0 & 0 & 1 & 0 \\ 0 & 0 & 1 & 0 \end{bmatrix}$$

then $$A \cdot B^T = \begin{bmatrix} 1 & 0 & 0 & 1 & 1 & 0 & 0 \\ 1 & 1 & 0 & 0 & 0 & 1 & 1 \\ 0 & 1 & 1 & 0 & 1 & 0 & 0 \\ 0 & 0 & 1 & 1 & 0 & 0 & 0 \\ 0 & 0 & 0 & 0 & 0 & 1 & 1 \end{bmatrix} \cdot \begin{bmatrix} 1 & 0 & 0 & 1 \\ 1 & 0 & 0 & 1 \\ 0 & 1 & 0 & 1 \\ 0 & 1 & 0 & 1 \\ 1 & 1 & 0 & 0 \\ 0 & 0 & 1 & 0 \\ 0 & 0 & 1 & 0 \end{bmatrix}$$

$$= \begin{bmatrix} 1.1 + 0.1 + 0.0 + 1.0 + 1.1 + 0.0 + 0.0 & & & \\ = 1 + 1 = 0 \ (\text{mod } 2) & 0 & 0 & 0 \\ 0 & 0 & 0 & 0 \\ 0 & 0 & 0 & 0 \\ 0 & 0 & 0 & 0 \\ 0 & 0 & 0 & 0 \end{bmatrix}$$

$$= \begin{bmatrix} 0 & 0 & 0 & 0 \\ 0 & 0 & 0 & 0 \\ 0 & 0 & 0 & 0 \\ 0 & 0 & 0 & 0 \\ 0 & 0 & 0 & 0 \end{bmatrix} \pmod 2$$

12.3.2 Fundamental Circuit Matrix (B_f)

A submatrix $B_f(G)$ of circuit matrix $B(G)$ in which all rows corresponds to fundamental circuits with respect to a spanning free in G of graph G, is called a fundamental circuit matrix B_f.

Consider the graph G and spanning tree T (in bold lines) given in **Fig. 12.3**(*a*).

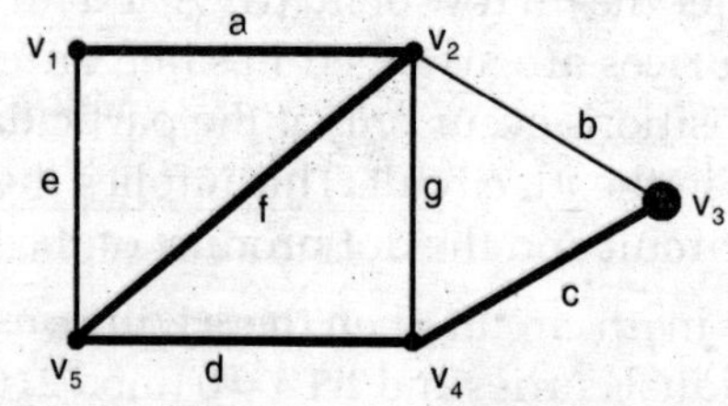

Fig. 12.3 (*a*) Graph G and spanning tree T

The fundamental circuits with respect to spanning tree T are (a, f, e), (f, d, g), (f, d, c, b) and the matrix $B_f(G)$ with respect to these fundamental circuits is given in **Fig. 12.3**(*b*).

$$B_f(G) = \begin{array}{c} \\ 1 \\ 2 \\ 3 \end{array} \begin{array}{c} \begin{array}{ccccccc} e & g & b & a & f & d & c \end{array} \\ \left[\begin{array}{ccc:cccc} 1 & 0 & 0 & 1 & 1 & 0 & 0 \\ 0 & 1 & 0 & 0 & 1 & 1 & 0 \\ 0 & 0 & 1 & 0 & 1 & 1 & 1 \end{array}\right] \end{array}$$

Fig. 12.3 (*b*) Fundamental circuit matrix of G

If n is the number of vertices and e be the number of edges in connected graph g then B_f is an $(e - n + 1)$ by e matrix. The number of fundamental circuits in G is equal to the number of chords and it is $(e - n + 1)$ because each fundamental circuit is produced by one chord.

The column of matrix B_f are arranged such that all the $e - n + 1$ chords corresponds to the first $e - n + 1$ columns similarly the rows are arranged such that the first row corresponds to the fundamental circuit made by the chord in the first column, second row to the fundamental circuit made by the second and so on.

So a matrix B_f is arranged as

$$B_f = [I_\mu \;\; B_t]$$

where I_μ is the identify matrix of order $\mu = e - n + 1$ and B_t is the remaining $(e - n + 1)$ by $(n - 1)$ submatrix, corresponding to the branches of the spanning tree T in G.

Each row of matrix B_f is a vector and it is clear that the vectors are linearly independent. Hence the rank of matrix $B_f = \mu = e - n + 1$.

12.3.3 Rank of Matrix B

Since B_f is a sub-matrix of matrix B

$\Rightarrow$ rank of $B \geq$ rank of $B_f = \mu = e - n + 1$

If A is the incidence matrix of the graph G, then by Theorem 12.2 we have

$$A \cdot B^T = 0 \pmod 2.$$

Thus,

$$\text{rank}(A) + \text{rank of } B \leq e$$

$\Rightarrow$ $\text{rank}(B) \leq e - \text{rank of }(A)$

$\Rightarrow$ $\text{rank}(B) \leq e - (n - 1)$ (rank of $A = n - 1$)

$\Rightarrow$ $\text{rank of }(B) \leq e - n + 1$

But $\text{rank of }(B) \geq e - n + 1$

Therefore, $\text{rank of } B = e - n + 1$

12.4 CUT-SET MATRIX

A cut-set matrix $C = [c_{ij}]$ with respect to a cut-set in the graph G is defined as:

$$c_{ij} = \begin{cases} 1, & \text{if ith cu - set contains } j\text{th edge} \\ 0, & \text{otherwise} \end{cases}$$

The rows and columns of cut-set matrix corresponds to the cut-sets and edges of the graph G respectively.

For example, the its cut-sets and cut-set matrix are shown in **Fig. 12.4**.

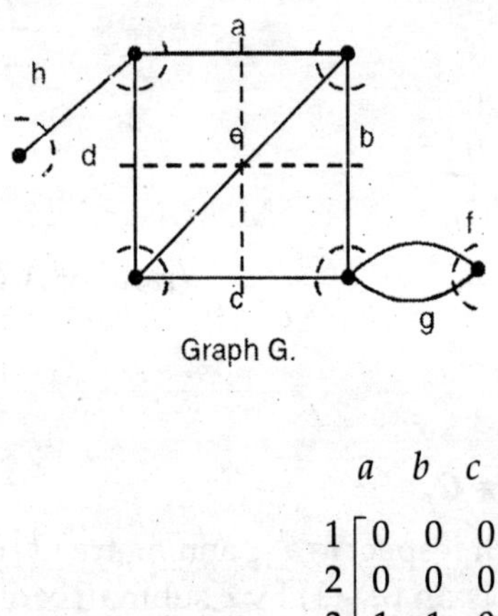

Graph G.

Cut-sets

$C_1 = \{h\}$
$C_2 = \{f, g\}$
$C_3 = \{a, e, b\}$
$C_4 = \{d, e, c\}$
$C_5 = \{d, e, b\}$
$C_6 = \{a, e, c\}$
$C_7 = \{a, d\}$
$C_8 = \{b, c\}$

$C = [C_{ij}] =$

	a	b	c	d	e	f	g	h
1	0	0	0	0	0	0	0	1
2	0	0	0	0	0	1	1	0
3	1	1	0	0	1	0	0	0
4	0	0	1	1	1	0	0	0
5	0	1	0	1	1	0	0	0
6	1	0	1	0	1	0	0	0
7	1	0	0	1	0	0	0	0
8	0	1	1	0	0	0	0	0

Fig. 12.4. Cut-set matrix of graph G with aspect to eight cut-sets in G

12.4.1 Observations

(*i*) A row with all 0's in matrix C corresponds to a self-loop in graph G.

(*ii*) Parallel edges in graph G produce identical columns in matrix C.

(*iii*) A permutation of rows or columns in a cut-set matrix C corresponds to a relabeling of the cut-sets and edges respectively.

(*iv*) If graph G is non-separable their every set of edges incident on a vertex v is cut-set. Thus every row of incidence matrix $A(G)$ is included as a row in the cut-set matrix C(G). *i.e.*, matrix C(G) contains matrix $A(G)$ for a non-separable graph.

Theorem 12.3. Let B and C be, respectively, the circuit matrix and cut-set matrix of a graph G whose columns are arranged using the same order of edges. Then every row of B is orthogonal to every row of C; that is

$$B \,.\, C^T = 0 = C \,.\, B^T \ (\text{mod } 2)$$

Proof: Consider a circuit T and a cut-set S in graph G. We know that there are even number of edges common between a circuit and cut-set of a graph.

Thus the number of 1's common in the ith circuit and jth cut-set is even and hence the sum of the products of the common entries will be $\underbrace{1.1 + 1.1 + \ldots + 1.1}_{\text{even time}}$ $= 0$ (mod 2)

Hence the theorem.

12.4.2 Rank of Matrix C

As we know that

$$B \,.\, C^T = 0 \quad (\text{mod } 2)$$

$\Rightarrow$ rank of B + rank C $\le e$

$\Rightarrow$ rank C $\le e - e + n - 1 \ldots (a)$

By (*iv*) observation of cut-set matrix

$$\text{rank of C(G)} \ge \text{rank of } A(G)$$

$$\Rightarrow \qquad \text{rank C(G)} \ge n - 1 \qquad\qquad (\text{rank of } A(G) = n - 1)$$

by equations (*a*) and (*b*)

$$\text{rank of C(G)} = n - 1$$

12.4.3 Fundamental Cut-set Matrix C_f

A fundamental cut-set matrix C_f (with respect to a spanning tree T) of concept of graph G with n-vertices and e-edges is an $(n - 1)$ by e submatrix of matrix C such that the rows in C_f corresponds to the set of fundamental cut-sets with respect to the given spanning tree. As we know that a fundamental cut-set contains exactly one branch of the spanning tree. So the matrix C_f can also be partitioned into two submatrices containing an identify matrix I_{n-1} of order $(n - 1)$.

i.e., $$C_f = \left[C_s \ \vdots \ I_{n-1} \right]$$

The columns of the identity matrix I_{n-1} corresponds to the $(n - 1)$ branches of the spanning tree and the columns of the submatrix C_s corresponds to the chord $(e - n + 1)$ of the graph G.

The fundamental matrix C_f of the graph G with respect to the spanning tree T (in bold lines) shown in **Fig. 12.5.**

Fundamental cut-sets are shown by dotted lines in the Graph G.

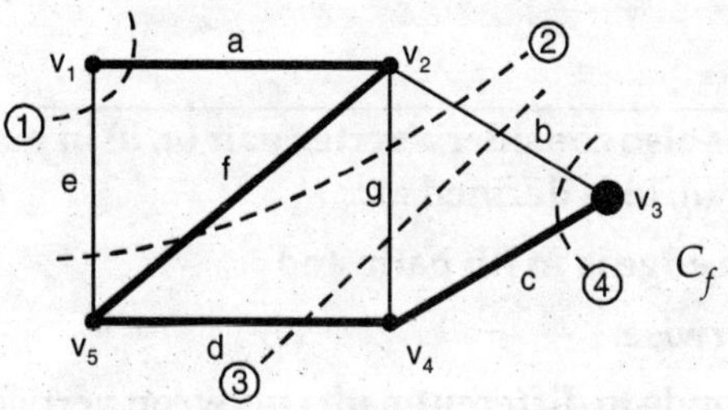

$$C_f = \begin{array}{c} \\ 1 \\ 2 \\ 3 \\ 4 \end{array} \overset{\begin{array}{ccccccc} e & g & b & a & f & d & c \end{array}}{\begin{bmatrix} 1 & 0 & 0 & 1 & 0 & 0 & 0 \\ 1 & 1 & 1 & 0 & 1 & 0 & 0 \\ 0 & 1 & 1 & 0 & 0 & 1 & 0 \\ 0 & 0 & 1 & 0 & 0 & 0 & 1 \end{bmatrix}}$$

Fig. 12.5. Fundamental cut-set matrix Cf

12.5 RELATIONSHIP AMONG A_R, B_F AND C_F

Let G be a graph. Consider the reduced incident matrix A_R, fundamental circuit matrix B_f and the fundamental cut-set matrix C_f of graph G.

Then we know that

$$B_f = [I_\mu \vdots B_t] \quad \text{and} \quad C_f = [C_s \vdots I_{n-1}]$$

where B_t denotes the submatrix corresponding to the branches of spanning tree t and C_s denotes the submatrix corresponding to the chords.

Consider a spanning tree T in graph G for both the matrices B_f and C_f and let the order of edges in both the matrices be same. Also the edges in both the matrices be same. Also the edges of the reduced incident matrix A_R of size $(n-1)$ by e are arranged in the same order as in B_f and C_f. Then consider the partition of matrix A_R into two submatrices A_c and A_t where the columns of A_t and A_c corresponds to the branches of the spanning tree T and the chords respectively.

i.e., $$A_R = [A_c : A_t]$$

We know that the columns of the three matrices are arranged in same order

$$\Rightarrow \quad A_R \cdot B_f^T = 0 \pmod 2$$

$$\Rightarrow \quad [A_c : A_t] \cdot \left[\frac{I_\mu}{B_t^T}\right] = 0 \pmod 2$$

$$\Rightarrow \quad A_c + A_t \cdot B_t^T = 0 \pmod 2$$

Since A_t is non-singular $\Rightarrow$ A_t^{-1} exists.

$$\Rightarrow \quad A_t^{-1} \cdot A_c = -B_t^T$$

or $$B_t^T = A_t^{-1} \cdot A_c \qquad \text{(since } -1 = 1 \text{ in mod 2 arithmetic)}$$

Similarly

$$C_f \cdot B_f^T = 0 \pmod 2$$

$$[C_s \mid I_{n-1}] \cdot \left[\frac{I_\mu}{B_t^T}\right] = 0 \pmod 2$$

$$\Rightarrow \quad C_s + B_t^T = 0 \pmod 2$$

$$\Rightarrow \quad C_s = -B_t^T$$

$$= B_t^T = A_t^{-1} . A_c$$

Hence $$B_t^T = C_s = A_t^{-1} . A_c$$

12.6 PATH MATRIX

Let us consider a graph G and let us also consider a vertex pair (*a*, *b*) in graph G. Then a path matrix $P(a, b) = [p_{ij}]$ for (*a*, *b*) is defined as:

$$p_{ij} = \begin{cases} 1, & \text{if } j\text{th edge is in } i\text{th path, and} \\ 0, & \text{otherwise.} \end{cases}$$

The rows in matrix *P*(*a*, *b*) corresponds to different paths between vertices *a* and *b* and the columns corresponds to the edges in *G*.

For example consider all paths between vertices *a* and *d* in **Fig. 12.6(*a*)**.

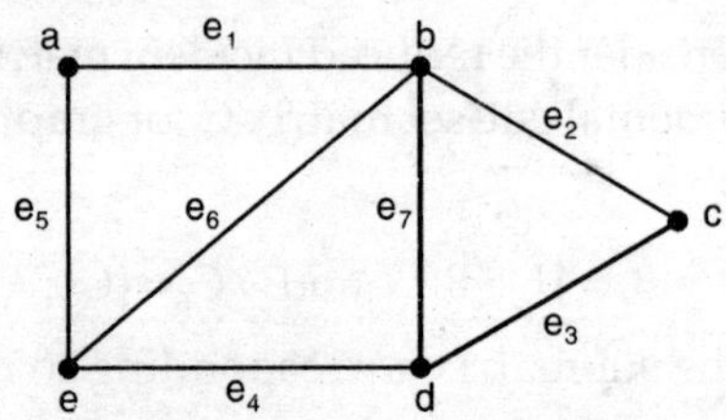

Fig. 12.6(*a*)

The four differents paths are (e_1, e_7), (e_1, e_6, e_4), (e_5, e_6, e_7) and (e_5, e_4) are numbered as (1), (2), (3) and (4) respectively.

Then
$$P(a, d) = \begin{array}{c} \\ 1 \\ 2 \\ 3 \\ 4 \end{array} \begin{array}{c} \begin{array}{ccccccc} e_1 & e_2 & e_3 & e_4 & e_5 & e_6 & e_7 \end{array} \\ \begin{bmatrix} 1 & 0 & 0 & 0 & 0 & 0 & 1 \\ 1 & 0 & 0 & 1 & 0 & 1 & 0 \\ 0 & 0 & 0 & 0 & 1 & 1 & 1 \\ 0 & 0 & 0 & 1 & 1 & 0 & 0 \end{bmatrix}_{4 \times 7} \end{array}$$

12.6.1 Observations

(*i*) The column of all 0's corresponds to an edge that does not belong to any path between the specified vertex pair (*a*, *b*).

(*ii*) The column of all 1's corresponds to an edge that lies in every path between a specified vertex pair (*a*, *b*).

(*iii*) The ring sum of any two rows of matrix *P*(*a*, *b*) corresponds to a circuit or an edge disjoint union of circuits.

12.7 ADJACENCY MATRIX

The adjacency matrix *M* of a graph *G* (without parallel edges) with *n*-vertices and *e*-edges is an *n* by *n* symmetric matrix defined as :

$$M(G) = [m_{ij}] = \begin{cases} 1, & \text{if } i\text{th vertex is adjacent to } j\text{th vertex, and} \\ 0, & \text{otherwise} \end{cases}$$

(*i.e.*, no edge between *i*th and *j*th vertices)

Consider the graph *G* and it's adjacency matrix *M*(*G*) in **Fig. 12.7**(*a*) and (*b*) respectively.

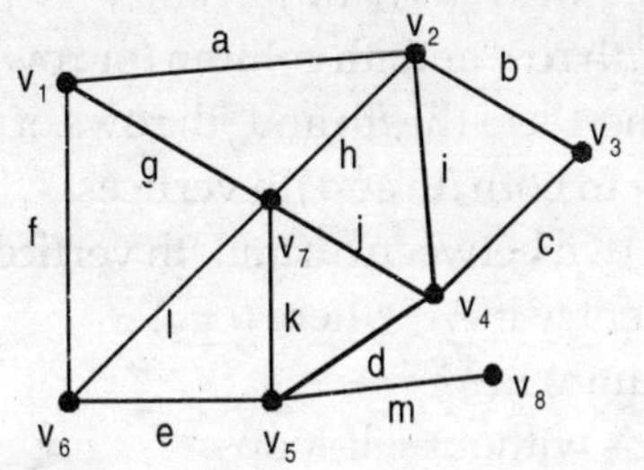

Fig. 12.7(*a*) Graph G

$$\begin{array}{c} \\ v_1 \\ v_2 \\ v_3 \\ v_4 \\ v_5 \\ v_6 \\ v_7 \\ v_8 \end{array}\begin{array}{c} \begin{array}{cccccccc} v_1 & v_2 & v_3 & v_4 & v_5 & v_6 & v_7 & v_8 \end{array} \\ \begin{bmatrix} 0 & 1 & 0 & 0 & 0 & 1 & 1 & 0 \\ 1 & 0 & 1 & 1 & 0 & 0 & 1 & 0 \\ 0 & 1 & 0 & 1 & 0 & 0 & 0 & 0 \\ 0 & 1 & 1 & 0 & 1 & 0 & 1 & 0 \\ 0 & 0 & 0 & 1 & 0 & 1 & 1 & 1 \\ 1 & 0 & 0 & 0 & 1 & 0 & 1 & 0 \\ 1 & 1 & 0 & 1 & 1 & 1 & 0 & 0 \\ 0 & 0 & 0 & 0 & 1 & 0 & 0 & 0 \end{bmatrix} \end{array}$$

Fig. 12.7(*b*) Adjacency matrix of G

12.7.1 Observations

(*i*) The principle diagonal contains all matrix 0's if *G* is without self-loops.

(*ii*) If *G* is without self-loop then the total number of 1's in a row or column gives the degree of the corresponding vertex. *For example,* the row corresponding to vertex v_7 has 5 one's and this is the degree of v_7 in graph *G*. (**Fig. 12.7**(*a*) and (*b*)).

(*iii*) A graph *G* is disconnected and contains two components g_1 and g_2 if its adjacency matrix *M*(*G*) can be partitioned as

$$M(G) = \left[\begin{array}{c|c} M(g_1) & 0 \\ \hline 0 & M(g_2) \end{array}\right]$$

where $M(g_1)$ and $M(g_2)$ are adjacency matrices of component g_1 and g_2 respectively.

12.7.2 Powers of Adjacency Matrix *M* of a Graph *G*

Consider a graph *G* and its adjacency matrix in **Fig. 12.8**(*a*) and (*b*) respectively.

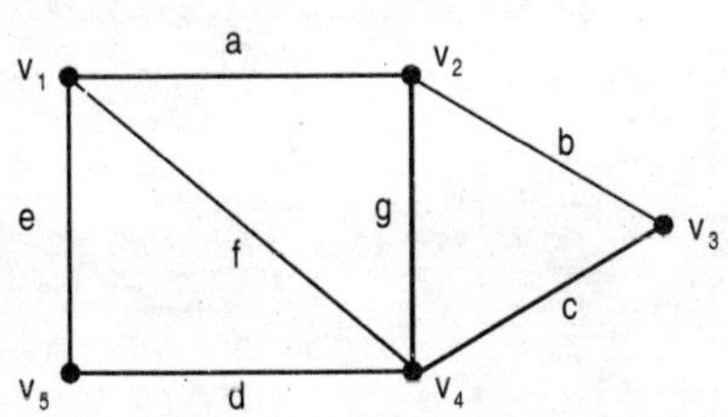

Fig. 12.8 (*a*) Graph G

$$M = \begin{array}{c} \\ v_1 \\ v_2 \\ v_3 \\ v_4 \\ v_5 \end{array}\begin{array}{c} \begin{array}{ccccc} v_1 & v_2 & v_3 & v_4 & v_5 \end{array} \\ \begin{bmatrix} 0 & 1 & 0 & 1 & 1 \\ 1 & 0 & 1 & 1 & 0 \\ 0 & 1 & 0 & 1 & 0 \\ 1 & 1 & 1 & 0 & 1 \\ 1 & 0 & 0 & 1 & 0 \end{bmatrix}_{5 \times 5} \end{array}$$

Fig. 12.8 (*b*) Adjacency Matrix

Let us multiply matrix $M_{5 \times 5}$ by itself. The result is another 5×5 matrix M^2 is given in **Fig. 12.8**(*c*)

$$M^2 = \begin{bmatrix} 3 & 1 & 2 & 2 & 1 \\ 1 & 3 & 1 & 2 & 2 \\ 2 & 1 & 2 & 1 & 1 \\ 2 & 2 & 1 & 4 & 1 \\ 1 & 2 & 1 & 1 & 2 \end{bmatrix}$$

Fig. 12.8(*c*)

Now, the value of an off-diagonal entry in M^2 (*i.e.*, the *ij*th entry ($i \neq j$) in M^2).
= number of 1's in the dot product of *i*th row and *j*th column (or row) of M
= number of common positions having 1's in the *i*th and *j*th rows of M.
= number of vertices that are adjacent to both *i*th and *j*th vertices.
= number of different paths of length two between *i*th and *j*th vertices.
And for the diagonal entries, *i.e.*, *ij*th entry in M^2 where ($i = j$)
= number of 1's in the *i*th row (or column) in M
= degree of corresponding vertex if G is without self-loops.
Thus, the *ij*th entry in X^r.
= dot product of *i*th row X^{r-1} and *j*th column (row) of X.
= number of different edge sequences of *r*-edges between *i*th and *j*th vertices.

Remark: If M is the adjacency matrix of graph G with *n*-vertices. Consider

$$X = M + M^2 + \ldots + M^{n-1} \text{ (in the ring of integers)}$$

Then G is connected iff there exist no entry in matrix X that is zero.

Consider the two adjacency matrices, $M(G_1)$ and $M(G_2)$ of two graphs G_1 and G_2.

$$M(G_1) = \begin{bmatrix} 0 & 1 & 0 & 1 \\ 1 & 0 & 1 & 1 \\ 0 & 1 & 0 & 1 \\ 1 & 1 & 1 & 0 \end{bmatrix} \quad \text{and} \quad M(G_2) = \begin{bmatrix} 1 & 0 & 0 & 0 \\ 0 & 1 & 0 & 0 \\ 0 & 0 & 1 & 0 \\ 0 & 0 & 0 & 1 \end{bmatrix}$$

then the matrices X_1 and X_2 for two graphs G_1 and G_2 are

$$X_1 = M(G_1) + M^2(G_1) + M^3(G_1) \quad \text{and}$$
$$X_2 = M(G_2) + M^2(G_2) + M^3(G_2)$$

$$X_1 = \begin{bmatrix} 2 & 4 & 2 & 5 \\ 5 & 4 & 5 & 5 \\ 2 & 5 & 2 & 5 \\ 5 & 5 & 5 & 4 \end{bmatrix}$$

and $$X_2 = \begin{bmatrix} 1 & 2 & 0 & 0 \\ 2 & 1 & 0 & 0 \\ 0 & 0 & 1 & 2 \\ 0 & 0 & 2 & 1 \end{bmatrix}$$

The matrix X_1 indicates that graph G_1 is connected as there is no zero in X_1. Whereas matrix X_2 has 0 as its entry so graph G_2 is disconnected.

The graphs drawn with respect to the adjacency matrices $M(G_1)$ and $M(G_2)$ are shown in **Fig. 12.9** (*a*) and (*b*).

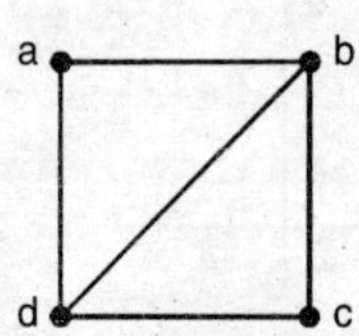

Fig. 12.9. (*a*) Graph G1

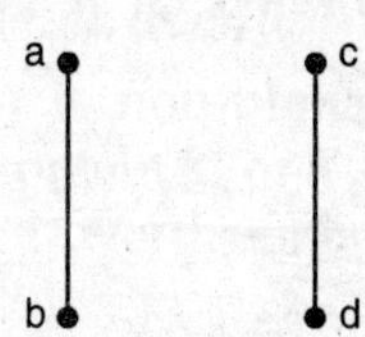

Fig. 12.9(*b*). Graph G2.

Clearly the graph G_1 with respect to incidence matrix $M(G_1)$ is a connected graph and graph G_2 with respect to incidence matrix $M(G_2)$ is a disconnected graph having two components.

SUMMARY

In this chapter we have illustrated how matrices are useful in studying graphs. Out of the several matrices associated with graphs, the incidence matrix A and the adjacency matrix X used to describe a simple graph G completely. Whereas the circuit matrix B and the cut-set matrix C display some important features of graphs. We have also describe the path matrix for graphs which contain less information than B and C does.

Also we investigated the submatrices A_R, B_f and C_f of A, B and C respectively to extract more features of a graph. We also established the relationship between the three submatrices.

EXERCISES

1. Write down the

(*i*) incidence matrix (*ii*) Adjacency matrix for the given graphs :

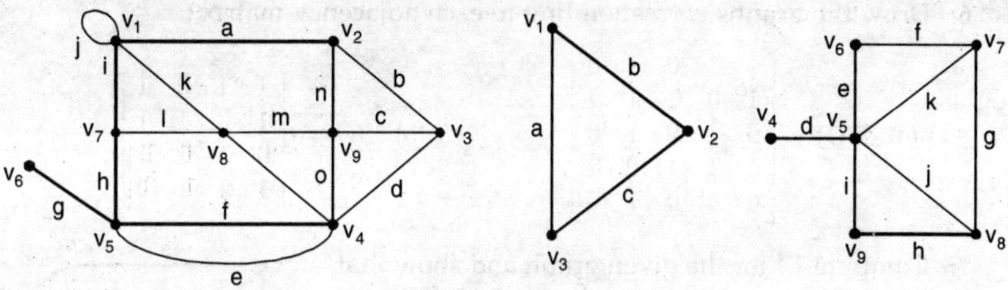

Fig. 12.10

2. Check whether the graphs represented by the adjacency matrices are connected.

(*i*) $X(G) = \begin{bmatrix} 1 & 1 & 0 & 1 \\ 0 & 1 & 0 & 1 \\ 1 & 0 & 1 & 0 \\ 0 & 1 & 0 & 1 \end{bmatrix}$ (*ii*) $X(G) = \begin{bmatrix} 0 & 0 & 1 & 0 \\ 0 & 0 & 1 & 1 \\ 1 & 1 & 0 & 0 \\ 0 & 1 & 0 & 0 \end{bmatrix}$

(*iii*) $X(G) = \begin{bmatrix} 1 & 0 & 1 & 0 \\ 0 & 0 & 1 & 0 \\ 0 & 1 & 0 & 1 \\ 0 & 1 & 0 & 1 \end{bmatrix}$ (*iv*) $X(G) = \begin{bmatrix} 1 & 1 & 1 & 1 \\ 0 & 1 & 1 & 0 \\ 1 & 0 & 1 & 1 \\ 1 & 0 & 0 & 1 \\ 0 & 1 & 1 & 1 \end{bmatrix}$

3. Consider the graph given below : write matrices A_f, B_f and C_f with respect to the spanning tree (a, b, d, c, g).

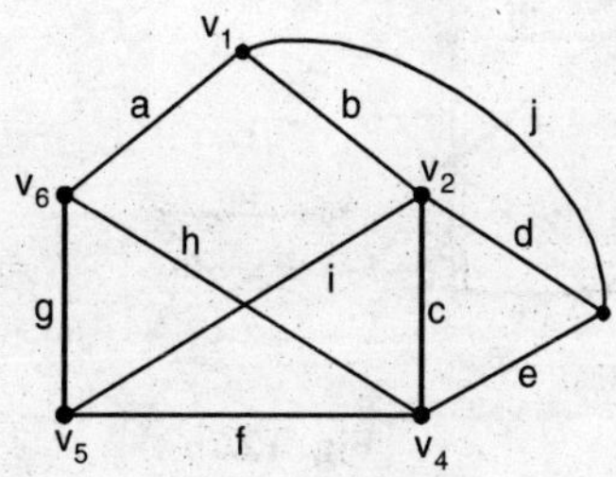

Fig. 12.11

4. Consider the graph given below: Compute $Y = X + X^2 + X^3 + X^4$

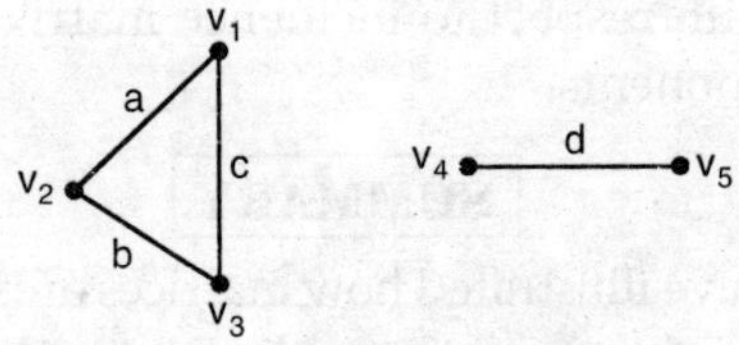

Fig. 12.12

and show that the given graph is disconnected where X is the adjacency matrix of the graph G.

5. Write the path matrix $P(v_1, v_4)$ for the given graph :

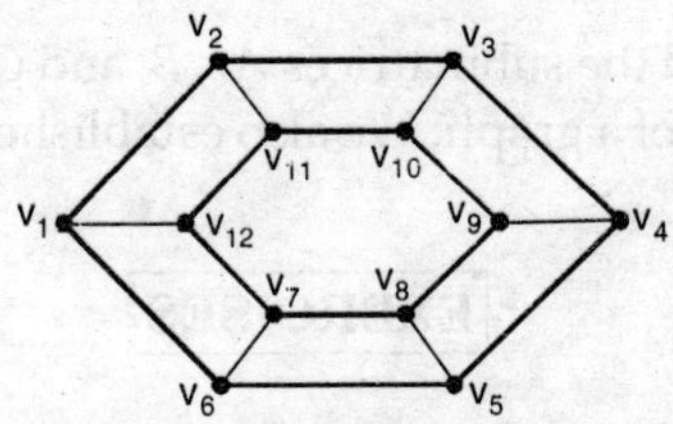

Fig. 12.13

6. Draw the graphs corresponding to each adjacency matrix :

(a) $X(G) = \begin{bmatrix} 1 & 1 & 0 & 1 & 0 \\ 1 & 0 & 1 & 1 & 1 \\ 0 & 1 & 0 & 1 & 0 \\ 1 & 1 & 1 & 0 & 1 \\ 0 & 1 & 0 & 1 & 0 \end{bmatrix}$ (b) $X(G) = \begin{bmatrix} 1 & 1 & 0 & 1 \\ 1 & 0 & 1 & 1 \\ 0 & 1 & 1 & 1 \\ 1 & 1 & 1 & 0 \end{bmatrix}$

7. Compute X^4 for the given graph and show that the ijth entry in X^2 is the number of different edge sequences of 2-edges between vertices v_i and v_j, where $1 \le i \le 5$ and $1 \le j \le 5$.

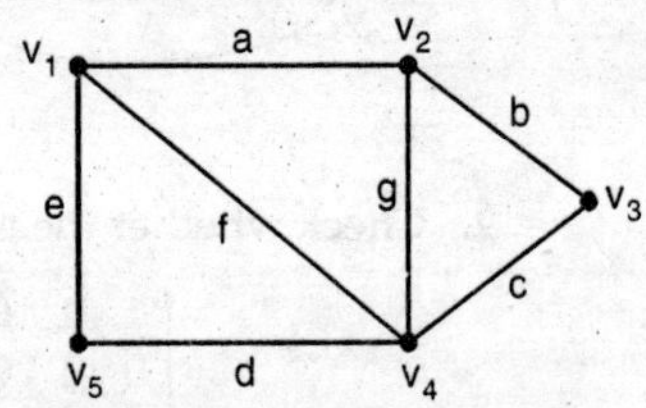

Fig. 12.14

8. Write the incidence matrices for the labeled simple graphs given below:

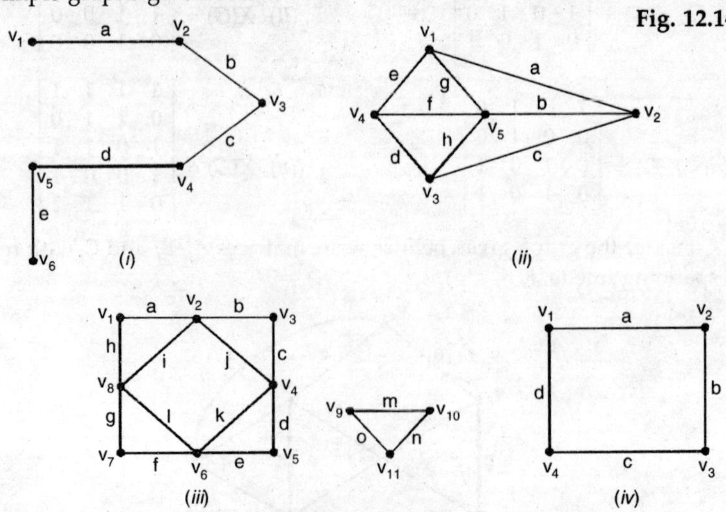

Fig. 12.15

9. Prove that reduced incidence matrix of a tree is non-singular.

[**Hint:** A graph with n vertices and $n-1$ edges is disconnected if it is not a tree. Thus the rank of the matrix A_f *i.e.*, reduced incidence matrix is $\leq n-1$. Hence the reduced incidence square matrix of order $(n-1)$ will not be non-singular].

10. Define (*i*) cut-set matrix (*ii*) fundamental cut-set matrix (*iii*) circuit matrix and (*iv*) fundamental circuit matrix. Give example in each case.

11. Show that the rank of incidence matrix A of a disconnected graph of n vertices, and k-components is $(n-k)$.

12. Characterize simple, self-dual graphs in terms of their circuit and cut-set matrices.

[**Hint:** Write the circuit and cut-set matrices of any self--dual graph].

13. Write the adjacency matrix $X(G)$ for the given graph.

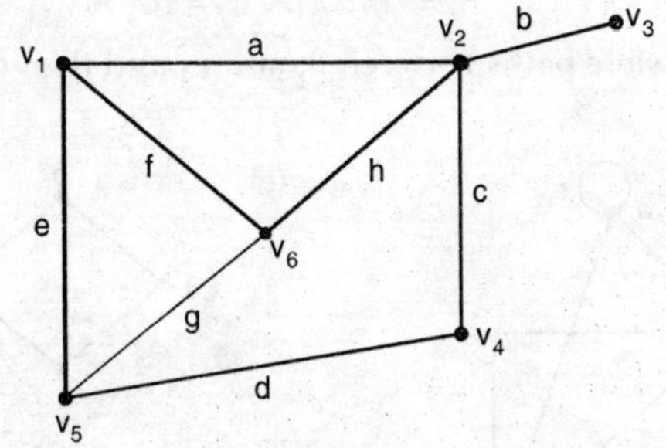

Fig. 12.16

Show that each diagonal entry in X^3 equals twice the number of triangles passing through the corresponding vertex.

14. Show that a graph G is bipartite iff the sum of each row of a circuit matrix is zero mod 2.

15. Show that the rank of the incidence matrix of the given graph G is less than order of graph G.

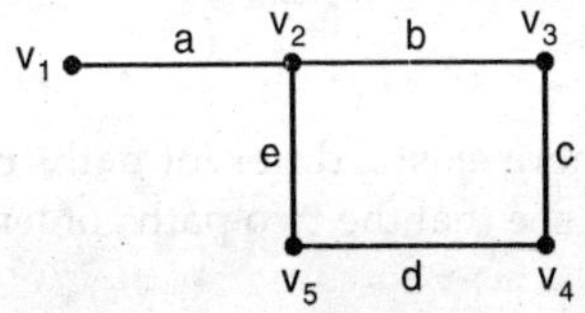

Fig. 12.17

16. Find the number of distinct walks (or edge-sequences) of length 3 from vertex v_2 to v_6 in graph G.

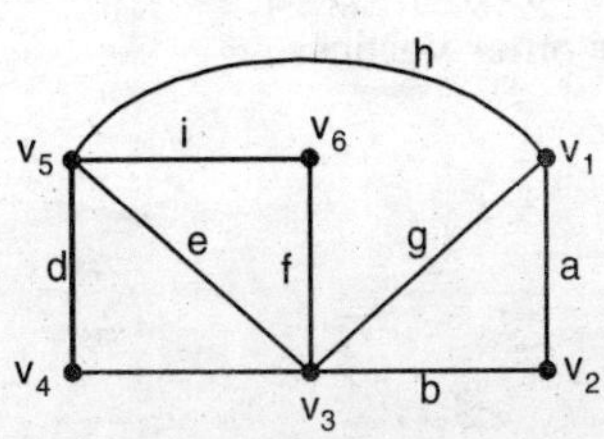

Graph G

Fig. 12.18

[**Hint:** Complete X^3]

ANSWERS

1. (*i*) Connected

(*ii*) Disconnected as $X(G) = \left[\begin{array}{c|c} 0 & X(G_1) \\ \hline X(G_2) & 0 \end{array}\right]$

(*iii*) Interchanging the first and fourth column of the given matrix we obtained

$$\begin{bmatrix} 0 & 0 & 1 & 1 \\ 0 & 0 & 1 & 0 \\ 1 & 1 & 0 & 0 \\ 1 & 1 & 0 & 0 \end{bmatrix} = \left[\begin{array}{c|c} 0 & X(G_1) \\ \hline X(G_2) & 0 \end{array}\right] \text{ So it is disconnected.}$$

(*iv*) Connected

3. **Hint:** Write $A_f = [A_C : A_s]$, $B_f = [I_C : {}_{Bf}]$, $C_f = [C_s : I_{n-1}]$.

5. **Hint:** List all possible paths between v_1 and v_4 and then construct the matrix.

6. (*a*)

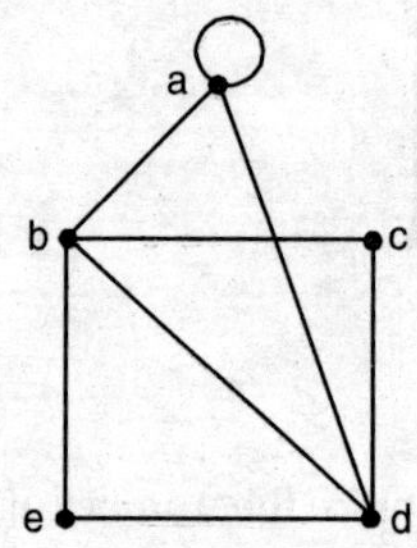

(*b*)

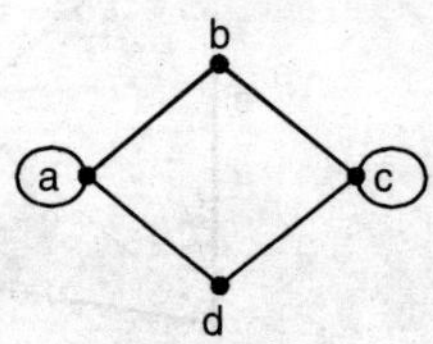

Fig. 12.19

7. $X_2 = \begin{bmatrix} 3 & 1 & 2 & 2 & 1 \\ 1 & 3 & 1 & 2 & 2 \\ 2 & 1 & 2 & 1 & 1 \\ 2 & 2 & 1 & 4 & 1 \\ 1 & 2 & 1 & 2 & 2 \end{bmatrix}$

$X_{24} = 2$ indicates that there exist 2 different paths of length 2 between v_1 and v_4. From the graph we can see that the two paths of length 2 between v_1 and v_4 is (*a*, *g*) and (*e*, *d*).

Similarly one can observe for other entries.

13. **Hint:** Compute X^3. One can see from the graph that there are 2 triangles passing through vertex v_1 as $v_1, v_2, v_6, v_1, v_6, v_5$. So $X_{11}{}^3 = 2, 2 = 4$ should be in the matrix X^3. Similarly verify for other vertices.

■■■

Chapter 13

COLOURING OF GRAPHS

13.1 INTRODUCTION

Suppose we want to assign time slots for final examinations in a university so that two courses with a common student have different slots. How many different time slots do we need?

We create a vertex for each exam with two vertices adjacent when the two examinations have a common student.

Since we are only interested in positioning the vertices (*i.e.*, calculate the minimum number of different time slots), so by colouring the vertices with different colours we can group them into different sets, *for example*, set of vertices painted by red colour, another set by blue, and so on.

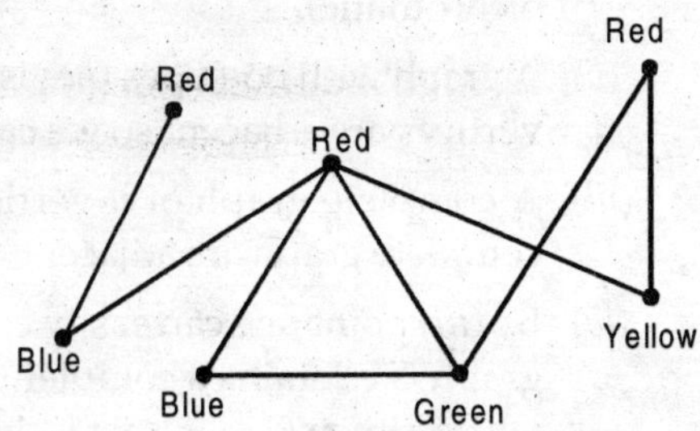

This is called partitioning problem. The colouring and partitioning problem can be performed on vertices or edges of the graph.

Another problem is to colour the regions of a map (map is a partition of the plane into connected regions) in case of a planar graph. In this chapter we are considering all these types of problems.

13.2 PROPER COLOURING

By proper colouring (or colouring) of a graph we simply mean to colour (or paint) all the vertices of the graph with colours such that no two adjacent vertices have the same colour.

A graph in which every vertex is painted according to proper colouring is called properly coloured graph.

For example, consider the proper colourings of a graph shown in **Fig. 13.1**.

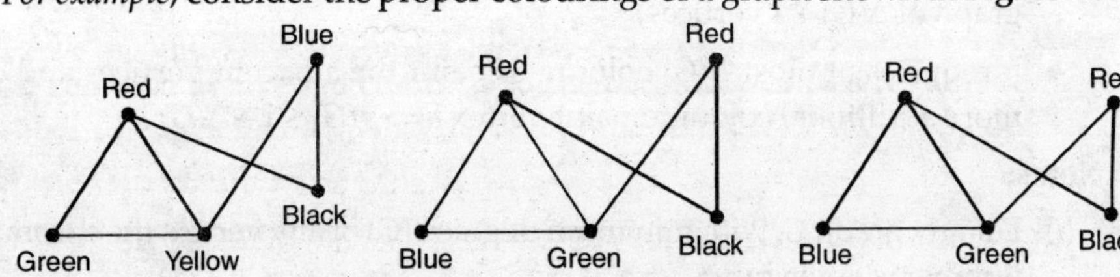

(*a*) (*b*) (*c*)

Fig. 13.1. Proper colouring of a graph.

13.3 CHROMATIC NUMBER

We are interest in finding cut a proper colouring of a graph G that requires minimum colour and this number is called the chromatic number of the graph denoted by $\chi(G)$.

13.3.1 Definition

A graph *G* is called *m*-chromatic if it requires exactly *m*-colours to paint it properly (*i.e.*, *G* cannot be painted properly with less than m-colors) and the number *m* is called chromatic number of graph *G*.

i.e., $\chi(G) = m$

For example: In **Fig. 13.1** three different proper colourings of a graph *G* is given. One can easily verify that the graph is 3-chromatic *i.e.*, $\chi(G). = 3$ as it requires at least 3-colours to paint it properly.

Notes :

(*i*) A graph consisting of one edge (or more) is at least 2-chromatic (or bichromatic).

(*ii*) A graph with no-edges (*i.e.*, isolated vertices) is 1-chromatic since no two vertices are adjacent so we can paint all the vertices with one colour.

(*iii*) A complete graph of *n*-vertices is *n*-chromatic, as all the vertices of a complete graph are adjacent.

(*iv*) By (*iii*) point one can easily conclude that if a graph *G* contains a complete graph of say *k*-vertices then *G* is at least *k*-chromatic.
For example,

The graph in **Fig. 13.2** contains a triangle (*i.e.*, a complete graph of a 3 vertices) so *G* is at least 3-chromatic.

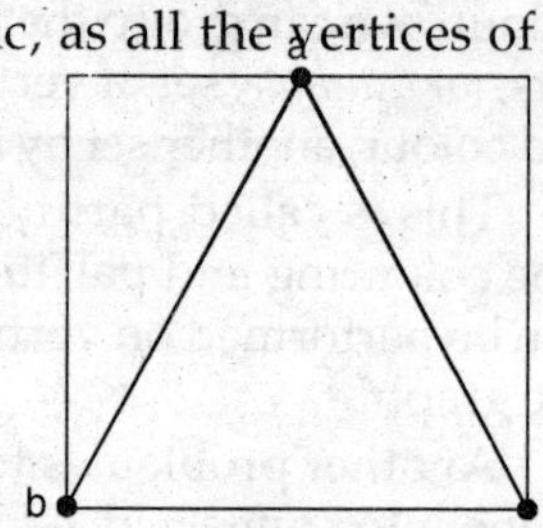

Fig. 13.2. Graph G

Theorem 13.1. For any graph *G* with maximum degree Δ of any vertex, chromatic number is defined as

$$\chi(G) \le 1 + \Delta(G)$$

Proof: Let *G* be a graph with *n*-vertices and let the degrees of all the vertices are at most $\Delta(G)$.

- each vertex *v* has at most $\Delta(G)$ adjacent vertices (*i.e.*, *G* has a complete graph of $\Delta(G) + 1$ vertices)
- it requires at most $\Delta(G)$ colours to paint the adjacent vertices and one more additional colour to paint vertex $v \Rightarrow \chi(G) \le 1 + \Delta(G)$.

Notes:

(*i*) For any graph *G*, with minimum degree $\delta(G)$ of any vertex, the chromatic number is given by

$$\chi(G) \ge \frac{n}{(n - \delta(G))}$$

where n is the total number of vertices in G.

(*ii*) If G has no complete graph of $(D(G) + 1)$ vertices then the chromatic number of graph G.

$$\chi(G) \geq \Delta(G).$$

Example 13.1. Find the chromatic number of the graph G shown in Fig. 13.3.

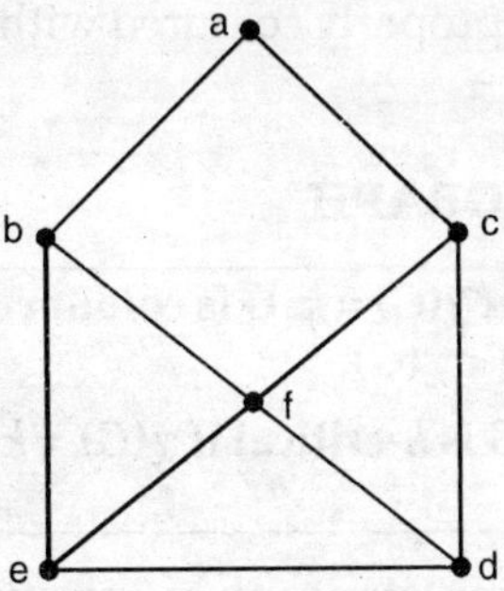

Fig. 13.3. Graph G

Solution: For the given graph G,

$$\Delta(G) = 4 \text{ (degree of vertex } f)$$

$$\Rightarrow \quad \chi(G) \leq 4 + 1 = 5 \qquad \text{...(1)}$$

and also $\quad \delta(G) = 2 \quad$ (degree of vertex a which is the minimum)

$$\chi(G) \geq \frac{n}{(n - \delta(G))}$$

$$\chi(G) \geq \frac{6}{(6-2)} = X(G) \geq \frac{3}{2} \qquad \text{...(2)}$$

But the graph G contains triangles

$$\Rightarrow \quad \chi(G) \geq 3 \qquad \text{...(3)}$$

By (1) and (3)

$$3 \leq \chi(G) \leq 5$$

Let us now start painting the vertices of the triangle, say b, f, e, by three colours red, blue and green respectively.

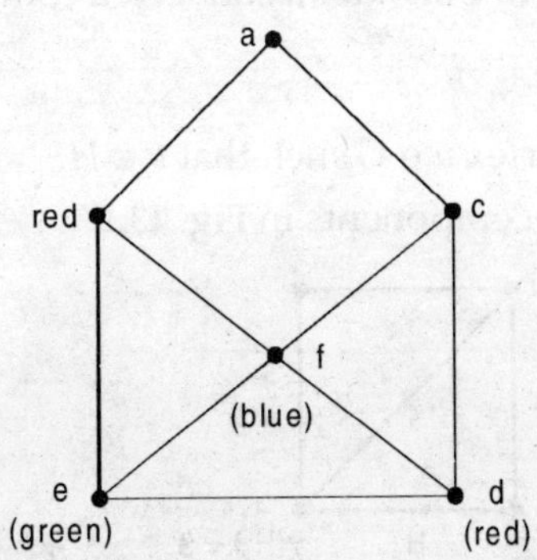

Fig. 13.3 (a)

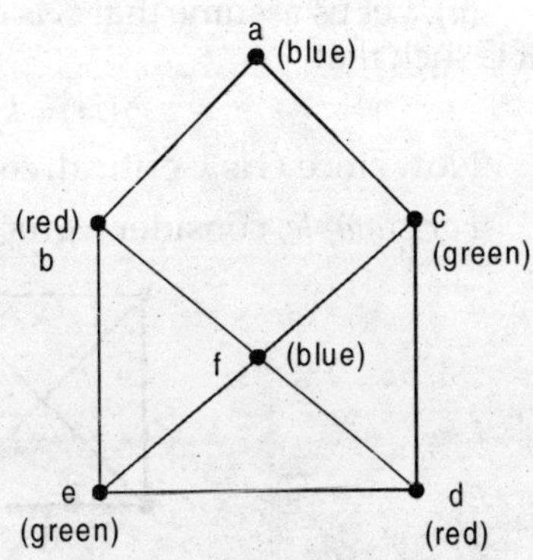

Fig. 13.3 (b)

Now since d is adjacent to f and e both so we cannot paint it by blue colour or green colour so only choice left is red. (See **Fig. 13.3**(*a*))

Vertex C is adjacent to d and f so only colour left for c is green because d and f are coloured with red and blue respectively.

Now vertex a is adjacent to vertices b and c, which are coloured by red and green respectively. So a can be coloured by blue colour. (See **Fig. 13.3** (*b*)).

Hence the graph G can be properly coloured with three vertices. So

$$\chi(G) = 3$$

13.4 COLOUR-CRITICAL GRAPH

A graph G is k-chromatic if $\chi(G) = k$. G is colour critical or k-critical if for every proper subgraph H of G, $\chi(H) < \chi(G)$.

In other words; A graph G is k-critical if $\chi(G) = k$ and $\chi(G - v) < \chi(G)$ for each vertex v of G.

For example, consider a complete graph of 3-vertices in **Fig. 13.4**(*a*).

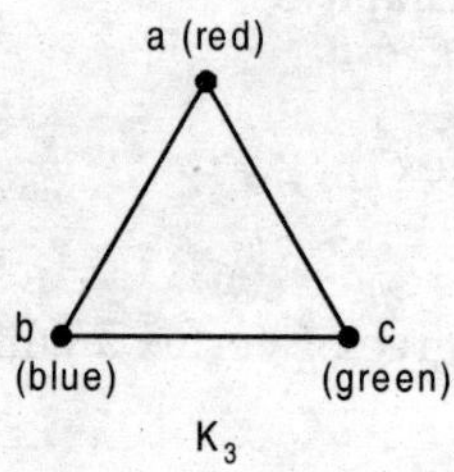

Fig. 13.4(a)

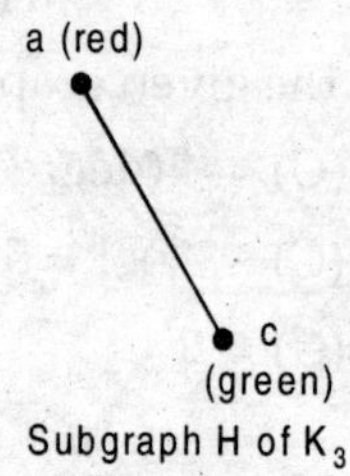

Fig. 13.4(b)

K_3 is 3-critical because $\chi(K_3) = 3$ and if we consider a proper subgraph of K_3 as H (given in **Fig. 13.4(*b*)** then $\chi(H) = 2$ (*i.e.*, H can be coloured properly by 2-colours).

$$\Rightarrow \qquad 2 = \chi(H) < \chi(G) = 3$$

Theorem 13.2. If graph G is k-critical then show that

(*a*) G is connected graph.

(*b*) every vertex v of graph G has at least $k - 1$ degree *i.e.*,

$$d(v) \geq k - 1 \quad \text{for all } v \in G.$$

(*c*) Graph G cannot be partitioned into subgraphs.

Proof: G is k critical $\Rightarrow \chi(G) = k$ and $\chi(G - v) < \chi(G)$ for each $v \in G$.

(*a*) Let us assume that G is disconnected. Consider a connected component H of G such that

$$\chi(H) = k$$

Now since G is k-critical, consider a vertex $v \in G$ such that $v \notin H$.

For example, consider a graph G with 2-components in **Fig. 13.5**.

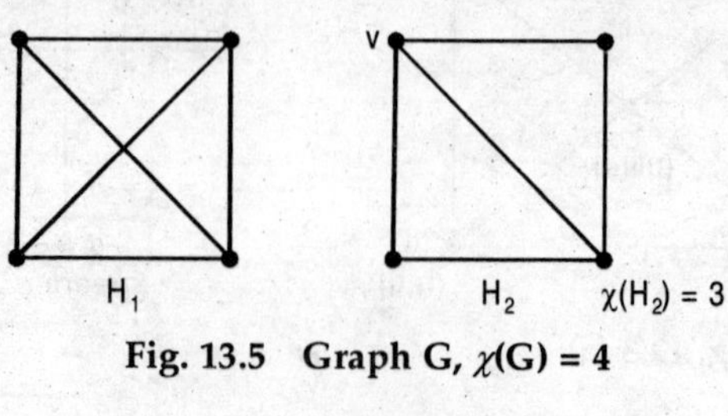

Fig. 13.5 Graph G, χ(G) = 4

In case of disconnected graph.

$$\chi(G) = \max\{\chi(H) \mid H \text{ is connected component of } G\}$$

Now clearly $\quad \chi(G - v) = 4 = \chi(H_1) = \chi(G)$

$\Rightarrow$ For each $\quad v \in G, \quad v \notin H \quad \chi(G - v) = \chi(G)$

which contradicts that G is k-critical.

Hence G must be connected.

(*b*) Since G is k-critical $\Rightarrow \chi(G - v) < k$

or $\quad \chi(G - v) \le k - 1$

Let us consider that $\deg(v) < k - 1;\ v \in G$

$\Rightarrow \quad \deg(v) \le k - 2$

$\Rightarrow$ The adjacent vertices of v can be coloured properly by at most $(k-2)$ colours (See **Fig. 13.5(*b*)**).

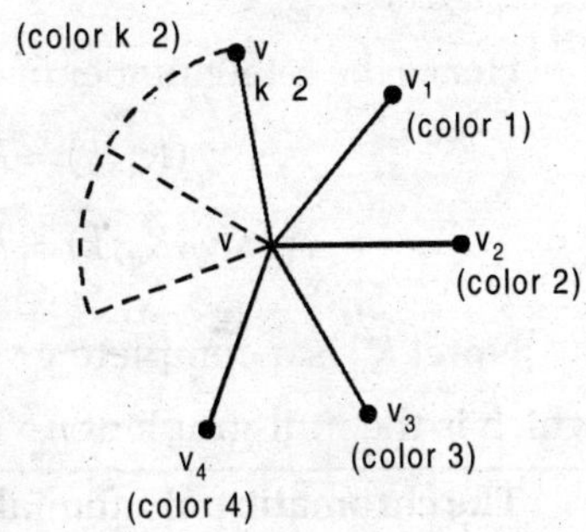

Fig. 13.5(b)

That is we need one extra colour apart from $(k - 2)$ colours to colour vertex v.

$\Rightarrow$ Total colour needed to colour the graph G

$$= (k - 2) + 1 = (k - 1)$$

$\Rightarrow \quad \chi(G) = k - 1$, but its given that

$\chi(G) = k$ as G is k-critical

$\Rightarrow \quad \deg(v) < k - 1$ is not possible.

$\Rightarrow \quad \deg(v) \ge k - 1$

(*c*) Consider the two subgraph H_1 and H_2 of graph G.

$\Rightarrow \quad G = H_1 \cup H_2$

and $\quad \chi(H_1) < k$

and $\quad \chi(H_2) < k$.

$\Rightarrow \quad \chi(H_1) \le k - 1$

and $\quad \chi(H_2) \le k - 1$

The vertices of the graph $H_c = H_1 \cap H_2$ (*i.e.*, the common vertices in H_1 and H_2) can be coloured with same colours in both the subgraphs H_1 and H_2.

$\Rightarrow \quad \chi(H_1 \cup H_2) \le k - 1$

or $\quad \chi(G) \le k - 1$

which contradicts that G is k-critical. Hence G cannot be partitioned.

Note: A complete graph of n-vertices in always n-critical.

13.5 CHROMATIC POLYNOMIAL

A graph G of n vertices can be coloured properly in different ways by a sufficiently large number of colours. This properly of a graph is expressed by means of a polynomial called **chromatic polynomial**.

Chromatic polynomial of a graph G of n-vertices is denoted as $\chi(G\,;k)$ gives the number of ways of properly colouring the graph, using at most k-colours.

For example; lets find out the chromatic polynomial for a null-graph of n-vertices.

Since each vertex can be coloured by using any of the k-colours.

i.e., there exist k-ways to colour a vertex v_1 similarly k-ways to colour vertex v_2, v_3, ..., v_n each.

Hence the total number of ways.

$$\chi(\overline{K_n};k) = k\,.\,k\,.\,k \ldots k \quad (n\text{-times})$$

$$\chi(\overline{K_n};k) = k^n$$

Note: K_n is a complete graph of n-vertices and $\overline{K_n}$ is the complement of K_n which is the null-graph of n-vertices.

The chromatic polynomial $\chi(G; k)$ of a graph G with n-vertices is defined as:

$$\chi(G;k) = \sum_{r=1}^{n} w_r \,.\, {}^kC_r$$

where w_r is the total number of ways of proper colouring G using exactly r-colours and ${}^kC_r$ gives the total number of ways to select r-colours from the given k-colours.

So $\chi(G;k) = w_1\,.\,{}^kC_1 + w_2\,.\,{}^kC_2 + \ldots + w_n\,.\,{}^kC_n$

or $\chi(G;k) = w_1\,.\,\dfrac{k}{1!} + w_2\,.\,\dfrac{k(k-1)}{2!} + \ldots + w_n\,\dfrac{k(k-1)\ldots(k-n+1)}{n!}$

Each w_r will be evaluated separately for the given graph G.

Consider a graph with a single edge, graph requires at least two colour for proper colouring which simply implies that one cannot colour the graph with a single colour and hence w_1 (*i.e.,* the number of ways to colour a graph using exactly one colour) is zero.

Similarly a graph G of n-vertex can be coloured properly by n-colours in n ! ways. As the first vertex can be coloured in n-ways, second can be in $(n-1)$ ways, third in $(n-2)$ ways and so on and in the last vertex would be a single choice. So

$$C_n = n(n-1)\,(n-2) \ldots 3.2.1 = n\,!$$

Theorem 13.3. If an n-vertex graph T is a tree then

$$\chi(T;k) = k(k-1)^{n-1}$$

k is the number of colours given to paint T.

Proof: Consider the tree of n-vertices given in **Fig. 13.6** choose a vertex $v \in T$ as a root. v can be coloured in k-ways.

If we extend a proper colouring to new vertices as we grow the tree from v, at each step only the colour of the parent is neglected and we are left with $k - 1$ choices out of k-colours (one neglected as the colour of parent).

$\Rightarrow$ root v can be coloured in k-ways and the $(n-1)$ vertices has $(k-1)$ choices left for each of the $(n-1)$ vertices.

$\Rightarrow \chi(T; k) = k \cdot \{(k-1) \cdot (k-1) \ldots (k-1)$
$\{n\text{-}1 \text{ times}\}$

$$\chi(T; k) = k \cdot (k-1)^{n-1}$$

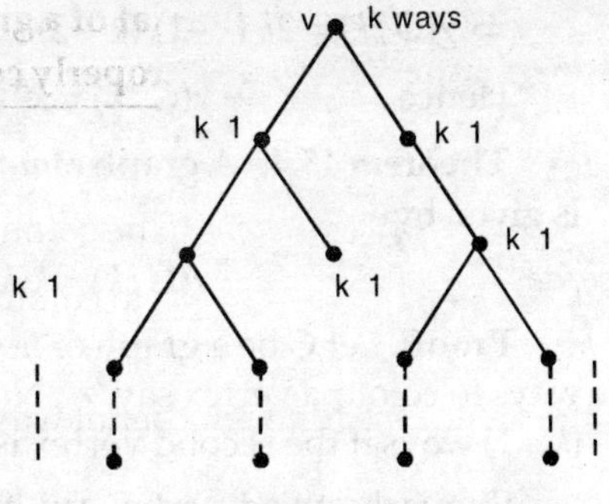

Fig. 13.6

Hence the theorem

Example 13.2. Find out the chromatic polynomial of the given graph.

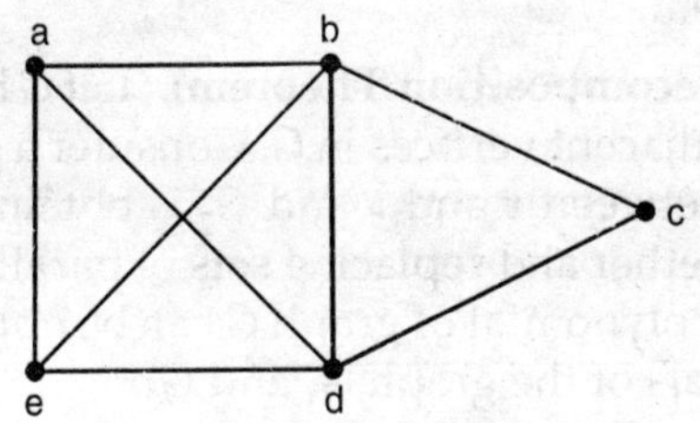

Fig. 13.7

Solution: As the given graph contains a complete graph of 4-vertices. So it requires at least 4-colour to colour it properly.

$$\Rightarrow \qquad C_1 = 0 = C_2 = C_3$$

Now to evaluate w_4; w_4 = number of ways to colour a graph with 4-colours.

Let the 4-colours be C_1, C_2, C_3 and C_4. Assign them to the vertices a, b, d and e in

$$4\,! = 24 \text{ ways.}$$

Then there will be two choices left for vertex c. (either colour of vertex a or vertex e).

Thus $\qquad w_4 = 24 \,.\, 2 = 48$

The total number of vertices in the graph = 5

$\Rightarrow \quad w_5 = 5\,!$ $\qquad$ (*i.e.*, at most 5-colours required to paint the graph properly)

So now

$$\chi(G; k) = w_1 \frac{k}{1!} + w_2 \frac{k(k-1)}{2!} + \ldots + w_n \frac{k(k-1)\ldots(k-n+1)}{n!}$$

$$= 0 + 0 + 0 + w_4 \cdot \frac{k(k-1)(k-2)(k-3)}{4!}$$

$$+ w_5 \cdot \frac{k(k-1)(k-2)(k-3)(k-4)}{5!}$$

$\Rightarrow \chi(G\,;k) = 2k\,(k-1)\,(k-2)\,(k-3) + k(k-1)\,(k-2)\,(k-3)\,(k-4)$

Hence $\quad \chi(G\,;k) = k^5 - 8k^4 + 23k^3 - 28k^2 + 12k$

Theorem 13.4. A graph of n-vertices is complete iff it's chromatic polynomial is given by

$$\chi(G\,;k) = k(k-1)\,(k-2)\,...(k-n+1)$$

Proof: Let G be a graph of n-vertices. Then with k-colours there are exactly k-ways to colour a vertex say 'v'. Now the second vertex can be coloured properly in $(k-1)$ ways if the second vertex is adjacent to vertex v.

Similarly, third vertex can be coloured properly in $(k-2)$ ways if the third vertex is adjacent to vertex v and the second vertex.

Hence the nth vertex can be coloured properly in $(k-n+1)$ ways if and only if n^{th} vertex is adjacent to previously selected $(n-1)$ vertices.

Thus the graph G with n-vertices is complete iff its chromatic polynomial is

$$\chi(G\,;k) = k(k-1)\,(k-2)\,...(k-n+1)$$

Hence the theorem.

Theorem 13.5 (Decomposition Theorem). Let G be a graph of n-vertices and x and y be two non-adjacent vertices in G. Consider a graph G_1 which is obtained by joining an edge between x and y and G_2 is obtained by fusing the two non-adjacent vertices together and replacing sets of parallel edges with single edges. Then the chromatic polynomial of graph G can be obtained by taking sum of the chromatic polynomials of the graph G_1 and G_2.

i.e., $\quad \chi(G; k) = \chi(G_1; k) + \chi(G_2; k)$

Obviously the number of vertices in G_2 are $(n-1)$.

Proof: x and y are two non-adjacent vertices in G. In the proper colouring of graph G there exist two cases (*i*) when x and y have same colour (*ii*) when x and y above different colour.

Case (*i*): When the two non-adjacent vertices (x and y) have same colour. Then we fuse them together in G and thus the proper colouring of G is the same as of graph G_2.

Case (*ii*): When x and y have different colour in G. The adding an edge between them does not affect the colouring of G. Thus the number of ways of properly colouring G such that x and y have different colour is equal to the number of ways of properly colouring graph G_1.

Thus $\quad \chi(G; k) = \chi(G_1; k) + \chi(G_2; k)$

Hence the theorem.

Example 13.3. Find out the chromatic polynomial of the graph given in Fig. 13.8.

Solution: Let us find the chromatic polynomial of the given graph by the help of Theorem 13.5. The procedure of obtaining graphs G_1 and G_2 from graph G is shown in **Fig. 13.8 (*a*)**.

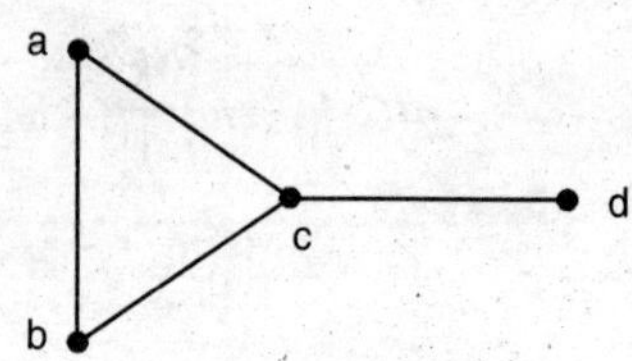

Fig. 13.8

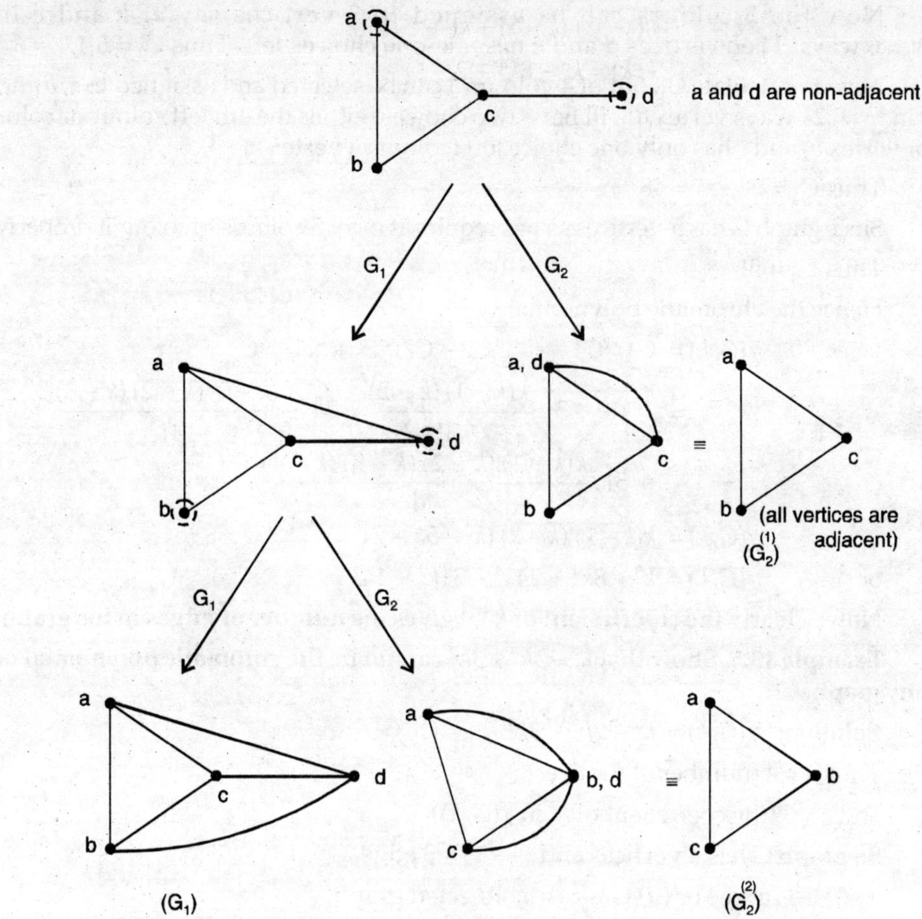

Fig. 13.9

The obtained graphs G_1, $G_2^{(1)}$, and $G_2^{(2)}$ are the complete graphs of 4 and 3-vertices *i.e.*, K_4 and K_3.

$$\chi(G;k) = \text{chromatic polynomial of } K_4 + \text{chromatic polynomial of } K_3 + \text{chromatic polynomial of } K_3$$

So $$\chi(G;k) = k(k-1)(k-2)(k-3) + 2k(k-1)(k-2)$$
$$= k(k-1)^2(k-2).$$

Example 13.4 Find the chromatic polynomial of graph G. *(U.P.T.U-2002-03)*

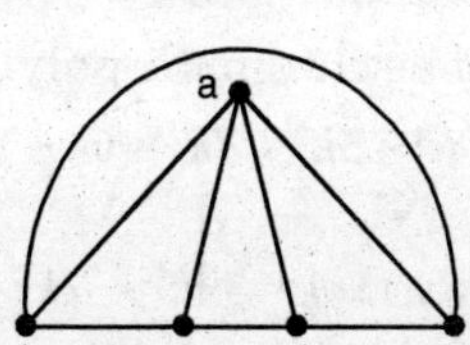

Fig. 13.10

Solution: Given graph G has a triangle so we require at least 3-colors to colour G properly.

$$\Rightarrow \qquad C_1 = C_2 = 0$$

Now the 3-colours can be assigned to 3-vertices say a, b and c in $3! = 6$ ways. Then vertices d and e has one-one choices left. Thus $C_3 = 6.1.1 = 6$

Now to calculate C_4, out of 4-colours 3 can be selected and assigned to a, b and c in $^4P_3 = 24$ ways vertex d will have two choices (either the 4th left colour or color or vertex b) and e has only one choice left (colour of vertex c).

Thus $C_4 = 24.2.1 = 48$

Since graph G has 5-vertices so we, require at most 5-colours to colour it properly

Thus $C_5 = 5!$

Hence the chromatic polynomial

$$\chi(G, k) = C1.\,^kC1 + C_2.\,^kC2 + C3.\,^kC_3 + \ldots C_n.\,^kC_n$$

$$= 0 + 0 + 3!.\frac{k(k-1)(k-2)}{3!} + 24.\frac{k(k-1)(k-2)(k-3)}{4!}$$

$$+ 5!.\frac{k(k-1)(k-2)(k-3)(k-4)}{5!}$$

$$\chi(G, k) = k(k-1)(k-2)(k^2 - 5k + 7)$$

or $$\chi(G, k) = k^5 - 8k^4 + 24k^3 - 31k^2 + 14k$$

Note: Clearly the **co-efficient** of k^{n-1} gives the **number of edges** in the graph.

Example 13.5. Show that $k^4 - 3k^3 + 3k^2$ cannot be the chromatic polynomial of any graph.

Solution: $\chi(G, k) = k^4 - 3k^3 + 3k^2$

Thus $n = 4$ (number of vertices)

and $e = 3$ (the coefficient of k^3 in $\chi(G, k)$)

So graph G has 4 vertices and $(4 - 1) = 3$ edges

$\Rightarrow$ (*i*) G is a tree or (*ii*) G is a disconnected graph.

(*i*) If G is a tree with $n = 4$ and $e = 3$

The chromatic polynomial is

$$\chi(G, k) = k(k-1)^{n-1} = k(k-1)^3 \neq k^4 - 3k^3 + 3k^2$$

(*ii*) If G is disconnected, $G = K_1 \cup K_3$

(*i.e.*, G is having two components, one is complete graph of 3-vertices and another is an isolated vertex)

So, the chromatic polynomial is

$$\chi(G, k) = k.\{k(k-1)(k-2)\} \neq k^4 - 3k^3 + 3k^2$$

Thus $k^4 - 3k^3 + 3k^2$ can not be chromatic polynomial of any graph.

Example 13.6. Show that $k^4 - 5k^3 + 7k^2 - 6k + 3$ is not a chromatic polynomial of any graph.

Solution: Given that $\chi(G, k) = k^4 - 5k^3 + 7k^2 - 6k + 3$

$\Rightarrow$ $n = 4, e = 5$

$\Rightarrow$ G is not a tree

So $G = K_4 - e$

So $\chi(G, k) = k(k-1)(k-2) \neq k^4 - 5k^3\ 7k^2 - 6k + 3$

Hence, the given polynomial can not be a chromatic polynomial of any graph.

Example 13.7. Show that a simple connected planar graph with 8 vertices and 13 edges cannot be bichromatic.

Solution: Let us suppose that G is bichromatic

$\Rightarrow$ *G* contains cycles of even length

$\Rightarrow$ every vertex in *G* has even degree.

Now by Euler's formula $r = e - n + 2 = 13 - 8 + 2 = 7$

That is the number of regions in $G = 7$

$= \text{number of vertices in } \overline{G}$ (dual of G)

number of edges in $\overline{G} = 13$

So in $\overline{G}$, $n' = 7$ and $e' = 13$

Since degree of each vertex in G is even and $\leq 2 \Rightarrow$ degree of each vertex in $\overline{G}$ is also even and greater than or equal of 4; as each region in G is bounded by 4 or more edges (because G contains circuits of even length).

$\Rightarrow \quad 2e' \geq 4n'$

or $\quad 2.13 \geq 4.7$

$\Rightarrow \quad 26 \geq 28$

which is a contradiction

Hence G can not be bichromatic.

13.6 FOUR COLOUR PROBLEM

Previously we have discussed the problem of colouring the vertices or edges of the graph. Let us now consider the problem of colouring properly the regions of a planar graph.

13.6.1 Definition

A planar graph *G* is said to be properly coloured if no two adjacent regions have same colour. If two regions share a common edge, they are adjacent.

We are interested in finding out the minimum number of colours that require for proper colouring of a planar graph. This leads us to the most famous **conjecture** in graph theory.

13.6.2 Conjecture

Every planar graph (or map) can be properly coloured with four colours.

Till now no mathematican has found a map which require more than 4-colours

Theorem 13.5. Five Colour Theorem: The vertices of every finite planar graph can be coloured properly with five colour (or fewer).

Proof: The proof of this theorem is based on the induction on the vertices of a planar graph. Since the vertices of all the planar graphs with 1, 2, 3, 4 or 5-vertices can be properly coloured by five or less colours.

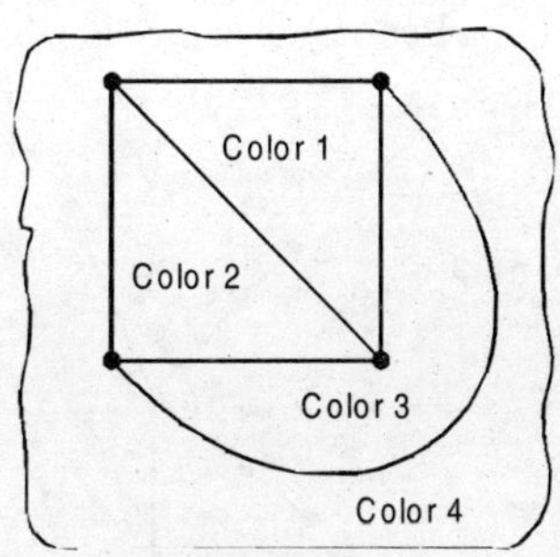

Fig. 13.11. Graph properly coloured with 4-colours.

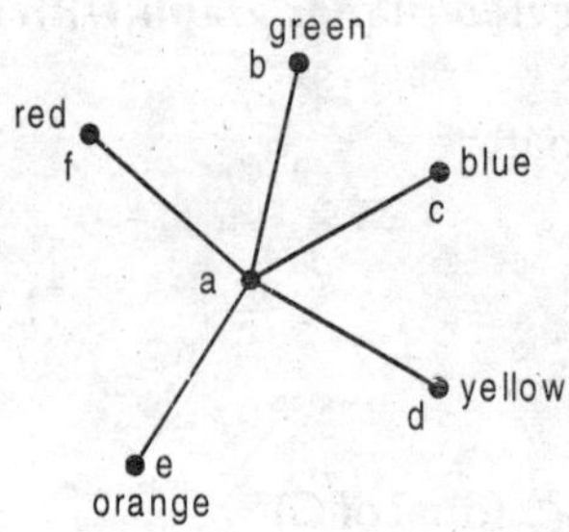

Fig. 13.12(a)

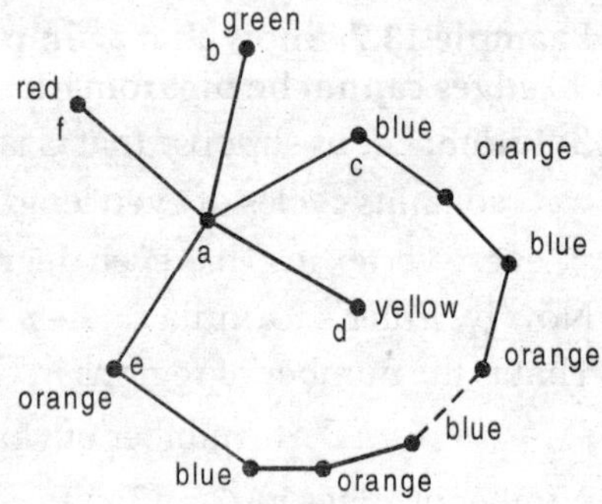

Fig. 13.12(b)

Let us assume that every planar graph with $n - 1$ vertices is properly colourable with five colours or fewer. So we have to show that there is no graph of n-vertices which require more than 5-colours for proper colouring.

Since G is a planar graph so it must have at least one vertex with degree 5 or less. Let a be that vertex in G (**Fig. 13.12(*a*)**).

The graph $G - a$ has $(n - 1)$ vertices and hence require at most 5-colours for proper colouring (by induction hypothesis).

Now we add in graph $(G - a)$ and all edges incident since degree of a is five and all the five colours have been used in colouring the adjacent vertices. (**Fig 13.12(*a*)**).

Now suppose there is a path between vertices c and e (or d and f) coloured alternately with blue and orange (**Fig. 13.12(*b*)**). Then there cannot exist a similar path between vertices b and d coloured alternately with green and yellow colours because graph G is planar and if such a path will exist it will intersect the path between c and e and will violate the condition of planar graph.

Thus we can colour vertex d with green colour and graph $(G - v)$ is still properly coloured. So we are left with a colour *i.e.*, yellow with which we can paint vertex v in G. Similar argument can be given by taking other paths (*i.e.*, between b and e).

Hence the theorem.

Theorem 13.16. (Six Colour Theorem): Every simple planar graph is six colourable.

Proof: Since G is planar, so there exist at least one vertex with degree 5 or less. Let a be that vertex. **Fig. 13.13(*a*)**.

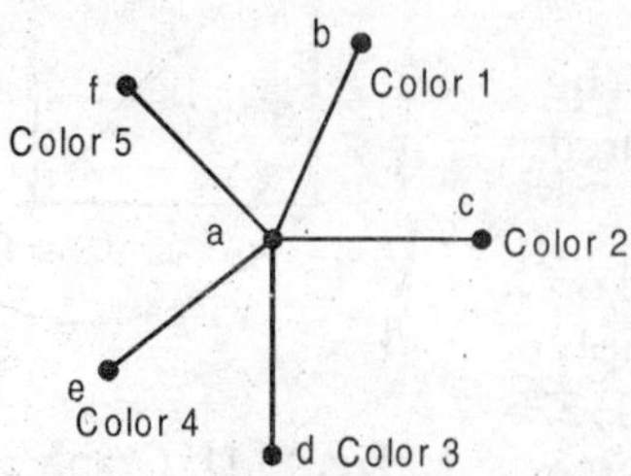

Fig. 13.13(a)

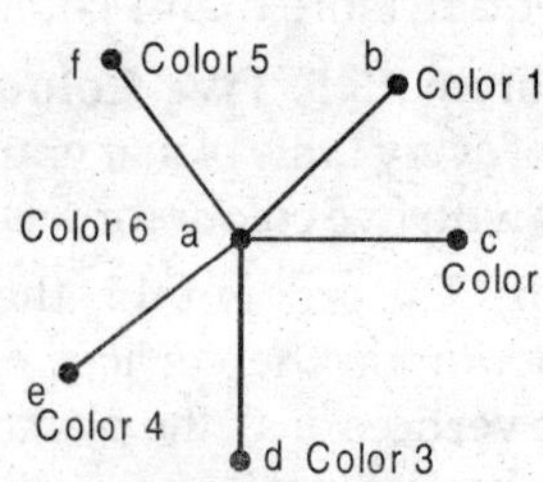

Fig. 13.13(b)

The adjacent vertices of a can be painted properly by using any of the five colours and vertex a can be painted by the left sixth colour. Keeping the graph properly coloured (**Fig. 13.13**(*b*)).

Hence the theorem.

SUMMARY

Firstly we have discussed the proper colouring of the graph which led us to the chromatic polynomial and chromatic partitioning of the vertices of the graph, which is based on the concept of an independent set of vertices.

On the other hand we have also discussed the proper colouring of the regions of planar graph, state of the most famous four colour problem and derived some useful result for colouring the vertices of planar graph. (*i.e.*, the five colour and six colour theorems).

EXERCISES

1. Find the chromatic number of the following graphs :

(*i*) K_n (*ii*) $K_{m,\,n}$; $m, n \geq 1$ (*iii*) C_n (*iv*) ω_n (*v*) $K_n - e$, where e is any edge of K_n

(*vi*) $\overline{K_n}$ (*i.e.*, complement of K_n)

(*vii*) P_n (*i.e.*, path graph of order n or ($n - e$, where $e \in C_n$)

2. Show that a tree is bichromatic.

3. Show that the edge connectivity of every n-critical graph G is at least $(n - 1)$.

4. Show that a simple planar graph G with 10 vertices and 17 edges cannot be bichromatic. (*i.e.*, $\chi(G) \neq 2$).

[**Hint:** Use Euler's formula to find r and apply the concept of duality].

5. Show that a simple connected planar graph G with $n = 8$ and $e = 13$ cannot be 2-coloured.

6. Determine the chromatic number of each of the following graphs :

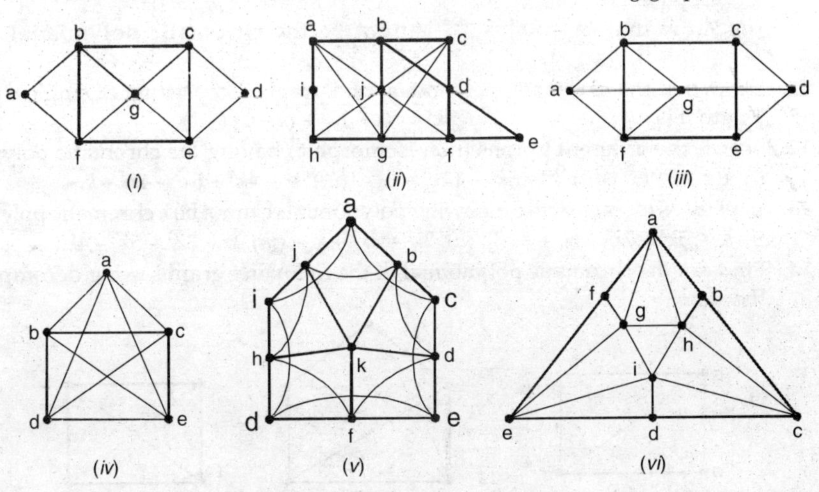

Fig. 13.14

7. Prove that any m-chromatic graph contains m-critical subgraph.

8. Compute the chromatic number $\chi(G)$ for each of the following graphs and also find $\chi(G)$-critical subgraph.

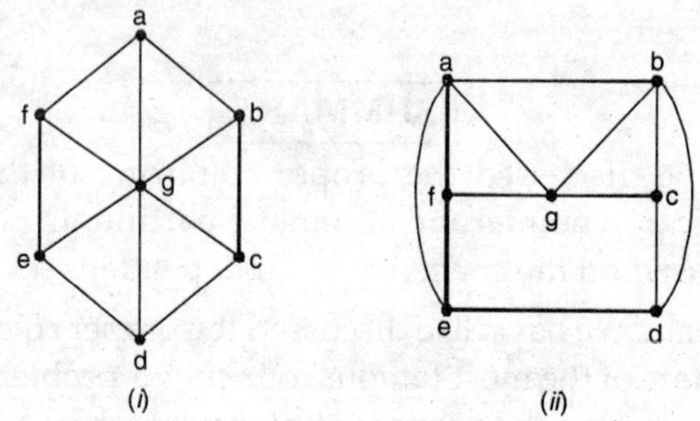

Fig. 13.15

9. Find the chromatic polynomial of the following graphs :

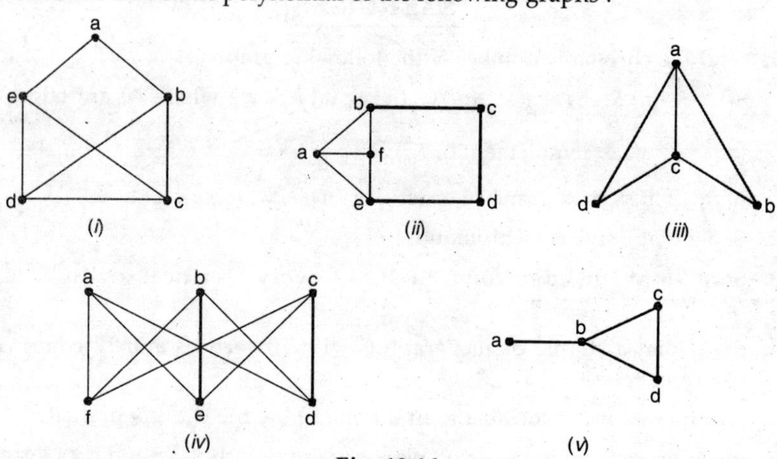

Fig. 13.16

10. (*i*) Show that $k^4 - 3k^3 + 5k^2 - 1$ cannot be the chromatic polynomial of a graph G.

    (*ii*) Show that $k^4 - 3k^3 + 3k^2$ cannot be the chromatic polynomial of any graph G.

11. Show that the chromatic polynomial of a graph G consisting of single circuit of length n is
$$P_n(k) = (k-1)^n + (k-1)(-1)^n$$

12. Sketch the different graphs (non-isomorphic) having the chromatic polynomial
(*i*) $k(k-1)^4$ (*ii*) $k^4 - 6k^3 + 11k^2 - 6k$ (*iii*) $k^5 - 4k^4 + 6k^3 - 4k^2 + k$.

13. Explain, why each of the following polynomials cannot be a chromatic polynomial
(*i*) $k^4 - 5k^3 + 7k^2 - 6k + 3$ (*ii*) $3k^3 - 4k^2 + k$ (*iii*) $k^4 - 3k^3 + 5k^2 - 4k$.

14. Find out the chromatic polynomial of the following graphs using decomposition theorem:

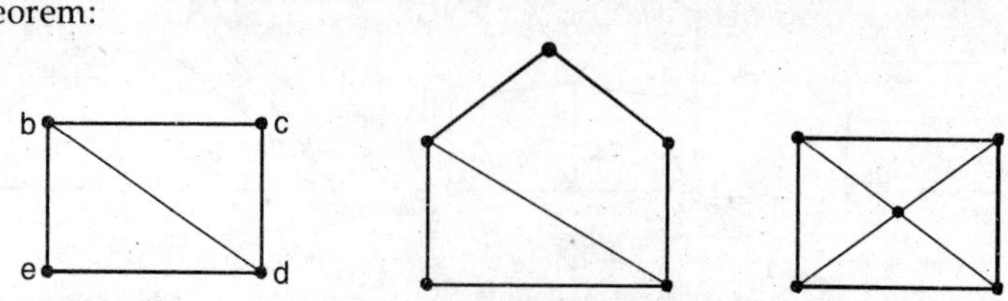

Fig. 13.17

15. Find the chromatic polynomial of $K_{2, n}$.

16. Show that a non-null graph G is 2-chromatic iff, for all odd k, every diagonal entry in matrix X^k is zero, where X is the adjacency matrix of the graph G.

ANSWERS

1. (*i*) n (*ii*) 2 (*iii*) 2 if n is even, 3 if n is odd (*iv*) 3 if n is even, 4 if n is odd.

(*v*) $n - 1$ (*vi*) one (*vii*) two

6. (*i*) 3 (*ii*) 3 (*iii*) 4 (*iv*) 5 (*v*) two (*vi*) three.

8. (*i*) $\chi(G) = 3$ (*ii*) $X(G) = 3$.

9. (*i*) $k(k-1)(k-2)(k^2-4k+5)$, (*iii*) $k(k-1)(k-2)^2$, (*v*) $k(k-1)^2(k-2)$.

12. (*i*)

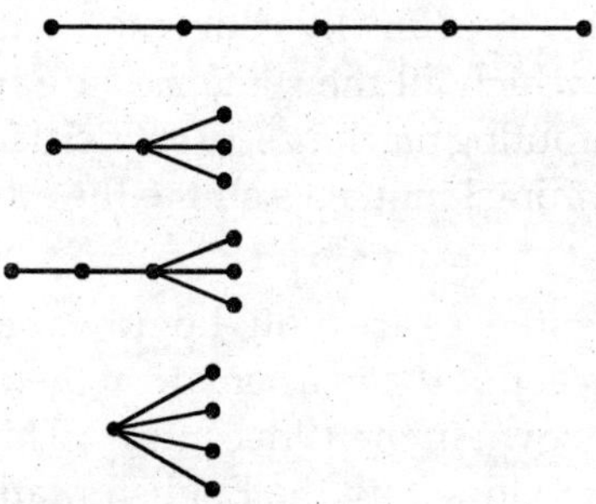

Fig. 13.18

(*ii*) $k^4 - 6k^3 + 11k^2 - 6k = k(k-1)(k-2)(k-3)$

It is a complete graph of 4-vertices. So G is

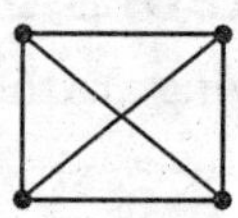

Fig. 13.19

14. (*i*) $k(k-1)(k-2)^2$

(*ii*) $k(k-1)(k-2)^2(k-3) + 2k(k-2)^2(k-2)$

15. $k(k-1)(k-2)^n + k(k-1)^n$

Chapter

14 PROPOSITIONAL CALCULUS AND FORMAL LOGIC

14.1. INTRODUCTION

Logic is the study of reasoning ; it is specifically concerned with whether reasoning is correct. It in the basis on which all the sentences are built. It has practical application to the design of computing machines, to artificial intelligence, to computer programming, to programming languages and to the other areas of computer science.

Logic is the basis on which all the sentences are built. Logic was extensively studied and developed in ancient Greece. But the mathematical theory of logic, called symbolic logic, only came into its own in the 19th century. This algebraic way of studying the arguments was developed by the English Mathematician George Boole (1815-1864). The main objective of logic is to formulate rules which may be helpful in taking decisions whether any argument or reasoning is "valid".

14.2. SYMBOLS AND NOTATIONS

Following are the symbols and notations which we will use very often in this chapter.

N	Set of natural numbers
R	Set of real numbers
$p \vee q$	p or q (p, q being statements)
$p \oplus q$	either p or q but not both
$p \wedge q$	p and q
$\sim p$	not p
$p \to q$	p implies q p only if q p is sufficient for q
$p \leftrightarrow q$	p if and only if q p is necessary and sufficient for q p implies and is implied by q
$p \Rightarrow q$	if p is true the q is true
$p \Leftrightarrow q$	p is true if and only if q is true
$p \equiv q$	p is eqvavalent to q

T	Tautalogy
F	Contradiction
$\therefore$	therefore
$\forall$	for all
$\exists$	there exists
$\exists!$	there exists one and only one
$P(X)$	Power set of X
CNF	Conjunctive normal form
DNF	disjunctive normal form
PCNF	Principle conjunctive normal form
PDNF	Principle Disjunctive normal form

14.3. PROPOSITIONS

Consider the sentence "The Taj Mahal is red". When you read this sentence, we can immediately decide whether it is ture or false. It would not happen that some people will say it is true and other will say it is false. Everybody would have the same opinion. So this sentence is either universally true or universally false but not both. Similarly "Delhi is the capital of India" is a declarative sentence which is either true or false but not both. In mathematical logic we call such sentences as **statements** or **propositions**. Therefore "**proposition** is a declarative sentence which is either universally true or universally false but not both". It is denoted by lower case latters p, q etc. If the proposition is universally true we give it a truth value T (or 1) otherwise F (or 0).

Example 14.1. Consider the following sentences and check whether the given sentences are proposition or not.

1. The president of India is a woman.
2. An elephant weights more than a human being.
3. If a stone is thrown in the air, it will fall to the ground.
4. Watch the movie.
5. How wonder ful !
6. What is your name ?

Here the first sentence is the declarative sentence which is universally false therefore is a proposition. Similarly, second sentence is also a proposition (think about it). The sentence 3rd is a proposition with the truth value T. But the 4th, 5th and 6th sentences are not even the declarative sentences and therefore none of them are propositions.

From the above example we conclude that the sentences that are not statements include questions, exclamations and expression of opinions are not proposition.

Example 14.2. Check whether the following sentences are propositions or not. Give reason also.

1. The equation $x^2 + 25$ has two solutions.
2. $3 + 3 = 6$
3. $x^2 > 0, x \in R$
4. If x and y are even numbers then the sum $x + y$ is also an even number.
5. $x + y > 0$

Solution : 1. It is a proposition being a declarative sentence which is universally true.

2, 3 and 4 are propositions [same reason as in (1)], but sentence given by 5 is not a proposition as it is neither universally true nor universally false, everything depends on the value or x and y.

14.4. LOGICAL CONNECTIVES

When we are talking to some one we never use simple sentences always. We use more complicated sentences which are formed by two or more simple sentences by using the words like 'and', 'or', if ... then etc. These words or phrases or connections or links are called logical connectives as in mathematics we use +, – ,*, / - There are five such connectives which are given as

14.4.1 Disjunction

The disjunction of two original statements p and q is a new compound statement written as $p \vee q$ and is nothing but p or q.

Consider a sentence "Rahul or Rachna went to see a movie". This sentence can be written as "Rahul went to see a movie or Rachana went to see a movie". So the given statement is actually made up of two simple said above statements by the connective called disjunction ($\vee$).

The truth value of the compound statements $p \vee q$ was depends on the truth values of the atomic statements therefore the truth values of the compound statement ($p \vee q$) are given by the following truth **Table 14.1.**

Table 14.1

p	q	$p \vee q$
T	T	T
T	F	T
F	T	T
F	F	F

You find your first number of rows in a Table

$$= (\text{number of propositions})^2$$

Since we have two propositions in this example, we know 2^2 or 4,so the number of rows are 4. (Similarly three propositions would be 8; four would be 16 and so on). The second column is found by having the first column. *For example,* we have two T's in the first column and therefore we assign those two T's one true value & one false value (in that order). We have two F's also so we assign those two F's one true valueand one false value (in that order). Now, how do we find p Ú q.

For the two statements p and q following cases will give the clarity of first two columns of the truth Table.

1. p is true, q is true
2. p is true, q is false
3. p is false, q is true
4. p is false, q is false.

That is for a single value of p there are two possibilities of q so consider the following figure

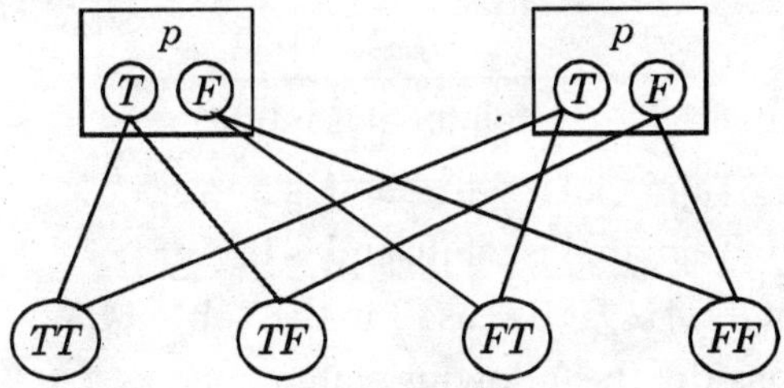

Fig. 14.1.

From the above truth Table it is clear that even if one of the p or q is true then the resultant compound statement is true otherwise false. Disjunction has the same properties as union. It is also called **inclusive or.**

Now we define **"exclusive or"** for two propositon p & q is denoted by $p \oplus q$ is true when exactly one of the proposition is true otherwise false that is $p \oplus q$ is true when either p is true or q is true not both. Again, the truth value of the compound statement. $p \oplus q$ depends on the truth value of the atomic statements, therefore for the truth value of $p \oplus q$ consider the following truth **Table 14.2.**

Table 14.2.

p	q	$p \oplus q$
T	T	F
T	F	T
F	T	T
F	F	F

Example 14.3. Let p be the statement "Sonmoy is an architect" and q is "Ashok is a draftman". Then $p \vee q$ is

"Sonmoy is an architect or Ashok is a draftman"

Example 14.4. Let A and B be sets. Then

$$A7 \cup B = \{x : x \in A \text{ or } x \in B\}$$
$$= \{x : x \in A \vee B\}$$

14.4.2. Conjunction

If p and q are two propositions then their conjunction is "p and q" denoted by $p \wedge q$ gives the new compound statement which is true only when both of them are true. If at least one of the statement is false then the compound statement is false.

Sometimes the symbols 'and', 'dot(·)', '&', are used to mean the conjunction. It is noTable that conjunction have the same properties as intersection.

The truth values of the compound statement '$p \wedge q$' are given by the **Table 14.3.**

Table 14.3.

p	q	$p \wedge q$
T	T	T
T	F	F
F	T	F
F	F	F

Example. 14.5. Let p : Tajmahal is white.

q : Tajmahal is white and is in Agra.

$p \wedge q$: Tajmahal is white and is in Agra.

The truth value of $p \wedge q$ is T. (By case 1 in the **Table 14.3**).

Example 14.6. Consider the following statements :

1. Mumbai is in Maharashtra and $2 + 2 = 4$
2. Mumbai is in U.P. and $2 + 2 = 4$
3. Bangalore is in Mysore and $2 + 2 = 4$
4. Bangalore is in Gujarat and $2 + 2 = 5$

The statements (1) and (3) are true and the statements (2) and (4) are false.

14.4.3 Negation (Not)

If p is a proposition, its negation is denoted by ~p gives the statement which is true only when p is false. Negation has the same properties as compliment operation in a set.

Its truth value is given by truth Table 14.4.

Table 14.4.

p	$\sim p$
T	F
F	T

Example 14.7. p : Ram is a good boy

then $\sim p$: Ram is not a good boy

Example 14.8. p : The sun is shining.

$\sim p$: The sun is not shining.

Example 14.9. $p : 8 + 1 = 5$

$\sim p : 8 + 1 \neq 5$

14.4.4. Conditional Operator [If... then]

If p and q are statements, the compound statement if p then q is denoted by $p \rightarrow q$, is called a conditional statement or implication. The statement p is called the antecedent or hypothesis and the statement q is called the consequent or conclusion.

The connective if ... then is denoted by the symbol $\rightarrow$.

Example 14.10. p : I will do my work.

q : I will go to see movie.

then $p \rightarrow q$ implies If p then q, *i. e.,*

If I will do my work then I will go to see movie is the resultant compound statement.

Truth values of $p \rightarrow q$ are given by the **Table 14.5.**

Table 14.5.

p	q	$p \rightarrow q$
T	T	T
T	F	F
F	T	T
F	F	T

By the truth Table it is clear that the compound statement $p \rightarrow q$ is false only if p is true and q is false.

Example 14.11. Let p & q be two statements, such that

p: Two triangles are congruent and

q: Two triangles are similar.

Then $p \rightarrow q$ is the statement :

If two triangles are congruent, then they are similar

or two triangles are congruent implies that they are similar

or two triangles are congruent only if they are similar

or the necessary condition for the two trianges to be similar is that they are congruent.

Example 14.12. Assume your father gave you the following promise :

'If you receive a grade of 95 or better in the final examination then you will get a car as reward". Your father has made a promise to you (placed a condition with you).

If you fulfill his condition you expect the conclusion (getting a car) to be forthcoming. Your final result has come to you. Has your father told the truth (kept the promise) or is he guilty of a falsehood?

Case I : Your final exam score was greater than 95 (Condition is true) and you receive a car (conclusion is true). The father told the truth.

Case II : Your final exam score was greater than 95 and you did not get a car. The father lied.

Case III : Your final exam score was less than 95 and you get a car. The father told the truth (perhaps your overall result was excellant).

Case IV : Your final exam score was less than 95 and you did not get a car. The father told the truth.

There are other forms in which a conditional statement may appear. For any conditional statement $p \to q$, three other related statements are as follows :

Statement : $p \to q$ If p, then q
Converse : $q \to p$ If q, then p
Inverse : $\sim p \to \sim q$ If not p, then not q
Contrapositive: $\sim q \to \sim p$ If not q, then not p

Example 14.13.

Statement : If it is raining then I get wet.
Negation : It is not the case that if it is raining then I get wet.
Or It is raining and I will not get wet
Converse : If I get wet then it is raining
Inverse : It it is not raining then will not get wet.
Contrapositive: I will not get wet if it is not raining.

Example 14.14.

Statement : If x is positive, then $x \neq 0$.
Negation : If x is positive and $x = 0$.
Converse : It $x \neq 0$ then x is positive.
Or It is not the case that if x is positive then $x \neq 0$.
Inverse : If x is negative, then $x = 0$.
Contrapositive: If $x = 0$ then x is negative.

The following is a truth Table for these related statements of a conditional statement $p \to q$.

Table 14.6.

p	q	$\sim p \to \sim q$	$q \to p$	$\sim q \to \sim p$	$p \to q$
T	T	T	T	T	T
T	F	T	T	F	F
F	T	F	F	T	T
F	F	T	T	T	T

14.4.5. Biconditional Operator (if and only if)

If p and q are propositions, the biconditional statement p if and only if q is denoted by $p \leftrightarrow q$ is true when p and q both have the same truth values.

$p \leftrightarrow q$ is same as $p \to q$ and $q \leftrightarrow p$ *i. e.*

$p \leftrightarrow q$ is the conjunction of $p \to q$ and $q \to p$

i.e. $p \leftrightarrow q$ is $(p \to q) \wedge (q \to p)$

The truth values of the statement $p \leftrightarrow q$ are given by the **Table 14.7.**

Table 14.7.

p	q	$p \to q$	$q \to p$	$(p \to q) \wedge (q \to p)$ or $p \leftrightarrow q$
T	T	T	T	T
T	F	F	T	F
F	T	T	F	F
F	F	T	T	T

Example 14.15. Compute the truth Table of the statement $(p \to q) \leftrightarrow (\sim q \to \sim p)$

p	q	$p \to q$	$\sim q$	$\sim p$	$\sim q \to \sim p$	$(p \to q) \leftrightarrow (\sim q \to \sim p)$
T	T	T	F	F	T	T
T	F	F	T	F	F	T
F	T	T	F	T	T	T
F	F	T	T	T	T	T

Precedance Rule

While dealing with operations on numbers, we have realised the need for applying the BODMAS Rule.

Similarly while calculating the truth value of compound propositions involving more than one connective, we have a similar rule which tells us which connective to apply first. This rule is named as "the rule of precedence". According to this rule the order of preference in which the connectives are applied in a formula of propositions that has no bracket is

(*i*) $\sim$

(*ii*) $\wedge$

(*iii*) $\vee$ and $\oplus$

(*iv*) $\to$ and $\leftrightarrow$

Note that $\vee$ and $\oplus$ *i. e.*, 'inclusive or' and 'exclusive or' are both third in order of preferce. Therefore we can apply either of them first with the same end result. Similarly, for position $\leftrightarrow$ and $\to$.

In order to clear this rule consider the following example.

Example 14.16. $p \to q \wedge \sim r \leftrightarrow r \oplus q$

Solution : To find the truth value of the following compound statement we use precedence Rule. Then we will write this statement as $(p \to (q \wedge (\sim r)) \leftrightarrow (r \oplus q)$.

Consider the truth Table

p	q	r	$\sim r$	$r \oplus q$	$q \wedge (\sim r)$	$p \to (q \wedge \sim r)$	$p \to (q \wedge \sim r) \leftrightarrow r \oplus q$
T	T	T	F	F	F	F	T
T	T	F	T	T	T	T	T
T	F	T	F	T	F	F	F
T	F	F	T	F	F	F	T
F	T	T	F	F	F	T	F
F	T	F	T	T	T	T	T
F	F	T	F	T	F	T	T
F	F	F	T	F	F	T	F

14.4.6 Tautologies and Contradiction

Consider the following composite statements and their truth Tables :

(i) $p \longleftrightarrow p$

p	$p \leftrightarrow p$
T	T
F	T

(ii) $p \vee \sim p$

p	$\sim p$	$p \vee \sim p$
T	F	T
F	T	T

(iii) $p \rightarrow p \vee q$

p	q	$p \vee q$	$p \rightarrow p \vee q$
T	T	T	T
T	F	T	T
F	T	T	T
F	F	F	T

In the above truth Tables we conclude that the statements $p \leftrightarrow p$, $p \vee \sim p$ and $p \rightarrow p \vee q$ are always true irrespective of the truth values of p and q. Such statements are called **Tautologies**. Thus **Tautology** is a composite statement which is always true irrespective of the truth value of the component statement.

A Contradiction is a composite statement which is always false irrespective of the truth values of the component Statements.

For example $r : p \wedge \sim p$

p	$\sim p$	$p \wedge \sim p$
T	F	F
F	T	F

$r : (p \vee q) \wedge (\sim p) \wedge (\sim q)$

p	q	$p \vee q$	$\sim p$	$\sim q$	r
T	T	T	F	F	F
T	F	T	F	T	F
F	T	T	T	F	F
F	F	F	T	T	F

It is clear that the negation of a tautology is a contradiction and vice-versa.

Example 14.17. Determine whether the following statements are tautologies:

1. $r : (p \wedge \sim q) \vee (q \wedge \sim p)$

2. $r : p \wedge q \rightarrow p$

3. $(p \rightarrow q) \vee (q \rightarrow p)$

Solution : 1.

p	q	$\sim p$	$\sim q$	$p \wedge \sim q$	$q \wedge \sim p$	r
T	T	F	F	F	F	F
T	F	F	T	T	F	T
F	T	T	F	F	T	T
F	F	T	T	F	F	F

The given compound statement is not a Tautology.

2.

p	q	$p \wedge q$	$p \wedge q \rightarrow p$
T	T	T	T
T	F	F	T
F	T	F	T
F	F	F	T

The given compound statement is Tautology.

3.

p	q	$p \rightarrow q$	$q \rightarrow p$	$(p \rightarrow q) \vee (q \rightarrow p)$
T	T	T	T	T
T	F	F	T	T
F	T	T	F	T
F	F	T	T	T

The given compound statement is Tautology.

Converse, Inverse and Contrapositive Implication

For two statements p and q if $p \Rightarrow q$ then, it's —

(*i*) Converse is given as $q \Rightarrow p$

(*ii*) Inverse is given as $\sim p \Rightarrow \sim q$

(*iii*) Contrapositive is given as $\sim q \Rightarrow \sim p$.

EXAMPLES

Example 14.18. Find converse, inverse and contiapositive of the followings :

(*i*) if $3x - 6 = 0$ then $x = 2$

(*ii*) if $3 > 4$ then $5 \neq 7$

Solution : (*i*) Let $\quad p \equiv (3x - 6 = 0)$

and $\quad q \equiv (x = 2)$

Then the given implication is $p \Rightarrow q$ (symbolicals) so converse is

$$q \Rightarrow p$$

i.e., $\quad (x = 2) \Rightarrow (3x - 6 = 0)$

Inverse is $\quad \sim p \Rightarrow \sim q$

i.e., $\quad \sim(3x - 6 = 0) \Rightarrow \sim(x = 2)$

or $\quad (3x - 6 \neq 0) \Rightarrow (x \neq 2)$

and contrapositive is;

$$\sim q \Rightarrow \sim p$$

i.e., $\quad \sim(x = 2) \Rightarrow \sim(3x - 6 = 0)$

or $\quad (x \neq 2) \Rightarrow (3x - 6 \neq 0)$

or if $\quad (x \neq 2)$ then $(3x - 6 \neq 0)$

(*ii*) if $\quad 3 > 4$ then $5 + 7$

Let $p \equiv (3 > 4)$ and $q \equiv (5 \neq 7)$

Then the given implications symbolically written as $p \Rightarrow q$, thus

(*a*) Converse : "if $3 \neq 7$ then $3 > 4$"

(*b*) Inverse : "if $3 \leq 4$ then $5 = 7$"

(*c*) Contrapositve : "if $5 = 7$ then $3 \leq 4$"

Example 14.19. Give converse, inverse and contrapositive of the following : "If teacher is absent, then some students do not complete their homework".

Solution : Let $p \equiv$ teacher is absent

$q \equiv$ some students do not complete their homework

Then symbolically it is written as $p \Rightarrow q$

Converse : $q \Rightarrow p$

if some students do not complete their homework then teacher is absent.

Inverse : $\sim p \Rightarrow \sim q$

if teacher is not absent than all students complete their homework.

Contrapositive : $\sim q \Rightarrow \sim p$

if all students complete their homework then teacher is not absent.

14.4.7 Logical Equivalence

Two prepositions *P(p, q, r, Λ, V, ...)* and *Q(p, q, r, Λ, V, ...)* are said to be logically equivalent if they both have same truth Tables. Symbolically we unite $P \equiv Q$. *For example, P(p → q)* is logically equivalent to it's compositive *Q(~ q → ~ p)*.

p	q	$p \to q$
T	T	T
T	F	F
F	T	T
F	F	T

$P \equiv Q$

$\sim p$	$\sim q$	$\sim q \to \sim p$
F	F	T
F	T	F
T	F	T
T	T	T

14.4.8 Logical Implication

A statement or preposition *P(p, q, r, Λ, V, ...)* logically implies *Q(p, q, r, Λ, V, ...)* if True *i.e.*, "*T*" value in the truth Table of *P* implies True "*T*" value in the truth Table of 8. Symbolically, we unite $P \Rightarrow Q$. For example; *P(p ↔ q) ⇒ Q (p → q)*.

p	q	$p \leftrightarrow q$
T	T	T
T	F	F
F	T	F
F	F	T

p	q	$p \to q$
T	T	T
T	F	F
F	T	T
F	F	T

Consider the truth Table of $(p \leftrightarrow q)$, the "*T*" value at 1st and 4th position correspondes to the truth value "*T*" at 1st and 4th position in the truth Table of $(p \to q)$. So we say that $(p \leftrightarrow q)$ logically implies $(p \to q)$.

14.5. THE LAWS OF LOGIC

Let T stand for tautology and F stand for contradiction and p is any propositions then the identity law is given by

$$p \vee T = T$$
$$p \wedge T = p$$
$$p \wedge F = \text{F}$$
$$p \vee F = p$$

Verification is given by the following truth Table

p	T	F	$p \vee T$	$p \wedge T$	$p \wedge F$	$p \vee F$
T	T	F	T	T	F	T
F	T	F	T	F	F	F

The 3rd, 4th, 5th and 6th columus give the result,

Similarly we can give all the basic law of logic which are given by the **Table 14.7.**

Table 14.7.

Basic Logical Laws

1.	$p \vee q \equiv q \vee p$ $p \wedge q \equiv q \wedge p$	Commutative law
2.	$(p \vee q) \vee r \equiv p \vee (q \vee r)$ $(p \wedge q) \wedge r \equiv p \wedge (q \wedge r)$	Associative law
3.	$p \wedge (q \vee r) \equiv (p \wedge q) \vee (p \wedge r)$ $p \vee (q \wedge r) \equiv (p \vee q) \wedge (p \vee r)$	Distributive laws
4.	$p \vee F \equiv p, p \wedge F = F$ $p \wedge T \equiv p, p \vee T = T$	Identity law
5.	$p \wedge \sim p \equiv F$ $p \vee \sim p \equiv T$	Negation law
6.	$p \vee p \equiv p$ $p \wedge p \equiv p$	Idempotent law
7.	$p \wedge (p \vee q) \equiv p$ $p \vee (p \wedge q) \equiv p$	Absorption law
8.	$\sim(p \vee q) \equiv (\sim p) \wedge (\sim q)$ $\sim(p \wedge q) \equiv (\sim p) \vee (\sim q)$	De Morgan's law
9.	$\sim(\sim p) \equiv p$	Involution law
10.	$p \rightarrow q \equiv \sim p \vee q$	
11.	$p \leftrightarrow q \equiv (p \wedge q) \vee (\sim p \wedge \sim q)$	

All the above results can be proved by the use of truth Table.

14.6. PREDICATES

Consider the declarative sentence "He is a good student". This sentence is not a proposition as it is not clear that whether it is true of false unless we know. 'who is he'. Such sentences are the examples of **propositional functions** or **predicates**. Thus the predicate in a variable x is a sentence $p(x)$ that becomes a proposition when we give a particular value to x. Let us take, "he is Ram", then the given sentence becomes "Ram is a good student" is a proposition as it will have a definite truth value either T or F.

Consider the another example $x + 2 = y$. It is again a proposition function unless we assign the values to x and y from a given set. Normally the propositional functions are denoted by $P(x)$.

Example 14.20. Let $p(x)$ be "$x^2 - 6x + 9 = 0$". Then $p(x)$ is a propositional function on the set of integers Z. $p(x)$ is true only when x is 3.

Let $p(x)$ be a propositional function on a set X. Then the set of all elements of $a \in X$ with due property that $p(a)$ is true is called the turn set T of $p(x)$. In other words,

$$T = \{a; a \in S, p(a) \text{ is true}\}$$

$$\text{or simply } T = \{a; p(a)\}$$

Example 14.21. Let $p(x)$ be "$x + 3 > 3$". Then the truth set of $p(x)$ on N, the set of natural numbers is N.

Most of the statements in mathematics and computer science use terms such as "for every" and "for some". *For example,* in mathematics we have the following theorem :

For every triangle T, the sum of the angles of T is equal to 180°.

In computer science, we have the theorem :

for some program P, the output of P is P itself.

Now we extend the logical system so that we can handle statements that includes "for every" and "for some".

14.7 QUANTIFIERS

Certain declarative sentences involves words that indicate quantify such as **all, same, none** or one. Since these words indicate quantity they are called quantifiers.

e.g.,
1. Some parallelogram are squares.
2. There are same real numbers that are not national numbers.

14.7.1 Universal Quantifier

The quantifier "all" is called universal quantifer denoted by $\forall x$.

14.7.2 Existential Quantifier

The quantifier "some" is the existential quantifier denoted by $\exists x$.

If given $\forall x F(x) \rightarrow$ for each x in the universe of discourse $F(x)$ is true. $\exists x, F(a) \Rightarrow$ there is at least one x in the universe of discourse such that $F(x)$ is true.

e.g.: 1. $M(x)$: "x is a mathematian", $R(x)$: "x is rational". All mathematians are rational"

$$\forall (x) M(x) \rightarrow R(x)$$

"some parallelograms are squares"

$\Rightarrow$ "there is at least one parallelogram that is square".

$$\exists\, x\,[P(x) \wedge S(x)]$$

2. Let $P(x)$: "x is a parallelogram"

$S(x)$: "x is a square".

Alternative ways to write universal quantifiers are as follows:

for every $x, p(x)$

or for all $x, p(x)$

or for any $x, p(x)$

i.e., the symbol $\forall$ may be read "for every", "for all", or "for any".

Therefore the statement $\forall x\, p(x)$ is called universal quantified statement.

Alternative ways to write existential quantifier are as follows :

for some $x, p(x)$

or for at least one $x, p(x)$

or there exists x such that, $p(x)$.

i.e., the symbol $\exists$ may be read "for some", "for at least one" or "there exists".

Thus the statement $\exists\, x\, p(x)$ is called existentially quantified statement.

We can say that the statement $(\exists\, x \in S)\, p(x)$ is equivalent to the statement that the truth set of $p(x)$ is non empty. That is $\{x; x \in S, p(x)\} \neq \phi$.

Quantifiers		
Statement	*When is it true ?*	*When is it false ?*
$\forall x\, p(x)$	$p(x)$ is true for every x	This is an x for which $P(x)$ is false.
$\exists x\, p(x)$	There is an x for which $p(x)$ is true	$p(x)$ is false for every x

Sentence	**Abbreviated Meaning**
$\forall\, x\, F(x)$	all true
$\exists x, F(x)$	at least one true
$\sim [\exists x, F(x)]$	none true
$\forall\, x, [\sim F(x)]$	all false
$\exists x, [\sim F(x)]$	at least one false
$\sim [\exists x, [\sim F(x))]$	none false
$\sim [\forall\, x, F(x)]$	not all true
$\sim [\forall\, x, [\sim F(x)]]$	not all false

$$\text{all true } [\forall x, F(x)] \equiv \{\sim \exists x (\sim F(x)\} \text{ none false}$$
$$\text{all false } [\forall x, (\sim F(x)] \equiv [\sim [\exists x, F(x)] \text{ none true}$$
$$\text{not all true } [\sim (\forall x, F(x)], \equiv [\exists x, (\sim F(x)] \text{ at least one false}$$
$$\text{not all false } [\sim \forall x, \{\sim F(x)] \equiv \exists x, F(x), \text{ at least one true}$$

EXAMPLES

Example 14.22. The statement "for some x, $x^2 - 6x + 9 = 0$" Can be written $\exists\, x\, (x^2 - 6x + 9 = 0)$.

This is a true statement since

$$\{x : x^2 - 6x + 9 = 0\} = \{2, 3\} \neq \phi$$

Example 14.23. The statement $(\exists\, x \in N)(3x + 9 = 6x + 2)$ is false. Since there is no integer x such that $3x + 9 = 6x + 2$, and therefore $\{x : 2x + 3 = 5x + 1\} = \phi$.

Example 14.24. The statement "for every natural number n, $n + 3 > 1$" can be written

$$\text{"}(\forall\, n \in N)(n + 5) > 1\text{"}$$

This is a true statement since

$\{\, n : n + 5 > 1\} = \{1, 2, 3, ...\} = N$

Example 14.25. The statement $(\forall n \in N)(n + 1 > 3)$ is false since

$\{\, n : n + 1 > 5\} = \{5, 6, 7, ...\} \neq N$

Example 14.26. The statement

for every real number x, $x^2 \geq 0$

is a universally quantified statement. The domain of discourse is the set of real numbers. The statement is true therefore have truth value T.

Example 14.27. To verify that the existentially quantified statement for some real number x, $\dfrac{1}{1+x^2} > 1$ is false.

Now we have to showe that the proposition $\dfrac{1}{1+x^2} > 1$ for some real number x is false is equivalent to prove $\dfrac{1}{1+x^2} \leq 1$ is true.

Now consider $\dfrac{1}{1+x^2} \leq 1$

Since $\quad x^2 \geq 0$

$$\Rightarrow \quad 1 + x^2 \geq 1$$

$$\Rightarrow \quad \frac{1}{1+x^2} \leq 1$$

$\therefore$ The statement $\dfrac{1}{1+x^2} \leq 1$ is true for every real number x.

Thus the statement $\dfrac{1}{1+x^2} > 1$ is false for every real number x.

Therefore, the result is proved.

Theorem 14.1. Generalized, De Morgan Laws for Logic

If p is a propositional function, each pair of propositions in (*a*) and (*b*) has the same truth values (*i.e.*, either both are true or both are false).

(*a*) $\overline{\forall x p(x)}$, $\exists x \overline{p(x)}$

(*b*) $\overline{\exists x p(x)}$, $\forall x \overline{p(x)}$

Proof : (*a*) Suppose that the proposition $\overline{\forall x p(x)}$ is true then the proposition $\forall x P(x)$ is false.

The proposition $\forall x p(x)$ is false if $P(x)$ is false for at least one x in the domain of discourse, *i.e.*, D. But if $p(x)$ is false for at least one x in D then $\overline{p(x)}$ is true for at least one x in D.

$\therefore$ the proposition $\exists x \overline{p(x)}$ is true.

Thus if the proposition $\overline{\forall x p(x)}$ is true then $\exists x \overline{p(x)}$ is true.

Similarly if the proposition $\overline{\forall x p(x)}$ is false then the proposition $\exists x \overline{p(x)}$ is false.

Part (*b*) is left for the students.

Example 14.28. Consider the proposition

"There is a criminal who has committed every crime".

In symbolic form

A : Set of criminals

B : Set of crimes

$(\exists\, x \in A)\ (\forall\, y \in B)$ (x has committed y)

$(\exists\, x \in A)\ (\forall y \in B)\ p(x,y)$

Now consider its negation

$\sim (\exists x \in A)\ (\forall y \in B)\ p(x, y)$

$\equiv (\forall x \in A) \sim(\forall y \in B)\ p(x,y)$

$\equiv (\forall x \in A)\ (\exists y \in B) \sim p(x,y)$

$\equiv (\forall x \in A)\ (\exists y \in B)$ (x has not committed y)

Example 14.29. Since $\sim(p \vee q) = \sim p \wedge \sim q$, so we have

$$\begin{aligned}\sim(\forall\, x\, p(x) \vee \exists\, y\, p(y)) &= \sim(\forall\, x\, p(x)) \wedge \sim (\exists\, y\, p(y)) \\ &= \exists\, x \sim p(x) \wedge \forall\, y \sim p(y)\end{aligned}$$

14.7.3 Nested Quantifiers

Two **quantifiers are nested** is one is within the scope of the other one, *for example :*

$$\forall x \exists y\ (x + y = 0).$$

We read quantifiers from left to right. Everything within the scope of a quantifier can be thought as a propositional function. *For example :*

$$\forall x \exists y\,(x + y = 0)$$

is the same thing as $\forall x Q(x)$ where $Q(x)$ is $\exists y\, P(x, y)$, where $P(x, y)$ is $x + y = 0$.

14.7.4 Order of Quantifiers

- It is important to note that the order of the quantitifers is important, unless all the quantifiers are universal quanitfiers or all are existential quantifiers.
- $\forall x \forall y\, P(x, y) \equiv \forall y \forall x\, P(x, y)$
- $\exists x \exists y\, P(x, y) \equiv \exists y \exists x\, P(x, y)$
- $\forall x \exists y\, P(x, y) \equiv \exists y A x\, P(x, y)$

14.7.5 Negating Nested Quantifiers

Statements involving nested quantifiers can be negated by successively applying the rules for negating statements involving a single quantifier. Here are the four cases involving two quantifiers.

$$\neg(\forall x\ \forall y\ P(x, y)) \equiv \exists x \neg(\forall y\ P(x, y)) \equiv \exists x\ \exists y\ \neg P(x, y)$$
$$\neg(\forall x\ \exists y\ P(x, y)) \equiv \exists x \neg(\exists y\ P(x, y)) \equiv \exists x\ \forall y\ \neg P(x, y)$$
$$\neg(\exists x\ \ \forall y\ P(x, y) \equiv \exists x \neg(\forall y\ P(x, y)) \equiv \forall x\ \exists y\ \neg P(x, y)$$
$$\neg(\exists x\ \ \exists y\ P(x, y) \equiv \forall x \neg(\exists y\ P(x, y)) \equiv \forall x\ \forall y\ \neg P(x, y)$$

14.7.6 Thinking of Quantification as Loops

In working with quantifications of more than one variable, it is sometimes helpful to think in term of nested loops.

Of course, if there are infinitely many elements in the universe of discourse of some variable, we cannot actually loop through all values. Nevertheless, this way of thinking is helpful in understanding nested quantifiers.

(*a*) $\forall x\ \forall y\, P(x, y)$

Assume that U is the universe of discourse and assume that the variables x and y belong to this universe of discourse.

- The proposition $\forall x\ \forall y\, P(x, y)$ is true when the propositional function $P(x, y)$ is true for each pair (x, y).
- The proposition $\forall x\ \forall y\, P(x, y)$ is false if there exists at least one pair (x, y) such that the propositional function $P(x, y)$ is false for that pair.

(*b*) $\forall x\ \exists y\, P(x, y)$

- The proposition $\forall x\ \exists y\, P(x, y)$ is true if for each x, there is at least one y for which the propositional function $P(x, y)$ is true.

- The proposition $\forall x\ \exists y\ P(x, y)$ is false if there exists at least one x such that the propositional function $P(x, y)$ is false for all y.

(c) $\exists x\ \forall y\ P(x, y)$

- The proposition $\exists x\ \forall y\ P(x, y)$ is true if there exists at least one x for which the propositional function $P(x, y)$ is true for all y.
- The proposition $\exists x\ \forall y\ P(x, y)$ is false if for all x, there is at least one y for which the propositional function $P(x, y)$ is false.

(d) $\exists x\ \exists y\ P(x, y)$

- The proposition $\exists x\ \exists y\ P(x, y)$ is true if there exists at least one pair (x, y) for which is the propositional function $P(x, y)$ is true.
- The proposition $\exists x\ \exists y\ P(x, y)$ is false if the propositional function $P(x, y)$ is false for all pairs (x, y).

14.7.7 Definition : Prenex Normal Form (PNF)

A statement is in **prenex normal form** if and only if it is of the form

$$Q_1x_1\ Q_2x_2 \ldots Q_kx_k\ P(x_1, x_2, \ldots, x_k)$$

where each Q_i, $i = 1, 2, \ldots, k$ is either the existential quantifier or the universal quantifier, and $P(x_1, x_2, \ldots, x_k)$ is a predicate involving no quantifiers.

For example, $\exists x \forall y\ (P(x, y) \wedge Q(y))$ is in prenex normal form, whereas $\exists x\ (P(x) \wedge \forall x Q(x))$ is not because the quantifiers do not all occur first.

14.8 NORMAL FORM

Let $A(P_1, P_2, \ldots P_n)$ be a statement formula where $P_1, P_2, \ldots . P_n$ are the atomic variables, if we consider all possible assignments of the truth values to $P_1, P_2, \ldots P_n$ and obtain the resulting truth values of the formula A, then we get a resulting truth Table for A..

If all values for A are true then A is identically true or a *tautology*.

If all values for A are false then A is identically false or a *contradiction*.

If A has the resulting truth value true for at least one combination of truth value assigned to $P_1, P_2, \ldots P_n$ then A is said to be *satisfiable*.

Decision Problem

The problem of determining in a finite number of steps whether a given statement formula is tautology or a contradiction or at least satisfiable is known as decision problem. The decision problem always has a solution. We use the word 'product' in place of 'conjunction'. We use the word 'sum' in place of 'disjunction'.

Elementary Product

A product (conjunction) of the variables or their negation in a formula is called Elementary product.

e.g., p, $\sim p$, $\sim p \wedge q$, $p \wedge q$, $\sim p \wedge q \wedge \sim p$, $p \wedge \sim p$, $q \wedge \sim p$ etc.

Elementary Sum

A sum (Disjunction) of the variable and their negation in a formula is called elementary sum.

For example, p_1, $\sim p$, $\sim p \vee p$, $\sim p \vee q$, $p \vee \sim q$, $p \vee q \vee \sim p$ etc.

Factor

Any part of an elementary sum or product which is itself an elementary sum or product is called a factor of the original elementary sum or product *e.g.*

$\sim p, p \wedge \sim p$ are the factors of $\sim p \wedge q \wedge p \wedge \sim p$,

14.8.1 Disjunctive Normal Form (DNF)

A formula which is equivalent to a given formula and which consists of "a sum of elementary products" is called a disjunctive normal form of the given formula. Disjunctive normal form of the given formula is not unique.

Example 14.30. Obtain *DNF* of

(*a*) $p \wedge (p \rightarrow q)$

(*b*) $\sim(p \vee q) \leftrightarrow (p \wedge q)$

Solution : (*a*) $p \wedge (p \rightarrow q) \equiv p \wedge (\sim p \vee q)$

$\equiv (p \wedge \sim p) \vee (p \wedge q)$ (By Distributive law)

is the required DNF.

(*b*) $\sim(p \vee q) \leftrightarrow (p \wedge q)$

$\equiv [\sim(p \vee \Box q) \wedge (p \wedge q)] \vee [(p \vee q) \wedge \sim(p \wedge q)]$

$\equiv [(\sim p \wedge \sim q) \wedge (p \wedge q)] \vee [(p \vee q) \wedge (\sim p \vee \sim q)]$

$\equiv (\sim p \wedge \sim q \wedge p \wedge q) \vee [(p \wedge \sim p) \vee (p \wedge \sim q) \vee (q \wedge \sim p) \vee (q \wedge \sim q)]$

By Distributive Law

$= (\sim p \wedge \sim q \wedge p \wedge q) \vee (p \wedge \sim p) \vee (p \wedge \sim q) \vee (q \wedge \sim p) \vee (q \wedge \sim q)$

It is clear from the above formula that DNF is identically false if every elementary product appearing in its DNF is identically false if at least one of them is true then the formula is true.

14.8.1.1 Minterms

Let p and q be two statement variables. All possible formulas which consists of conjunction of p (or its negations) and q (or its negation). None of the formula should contain both variable and its negation. These formulas are called minterms. No two minterms are equivalent.

For n variables $p_1, p_2, ...p_z$ there are 2^n such minterms. In particular for 2 varibles say p and q there are 2^2 such formulas $p \wedge q, p \wedge \sim q, \sim p \wedge q$, and $\sim p \wedge \sim q$.

$p \wedge \sim p$ is not a minterm.

One can find the minterms by using truth Table. Each row in the truth Table will give one minterm by using conjunction between the variables. *For example* consider the Table having two variables attached

x	y	minterms
T	T	$x \wedge y$
T	F	$x \wedge \sim y$
F	T	$\sim x \wedge y$
F	F	$\sim x \wedge \sim y$

An in case of truth Table there are 2^n rows for n number of variables, these rows gives each minterms. Similarly in the case of three variables.

x	y	z	minterms
T	T	T	$x \wedge y \wedge z$
T	T	F	$x \wedge y \wedge \sim z$
T	F	T	$x \wedge \sim y \wedge z$
T	F	F	$x \wedge \sim y \wedge \sim z$
F	T	T	$\sim x \wedge y \wedge z$
F	T	F	$\sim x \wedge y \wedge \sim z$
F	F	T	$\sim x \wedge \sim y \wedge z$
F	F	F	$\sim x \wedge \sim y \wedge \sim z$

14.8.1.2 Principal Disjunctive Normal Form (PDNF)

For a given formula, an equivalent formula consisting of **disjunction of minterms** only is known as principal disjunctive normal form or **"Sum of product canonical form"**.

Example 14.31. Write *PDNF* of the following :

(*i*) $(\sim p \vee \sim q) \rightarrow (p \leftrightarrow \sim q)$

(*ii*) $q \wedge (p \vee \sim q)$

Solution : (*i*) $\sim (p \wedge q) \rightarrow (p \wedge \sim q) \vee (\sim p \wedge q)$

$$\equiv \sim\sim(p \wedge q) \vee [(p \wedge \sim q) \vee (\sim p \wedge q)]$$

$$\equiv (p \wedge q) \vee (p \wedge \sim q) \vee (\sim p \wedge q)$$

Disjunction of minterms, therefore is the required PDNF.

(*ii*) $q \wedge (p \vee \sim q)$

$$\equiv (q \wedge p) \vee (q \wedge \sim q)$$

$$\equiv (q \wedge p) \vee F$$

$\equiv (q \wedge p)$ is the required form.

14.8.2 Conjunctive Normal Form (CNF)

A formula which is equivalent to a given formula which consists of a product of elementary sum is called conjunctive normal form of the given formula. For a given formula conjunctive normal form is not unique. For a given formula CNF is identically true if every elementary sum is identically true.

14.8.2.1 Principal Conjunctive Normal Form (PCNF)

To define the principal conjunctive normal form, we first define formula which are called maxterms.

14.8.2.2 Maxterms

For a given number of variables, the maxterms consists of disjunctions in which each variable or its negation but not both appears only once. Maxterms are the dual of minterms. One can find maxterms by using truth Table as in the case of minterms by replacing the sign conjunction ($\wedge$) by the disjunction ($\vee$).

For a given formula an equivalent formula consisting of conjunction of the maxterms only is known as Principle conjunctive normal form or 'product of sums canonical form. Method, for obtaining *PCNF* is similar to the one described for *PDNF*.

Example 14.32. Obtain the principle conjunctive normal form of

$$(\sim p \to r) \wedge (q \leftrightarrow p)$$
$$\equiv [\sim(\sim p) \vee r] \wedge [(q \to p) \wedge (p \to q)]$$
$$\equiv (p \vee r) \wedge [(\sim q \vee p) \wedge (\sim p \vee q)]$$
$$\equiv [(p \vee r) \vee F] \wedge [(\sim q \vee p) \vee F] \wedge [(\sim p \vee q) \vee F]$$
$$\equiv [(p \vee r) \vee (q \wedge \sim q)] \wedge [(\sim q \vee p) \vee (r \wedge \sim r)] \wedge [(\sim p \vee q) \vee (r \wedge \sim r)]$$
$$\equiv (p \vee r \vee q) \wedge (p \vee r \vee \sim q) \wedge (\sim q \vee p \vee r) \wedge (\sim q \vee p \vee \sim r) \wedge (\sim p \vee q \vee r) \wedge (\sim p \vee q \vee \sim r)$$
$$\equiv (p \vee q \vee r) \wedge (p \vee r \vee \sim q) \wedge (p \vee \sim q \vee \sim r) \wedge (\sim p \vee q \vee r) \wedge (\sim p \vee q \vee \sim r)$$

Example 14.33. Obtain the conjunctive normal form of

(i) $p \wedge (p \to q)$

(ii) $\sim(p \vee q) \leftrightarrow (p \wedge q)$

Solution : (*i*) $p \wedge (p \to q)$

$$p \wedge (\sim p \vee q)$$

is the required conjunctive normal form.

(*ii*)
$$\sim(p \vee q) \leftrightarrow (p \wedge q) \equiv (\sim(p \vee q) \to (p \wedge q)) \wedge ((p \wedge q) \to \sim(p \vee q))$$
$$\equiv ((p \vee q) \vee (p \wedge q)) \wedge (\sim(p \wedge q) \vee \sim(p \vee q))$$
$$\equiv (p \vee q) \wedge (p \vee q) \wedge ((\sim p \vee \sim q) \vee (\sim p \wedge \sim q))$$
$$= (p \vee q) \wedge (\sim p \vee \sim q) \wedge (\sim p \vee \sim q)$$
$$= (p \vee q) \wedge (\sim p \vee \sim q)$$

Example 14.34. Obtain the Principle Conjunctive Normal (PCNF) form of

(i) $\sim (p \vee q)$

(ii) $\sim (p \to q)$

Solution :
$$\sim (p \vee q) \equiv \sim p \wedge \sim q$$
$$\equiv (\sim p \vee F) \wedge (\sim q \vee F)$$
$$\equiv (\sim p \vee (q \wedge \sim q)) \wedge (\sim q \vee (p \wedge \sim p))$$
$$\equiv (\sim p \vee q) \wedge (\sim p \vee \sim q) \wedge (\sim q \vee p) \wedge (\sim q \vee \sim p)$$
$$\equiv (\sim p \vee q) \wedge (\sim p \vee \sim q) \wedge (\sim q \vee p)$$

Example 14.35. Put the function $f = [(x \wedge y')' \vee z'] \wedge (x' \vee z)'$ in the conjunctive normal form (CNF).

Solution: $((x \wedge y')' \vee z') \wedge (x' \vee z)'$

$$= (x' \vee y \vee z') \wedge (x \wedge z')$$
$$= (x \wedge z') \wedge (x' \vee y \vee z')$$
$$= (x \wedge z' \wedge x') \vee (x \wedge z' \wedge y) \vee (x \wedge z \wedge z')$$

Example 14.36. Put the function $f = x \wedge (y \vee z)$ in the conjunctive normal form.

Solution : The given function is already in conjunctive normal form.

Example 14.37. Find the *CN* form of the function $f = [x \wedge (y' \vee z)] \vee z'$ and then find its *DN* form from it.

Solution: $f = (x \wedge (y' \vee z)) \vee z'$

$$= (x \vee z') \wedge [(y' \vee z) \vee z']$$
$$= (x \vee z') \wedge (y' \vee (z \vee z'))$$
$$= (x \vee z') \wedge (y' \vee T)$$
$$= (x \vee z') \wedge T$$
$$= x \vee z'$$
$$= (x \vee z') \vee F$$
$$= (x \vee z') \vee (y \wedge y')$$
$$= (x \vee z' \vee y) \wedge (x \vee y' \vee z') \text{ is the } CN \text{ form}$$

Now consider CN form of the function f

$$f(x, y, z) = (x \vee z' \vee y)\,(x \vee y \vee z')$$

As $f = ((f))' = [[(x \vee y \vee z') \wedge (x \vee y' \vee z')]']'$

$$= [(x \vee y \vee z') \vee (x \vee y' \vee z')']'$$
$$= [(x' \wedge y' \wedge z) \vee (x' \wedge y \wedge z)]'$$
$$= (x' \wedge z)'$$
$$= x \vee z'$$
$$= (x \wedge T) \vee (z' \wedge T)$$
$$= [x \wedge (y \vee y')] \vee [z' \wedge (y \vee y')$$
$$= (x \wedge y) \vee (x \wedge y') \vee (z' \wedge y) \wedge (z' \wedge y')$$

is the required DN fom.

SOLVED PROBLEMS

Problem 14. Which of the following are logically equivalent to $\forall x\, \exists y p(x, y)$? Explain.

1. $\exists x\, \forall y\, \overline{P(x, y)}$ 2. $\forall x\, \forall y\, \overline{P(x, y)}$

Solution: 1. $\overline{\forall x\, \exists y\, P(x, y)} \equiv \exists x\, \forall y\, \overline{P(x, y)}$

To show this more clearly : $\overline{\forall x\, \exists y\, P(x, y)}$ By DeMorgan's Laws

$$\equiv \exists x\ \overline{(\exists y\ P(x, y)}$$

$$\equiv \exists x\ \forall y\ \overline{P(x, y)}$$

Thus the given statement is logically equivalent, (*i.e.,*)

$$\overline{\forall x\ \exists y\ P(x, y)} \equiv \forall x\ \forall y\ \overline{P(x, y)}$$

But $\qquad \forall x\ \forall y\ \overline{P(x, y)} \not\equiv \exists x\ \forall y\ \overline{P(x, y)}$

So the statement is not logically equivalent.

Note : Many people claimed that the two statements were logically equivalent by setting $P(x, y) = x < y$ the domain D of all integers. Doing this would get a truth value of false for each statement, but this does not necessarily mean that the two statements are logically equivalent :

Let $\qquad D$ = all integers.

$$P(x, y) = x < y$$

$\overline{\forall x\, \exists y\, P(x, y)} \equiv \exists x\ \forall y\ \overline{P(x, y)}$ Means that there exists $x \in D$ such that for all $y \in D$.

$x \geq y$. This is of course is false since we can always choose $y = x + 1$. Now the second statement says $\forall x\ \overline{\exists y\ P(x, y)} \equiv \forall x\ \forall y\ \overline{P(x, y)}$ which claims that for every x, for every y, $x \geq y$, which is also false. However the statements

1. There exists an x such that for all y, $x\,^3\,y$, and
2. for every x, for every y, $x \geq y$ are not equivalent, although they are both false.

Further more, we can show that (2) implies (1) but is not equal to (1).

Problem 14.2s. Give the meaning of easch statement given below. Clarify each statement by rephrasing it.

1. Economist Robert J. Samuelson : Every environmental problem is not a tragedy.
2. Headline over a Martha Stewart Column : All Lampshades can't be cleaned.
3. From Newsweek: Formal investigations are a sound practice in the right circumstances, but every circumstance is not right.

Solution:

1. *As stated:* No evnrionment problem is a tragedy.

   Intended: Not every environmental problem is a tragedy.
2. *As stated*: Every lampshade cannot be cleaned or no lampshade can be cleaned.

Intended: Not every lampshade can be cleaned.

3. *As stated*: No circumstance is right. Formal investigation are never a sound practice.

   Intended: Not every circumstance is right, Investigations sometimes sound.

Give an argument using rules of inference to show that the conclusion follows from the hypothesis.

EXERCISES

1. Give two propositions each, the truth values of which are T and F respectively. Also give two examples of the sentances that are not propositions.

2. Give the truth value of the conjunction of the following propositions.

(*i*) "$3 \div 5 = 2$", "Ram is in Delhi".

(*ii*) "$5 + 7 = 12$", $x^2 \geq 0 \ \forall\ x$

(*iii*) "$x^2 - 3x + 5 = 0$ has two solutions". "$9 \geq 7$".

3. Write down two propositions p and q for wihch $q \rightarrow p$ is true but $p \leftrightarrow q$ is false.

4. Write down the truth Table of $p \rightarrow q \wedge \sim r \leftrightarrow r \wedge q$

5. Form the truth Table of $p \wedge q \vee \sim r$ and $(p \wedge q) \vee (\sim r)$ and explain where differ.

6. How would you bracket the following formulae to correctly interpret them?

(*i*) $\sim p \vee q$

(*ii*) $\sim p \rightarrow \sim qs$

(*iii*) $p \rightarrow q \leftrightarrow \sim p \vee q$

(*iv*) $p \oplus q \wedge \mathrm{r} \rightarrow \sim p \vee q \leftrightarrow p \wedge r$

7. Use the laws given in this section to show that $\sim(\sim p \wedge q) \wedge (p \vee q) \equiv p$

8. Write down suiTable mathematical statements that can be represented by the following symbolic propositions. Also write down their negations. Give the truth value of the given propositions.

(*i*) $(\forall\ x)\ (\exists\ y) p$

(*ii*) $(x \geq 0) \rightarrow (\exists\ y)\ (y^2 = x)$

(*iii*) $(\exists\ x)\ (\exists!\ y)\ (x\ y = 0)$

(*iv*) $\sim(\exists x)\ (\exists!\ y)\ (x + y = 0)$

9. Obtain the disjunctive normal form of the Boolean expression

(*i*) $f(x_1, x_2, x_3) = (x_1' \wedge x_2) \vee (x_1 \wedge x_3)$

(*ii*) $= (x_1 \vee x_2)' \vee (x_1' \wedge x_3)$

10. Obtain the CNF of the Boolean expression

(*i*) $f(x_1, x_2, x_3) = (x_1 \wedge x_2)' \vee (x'_1 \wedge x_3)$

(*ii*) $f(x_1, x_2, x_3) = (x_1 \wedge x_2') \vee (x_1' \wedge x_3')'$

11. Reduce the following Boolean expression to a simple form.

(*i*) $f(x_1, x_2) = (x_1 \wedge x_2) \wedge (x_1 \wedge x_2')$

(*ii*) $f(x_1, x_2, x_3) = (x_1 \wedge x_2) \vee (x_1 \wedge x_2' \wedge x_3) \vee (x_1 \wedge x_3)$

(*iii*) $f(x_1, x_2, x_3) = (x_1 \wedge x_2) \vee ((x_1 \wedge x_2) \wedge x_3) \vee (x_2 \wedge x_3)$

12. Check whether the following sentences are statements or not.

(*i*) Come in !

(*ii*) What are you doing?

(*iii*) 7 is a lucky number.

(*iv*) This statement you are reading is false.

13. State the following argument in symbolic form and test its validity.

"If the parcel is not propely addressed or is too large, then the Post office will not accept it. The parcel is not too large. If Ram wrote the address on the parcel, then it is properly addressed. Hence if Ram wrote the address the Post Office will accept the parcel. " ***(End-term Exam., GGSIPU, 2001)***

14. Prove the following implication :

$P \rightarrow Q \cdot Q \rightarrow S \cdot P \vee Q \Rightarrow S \vee R$ ***(End-term Exam., GGSIPU, 2001)***

15. Determine whether the following wffs are tautology, a contradiction.

(*i*) $(P \rightarrow r) \vee (r \rightarrow \urcorner q) \wedge \urcorner r$

(*ii*) $((P \vee q \rightarrow r) \wedge p) \rightarrow (r \rightarrow p)$ ***(End-term Exam., GGSIPU, 2001)***

16. (*a*) Show that the following arguments is a fallacy :

$P \rightarrow q, \neg p \vdash \neg q$ where symbol $\neg$ represents not (~).

(*b*) Write the Converse, Inverse and Contrapositive of the implication :

"If it is rose, then it is red"

(*c*) Negate the statement:

"Every student has at least one Course where the teacher is an Asstt. Professor." ***(Terminal Exam., GGSIPU, 2001)***

17. (*a*) Show that

$$\neg P \wedge (\neg q \wedge r) \vee (q \wedge r) \vee (p \wedge r) \equiv r$$

(*b*) Obtain the conjunctive Normal form of $(P \rightarrow Q) \wedge (Q \vee (P \wedge R))$ and determine whether or not it is a tautology.

(*c*) Show that $\neg (P \wedge Q)$ follows from $\neg P \vee \neg Q$ using poof by contradiction. ***(Terminal Exam., GGSIPU, 2001)***

18. Use resolution principle in testing the validity of the following argument:-

"If the musicians sing the songs, then crowds will come, if the ticket prices are not too high.

If the musicians sing the songs, then the ticket prices are not too high. Therefore, if the musicians sing the song, then crowds will come" ***(End-Term Exam., GGSIPU, 2002)***

19. (*a*) Write negetion, converse, inverse and contra positive statements of the following implication:

"If it is cold, then she wears a coat"

(*b*) Prove the De Morgan's Rule in Predicate logic

$$\sim(\forall\, x\, P(x)) \Leftrightarrow \exists x \sim P(x)$$

Also negate the statement" Every locality in New Delhi is clean"

(*c*) Define conjunctive Normal Form (CNF) and Disjunctive Normal Form (DNF) in prepositional logic Also convert $P \leftrightarrow Q$ into DNF. ***(End-term Exam., GGSIPU, 2002)***

20. (*a*) Show tha $\sim(P \wedge Q)$ logically follows from $\sim P \wedge \sim Q$ without using Truth Tables.

(*b*) If $P \to (Q \to R)$ is true then prove that $Q \to (P \to R)$ is true.

(End Term Exam., GGSIPU, 2001)

ANSWERS

1. $p : x^2 < 0 \ \forall \ x \in R$" , the truth value of p is F

q : Taj Mahal is in Agra", the truth value of q is T.

Example of the sentences that are not propositions.

1. Do work.
2. $x + y \geq 9 \ \forall x, y \in R,$

2. (*i*) Let $p : 3 \div 5 = 2$

q : Ram is in Delhi

The truth value of p is F

The truth value of q is T

Therefore truth value of $p \wedge q$ = F

(*ii*) Let $p : 5 + 7 = 12$

$q : x^2 \geq 0 \ \forall \ x$

the truth value of p is T

the truth value of q is T

Therefore truth value of $p \wedge q$ is T

(*iii*) Let $p : x^2 - 3x + 5 = 0$ has two solutions.

$q : 9 \geq 7$

the truth value p is T

the truth value q is T

the truth value $p \wedge q$ is T.

5. $(p \to (q \wedge (\sim r)) \leftrightarrow (r \wedge q)$

p	q	r	$\sim r$	$q \wedge (\sim r)$	$p \to (q \wedge \sim r)$	$r \wedge q$	$LHS \leftrightarrow RHS$
T	T	T	F	F	F	T	F
T	T	F	T	T	T	F	F
T	F	T	F	F	F	F	T
T	F	F	T	F	F	F	T
F	T	T	F	F	T	T	T
F	T	F	T	T	T	F	F
F	F	T	F	F	T	F	F
F	F	F	T	F	T	F	F

6. (*i*) $(\sim p) \vee q$

(*ii*) $(\sim p) \to (\sim q)$

(*iii*) $(p \to q) \leftrightarrow ((\sim p) \vee q))$

(*iv*) $((p \oplus q) \wedge r) \to ((\sim p) \vee q) \leftrightarrow (p \wedge r)$

7. $\sim(\sim p \wedge q) \wedge (p \vee q)$

$\equiv (p \vee \sim q) \wedge (p \vee q)$

$\equiv p \vee (\sim q \wedge q)$

$\equiv p \vee 0$

$\equiv p$

8. (*i*) True

(*ii*) True

(*iii*) True

(*iv*) is equivalent to $(\forall\ x \in R)\ (\sim (\exists!\ y \in R)\ (x + y = 0))$ *i.e.*, for every x there is no unique y such that $x + y = 0$

This is clearly false because for every x these is a unique $y(= -x)$ such that $x + y = 0$

12. All the given sentences are not propositions as

(*i*) Exclamation

(*ii*) Question

(*iii*) and (*iv*) are expressions of opinion.

■■■

Chapter

15 METHODS OF PROOF

15.1. INTRODUCTION

In the previous chapter we studied about statements and their truth values. In this chapter we will discuss the way in which statements can be linked to form a logically valid argument. we will talk about what a theorem is and what constitutes a mathematical acceptable proof ?

The statement that is proved to be true is called a ***theorem.*** Special kinds of theorems are referred to as ***lemmas*** and ***corollaries***. A ***lemma*** is a theorem that is usually not too interesting in its own right but is useful in proving another theorem. A ***corollary*** is a theorem that follows quickly from another theorem. A ***conjecture*** is a statement whose truth value is unknown. When a proof of conjecture is found, the conjecture becomes a theorem. Many times conjectures are shown to be false, so they are not theorems. An argument that establishes the truth of the theorem is called a ***proof*** and ***logic*** is a tool for the analysis of proofs.

15.2. METHODS OF PROOF

Consider an implication $P \Rightarrow Q$ where P and Q are compound statements involving any number of propositional variables. That is $P \equiv (p_1, p_2, \ldots p_n)$ and $Q \equiv (q_1, I_2, \ldots q_m)$. If $P \Rightarrow Q$ is a tatuology we say Q logically follows from P.

Suppose $(p_1 \wedge p_2 \wedge p_n) \Rightarrow q$ is a tautology then this implication is true regardless of the truth values of any of its components. Here we say that q logically follows from $p_1, p_2, \ldots p_n$. This can also be written as

$$\begin{array}{l} p_1 \\ p_2 \\ \vdots \\ p_n \\ \hline \therefore \quad q \end{array} \qquad \text{or,} \qquad p_1, p_2, \ldots. p_n \vdash q$$

where the symbol $\therefore$ or $\vdash$ means therefore. That is if p_1 is true, p_2 is true, ...p_n is true then q is true. Virtually all mathematical theorom or arguments are camposed of implications of $(p_1 \wedge p_2 \wedge, \ldots \wedge p_n) \Rightarrow q$ type. Here $p_1, p_2, \ldots p_n$ are called premises or hypothesis and q is called the conclusion. Thus to "prove the theorem" we means to show that the implication $(p_1 \wedge p_2 \wedge \ldots \wedge p_n)$

Note. The proof does not show that the conclusion is true but simply shows that the conclusion will be true if all the premises are true. Hence mathematical proofs often begin with the statement "suppose that $p_1, p_2, \ldots p_n$ are true and conclude with the statement" therefore, q is ture"

15.2.1. Truth Tables

Consider an argument

$$p_1, p_2, \ldots\, p_n \vdash q$$

To show that the argument is valid, we can also make use of truth tables. The argument is valid if p_1 is true, p_2 is true and so on p_n is true logically implies that q is true. Which can be represented as $(p_1 \wedge p_2 \wedge \ldots p_n) \to q$ in a truth table and if it is a tautology, we say the argument is valid otherwise the argument is fallacy or not valid. The validity of the arguments based on tautologies depends only on the form of the statements involved and not on the truth values of the variables they contain and hence these argumetns are also knwon as *rules of inference.* In other words a rule of inference is universally valid.

Example 15.1 : Check the validity of the argument. If f drive to work, I will arrive tired. I drive to work. Therefore, I will arrive tired". by making use of truth table.

Solution: Let p : f drive to work

and q : f arrive tired.

Then the argument is represented as

$$p \to q, p \vdash q$$

or

$$p \to q$$

$$p$$

$$\therefore q$$

To show that this argument is valid we need to construct the truth table for

$$[(p \to q) \wedge p] \to q$$

p	q	$p \to q$	$(p \to q) \wedge p$	$(p \to q) \wedge p \to q$
T	T	T	T	T
T	F	F	F	T
F	T	T	F	T
F	F	T	F	T

Since it is a tautology, the argument is universally valid, and so is a rule of inference, called modes panes of law of detachment.

Table 15.1. Rules of Inference

	Rule of Inference	Tautology	Name of the Rule
1.	$\dfrac{p}{\therefore p \vee q}$	$p \to (p \vee q)$	Addition
2.	$\dfrac{p \wedge q}{\therefore p}$	$(p \wedge q) \to p$	Simplification
3.	ps $\dfrac{p \to q}{\therefore q}$	$[p \wedge (p \to q)] \to q$	Modes pones
4.	$\sim q$ $\dfrac{p \to q}{\therefore \sim p}$	$[(\sim q) \wedge (p \to q)] \to \sim p$	Modus tollens
5.	$p \to q$ $\dfrac{q \to r}{\therefore p \to r}$	$[(p \to q) \wedge (q \to r)] \to (p \to r)$	Hypothetical Syllogism
6.	$p \vee q$ $\dfrac{\sim p}{\therefore q}$	$(p \vee q) \wedge (\sim p) \to q$	Disjunctive Syllogism.

Now we will discuss some of the rules given in the table.

15.2.2. Law of Detachment (Modes Pones)

'Modes pones' is a latin term which means 'Method of affirmation'.

It is a tautology given in the **Table 15.2.**

$$[(p \to q) \wedge p] \Rightarrow q$$

Table 15.2.

p	q	$p \to q$	$(p \to q) \wedge p$	$[(p \to q) \wedge p] \to q$
T	T	T	T	T
T	F	F	F	T
F	T	T	F	T
F	F	T	F	T

Modes pones states that *if both an implications and its hypothesis are known to be true, then the conclusion of this implication is true.*

This form of valid argument is called the law of detachment (Modes pones) because of conclusion q is detached from a premise $(p \to q)$.

Example 15.2. Suppose that the implication 'If it rains today, then we will have rain bath' and its hypothesis is 'it is raining today,' are true. Then by Modes pones, it follows that the conclusion of the implication, "We will have rain bath," is true, as given below.

p: It rains today.

q: We will have rain bath.

Then premises $(p \to q)$ and hypothesis is p

$\therefore [(p \to q) \wedge p] \to q$ by Modes Pones

Example 15.3. Consider the following argument

If I work hard, then I will get first position in the examination.

I work hard.

Therefore, I get first position in the examination.

This is a **valid argument** by Modus pones as

If p : I work hard

q : I get first position in the examination

In the symbolic form

$$\begin{array}{l} p \to q \\ p \\ \hline \therefore \quad q \end{array}$$

That is $[(p \to q) \wedge p] \Rightarrow q$ is a Tautology.

Example 15.4. In the **example 15.1(*b*)** the following argument is not a valid argument.

If I work hard, then I will get first position in the examination

I get first position in the examination.

Therefore, I work hard.

p : I work hard

q : I will get first position in the examination

In symbolic form

$$\begin{array}{l} p \to q \\ q \\ \hline \therefore \quad p \\ \hline \end{array}$$

$(p \to q) \wedge q \to p$ is not a tautology given below

p	q	$p \to q$	$(p \to q) \wedge q$	$(p \to q) \wedge q \to p$
T	T	T	T	T
T	F	F	F	T
F	T	T	T	F
F	F	T	T	F

Example 15.5. State which rule of inference is the basis of the following argument,' "It is below freezing now. Therefore, it is either below freezing or raining now".

Let p : It is below freezing now.

q : It is raining now.

therefore $p \to p \vee q$

Here **"addition"** is the **rule of inference.**

Example 15.6. State which rule of inference is used in the argument :

If it rains today then we will not have a play today. If we do not have a play today, then we will have a play tomorrow. Therfore if it rains today then we will have a play tomorrow.

p : It rains today

q : we will not have a play today

r : we will have a play tomorrow.

Hypothesis $(p \to q) \wedge (q \to r) \Rightarrow (p \to r)$

Here we use the rule of inference **'Hypothetical syllogism.'**

15.2.3. Law of Contra Positive (or Modus Tollens)

Modes tollens means method of **Denial**. It is given by tautology

$[(p \to q) \wedge \sim q] \to \sim p$

Consider the truth **Table 15.3.**

Table 15.3.

p	q	$p \to q$	$\sim q$	$(p \to q \wedge \sim q$	$(p \to q) \wedge \sim q \Rightarrow \sim p$
T	T	T	F	F	T
T	F	F	T	F	T
F	T	T	F	F	T
F	F	T	T	T	T

15.2.4. Disjunctive Syllogism

It is given by the tautology $[(p \vee q) \wedge \sim p] \Rightarrow q$

Consider the example

Example 15.7. Either Ram is a good boy or Rahul is a good boy. Ram is not a good boy therefore Rahul is a good boy.

Let p : Ram is a good boy

q : Rahul is a good boy

$[(p \vee q) \wedge (\sim p)] \to q$ is a tautology given as

p	q	$\sim p$	$(p \vee q)$	$(p \vee q) \wedge (\sim p)$	$[(p \vee q) \wedge (\sim p)] \Rightarrow q$
T	T	F	T	F	T
T	F	F	T	F	T
F	T	T	T	T	T
F	F	T	F	F	T

Therefore, Disjunctive syllogism is a valid argument.

15.2.5. Hypothetical Syllogism or Law of Transitivity

It is one of the most useful rule of inference and is given by the tautology

It is also known as **Chain rule.**

$[(p \to q) \wedge (q \to r)] \Rightarrow p \to r$

Consider the truth **Table 15.4.**

Table 15.4.

p	q	r	$p \to q$	$q \to r$	$(p \to q) \wedge (q \to r)$	$p \to r \wedge$	$(p \to q) \wedge (q \to r) \Rightarrow p \to r$
T	T	T	T	T	T	T	T
T	T	F	T	F	F	F	T
T	F	T	F	T	F	T	T
T	F	F	F	T	F	F	T
F	T	T	T	T	T	T	T
F	T	F	T	F	F	T	T
F	F	T	T	T	T	T	T
F	F	F	T	T	T	T	T

Therefore hypothetical syllogism is a valid argument.

Example 15.8. If I will do my work, I will go to see movie.

If I will go to see movie, I will have popcorns.

Therefore if I will do my work, I will have popcorns.

Let p : I will do my work.

q : I will go to see movie.

r : I will have popcorn.

$p \to q$

$q \to r$

$\Rightarrow p \to r$.

$(p \to q) \wedge (q \to r) \Rightarrow (p \to r)$

We observe that validity of an argument depends on its form but not on its contents. *For example,* consider the following arguments.

Example 15.9. If a polygon is a square than it has 4 sides.

If a polygon has 4 sides, then it has 3 angles.

Therefore, if a polygon is a square, then it has 3 angles.

Here the conclusion is false, but the argument is valid since it is of the form **Hypothetical Syllogism**.

p : Polygon is a square

q : Polygon has 4 sides

$\therefore$ r : Polygon has 3 angles

$(p \to q) \wedge (q \to r) \Rightarrow (p \to r)$

Example 15.10. Check the validity of the following argument "If a polygon is square, it has 3 sides. If a polygon has 3 sides, then it has 4 angles. Therefore if a polygon is a square then it has 4 angles.

Solution: Let p: A polygon is square

q : A polygon has 3 sides

r : A polygon has 4 angles.

then the argument is symbollic form can be represented as.

$p \to q$

$q \to r$

$\therefore p \to r$

which is the law of transitivity (rule of inference) and its truth table is shown be low:

Table 15.5.

p	q	r	$p \to q$	$q \to r$	$(p \to q) \wedge (q \to r)$	$p \to r \wedge$	$(p \to q) \wedge (q \to r) \Rightarrow p \to r$
T	T	T	T	T	T	T	T
T	T	F	T	F	F	F	T
T	F	T	F	T	F	T	T
T	F	F	F	T	F	F	T
F	T	T	T	T	T	T	T
F	T	F	T	F	F	T	T
F	F	T	T	T	T	T	T
F	F	F	T	T	T	T	T

Here we can see that the argument is valid although the premises or hypothesis are false. Thus, the validity of arguments depend on its form and not on its contents.

15.2.6. Direct Proof

The direct proof is based entirely on modes pones.

A direct proof of $p \Rightarrow q$ is a logical valid argument in which we start with the assumption that hypothesis is true, that is, p is true.

A direct proof assumes that $p(x_1, x_2, ... x_n)$ is true and then using $p(x_1, x_2, ... x_n)$ as well as other axioms, definitions and previousely derived theorems shows directly that $q(x_1, x_2, ... x_n)$ is true.

Example 15.11. Give a direct proof of the statement,'

"The product of two odd integers is odd."

Let x and y be two odd integers then

$$x = 2n + 1 \quad \text{for } n, m \in Z$$

$$y = 2m + 1$$

$$\begin{aligned} xy &= (2n + 1)(2m + 1) \\ &= 4mn + 2n + 2m + 1 \\ &= 2(mn + n + m) + 1 \\ &= 2p + 1 \quad p \in Z \end{aligned}$$

is an odd integer

$\therefore$ The result follows.

Example 15.12. Give a direct proof of the statement,

"The square of an even integer is an even integer".

Let x be an even integer then for some $n \in Z$

$$x = 2n$$

Squaring on both sides

$$x^2 = (2n)^2 = 4n^2 = 2(2n)^2$$
$$= 2p \quad \text{where } p = 2n^2 \in Z$$
$$= \text{even}$$

$\therefore$ The result follows.

Example 15.13. If $x \in R$ (set of real numbers) such that $x^2 = K$ then either $x = \sqrt{K}$ or $x = -\sqrt{K}$.

Let p and q are two propositions

$p : x \in R$ such that $x^2 = K$ and

$q : x = \sqrt{K}$ or $x = -\sqrt{K}$

Now $x^2 = K$

$$\Rightarrow \sqrt{x^2} = \pm\sqrt{K}$$

Therefore, p is true and $(p \Rightarrow q)$ is true, conclude that q is true by Rule of "***Modus Pones***".

Example 15.14. Give a direct proof of the following statement.

For all real numbers x, x_1, x_2 and y,

If $\quad x = \max\{x_1, x_2\}$

and $\quad y \geq x$

Then $\quad y \geq x_1$ and $y \geq x_2$

Now given $\quad y \geq x$

$\Rightarrow y \geq \max\{x_1, x_2\}$

Case I : If $\max\{x_1, x_2\} = x_1$

$\Rightarrow x_1 \geq x_2$

$\Rightarrow y \geq x_1$ as well as $y \geq x_2$

Case II : If $\max\{x_1, x_2\} = x_2$

$\Rightarrow x_2 \geq x_1$

$\Rightarrow y \geq x_2$ as well as $y \geq x_1$

Therefore $y \geq x_1$ and $y \geq x_2$.

15.2.7. Indirect Methods

In the direct method we prove the argument $p \Rightarrow q$ by directly using that p is true (*i.e.*, hypothesis is true) then proved that q is true.

Now we will discuss an important method of proof called indirect method of proof a which follows from the tautlogy $(p \rightarrow q) \Leftrightarrow (nq \rightarrow \sim p)$

p	q	$p \rightarrow q$	$\sim q \rightarrow np$	$(p \rightarrow q) \Leftrightarrow (\sim q \rightarrow \sim p)$
T	T	T	T	T
T	F	F	F	T
F	T	T	T	T
F	F	T	T	T

we can also see that $(p \rightarrow q)$ is logically equivalent to its contrapositive $(nq \rightarrow np)$ (see the truth table).

Thue to prove $(p \rightarrow q)$ in directly, we will prove its contrapositive. That is we assume q is false (the statement nq is true) and show that p is then false (the statement np is ture). This indirect method of proofs is also known as method of contrapositive.

15.2.8 Proof by Contrapositive

The method of contrapositive says $p \Rightarrow q$ is logicaly equivalent to its contrapositive $(\sim q \Rightarrow \sim p)$ *i.e.,*

$$(p \Rightarrow q) \equiv (\sim q \Rightarrow \sim p)$$

Consider the following truth Table 15.6.

Table 15.6.

p	q	$p \rightarrow q$	$\sim p$	$\sim q$	$\sim q \rightarrow \sim p$
T	T	T	F	F	T
T	F	F	F	T	F
F	T	T	T	F	T
F	F	T	T	T	T

From the third and sixth column both $p \rightarrow q$ and $\sim q \rightarrow \sim p$ are equivalent.

In other words, in order to prove $p \Rightarrow q$ we assume that q is false and then show that p is false.

Example 15.15. Prove that "If $x, y \in Z$ (set of integer) such that xy is odd then both x and y are odd, by proving its contrapositive.

As x and y both are integers then their product is also an integer *i.e.,* xy is an integer. Now

P: xy is odd.

q: x and y are odd.

Now we need to show

$$p \Rightarrow q$$

By using method of contrapositive we will prove $\sim q \Rightarrow \sim p$

Consider $\sim q$, *i.e.,* both x and y are not odd

Case I : x and y both are even

$\Rightarrow$ $x = 2n$ and $y = 2m$ for some $m, n \in Z$

$\Rightarrow$ $xy = (2n)(2m)$

$= 4nm$

$= 2(2mn)$

$= 2p$ where $p = 2mn \in Z$

$\Rightarrow xy$ is even

$\Rightarrow \sim p$ is true.

$\Rightarrow \sim q \Rightarrow \sim p$

Case II : One of x and y is even

Let x is even and y is odd

$\Rightarrow$ $x = 2n$ and $y = 2m + 1$ for some integer $m, n \in Z$

$\Rightarrow$ $xy = (2n)(2m + 1)$

$= 2q$ where $q = n(2m + 1) \in Z$

$\Rightarrow$ xy is even

$\Rightarrow$ $\sim p$ is true

$\Rightarrow$ $\sim q \Rightarrow \sim p$

Therefore, by method of contrapositives the result holds.

Example 15.16. Prove that "If x and y are two integers such that x is even and y is odd then xy is even.

We will prove the result by direct method. By the given hypothesis

$p : x$ is even & y is odd.

$q : xy$ is even

$\Rightarrow$ $x = 2n$ and $y = 2m + 1$ for some $m, n \in Z$

$\Rightarrow$ $xy = (2n)(2m + 1)$

$= 4nm + 2n$

$= 2(2nm + n)$

$= 2p$ where $p = (2mn + n) \in Z$

$=$ even

Therefore xy is even.

Example 15.17. Prove that statement

"If x is an integer and x^2 is even then x is also even",

by proving its contrapositive.

Let p : x is an intger and x^2 is an even intger.

q : x is an even intger.

We need to prove $p \Rightarrow q$

By using contrapositive

$\sim q \Rightarrow \sim p.$

$\sim q$ says x is odd integer.

$\Rightarrow x = 2n + 1, n \in Z$

squaring on both sides

$$\begin{aligned} x^2 &= (2n+1)^2 \\ &= 4n^2 + 1 + 4n \\ &= (4n^2 + 4n) + 1 \\ &= 2(2n^2 + 2n) + 1 \\ &= 2p + 1; \quad p = 2n^2 + 2n \in Z \end{aligned}$$

$\Rightarrow$ x^2 is odd integer

Therefore $(\sim p)$ holds.

Therefore result is true by the method of contrapositive.

Example 15.18. Give an indirect proof of the theorem "If $3n + 2$ is odd, then n is odd."

$p : 3n + 2$ is odd

$q : n$ is odd

We need to prove $p \Rightarrow q$

Consider the contrapositive.

$\sim q \Rightarrow \sim p.$

$\sim q : n$ is even

$\Rightarrow n = 2k \quad , \quad k \in Z$

Consider

$$\begin{aligned} 3n + 2 &= 3(2k) + 2 \\ &= 6p + 2 \\ &= 2(3k + 1) \\ &= \text{even} \end{aligned}$$

$\Rightarrow 3n + 2$ is even

$\Rightarrow \sim p$ holds

Therefore by method of contrapositive the result holds.

This result (**Example 15.19**) can also be proved by another method called method of contradiction given as :

15.2.9 Method of Contradiction

This method is also called **reductio ad absurdum** in latin because it believes on reducing a given assumption to an absurdity.

For the argument $p \Rightarrow q$ we need to prove that if p is true then q is true. In this method we start by assuming that q is false (*i.e.*, $\sim q$ is true). Then by a logical argument we arrive at a situation where a statement is true as well as false that is we reach a contradiction $r \wedge \sim r$ for some statement r. This means that the truth of $\sim q$ implies a contradiction, a statement that is always false. This can only happen when $\sim q$ is false also. Therefore q must be true.

Now consider the **Example 15.17** again

we need to prove the theorem that

"If $3n + 2$ is odd, then n is odd".

$p : 3n + 2$ in odd

$q : n$ is odd

Now we use method of contradiction and we will prove the result.

Let $\sim q$ is true that is n is even

then $n = 2k$ for some integer k.

It follows that $3n + 2 = 3(2k) + 2 = 6k + 2 = 2(3k + 1)$

$\Rightarrow 3n + 2$ is even (since it is a multiple of 2)

Since the negation of the conclusion of the implication implies that the hypothesis is false, therefore the original implication is true.

Example 15.19. Show that $\sqrt{3}$ is irrational.

Let us try and prove the given statement by contradiction. For this we begin by assuming that $\sqrt{3}$ is rational. This means that there exists positive integers x and y

Such that $\sqrt{3} = \frac{x}{y}$, where x and y are relatively prime.

This implies $x = \sqrt{3}\, y$

$\Rightarrow x^2 = 3y^2$

$\Rightarrow 3 | x^2$ (3 divides x^2)

$\Rightarrow 3 | x$ (3 divides x) ...(1)

Therefore, by definition, $x = 3p$ for some $p \in Z$

$\therefore \quad x^2 = 9p^2$

But $\quad x^2 = 3y^2$ also

So $\quad 9p^2 = 3y^2$

$\Rightarrow 3p^2 = y^2$

$\Rightarrow 3 | y^2$ (3 divides y^2)

$\Rightarrow 3 | y$ ...(2)

(1) and (2) implies that 3 divides both x and y which contradicts our assumption at x and y are relatively prime (that have no common factor).

Therefoere we conclude that our assumption is wrong that $\sqrt{3}$ is rational

So $\sqrt{3}$ is irrational.

Example 15.20. Use the method of proof by contradiction to show that for $x \in R$, if $x^3 + 4x = 0$ then $x = 0$.

Let us assume that $x^3 + 4x = 0$ and $x \neq 0$

Then $x(x^2+4x) = 0$ and $x = 0$.

$\Rightarrow x^2+4 = 0$

$\Rightarrow x^2 = -4$

But $x \in R$ and $x^2 = -4$ *Not Possible.*

Therefore our assumption is false.

Therefore the given statement is true.

Example 15.21. Prove by method of contradiction the following statement :

"For all real numbers x and y, if $x+y \geq 2$ then either $x \geq 1$ or $y \geq 1$".

Solution : Let us assume that for $x, y \in R$,

$x+y \geq 2$ and the conclusion is not true.

Case I. $x < 1$ and $y < 1$

Adding both the inequalities

$x+y < 1$ Not possible

Case II. $x \geq 1$ and $y \geq 1$

Adding both the inequalities

$x+y \geq 2$

which is true.

Case III. $x \geq 1$ and $y < 1$

Now consider the given inequality

$\Rightarrow x+y \geq 2$

$\Rightarrow x \geq 2-y$ ...(1)

As $y < 1$

$\Rightarrow -y > -1$

$\Rightarrow 2-y > 2-1$

$\Rightarrow 2-y > 1$

$\therefore$ (1) $\Rightarrow x \geq 1$ which is true

Therefore, the result is true for either $x \geq 1$ or $y \geq 1$.

Example 15.22. From the following statement

"If n is an integer not divisible by 3, then $n^2 \equiv 1 \pmod 3$".

Let p : n is an integer not divisible by 3

$q : n^2 \equiv 1 \pmod 3$.

Now $p \equiv p_1 \vee p_2$ where $p_1 : n \equiv 1 \pmod 3$ " and

$p_2 : n \equiv 2 \pmod 3$

Hence to show that $p \to q$ it can be shown that $p_1 \to q$ and $p_2 \to q$. It is easy to give direct proof of these two implications.

Case I. Let us suppose that p_1 is true.

$\Rightarrow n \equiv 1 \pmod 3$

$$\Rightarrow 3 \mid (n-1) \qquad [3 \text{ divides } (n-1]$$

$$\Rightarrow n-1 = 3k, k \in Z$$

$$\Rightarrow n = 3k+1$$

By squaring

$$n^2 = (3k+1)^2$$

$$= 9k^2 + 6k + 1$$

$$= 3(3k^2 + 2k) + 1$$

$$\Rightarrow n^2 \equiv 1 (\text{mod } 3)$$

Therefore $p_1 \Rightarrow q$ is true

Case II : Let us suppose that p_2 is true.

$$\Rightarrow n \equiv 2 \ (\text{mod}3)$$

$$\Rightarrow 3 \mid (n-2) \qquad [3 \text{ divides } (n-2)]$$

$$\Rightarrow n-2 = 3r \qquad r \in Z$$

$$\Rightarrow n = 3r+2$$

By squaring

$$n^2 = (3r+2)^2$$

$$= 9r^2 + 12r + 4$$

$$= 3(3r^2 + 4r + 1) + 1$$

Hence $n^2 \equiv 1(\text{mod } 3)$.

Therefore $p_2 \rightarrow q$ is true.

Since it has been shown that $p_1 \rightarrow q$ and $p_2 \rightarrow q$ are true, it can be concluded that $(p_1 \vee p_2) \rightarrow q$ is true.

As $p \equiv p_1 \vee p_2$

$\Rightarrow p \rightarrow q$ is true.

Therefore the result follows.

15.2.10 Proof by Counter-example

Usually indirect technique to prove a statement which claims that a property holds for all object of a certain type, we must use steps that are valid for all objects of that type. These steps do not make references to any particular object but to disprove such a statement, we need only to show one particular object or set of objects for which the claim fails. The object for which the claim fails is reffered as counter example.

Example 15.23. Prove or disprove the statement that if a and b are real numbers. Then ($a^2 = b^2$) if and only if ($a = b$).

Solution: To disprove this statement, we need only to find one example for which the is false.

Since $(-2)^2 = (2)^2 = 4$ but $-2 \neq 2$.

Thus, the reslut is false

Example 15. 24. Prove or disprove: "for all prime numbers, *p*, 2*p* + 1 is prime".

Solution: Since, the statement is an assertion for all primes, to disprove it we only need to find a prime *p* for which $(2p + 1)$ is not a prime.

Clearly for $p = 2, 2p + 1 = S$ is a prime

for $p = 3, 2p + 1 = 7$ is again a prime

for $p = 5; 2p + 1 = 11$ is also a prime

but for $p = 7, 2p + 1 = 15 = 3.5$ is not a prime.

Hence we have obteined acounter example to the statement.

Example 15.25. Check the validity of the following argument. "If I am elected, the streets will be clean. If the sunshines, the republocrates will vote. If the republocrats vote, & will be elected. The sun is not shining. Thus, the streets will not be clan."

Solution : Let

p: *f* am elected

q: the streets will be clean

r: sunshines

s: the repubocrates will vote

Now, we will represent the sentences into symbolic form. Then the argument is

$$\text{Arguments} \vdash \text{Conclusion}$$

$$p \to q; r \to s; s \longrightarrow p; \sim r \vdash \sim q$$

Direct Proof : Given

$p \to q$

$r \to s, s \to p$

$\Rightarrow \quad r \to p$ (by law of transitivity)

Now $r \to p$ and $p \to q$

$\Rightarrow \quad r \to q$ (by law of transitivity)

$r \to q \equiv \sim q \to \sim r$

So $\left.\begin{matrix}\sim q \to \sim r \\ \sim r\end{matrix}\right\} \Rightarrow \sim q$

Thus argument is not valid.

Indirect Proof

Let the given argument is not valid.

$\Rightarrow \quad \sim q$ is false $\Rightarrow q$ is true

now $p \to q$ and $s \to p$

$\Rightarrow \quad s \to q$ (by transitive law)

Again $r \to s, s \to q \Rightarrow r \to q$

$r \to q$ and q is true

$\Rightarrow \quad r$ is not true

$\Rightarrow \quad \sim r$ is true

⇒ all the argument (premises) are valid.

Hence our assumption is the *i.e.*, q is true.

⇒ $\sim q$ is false and thus argument is not valid.

15.3 RULES OF INFERENCE FOR QUANTIFIED PROPOSITIONS

1. $\forall x,\ P(x) \vdash P(x)$ for all c
2. $P(x)$, all $c \vdash \forall x,\ P(x)$
3. $\exists x, P(x) \vdash P(c)$ for some c
4. $P(x)$, for some $c \vdash \exists x, P(x)$

Consider the Argument

Lions are dangerous animals. There are lions. Therefore, there are dangerous animals.

$L(x)$: x is a lion

$D(x)$: x is a dangerous animal

$$\forall x,\ [L(x) \rightarrow D(x)]$$

$$\mathbf{\exists x, L(x)}$$

$$\therefore \exists x, D(x)$$

Proof :

	1. $\exists x, L(x)$	Premise 2
⇒	2. $L(a)$	Step 1 & Rule 3
	3. $\forall x\ [L(x) \rightarrow D(x)]$	Premises 1
	4. $L(a) \rightarrow D(a)$	Step 3, Rule 2
⇒	5. $D(a)$	(Step 2 & 4 and by claw of detouchment)
⇒	$\exists x, D(x)$	

Starting for $\exists x$, so ending with $\exists x$.

15.4 MATHEMATICAL INDUCTION

In this section we will examine mathematical induction, a technique for proving propositions over the positive integers. What does this principle really says? It says that if you can walk a few steps, say m steps, and if at each stage you can walk one more step, then you can walk any distance. Mathematical induction is used extensively to prove results about a large variety of discrete objects. *For example,* it is used to prove results about the complexity of algorithms, the correctness of certain types of computer programe, theorems about graphs and trees, as well as wide range of identities and inequalities.

An Analogy : Mathematical induction is often useful in overcoming a problem such as this one. A proof by mathematical induction similar to knocking over a row or closely spaced dominos that are standing on end. To knock over the six dominos in **Fig. 15.1** all you need to do is push domino 1 to right. To be assured that they all will be knocked over, some work must be done ahead of time. The dominos must be positioned so that if any domino is pushed to the right, it will push the next domino in the line.

Now imagine the propositions $p(1)$, $p(2)$, $p(3)$, ...to be an infinite line of dominos. Let's see if these propositions are in the same formation as the dominos were. First, we will focus on one specific point of the line : $p(99)$ and $p(100)$. We are not going to prove that either of these propositions is true, just that the truth of $p(99)$ implies the turth of $p(100)$. In terms of our analogy, if $p(99)$ is knocked over, it will knock over $p(100)$.

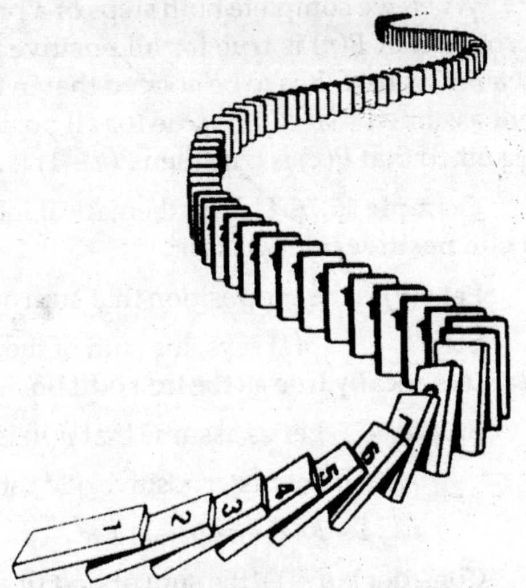

Fig. 15.1. Illustrating how mathematical induction works using dominos

In proving $p(99) \Rightarrow p(100)$, we will use $p(99)$ as our premise. Consider the result "sum of the positive integers 1 to 100 is $\frac{100(100+1)}{2}$. We start by observing that the sum of the positive integers from 1 to 100 is $(1 + 2 + ... + 99) + 100$. That is, the sum of the positive integers from 1 to 100 equals the sum of the first ninety nine and 100. Now we apply our premise

$$p(99) : 1 + 2 + ... 99 = \frac{99(99+1)}{2}$$

then we consider

$$\begin{aligned} p(100) &= 1 + 2 + ... + 99 + 100 \\ &= (1 + 2 + ... + 99) + 100 \\ &= \frac{99}{2}(99 + 1) + 100 \\ &= 99 \times \frac{100}{2} + 100 = \frac{(100)(101)}{2} \\ &= \frac{100\,(100+1)}{2} \end{aligned}$$

and we get the success is proving the required result.

The validity of mathematical induction follows from the following fundamental axiom about the set of integers.

(a) The well ordering property : Every nonempty set of non-negative integers has a least element.

> **Definition : Mathematical Induction**
>
> **A proof by mathematical induction that $P(n)$ is true for every positive integer n consists of two steps :**
>
> **1. Basic step : The proposition $p(1)$ is shown to be true.**
>
> **2. Inductive Step : The implication $P(n) \rightarrow P(n+1)$ is shown to be true for every positive integer n.**

When we complete both steps of a proof by mathematical induction, we have proved that $P(n)$ is true for all positive integers n, that is, we have shown than $\forall n\, P(n)$ is true. It is to be noticed that in the proof by mathematical induction it is not assumed that $P(n)$ is true for all positive integers! It is only shown that if it is assumed that $P(n)$ is true then $P(n + 1)$ is also true.

Example 15.26. Use mathematical induction to prove that the **sum of the first n odd positive integers is n^2.**

Let $p(n)$ is the proposition that sum of the first n odd positive integers is n^2.

Step 1. $p(1)$ says, the sum of the first one odd positive integer is 1^2 which isautomatically true as the first odd positive integer is 1.

Step 2. Let us assume that $p(n)$ is true

i.e., sum of first n positive odd integer is n^2

i.e., $1 + 3 + 5 + ... (2n - 1) = n^2$.

Consider $p(n + 1)$, the sum of first $(n + 1)$ positive odd integers ;

$$\begin{aligned} 1 + 3 + 5 + ... + (2n - 1) + (2n + 1) &= n^2 + 2n + 1 \\ &= (n + 1)^2 \end{aligned}$$

This implies that $p(n + 1)$ is true.

Therefore by the principle of mathematical induction the result is true.

Example 15.27. Prove the following result by the method of mathematical induction.

For all $n \geq 1$, $n^3 + 2n$ is a multiple of 3.

Let $p(n) : n^3 + 2n$ is a multiple of 3, $n \geq 1$

Basic step : $p(1) : 1^3 + 2.1 = 3$ is a multiple of 3

$\therefore p(1)$ is true.

Induction step : Assume that $p(n)$ is ture *i.e.,* $n \geq 1$ and $n^3 + 2n$ is a multiple of 3.

Consider $p(n + 1)$.

$$\begin{aligned} (n + 1)^3 + 2(n + 1) &= n^3 + 3n^2 + 3n + 1 + 2n + 2 \\ &= (n^3 + 2n) + 3n^2 + 3n + 3 \\ &= (n^3 + 2n) + 3(n^2 + n + 1) \end{aligned}$$

Now $n^3 + 2n$ is a multiple of 3 as $p(n)$ is true (By our hypothesis)

$(n + 1)^3 + 2(n + 1)$ is the sum of two multiples of 3; therefore, it is also a multiple of 3.

Thus by Principle of mathematical induction the result is true $\forall n$.

Example 15.28. Use mathematical induction to prove that **$n^3 - n$ is divisible by 3 whenever n is a positive integer.**

Let $p(n) : (n^3 - n)$ is divisible by 3, $n \in Z^+$ (set of positive integers)

Basic Step :

$p(1) = (1^3 - 1) = 0$ is divisible by 3

Therefore, $p(1)$ is true.

$p(2) = 2^3 - 2 = 6$ is divisible by 3

Therefore, $p(2)$ is true.

Induction Step :

Now let us assume that $p(n)$ is true that is $(n^3 - n)$ is divisible by 3. We need to show $p(n + 1)$ is true, that is $[(n + 1)^2 - (n + 1)]$ is divisible by 3.

As $(n^3 - n)$ is divisible by 3 then by division algorithm there exists a positive integer k_1, such that

$$n^3 - n = 3k_1$$

Now consider $p(n + 1)$

$$\begin{aligned}(n + 1)^3 - (n + 1) &= (n^3 + 3n^2 + 3n + 1) - (n + 1)\\ &= n^3 + 3n^2 + 3n - n\\ &= (n^3 - n) + 3(n^2 + n)\end{aligned}$$

Now $(n^3 - n)$ is divisible by 3 as $p(n)$ is true.

$(n + 1)^3 - (n + 1)$ is the sum of two multiplies of 3, that is

$$\begin{aligned}(n + 1)^3 - (n + 1) &= 3k_1 + 3k_2 && k_1 = n^3 - n \in Z\\ &= 3(k_1 + k_2) && k_2 = n^3 + n \in Z\end{aligned}$$

$\Rightarrow ((n + 1)^3 - (n + 1))$ is divisible by 3 therefore by principle of mathematical induction the result is true for $\forall n$.

Example 15.29. Use mathematical induction to prove that

$$\mathbf{1 + 2 + 2^2 + ... + 2^n = 2^{n+1} - 1}$$

for all non-negative integer *n*.

Solution : Let $p(n)$ be the predicate

$$1 + 2 + 2^2 + ... + 2^n = 2^{n+1} - 1$$

Basic Step : $p(0)$

$$1 = 2^1 - 1 = 1$$

$\therefore$ $p(0)$ is true.

Induction Step : Let us assume that the result is true for $n = k$, that is, $p(k)$ is true that is

$$1 + 2 + 2^2 + ... + 2k = 2^{k+1} - 1 \text{ is true.}$$

Now to prove that $p(k + 1)$ is true. *i.e.*,

$$1 + 2 + 2^2 + 2^3 + ... + 2^k + 2^{k+1} = 2^{k+2} - 1$$

Consider $p(k + 1)$

$$\begin{aligned}1 + 2 + 2^2 + .. + 2^k &= 2^{k+1} - 1\\ 1 + 2 + 2^2 + ... + 2^k + 2^{k+1} &= 2^{k+1} - 1 + 2^{k+1}\\ &= 2.\ 2^{k+1} - 1\\ &= 2^{k+2} - 1\\ &= 2^{(k+1)+1} - 1\end{aligned}$$

$\therefore p(k+1)$ is true.

Therefore by principle of mathematical induction the result is true $\forall n$.

Example 15.30. Use mathematical induction to prove that

$$1^2 + 2^2 + 3^2 + ... + n^2 = \frac{n}{6}(n+1)(2n+1) \ \forall \ n \in N$$

Solution : Let $p(n)$ is the predicate

$$1^2 + 2^2 + ... + n^2 = \frac{n}{6}(n+1)(2n+1)$$

Since we want to prove it for every $n \in N$, we take $n = 1$

Basic Step : $1^2 = 1 = \frac{1}{6}(1+1)(2.1+1) = \frac{2.3}{6} = 1$

Therefore $p(1)$ is true.

Induction Step : Let for an arbitracy $k \in N$, $p(k)$ is true

i.e., $1^2 + 2^2 + ... + k^2 = \frac{1}{6}k(k+1)(2k+1)$ is true.

Consider $p(k+1)$

$$\begin{aligned} p(k+1) &= 1^2 + 2^2 + ... + k^2 + (k+1)^2 \\ &= \frac{k(k+1)}{6}(2k+1) + (k+1)^2 \\ &= \frac{(k+1)[k(2k+1) + 6(k+1)]}{6} \\ &= \frac{1(k+1)[2k^2 + k + 6k + 6]}{6} \\ &= \frac{1}{6}(k+1)[2k^2 + 3k + 4k + 6] \\ &= \frac{1}{6}(k+1)(k+2)(2k+3) \end{aligned}$$

Therefore the result is true for $p(k+1)$.

As both the conditions of the principle of mathematical induction hold. Therefore $p(n)$ is true for every $n \in N$.

Example 15.31. Use mathematical induction to prove that $2^n < n!$ for every positive integer n with $n \geq 4$.

Solution. Let $p(n)$ be the predicate

$$2^n < n!$$

Basic step : $p(4)$, $2^4 = 16$.

$$4! = 4 \times 3 \times 2 \times 1 = 24$$

$\therefore 2^4 < 4!$

$\therefore p(4)$ is true.

Induction step : Let us assume that $p(k)$ is true

$$\Rightarrow 2^k < k!\,,\, k \geq 4$$

Consider $p(k+1)$

$$2^{k+1} = 2^k.2 < 2.\,k!$$

As $k \geq 4$

$k + 1 \geq 4 > 2$

$(k+1) > 2$

$\therefore 2^{k+1} < (k+1)\, k!$

$= (k+1)\,!$

$\Rightarrow 2^{k+1} < (k+1)!$

$\therefore p(k+1)$ is true.

Therefore by principle of mathematical induction the result is true $\forall n$.

Example 15.32 Show that $2^n > n^3$ for $n \geq 10$

Solution : Let $p(n)$ be the predicate $2^n > n^3$.

Basic Step : $p(10)$?

$2^{10} = 1024$

$10^3 = 1000$

$\therefore 2^{10} > 10^3$

$\therefore p(1)$ is true.

Induction Step : Let us assume that $p(k)$ is true for an arbitrary $k \geq 10$.

Therefore $2^k > k^3$

Consider $p(k+1)$

$2^{k+1} = 2.2^k > 2.k^3$ By our assumption

$$> \left(1 + \frac{1}{10}\right).k^3 \qquad ...(1)$$

As $$2 > \left(1 + \frac{1}{10}\right)^3$$

As $k \geq 10 \Rightarrow \dfrac{1}{10} \geq \dfrac{1}{k}$

$$1 + \frac{1}{10} \geq 1 + \frac{1}{k}$$

$$\Rightarrow \left(1 + \frac{1}{10}\right)^3 \geq \left(1 + \frac{1}{k}\right)^3$$

$$\Rightarrow \left(1 + \frac{1}{10}\right)^3 k^3 \geq \left(1 + \frac{1}{k}\right)^3 k^3$$

$\therefore$ (1) $$\Rightarrow 2^{k+1} \geq \left(1 + \frac{1}{k}\right)^3 k^3$$

$$= (k+1)^3$$

Thus $p(k+1)$ is true if $p(k)$ is true for $k \geq 10$.

Therefore by principle of Mathematical Induction the result is true.

Example 15.33. Use principle of mathematical induction to show that for any integer $n > 1$, $\frac{1}{\sqrt{1}} + \frac{1}{\sqrt{2}} + \ldots + \frac{1}{\sqrt{n}} > \sqrt{n}$.

Solution : Let $p(n)$ be the predicate

$$\frac{1}{\sqrt{1}} + \frac{1}{\sqrt{2}} + \ldots + \frac{1}{\sqrt{n}} > \sqrt{n}$$

Basic Step : $p(2)$

$$\frac{1}{\sqrt{1}} + \frac{1}{\sqrt{2}} > \sqrt{2}.$$

$$1 + \frac{\sqrt{2}}{2} > \sqrt{2}$$

Therefore $p(1)$ is true

Induction Step : Now let us assume that $p(k)$ is true for some $k \geq 2$. Then $p(k+1)$

$$\frac{1}{\sqrt{1}} + \frac{1}{\sqrt{2}} + \ldots + \frac{1}{\sqrt{k}} + \frac{1}{\sqrt{k+1}}$$

$$> \frac{\sqrt{k(k+1)} + 1}{\sqrt{k+1}} \qquad \text{[as } p(k) \text{ is true]}$$

As $\quad k + 1 > k \quad , \; k > 1$

$$\sqrt{k+1} > \sqrt{k}$$

$$\sqrt{k}\,\sqrt{k+1} > k$$

$$\therefore \quad \frac{1}{\sqrt{1}} + \frac{1}{\sqrt{2}} + \ldots + \frac{1}{\sqrt{k+1}} > \frac{k+1}{\sqrt{k+1}}$$

$$= \sqrt{k+1}$$

Hence $p(k+1)$ is true

therefore $p(n)$ is true $\forall n > 1$.

Example 15.34. Use mathematical induction to prove that

$$1 + \frac{1}{4} + \frac{1}{9} + \ldots + \frac{1}{n^2} \leq 2 - \frac{1}{n} \quad \forall n \in N.$$

Solution : Let $p(n)$ be the predicate

$$1 + \frac{1}{4} + \frac{1}{9} + \ldots + \frac{1}{n^2} \leq 2 - \frac{1}{n} \quad \forall n \in N.$$

Basic Step : $p(1)$

$$1 \leq 2 - \frac{1}{1} = 1$$

$\therefore$ $p(1)$ is true.

Induction Step : Let us assume that $p(k)$ is true for $k \in N$.

$$\therefore \quad 1 + \frac{1}{4} + \frac{1}{9} + \ldots + \frac{1}{k^2} \leq 2 - \frac{1}{k}$$

Consider $p(k+1)$

$$1+\frac{1}{4}+\frac{1}{9}+....+\frac{1}{k^2}+\frac{1}{(k+1)^2} \le \left(2-\frac{1}{k}\right)+\frac{1}{(k+1)^2} \qquad ...(i)$$

(As $p(k)$ is true)

Now $$\left(2-\frac{1}{k}\right)+\frac{1}{(k+1)^2}+\frac{1}{k+1}-\frac{1}{k+1}$$

$$= 2-\frac{1}{k+1}+\left[\frac{1}{(k+1)^2}+\frac{1}{k+1}-\frac{1}{k}\right]$$

$$= 2-\frac{1}{k+1}+\left[\frac{k+k(k+1)-(k+1)^2}{k(k+1)^2}\right]$$

$$= 2-\frac{1}{k+1}+\left(\frac{k^2+2k-k^2-2k-1}{k(k+1)^2}\right)$$

$$= 2-\frac{1}{k+1}-\frac{1}{k(k+1)^2}$$

$$= 2-\frac{1}{k+1}+\text{negative}$$

$$\le 2-\frac{1}{k+1}$$

Therefore, $1+\frac{1}{4}+\frac{1}{9}+...+\frac{1}{(k+1)^2} \le 2-\frac{1}{k+1}$.

Therefore $p(k+1)$ is true.

Thus by principle of mathematical induction $p(n)$ is true $\forall n \in N$.

15.5 VARIANTS OF MATHEMATICAL INDUCTION

In the mathematical induction we just studied, the constraint on n is positive integer. In fact, we can have variants:

- in which n is a non-negative integer;
- or, in which n is a positive integer $\ge m$

To deal with the above situations, all we need is:

- Adjust the basis step, so that it considers $n = 0$ or $n = m$ instead of $n = 1$.
- Adjust the inductive step, so that $P(k) \to P(k+1)$ is proved for all non-negative integer k or all integer $k \ge m$.

Example: 15.35. *Suppose that, for a finite set S, $|S| = n$. Show that each $|P(S)| = 2^n$. Here P(S) stands for power set of S.*

Solution: Note that we cannot consider $n = 1$ in the basis step!

- Because S could be the empty set and thus n could be 0. That means, we have to make sure the above statement is true for all *non-negative* integer n (not just all positive integer n).

- If we consider $n = 1$ in the basis step, then the entire proof ignores the possibility of $n = 0$.
- Similarly, when we do the inductive step, we cannot just prove it for all positive integer k. We should prove it for all non-negative integer k.

Thus

Basis step: When $n = 0$, S is the empty set. Hence, $P(S) = \{\Phi\}$, which means $|P(S)| = 1 = 2^0$.

Inductive step: Assume that, for all S such that $|S| = k$ (where k is a non-negative integer), $|P(S)| = 2^k$. Now we show that, for all S' such that $|S'| = k = 1$, $|P(S')| = 2^{k+1}$.

Clearly, all S' such that $|S'| = k + 1$ can be written as $S' = S \cup \{a\}$, where $|S| = k$ and a is not in S. To count $|P(S')|$, *i.e.*, the number of subsets of S', we only need to count: (*a*) $|P(S)|$, *i.e.*, the number of subsets of S; (*b*) the number of subsets of S' that contains a.

Example for Variant

By the inductive assumption, we know that $|P(S)| = 2^k$.

What is the number of subsets containing a? We note that each subset containing 'a' uniquely corresponds to a subset not containing '*a*' (by eliminating '*a*' from the subset). Hence, this number is also $|P(S)| = 2^k$.

We sum up these two numbers and get that $|P(S')| = 2^k = 2^{k+1}$.

15.6 MATHEMATICAL INDUCTION WITH NON-ZERO BASE CASE

In some problems, we are interested in showing some proposition is true for integers greater than or equal to same lower bound say b. In general, base case in induction does not have to be 0. If base case is same integer b induction proves proposition for $n = b, b + 1, b + 2$, and iduction does not say anything about $n = 0, 1, 2, \ldots b - 1$.

Thus to prove by induction that for all $n \geq b$ $P(n)$ holds the procedure is as follows:

(*a*) Base Case: Prove $P(b)$.

(*b*) Inductive Case: Assume inductive hyptothesis $P(k)$ for any $k \geq b$ and prove $P(k + 1)$.

or

Assume inductive hypothesis $P(k - 1)$ for $k > b$ and prove $P(k)$.

Example15.36 : *We have lots of Rs. 3 and Rs. 5 stamps. Show that any amount of money that is at least Rs. 8 can be made using Rs. 3 and Rs. 5 stamps.*

i.e., for all $n \geq 8$

$P(n)$ *: n can be made using Rs. 3 and Rs. 5 stamps.*

Solution : We need to prove that

$$P(n) = b \times 3 + c \times 5 \text{ for same } +ve \text{ integers } b \text{ and } c.$$

Base Case : $P(8)$ and $P(9)$ are true

$$8 = 1 \times 3 + 1 \times 5$$

$$9 = 3 \times 3 + 0 \times 5$$

Inductive Step : Let us assume that $P(n)$ is true for $n \geq 9$. Now we need to prove it for $(n + 1)$.

Case 1: When $n \geq 9$ and there is no stamp of Rs. 5 in sum n, there will be at least 3 stamps of Rs. 3. Take three of then out and put in two stamps of Rs. 5 and hence the sum add up to $(n + 1)$.

For *e.g.*, $\Rightarrow$ $9 = 3 \times 3$ here $b = 3$ and $c = 0$

and $9 + 10 = 0 \times 3 + 2 \times 5$

here $b = 0$ and $c = 2$

Case II: Let the sum n contains a single stamp of Rs. 5. Take out the stamp of Rs. 5 and add two stamps of Rs. 3 and hence the sum be comes $(n + 1)$.

For *e.g.*, $\Rightarrow$ $11 = 2 \times 3 + 1 \times 5$ here $b = 2$

and $11 + 1 = 4 \times 3 + 0 \times 5$ and $c = 1$

here $b = 4$ and $c = 0$

Hence $P(n)$ holds for all $n \geq 8$.

15.7 PRINCIPLE OF STRONG MATHEMATICAL INDUCTION

Now let us look at the situation where the principle of mathematical induction does not work. Consider the Fibonacci Sequence of numbers 1, 1, 2, 3, 5, 8, 13, ... Each term in the sequence, from the third term on, is obtained by adding the previous two terms. So, if a_n is the nth term, then $a_1 = 1, a_2 = 1$ and $a_n = a_{n-1} + a_{n-2}\ \forall\, n \geq 3$.

Suppose we want to show that $a_n < 2^n\ \forall\, n \in N$

Using the principle of mathematical induction.

We use the predicate $p(n) : a_n < 2^n$.

Basic step : $p(1)$ is true.

Inductive Step : Let us assume that $p(k)$ *i.e.*, $a_k < 2^k$ is true for $k \in N$. we need to prove that $p(k + 1)$ is true.

$p(k + 1)$:

$a_{k+1} < 2^{k+1}$

$a_k + a_{k-1} < 2^{k+1}$

We know by our assumption that $a_k < 2^k$. but we don't know about a_{k-1}. So we cannot apply the principle of mathematical induction. In such a situation, a stronger, more powerful version of the principle of mathematical induction is to apply is given as.

Principle of Strong Mathematical Induction

***p(n)* is true for all positive integer *n* if**

1. Basic step : The proposition *p*(1) is shown to be true.

2. Induction step : $[p(1) \wedge p(2) \wedge p(3) \wedge .... \wedge P(n)] \rightarrow p(n+1)$ is true for every positive integer *n*.

Let us now go back to the Fibonacei sequence.

To use the strong form of the principle of mathematical induction, We take $n = 1$.

We have seen that $p(1)$ is true.

We also need to see if $p(2)$ is true.

This is because we have to use the relation

$a_n = a_{n-1} + a_{n-2}$ which is valid for $n \geq 3$.

Now we know $p(1)$ and $p(2)$ both are true then we go to next step. In step 2. we assume that $p(n)$ is true for $1 \leq n \leq k, k \geq 0$.

i.e., $a_n < 2^n \qquad 1 \leq n \leq k$

$p(k+1)$:

$$a^{k+1} = a_k + a_{k-1} \quad \text{[By Fibonacei sequence]}$$
$$< 2^k + 2^{k-1} \quad \text{(by our assumption)}$$
$$= 2^{k-1}(2+1)$$
$$< 2^{k-1}.2^2$$
$$= 2^k + 1$$

$\therefore p(k+1)$ is true.

Therefore $p(n)$ is true for all n.

From the above example it is clear that in case of strong mathematical induction we ofter need to check Basic Step for more than one value of n.

Both the form of mathematical induction is actually equivalent. This is because that one can be obtained from other. So we can use either form of mathematical induction.

Example 15.37. Show that if *n* is an integer greater than 1, then *n* can be written as the product of primes.

Solution:

Let $p(n)$: If n is an integer greater than 1, then n is a product of primes.

Basic Step : $p(2)$: 2 is an integer greater than is 1 and 2 itself is a prime.

$p(3)$: 3 again it self is a prime & is greater than 1.

$p(4)$: $4 = 2 \times 2$ is the product of primes.

Induction Step :

Assume that $p(k)$ is true for all positive integers k with $k \leq n$.

We need to show that $p(n+1)$ is true under this assumption.

Case I. $(n+1)$ is a prime. Then $p(n+1)$ is true automatically.

Case II. $(n+1)$ is not a prime *i.e.*, $(n+1)$ is a composite and can be written as product of two positive integers. *i.e.*, $(n+1) = ab$ a and b are positive integer with. $2 \le a \le b < n+1$

By our hypothesis a and b can be written as product of primes *i.e.*,

$$a = p_1p_2 \ldots p_r \qquad p_i \text{ are prime} \qquad 1 \le i \le r$$

$$b = q_1q_2 \ldots q_s \qquad q_j \text{ are prime} \qquad 1 \le j \le s$$

$$\therefore \quad n+1 = ab$$

$$= (p_1p_2 \cdots p_r)(q_1q_2 \cdots q_s)$$

That is, $(n+1)$ can be written as product of prime.

Thus $p(n+1)$ is true.

Therefore, by principle of **strong mathematical induction** the result is true.

Example 15.38. To prove that $\sum_{i=1}^{n} \frac{1}{\sqrt{i}} \le 2\sqrt{n} - 1 \;\; \forall \quad n \in N.$

Which form of the principle of mathematical induction whould you use and why ? Also prove the inequality.

Solution : Here we will use the weak form of principle of mathematical induction as the assumption $p(k)$ is true is enough to prove that $p(k+1)$ is true. We don't need to use $p(1), p(2), \ldots p(K-1)$ also.

Basic Step : $p(1) = 1 \le 2 - 1 = 1$

$\therefore p(1)$ is true.

Induction Step : Let us assume that $p(k)$ is true for some $k \in N$

Then $\frac{1}{\sqrt{1}} + \frac{1}{\sqrt{2}} + \ldots + \frac{1}{\sqrt{k}} < 2\sqrt{k} - 1$

Consider,

$p(k+1)$?

$$\frac{1}{\sqrt{1}} + \frac{1}{\sqrt{2}} + \ldots + \frac{1}{\sqrt{k}} + \frac{1}{\sqrt{k+1}} < \left(2\sqrt{k} - 1\right) + \frac{1}{\sqrt{k+1}} < \left(2\sqrt{k} - 1\right) + \frac{1}{\sqrt{k+1}}$$

$$= \frac{2\sqrt{k}\sqrt{k+1} - \sqrt{k+1} + 1}{\sqrt{k+1}} = \frac{\sqrt{k}\left(2\sqrt{k+1} - \sqrt{k}\right)}{(k+1)}$$

As $\quad k+1 > k$

$$\Rightarrow \sqrt{k+1} > \sqrt{k}$$

$$\Rightarrow \frac{\sqrt{k}}{\sqrt{k+1}} < 1$$

$$\therefore \quad 1 + \frac{1}{\sqrt{2}} + \frac{1}{\sqrt{3}} + \ldots + \frac{1}{\sqrt{k+1}} < 2\sqrt{k+1} - \sqrt{k}$$

$$< 2\sqrt{k+1} - 1$$

$\therefore p(k+1)$ is true.

Therefore by principle of mathematical induction the result is true $\forall n$.

Example 15.39. Use mathematical induction to prove that the sum of the first *n* odd positive integers is n^2.

Solution : Let $P(n)$ denote the proposition that the sum of the first n odd positive integers is n^2. We must first complete the basic step; that is, we must show that $P(1)$ is true. Then we must carry out the inductive step; that is, we must show that $P(n + 1)$ is true when $P(n)$ is assumed to be true.

Basic step : P(1) states that the sum of the first one odd positive integers is 1^2. This is true as the sum of the first odd positive integer is 1.

Inductive Step : To complete the inductive step we must show that the proposition $P(n) \to P(n+1)$ is true for every positive integer n. To do this, suppose that $P(n)$ is true for a positive integer n; that is,

$$1 + 3 + 5 + ... + (2n - 1) = n^2$$

Note : That the nth odd positive integer is $(2n - 1)$, since this integer is obtained by adding 2 a total of $n - 1$ times to 1. We must show that $P(n + 1)$ is true, assuming that $P(n)$ is true. Note that $P(n + 1)$ is the statement that

$$1 + 3 + 5 + ... + (2n - 1) + (2n + 1) = (n + 1)^2$$

LHS $\Rightarrow 1 + 3 + 5 + ... + (2n - 1) + (2n + 1)$

$= n^2 + 2n + 1$ [By our hypothesis]

$= (n + 1)^2$

Therefore, the result is true for (n + 1) also. Hence, by principle of mathematical Induction the result is true.

SOLVED PROBLEMS

Problem 15.1: Give an argument using rules of inference to show that the conclusion follows from the hypothesis.

1. Hypotheses if Srishti can sing or kshitij can play, then I'll buy the compact disc. Srishti can sing. I'll buy the compact disc player. Conclusion : I'll buy the compact disc and the compact disc player.

Solution: p : Srishti can sing

q : Kshitij can play

r : I'll buy the compact disc

s : I'll buy the compact disc player

$$(p \vee q) \to r$$
$$p$$
$$s$$
$$\therefore (r \wedge s)$$

Explanation: S is a hypothesis, so it holds. $((p \vee q) \to r) \wedge p$ implies r by Modus pones.

2. Hypothesis : Ken, a member of the Titans, can hit the ball a longway. Everyone who can hit the ball a long way can make a lot of money. Conclusion : Some member of the Titans can make a lot of money.

Solution. p : Member of the Titan
q : Can hit the ball a long way
r : Make lot of money

$$p \to q$$
$$q \to r$$
$$\therefore \quad p \to r \qquad \text{By transitive law}$$

3. Hypothesis: Everyone in the discrete mathematics class loves proofs. Someone in the discrete mathematics class has never taken calculus. Conclusion: Someone who loves proofs has never taken calculus :

Solution: Let D be the discrete math class.

Let $P(x)$: x loves proofs.
$Q(x)$: x has never taken calculus

Then $\forall x, P(x) \quad (\forall x \in D$, that is $P(x))$
$\exists x, Q(x)$
$\exists x\,(Q(x) \wedge P(x))$,

so someone loves proofs and have never taken calculus.

$\exists x \in D, P(x)$ is proven.

Mathematical Induction

Problem 15.2. Prove that :

$$\frac{1}{2^2-1}+\frac{1}{3^2-1}+\ldots\ldots+\frac{1}{(n+1)^2-1}=\frac{3}{4}-\frac{1}{2(n+1)}-\frac{1}{2(n+2)}$$

i.e.,

$$\sum_{i=1}^{n}\frac{1}{(i+1)^2-1}=\frac{3}{4}-\frac{1}{2(n+1)}-\frac{1}{2(n+2)}$$

Solution: Basic step: $\displaystyle\sum_{i=1}^{1}\frac{1}{(i+1)^2-1}-\frac{1}{2^2-1}=\frac{1}{3}$

RHS. $\displaystyle\frac{3}{4}-\frac{1}{4}-\frac{1}{6}=\frac{1}{3}$

Thus the result in true for $n = 1$.

Inductive step : Let $k \in D$ be given. Assume $P(k)$ is true *i.e.,*

$$p(k)=\sum_{i=1}^{k}\frac{1}{(i+1)^2-1}=\frac{3}{4}-\frac{1}{2(k+1)}-\frac{1}{2(k+2)}$$

$$\therefore \quad p(k+1)=\sum_{i=1}^{k+1}\frac{1}{(i+1)^2-1}$$

$$=\sum_{i=1}^{k}\frac{1}{(i+1)^2-1}+\frac{1}{(k+2)^2-1}$$

$$=\frac{3}{4}-\frac{1}{2(k+1)}-\frac{1}{2(k+2)}+\frac{1}{(k+2)^2-1}$$

$$= \frac{3}{4} - \frac{1}{2(k+1)} - \frac{1}{2(k+2)} + \frac{1}{(k+1)(k+3)}$$

$$= \frac{3}{4} - \frac{1}{2(k+2)} - \frac{2-(k+3)}{2(k+1)(k+3)}$$

$$= \frac{3}{4} - \frac{1}{2(k+2)} + \frac{(-k-1)}{2(k+1)(k+3)}$$

$$= \frac{3}{4} - \frac{1}{2(k+2)} + \frac{1}{2(k+3)}$$

As $k \in D$ was arbitrary, the inductive step, $p(k) \to p(k+1)$, holds and along with the base case proves $\forall n \in D, p(n)$.

Problem 15.3. Prove by principle of mathematical induction

$$1^2 - 2^2 + 3^2 - ......+(-1)^{n+1} n^2 = \frac{(-1)^{n+1} n(n+1)}{2}$$

Solution: Let $p(n) = 1^2 - 2^2 + 3^2 + ...+(-1)^{n+1} n^2 = \frac{(-1)^{n+1} n(n+1)}{2}$

$$= \sum_{i=1}^{n} (-1)^{i+1} i^2 = \frac{(-1)^{n+1} n(n+1)}{2}$$

Basic step: $$p(1) = \sum_{i=1}^{1} (-1)^{i+1} i^2 = (-1)^2 i^2 = 1$$

RHS. $$\frac{(-1)^2 \, 1(1+1)}{2} = 1$$

Inductive step: Let us assume that $p(k)$ is true, that is

$$\sum_{i=1}^{k} (-1)^{i+1} i^2 = \frac{(-1)^{k+1} k(k+1)}{2}$$

We will show that $p(k+1)$ is true.

$$\sum_{i=1}^{k+1} (-1)^{i+1} i^2 = \sum_{i=1}^{k} (-1)^{i+1} i^2 + (-1)^{k+2} (k+1)^2$$

$$= \frac{(-1)^{k+1} k(k+1)}{2} + (-1)^{k+2} (k+1)^2$$

$$= (-1)^{k+1} (k+1) \left[\frac{k}{2} - (k+1)\right]$$

$$= (-1)^{k+1} (k+1) \left[\frac{-k-2}{2}\right]$$

$$= \frac{(-1)^{k+2}(k+1)(k+2)}{2}$$

As $k \in D$ is arbitrary, the inductive step is proven for all k and along with the basic case the result is true for all k.

Problem 15.4. Use mathematical induction to show that if $E_1, E_2, \ldots, E_n$ are events, then

$$P(E_1 \cup E_2 \cup \ldots \cup E_n) \leq \sum_{i=1}^{n} P(E_i)$$

Solution : Basic step for $n = 1$. The result is true.

Inductive step. Let the result is true for $n = k$, *i.e.*,

$$P(E_1 \cup E_2 \cup \ldots \cup E_k) \leq \sum_{i=1}^{k} P(E_i)$$

Then $\quad P(E_1 \cup E_2 \cup \ldots \cup E_k \cup E_{k+1})$

$$= P(E_1 \cup \ldots \cup E_k) + P(E_{k+1}) - P((E_1 \cup E_2 \ldots \cup E_k) \cap E_{k+1})$$

$$\leq \sum_{i=1}^{k} P(E_i) + P(E_{k+1}) - P(E_1 \cup \ldots E_k) \cap E_{k+1}) \leq \sum_{i=1}^{k+1} P(E_i)$$

As k is arbitrary thus the result is true for all n.

Problem 15.5. Find the sum

$$\mathbf{1.2 + 2.3 + \ldots + (n-1)n}$$

Solution : The sum $1.2 + 2.3 + \ldots + (n-1).n$

$$= 2\left[\sum_{i=2}^{n} C(i,2)\right]$$

Since each term is $C(i, 2) . 2$ and $C(i,2) = \dfrac{i(i-1)}{2}$

So using $\quad \displaystyle\sum_{1=k}^{n} C(i,k) = C(n+1, k+1)$

We have $\quad \displaystyle\sum_{i=2}^{n} C(i,2) = C(n+1, 3)$

So the required sum $= 2\,C(n+1, 3)$

$$= 2\,\frac{(n+1)(n)(n-1)}{3!}$$

$$= \frac{n(n+1)(n-1)}{3}$$

EXERCISE 1

Convert each of the following arguments into the language of symbols and check their validity also.

1. Either the moon is cool or oxygen is a metal.

The moon is cool.

Therefore, oxygen is metal.

2. Either Srishti will play or Ribhav will practise karate.

If Ribhave practises Karate, then Srishti studies.

Srishti does not study.

Therefore, Ribhav will practise Karte.

3. If Madhu is a 'President', she will be Head Girls in the School. If Madhu is Head Girl in the school; she wiil decide for annual function in the school.

Therefore, if Madhu is a 'President', she will decide for annual functioin in the school.

4. Construct an argument using rules of inference to show that the hypotheses Raj works hard," "If Raj works hard, then he is a dull boy," and "If Raj is a dull boy, then he will not get the job" imply the conclusion "Raj will not get the job."

5. What rules of inference are used in the following famous argument? "All men are mortal. Socrates is a man. Therefore, Socrates is mortal."

6. Show that $\sqrt{5}$ is irrational.

7. What rules of inference are used in the following argument? "No man is an island. Manhattan is an island. Therefore, Manhattan is not a man."

8. Write down the contrapositive of the statement "If f is a 1 – 1 function from a finite set X into itself, then f must be surjective".

9. Prove that the square of an even number is an even number using

(*a*) a direct proof

(*b*) an indirect proof

(*c*) a proof by contradiction

10. Prove that if n is integer and $5n + 4$ is even, then n is even using.

(*a*) an indirect proof

(*b*) a proof by contradiction

11. Prove that the sum of two odd integers is even

12. Prove that the sum of two rational numbers is rational.

13. Prove that if x and y are real numbers, then $\max(x, y) + \min(x, y) = x + y$.

(**Hint :** Use a proof by cases, with the two cases corresponding to $x \geq y$ and $x < y$, respectively).

14. Use a proof by cases to show that min (a, min (b, c)) = min (min(a, b), c) whenever a, b and c are real numbers.

15. Prove or disprove that if m and n are integers such that $mn = 1$, then either $m = 1$ and $n = 1$, or else $m = -1$ and $n = -1$.

16. Prove that $m^2 = n^2$ if and only if $m = n$ or $m = -n$.

EXERCISE 2

1. Use mathematical induction to prove that

$$1 + \frac{1}{4} + \frac{1}{9} + \dots + \frac{1}{n^2} \leq 2 - \frac{1}{n} \quad \forall n \in N$$

2. Use induction to prove that any integer $n \geq 2$ is either a prime or a product of primes.

3. Use the method of proof by contradiction to show that

for $x \in R$, if $x^3 + 4x = 0$ then $x = 0$

4. Show that $1^3 + 2^3 + \dots + n^3 = [n(n+1)/2]^2$ whenever n is a positive integer.

5. Prove that $1^2 + 3^2 + 5^2 + \dots + (2n + 1)^2 = (n + 1)(2n + 1)(2n + 3)/3$ whenever n is a non-negative integar.

6. Prove that $1 \cdot 1! + 2 \cdot 2! + \dots + n \cdot n! = (n + 1)! - 1$ whenever n is a positive integar.

7. Prove that $3n < n!$ whenever n is a positive integer greater than 6.

8. Show that $2n > n^2$ whenever n is an integer greater than 4.

9. Prove using mathematical induction that

$$1 \cdot 2 + 2 \cdot 3 + \dots\dots + n(n + 1) = n(n + 1)(n + 2)/3$$

whenever n is a positive integer.

10. Use mathematical induction to show that 3 divides $n^3 + 2n$ whenever n is a non-negative integer.

11. Use mathematical induction to show that 5 divides $n^5 - n$ whenever n is a non-negative integer.

12. Use mathematical induction to prove that if $A_1, A_2, \dots, A_n$ are subsets of a universal set U, then

$$\overline{\bigcup_{k=1}^{n} A_k} \subseteq \bigcap_{k=1}^{n} \bar{A}_k$$

13. Prove that if $A_1, A_2, \dots\dots, A_n$ and $B_1, B_2, \dots, B_n$ are sets such that $A_k \subseteq B_k$ for $k = 1, 2, \dots, n$, then

(a) $$\bigcup_{k=1}^{n} A_k \subseteq \bigcup_{k=1}^{n} B_k$$

(b) $$\bigcap_{k=1}^{n} A_k \subseteq \bigcap_{k=1}^{n} B_k$$

14. Prove that

$$1 + \frac{1}{\sqrt{2}} + \frac{1}{\sqrt{3}} + \dots + \frac{1}{\sqrt{n}} > 2\left(\sqrt{n+1} - 1\right)$$

15. Prove by induction :

$1 + 2^3 + \dots + n^3 = [n(n + 1)/2]^2$ for $n \geq 1$

16. Prove by induction :

$1/(1 \cdot 2) + 1/(2 \cdot 3) + \dots + 1/[n(n + 1)] = n/(n + 1)$ for $n \leq 1$

17. Prove by induction :

$$n^2 > n + 1 \text{ for } n \leq 2$$

18. Prove by induction for $n \leq 1$:

$1 + 3^2 + 5^2 + ... + (2n + 1)^2 = (n + 1)(2n + 1)(2n + 3)/3$

19. (*a*) Prove by induction that $2n + 1 \leq 2^n$ for $n \leq 3$.

(*b*) Use the results of part (*a*) to prove by induction that $n^2 \geq 2^n$ for $n \leq 4$.

20. Let x be a real number such that $x \leq 1 + nx$.

$$(1 + x)^n \leq 1 + nx.$$

21. For $n \leq 1$, prove that $(x - y)$ is a factor of $(x^n - y^n)$. Hint : Add $0 = x^n y - x^n y$ to the term x^{n+1} and factor.

22. Prove by induction that, for any collection of n sets, $A_1, A_2, ..., A_n$, we have

$$(A_1 \cap A_2 \cap ... \cap A_n)' = A_1' \cup A_2' \cup ... \cup A'_n.$$

23. Applying the principle of mathematical induction, show that

(*i*) $1 + a + a^2 + ... + a^n = \dfrac{1 - a^{n+1}}{1 - a} \quad (n \neq 1)$

(*ii*) $1 + 2 + 3 + ... + n = \dfrac{n(n + 1)}{2}$

(*iii*) $1 + 2 + 2^2 + 2^3 + ... + 2^n n = 2^{n+1} - 1$

(*iv*) $2 + 2^2 + 2^3 + ... + 2^n = 2^{n+1} - 2$

(*v*) $1^2 + 3^2 + 5^2 + ... + (2^n - 1)^2 = \dfrac{n(2n - 1)(2n + 1)}{3}$

(*vi*) $2^n < n! \qquad (n \geq 4)$

(*vii*) $1.1! + 2.2! + 3.3! + ... n.n! = (n + 1)! - 1$, where (factorial)

$n! = 1.2.3. .... n.$

(*viii*) $1.2 + 2.3 + ... + n(n + 1) = \dfrac{n(n + 1)(n + 2)}{3}$

(*ix*) $1^3 + 2^3 + ... + n^3 = (1 + 2 + 3 + ... + n)^2$.

(*x*) $\dfrac{1^2}{1.3} + \dfrac{2^2}{3.5} + ... + \dfrac{n^2}{(2n - 1)(2n + 1)} = \dfrac{n(n + 1)}{2(2n + 1)}$

EXERCISE 3

1. (*a*) Show that the following proposition is a tautology

(*i*) $[(P \Rightarrow q) \wedge (q \Rightarrow r)] \Rightarrow (P \Rightarrow r)$

(*ii*) $\sim(P \vee q) \vee [(\sim P) \wedge q] \vee P$

(*b*) Establish the equivalence

$P \Rightarrow q \vee r \equiv P \wedge \sim P \Rightarrow r \equiv P \wedge \sim r \Rightarrow q$

Hence rewrite the following sentence in two different ways.

If n is prime; then n is odd or n is 2.

(*c*) Using laws and algebra of propositions, show that

(*i*) $(P \wedge q) \vee (P \wedge \sim q) \equiv P$

(*ii*) $(P \Rightarrow q) \wedge (r \Rightarrow q) \equiv (P \vee r) \Rightarrow q$

(*iii*) $\sim (P \vee (\sim P \wedge q)) \equiv \sim P \wedge q$

2. (*a*) Check the validity of the following argument.

   If I try hard and I have a talent, then I will become a scientist.

   If I become scientist then I will be happy. Therefore, if I will not try hard or I do not have talent.

   (*b*) Show that each of the following inference is fallacy.

   (*i*) If today is Anita's birthday, then today is February 16. Today is Feb. 16. Hence today is Anita's birthday.

   (*ii*) If the client is guilty, then he was at the scene of the crime. The client was at the scene of the crime. Hence the client is guilty.

   (*c*) Check the validity of the following arguments

   If P and q are true and r is false, find the truth values of the following

   (*i*) $(P \wedge \sim q) \Rightarrow r$ (*ii*) $(P \Leftrightarrow r) \Rightarrow r$

   (*iii*) $(P \vee q) \Leftrightarrow (P \Rightarrow \sim r)$

3. Show that the proposition $P \wedge - q$ and $(P \cup q) \wedge (-p) \wedge (-q)$ are contradiction.

4. Show that t is valid conclusion from the given promises

   $- p \wedge q, r \Rightarrow p, - r \Rightarrow s$ and $s \Rightarrow t$

5. (*a*) Define min terms and max terms obtained the disjunctive normal form (join of min terms) of the following boolean expression

   $(x_1 \wedge x_2^c) \vee (x_2 \wedge x_3^c) \vee (x_3 \wedge x_1^c)$

   where superscript '*c*' represents complement.

   (*b*) Determine the validity of the following argument with explanation :

   "If wages increase, then there will be inflection. The cost of living will not increase if there is no inflation wages will increase, therefore, the cost of living will not increase."

   (*c*) Without truth table prove that following is a tautology $P \wedge (P \vee Q) \rightarrow Q$.

6. Prove that following are all equivalence

   (*a*) $(p \wedge q) \vee (\sim P \wedge q) \Leftrightarrow q$

   (*b*) $p \rightarrow q \Leftrightarrow \sim q \rightarrow \sim p$

   (*c*) $p \vee q \Leftrightarrow q \vee p$

7. What common relation on Z are the transitive closures of the following relations:

   (*a*) aSb if and only if $a + 1 = a$

   (*b*) aRb if and only if $|a - b| = 2$

8. (*a*) Show that

   $((P \vee Q) \wedge \sim(\sim P \wedge (\sim Q \vee \sim R))) \vee (\sim P \wedge \sim Q) \text{ v } (\sim P \wedge \sim R)$ is a Tautology.

   (*b*) Prove that $(P \rightarrow Q) \Leftrightarrow (\sim P \vee Q)$.

9. Show that the truth values of following formulas are independent of their components.

   (*a*) $(P \wedge (P \rightarrow Q)) \rightarrow Q$

   (*b*) Check whether this statement $P \rightarrow Q, \sim P \mid\!\!-\!\!- \sim Q$ is fallacy of a valid argument.

ANSWERS

EXERCISE 1

1. p : The moon is cool

   q : Oxygen is a metal

   $p \vee q$

   p

   $\therefore \quad q$

   Argument is $(p \vee q) \wedge p \Rightarrow q$

   The argument is true if $(p \vee q.) \wedge p \rightarrow q$ is a tautology.

p	q	$p \vee q$	$(p \vee q) \wedge p$	$(p \vee q) \wedge p \rightarrow q$
T	T	T	T	T
T	F	T	T	F
F	T	T	F	T
F	F	F	F	T

   As the last column is not a tautology, Therefore it is not a valid argument.

2. p : Srishti will play

   q : Ribhav will practise Karate

   r : Srishti Studies

   $p \quad \vee q$

   $p \quad \rightarrow r$

   $\sim r$

   $\therefore \quad q$

   The given argument is $(p \vee q) \wedge (q \rightarrow r) \wedge (\sim r) \Rightarrow q$.

	p	q	r	$p \vee q$	$q \rightarrow r$	$\sim r$ $\wedge (\sim r)$	$(p \vee q) \wedge (q \rightarrow r)$ $\wedge (\sim r) \rightarrow q$	$(p \vee q) \wedge (q \rightarrow r)$
T	T	T	T	T	F	F	T	
T	T	F	T	F	T	F	T	
T	F	T	T	T	F	F	T	
T	F	F	T	T	T	T	T	F
F	T	T	T	T	F	F	T	
F	T	F	T	F	T	F	T	
F	F	T	F	T	F	F	T	
F	F	F	F	T	T	F	T	

   Not a valid argument.

3. (a) p : Kangaroos live in Australia.

   q : Kangaroos are Marsupials.

   $p \wedge q$

   $\therefore \quad p$

Argument : $(p \wedge q) \rightarrow p$

The Rule of Inference is **"SIMLIFICATION"**

(*b*) p : It is hotter than 100 degrees today.

q : The pollution is dangerous.

$p \vee q$

$\sim p$

$\therefore \quad q$

Argument : $(p \vee q) \wedge (\sim p) \rightarrow q$

The rule of Inference is **"DISJUNCTIVE SYLLOGISM".**

(*c*) p : Lisa is an excellent swimmer.

q : Lisa can work as a lifeguard.

p

$p \rightarrow q$

$\therefore \quad q$

Argument : $[p \wedge (p \rightarrow q)] \rightarrow q$

The rule of Inference is **"MODUS PONES".**

(*d*) p : Rahul will work at a computer company this summer.

q : Rahul will be a beach bum.

p

$\therefore \quad p \vee q$

Argument : $p \rightarrow p \vee q$

The rule of Inference is **"ADDITION".**

(*e*) p : Raj work all night on this homework.

q : I can answer all the exercise

r : I will get better understanding

$p \rightarrow q$

$q \rightarrow r$

$\therefore \quad p \rightarrow r$

Argument : $(p \rightarrow q) \wedge (q \rightarrow r) \Rightarrow (p \rightarrow r)$

The rule of Inference is **"HYPOTHETICAL SYLLOGISM".**

4. p : Madhu is a 'President'.

q : Madhu will be Head Girl in the School.

r : Madhu will decide for annual function in the school

$p \rightarrow q$

$q \rightarrow r$

$\therefore \quad p \rightarrow r$

Here Argument is $(p \rightarrow q) \wedge (q \rightarrow r) \Rightarrow (p \rightarrow r)$

By the rule of **"Hypothetical Syllogism"** the given argument is a valid argument.

5. p : Raj works hard.

q : Raj is a dull boy.

r : Raj will not get the job.

p

$p \to q$

$q \to r$

$\therefore \quad r$

Here Argument is $p \wedge (p \to q) \wedge (q \to r) \Rightarrow r$

p	q	r	$p \to q$	$q \to r$	$p \wedge (p \to q) \wedge (q \to r)$	$p \wedge (p \to q) \wedge (q \to r) \to r$
T	T	T	T	T	T	T
T	T	F	T	F	F	T
T	F	T	F	T	F	T
T	F	F	F	T	F	T
F	T	T	T	T	F	T
F	T	F	T	F	F	T
F	F	T	T	T	F	T
F	F	F	T	T	F	T

Valid argument.

6. p : Socrate

q : Man

r : Mortal

$p \to q$: socrate is a man.

$q \to r$: All men are mortal

$p \to r$: Socrates is mortal

Argument : $(p \to q) \wedge (q \to r) \Rightarrow p \to r$

Therefore, by Rule "**Hypothetical Syllogism**", the argument is valid.

■■■

Chapter

16 FINITE STATE MACHINES

16.1 INTRODUCTION

A Digital computer may be viewed as a machine which has the following properties – At each step in time, machine is in a certain "internal state", machine reads some "input", machine prints some "output" which depends only on its internal state and the input and then, possibly, machine changes its internal state. There are various ways to formalize the notion of a machine and there exists a hierarchy of such machines.

In this chapter we define strings, alphabets, operations on strings, the concept of finite state machines and the language and grammer accepted by these finite state machines.

16.2 ALPHABETS

Alphabet is a finite non-empty set of characters or symbols usually denoted by Σ or A and the elements of Σ are called letters.

For example $\Sigma = \{0, 1\}$ is a set of binary alphabets

$\Sigma = \{a, b, c\}$ is a set of alphabets of all lower case english letters

16.2.1 String Over Alphabets

A string or word say w on a set of alphabet Σ is a finite sequence of its elements.

For example w_1 = aabba and w_2 = bbaba are strings on $\Sigma = \{a, b\}$.

The empty sequence of letters is also considered as a string on Σ, called the empty string. An empty string is usually denoted by λ or $\in$.

The set of all strings on Σ(or A) is denoted by Σ^* (or A^*).

16.2.2 Length of String

The length of a string w is denoted as $|w|$ and defined as "the number of positions for characters that it cartains".

For example if w =010011 over $\Sigma = \{0, 1\}$

Then $|w| = 6$

Note : The length of empty string is zero. That is $|\lambda|$ or $|\Sigma| = 0$. if λ or Σ denotes the empty string.

16.2.3 Reverse of a String

Consider a string w over Σ. Then the reverse of w can be obtained by writting the elements of w in reverse order and is denoted as w^R.

For examples (*i*) if w = *ababab* over $\Sigma = \{a,b\}$ then w^R = *bababa* over $\Sigma = \{a, b\}$.

(*ii*) w = 100110 then w^R = 011001 over $\Sigma = \{0, 1\}$

16.2.4 Concatenation of Strings

Consider two string w_1 and w_2 on alphabet Σ (or A). The concatenation of w_1 and w_2 written as, $w_1 w_2$ is the string obtained by writing down the letters of string w_1 followed by letters of string w_2.

For examples (*i*) if $u = ababb$ and $v = bbaa$ over $\Sigma = \{a, b\}$ then $uv = ababbbbaa$ is the concatentation of u and v over $\Sigma = \{a, b\}$.

(*ii*) $w_1 = 011001$ and $w_2 = 11001$ over $\Sigma = \{0, 1\}$. Then $w_1 w_2 = 01100111001$ and $w_2 w_1 = 11001011001$ Clearly $w_1 w_2 \neq w_2 w_1$.

16.2.5 Below are some specific properties of string concatenation

1. Concatenation on *a* set Σ^*, is associative since each u, v, w in Σ^*, $u(vw) = (uv)w$

   For Example let $\Sigma = \{0, 1\}$ the $\Sigma^* = \{0, 1\}^*$ represents all strings of 0 and 1 including the null string.

   let $u = 0101 \in \Sigma^*$

   $v = 11001 \in \Sigma$

   and $w = 0011 \in \Sigma^*$

Then $u(vw) = 0101(110010011) = 0101110010011$ and $(uv)w = (010111001)\,0011 = 0101110010011$ clearly $u(vw) = (uv)w$.

2. The set Σ^* has an identity element as null string λ or $\in$ with respect to binary operation of concatenation as :

   $w\in = \in w = w$ for all $w \in \Sigma^*$

   For example : $\Sigma^* = \{0, 1\}^*$

   $w = 11001$

   Then $w\in = \in w = \in 11001 = 11001 = w$

3. Σ^*has left and right cancelleation. For u, v, w,. *in* Σ^*

   $u\,v = uw \Rightarrow v = w$ (left cancellation)

   and $vu = wu \Rightarrow v = w$ (right canecllation)

4. For $u, v \in \Sigma^*$, we have

   $|vu| = |u| + |v|$

   Where $|u|$, $|v|$ and $|uv|$ represents the length of strings u, v and uv respectively.

16.3 SOME MORE OPERATIONS ON STRINGS

In this section we define some more operations on strings over alphabet set Σ.

16.3.1 Transpose Operation

For any string $w \in \Sigma^*$ and an element say $a \in \Sigma$

$$(wa)^T = a(w)^T$$

For example $(01100\underline{1})^T$, is $\underline{1}00110$ where $w = 01100 \in \Sigma^*$ and $1 \in \Sigma = \{0, 1\}$.

16.3.2 Palindrome

A string which remains the same whether written forward or backward is called a *palindrome.*

For example (*i*) Malayalam is a palindrome.

(*ii*) 110010011 is a palindrome over $\Sigma = \{0, 1\}$. Clearly, an even length polindrome can be obtained by concatenation of a string and its transpose.

For example if $w = abab$ then $w^T = baba$ and $w.w^T = ababbaba$ is a palindrome of even length.

16.3.2.1 Definition

(*a*) $\in$ is a palindrome (*b*) If a is any symbol, the string a is a palindrome (*c*) If a is any symobl and x is a palindrome, then axa is a palindrome (*d*) Nothing is a palindrome unless it follows from (*a*) – (*c*).

16.3.3 Prefix and Suffix of String

A prefix of a string is a substring of leading symbols of that string. That is u is a prefix of a string w if there exists w' in Σ^* such that $w = u.\ w'$. Then we write $u < w$.

For example, the string *abc* has four prefixes as $\in$, *a*, *ab*, and *abc*.

A suffix of a string is a substring of trailing symbols of that string. That is u is a suffix of w if there exists $w' \in \Sigma^*$ *such that*

$$w = w'u$$

For example, the string *abc* has four suffixes, as $\in$, *c*, *bc* and *abc*.

16.3.4 Levi s Theorem

Let u, v, w and $x \in \Sigma^*$ and let $uv = wx$. Then

(*a*) There exists a unique string $y \in \Sigma^*$ such that $y = wy$ and $x = yv$ if $|u| > |w|$

(*b*) $u = w, x = v$ That is $y = \in$ if $|u| = |w|$

(*c*) There exists a unique string $z \in \Sigma^*$ such that $w = uz$ and $v = zx$ if $|u| < |w|$

Proof : The proof of this theorem is given with the help of diagrams

Case A

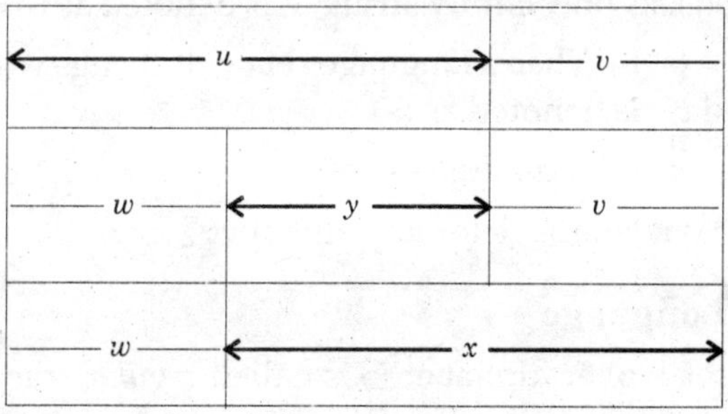

$|u| > |w|;\ u = wy$ and $x = v$

Case B

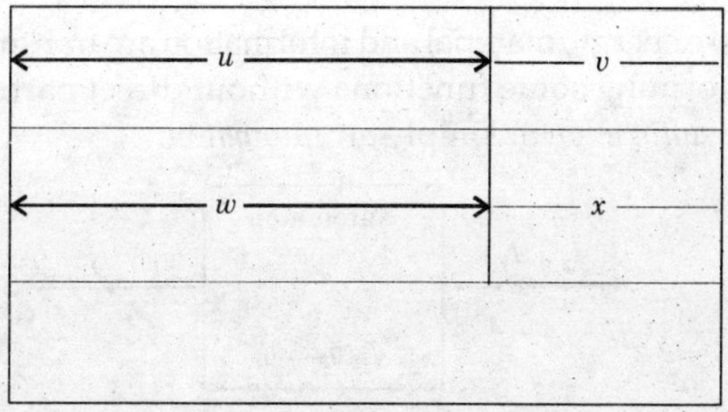

$|u| = |w|;\ u = w$ and $x = v$

Case C

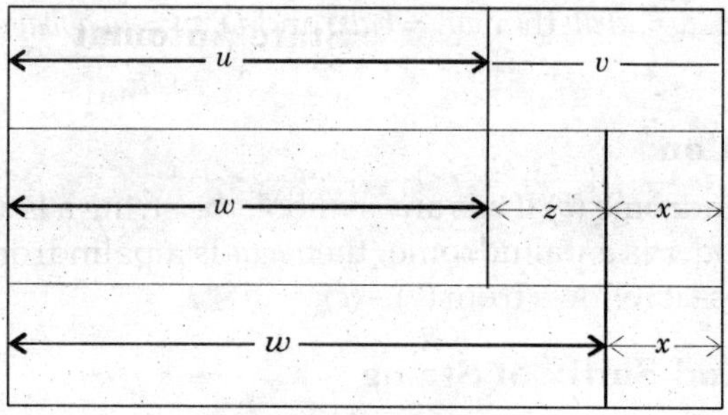

$|u|<|w|$; $w = iz$ and $v = zx$

16.4 POWERS OF AN ALPHABET

Let Σ be the set of alphabet Then Σ^n is the set of all strings of length *n* over alphabet set Σ.

For example (*i*) If $\Sigma = \{0, 1\}$, then $\Sigma^2 = \{00, 11, 01, 10\}$ represents the set of all strings of length 2 over = Σ {0, 1}

(*ii*) $\Sigma^3 = \{000, 111, 010, 101, \text{--}\}$ represents the set of all strings of length 3 over alphabet set $\Sigma = \{0, 1\}$

16.5 LANGUAGES

Let Σ be a set of alphabet and Σ^* is the set of all strings over Σ. Then a subset *L* of Σ^* is called *language* over Σ.

Note : (*i*) If L = ϕ, then language set is called *empty language.* That is there is no string in *L*.

(*ii*) If *L* consists only empty string, it is denoted as $L = \{\in\}$

(*iii*) Let $\Sigma = \{a, b\}$. Then a language set of all strings with equal number of *a's* and *b's* is denoted as :

$L = \{\in, ab, aabb, ba, abab, \text{----}\}$

(*iv*) $L = \Sigma^*$ is a language for any alphabet Σ.

16.5.1 Regular Language

A language set *L* over alphabet Σ is called *regular language* if some finite machine (finite automaton) recognize it.

16.6 FINITE STATE AUTOMATA

A system where energy, material and information are transmitted, transformed and used for performing some functions without direct participation of man is called a *finite state automaton* or simply an *automaton.*

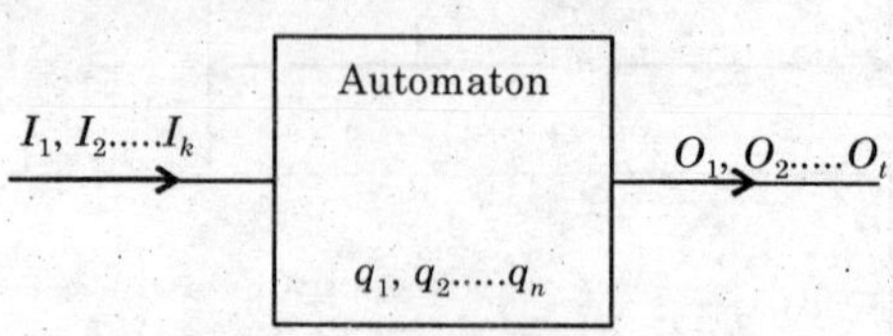

Fig. 16.1. An Automaton

16.6.1 Characteristics of a finite State Automaton

The five characteristics of an automaton are defind as :

(i) **Input :** The values $I_1, I_2, ...,$ applied to the input side of the automaton takes a finite number of fixed values from the input alphabet Σ.

(ii) **Output :** Each of O_1 O_2, ... take finite numbers of fixed values from an output $O, O_1, O_2 ... O_t$ are the output of the automaton.

(iii) **States :** At any instance of time the automaton can be in one of the states q_1, q_2, ... q_n.

(iv) **State relation :** At any instance, the next state of an automaton is determined by the help of present or current state and the present input to the automaton.

(v) **Output relation :** As we know that at any instance automaton is in same state. The output of an automaton depends either on state or on both the input and the state.

An automaton in which output depends only on state is called an automaton without memory where as when output depends on both state and input, automaton is called with memory.

Do You Know
On reding an input symbol, the automaton moves to a next state which is given by the state relation.

16.6.2 Formal Description of an Automata

A Deterministic (DFA) finite state automaton or simply an automaton say M can be represented by a 5–tuple $(Q, \Sigma, \delta, q_0, F)$ where

(i) Q is a finite conempty set of states, *i.e.*; $Q = (q_0, q_1, ... q_n)$

Where $q_0, q_1, ... q_n$ are the *n*- different state of the automaton.

(ii) Σ is a finite non-empty set of input symbols called input alphabet

(iii) δ is a transition function defined from $Q \times \sum \xrightarrow{\text{in to}} Q$. The change of states during the transition is given by this function δ. Usually the transition function is represented in a tabular form called transition table or as a digraph called transition diagram.

(iv) $q_0 \in Q$ is the initial state of the automaton

(v) $F \subseteq Q$ is a non-empty finite set of final states.

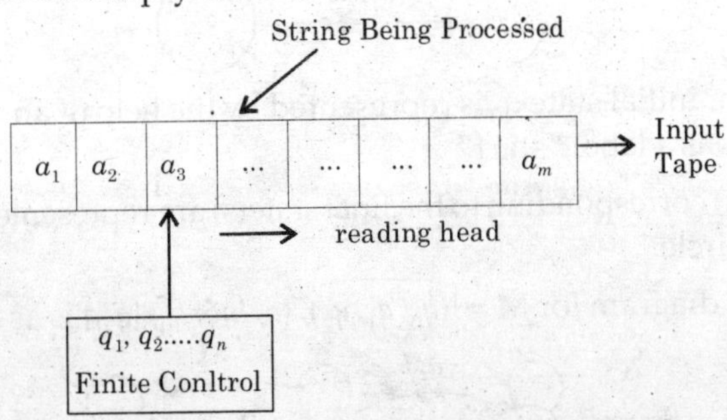

Fig. 16.2. B lock diagram of Fiuite Automata.

Note: In DFA, there is a unique state change with respect to a single input symbol.

16.6.3 The Major Compneuts of Finite Automaton are as Follows

(*i*) **Input Tape :** An input tape is divided into number of squares, each square contains a single input symbol from the input alphabet Σ. The end squares of an input tape contains markers. The left...end square contains and end–marker ¢ and the right ... end square contains an end–marker \$. The left-to-right sequence of symbols between the end-markers is the input string to be processed. An input tape is said to be of infinite length if it does not contains end-markers.

(*ii*) **Reading Head :** A pointer which moves from left-to-right -(or right-to-left) and reads (or examines) only one square of input tape at a time is called the reading head. Intially, it is placed at the left most square of the tape and the moves towards the right side only.

(*iii*) **Finite Control :** Finite control is the main part of an automaton. It contains all the transition states of set Q. The input to the finite control is usually a symbol from input tape which is under the reding head or present state of the automaton M. Output will be the movement of reading head along the tape to the next square. The next state of the automaton is given by $\delta(q, a)$ where $q \in Q$ and $a \in \Sigma$.

16.7 TRANSITION FUNCTION

There are two ways to represent. The transition function δ of the finite automation.

16.7.1 Transition Diagrams

The transition diagram for δ in automaton $M = (Q, \Sigma, \delta, q_0, F)$ is a directed graph defined as follows :

(a)There is a node with respect to each state $q \in Q$.

(b) For each state $q \in Q$ and each input symbol $a \in \Sigma$, let $(q, a) = q_1$. Then the transition diagram has a directed edge from node (or vetex) q to node (or vertex) q_1 with label 'a'. If there are several input symbols that cause transition from q to q_1 then there will be only one edge (directed) from q to q_1 labeled by the list of all those input symbols.

For example if $\delta(q_1, a) = q_2$ and $\delta(q_1, b) = q_2$ Then the the transition diagram will be

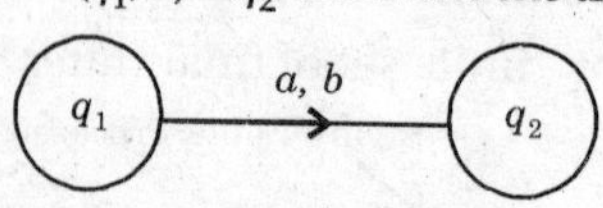

(c) The node for initial state q_0 is represented by the help of an arrow, pointing fowards it and it has a label "start"

(d) The node (s) corresponding to the final state(s) are represented by concentric circles or double circle.

The transition diagram for $M = \{(q_0, q_1, q_2), (a, b)\delta, q_0, \{q_2\}\}$ is as follwos.

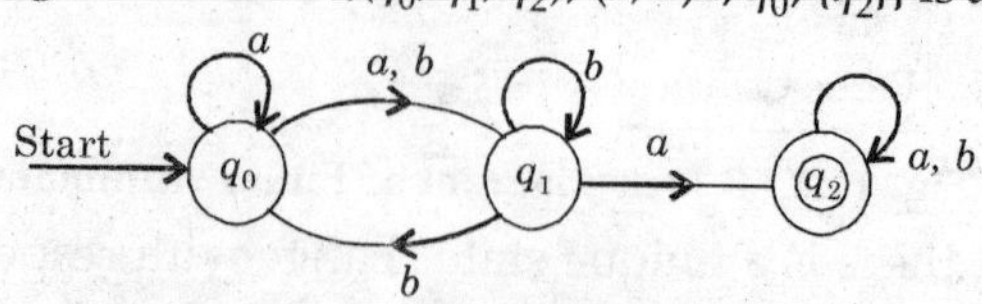

Fig. 16.3. Transition Diagram δ for M

16.7.2 Transition Tables

A transition table is a tabular represention of functions like δ that takes two arguments and returns a value. In the transition table, the rows corresponds to states and columns corresponds to input symbols. The intial state is represented by an arrow pointing towards it and final state is represented in a circle.

For example the transition table of machine M is as follows,

$M = (\{q_0, q_1, q_2\}\{a, b\}\ \delta, q_0, \{q_2\})$

δ	a	b
→ q_0	q_2	q_0
q_1	q_1	q_2
(q_2)	q_0	q_1

Fig. 16.4. Transition table for δ.

16.7.3 Properties of Transition Function

Property (1) : $\delta(q, \in) = q$. That is, if we are in state q and read no inputs, then we are still in state q.

Property (2) : For all string x and an input symbol a

$$\delta(q, x\,a) = \delta(\delta(q, x), a)$$

and

$$\delta(q, ax) = \delta(\delta(q, a), x)$$

To compute $\delta(q, xa)$ first compute $\delta(q, x)$ the state in which thet the automaton is after processing string x. Suppose it is q'. Then $\delta(\delta(q,x),a)$ is what we get by making a transition from state q' on input a. That is

$$\delta(q, xa) = \delta(q', a).$$

16.8 HOW AN AUTOMATON PROCESSES STRINGS

The "language" of an automaton is the set of all strings accepted by the automaton let $w = a_1, a_2, a_3 .. a_n....$ be a string of input symobls. The automaton is in the start state or initial state q_0 let us define the transition function δ as $\delta(q_0, q_1) = q_1$, to find the state that the automaton enters after processing the very first input symbol a_1. Now we process the next input symbol a_2 by evaluating

$\delta(q_1, a_2)$; suppose this state is q_2

That is $\delta(q_1, a_2) = q_2$

Continuing in this manner we have

$\delta(q_{i-1}, a_i) = q_i$ for all i

Thus, for $x = a_1 a_2 ... a_n$

$\delta(q_0, q_1, q_2 ... q_{n-1}), q_n) = \delta(q_{n-1}, a_n) = q_n$.

If $q_n \in F$(set of final states) then input $x = a_1 a_2 ...a_n$ is *accepted* by the automaton and if $q_n \notin F$, then the string is rejected.

> **Do You Know**
>
> A final stat $q \in F$ is also known as accepting state.

16.8.1 Definition

A string x is accepted by an automaton $M = (Q, ,\Sigma, \delta, q_0, F)$ if $\delta(q_0, x) = q$ for some $q \in F$.

Example 16.1. *Given a DFA* $M = (\{q_0, q_1, q_2, q_3\}, \{a, b\}, \delta, q_0, \{q_0\})$. *The transition diagram for the DFA is given below. Check whether strings (i) aababa and (ii) bababa are accepter by the DFA*

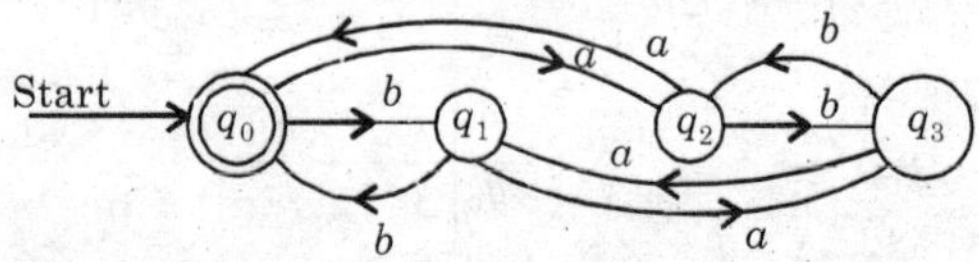

Solution : For string (i) *aababa*

$$\delta(q_0, \underset{\uparrow}{a}\, ababa) \vdash \delta(q_2, \underset{\uparrow}{a}\, b\, a\, ba) \vdash \delta(q_0, \underset{\uparrow}{b}\, a\, ba) \vdash \rightarrow$$
$$\rightarrow \delta(q_1, \underset{\uparrow}{a}\, ba) \vdash \delta(q_3, \underset{\uparrow}{b}a) \vdash \delta(q_2, \underset{\uparrow}{a}) \vdash \delta(q_0, \underset{\uparrow}{\in}) = q_0 \in F$$

(The symbol $\overline{\uparrow}$ indicates the current input symbol being processed by the DFA)

Since $q_0 \in F$ the string *aababa* is accepted by the DFA.

Similary for spting (*ii*) *bababa*

$$\delta(q_0, \underset{\uparrow}{b}ababa) \vdash \delta\,(q_1, \underset{\uparrow}{a}baba) \quad \delta\,(q_3, \underset{\uparrow}{b}aba) \vdash \delta(q_2, \underset{\uparrow}{a}ba)$$
$$\vdash \delta(q_0, \underset{\uparrow}{b}a) \vdash \delta\,(q_1, \underset{\uparrow}{a}) \vdash \delta\,(q_3, \underset{\uparrow}{\in}) = q_3 \notin F$$

Thus string not accepted.

16.8.2 The Language of a DFA

The language $L(M)$ of a DFA, $M = (q, \Sigma, \delta, q_0, F)$ is defined as

$$L(M) = \{w \mid \delta(q_0, w) \in F\}$$

That is, the language $L(M)$ of DFA M is the set of strings w that take the initial state q_0 to one of the final states.

If language 'L' over input alphabet Σ is $L(M)$ then we say that L is a *regular language.***

Example 16.2. *Design a DFA that acceptes language* $L = \{w \mid w$ *has both an even number of a's and an even number of b's*$\}$ *over* $\Sigma = (a, b)$

Solution : By the definition of L it is clear that the DFA has to count and remember the number of a's and the number of b's and that to in modulo 2. (Because both a and b has to be even in numbers). Thus, there exists four states for this counting.

** A language L over input alphabet Σ is the set of strings w such that $w \in \Sigma^*$. That is if Σ is an alphabet set and $L \subseteq \Sigma^*$, then L is a language over Σ. *For example* $L = \{\in, 0, 1, 01, 10, 001, 100..\}$ is a language over $\Sigma = \{0, 1\}$.

State q_0 : Both the number of a's and number of b's seen so far are even.

State q_1 : The number of a's seen so far is even and the number of b's seen so far is odd.

State q_2 : The number of a's seen so far is odd and the number of b's seen so far is even

State q_3 : Both the number of a's and the number of b's seen so for are odd.

Thus, state q_0 is both the initial as well as the only final state.Initial because before processing any input the numbers of a's and b's seen so for are zero that is even and since it describes exactly the candition for string of a's and b's to be in language (that is accepted by DFA) it is the only final state.

Thus, the DFA is as follows :

$$m = (\{q_0, q, q_2, q_3\}, \{a, b\}, \delta, q_0, \{q_0\})$$

The transition diagram for δ is as follows :

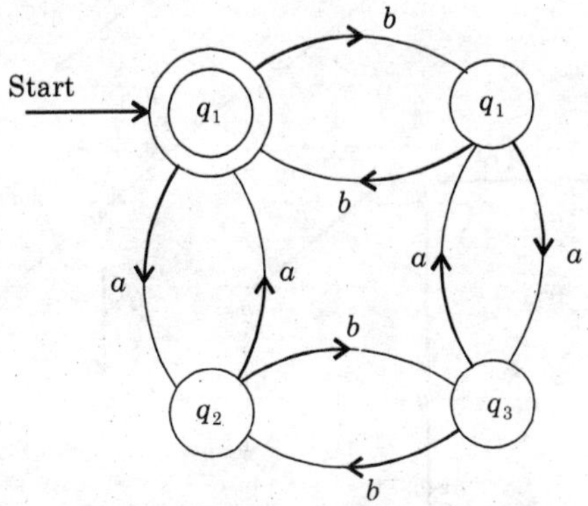

Fig. 16.5. Transition diagram for the DFA of example.

16.9 NON-DETERMINISTIC FINITE AUTOMATA

A finite automata which has the power to be in several states with respect to same input symbol at once is known as "Non-Deterministic Finite Automata" (NDFA or NFA) A transition diagram for an NFA is shown in **Fig. 16.6.**

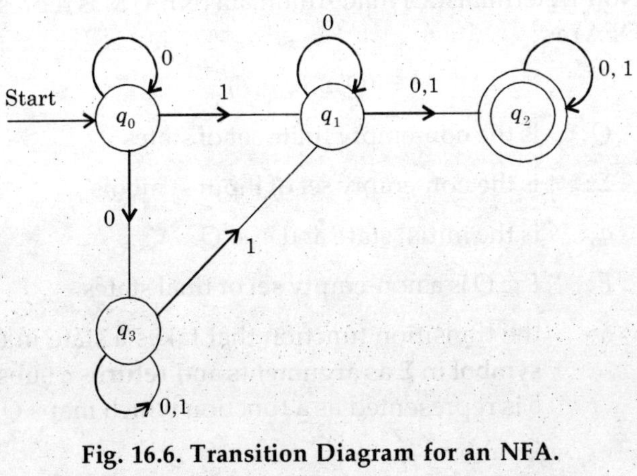

Fig. 16.6. Transition Diagram for an NFA.

Observe that there are two edges labeled 0 out of state q_0, one going to q_3 and the other going back to q_0. Similarly, from state q_1, there are again two edges labeled 0, one going to q_2 and the other going back to q_1. From q_3 again these are two edges labeled 1, one going to q_1 and the other back to q_3.

> ***Do You Know***
>
> DFA is a special case of NFA in which for each state there is a unique transition on each symbol.

An input sequence say $a_1, a_2, \ldots a_m$ is accepted by NFA if there exists a sequence of transitions, corresponding to the input sequence, that leads from the initial state to some final state (as in the case of DFA). The processing of string 0101 for the NFA given in **Fig. 16.6** is shown in Fig. 16.7.

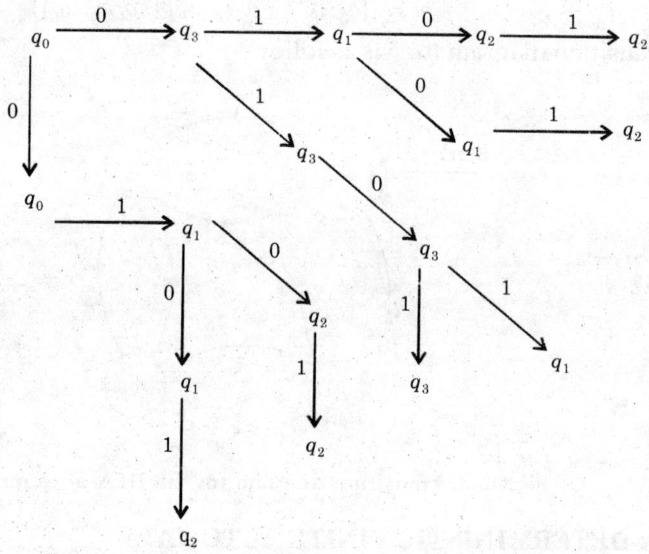

Fig. 16.7. The states an NFA is in during the processing of input spring 0101.

16.9.1 Formal Definition of NFA

An Non-Deterministic Finite Automata (NFA) M is represented by 5-type (just like of DFA) as :

$$M = (Q, \Sigma, \delta, q_0, F)$$

where Q: is the non-empty finite set of states

Σ: is the non-empty set of input symbols.

q_0: is the initial state and $q_0 \in Q$.

F: $F \subseteq Q$ is a non-empty set of final states.

δ: the transition function that takes a state in Q and an input symbol in Σ as arguments and returns a subset of Q. That is δ is represented as a function which maps $Q \times \Sigma$ into 2^Q.

Thus, according to the definition of NFA, the NFA of **Fig. 16.6** is special formally as:

$$M = (\{q_0, q_1, q_2, q_3\}, \{0, 1\}, \delta, q_0, \{q_2\})$$

where the transition function δ is given by the transition table of **Fig. 16.8.**

> ***Do You Know***
>
> 2^Q is the power set of Q which gives all possible subsets of set Q.

	0	1
q_0	$\{q_0, q_3\}$	$\{q_1\}$
q_1	$\{q_1, q_2\}$	$\{q_2\}$
q_2	$\{q_2\}$	$\{q_2\}$
q_3	$\{q_3\}$	$\{q_1, q_3\}$

Fig. 16.8.

16.10 THE EXTENDED TRANSITION FUNCTION

The function δ can be extended to a function $\hat{\delta}$ mapping $Q \times \Sigma^*$ to 2^Q and reflecting sequence of inputs as follows :

***(i)* Basis :** $\hat{\delta}(q, \varepsilon) = \{q\}$, which means without reading any input symbol, we are only in the state we began in.

***(ii)* Induction :** Let $w = x_a$, where x is a string of input symbols and a is the final symbol of w suppose that

$$\hat{\delta}(q,x) = \{p_1, p_2, \dots p_k\}$$

where $p_1, p_2, \dots p_k \in Q$

Let $\bigcup_{i=1}^{k} \delta(p_i, a) = \{r_1, r_2, \dots r_m\}$

Then $\hat{\delta}(q, w)$ or $\hat{\delta}(q, x_a) = \{r_1, r_2, \dots r_m\}$

That is we compute $\hat{\delta}(q, w)$ by first computing $\hat{\delta}(q, x)$ and by then following any transition from any of these states that is labeled a.

Example 16.3. *Use $\hat{\delta}$ to describe the processing of string a a b a b for the given NFA.*

$$M = (\{q_0, q_1, q_2\}, \{a, b\}, \delta, q_0, \{q_2\})$$

δ is given as

	a	b
$\rightarrow q_0$	$\{q_0, q_1\}$	$\{q_0\}$
q_1	ϕ	$\{q_2\}$
((q_2))	ϕ	ϕ

> ***Do You Know***
>
> For an input symbol a
>
> $\hat{\delta}(q, a) = \delta(q, a)$

Solution : The processing of string *a a b* is defined by following steps.

(i) $\hat{\delta}(q_0, \in) = \{q_0\}$

(ii) $\hat{\delta}(q_0, a) = \{q_0, q_1\}$

(iii) $\hat{\delta}(q_0, aa) = \delta(\delta(q_0, a), a) = \delta(\{q_0, q_1\} a) = \delta(q_0, q\} \cup \delta(q_1, q)$

$= \{q_0, q_1\} \cup \{\phi\} = \{q_0, q_1\}$

(iv) $\hat{\delta}(q_0, aab) = \delta(q_0, b) \cup \delta(q_1, b) = \{q_0\} \cup \{q_2\} = \{q_0, q_2\}$

Step (*i*) is the basic rule. We obtained step (*ii*) by applying δ to the initial state, q_0, that is in the previous set and obtain $\{q_0, q_1\}$ as the result. Step (*iii*) is obtained by taking the union over the two states in the previous set of what we get when we apply δ to them with input *a*. Similarly we obtained step (*iv*).

16.11 THE LANGUAGE OF NFA

The language *L* (*M*) of an NFA *M* is the set of strings *w* in Σ^* such that the subset of set *Q* obtained after processing string *w* contain at least one final state from set *F* (*i.e.*, the set of final states).

In other words

$$L(M) = \{w | \hat{\delta}(q_0, w) \cap F \neq \phi\}$$

or $$L(M) = \{w | \hat{\delta}(q_0, w) = \{r_1, r_2, ... r_m\} \text{ and } F \subseteq \{r_1, r_2, ... r_m\}$$

16.12 THE EQUIVALENCE OF DFAS AND NFAS

Since every DFA is an NFA, the language that can be described by some NFA can also be described by some DFA. Thus, NFA accepts regular language as DFA does. Moreover, two automaton say M_1 & M_2 are said to be equivalent, if they both accepts same set of languages *i.e.*, $L(M_1) = L(M_2)$. So if a language is identified or recognized by an NFA, then there exists a DFA which also recognize the same language.

16.12.1 Theorem

For every non-deterministic finite automaton there exists an equivalent deterministic finite automation. That is; If *L* be a set accepted by an NFA, then there exists a DFA which also accepts the same set *L*.

Proof

The proof that DFA accepts the same language set *L* as NFA does, involves an important "construction" called the **subset construction** because it involves constructing all subsets of the set of states of the NFA.

This subset construction starts from NFA say $M = (Q, \Sigma, \delta, q_0, F)$. Which accepts the language set *L*. Define a Deterministic Finite Automaton (DFA) say

$$M_D = (Q_D, \Sigma, \delta_D, q_{0_D}, F_D)$$

The components of M_D are constructed as follows :

(*a*) Q_D is the set of subsets of Q *i.e.*, the power set. Thus, if Q has n-states Q_D will have 2^n states in it. so $Q_D = 2^Q$.

(*b*) F_D is the set of all subsets in Q_D containing a final state of automaton M.

(*c*) q_{0_D} is the initial state of DFA M_D and $q_{0_D} = [q_0]$. Since an element of Q_D will be denoted by $[q_1, q_2, \dots q_n]$, where $q_1, q_2, \dots q_n \in Q$, which means corresponding to a set of states $q_1, q_2, \dots q_n$ of NFA we have a single state $[q_1, q_2, \dots q_n]$ in DFA. Thus, $q_{0_D} = [q_0]$.

(*d*) Now we define δ_D as $([q_1, q_2, \dots q_i], a) = [p_1, p_2, \dots p_j]$

if and only if $\delta(\{q_1, q_2, \dots q_i\}, a) = \{p_1, p_2, \dots p_j\}$

That is, δ_D is applied to an element $[q_1, q_2, \dots q_i]$ of Q_D is computed by applying δ to each state of Q represented by $\{q_1, q_2, \dots q_i\}$.

$$\text{i.e.,}\quad \delta_D([q_1, q_2, \dots q_i], a) = \delta(q_1, a) \cup \delta(q_2, a) \cup \dots \cup \delta(q_i, a)$$

$$= [p_1, p_2, .. p_j]$$

This new set of states *i.e.*, $\{p_1, p_2, \dots p_j\}$ has a representative of $\{p_1, p_2, \dots p_j\}$ in Q_D.

On the input string w, it is easy to show by induction that

$$\delta_D(q_{0_D}, w) = [q_1, q_2, \dots q_k]$$

if and only if $\qquad \delta(q_0, w) = \{q_1, q_2, \dots q_k\}$

Basis

For $|w| = 0$, the result is trivial. Since $q_{0_D} = [q_0]$ and x must be $\in$.

Induction

Let us suppose the result is true for strings of length m or less consider a string $w = xa, a \in \Sigma$, such that $|w| = m + 1$. That is $|x| = m$ and $|a| = 1$, which gives $|w| = |xa| = m + 1$.

Then $\qquad \delta_D(q_{0_D}, xa) = \delta_D(\delta_D(q_{0_D}, x), q)$

Since $|x| = m$, by induction hypothesis.

$$\delta_D(q_{0_D}, x) = [p_1, p_2, \dots p_j]$$

if and only if $\qquad \delta(q_0, x) = \{p_1, p_2, \dots p_j\}$

And by the definition of δ_D

$$\delta_D([p_1, p_2, \dots p_j], a) = [r_1, r_2, \dots r_k]$$

if and only if $\delta(\{p_1, p_2, \dots p_j\}, a) = \{r_1, r_2, \dots r_k\}$

Thus $\qquad \delta_D(q_{0_D}, xa) = [r_1, r_2, \dots r_k]$

if and only if $\delta(q_0, xa) = \{r_1, r_2, ... r_k\}$

Now $\delta_D(q_{0_D}, w)$ is in F_D exactly when $\delta(q_0, w)$ contains a state of Q that is in F. This completes the proof and hence $L(M) = L(M_D)$.

Example 16.4. *Construct a DFA equivalent to the given NFA*

$$M = (\{q_0, q_1, q_2\}, \{a, b\}, \delta, q_0, \{q_2\})$$

where transition function δ is given in tabular form as

	a	b
q_0	(q_0, q_1)	q_0
q_1	—	q_2
q_2	—	—

Solution : For DFA M_D,

(*i*) Q_D : the states are subset of (q_0, q_1, q_2). There are $2^3 = 8$ subsets as $(\phi, [q_0], [q_1], [q_2], [q_0, q_1], [q_0, q_2], [q_1, q_2], [q_0, q_1, q_2]$.

(*ii*) $[q_0]$ is the initial state.

(*iii*) $F_D = ([q_2], [q_0, q_2], [q_1, q_2], [q_0, q_1, q_2]$ are the final states as all these states contains the final state of NFA *i.e.*, state q_2.

Construction of δ_D

Clearly the initial state of DFA is $[q_0]$

so

(*i*) $\delta_D([q_0], a) = [q_0, q_1]$ and $\delta_D([q_0], b) = [q_0]$

as $\delta(q_0, a) = \{q_0, q_1\}$

Now one of the two sets we computed is old and that set is $[q_0]$. The new set we obtained is $[q_0, q_1]$, so we must compute its transactions.

(*ii*) $\delta_D([q_0, q_1], a) = \delta(q_0, a) \cup \delta(q_1, a)$

$= [q_0, q_1] \cup \phi = [q_0, q_1]$

and $\delta_D([q_0, q_1], b) = \delta(q_0, b) \cup \delta(q_1, b) = [q_0] \cup [q_2] = [q_0, q_2]$

Now we compute transactions for the new state obtained by above computations and it is $[q_0, q_2]$.

(*iii*) $\delta_D([q_0, q_2], a) = \delta(q_0, a) \cup \delta(q_2, a) = [q_0, q_1]$

and $\delta_D([q_0, q_2], a) = \delta(q_0, a) \cup \delta(q_2, b) = [q_0]$

and $\delta_D([q_0, q_2], a) = \delta(q_0, b) \cup \delta(q_2, b) = [q_0]$ q

We have not obtained any new states. Thus, the subset construction has converged. That is we know all the accessible states and their transactions. The δ_D for DFA is given in Fig.

	a	b
$[q_0]$	$[q_0, q_1]$	$[q_0]$
$[q_0, q_1]$	$[q_0, q_1]$	$[q_0, q_2]$
$[q_0, q_2]$	$[q_0, q_1]$	$[q_0]$

Fig. 16.9. δ_D for DFA M_D.

The diagram of δ_D is also given below :

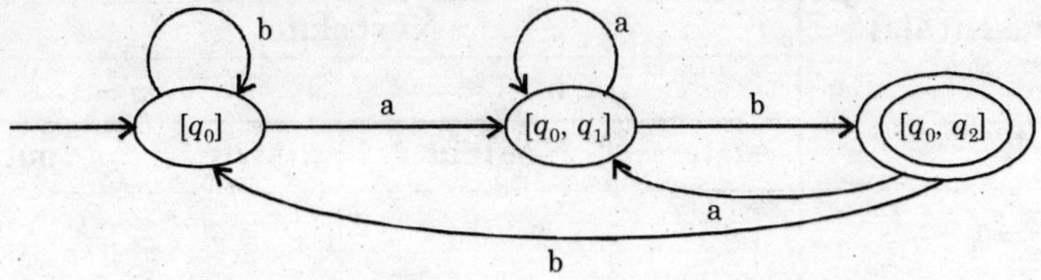

Fig. 16.10. The DFA M_D corresponding to NFA M.

16.13 FINITE AUTOMATON WITH OUTPUT : MEALY AND MOORE MACHINES

We have studied automaton which accepts or rejects a string. Now we will consider automaton in which outputs can be choosen from some other alphabet. In general, the output $0(t)$ of the automaton is a function of the present state $q(t)$ and the present input $x(t)$ *i.e.,*

$$0(t) = f(q\,(t), x(t))$$

where f is known as the output function

The automaton, where the output $0(t)$ depends on both present state and input is called, **Mealy Machine** whereas the automaton where the output $0(t)$ depends only on the present state of the automaton that is,

$$0(t) = f(q\,(f))$$

is called **Moore Machine.**

16.13.1 Moore Machine

A moore machine is represented by six – tuple $(Q, \Sigma, \Delta, \delta, f, q_0)$, where

(*i*) Q is a non-empty finite set of states.

(*ii*) Σ is a non-empty finite set of inputs.

(*iii*) Δ is the output alphabet.

(*iv*) δ is the transition function $\Sigma \times Q$ into Q.

(*v*) f is the output function on mapping Q into Δ, *i.e.* $f : Q \rightarrow \Delta$ and

(*vi*) q_0 is the initial state.

16.13.2 Mealy Machine

A mealy machine is represented by a six – tuple $(Q, \Sigma, \Delta, \delta, f, q_0)$

All the symbols except f have the same meaning as in moore machine.

Here f is the output function which maps $\Sigma \times Q$ into Δ, *i.e.,* $f : Q \times \Sigma \rightarrow \Delta$

Table 16.1 and **Table 16.2** represent the moore and mealy machines respectively present.

Table 16.1. Moore Machine.

Present state	Next state		Output (f)
	$a = 0$	$a = 1$	
$\rightarrow q_0$	q_1	q_3	0
q_1	q_1	q_0	1
q_2	q_0	q_1	1
q_3	q_2	q_3	0

Table 16.2. Mealy Machine.

Present state	Next state			
	$a = 0$		$a = 1$	
	state	output	state	output
$\rightarrow q_0$	q_2	0	q_0	0
q_1	q_3	1	q_2	0
q_2	q_0	1	q_1	1
q_3	q_1	0	q_3	1

The output strings obtained from both the tables for a given input string 0101 are 010100 and 0111 respectively. The transition of states in cases of moore machine is given as.

$$\begin{array}{l}\text{output} \rightarrow \\ \text{input} \rightarrow\end{array} \longrightarrow q_0 \xrightarrow[0]{0} q_1 \xrightarrow[1]{1} q_0 \xrightarrow[0]{0} q_1 \xrightarrow[1]{1} q_0 \xrightarrow{0} \vdash 01010$$

giving output 01010. For the input string $\in$, the output in case of moore machine is

$$f(q_0) = 0$$

That's why the length of output string is one more than the length of input string in case of moore machine whereas it remains as it is in case of mealy machine.

The transition of states incase of mealy machine is given as

$$q_0 \xrightarrow[0]{0} q_2 \xrightarrow[1]{1} q_1 \xrightarrow[1]{1} q_3 \xrightarrow[1]{1} q_3 \vdash 0111$$

giving output 0111.

The output of an empty string $\in$ in mealy machine will remain $\in$ only as we get an output only on the application of an input symbol.

Note : A finite automaton can be converted into a moore machine by introducing $\Delta = \{0, 1\}$ and defining $f(q) = 1$ if $q \in F$ and $f(q) = 0$ if $q \neq F$.

16.14 MINIMIZATION OF FINITE AUTOMATA

We know that a DFA recognize a unique language but its converge is not true. So for a given language there exists more than one DFA that accepts it. All these DEAs are called, equivalent automaton.

So here we construct an automaton with minimum number of states equivalent to a given automaton M. Thus we define some relations in Q.

16.14.1 Equivalent States

Two states say q_i and q_j are said to be equivalant if both (q_i, w) and $\delta(q_j, w)$ are final states or both of them are non-final states for all $w \in \Sigma$

* In Case of mealy and moore machines the transitions are shown as $q_i \xrightarrow[\text{input}]{\text{output}} q_j$

16.14.2 *m*-Equivalent states

Two states q_i and q_j are said to be *m*-equivalent if for all strings *w* of lenght *m* or less, both $\delta(q_i, w)$ and $\delta(q_j, w)$ are final or both non final states.

Note : Any two final states or non-final states are 0-equivalent

The above **equivalent relations** are defined on set of states *Q* and thus, the relations partitions *Q*. Into disjoint equivalent classes. Let *S* and S_n be the partitions of *Q*. The elements of S_n are *n*-equivalent classes*

Further if q_i and q_j are $(n + 1)$ equivalent, then we say that they are *n*–equivalent

Continuing in this way, at the end we get $S_n = S_{n+1}$ for some *n*, where S_n is the set of *n*-equivalent classes.

16.14.3 Construction of Minimum Automata

The construction of minimized finite automata is based upon a result which says – Two states q_i and q_j are $(n + 1)$ equivalent if (*i*) they are *n* – equivalent and (*ii*) $\delta(q_i, a)$ and $\delta(q_j, a)$ are also *n*-equivalent for every $a \in \Sigma$.

Step I : Firstly we construct the 0–equivalent classes S_0 as $S_0 = (Q_1^0, Q_2^0)$ where Q_1^0 is the set of all final states any two final states are o-equivalent and $Q_2^0 = Q - Q_1^0$, where *Q* is the set of states.

Step II : (Construction of S_{n+1} from S_n) Let Q_i^n be any subset in S_n. If q_1 and q_2 are in Q_i^n then they are $(n + 1)$ equivalent provide $\delta(q_1, a)$ and $\delta(q_2, a)$ are *n* – equivalent for all $a \in \Sigma$.

⇒ So, find out whether $\delta(q_1, a)$ and $\delta(q_2, a)$ are in same equivalent class in S_n for every $a \in \Sigma$. If yes, then q_1 and q_2 are $(n + 1)$– equivalent.

⇒ Q_i^n is further divided into $(n + 1)$ – equivalent dasses. Thus, repeat the above for every Q_i^n in S_n to get all the elements of S_{n+1}.

Step III: Construct S_n for all $n = 1, 2, \ldots$ until $S_n = S_{n+1}$

Step IV : Thus, the equivalence classes obtained in step III are the required states in minimum state automaton. The state table is obtained by replacing a state *q* by the corresponding equivalence class $[q]$.

Example 16.5. *Construct the minimium state automaton equivalent to the tranisition table given in figure*

	Input	
State	*a*	*b*
$\rightarrow q_0$	q_1	q_0
q_1	q_0	q_2
q_2	q_3	q_1
(q_3)	q_3	q_0
q_4	q_3	q_5
q_5	q_6	q_4
q_6	q_5	q_6
q_7	q_6	q_3

Fig. 16.11. FA of Example

* if q_i and q_j are *n*-equivalent for all $n \geq 0$, then we say that they are equivalent.

Solution : (*i*) First of all let us construct $S_0 = (Q_1^0, Q_2^0)$

$Q_1^0 = \{q_3\}$ as q_3 is the final state and

$Q_2^0 = Q - \{q_3\} = \{q_0, q_1, q_2, q_3, q_4, q_5, q_6, q_7\}$

Thus $S_0 = (\{q_3\}, \{q_0, q_1, q_2, q_4, q_5, q_6, q_6\})$

As Q_1^0 contains only one state *i.e.*, q_3 it cannot be partitioned further. $Q_1^1 = \{q_3\}$.

(*ii*) Consider q_0 and $q_1 \in Q_2^0$.

Now, $s(q_0, a) = q_1\ d(q_1, a) = q_0$

and $d(q_0, b) = q_0,\ d(q_1, b) = q_2$

The entries under a column corresponding to q_0 and q_1 are q_1 are q_1, q_0. That is they all lie in Q_2^0. The entries under b–column are q_0 and q_2 and they also lie in Q_2^0. Thus, q_0 and q_1 are 1-equivalent (*i.e.*, $q_0 \equiv q_1$)

Now let us take q_1 and q_2 and cheek their equivalence with same procedure and it shows that q_1 and q_2 are not-equivalent as the entries under a-column corresponding to q_1 and q_2 are q_0 and q_3 respectively. $q_0 \in Q_2^0$ and $q_3 \in Q_1^0$. Thus, $q_1 \not\equiv q_2$ (*i.e.*, both q_0 and q_3 are not in same grup).

Similarly q_1 and q_4 and q_1 and q_7 are not 1-equivalent and $q_0, q_1\ q_5$ and q_6 are 1-equivalent

Thus, $S_1 = (\{q_3\}\ \{q_0, q_1, q_5, q_6\}, \{q_2, q_4\}, \{q_7\})$

Here, $Q_1' = \{q_3\}$ and $Q_2' = (\{q_0, q_1, q_5, q_6\}, \{q_3, q_4\}\ \{q_7\})$

(*iii*) $\{q_7\}$ being single state cannot be partitioned further and it will be there in S_2.

Now for $\{q_1, q_4\}$ consider the entries under a-column corresponding to q_2 and q_4 and they are q_3 and q_3 respectively and lies in the same group class in S_1. Now the entries under b-column corresponding to q_2 and q_4 are q_1 and q_5. These also lies in same equivalence set in S_1. This q_2 and q_4 are 2-equivalent and $\{q_2, q_4\}$ will be there in S_2

(*iv*) Now we cheek the 2-equivalence in the set $\{q_0, q_1, q_5, q_6\}$. q_0 and q_1 are not 2-equivalent as the entries under b-column corresponding to q_0 and q_1 are q_0 and q_2. $q_0 \in \{q_0, q_1, q_5, q_6\}$ and $q_2 \in \{q_2, q_4\}$ *i.e.*, both lies in different equivalence sets.

Similarly q_0 and q_5 are also not 2-equivalent where as q_0 and q_6 are 2-equivalent and q_1 and q_5 are two equivalent.

Thus, the set $\{q_0, q_1, q_5, q_6\}$ is partioned into two subsets as $(\{q_0, q_6\}, \{q_1, q_5\})$

So $S_2 = (\{q_3\}\ \{q_0, q_6\}\ \{q_1, q_5\}, \{q_2, q_5\}, \{q_7\})$

(*v*) Now q_0 is 3-equivalent to q_6. q_1 is 3-equivalent to q_5 and q_2 is 3-equivalent to q_4 and hence (check yourself)

$S_3 = S_2 = (\{q_3\}\ \{q_0, q_6\}\ \{q_1, q_5\}\ \{q_2, q_4\}\ \{q_7\})$

gives the equivalence classes. The minimum state automaton is

$$m' = (Q', \Sigma = (a, b), \delta', q_0', F')$$

Where $$Q' = ([q_3], [q_0, q_6], [q_1, q_5], [q_2, q_4], [q_7])$$

$$q_0' = [q_0, q_6]\ F' = [q_3].$$

and δ^1 is given in the following table

State	Input a	b
$[q_0, q_6]$	$[q_1, q_5]$	$[q_0, q_6]$
$[q_1, q_5]$	$[q_0, q_6]$	$[q_2, q_4]$
$[q_2, q_4]$	$[q_3]$	$[q_1, q_5]$
$[q_3]$	$[q_3]$	$[q_0, q_6]$
$[q_7]$	$[q_0, q_6]$	$[q_3]$

The transition diagram (or the minimum state automaton) is represented as :

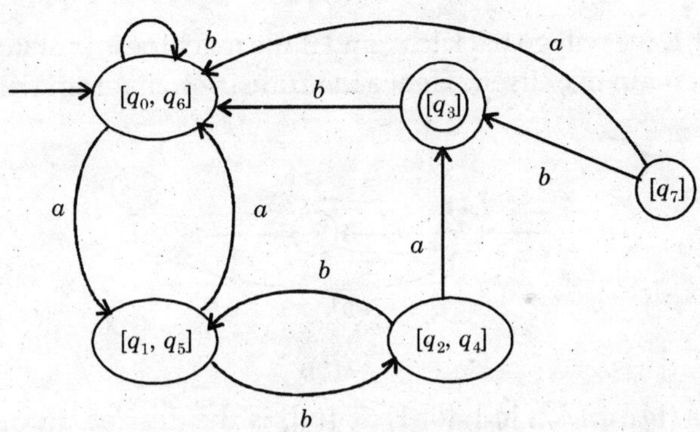

Fig. 16.12. Minimum state automaton of the above example

Example 16.6. Let $\Sigma = \{a, b\}$. Constant an automaton 19 which accept all strings over $\{a, b\}$ where the number of b's is divisible by 3.

Solution: Since the number of $b'x$ is divisiable by 3, we require 3 states as q_0, q_1, q_2 where q_0 is the initial as well as final state. For any a, the machines remains in the same state. Hence the transition diagram will be

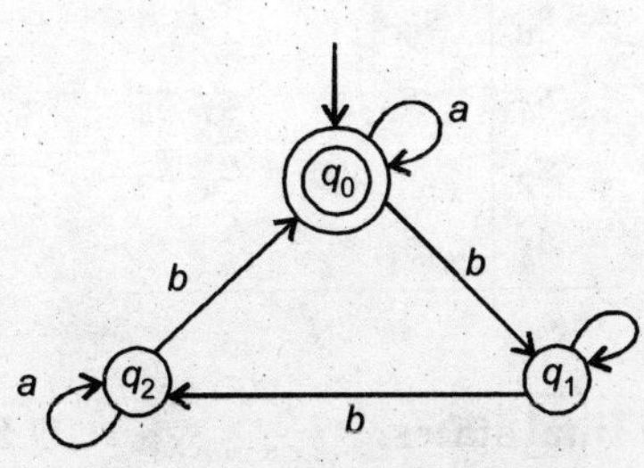

Fig. 16.13

Example 16.7. Construct an automaton M over $\Sigma = \{0, 1\}$ which acepts all strings over Σ which sterts with 0 and end with 1.

Solution : The diagram is shown below:

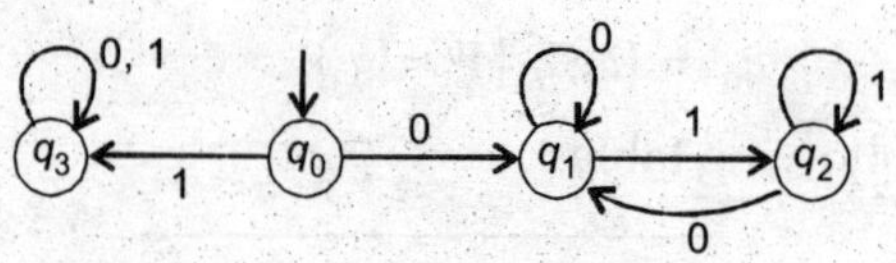

Fig. 16. 14

Example 16.8. Let Σ {0, 1}. Construct an automaton M which will accept precisely those words from Σ which end in two 0's. [U.P.T.U. 2002-03]B.Tech.

Solution : Since O^2 is aceepted, but not $\wedge$ or 0, we need three states q_0, q_1 and q_2. q_0 is the initial and q_2 is the final state. The transition between these three is shown below.

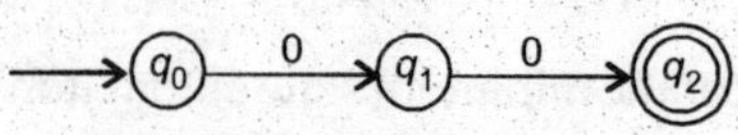

(*a*)

For input 1, we will go back to q_0 and if the machine is in state q_2 receives input as O, if will remain in q_2 itself. Hence the transition diagram will be.

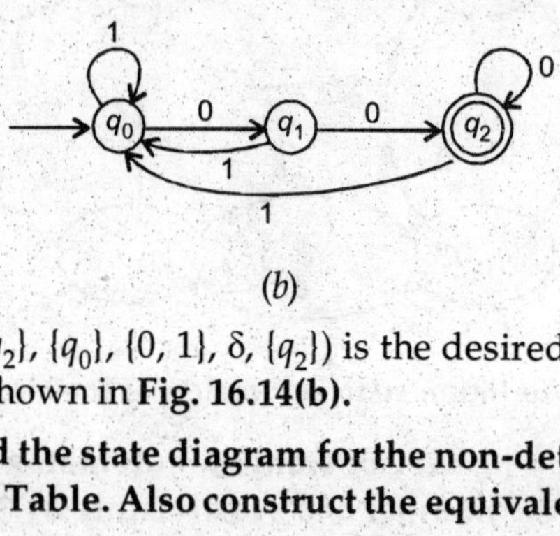

(*b*)

Thus $M = (\{q_0, q_1, q_2\}, \{q_0\}, \{0, 1\}, \delta, \{q_2\})$ is the desired automaton and δ, the transition function is shown in **Fig. 16.14(b).**

Example 16.9. Find the state diagram for the non-deterministic finite state automaton withe state Table. Also construct the equivalent DFA.

	f	
State	**0**	**1**
$\rightarrow s_0$	s_0, s_1	s_3
s_1	s_0	s_1, s_3
s_2		s_0, s_2
s_3	s_0, s_1, s_2	s_1

where s_2 and s_3 are the final states. [U.P.T.U. B.Tech 2009-10]

Solution : The transition diagram is shown next page

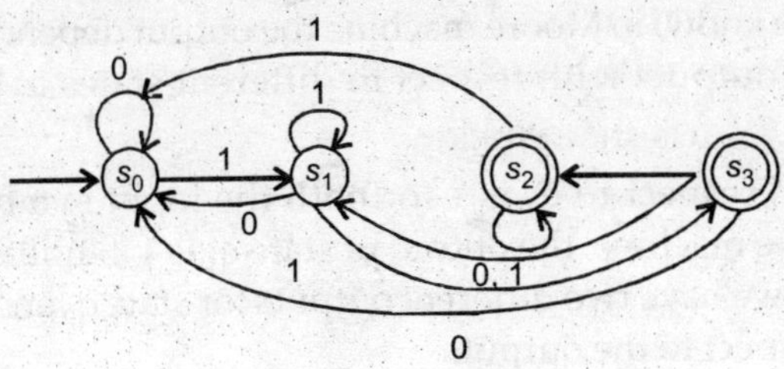

Fig. 16.15 (a)

The transition Table for the equivalent DFA is shown below in **Fig. 16.15 (*b*).**

δ_D:

State	*input* 0	1
$[s_0]$	$[s_0, s_1]$	$[s_3]$
$[s_3]$	$[s_0]$	$[s_1, s_3]$
$[s_0, s_1]$	$[s_0, s_1]$	$[s_1, s_3]$
$[s_1, s_3]$	$[s_0, s_1, s_2]$	$[s_1, s_3]$
$[s_0, s_1, s_3]$	$[s_0, s_1]$	$[s_0, s_1, s_2, s_3]$
$[s_0, s_1, s_2, s_3]$	$[s_0, s_1, s_2]$	$[s_0, s_1, s_2, s_3]$

Clealry :

$$\begin{aligned}\delta_D([s_0, s_1], 0) &= d(s_0, 0) \cup \delta(s_1, 0)\\ &= (s_0, s_1) \cup (s_0)\\ &= [s_0, s_1]\end{aligned}$$

Simillarly.

$$\begin{aligned}\delta_D([s_1, s_3], 0) &= \delta(s_1, 0) \cup \delta(s_3, 0) = (s_0, s_1) \cup (s_0, s_1, s_2)\\ &= [s_0, s_1, s_2]\\ \delta_D([s_1, s_3], 1) &= \delta(s_1, 1) \cup \delta(s_3, 1) = (s_3) \cup (s_1) = [s_1, s_3]\end{aligned}$$

and

$$\begin{aligned}\delta_D([s_D, s_1, s_2], 0) &= \delta(s_0, 0) \cup \delta(s_1, 0) \cup \delta(s_2, 0)\\ &= (s_0, s_1) \cup (s_0) \cup \varnothing = [s_0, s_1]\end{aligned}$$

and so on.

Example 16.10. Construct a Moore Machine equivalent to Mealy Mechine M gi ven belwo in Table

Present state	**Next state** $a = 0$ **state**	**output**	$a = 1$ **state**	**output**
$\rightarrow q_1$	q_1	1	q_1	0
q_2	q_2	1	q_2	1
q_3	q_3	1	q_3	1
q_4	q_4	0	q_4	1

Fig. 16.16 (a)

Solution : As we know, in Moore machine the output depends only at present state. We check the outputs with respect to different states in the given Mealy machine.

The output with respect to q_1 is 1 for both the input symbols. Thus in the corresponding Moore machine, the output of state q_1 is 1. Simillarly the output of state q_u is also 1. But we have two different outputs for state q_2 and q_3. Partition the states into 2 with respect to the outputs.

That is

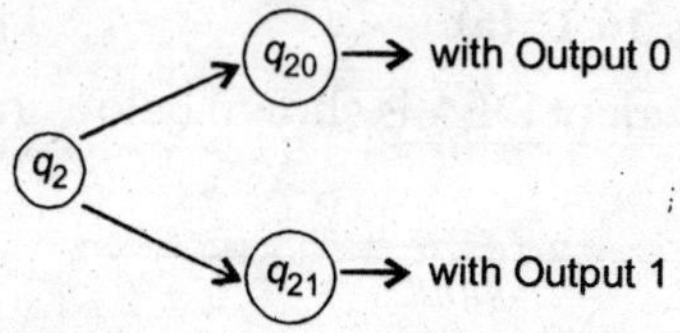

Simillarly

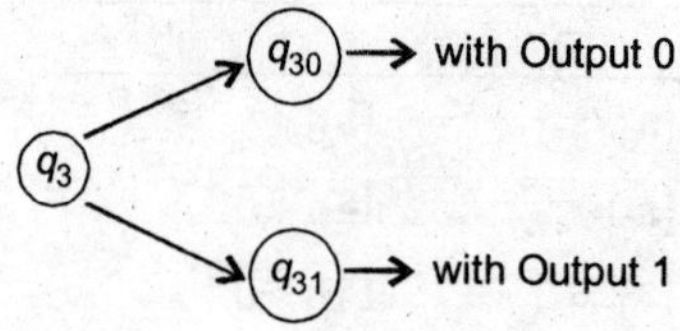

Thus, the new Mealy machine will have 6 states and the table is shown below.

Present state	Next state			
	a = 0		a = 1	
	state	output	state	output
→ q_1	q_1	1	q_{20}	0
q_{20}	q_4	1	q_4	1
q_{21}	q_4	1	q_4	1
q_{30}	q_{21}	1	q_{31}	1
q_{31}	q_{21}	1	q_{31}	1
q_4	q_{30}	0	q_1	1

Fig. 16.16 (b)

How, there is no state with two different outputs. That is, output is with respect to state only and input symbols have no significance over the output. Thus the required Moore Machine is as follows:

Present state	Next state		output
	a = 0	a = 1	
→ q_1	q_1	q_{20}	1
q_{20}	q_4	q_4	0
q_{21}	q_4	q_4	1
q_{30}	q_{21}	q_{31}	0
q_{31}	q_{21}	q_{31}	1
q_4	q_{30}	q_1	1

Fig. 16. 16 (c)

One can check that for an input-string w = 111, both the machines in **Fig. 16.16 (*a*)** and **Fig. 16 16 (*c*)** have some output.

Example 16.11. Change the given Moore machine into mealy machine:

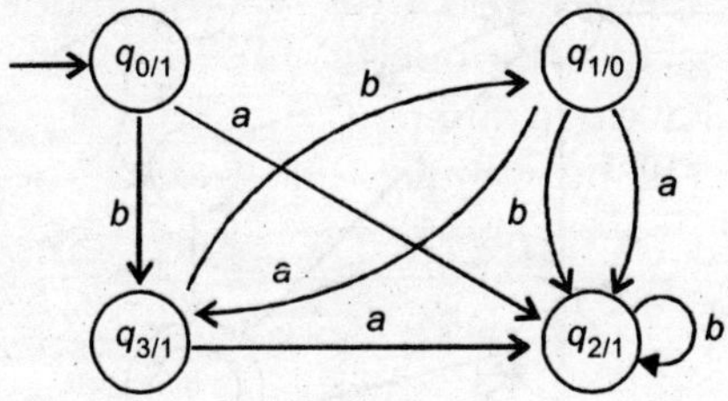

Solution : The transition Table with respect to given transition diagram is shown below.

Present state	Next state $I = a$	Next state $I = b$	output
→ q_0	q_2	q_3	1
q_1	q_3	q_2	0
q_2	q_1	q_2	1
q_3	q_2	q_1	1

Moore Machine Fig. 16. 17 (a)

The equaivalent mealy machine in shown in

Present state	Next state			
	$I = a$		$I = b$	
	state	output	state	output
→ q_0	q_2	1	q_3	1
q_1	q_3	1	q_2	1
q_2	q_1	1	q_2	1
q_3	q_2	1	q_1	0

Fig. 16. 17 (a)

EXERCISES

1. Define deterministic finite automata. Construct an NDFA that accepts the set {*ab, ba*} and use it to find a deterministic automaton accepting the same set
2. Construct an automaton M which accepts all the words on $\Sigma = \{a, b, c\}$ ending with bc.
3. The transition diagram of a non-deterministic finite automation M is given below, construct a DFA equivalent to M.

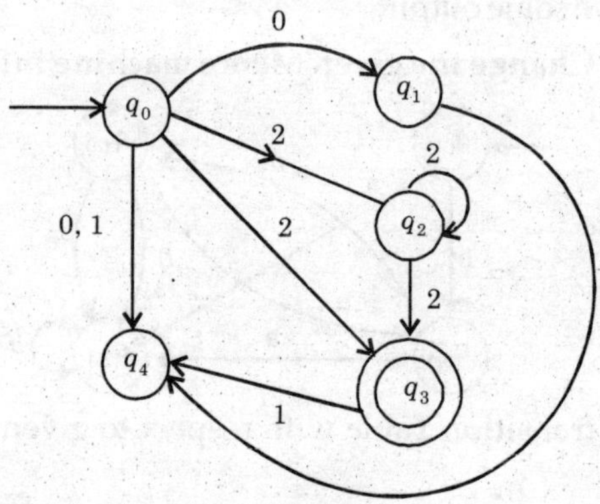

4. Describe the string w in the language L accepted by the automaton M in the folowing diagram :

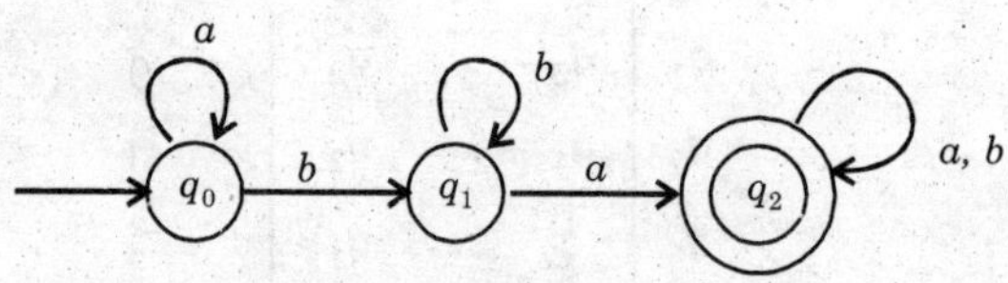

5. Construct a mealy machine which is equivalent to the more machnie given in Table.

Table 16.1. Moore Machine

Present state	Next State $a = 0$	$a = 1$	Output
$\rightarrow q_0$	q_1	q_2	1
q_1	q_3	q_1	0
q_2	q_2	q_1	0
q_3	q_0	q_3	1

6. Construct a more machine equivalent to the mealy machine M given in the Table:

Table 16.2. Mealy Machine

Present state	Next State $a = 0$ State	Output	$a = 1$ State	Output
$\rightarrow q_1$	q_1	1	q_2	0
q_2	q_4	0	q_4	1
q_3	q_2	1	q_3	0
q_4	q_3	0	q_1	1

7. Construct a minimum state automata equivalent to a given automaton whose transition table is given below.

States	Input 0	Input 1
$\rightarrow q_0$	q_0	q_3
q_1	q_2	q_5
q_2	q_3	q_4
q_3	q_0	q_5
q_4	q_0	q_6
q_5	q_1	q_4
q_6	q_1	q_3

ANSWERS

1.

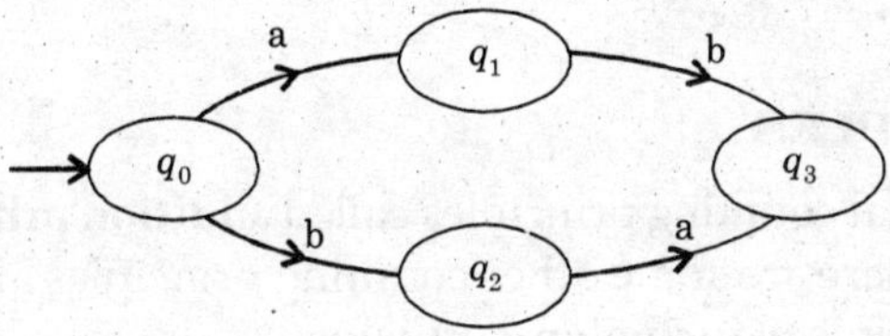

$M = (q_0, q_1, q_2, q_3\} \Sigma, \delta, q_0, \{q_3\})$
where $\Sigma = (a, b)$

■■■

Chapter

17 COMBINATORICS

17.1 INTRODUCTION

The field of combinatorial mathematics deals with the problem of counting or enumeration. The term combination deals with problem of selection of objects and the problem of number of ways of arrangements of objects comes under permutations. Two types of problems related to addition and multiplication based on addition and multiplication principles respectively have great importance in combinatorics.

17.2 COUNTING PRINCIPLES

There are two fundamental counting principles called **addition principle** and **multiplication principle**. There are some other counting principles also such as **bijection principle** and principle of **inclusion-exclusion.**

17.2.1 Addition Principle (or Sum Rule)

Suppose that *A* and *B* are two mutually exclusive (or disjoint) events ; that is, they never occur together. Also suppose that *A* occurs in *m* ways and *B* in *n* ways. Then *A* or *B* can occur in $m + n$ ways.

This rule can be extended for then two disjoint events also. If $E_1, E_2, ..., E_m$ are m disjoint events and E_1 can happen in n_1, ways, E_2 can happen in n_2 ways and so on. Then E_1 or E_2 or ... or E_n can happen in $n_1 + n_2 + ... + n_m$ ways.

Example 17.1. In how many ways can we obtain a sum of 4 or 6 when two distinguishable dice are rolled?

Solution: Possibility of obtaining a sum of (1, 3), (2, 2) and (3, 1). So $m = 3$.

Possibility of obtaining a sum of 6 are (2, 4), (3, 3), (1, 5), (5, 1), (4, 2) *i.e.*, $n = 5$.

So total number of possibilities to get sum of 4 or 6 = 5 + 3 = 8.

Note: Here the two dice are distinguishable, thus in the ordered pair of outcome (a, b), a indicates the outcome on dice 1 and b indicates the outcome on dice 2.

Clearly $(a, b) \neq (b, a)$

But if the two dice are indistinguishable then $(a, b) = (b, a)$ and in that case the total number of ways to obtain a sum of 4 or 6 is $m + n = 2 + 3 = 5$.

where $m \Rightarrow$ getting a sum of 4 (*i.e.*, (1, 3) = (3, 1) and (2, 2))

$n \Rightarrow$ getting a sum of 6 (*i.e.*, (1, 5) = (5, 1), (4, 2) = (2, 4) and (3, 3)).

Example 17.2. A computer password consists of an alphabet followed by 3 or 4 digits. Find (*a*) the total number of passwords that can be created, and (*b*) the number of passwords in which no digit repeats.

Solution : (*a*) Total number of passwords = number of passwords having 4 – characters (*i.e.*, one alphabet and 3 digits) + number of passwords having 5-characters (*i.e.*, an alphabet and 4 digits).

$$= (26) . (10) . (10) . (10) + (26) . (10) . (10) . (10) . (10)$$

$$= 2,86,000.$$

(*b*) When no digit repeats then the number of 4-character passwords is

$$= (26) . (10) . (9) . (8) = 18,720.$$

and the number of 5-character passwords is

$$= (26) . (10) . (9) . (8) . (7) = 1,31,040.$$

Thus a total of 1,49,760 passwords are there.

17.2.2 Multiplication Principle (or Product Rule)

Suppose that an event *X* can be decomposed into two stages, *A* and *B*. Let stage *A* occur in *m* ways and suppose that these stages are unrelated, in the sense that stage *B* occurs in *n* ways regardless of the outcome of degree *A*. Then event *X* occurs (*i.e.*, Both the stages *A* and *B* occurs) in $m \times n$ ways.

This rule is applicable even if event *X* can be decomposed in more than two stages $A_1, A_2, ..., A_m$ such that A_1 occurs in n_1 ways, A_2 occurs in n_2 ways and so on and finally A_m occurs in A_m ways. Then event *X* occurs in $n_1 \times n_2 \times ... \times n_m$ ways.

Example 17.3. Find the number of ways in which we can post 5 letters in 10 letter boxes.

Solution: We can post the first letter in 10 ways. Similarly the second letter in 10 ways, third letter in 10 ways and so on.

So let *X* be the event of posting letter is broken into 5 stages A_1, A_2, A_3, A_4 and A_5 where A_i is the event of posting in letter into 10 letter boxes.

So A_1 can occur in 10 ways *i.e.*, $n_1 = 10$

A_2 can occur in 10 ways *i.e.*, $n_2 = 10$

A_3 can occur in 10 ways *i.e.*, $n_3 = 10$

and similarly $n_4 = n_5 = 10$

So event *X* can occur in $10 \times 10 \times 10 \times 10 \times 10$

$$= 10^5 \text{ ways.}$$

Example 17.4. Three travellers arrives at a town where there are four hotels, in how many ways can they take up their quarters, each at a different hotel?

Solution: The first traveller has choice of four hotels, and when he has made his selection in any one way, the second traveller has a choice of 3; therefore the first two can make their choice in 4×3 ways ; and with any one such choice the third traveller can select his hotel in 2 ways; hence the required number of ways is $4 \times 3 \times 2 = 24$.

Example 17.5. How many ways are there to select 2 cards from a deck of 52 such that:

(*i*) The first card is a diamond and second card is a club.

(*ii*) The first card is a queen and second is not a king.

(*iii*) The first card is a club and second is queen.

Solution: We know that out of 52 cards there are 13 cards of diamond, 13 of heart, 13 of club and 13 of. In total there are 4 kings, 4 queens and 4 aces in a deek of 52 cards.

(*i*) A diamond from 13 can be chosen in 13 ways, and a club can be selected in 13 ways. Thus both the cards can be selected in 169 ways.

(*ii*) First card is queen, which can be selected in 4 ways (as there are 4 queens in 52 cards). The second card is selected from 52 – 1 = 51 cards (because replacement of selected cards in not allowed). Such that the card selected is not a king. We know that there are 4 kings in 51 cards, so the second card which is not a king can be selected in 51 – 4 = 47 ways. Thus the two cards can be selected in 4.47 = 188 ways.

(*iii*) First card of club can be selected in 13 ways and second card of queen can be selected in 4 ways. Thus the 2 cards can be selected in 13 = 52 ways.

Example 17.6. Consider three different cities say *P*, *Q* and *R*. There are 4 railway tracks between *P* and *Q*, 3 railway tracks between *Q* and *R* and 2 different tracks that go directly from *A* to *C*. Find out ,

(*i*) Total number of ways to go from *P* to *R* altogether (*i.e.*, *P* to *R* via *Q* and *P* to *R* directly).

(*ii*) Total number of ways to go from *P* to *R* and then back to *P*.

(*iii*) Total number of ways that go from *P* to *R* via *Q* and return directly from *R* to *P*.

(*iv*) Total number of ways to leave *Q* if a person is in city *Q*.

Solution: The situation is defined below:

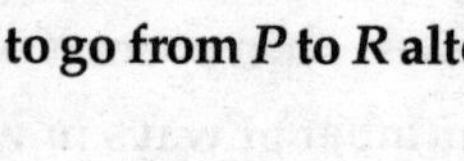

(*i*) Number of ways to go from *P* to *R* = *P* to *R* via *Q* + *P* to *R* directly

= (*P* to *Q*) . (*Q* to *R*) + (*P* to *R* directly)

= 4 . 3 + 2 = 14 ways.

As there are 4 ways to go from *P* to *Q* and 3 ways to go from *Q* to *R*, giving 4 . 3 = 12 ways to go from *P* to *R*. Parallely there are 2 ways to go from *P* to *R* directly. Thus becomes 12 + 2 = 14 ways altogether.

(*ii*) As we know that there are 14 ways to go from *P* to *R* and hence 14 ways to come back from *R* to *P*. Thus giving a total of 14.14 = 14^2 ways.

(*iii*) Total number of ways to go from *P* to *R* via *Q* are 4.3 = 12 ways.

Total number of ways to come to P from R directly are 2. So the total number of ways to go from P to R via Q and return directly form R to P are $12.2 = 24$ ways.

(*iv*) There are 4 routes from Q to P and 3 routes from Q to R, Thus giving $4 + 3 = 7$ ways to leave station Q.

17.2.3 Factorial

Consider the product of all positive integers from 1 to n as

$$1.2.3 \ldots (n-2)(n-1).n$$

This product is denoted as $n\,!$ and is read as "n factorial".

That is, $n\,!$ or $\underline{|n} = n.(n-1).(n-2) \ldots, 3.2.1.$

Notes : (*i*) Factorial is defined for positive integers only

(*ii*) $0\,! = 1$ and $1\,! = 1$

(*iii*) In general $n\,! = n.(n-1)\,!$

17.2.4 Bijection Principle

To count the number of elements in a very large set is not so easy. Let us consider a set X and we wish to count the number of elements in X *i.e.*, $n(X)$. Consider another set Y and $n(Y)$ is known to us. Now if there exist a one-to-one and onto mapping from X to Y then $n(X) = n(Y)$.

Thus **the bijection principle says that if X and Y are two finite sets such that there is a one-one-onto mapping exist between X and Y then $n(X) = n(Y)$.**

17.2.5 Principle of Inclusion Exclusion

Consider the two finite non-empty sets X and Y, then the number of elements in the two sets is

$$n(X \cup Y) = n(X) + n(Y) - n(X \cap Y)$$

or $$|X \cup Y| = |X| + |Y| - |X \cap Y|$$

Let us prove the principle using Vien diagram (Fig. 17.1)

Clearly $$X \cup Y = (X - Y) \cup (Y - X) \cup (X \cap Y)$$

or $$|X \cup Y| = |(X - Y)| + |(Y - X)| + |(X \cap Y)| \quad \ldots (1)$$

and $$|X| = |(X - Y)| + |(X \cap Y)| \quad \ldots (2)$$

$$|Y| = |(Y - X)| + |(X \cap Y)| \quad \ldots (3)$$

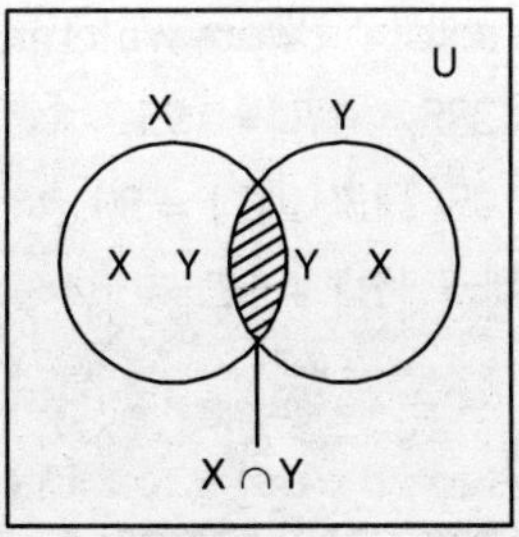

Fig. 17.1

Substitute the value of $|X-Y|$ and $|Y-X|$ in equation (1) from equations (2) and (3), thus

$$|X \cup Y| = |X| - |X \cap Y| + |Y| - |X \cap Y| + |X \cap Y|$$

$$|X \cup Y| = |X| + |Y| - |X \cap Y| \quad \textbf{Proved.}$$

The principle of Inclusion-Exclusion for 3-non-empty finite sets X, Y and Z is

$$|(X \cup Y \cup Z)| = |X| + |Y| + |Z| - |X \cap Y| - |X \cap Z| - |(Y-Z)| + |X \cap Y \cap Z|$$

Example 17.7. In an examination 65% of the total students passed in science, 70% of the total students passed in mathematics and 50% of the total students passed in science and mathematics both. If 90 students failed in both science and mathematics, calculate the total number of students appeared in the examination.

Solution: Let set A be the set of students passed in science and B be the set of students passed in mathematics then

$$|A| = 65\%, \quad |B| = 70\% \text{ and } |A \cap B| = 50\% \text{ then}$$

$$|A \cup B| = 65\% + 70\% - 50\% = 85\%.$$

That is 85% of the total students passed the examination.

So failed students are 100% – 85% = 15%. That is, 15 out of 100 failed, now if 90 failed then strength is $= \frac{100 \times 90}{15} 600$.

So the total number of students are 600.

Example 17.8. A survey of 500 T.V. viewers gives the following data:

(*i*) 285 viewers watch football.

(*ii*) 195 watch hockey.

(*iii*) 115 watch basketball.

(*iv*) 45 watch football and basketball.

(*v*) 70 watch football and hockey.

(*vi*) 50 watch hockey and basketball.

(*vii*) 50 viewers do not watch any of these three.

Calculate the number of viewers who watch all the three sports. Also calculate the number of viewers watching only one sport.

Solution: Let A be the set of viewers watching football, B bet the set of viewers watching hockey and C be the setof viewers watching basketball. Given that,

$$|A| = 285, \quad |B| = 195, \quad |C| = 115, \quad |A \cup B| = 70$$

$$|A \cup C| = 45 \quad |B \cup C| = 50.$$

then by $|A \cup B \cup C| = |A| + |B| + |C| - |A \cup B| - |A \cup C| - |B \cup C| + |A \cup B \cup C|$

Clearly,

$$|A \cup B \cup C| = 500 - 50 = 450 \text{ (50 do not watch any sport)}$$

So $450 - 285 - 195 - 115 + 70 + 45 + 50 = |A \cap B \cap C|$

$\Rightarrow$ $|A \cap B \cap C| = 20$

That is 20 people watch all the three sports.

(*ii*) Now calculate the number of people watching exactly one sport.

That is, only football or only hockey or only basketball.

So only A + only B + only C

$$= |(A \cup B \cup C)| - [|(A \cap B)| + |(A \cap C)| + |(B \cap C)|] + [2|(A \cap B \cap C)|]$$

$$= (500 - 50) - [70 + 50 + 45] + 2.20 = 325$$

Note: Another form of Inclusion Exclusion is

$$|A \cup B \cup C| = \text{only } (A) + \text{only } (B) + \text{only } (C) + [(A \cap B) + (B \cap C) + (A \cap C)] - 2[(A \cap B \cap C)]$$

17.2.6 Pigeonhole Principle

If there are *n* pigeonholes and more than *n* pigeons then at least one pigeonhole will be occupied by at least two pigeons.

Example 17.9. If there are 27 or more students in a class then the name of atleast two students' will begin with the same letter. As there are 26 letters in English alphabet.

Example 17.10. Suppose there are some red gloves, some blue gloves and some black gloves in a box. What is the least number of gloves that one should grab from the box (without looking at the contents) to be sure of getting a matching pair?

Solution : Let the three colours be the 3 pigeonholes. *i.e.*, $n = 3$.

Therefore, if one grabs $n + 1 = 4$ gloves (pigeons) atleast 2 of them will share same colour.

Example 17.11. Consider a tournament of *n*-players in which each of the *n*-players, plays against every other player. Also each player wins at least once. Show that there are at least 2 players having the same number of wins.

Solution : If each of n-players plays against every other player, then the number of wins for the player is at least 1 and at most $(n - 1)$.

These $(n - 1)$ wins corresponds to $(n - 1)$ pigeonholes and n-players corresponds to n-pigeons. So according to pigeonhole principle at least 2 pigeons (players) will share at least one pigeonhole (win). That is at least 2 players will have the same number of wins.

Extended Pigeonhole Principle

If there are *n* pigeonholes ($m \geq n$), then at least one pigeonhole contains at least ($p + 1$) pigeons. Where $p = [(m - 1)/n]$

Proof : If every pigeonhole contains at most p pigeons,

then the number of pigeons $\leq np \leq n \,.\, (m - 1)/n = m - 1 < m$.

which is a contradiction.

Hence the theorem.

Example 17.12. If seven colours are used to paint 50 cars, then show that at least 8 cars will have same colour.

Solution : Let the seven colours be 7 pigeonholes, *i.e.*, $n = 7$ and 50 cars be pigeons, *i.e.*, $m = 50$.

by extended pigeonhole principle at least $\left(\left[\frac{m-1}{n}\right]+1\right)$ cars will have same colour.

So number of cars having same colour $= \left[\frac{50-1}{7}\right] + 1 = 7 + 1 = 8.$

Example : Suppose there are 26 persons and 7 cars to transport them. Then $p = \left(\left[\frac{26-1}{7}\right]\right) = 3.$

Therefore, at least 1 car must have $p + 1$ *i.e.*, 4 or more persons.

17.3 PERMUTATION AND COMBINATION

Consider a set of n-distinct objects $\{x_1, x_2, ..., x_n,\}$ from which we have to take a sample of $r, 1 \le r \le n,$ objects. How many ways are there to do the task? However, the problem in not a complete one, in particular the phrase 'take a sample' is somewhat ambiguous.

(*a*) Is the order in which we take the sample important ? *For example,* if $r = 2$, do we consider first picking x_1 and then x_2 to be different from picking x_2 first and then x_1.

(*b*) Do we allow repetition of elements ? For instance, do we allow a sample to consists of two or three x_i's ?

Each of these questions has two answers. Thus the product rule says that the original problem is made up of four separate problems. That is the main event X is composed of four events. E_1, E_2, E_3 and E_4 as:

(*i*) How many such samples are there if the order is important and we allow repetitions ?

(*ii*) How many samples are there if the order is important but we don't allow repetitions ?

(*iii*) How many samples are there in which the order is not important and repetition is allowed ?

(*iv*) How many sample are there in which order is not important and also repetition is not allowed ?

If the order is important, then we say that our sample consists of **arrangements.** If the order is not important then we have **detections.** If repetition is allowed, then we are using **replacement**. An arrangement of a r-objects with replacement is called an r-sequence (or simply sequence).

For example, vehicle numbers are example of sequence.

17.3.1 Definition

An r-permutation (or simply permutation) is an arrangement without replacement.

Thus, the permutations which can be made by taking the letters $x, y, z\ t$ two at a time are twelve in number, namely,

$$xy, xz, xt, bz, bt, zt$$

$$yx, zx, tx, zy, tb, tz$$

each of these presenting a different arrangement of two letters. The permutation of n-objects taken r at a time is denoted by ${}^nP_r$ or $P(n, r)$.

A selection of r-objects with replacement is called an *r-multiset*. *For example,* the word MATHEMATICS is a multi-set with 2 M's, 2 A's, 2 T's, 1 H, 1 E, 1 I, 1 C and 1 S.

17.3.2 Definition

An r-combination (or simply combination) is a selection of r-objects without replacement.

Thus, the *combination* which can be made by taking the four letters x, y, z and t two at a time are six in numbers, namely,

$$xy,\ xz,\ xt,\ yz,\ yt \text{ and } zt$$

each of these presenting a different *selection* of two letters.

The combination of n-objects taken r at a time is denoted by ${}^nC_r$ or $C(n, r)$.

Theorem 17.1: The number of permutations of n distinct objects taken $r\ (0 \le r \le n)$ at a time is given by

$${}^nP_r = \frac{n!}{(n-r)!}$$

Proof: Consider the problem of arranging n-objects taken r at a time as filling r places from n-objects

Let there be r-empty places, as

Places: [1] [2] [3] ... [r]

Out of n-objects, there are n choices to fill place 1. Since repetition is not allowed so choices for second place are $(n-1)$, for third place are $(n-2)$ and so on. Thus choices left for the rth place are $(n-(r-1))$. **(Fig. 17.2)**

[1] [2] [3] ... [r] : Places

n   n 1   n 1   n (r 1) : Choices

Fig. 17.2

Hence the r-places can be filled in

$$n.(n-1).(n-2).\ldots.(n-(r-1)) \text{ ways.}$$

$$\Rightarrow \qquad {}^nP_r = n.(n-1).(n-2)\ldots(n-(r-1))$$

$$= \frac{n.(n-1).(n-2).....(n-(r-1).(n-r)!}{(n-r)!}$$

$$= \frac{n.(n-1).(n-2).....(n-(r-1).(n-r)...3.2.1}{(n-r)!}$$

Thus, $\quad {}^nP_r = \dfrac{n!}{(n-r)!}$

Notes : (*i*) We take all n-objects at a time then

$${}^nP_r = n(n-1)\,(n-2)\ldots 3\,.\,2\,.\,1 = n\,!$$

which is equivalent to fill n-places with n-objects without replacement.

(*ii*) When repetition of objects is allowed then each place has n-choices. Thus the total number of ways to fill r-places with n-objects, *with replacement*, are $n\,.\,n\,.\,n \ldots n$ (r-times).

That is $\quad {}^nP_r = n^r\ (r \geq 0)$

Example 17.13. How many different numbers can be formed by using six out of the nine digits 1, 2, 3, ..., 9 ?

Solution: Here we have a different objects (or things) and we have to find the number of permutations of them taken 6 at a time ;

$\therefore$ the required result $= {}^9P_6$

$$= \frac{9!}{(9-6)!} = \frac{9!}{3!} = \frac{9.8.7.6.5.4.\cancel{3!}}{\cancel{3!}}$$

$$= 9\,.\,8\,.\,7\,.\,6\,.\,5\,.\,4 = 60480.$$

Example 17.14. How many different arrangements can be made by taking 5 of the letters of the word EQUATION ?

Solution: Total number of letters in EQUATION = 8. Out of 8 we have to take 5 at a time. So the required result is

$${}^8P_5 = \frac{8!}{(8-5)!} = \frac{8!}{3!}$$

$$= 8\,.\,7\,.\,6\,.\,5\,.\,4 = 6720.$$

Example 17.15. How many permutations can be made out of the letters of the word TRIANGLE ? How many of these will begin with *T* and end with *E*.

Solution: Number of permutations of 8 letters of TRIANGLE taken 8 at a time is ${}^8P_8 = 8\,! = 40320.$

Now consider the problem as filling 8 position by 8-objects such that *T* fills the first place and *E* fills the last place.

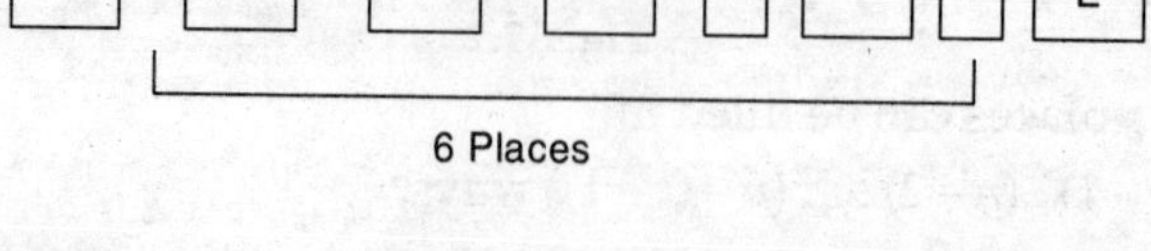

Now the remaining 6 places can be filled by 8 – 2 (T and E) = 6 objects in ${}^6P_6$ ways.

That is $\quad {}^6P_6 = 6\,! = 720$ ways

So there are 720 permutations starting with T and ending with E.

Example 17.16. If $4\ {}^nP_3 = 5^{n-1}P_3$, calculate n.

Solution: $\quad 4\ {}^nP_3 = 5^{n-1}P_3 \quad$ (given)

We know that $\quad {}^nP_r = \dfrac{n\,!}{(n-r)\,!}$

So $$4 \cdot \frac{n\,!}{n-3\,!} = 5\,\frac{(n-1)\,!}{(n-1-3)\,!}$$

$$\Rightarrow \quad \frac{4 \cdot n\,!}{(n-3)\,!} = \frac{5(n-1)\,!}{(n-4)\,!}$$

or $$\frac{4 \cdot n(n-1)\,!}{(n-3)(n-4)\,!} = \frac{5(n-1)\,!}{(n-4)\,!}$$

or $$4n = 5(n-3)$$

or $$4n = 5n - 15$$

or $$n = 15 \quad \textbf{Ans.}$$

Example 17.17. One wants to arrange 5 balls Red, Yellow, Green, Blue and Orange in a row such that red ball always occupies second position and such that the blue and yellow balls are always adjacent to each other.

Solution: The problem is to arrange 5 places with 5-balls as :

1	2	3	4	5
	R			

Red ball is at second position always blue and yellow balls are always adjacent. The possible arrangements are shown below :

G	R	B	Y	O	O	R	B	Y	G
G	R	O	B	Y	O	R	G	B	Y
G	R	Y	B	O	O	R	Y	B	G
G	R	O	Y	B	O	R	G	Y	B

In each arrangement the two balls blue and yellow are adjacent.

So there are 8 such arrangements are possible.

Example 17.18. How many number plates of length 5 can be made for vehicles such that

(*a*) Each plate contains two different letters followed by three different digits.

(*b*) The first digit cannot be zero.

Solution: (*a*) There are 26 ways to fill the very first position and 25 ways for the second position. Thus giving 26.25 ways to fill the first two positions by different letters. That is $^{26}P_2$. Out of the 10 digits, 0, 1, 2, 3, 4, 5, ..., 9. There are 10 ways to fill the third position, 9 to fill the fourth position and 8 to fill the fifth position. So that 10 . 9 . 8 ways to fill the last three positions of the number plate with three different digits. That is $^{10}P_3$.

Hence the total number of ways to solve the problem are 26 . 25 . 10 . 9. 8 = $^{26}P_2 \, . \, ^{10}P_3 = 468000$.

(*b*) '0' cannot be the first digit. If 0 is the first digit (that is 0 is at third place) then the next two places can be filled by remaining 9 digits in $^9P_2$ ways. So there are $^9P_2$ permutations exist for which digit 0 is the very first digit. So the total number of permutations where 0 is not the first digit are

$$^{26}P_2 \, . \, ^{10}P_3 - {}^9P_2 = 468000 - 72$$
$$= 467928.$$

17.3.3 Circular Permutation

Suppose *n* objects or persons ($x_1, x_2, ..., x_n$) are to be arranged around a circular table. There are (*n* – 1) ! such permutations, consider the following circular arrangements.

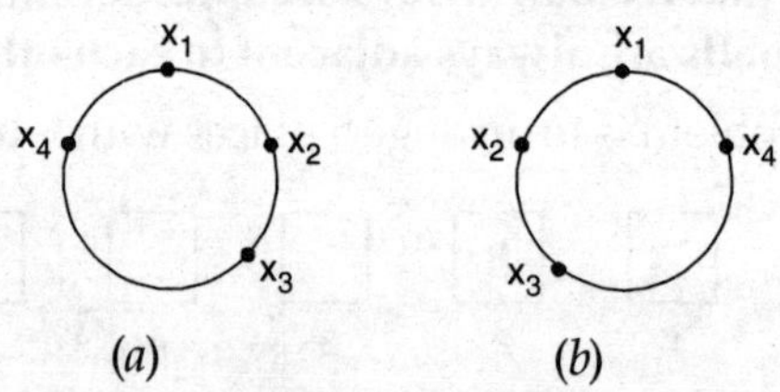

Fig. 17.3

In **Fig. 17.3**(*a*) the order of arrangement is **clockwise** whereas in **Fig. 17.3**(*b*) the order is **anticlockwise**. There are two different arrangements. When the distribution is made between the two arrangements of *n* different objects around a circle, then the number of arrangements are $(n-1)$! Whereas if no distinction is made between the clockwise and anticlockwise of *n* different objects around a circle, then the number of arrangements is $\frac{1}{2}(n-1)!$.

Example 17.19. Find the number of ways in which (*i*) five person can sit around a table (*ii*) five persons can sit around a table such that two persons insist on sitting next to one another.

Solution: (*i*) Five persons can sit around a table in $(5-1)! = 4! = 24$ ways.

(*ii*) Out of 5, two persons wants to sit next to each other, so treating the two person as one. There are 4 persons can be arranged in 3 ! ways around a circular table. Two persons can be arranged in 2 ! ways among themselves. Hence total number of ways are 3 ! . 2 ! = 12 ways.

Theorem 17.2. The number of arrangements of a multi-set containing n-objects, taken all at a time, when p of the objects are exactly alike of one kind. q of them exactly alike of another kind, r of them exactly alike of a third kind, and the rest all different, is given by

$$^{n}P_{p,q,r} \text{ or } P(n, p, q) = \frac{n!}{p!.q!.r!}$$

Proof: Out of n-objects, let p *of* them be a, q of them to be b and r of them tobe c and rest to be unlike.

Let x be the required number of permutations; then if the objects a were replaced by p unlike objects different from any of the rest, from *anyone* of the x permutations, without altering the position of any of the remaining objects one can form p ! new permutations. If this change can be made in each of the x permutations, then we should obtain $x . p!$ permutations.

Similarly if the q objects b were replaced by q unlike objects say $b_1, b_2, ..., b_q$, the number of permutations would be

$$x . p! . q!$$

and by replacing the r-objects c by r unlike objects. Then the final permutation should be

But all the n-objects are different and therefore admit of n ! permutations among themselves.

Hence $\quad x . p! . q! . r! = n!$

$$\Rightarrow \qquad x = \frac{n!}{p!.q!.r!} = P(n; p, q, r)$$

Note: The other theorem can be generated, as "If out of n-objects, n_1 are alike, n_2 are alike of the first kind.

Second kind and ... n_r are alike of the rth kind then the number of arrangements taken n of the objects at a time is given by

$$P(n; n_1, n_2, ..., n_r) = \frac{n!}{n_1!.n_2!...n_r!}$$

where $\qquad n_1 + n_2 + ... + n_r \leq n$

Example 17.20. How many different permutations can be made out of the letters of the word "ASSASSINATION" taken all together ?

Solution: We have a total of 13 letters, out of which 4 are S, 3 are A, 2 are I and 2 are N, 1 is T and 1 is 0.

$$\text{So the number of permutations} = \frac{13!}{4!.3!.2!.2!.1!.1!}$$

$$= 10810800$$

Example 17.21. How many numbers can be formed with the digits 1, 2, 3, 4, 3, 2, 1, so that the odd digits always occupy the odd places ?

Solution: There are a total of 7 places, out of which 4 are odd and 3 are even. Thus the 4 odd places can be occupied by 4 odd digits 1, 3, 3, 1 in $\frac{4!}{2!.2!}$ ways. As there are two 1*s* and two 3*s*.

So the remaining 3-even places can be occupied by the 3-even numbers 2, 4, 2, having two 2*s* in $\frac{3!}{2!}$ ways.

Hence the total number of permutations are $\frac{4!}{2!.2!} \times \frac{3!}{2!} = 18$

Example 17.22. How many anagrams can be made by using the letters of the ENGLAND word such that,

(*a*) Anagrams begin and end with vowel.

(*b*) All the vowels come together.

Solution: (*a*) We have 2 vowels and 5 consonants in which 2 consonants are alike. The first place can be filled in 2 ways and the last in one way. The rest of the 5 places can be filled by 5-constants in which 2 consonants are alike in $\frac{5!}{2!} = 120$.

(*b*) Assume that the vowels (*E*, *A*) is a single letter. Then the letters (*E*, *A*), *N*, *G*, *L*, *N*, *D* can arranged in $\frac{6!}{2!}$ ways. Also *E*, *A* can be arranged among themselves in 2 ! = 2 ways.

Hence the total number of anagrams = $\frac{6!}{2!} . 2 = 6!$ ways.

Example 17.23. In how many ways can be arrange 15 books on a shelf, where out of 15, 7 books are of Hindi 6 are of history and 2 are of science. Such that

(*a*) No distribution is to be made in the books of the same subjects ?

(*b*) All books of the same subject are together ?

Solution: (*a*) $n = 15$, $p = 7$, $q = 6$ and $r = 2$, no distinction is required so total number of such permutations are $\frac{15!}{7!.6!.2!} = 180180$.

(*b*) All books of same subject are together, then 7 books of Hindi can be arranged in 7 ! ways, 6 books of history can be arranged in 6 ! ways and science books can be arranged in 2 ! ways. This, altogether 7 ! . 6 ! . 2 ! = 7257600 ways.

Example 17.24. A student is to answer 8 out of 10 questions on an examination. How many choices are there ? How many choices exist if

(*i*) He must answer the first two questions ?

(*ii*) He must answer the first or second questions but not the both ?

(iii) He answer exactly four out of the first five six questions ?

(iv) He must answer at least four of the first six questions ?

Solution: Total choices are ${}^{10}C_8 = 45$

(i) If he must answer the first two questions. Then the left 6 questions can be selected from the remaining 8 questions in ${}^8C_6 = 28$ ways.

(ii) If he selects the first question, then he cannot answer the second one. So in that case out of 10 – 1 – 1 = 8 questions, 8 – 1 = 7 question can be selected in ${}^8C_7 = 8$ ways.

Similarly, if he answer the second question, he cannot answer the first one. Thus having a choice of ${}^8C_7 = 8$ ways.

Hence the total number of choices are 8 + 8 = 16.

(iii) If he answer exactly 4 out of first 5 questions, selection can be made in ${}^5C_4 = 5$ ways. The rest of the 4 questions can be answered out of the remaining 5 questions in ${}^5C_4 = 5$ ways.

Hence the total number of ways are 5.5 = 25.

(iv) If he can answer at least four out of the first six questions (That is he answer 4 out of 6, 5 out of 6 or 6 out of 6) in ${}^6C_4$, ${}^6C_5$ or ${}^6C_6$ ways.

If he answer 4 out of 6 questions in ${}^6C_4$ ways, then the remaining 4 questions can be selected out of the remaining 4 question in ${}^4C_4 = 1$ way.

Thus the total number of ways are $= {}^6C_4 . {}^4C_4 = 15$.

Similarly, 5 out of 6 and 3 out of 4 questions can be selected in ${}^6C_5 . {}^4C_3 = 24$ ways.

an 6 out of 6 and 2 out of 4 remaining questions can be selected in ${}^6C_6 . {}^4C_2 = 6$ ways.

Thus, total number of ways are 15 + 24 + 6 = 45.

17.3.4 Restricted Permutation

The number of permutations of *n* different objects (or elements) taken *r*-at a time in which *k* particular objects are (*i*) never included will be $P(n-k, r)$ or ${}^{n-k}P_r$ and (*ii*) always included will be $P(n-k, r-k) . P(r, k)$ where $n \geq k$ and $r \geq k$.

Example 17.25. How many words of three letters can be formed with the letters of the word BEAUTY if

(i) *B, U* and *T* are not to be included.

(ii) *B, U* and *T* are to be included always.

Solution: We are given with 6 letters out of which we are to take 3 letters after keeping *B, U* and *T* aside. That is $k = 3$. $n = 6$ and $r = 3$.

Then the total number of permutations are $P(n - k, r) = P(6 - 3, 3) = P(3, 3) = 6$ ways.

(*iii*) We are interested in including 3 letters *B*, *U* and *T* for the three places out of 3 places of three lettered words which can be done in $3! = 6$ ways. By the other formula $P(n-k, r-k) . P(r . k)$

$$= P(6-3, 3-3) . P(3, 3)$$
$$= P(3, 0) . P(3, 3)$$
$$= 1 . 3! = 6 \text{ ways}$$

Example 17.26. *m* men and *n* women are to be seated in a row so that no two women sit together. If $m > n$, show that the number of ways in which they can be seated is

$$\frac{m!(m+1)!}{(m-n+1)!}$$

Solution: Let the *m* men take their seats. They can sit in ${}^mP_m$ ways.

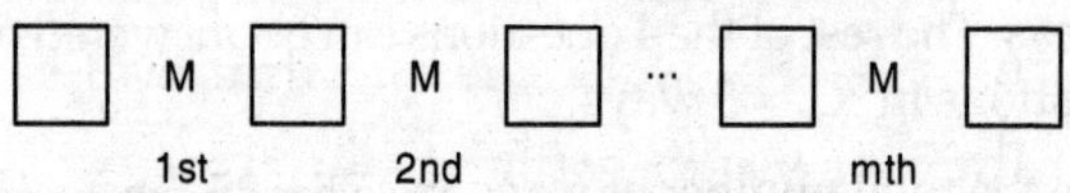

Since no two women are to sit together, we have $(m+1)$ places for the women, shown by □. Thus, the women can take their seats in ${}^{m+1}P_n$ ways. Hence the total number of ways in which *m* men and *n* women can take their seats subject to the given condition is

$$({}^mP_m) . ({}^{m+1}P_n) = \frac{m!}{0!} . \frac{(m+1)!}{(m-n+1)!} = \frac{m!(m+1)!}{(m-n+1)!}$$

Example 17.27. In how many ways can 7 boys and 5 girls be seated together in a row, so that no two girls are together ?

Solution: 7 boys can take their seats in ${}^7P_7 = 7!$ ways. Since no two girls are to sit together, we have $(7+1) = 8$ places for the girls. Thus the girls can take their seats in ${}^8P_5$ ways.

Hence the total number of ways are

$$7! . {}^8P_5 = \frac{7! . 8!}{(8-5)!} = \frac{7! . 8!}{3!} \; 33868800.$$

Example 17.28. A tea party is arranged for 2 m people along two sides of a long table with *m* chairs on each side. *r* men wish to sit on one particular side and *s* on the other, where $r, s \le m$. In how many ways can they be seated ?

Solution: We can arrange *r* persons on *m* chairs on a particular side in ${}^mP_r$ ways and persons on *m* chairs on the other side in ${}^mP_s$ ways. We can arrange $(2m - r - s)$ remaining persons on the remaining $(2m - r - s)$ chairs in ${}^{2m-r-s}P_{2m-r-s}$ ways.

Thus the number of ways of arranging the persons subject to the given conditions is

$$({}^mP_r) . ({}^mP_s) . ({}^{2m-r-s}P_{2m-r-s}).$$

Example 17.29. A family consists of a grandfather, m sons and daughter and $2n$ grandchildren. They are to be seated in a row for dinner. The grandchildren wish to occupy the n seats at each end and the grandfather refuses to have a grandchild on either side of him. In how many ways can be family be made to sit?

Solution: The total number of seats required at the table is $1 + m + 2n$. The children can occupy the n seats on either side of the table in $({}^{2n}P_{2n})$ ways. The grandfather can occupy a seat in $({}^{m+1}P_1)$ ways and the remaining seats can be occupied in $({}^{2n}P_{2n})$ ways. Hence the total number of ways are

$$({}^{2n}P_{2n}) \cdot ({}^{m+1}P_1) \cdot ({}^{m}P_{m}) = 2n\,! \,.\, m\,!\,(m+1).$$

Example 17.30. A question paper is split into two parts – Part A and Part B contains 5 questions and Part B has 4. Each question in Part A has an alternative. A student has to attempt at least one question from each part. Find the number of ways in which the student can attempt the question paper.

Solution: For each question in Part A, the student has 3 choices :

(*a*) Student does not attempt the question;

(*b*) Student attempts the first part of the question; and

(*c*) Student attempts the alternative part of the question.

Therefore, the total number of choices is 3^5, (3 choices for each of the 5 questions in Part A). But this includes a choice in which the student does not attempt any question in Part A.

Thus the total number of choices such that the student attempt at least one question from Part A is

$$3^5 - 1 = 243 - 1 = 242.$$

Similarly, Part B can be solved in $2^4 - 1 = 15$ ways. Hence the number of ways in which the student can attempt paper is $(242) \,.\, (15) = 3630$.

Theorem 17.3. The total number of combinations of n dissimilar things taken r at a time is ${}^nC_r = \dfrac{n!}{r!\,.(n-r)!}$

Proof: Let ${}^nC_r$ denote the required number of combinations. Then each of these combinations consists of a group of r dissimilar things which can be arranged among themselves in $r\,!$ ways.

Hence ${}^nC_r \,.\, r\,!$ is equal to the number of arrangements of n things taken r at a time,

That is $${}^nC_r \times r\,! = {}^nP_r$$

$${}^nC_r = \frac{{}^nP_r}{r\,!} = \frac{n\,!}{(n-r)\,!\,.\,r\,!}$$

Hence $${}^nC_r = \frac{n\,!}{(n-r)\,!\,.\,r\,!}$$

Notes : (*i*) If $r = n$, then $^nC_r = \dfrac{n!}{(n-n)!.n!} = 1$

(*ii*) If $r = 0$, then $^nC_0 = \dfrac{n!}{(n-0)!.0!} = 1$

Example 17.31. In how many ways can a selection of 5 books out of 12 books be made, (*a*) when one specified book is always included, (*b*) when one specified book is always excluded ?

Solution: (*a*) When a specified book is always included in every selection, we have only to choose 4 books out of the remaining 11.

Hence the number of ways $^{11}C_4 = \dfrac{11!}{4!.5!} = 330.$

(*b*) Since a specified book is always to be excluded, we have to select 5 books out of the remaining 11 books.

Hence the number of ways $= {}^{11}C_5$.

$$= \frac{11\times 10\times 9\times 8\times 7}{1\times 2\times 3\times 4\times 5} = 462$$

Example 17.32. Prove that $^nC_r = {}^{n-1}C_r + {}^{n-1}C_{r-1}$

Solution: Consider R.H.S.

$$^{n-1}C_r + {}^{n-1}C_{r-1}$$

$$= \frac{(n-1)!}{r!.(n-1-r)!} + \frac{(n-1)!}{(r-1)!.(n-r)!}$$

$$= (n-1)!\left\{\frac{1}{r!.(n-1-r)!} + \frac{1}{(r-1)!.(n-r).(n-1-r)!}\right\}$$

$$= \frac{(n-1)!}{(r-1)!.(n-r-1)!}\left\{\frac{1}{r} + \frac{1}{(n-r)}\right\}$$

$$= \frac{(n-1)!}{(r-1)!.(n-r-1)!}\left\{\frac{n-r+r}{(n-r).r}\right\}$$

$$= \frac{n.(n-1)!}{r.(r-1)!.(n-r).(n-r-1)!}$$

$$= \frac{n!}{r!.(n-r)!} = {}^nC_r \text{ (L.H.S.)}.$$

Example 17.33. A collection of 100 pens contains 8 defective ones.

(*a*) In how many ways can a sample of 10 pens be selected ?

(*b*) In how many ways can a sample of 10 pens be selected which contain 7 good pens and 3 defective ones ?

(*c*) In how many ways can the sample contains 10 pens such that either the sample contains 7 good pens and 3 defective ones or 5 good ones and 5 defective pens ?

Solution: (*a*) Out of 100 pens, 10 pens can be selected in

$$^{100}C_{10} = \frac{(100)!}{10!.90!} = C(100, 10).$$

(*b*) A sample of 10 and out of 100, such that sample contains 7 good pens and 3 defective ones in

$$^{92}C_7 . {}^8C_3 \text{ ways}$$

Since out of 100 pens 8 are defective, so 92 pens are good and 7 can be selected in $^{92}C_7$ ways. Out of 8 defective pens 3 can be selected in $^8C_3$ ways. Thus, the total becomes $^{92}C_7 . {}^8C_3$.

(*c*) A sample containing 7 good pens and 3 defective ones can be selected in $(^{92}C_7) . (^8C_3)$ ways. Another sample of 5 good pens and 5 defective ones can be selected in $(^{92}C_5) . (^8C_5)$ ways.

Hence the total number a ways to selected either of the samples is $(^{92}C_7) . (^8C_3) + (^{92}C_5) . (^8C_5)$.

Example 17.34. Prove that $^nC_r = {}^nC_{n-r}$.

Solution: R.H.S.,

$$^nC_{n-r} = \frac{n!}{(n-r)!.(n-n+r)!} = \frac{n!}{(n-r)!.r!}$$
$$= {}^nC_r; \quad \text{L.H.S.}$$

Note: Since $^nC_r = {}^nC_{n-r}$, so these two are called complementary combinations.

Example 17.35. Out of 7 consonants and 4 vowels, how many words can be made each containing 3 consonants and 2 vowels.

Solution: The number of ways of choosing the three consonants in $^7C_3$ and the number of ways of choosing the 2 vowels in $^4C_2$; and since each of the first groups can be associated with each of the second, the number of combined groups, each containing 3 consonants and 2 vowels is, $^7C_3 \times {}^4C_2$.

Further, each of these groups contains 5 letters, which may be arranged among themselves in 5 ! ways. Hence the required number of words.

$$= (^7C_3) . (^4C_3) . (5!)$$

$$= \left(\frac{7!}{3!.4!}\right) . \left(\frac{4!}{3!.1!}\right) . (5!) = 5 \times 7! = 25200$$

17.3.5 Restricted Combinations

The number of combinations of *n*-objects taken *r* at a time in which *k* special objects are (*i*) always included in the selection is $C(n-k, r-k)$ and (*ii*) never included in the selection is $C(n-k, r)$, where $k \le n$.

This selection can be made in two ways. When *k* particular objects are included in the selection, then in our selection of *r*-objects out of *n*-objects we keep the *k*-objects aside and now the rest of the $(r-k)$ objects can be selected out of $(n-k)$ objects in

$$^{n-k}C_{r-k} \equiv C(n-k, r-k) \text{ ways.}$$

whereas, if we don't want to include the particular objects, then the r-objects can be selected from the remaining $(n-k)$ objects in

$$^{n-k}C_r \equiv C(n-k, r) \text{ ways}$$

as we don't want to include k-particular object in our selection. So we discard them from the whole set of n-objects and hence the total r-objects can be selected in $C(n-k, r)$ ways out of $(n-k)$ objects.

Example 17.36. From a group of 25 people, 10 are to be selected for a committee. There are 3 people say *A, B* and *C* who decided that either all of three will join the committee or none of them will join. Calculate the number of such combinations if (*a*) three people *A, B* and *C* are included and (*b*) *A, B* and *C* are not include.

Solution: (*a*) Out of 25 people if 3 (*i.e.*, *A*, *B* and *C*) are included then the remaining people are $25-3=22$ and now we require to select only 7 people as are already there in the committee of 10.

So 7 people from 22 can be selected in $^{22}C_7$ ways.

Alter: Clearly $n = 25, \quad r = 10 \quad \text{and} \quad k = 3.$

then the selection can be made in

$$C(n-k, r-k) = C(25-3, 10-3) = C(22, 7) \text{ ways.}$$

(*b*) Here *A, B* and *C* are not included. That is 3-particular objects are not included in selecting 10 objects from a set of 25.

Clearly, $n = 25, \quad k = 3 \quad \text{and} \quad r = 10.$

k is not included and the selection can be made in

$$C(n-k, r) = C(25-3, 10) = C(22, 10) \text{ ways.}$$

17.3.6 Other Combinations

(*i*) The number of ways in which $(m+n)$ objects or things can be divided into two groups containing m and n things respectively is

$$\frac{(m+n)!}{m!.n!} = C(m+n, m)$$

Note: If $m = n$, the groups are equal, and in this case the number of different ways of sub-division is $\dfrac{2m!}{m!.m!.2!}$; for in any one way it is possible to interchange the two groups without obtaining a new distribution.

(*ii*) The number of ways in which $m+n+p$ things can be divided into three groups containing m, n and p things respectively is given by

$$\frac{(m+n+p)!}{m!.n!.p!}$$

Note: If $m = n = p$ then the obtain $\dfrac{3m!}{m!.m!.m!.3!}$ ways of distributing the total $3m$ things into 3 equal groups.

Similarly, $m_1 + m_2 + m_3 + ... + m_n$ things can be divided into n groups containing $m_1, m_2, ..., m_n$ things respectively in

$$\frac{(m_1 + m_2 + ... + m_n)!}{m_1!.m_2!...m_n!} \text{ ways}$$

and again if $m_1 = m_2 = ... = m_n = n$, that is all the n groups are equal then the total ways are

$$\frac{nm!}{(m!.m!...m!)\,n!} = \frac{(nm)!}{n(m!).n!}$$

(*iii*) The total number of combinations of $(m_1 + m_2 + m_3 + ...)$ objects in which m_1 are alike, m_2 are alike, and so on, taken any number at a time is

$\{(m_1 + 1)(m_2 + 1) ... (m_k + 1) ...\} - 1.$

(*iv*) The total number of ways in which it is possible to make a selection by taking some or all of n things is $2^n - 1$.

Example 17.37. Find the number of ways in which 15 recruits can be divided into 3 equal groups. Also find out the number of ways in which they can be drafted into three different regiments; five into each.

Solution: 15 has to be divided into 3 equal groups. That is each group will have 5 recruits.

Hence the total number of ways are

$$\frac{15!}{5!.5!.5!.3!} = 126126$$

Now we want to find out the number of ways also in which they can be drafted into three different regiments such that 5 recruits in each regiment *i.e.*, out of 15 objects, 5 objects are of one kind, 5 of second kind and 5 of third kind.

Hence total number of ways are $\dfrac{n!}{p!.q!.r!}$

$$= \frac{15!}{5!.5!.5!} = 756756$$

Note: The formulae for *permutation* should not be used until the suitable *selections* required by the question have been made.

Example 17.38. In how many ways can 12 things be divided equally among 4 persons.

Solution: Here 12 things have to be divided equally, among 4 persons, that is 3 things per person. That is $m = 3, n = 4$.

Thus the number of ways are $\dfrac{(nm)!}{(m!)^n.n!} = \dfrac{12!}{(3!)^4.4!}$

$$= \frac{(12)!}{(3!)^4.4!} = 369600$$

Example 17.39. Find the number of ways in which (*i*) a selection (*ii*) an arrangement of four letters can be made from the letters of the word *proportion*.

Solution: (*i*) There are letters of six different sorts, namely *o, o, o; p, p: r, r: t: i* and *n*.

In finding groups of four these may be selected as follows:

(*a*) Three letters alike, one different.

(*b*) Two letters alike, two others alike.

(*c*) Two letters alike, the other two different

(*d*) All four letters are different.

(*a*) For each of the five letters, *p, r, t, i* and *n*, can be taken with the single group of three alike letters *o*. Hence the total number of ways for selection this case is 5.

(*b*) Here the selection can be made in $C(3, 2)$ ways; for we have to choose two the out of 3 pairs *o, o; p . p; r r*. This gives a selections.

(*c*) Selection can be made in ${}^3C_1 \times {}^5C_2$; for we have to choose one of the three pairs *o, o ; p, p ; r, r* and then the two from the remaining 5 letters. This gives 30 selections.

(*d*) When all four are different, the selection can be made in ${}^6C_4 = 15$ ways.

Thus the total number of selections is $5 + 3 + 30 + 15 = 53$.

(*ii*) In finding different arrangements of 4 letters we have to permute in all possible ways each of the foregoing groups.

(*a*) In 4 letters word, three are alike and one is different, total number of permutations is

$$\frac{4!}{3!.1!} = 4.$$

Selection can be made in 5 ways. So that number of arrangements is $5 \times 4 = 20$.

(*b*) Here $n = 4, p = 2$ and $q = 2$, thus, the total number of arrangements is

$$3 \times \frac{4!}{2!.2!} = 18.$$

(*c*) $n = 4, p = 2, q = 1, r = 1$, thus, the total number of arrangements is

$$30 \times \frac{4!}{2!} = 360$$

(*d*) $n = 4, p = q = r = t = 1$ (*i.e.*, all are different). So total number of arrangements in this case is $15 \times 4\,! = 360$.

Hence the total number of arrangements is

$$20 + 18 + 360 + 360 = 758.$$

Example 17.40. A man has 5 friends ; in how many ways may be invite one or more of them for lunch ?

Solution: Out of n, some or all may be selected in $2^n - 1$ ways. So the total number of ways of inviting some or all 5 friends on lunch is $2^5 - 1 = 31$.

The result can be verified in following manner. The guests may be invited singly, in two's, three's ... ; therefore, the number of selections

$$= 5C_1 + 5C_2 + 5C_3 + 5C_4 + 5C_5 = 5 + 10 + 10 + 5 + 1 = 31.$$

17.4 GENERATING FUNCTION

In this section we introduce an alternative way to represent a sequence of numbers called generating function. Generating functions are very useful technique for solving recurrence relation which we will discuss in next section and also important for solving counting problems.

17.4.1 Definition

For the sequence $(a_0, a_1, a_2, ..., a_n)$ define a series

$$G(x) = a_0 + a_1x + a_2x^2 + ... + a_nx^n + ... = \sum_{n=0}^{\infty} a_nx^n$$

This infinite series is called generating function for the numeric functions $(a_0, a_1, a_2, ..., a_n)$.

For example, $G(x) = 1 + 2x + 4x^2 + 6x^3 + ...$ is a generating function for the sequence $(1, 2, 4, 6, ...)$.

Example 17.41. Find the generating function for the numeric functions (1, 2, 4, 8, ...).

Solution: $a_0 = 1, = 2^0, a_1 = 2 = 2^1, a_2 = 4, = 2^2 \quad a_3 = 8 = 2^3 ...$

So $$G(x) = a_0 + a_1x + a_2x^2 + ... + a_n x^n + ... = \sum_{n=0}^{\infty} 2^nx^n.$$

which is the required generating function

17.5 CLOSED TERM EXPRESSION FOR REPRESENTING GENERATING FUNCTION

Closed form expression is a compact way to represent a sequence and hence a generating function.

For example, the generating function

$$G(x) = 2^0x^0 + 2^1x + 2^2x^2 + ... + 2^nx^n + ...$$

$$= \sum_{n=0}^{\infty} 2^nx^n$$

For the sequence $(2^0, 2^1, 2^2, ..., 2^n, ...)$ is represented in compact from as

$$G(x) = 2^0 + 2^1x + 2^2x^2 + ... + 2^nx^n + ...$$

$$= 1 + 2x + 2^2x^2 + ... + 2^nx^n + ...$$

$$= \frac{1}{1-2x}$$

Notes : (*i*) The sum of infinite G.P.

$$a_0 + a_1x + a_2x^2 + a_3x^3 + \dots + a_nx^n + \dots$$

is $\dfrac{a_0}{1-r}$ if $r < 1$ and $\dfrac{a_0}{r-1}$ if $r > 1$

(*ii*) The sum of finite G.P.

$$a_0 + a_1x + a_2x^2 + \dots + a_nx^n$$

is $\dfrac{a(1-r^n)}{(1-r)}$ if $r < 1$ and $\dfrac{a(r^n-1)}{r-1}$ if $r > 1$

where, $r = \dfrac{a_1}{a_0} = \dfrac{a_2}{a_1} = \dots = \dfrac{a_n}{a_{n-1}}$

Example 17.42. Find the generating function for the sequence 1, – 1, 1, – 1,...

Solution: Clearly $a_0 = 1,\ \ a_1 = -1,\ \ a_2 = 1,\ \ a_3 = -1 \dots$

hence $a_n = (-1)^n$

So $G(x) = 1 . + (-1) . x + 1 . x^2 + (-1) . x^3 + \dots$

$= 1 - x + x^2 - x^3 + \dots$

which is an infinite G.P. with

$$r = \frac{a_1}{a_0} = \frac{a_2}{a_1} = \dots = \frac{a_n}{a_{n-1}} = -1.$$

So $G(x) = 1 - x + x^2 - x^3 + \dots = \dfrac{1}{1+x}$

Hence the generating function for the sequence (1, – 1, 1, – 1, ...) is

$$G(x) = \frac{1}{1+x}$$

Some of the closed form expressions for the generating functions and the corresponding sequences are given below:

Sequence	*Generating function*
$(-1)^n$	$\dfrac{1}{1+x}$
1	$\dfrac{1}{1-x}$
a^n	$\dfrac{1}{1-ax}$
$(-a)^n$	$\dfrac{1}{1+ax}$
n	$\dfrac{x}{(1-x)^2}$
$n+1$	$\dfrac{1}{(1-x)^2}$

n^2	$\frac{(1+x)x}{(1-x)^3}$
na^n	$\frac{ax}{(1-ax)^2}$
$(n+1)\,a^n$	$\frac{1}{(1-ax)^2}$
$\frac{1}{n!}$	e^x
$n(n+1)$	$\frac{2x}{(1-x)^3}$
$C(m, n)$	$(1+x)^m$
$(C(m-1+n, n))$	$\frac{1}{(1-x)^m}$

Example 17.43. Find the generating function for the sequence whose *n*th term is $a_n = n$.

Solution: Given that $a_n = n$.

So $$a_0 = 0, \quad a_1 = 1, \quad a_2 = 2, \quad a_3 = 3, \ldots$$

Thus $$G(x) = 0 + 1\,.\,x + 2x^2 + 3x^3 + \ldots + nx^n + \ldots$$

$$= x + 2x^2 + 3x^3 + \ldots + nx^n \qquad \ldots(1)$$

Multiplying equation (1) by x

$$x\,.\,G(x) = x^2 + 2x^3 + 3x^4 + \ldots \qquad \ldots(2)$$

Substract equations (2) from equation (1)

$$G(x) = x + 2x^2 + 3x^3 + \ldots$$

$$-x\,.\,G(x) = -x^2 \pm 2x^3 \pm \ldots$$

$$G(x)\{(1-x)\} = x + x^2 + x^3 + \ldots$$

$$G(x)\{(1-x)\} = x(1 + x + x^2 + \ldots) = \frac{x}{1-x}$$

$$\Rightarrow \qquad G(x) = \frac{x}{(1-x)^2}$$

Example 17.44. Find the generating function for the numeric function

$$a_n = 2^n, \qquad n \geq 0$$

Solution: Given that $a_n = 2^n, \quad n \geq 0.$

So $$a_0 = 2^0, a_1 = 2^1, a_2 = 2^2, \ldots$$

Thus $$G(x) = 2^0 + 2^1x + 2^2x^2 + \ldots + 2^nx^n + \ldots$$

$$= \sum_{n=0}^{\infty} 2^nx^n = \frac{1}{1-2x}$$

Hence the required generating function

$$G(x) = \frac{1}{1-2x}$$

Example 17.45. Find the generating function of the numeric function

$$a_n = C(m, n), \quad n \geq 0$$

Solution: $a_0 = C(m, 0), \quad a_1 = C(m, 1), \quad a_2 = C(m, 2), \ldots$

So $G(x) = C(m, 0) . C(m, 1) . x + C(m, 2) . x^2 + \ldots$

$$= 1 + \frac{m!}{(m-1)!} . x + \frac{m!}{(m-2)! . 2!} x^2 + \ldots$$

$$= 1 + mx + \frac{m(m-1)}{2!} x^2 + \ldots$$

$\therefore$ $G(x) = (1 + x)^m$

Example 17.46. Find the generating function for the numeric function

(i) $a_n = \frac{1}{(n+1)!}, \quad n \geq 0$ **(ii)** $a_n = C(n+3, n), \quad n \geq 0.$

Solution: (*i*) Given that $a_n = \frac{1}{(n+1)!}, \quad n \geq 0$

$$a_0 = \frac{1}{1!}, \quad a_1 = \frac{1}{2!}, \quad a_2 = \frac{1}{3!}, \ldots$$

So $G(x) = \sum_{n=0}^{\infty} a_n x^n = \frac{1}{1!} + \frac{1}{2!} x + \frac{1}{3!} x^2 + \ldots$

$$= \frac{1}{x} \left[-1 + 1 + \frac{x}{1!} + \frac{x^2}{2!} + \frac{x^3}{3!} + \ldots \right]$$

$$= \frac{1}{x} [-1 + e^x] = \frac{e^x - 1}{x}$$

Hence $G(x) = \frac{e^x - 1}{x}$

(*ii*) $a_n = C(n+3, n), \quad n \geq 0$

$$G(x) = \sum_{n=0}^{\infty} a_n x^n$$

$$= \sum_{n=0}^{\infty} C(n+3, n) . x^n$$

$$= C(3, 0) . x^0 + C(4, 1) x + C(5, 2) x^2 + \ldots$$

$$= 1 + 4x + 10x^2 + \ldots$$

$$= (1-x)^{-4}$$

$$= \frac{1}{(1-x)^4}, \quad |x| < 1.$$

Thus, the require generating function

$$G(sx) = \frac{1}{(1-x)^4}, \quad |x| < 1.$$

Example 17.47. Determine the numeric function for the corresponding generating functions.

(a) $G(x) = \dfrac{9}{1-5x}$ **(b)** $G(x) = \dfrac{10}{1-x} + \dfrac{12}{2-x}$

(c) $G(x) = \dfrac{6-29x}{1-11x+30x^2}$ **(d)** $G(x) = (1+x)^n + (1-x)^n$

Solution: $G(x) = \dfrac{9}{1-5x}$

$$\sum_{n=0}^{\infty} a_n x^n = 9 \cdot \frac{1}{1-5x} = 9 \cdot ((1-5x)^{-1})$$

$$= 9 \cdot (1 + 5x + (5x)^2 + (5x)^3 + \ldots)$$

$$= 9 \cdot \sum_{n=0}^{\infty} 5^n x^n$$

Hence the required numeric function is

$$a_n = 9 \cdot 5^n, \quad n \geq 0$$

(*b*) $G(x) = \dfrac{10}{1-x} + \dfrac{12}{2-x}$

$$\sum_{n=0}^{\infty} a_n x^n = 10 \cdot (1-x)^{-1} + 12 \cdot (2-x)^{-1}$$

$$= 10 \cdot [1 + x + x^2 + x^3 + \ldots] + \frac{12}{2}\left(1 - \frac{x}{2}\right)^{-1}$$

$$= 10\,[1 + x + x^2 + \ldots] + 6\left[1 + \frac{x}{2} + \left(\frac{x}{2}\right)^2 + \left(\frac{x}{2}\right)^3 + \ldots\right]$$

$$= 10 \cdot \sum_{n=0}^{\infty} 1^n \cdot x^n + 6 \cdot \sum_{n=0}^{\infty} \left(\frac{1}{2}\right)^n x^n$$

$$\therefore \quad a_n = 10 \cdot (1)^n + 6 \cdot \left(\frac{1}{2}\right)^n, \quad n \geq 0$$

(*c*) $$G(x) = \frac{6-29x}{1-11x+30x^2} = \frac{6-29x}{(1-5x)(1-6x)} \qquad \ldots(1)$$

Using partial fraction of (1)

$$\frac{6-29x}{(1-5x)(1-6x)} = \frac{A}{(1-5x)} + \frac{B}{(1-6x)}$$

Multiplying both sides by $(1-5x).(1-6x)$, we obtain

$$6-29x = A(1-6x) + B(1-5x) \qquad ...(2)$$

Put $\qquad x = \frac{1}{5}$ in equation (2)

$$A = -1$$

and put $x = \frac{1}{6}$ in equation (2)

$$B = 7$$

Thus $\qquad \frac{6-29x}{(1-5x)(1-6x)} = \frac{-1}{(1-5x)} + \frac{7}{(1-6x)}$

Now from equation (*i*) we obtain

$$G(x) = \frac{6-29x}{(1-5x)(1-6x)} = \frac{-1}{(1-5x)} + \frac{7}{(1-6x)}$$

$$\therefore \qquad \sum_{n=0}^{\infty} a_n x^n = -1 \sum_{n=0}^{\infty} 5^n x^n + 7 . \sum_{n=0}^{\infty} 6^n x^n.$$

Thus $\qquad a_n = -1 . 5^n + 7.6^n$

or $\qquad a_n = -1 . (5)^n + 7.(6)^n, \quad n \geq 0$

(*d*) $\qquad G(x) = (1+x)^m + (1-x)^{+m}$

$$\sum_{n=0}^{\infty} a_n x^n = \left[1 + m \cdot x + \frac{m(m+1)}{2!} x^2 + \frac{m(m-1)(m-2)}{3!} x^3 + ...\right]$$

$$+ \left[1 - m \cdot x + \frac{m(m-1)}{2!} x^2 + \frac{m(m-1)(m-2)}{3!} x^3 + ...\right]$$

$$= \left[\binom{m}{0} + \binom{m}{1} x + \binom{m}{2} x^2 + ...\right] + \left[\binom{m}{0} - \binom{m}{1} x + \binom{m}{2} x^2 + ...\right]$$

$$= \sum_{n=0}^{\infty} C(m, n) . x^n + \sum_{n=0}^{\infty} (-1)^n C(m, n) . x^n$$

Thus $\qquad a_n = C(m, n) + (-1)^n C(m, n)$

$$a_n = C(m, n) [1 + (-1)^n], \quad n \geq 0$$

Example 17.48. Find the number of solutions of $m_1 + m_2 + m_3 = 12$ where m_1, m_2, m_3 are non-negative integers such that

$$1 \leq m_1 \leq 4, \quad 3 \leq m_2 \leq 5, \quad 4 \leq m_3 \leq 6.$$

Solution: The number of solutions of the given equation is the coefficient of x^{12} in the expression.

$$
\begin{aligned}
& (x + x^2 + x^3 + x^4) \,.\, (x^3 + x^4 + x^5) \,.\, (x^4 + x^5 + x^6) \\
&= x(1 + x + x^2 + x^3) \,.\, x^3(1 + x + x^2) \,.\, x^4(1 + x + x^2) \\
&= x^8[(1 + x + x^2)^2 \,.\, (1 + x + x^2 + x^3)] \\
&= x^8[(1 + 2x + 3x^2 + 2x^3 + x^4) \cdot (1 + x + x^2 + x^3)] \\
&= x^8[(1 + 3x + 6x^2 + 8x^3 + 8x^4 + 6x^5 + 3x^6 + x^7)] \\
&= x^8 + 3x^9 + 6x^{10} + 8x^{11} + 8x^{12} + 6x^{13} + 3x^{14} + x^{15}
\end{aligned}
$$

So, the coefficient of x^{12} is 8. Hence there are 8 solutions.

17.6 RECURRENCE RELATION

For a numeric function $\{a_0, a_1, a_r, ...\}$ a_n equation which relates a_r to one or more a_i $(i < r)$ is called a recurrence relation. A recurrence relation is also called a difference equation.

For example,

$$a_r = 2a_{r-1}$$

$$a_r = a_{r-1} - a_{r-2}$$

are recurrence relations.

The numeric function is also referred as the solution of the recurrence relation.

17.6.1 Order and Degree of the Recurrence Relation

The difference between the highest and lowest subscripts of a_r or $f(x)$ or y_n is called the order of the recurrence relation.

For example, the recurrence relation

$a_n = a_{n-1} - a_{n-2}$ is of order 2 (*i.e.*, $n - (n - 2) = 2$)

The highest power of a_r or y_n or $f(x)$ is called the degree of the recurrence relation or difference equation.

For example, recurrence relation $a_r = 3a_{r-2}$ is of degree 1 and called linear recurrence relation, whereas the recurrence relation

$y_n^4 - 3y_{n-1}^3 + 2y_n = f(n)$ is of degree 4.

17.6.2 Linear Recurrence Relation with Constant Coefficient

If the degree of the recurrence relation is one, then relation is called linear recurrence relation.

In general,

$$A_0 a_n + A_1 a_{n-1} + A_2 a_{n-2} + ... + A_n a_0 = f(n).$$

is a linear recurrence relation where $A_0, A_1 ... A_n$ are constant and $f(n)$ is a function which satisfies the equation.

if $f(n) = 0$, recurrence relation is called **linear homogeneous recurrence relation,** and if $f(n) \neq 0$ the recurrence relation is called **linear non-homogeneous** recurrence relation.

17.6.3 Solution of Linear Homogeneous Recurrence Relation

Consider a liner homogeneous recurrence relation

$$A_0 a_n + A_1 a_{n-1} + A_2 a_{n-2} + \dots + A_n a_0 = 0. \quad \text{...(1)}$$

The solution of equation (1) is of the term

$$a_n = C\,\alpha^n$$

where a is the characteristic root of equation (*i*) and *A* is the constant. Substitute the value of *an in* equation (*i*) we have,

$$A_0(C\,.\,\alpha^n) + A_1(C\,.\,\alpha^{n-1}) + \dots + A_n(C\,.\,\alpha^0) = 0$$

or

$$A_0\,\alpha^n + A_1\,\alpha^{n-1} + A_2\,\alpha^{n-2} + \dots + A_n = 0 \quad \text{...(2)}$$

equation (2) is called the characteristic equation of the recurrence relation of equation (1).

If there are *m* roots $\alpha_1, \alpha_2, \dots, \alpha_m$ of the equation (1) then these exist four cases for the solution:

Case I: If all the *n*-roots are distinct, then

$$a_n = C_1\,\alpha_1{}^n + C_2\,\alpha_2{}^n + \dots + C_m\,\alpha_m{}^n.$$

Case II: If all the roots are repeated and repeated and real

i.e., $\alpha_1 = \alpha_2 = \alpha_3 = \dots = \alpha_m$ (say)

then $a_n = (C_1 + C_2 n + C_3 n^2 + \dots + C_m n^m)\,\alpha_1{}^n.$

For example, if $\alpha_1 = \alpha_2, \alpha_3, \alpha_4$ are the roots of the recurrence relations then its homogeneous solution is

$$a_n = (C_1 + C_2 n)\,.\,\alpha_1{}^n + C_3\,\alpha_3{}^n + C_4\,\alpha_4{}^n$$

Case III: If the roots of the characteristic equation are imaginary say $\alpha + i\beta$ is the root of the characteristic equation, where α and β are real numbers, then

$$a_n = C_1\,.\,(\alpha + i\beta)^n + C_2\,.\,(\alpha - i\beta)^n$$

Case IV: If the imaginary roots are repeated then

$$a_n = (C_1 + C_2 n)\,(\alpha + i\beta)^n + (C_3 + C_4 n)\,(\alpha - i\beta)^n$$

Example 17.49. There are *r* lines drawn in a plane such that no 2 lines are parallel and no. 3 lines are concurrent. If the plane is thereby divided into *f*(*r*) regions, find a recurrence relation for *f*(*r*).

Solution : The number of regions formed by $(r-1)$ is $f(r-1)$.

By hypothesis the rth line meets the other $(r-1)$ lines in $(r-1)$ distinct points. So these $(r-1)$ lines divide the rth line into *r* parts. Each such part divides a farmer region into 2 subregions and thus the rth line creates *r* more regions. So,

$$f(r) = f(r-1) + r$$

with initial condition $f(1) = 2$, *i.e.*, number of regions formed by one line are two.

Example 17.50. (The Tower of Hanoi) There are *r* circular disks of decreasing radii. Each disk has a hole at the center and 3 pegs (marked *X*, *Y* and *Z*) fixed vertically on a table so that the distance between the feet of any 2 of them is greater than the diameter of the largest disk. Initially these disks are slipped onto peg *X* with the largest disk at the bottom and the others on the top of this, in

decreasing order of size. A legal move is defined as the transfer of the top disk from 1 of the 3 pegs to the top of the stack on 1 of the other 2 pegs where it rests on a larger disk. Let *f*(*r*) be the number of legal moves needed to transfer all *r* disks from peg *X* to another peg. Find the recurrence relation for *f*(*r*) and solve it.

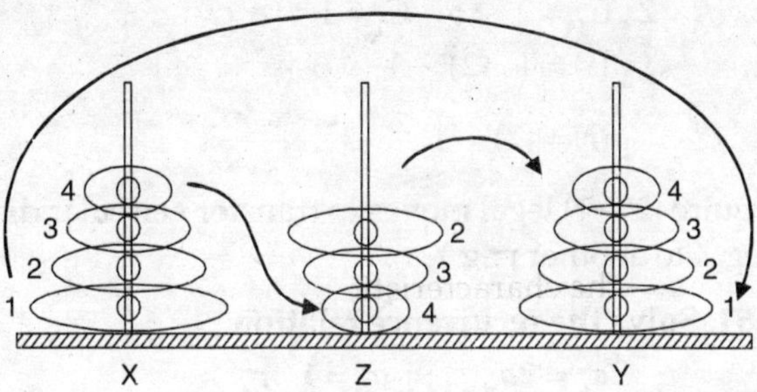

Fig. 17.4.

Solution. Let's fix the larger disk say disk (1) at the bottom of peg X. Now the total number of legal moves required to transfer the other $(r-1)$ disks to peg Z are $f(r-1)$.

We can transfer the larger disk (1) from peg X to peg Y in one legal move and then the other $(r-1)$ disks from peg Z to peg Y can be transferred into $f(r-1)$ legal moves. Thus the total number of legal moves, $f(r)$ required to transfer the r disks from peg X to peg Y are

$$f(r) = f(r-1) \text{ (legal moves from } X \text{ to } Z) + 1 \text{ (legal move from } X \text{ to } Y) + f(r-1) \text{ (legal move from } Z \text{ to } Y)$$

$$\therefore \quad f(r) = 2f(r-1) + 1$$

with the initial condition $f(1) = 1$

or $\quad f(r) - 2f(r-1) = 1, \quad f(1) = 1$

It's homogeneous solution is given as

$$f_{(r)}{}^{H} = C_1(2)^r \quad \text{...(1)}$$

Let $f_{(r)}^{P} = A_0$ be it's particular solution. (Since the RHS of the recurrence relation is a constant, so particular solution will be a constant).

So $\quad f_{(r)}^{P} - 2f_{(r-1)}^{P} = 1$

$$A_0 - 2A_0 = 1$$

$\Rightarrow \quad A_0 = -1$

Thus $\quad f_{(r)}^{P} = -1 \quad$ ...(2)

Hence the total solution

$$f(r) = f_{(r)}^{H} + f_{(r)}^{P}$$

$\Rightarrow \quad f(r) = C_1(2)^r + (-1)$ ...(3)

Given that $\quad f(1) = 1$

So $\quad 1 = C_1(2)^1 + (-1)$

$$2 . C_1 = 2 \Rightarrow C_1 = 1$$

So $\quad f(r) = 1 . (2)^r - 1$ (From (3))

or $\quad f(r) = (2)^r - 1$

Hence we require $(2^r - 1)$ legal moves to transfer r circular disks of decreasing radii from one peg X to another peg Y.

Example 17.51. Solve the recurrence relation

$$a_n = 2a_{n-1}; \quad a_0 = 1$$

Solution: Let $A . \alpha^n$ be the solution of the given recurrence relation

then $\quad A . \alpha^n = 2A . \alpha^{n-1}$

or $\quad A(\alpha^n - 2\alpha^{n-1}) = 0$

or $\quad \alpha^n - 2\alpha^{n-1} = 0$

or $\quad \alpha^{n-1}(\alpha - 2) = 0$

$$\alpha^{n-1} = 0 \quad \text{or} \quad \alpha - 2 = 0$$

We know that $\alpha^{n-1} \neq 0$ so $\alpha - 2 = 0$ which is the characteristic equation of the recurrence relation.

Clearly, $\alpha = 2$, *i.e.*, $\alpha = 2$ is the root of the characteristic equation.

So $\quad a_n = A . (2)^n$, given $a_0 = 1$.

Put $n = 0$ the solution we have

$$a_0 = A . (2)^0,$$

or $\quad 1 = A . (2)^0 \Rightarrow A = 1$

Hence the desired solution is

$$a_n = 1 . (2)^n$$

Example 17.52. Find the solution of the recurrence relation

$$S_k - 5S_{k-1} + 6S_{k-2} = 0.$$

Solution: Given that

$$S_k - 5S_{k-1} + 6S_{k-2} = 0. \quad \text{...(1)}$$

so the characteristic equation of equation (1) is

$$\alpha^2 - 5\alpha + 6 = 0$$

or $\quad (\alpha - 3) . (\alpha - 2) = 0$

$\Rightarrow \quad \alpha_1 = 3, \quad \alpha_2 = 2.$

are the two roots of the equation.

So the solution is

$$a_n = C_1(\alpha_1)^n + C_2(\alpha_2)^n$$

$$a_n = C_1(3)^n + C_2(2)^n.$$

Example 17.53. Solve the equation $a_n + 6a_{n-1} + 12a_{n-2} + 8a_{n-3} = 0$.

Solution: The characteristic equation of the given recurrence relation is

$$\alpha^3 + 6\alpha^2 + 12\alpha + 8 = 0$$

the roots of the equation is

$$\alpha_1 = -2, \quad \alpha_2 = 2, \quad \alpha_3 = -2.$$

Clearly, the roots are repeated thrice, so the general solution is

$$a_n = (C_1 + C_2\, n + C_2\, n^2) \,.\, (-2)^n.$$

Example 17.54. Solve the recurrence relation

$$a_n + 5a_{n-1} + 5a_{n-2} = 0; \quad a_0 = 0, a_1 = 2\sqrt{5}\,.$$

Solution: The characteristic equation of the given recurrence relation is

$$\alpha^2 - 5 \cdot \alpha + 5 = 0.$$

$$\alpha = \frac{-5 \pm \sqrt{25 - 4.5}}{2} = \frac{-5 \pm \sqrt{5}}{2}$$

$$\Rightarrow \qquad \alpha_1 = \frac{-5 + \sqrt{5}}{2}, \quad \alpha_2 = \frac{-5 - \sqrt{5}}{2}.$$

So, $$a_n = C_1\left(\frac{-5 + \sqrt{5}}{2}\right)^n + C_2\left(\frac{-5 - \sqrt{5}}{2}\right)^n$$

Initial conditions are $a_0 = 0, a_1 = 2\sqrt{5}$

put $n = 0$ and $n = 1$ in the general solution.

$$a_0 = 0 = C_1\left(\frac{-5 + \sqrt{5}}{2}\right) + C_2\left(\frac{-5 - \sqrt{5}}{2}\right)$$

$$\Rightarrow \qquad C_1 + C_2 = 0 \qquad ...(1)$$

$$a_1 = 2\sqrt{5} = \left(\frac{-5 + \sqrt{5}}{2}\right) \cdot C_1 + C_2\left(\frac{-5 - \sqrt{5}}{2}\right)$$

or $$-\frac{5}{2}(C_1 + C_2) + \frac{\sqrt{5}}{2}(C_1 - C_2) = 2\sqrt{5} \qquad ...(2)$$

By equation (1), put $C_1 + C_2 = 0$ in (2)

$$\Rightarrow \qquad \frac{\sqrt{5}}{2}(C_1 - C_2) = 2\sqrt{5}$$

$$\Rightarrow \qquad C_1 - C_2 = 4 \qquad ...(3)$$

By equations (1) and (3)

$$C_1 = 2, \quad C_2 = -2$$

Hence the solution of the recurrence relation is

$$a_n = 2\left(\frac{-5+\sqrt{5}}{2}\right)^n - 2\left(\frac{-5-\sqrt{5}}{2}\right)^n$$

Example 17.55. Find the solution of the recurrence relation

$$S_{k+4} + 2S_{k+3} + 3S_{k+2} + 2S_{k+1} + S_k = 0$$

Solution: The characteristic equation is

$$\alpha^4 + 2\alpha^3 + 3\alpha^2 + 2\alpha + 1 = 0$$

or $$(\alpha^2 + \alpha + 1) . (\alpha^2 + \alpha + 1) = 0$$

$$\Rightarrow \quad \alpha_1 = \alpha_2 = \frac{-1+i\sqrt{3}}{2}, \quad \alpha_3 = \alpha_4 = \frac{-1-i\sqrt{3}}{2}$$

Therefore, the solution of the homogeneous recurrence relation is

$$S_k = (C_1 + C_2 k) . \left(\frac{-1+i\sqrt{3}}{2}\right)^k + (C_3 + C_4 k)\left(\frac{-1-i\sqrt{3}}{2}\right)^k$$

Example 17.56. Solve the recurrence relation

$$a_n - 7a_{n-1} + 16a_{n-2} - 12a_{n-3} = 0$$

and find the particular solution. Given that $a_0 = 1, a_1 = 4, a_2 = 8$.

Solution: The characteristic equation is

$$\alpha^3 - 7\alpha^2 + 16\alpha - 12 = 0$$

the roots are

$$\alpha_1 = \alpha_2 = 2 \quad \text{and} \quad \alpha_3 = 3$$

So the general solution is

$$a_n = (C_1 + C_2 n)\, 2^n + C_3(3)^n \qquad ...(1)$$

put $n = 0$, $n = 1$ and $n = 2$ in equation (1)

$$C_1 + C_3 = 1.$$
$$(C_1 + C_2) . 2 + 3C_3 = 4$$
$$(C_1 + 2C_2) . 4 + 9C_3 = 8$$

On solving we have

$$C_1 = 5, \quad C_2 = 3 \quad \text{and} \quad C_3 = -4$$

Thus the required particular solution is

$$a_n = (5 + 3n)\,(2)^n - 4 . (3)^n.$$

17.6.4 Solution of Non-homogeneous Recurrence Relation

Consider the recurrence relation

$$A_0\, a_n + A_1\, a_{n-1} + ... + A_n\, a_0 = f(n) \qquad ...(1)$$

where $f(n)$ is a function satisfying the equation.

Step I: Find the homogeneous solution $a_n^{(H)}$ of the given recurrence relation by keeping $f(n) = 0$.

Step II: Find the particular solution $a_n^{(P)}$ of the given recurrence relation with respect to the given function $f(n)$ by making suitable gives.

Step III: Substitute the guess of the particular solution in the equation (*i*) and find out the undermined coefficients.

Step IV: Find out the general solution of the recurrence relation as

$$a_n = a_n^{(H)} + a_n^{(P)}$$

Step V: Find out the constants in the general solution by using the given initial conditions.

Guess of the Particular Solution With Respect to $f(n)$

The function $f(n)$ can be of two types:

(*i*) A polynomial, say $C_0 + C_1 n + C_2 n^2 + ... + C_k n^k$.

(*ii*) An exponential function say $f(n) = C.(t)^n$

Case I: If $f(n)$ is a polynomial of degree k

$$\text{say } f(n) = C_0 + C_1 n + C_2 n^2 + ... + C_k n^k.$$

then the particular solution is also a polynomial of degree k say

$$a_n^{(P)} = D_0 + D_1^n + ... + D_k n^k.$$

where $D_0, D_1, ..., D_k$ are the undermined co-efficients.

Case II: If $\quad f(n) = C.(t)^n$, then

$$a_n^{(P)} = D.(t)^n$$

where is the undermined co-efficient.

(*i*) If it is the root of the characteristic equation with multiplicity m *i.e.*,

$$\alpha_1 = \alpha_2 = ... = \alpha_m = t$$

then $\quad a_n^{(P)} = D.n^m.(t)^n$

(*ii*) If $\quad t = 1$, then $a_n^{(P)} = D.(1)^n$

and if 1 is the root of the characteristic equation with multiplicity m then

$$a_n^{(P)} = D.n^m.(1)^n$$

Case III: If $f(n)$ is a combination of both polynomial and exponential, *i.e.*,

(*a*) $\quad f(n) = (C_0 + C_1 n + ... + C_k n^k).(t)^n$

(*b*) $\quad f(n) = k(t)^n + (C_0 + C_1 n + C_2 n^2 + ... + C_k n^k)$

Then,

(*a*) (*i*) $\quad a_n^{(P)} = (D_0 + D_1 n + ... + D_k n^k).(t)^n$

(*ii*) $\quad a_n^{(P)} = n^m (D_0 + D_1 n + ... + D_k n^k).(t)^n$

If t is the root of the characteristic equation with multiplicity m.

(*b*) (*i*) $\quad a_n^{(P)} = D.(t)^n + (D_0 + D_1 n + ... + D_k n^k)$

(*ii*) $\quad a_n^{(P)} = D.n^m (t)^n + (D_0 + D_1 n + ... + D_k n^k)$

If t is the root of the characteristic equation with multiplicity m.

Example 17.57. Solve the recurrence relation

$$a_n - 5a_{n-1} + 6a_{n-2} = 9\,, \quad a_0 = 0, \quad a_1 = 1.$$

Solution: The corresponding homogeneou recurrence relation is

$$a_n - 5a_{n-1} + 6a_{n-2} = 0$$

the characteristic equation is

$$\alpha^2 - 5\alpha + 6 = 0$$

$$(\alpha - 3)(\alpha - 2) = 0$$

$$\alpha_1 = 3, \quad \alpha_2 = 2.$$

So the homogeneous solution is

$$a_n^{(H)} = C_1(3)^n + C_2(2)^n \qquad ...(1)$$

Particular Solution: $f(n) = 9$

i.e. a polynomial of degree 0 or a constant.

So $\quad a_n^{(P)} = D$ (a polynomial of degree 0 or constant).

Substitute $\quad a_n^{(P)} = D$ in the given recurrence relation, we have

$$D - 5D + 6D = 9$$

$$\Rightarrow \qquad 2D = 9 \Rightarrow D = \frac{9}{2}$$

So $$a_n^{(P)} = \frac{9}{2} \qquad ...(2)$$

So the general solution

$$a_n = a_n^{(P)} + a_n^{(H)}$$

$$a_n = C_1(3)^n + C_2(2)^n + \frac{9}{2}$$

To find C_1 and C_2 put $n = 0$ and $n = 1$ in the general solution, then

$$C_1 + C_2 + \frac{9}{2} = 0 \Rightarrow C_1 + C_2 = -\frac{9}{2} \qquad ...(1)$$

and $$3C_1 + 2C_2 = -\frac{7}{2} \qquad ...(2)$$

On solving (1) and (2) we have

$$C_1 = \frac{11}{2} \quad \text{and} \quad C_2 = -10$$

Hence $$a_n = \frac{11}{2}(3)^n - 10(2)^n + \frac{9}{2}$$

Example 17.58. Solve $a_n - 7a_{n-1} + 12a_{n-2} = n \cdot 4^n$.

Solution: The homogeneous recurrence relation is

$$a_n - 7a_{n-1} + 12a_{n-2} = 0$$

the characteristic equation is

$$\alpha^2 - 7\alpha + 12 = 0$$

$$(\alpha - 4)(\alpha - 3) = 0$$

$$\Rightarrow \quad \alpha_1 = 4, \quad \alpha_2 = 3$$

So the homogeneous solution is

$$a_n^{H} = C_1(4)^n + C_2(3)^n$$

Particular Solution: $f(n) = n \,.\, (4)^n$

Clearly $f(n)$ is a combination of a polynomial of degree 1 and exponential function $(4)^n$.

So $$a_n^{(P)} = (D_0 + D_1 n) \,.\, (4)^n$$

But one can see that 4 is the root of the characteristic equation with multiplicity $m = 1$. So the new particular solution.

$$a_n^{(P)} = n \,.\, (D_0 + D_1 n)(4)^n \qquad ...(1)$$

put the value of $a_n^{(P)}$ in the non-homogeneous recurrence relation, so

$$n \,.\, (D_0 + D_1 n)(4)^n - 7\{(n-1)(D_0 + (n-1) D_1) \,.\, (4)^{n-1}\}$$
$$+ 12\{(n-2)(D_0 + (n-2) D_1) \,.\, (4)^{n-2}\} = n \,.\, (4)^n$$

or $$(n D_0 + n^2 D_1) - \frac{7}{4}\{n(D_0 - 2D_1) + (D_1 - D_0) + n^2 D_1\} + \frac{12}{16}\{n(D_0 - 4D_1)$$
$$+ \{(4D_1 - 2D_0) + n^2 D_1\} = n$$

or $$\frac{n}{2} D_1 + n^2 \,.\, (0) + \left(\frac{D_0}{4} + \frac{5D_1}{4}\right) = n$$

Comparing the two sides, we have

$$\frac{D_1}{2} = 1 \;\Rightarrow\; D_1 = 2$$

and $$\frac{D_0}{4} + \frac{5D_1}{4} = 0 \;\Rightarrow\; D_0 = -10$$

So the solution is,

$$a_n^{(P)} = n \,.\, (-10 + 2n) \,.\, (4)^n \qquad ...(2)$$

Hence $$a_n = a_n^{(P)} + a_n^{(H)}$$

$$a_n = C_1(4)^n + C_2(3)^n + n(2n - 10) \,.\, (4)^n$$

Example 17.59. Solve, $a_n - 4a_{n-1} + 4a_{n-2} = n + 3^n$

Solution: The homogeneous recurrence relation is

$$a_n - 4a_{n-1} + 4a_{n-2} = 0$$

the characteristic equation is

$$\alpha^2 - 4\alpha + 4 = 0$$

$$\Rightarrow \quad \alpha_1 = \alpha_2 = 2.$$

So $$a_n^{(H)} = (C_1 + C_2 n) \cdot 2^n \qquad \text{...(1)}$$

Given $$f(n) = n + 3^n$$

n is a polynomial of degree one and 3^n the exponential function and clearly 3 is not a root of the characteristic equation.

So $$a_n^{(P)} = (d_0 + d_1 n) + d_2 (3)^n \qquad \text{...(2)}$$

Substitute the value of a_n^P is the recurrence relation, we have

$$\{(d_0 + d_1 n) + d_2 (3)^n\} - 4\{(d_0 + d_1 (n-1)) + d_2 (3)^{n-1}\} + 4\{(d_0 + d_1 (n-2)) + d_2 (3)^{n-2}\} = n + (3)^n$$

$$(3)^n \cdot \left(\frac{d_2}{9}\right) + nd_1 + (d_0 - 4d_1) = n + 3^n$$

Comparing both the sides we have

$$d_1 = 1, \quad \frac{d_2}{9} = 1 \Rightarrow d_2 = 9$$

and $$d_0 - 4d_1 = 0 \Rightarrow d_0 = 4$$

So the particular solution $a_n^{(P)} = (4 + n) + 9 \cdot (3)^n$

or $$a_n^{(P)} = (4 + n) + 3^{(n+2)}$$

Thus the general solution

$$a_n = a_n^{(H)} + a_n^{(P)} \text{ is}$$

$$a_n = (C_1 + n \cdot C_2)\, 2^n + (4 + n) + 3^{n+2}.$$

17.6.5 Solution of Recurrence Relation by the Method of Operators

Consider a shifting operator E such that

$$f(x + n) = E^n \cdot f(x)$$

or $$a_{n+m} = E^m \cdot a_n$$

Now, let us consider the recurrence relation as

$$A_0\, a_{n+m} + A\, a_{n+m-1} + A_2\, a_{n+m-2} + \ldots + A_n\, a_n = f(n).$$

which can be written with the help of shift operator as

$$A_0\, E^m a_n + A_1 \cdot E^{m-1} a_n + A_2 \cdot E^{m-2} a_n + \ldots + A_n\, E^0 a_n = f(n)$$

$$\Rightarrow (A_0\, E^m + A_1 \cdot E^{m-1} + A_2 \cdot E^{m-2} + \ldots + A_n\, E^0) \cdot a_n = f(n)$$

or $$\phi(E) \cdot a_n = f(n)$$

where $$\phi(E) = A_0\, E^m + A_1\, E^{m-1} + \ldots + A_n$$

Thus the particular, solution is given by

$$a_n^{(P)} = \frac{f(n)}{\phi(E)}$$

whereas, the homogeneous solution can be obtained by the characteristic polynomial method.

Let us Consider the Different Types of $f(n)$:

(*i*) If $$f(n) = C \,.\, b^n$$

then $$a_n^{(P)} = \frac{C}{\phi(b)} \,.\, b^n \quad \text{provided } \phi(b) \neq 0.$$

If $\phi(b) = 0$, then $(E - b)^m a_n = f(n) = C\, b^n$.

i.e., b is the root of the characteristic equation with multiplicity m.

$$\Rightarrow \quad a_n = \frac{C \,.\, \{n(n-1)(n-2) \ldots (n-(m-1))\}}{m!} \,.\, b^{n-m},\ m \geq 0.$$

(*ii*) When $f(n) = \sin kn$, where k is a constant.

Thus $$a_n^{(P)} = \frac{1}{\phi(E)} \,.\, \sin kn = \frac{1}{\phi(E)} \left[\frac{e^{ikn} - e^{-ikn}}{2!}\right]$$

where $e^{i\theta} = \cos\theta + i \sin\theta$ (Demoiver's Theorem)

So $$a_n^{(P)} = \frac{1}{2!} \left[\frac{1}{\phi(E)} \,.\, e^{ikn} - \frac{1}{\phi(E)} \,.\, e^{-ikn}\right]$$

which can be solved by using case (*i*)

i.e., consider $e^{ik} = b$ then

when $f(n) = \cos kn$, then

$$a_n^{(P)} = \frac{1}{\phi(E)} \,.\, \cos kn = \frac{1}{\phi(E)} \,.\, \left[\frac{e^{ikn} + e^{-ikn}}{2!}\right]$$

So $$a_n^{(P)} = \frac{1}{2} \left[\frac{1}{\phi(E)} \cdot (b)^n + \frac{1}{\phi(E)} \cdot (b^{-1})\right]$$

where $b = e^{ik}$

(*iii*) If $f(n) = p(n)$, where $p(n)$ is a polynomial degree n.

then $$a_n^{(P)} = \frac{1}{\phi(E)} \,.\, p(n)$$

$$= \frac{1}{\phi(1+\Delta)} \,.\, p(n) \qquad \{E = 1 + \Delta\}$$

$$= (A_0 + A_1\Delta + \ldots + A_n \Delta^n) \,.\, p(n) \text{ (using Binomial theorem)}$$

Clearly if degree of polynomial $p(n)$ is m then

$$\Delta^{m+1}\, p(n) = 0$$

That is the difference greater than the degree of the polynomial $p(n)$ will always be zero.

(*iv*) If $f(n) = k^n \cdot p(n)$, then the particular integral.

$$a_n^{(p)} = k^n \cdot \frac{1}{\phi(kE)} \cdot p(n)$$

which is nothing but case (*iii*).

Example 17.60. Solve the recurrence relation

$$a_{n+2} - 5a_{n+1} + 6a_n = 4^n$$

Solution: Given relation is:

$$a_{n+2} - 5a_{n+1} + 6a_n = 4^n$$

Then the corresponding homogeneous equation as

$$a_{n+2} - 5a_{n+1} + 6a_n = 0$$

The characteristic equation is

$$\alpha^2 - 5\alpha + 6 = 0$$

$$\Rightarrow \quad \alpha_1 = 2, \quad \alpha_2 = 3.$$

Thus, the homogeneous soluton

$$a_n^{(H)} = C_1 (2)^n + C_2 (3)^n \qquad \text{...(1)}$$

Now $f(n) = 4^n$.

So the particular soluton is

$$a_n^{(H)} = \frac{1}{\phi(E)} \cdot a^n = \frac{1}{\phi(E)} \cdot 4^n$$

$$= \frac{1}{(E^2 - 5E + 6)} \cdot 4^n$$

Clearly $\phi(4) = 4^2 - 5.4 + 6 = 2 \neq 0.$

So $$a_n^{(P)} = \frac{1}{\phi(4)} \cdot 4^n = \frac{1}{(4^2 - 5.4 + 6)} \cdot 4^n$$

$$a_n^{(P)} = \frac{4^n}{2} \qquad \text{...(2)}$$

Thus, the general solution

$$a_n = a_n^{(H)} + a_n^{(P)}$$

$$a_n = C_1(2)^n + C_2(3)^n + \frac{1}{2} \cdot 4^n$$

Example 17.61. Solve the given recurrence relation

$$a_{n+2} + k^2 a_n = \cos kn$$

Solution: Given recurrence relation is:

$$a_{n+2} + k^2 a_n = \cos kn$$

The corresponding homogeneous recurren ce relation is

$$a_{n+2} + k^2 a_n = 0$$

So, the characteristic equation is

$$\alpha^2 + k^2 = 0$$

$$\Rightarrow \qquad \alpha = \pm ik.$$

Consider the polar form

$$\Rightarrow \quad r . \cos\theta \pm i\, r \sin\theta = 0 \pm ik$$

$$\Rightarrow \qquad \theta = \tan^{-1}\left(\frac{k}{0}\right) = \tan^{-1}(\alpha) = \frac{\pi}{2}$$

and $\qquad r^2 = k^2 \quad \Rightarrow \quad r = k.$

Hence the homogenousm solution is given by

$$a_n^{(H)} = k^n \left[A_1 \cos n\frac{\pi}{2} + A_2 \sin n\frac{\pi}{2}\right] \qquad \text{...(1)}$$

Now for particular solution

$$f(n) = \cos kn$$

$$a_n^{(P)} = \frac{1}{\phi(E)} . \cos kn = \frac{1}{E^2 + k^2} \cos kn$$

$$= \frac{1}{E^2+k^2}\left[\frac{e^{ikn} + e^{-ikn}}{2}\right] = \frac{1}{2}\left[\frac{1}{E^2+k^2} . e^{ikn} + \frac{1}{E^2+k^2} . e^{-ikn}\right]$$

$$= \frac{1}{2}\left[\frac{1}{(e^{ik})^2 + k^2} . e^{ikn} + \frac{1}{(e^{-ik})^2 + k^2} . e^{-ikn}\right]$$

$$= \frac{1}{2}\left[\frac{e^{ikn}(e^{-2ik} + k^2) + e^{-ikn} . (e^{2ik} + k^2)}{(e^{2ik} + k^2) . (e^{-2ik} + k^2)}\right]$$

$$= \frac{1}{2}\left[\frac{e^{ik(n-2)} + k^2(e^{ikn} + e^{-ikn}) + e^{ik(2-n)}}{1 + k^2(e^{2ik} + e^{-2ik}) + k^4}\right]$$

$$= \frac{1}{2}\left[\frac{e^{ik(n-2)} + 2k^2 \cos kn + e^{ik(2-n)}}{1 + 2k^2 \cos 2k + k^4}\right]$$

$$= \frac{1}{2}\left[\frac{\cos k(n-2) + i \sin k(n-2) + 2k^2 \cos kn + \cos k(2-n) + i \sin k(n-2)}{1 + 2k^2 \cos 2k + k^4}\right]$$

$$= \frac{1}{2}\left[\frac{2\cos k(n-2) + 2k^2 \cos kn}{1 + 2k^2 \cos 2k + k^4}\right] \qquad (\text{as } \cos(-\theta) = \cos\theta \ \sin(-\theta) = -\sin\theta)$$

$$a_n^{(P)} = \left[\frac{\cos k(n-2) + k^2 \cos kn}{1 + 2k^2 \cos 2k + k^4}\right] \quad \text{... (ii)}$$

Thus, the general solution $a_n = a_n^{(H)} + a_n^{(P)}$

So $$a_n = k^n \left\{A_1 \cos\frac{\pi}{2} + A_2 \sin n\frac{\pi}{2}\right\} + \left\{\frac{\cos k(n-2) + k^2 \cos kn}{1 + 2k^2 \cos 2k + k^4}\right\}$$

Example 17.62. Solve the following difference equation (recurrence relation)

$$\boldsymbol{a_{r+2} - 2a_{r+1} + a_r = r^2 \cdot 2^r}$$

Solution: Given difference equation is

$$a_{r+2} - 2a_{r+1} + a_r = r^2 . 2^r$$

So the corresponding homogeneous equation is

$$a_{r+2} - 2a_{r+1} + a_r = 0$$

The characteristic equation is

$$\alpha^2 - 2\alpha + 1 = 0$$

$$\Rightarrow \quad \alpha = 1, 1$$

So the homogeneous solution is

$$a_n^{(H)} = (C_1 + C_2 r) . (1)^r$$

Now $f(r) = r^2 . 2^r$; then the particular solution is

$$a_n^{(P)} = \frac{1}{\phi(E)} r^2 . 2^r$$

$$= \frac{1}{(E-1)^2} . 2^r . r^2 = 2^r \frac{1}{(2E-1)^2} . r^2$$

$$= 2^r . \frac{1}{(2(1+\Delta)-1)^2} . r^2 = 2^r \frac{1}{(2\Delta+1)^2} . r^2$$

$$= 2^r (1 + 2\Delta)^2 . r^2$$

$$= 2^r (1 - 4\Delta + 12\Delta^2 ...) . r^2$$

$$= 2^r (r^2 - 4\Delta . r^2 + 12\Delta^2 . r^2 ...)$$

$$= 2^r \{r^2 - 4 . 2r + 12.2\}$$

$$= 2^r \{r^2 - 8r + 24\}$$

Thus $$a_n^{(P)} = 2^r . (r^2 - 8r + 24\}$$

So $$a_n = a_n^{(H)} + a_n^{(P)}$$

$$a_n = (C_1 + C_2 r) (1)^r + 2^r . (r^2 - 8r + 24)$$

Note: Operator Δ is analogous with differential and operator Δ^{-1} is analogous with integral.

17.6.6 Solving Recurrence Relation Using Generating Functions

Consider the recurrence relations as :

$A_0 a_{n+k} + A_1 a_{n+k-1} + A_2 a_{n+k} + \ldots + A_n a_n = f(n)$ ...(1)

where $A_0, A_1 \ldots A_n$ are constants.

We know that

$\sum_{n=0}^{\infty} a_n x^n = a_0 + a_1 x + a_2 x^2 + \ldots + a_n x^n + \ldots + \infty$ is the generating function of the sequence $(a_0, a_1, a_2, \ldots, a_n \ldots)$

Let us now define a procedure to solve the recurrence relation :

Step I: Multiply equation (*i*) by x^n and take summation from $n = 0$ to ∞. Thus

$$A_0 \sum_{n=0}^{\infty} a_{n+k}\, x^n + A_1 \sum_{n=0}^{\infty} a_{n+k-1} \cdot x^n + A_n \sum_{n=0}^{\infty} a_n x^n = \sum_{n=0}^{\infty} f(n) \cdot x^n$$

We know that $\sum_{n=0}^{\infty} a_n x^n = G(x)$

Now consider

$$\sum_{n=0}^{\infty} a_{n+k}\, x^n = a_k x^0 + a_{k+1} x^0 + a_{k+2} x^2 + \ldots \infty$$

$$= \frac{G(x) - (a_0 + a_1 x + \ldots + a_{k-1} x^{k-1})}{x^k}$$

Similarly,

$$\sum_{n=0}^{\infty} a_{n+k-1} \cdot x^n = a_{k-1} x^0 + a_k x + a_{k-1} x^2 + \ldots$$

$$= \frac{G(x) - (a_0 + a_1 x + \ldots + a_{k-2} x^{k-2})}{x^{k-1}}$$

and so on.

Step II: Collect the coefficients of $G(x)$, $a_0, a_1, \ldots, a_n$.

Step III: Simplify $\sum_{n=0}^{\infty} f(n) \cdot x^n$

Step IV: Then $G(x)$ can be obtained as

$$\sum_{n=0}^{\infty} a_n x^n = G(x) = \frac{\sum_{n=0}^{\infty} f(n)}{(\alpha^k + \alpha^{k-1} + \ldots + \alpha^0)} + A_0 \,.\, \text{terms} + A_1 \,.\, \text{terms}$$

$$+ \ldots + A_n \,.\, \text{terms} \quad \ldots(ii)$$

Solve equation (2) and compare the coefficients of x^n from both the sides to obtain a_n., *i.e.*, the nth term of the sequence or solution of the recurrence relation.

Example 17.63. Solve the homogeneous recurrence relation

$$a_{n+2} - 7a_{n+1} + 12a_n = 0, \qquad n \geq 0$$

Using generating function.

Solution: Given recurrence relation is

$$a_{n+2} - 7a_{n+1} + 12a_n = 0$$

Multiplying both the sides by x^n and summing from $n = 0$ to ∞. We have

$$\sum_{n=0}^{\infty} a_{n+2}\, x^n - \sum_{n=0}^{\infty} a_{n+1} \cdot x^n + 12 \sum_{n=0}^{\infty} a_n\, x^n = 0$$

We know that $\quad G(x) = \sum_{n=0}^{\infty} a_n\, x^n$

Thus the recurrence relation becomes,

$$\frac{G(x) - a_0 - a_1}{x^2} - 7\left(\frac{G(x) - a_0}{x}\right) + 12\, G(x) = 0$$

or $\quad G(x) - a_0 - a_1\, x - 7x\,(G(x) + a_0) + 12x\,.\, G(x) = 0 \qquad$ ...(2)

$$G(x)\,(12x^2 - 7x + 1) + a_0\,(7x - 1) - a_1 x = 0$$

$\Rightarrow \quad G(x) = (1 - 4x)\,(1 - 3x) = a_1 x - a_0\,(7x - 1)$

$\Rightarrow \quad G(x) = \dfrac{a_0(1-7x)+a_1 x}{(1-4x)(1-3x)} = \dfrac{A}{(1-4x)} + \dfrac{B}{(1-3x)}$

So $\quad \sum_{n=0}^{\infty} a_n\, x^n = A\,.\sum_{n=0}^{\infty} 4^n\, x^n + B\,.\sum_{n=0}^{\infty} 3^n\, x^n \quad \because \left\{\dfrac{1}{(1-a\cdot x)} = \sum_{n=0}^{\infty} a^n\, x^n\right\}$

Taking co-efficients of x^n from both the sides

$\therefore \quad a_n = A\,.\,4^n + B\,.\,3^n$

is the required solution.

Note: If the initial conditions *i.e.*, a_n's for $n = 0, 1, 2, \ldots.k$ substitute the values of the a_n's in the equation formed using generating function $G(x)$ and then solve the equation with the help of generating function.

Example 17.64. Solve the recurrence relation using generating function

$$a_n - 4a_{n-1} + 4a_{n-2} = 0, \qquad n \geq 2$$

Given $a_0 = 1, a_1 = 2$

Solution: Multiply both the sides of the given recurrence equation by x^n and summing from $n = 2$ to ∞, as equation is valid for $n \geq 2$.

Thus, the equation becomes

$$\sum_{n=2}^{\infty} a_n x^n - 4\sum_{n=2}^{\infty} a_{n-1} x^n + 4\sum_{n=2}^{\infty} a_{n-2} x^n = 0 \qquad ...(1)$$

$$(a_2x^2 + a_3x^3 + ...) - 4\,(a_1x^2 + a_2x^3 + ...) + 4\,(a_0x^2 + a_1x^3 + ...)$$

$$(G(x) - a_0 - a_1x) - 4x(G(x) - a_0) + 4x^2\,.\,G(x) = 0$$

as $\quad G(x) = \sum_{n=0}^{\infty} a_n x^n = a_0 + a_1x + a_2x^2 + ... + a_n x^n + ...$

So $\quad G(x)\,\{1 - 4x + 4x^2\} + a_0(4x - 1) - a_1x = 0$

or $\quad G(x)\,.\,\{1 - 4x + 4x^2\} = a_0(1 - 4x) + a_1x$

or $\quad G(x)\,.\,(1 - 2x)^2 = a_0(1 - 4x) + a_1x \qquad ...(2)$

Substitute the given initial conditions *i.e.*, $a_0 = 1$ and $a_1 = 2$ in equation (2), we have

$$G(x)\,.\,(1 - 2x)^2 = (1 - 4x) + 2x = 1 - 2x$$

$$G(x) = \frac{1 - 2x}{(1 - 2x)^2} = \frac{1}{1 - 2x}$$

Thus $\quad G(x) = \sum_{n=0}^{\infty} a_n x^n$

$$G(x) = \frac{1}{1 - 2x} = \sum_{n=0}^{\infty} 2^n x^n$$

So $\quad a_n = 2^n$

Example 17.65. Solve the non-homogeneous recurrence relation

$$\boldsymbol{a_n - 5a_{n-1} + 6a_{n-2} = 3^n, \qquad n \geq 2}$$

given that $\quad \boldsymbol{a_0 = 0, \quad a_1 = 2}$

Solution: Multiply both the sides of the equation by x^n and summing from $n =$ 2 to ∞, we have

$$\sum_{n=2}^{\infty} a_n x^n - 5\sum_{n=2}^{\infty} a_{n-1} x^n + 6\sum_{n=2}^{\infty} a_{n-2} x^n = \sum_{n=2}^{\infty} 3^n x^n$$

$$(G(x) - a_0 - a_1x) - 5x(G(x) - a_0) + 6x^2\,.\,G(x) = \sum_{n=2}^{\infty} 3^n x^n$$

$$G(x)\,.\,(1 - 5x + 6x^2) + a_0(5x - 1) - a_1x = 3^2x^2 \sum_{n=2}^{\infty} 3^{n-2} x^{n-2}$$

or $\quad G(x)\,.\,(1 - 5x + 6x^2) + 0\,.\,(5x - 1) - 2x = 3^2x^2\,.\,\sum_{n=2}^{\infty} 3^{n-2} x^{n-2}$

$$(a_0 = 0, a_1 = 2)$$

or $$G(x).(1-2x)(1-3x) = 9x^2.\sum_{k=0}^{\infty} 3^k x^k + 2x \quad \text{(substituting } n-2 \equiv k\text{)}$$

or $$G(x).(1-2x)(1-3x) = \frac{9x^2}{1-3x} + 2x = \frac{9x^2+2x-6x^2}{(1-3x)} = \frac{3x^2+2x}{(1-3x)}$$

or $$G(x) = \frac{3x^2+2x}{(1-3x)^2(1-2x)}$$

$$G(x) = \frac{-10}{(1-3x)} + \frac{3}{(1-3x)^2} + \frac{7}{(1-2x)}$$

So, $$\sum_{n=0}^{\infty} a_n x^n = -10\sum_{n=0}^{\infty} 3^n x^n + 3.\sum_{n=0}^{\infty}(n+1).3^n x^n + 7\sum_{n=0}^{\infty} 2^n x^n$$

Hence

$$a_n = -10.3^n + 3(n+1).3^n + 7.2^n$$
$$= 3^n(3n-7) + 7.2^n$$

or $$a_n = n.3^{n+1} - 7.3^n + 7.2^n$$

or $$a_n = n.3^{n+1} + 7(2^n - 3^n)$$

Example 17.66. Solve the recurrence relation using generating function

$$\boldsymbol{a_n - 4a_{n-2} = n+1, \quad a_0 = 0, \quad a_1 = 1 \qquad n \geq 2.}$$

Solution: Multiply the given recurrence relation by x^n and summing from $n = 2$ to ∞. Thus, we have

$$\sum_{n=2}^{\infty} a_n x^n - 4\sum_{n=2}^{\infty} a_{n-2} x^n = \sum_{n=2}^{\infty}(n+1).x^n$$

or $$(G(x) - a_0 - a_1 x) - 4x^2(G(x) = \sum_{n=2}^{\infty} n.x^n + \sum_{n=2}^{\infty} x^n$$

$$G(x)(1-4x^2) - a_0 - a_1 x = \sum_{n=2}^{\infty} n.x^n + x^2.\sum_{n=2}^{\infty} x^{n-2}$$

Substitute $a_0 = 0$ and $a_1 = 1$, we have

$$G(x)(1-4x^2) - x = \sum_{n=2}^{\infty} n.x^n + x^2.\frac{1}{1-x}$$

or $$G(x)(1-4x^2) = \sum_{n=2}^{\infty} n.x^n + \left(\frac{x^2}{1-x} + x\right) \qquad ...(1)$$

Now let us simplify the term $\sum_{n=2}^{\infty} n.x^n$ in equation (1)

Consider $$\sum_{n=2}^{\infty} n.x^n = A(x)$$

$$\Rightarrow \qquad A(x) = \sum_{n=2}^{\infty} n \, . \, x^n = 2x^2 + 3x^3 + 4x^4 + \ldots + \infty \qquad \ldots(2)$$

Multiply equation (2) by x, we have

$$x \, . \, A(x) = 2x^3 + 3x^4 + 4x^5 + \ldots + \infty \qquad \ldots(3)$$

Substract equation (3) from equation (2), we have

$$A(x)\,(1-x) = 2x^2 + x^3 + x^4 + \ldots + \infty$$

$$= 2x^2 + x^3\,(1 + x + x^2 + \ldots + \infty)$$

$$= 2x^2 + x^3 \, . \, \frac{1}{1-x}$$

$$= \frac{2x^2 - 2x^3 + x^3}{1-x} = \frac{2x^2 - x^3}{(1-x)}$$

So $$A(x) = \sum_{n=2}^{\infty} n\,x^n = \frac{2x^2 - x^3}{(1-x)^2}$$

Substitute the value of $A(x)$ in equation (1).

Thus we have

$$G(x)\,(1-4x^2) = \frac{2x^2 - x^3}{(1-x)^2} + \left(\frac{x^2}{1-x} + x\right)$$

$$\Rightarrow \quad G(x)\,(1-(2x)^2) = \frac{2x^2 - x^3}{(1-x)^2} + \left(\frac{x^2 + x - x^2}{1-x}\right)$$

$$= \frac{2x^2 - x^3 + x - x^2}{(1-x)^2} = \frac{x + x^2 + x^3}{(1-x)^2}$$

So $$G(x) = \frac{x + x^2 - x^3}{(1-x)^2(1-(2x)^2)}$$

$$= \frac{x + x^2 - x^3}{(1-x)^2(1-2x)(1+2x)}$$

Using partial fraction, we have

$$G(x) = \frac{1}{3}\cdot\frac{1}{(1-x)} - \frac{2}{9}\cdot\frac{1}{(1-x)^2} - \frac{1}{4}\cdot\frac{1}{(1-2x)} + \frac{5}{36}\cdot\left(\frac{1}{1+2x}\right)$$

So, $$\sum_{n=0}^{\infty} a_n x^n = \frac{1}{3}\sum_{n=0}^{\infty}(1)^n x^n - \frac{2}{9}\sum_{n=0}^{\infty}(n+1)\cdot(1)^n x^n - \frac{1}{4}\sum_{n=0}^{\infty}(-1)^n 2^n \cdot x^n + \frac{5}{36}\sum 2^n \cdot x^n$$

Thus the solution of the recurrence relation is

$$a_n = \frac{1}{3}(1)^n - \frac{2}{9}(n+1)\,(1)^n - \frac{1}{4}(-1)^n(2)^n + \frac{5}{36}(2)^n$$

POLYAS COUNTING THEOREM

1. Permutation

On a finite set A, a one-to-one maping from A onto itself is called a **permutation**. For example consider set $A = \{1, 2, 3, 4\}$, then a permutation

$$f_1 = \begin{pmatrix} 1 & 2 & 3 & 4 \\ 2 & 3 & 1 & 4 \end{pmatrix}$$

takes 1 into 2, 2 into 3, 3 into 1 and 4 into 4.

i.e.,

$$f_1(1) = 2, f_2(2) = 3,$$
$$f_1(3) = 1 \text{ and } f_1(4) = 4$$

2. Degree of a Permutation

Degree of a permutation is the number of elements in the set.

3. Representation of a Permutation

Fig. 17.5 is digraph of the permutation f_1.

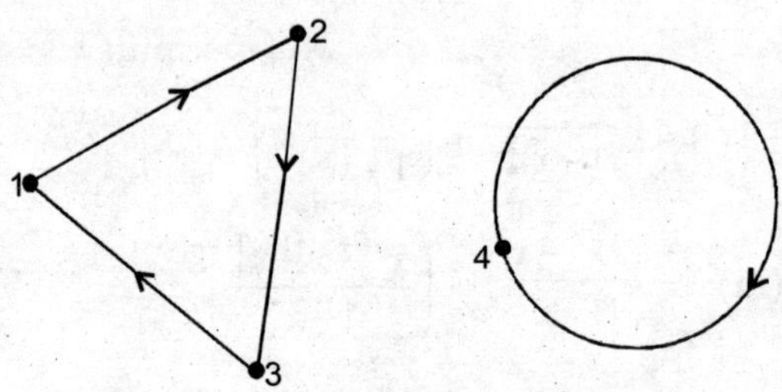

Fig. 17.5. Digraph of permutation f_1.

This permutation can also be represented as (1 2 3) (4). Where (1 2 3) and (4) are called **cycles** of permutations. This type of representation of a permutation is called Bcyclic **representation**. The numbers of edges (or elements) in a cycle of permutation is called the **length** of the cycle in the permutation. The cycles (1 2 3) and (4) of permutation f_1 are of length 3 and 1 respectively.

17.6.7 Definition (1)

Permutation f of degree m is said to be of type $(l_1, l_2, ..., l_m)$ if permutation f has l_i cycles of length $i = 1, 2, ..., m$.

Example 17.67. Consider a permutation $f = (1\ 3\ 5)\ (2\ 6)\ (4\ 7)$, f has cycles of length 3, 2 and 2 respectively. That is, it has no cycles of length 1, 4, 5, 6 and 7. so f is of type (0, 1, 0, 0, 0, 0).

Because l_1 = cycles of length 1 in $f = 0$

l_2 = cycles of length 2 in $f = 2$

l_3 = cycles of length 3 in $f = 1$

and so on.

Clearly $$1 \cdot l_1 + 2 \cdot l_2 + 3 \cdot l_3 + ... + m \cdot l_m = m$$

17.6.8 Definition (2)

Another way of representing the type of permutation f. Let us introduce m dummy variables, say $y_1, y_2, ..., y_m$.

Then the cycle structure of permutation f is given as $y_1^{l_2}\ y_2^{l_2} \cdots y_m^{l_m}$.

Hence the permutation (1 3 5) (2 6) (4 7) in the above example can be represented as

$$y_1^0\ y_2^2\ y_3^1\ y_4^0\ y_5^0\ y_6^0\ y_7^0 = y_2^2\ y_3$$

This is clear that any two distinct permutations may have same cycle structures. Consider another permutation as $(a\ b\ c)\ (d\ f)\ (e\ g)$.

The cycle structures of this permutation is also $y_2^2\ y_3^1$.

17.6.9 Definition (3)

Let us consider a set A with m elements. Then there exist a total of $m!$ permutations of set A, including the identify permutation.

A permutation which takes every element of set A into itself, is called **identity permutation.**

For example, on set $A = (a, b, c)$, permutation

$$f = \begin{pmatrix} a & b & c \\ a & b & c \end{pmatrix} \text{ is the identity permutation.}$$

Example 17.67. Let $A = (a, b, c)$ be a set with three elements. Then there will be a total $3! = 6$ permutations on A. The permutations are:

$$\begin{pmatrix} a & b & c \\ a & b & c \end{pmatrix}, \begin{pmatrix} a & b & c \\ a & c & b \end{pmatrix}, \begin{pmatrix} a & b & c \\ b & a & c \end{pmatrix}, \begin{pmatrix} a & b & c \\ c & b & a \end{pmatrix}, \begin{pmatrix} a & b & c \\ c & a & b \end{pmatrix}, \begin{pmatrix} a & b & c \\ b & c & a \end{pmatrix}$$

These six permutation can also be represented as

$(a)\ (b)\ (c)$, $(a)\ (b\ c)$, $(a\ b)\ (c)$, $(a\ c)\ (b)$, $(a\ c\ b)$ and $(a\ b\ c)$

Then their cycle structures, respectively are

$$y_1^3,\ y_1,\ y_2, y_1,\ y_2,\ y_1 y_2,\ y_3,\ y_3$$

Because in the first permutation, no cycles of length 2 and 3 and 3 cycles of length one. In second permutation one cycle of length one, one cycle of length 2 and no cycle of length 3 and so on.

17.6.10 Composition of Permutations

Let us take two permutations f_1 and f_2 on a set $S = (a\ b\ c\ d)$ as

$$f_1 = \begin{pmatrix} a & b & c & d \\ b & a & d & c \end{pmatrix} f_2 = \begin{pmatrix} a & b & c & d \\ c & d & a & b \end{pmatrix}$$

A composition $f_2\ f_1$ of two permutations is another permutation on set S defined as:

$$f_2 f_1(x) = f_2(f_1(x)),\ \forall\ x \in S$$

So $f_2 f_1$ for the given set $S = (a, b, c, d)$

is

$$f_2 f_1(a) = f_2(f_1(a)) = f_2(b) = d$$
$$f_2 f_1(b) = f_2(f_1(b)) = f_2(a) = c$$
$$f_2 f_1(c) = f_2(f_1(c)) = f_2(d) = b$$
$$f_2 f_1(d) = f_2(f_1(d)) = f_2(c) = a$$

So
$$f_2 f_1 = \begin{pmatrix} a & b & c & d \\ d & c & b & a \end{pmatrix}$$

17.7 PERMUTATION GROUP

Consider a set $A = (a_1, a_2, ..., a_m)$ then a set of all permutations (*i.e.*, m permutations). $S = (f_1, f_2, ..., f_m)$ on set A, forms a group under composition of two permutations as operation if

(*i*) S is closed with respect to composition of permutations.

(*ii*) Composition of permutation is associative. That is

$$f_1 \cdot (f_2 \cdot f_3) = (f_1 f_2) \cdot f_3$$

(*iii*) Identity exist. That is; exist a permutation

$$I \in S \text{ such that}$$
$$f_1 \cdot I = f_1 = I \cdot f_1$$

I is the identity permutation.

(*iv*) Inverse exist. That is; there exist a permutation $f^{-1} \in S$ for all $f \in S$ such that

$$f \cdot f^{-1} = I = f^{-1} \cdot f$$

The number of permutations m in a permutation group is called the **order** of the group and the number of elements in the set S on which the permutation are acting is called the **degree** of permutation group.

Example 17.69. Show that the set of all the six permutations on a set $A = (a, b, c)$ of 3 elements forms a permutation group with respect to composition of permutations as operation.

Solution: The six permutations are :

$$(a)\ (b)\ (c), (a)\ (bc), (ab)\ (c), (ac)\ (b)\ (abc) \text{ and } (acb)$$

Let $S = \{(a)\ (b)\ (c), (a)\ (bc), (ab)\ (c), (ac)\ (b)\ (abc), (acb)\}$

(*i*) We know that, the composition of two permutations is again a permutation so if $f_1 \in S$ and $f_2 \in S$.

then $f_2 f_1 \in S.$

Hence set S is closed.

(*ii*) Let $f_1 = (a)\ (bc), f_2 = (ab)\ (c)$ and $f_3 = (abc)$

then show that $f_1(f_2 f_3) = (f_1 f_2) f_3$

So
$$f_1(f_2 f_3) = \begin{pmatrix} a & b & c \\ a & c & b \end{pmatrix} \cdot \left\{ \begin{pmatrix} a & b & c \\ b & a & c \end{pmatrix} \cdot \begin{pmatrix} a & b & c \\ b & c & a \end{pmatrix} \right\}$$

$$= \begin{pmatrix} a & b & c \\ a & c & b \end{pmatrix} \cdot \left\{ \begin{pmatrix} a & b & c \\ c & b & a \end{pmatrix} \right\}$$

$$= \begin{pmatrix} a & b & c \\ c & a & b \end{pmatrix} \quad ...(1)$$

and $$(f_1 f_2) f_3 = \left\{ \begin{pmatrix} a & b & c \\ a & c & b \end{pmatrix} \cdot \begin{pmatrix} a & b & c \\ b & a & c \end{pmatrix} \right\} \cdot \begin{pmatrix} a & b & c \\ b & c & a \end{pmatrix}$$

$$= \left\{ \begin{pmatrix} a & b & c \\ b & c & a \end{pmatrix} \right\} \cdot \begin{pmatrix} a & b & c \\ b & c & a \end{pmatrix}$$

$$= \begin{pmatrix} a & b & c \\ c & a & b \end{pmatrix} \quad ...(2)$$

Clearly $f_1 (f_2 \cdot f_3) = (f_1 \cdot f_2) \cdot f_3$

Hence associative

(iii) Since the permutation $f_1 = \begin{pmatrix} a & b & c \\ a & b & c \end{pmatrix}$ is the identity permutation.

So identity eixst in S

i.e., for all $f_i \in S \;\; f_i \cdot I = f_i = I \cdot f_i$

(iv) The inverse of $(a)\,(b)\,(c)$ (*i.e.*, identity permutation) is identity permutation itself. The inverse of $(a)\,(bc)$ is $(a)\,(bc)$ itself as

$$\{(a)\,(bc)\} \cdot \{(a)\,(bc)\} = (a)\,(b)\,(c) \quad (\textit{i.e.}, \text{the identity permutation})$$

Similarly the inverses of other permutations also exist.

Hence set of six permutations over a set $A = (a, b, c)$ forms a group of order 6 and degree 3.

Thus, the set of all $m!$ permutations on a set A of m elements forms a permutation group of **order** $m!$ and **degree** m. This type of permutation group is called the **full symmetric group**, S_k. Hence the permutation group in example 17.7 is the full symmetric group S_3.

17.7.1 Cyclic Index of Permutation Group

For a permutation group S the expression obtained by dividing the sum of all cycle structures of all m permutations by m is called **cyclic index** of group P, denoted as $Z(P)$.

For example, the cyclic index of full symmetric group S_3, according to **example 17.7** is.

$$Z(S_3) = \frac{1}{3}(y_1^3 + y_1 y_2 + y_1 y_2 + y_1 y_2 + y_3 + y_3)$$

$$\Rightarrow \qquad \boxed{Z(S_3) = \frac{1}{3}\,(y_1^3 + 3y_1y_2 + 2y_3)}$$

Similarly the cyclic Index of S_4 full symmetric group of order 24 and degree 4 is

$$Z(S_4) \;=\; \frac{1}{24}(y_1^4 + 6\cdot y_1^2 y_2 + 8\cdot y_1 y_3 + 3\cdot y_2^2 + 6\cdot y_4)$$

17.7.2 Cyclic Index of Pair Group

Consider a graph G with n-vertices and $\frac{n(n-1)}{2}$ unordered pair of vertices. Permutation over the set of n-vertices of graph G also permutes the $\frac{n(n-1)}{2}$ unordered pair of vertices.

For example, consider the set $V = \{a, b, c\}$ of 4-vertices of a graph G then a permutation

$$\alpha_1 \;=\; \begin{pmatrix} a & b & c & d \\ d & a & c & b \end{pmatrix}$$

On the vertices a, b, c and d induces the permutation on the $\frac{4(4-1)}{2} = 6$ unordered vertex pairs. **(Fig. 17.6)**

$$\alpha_2 = \begin{pmatrix} ab & ac & ad & bc & bd & cd \\ da & dc & db & ac & ab & cb \end{pmatrix}$$ is a permutation on 6-unordered pairs.

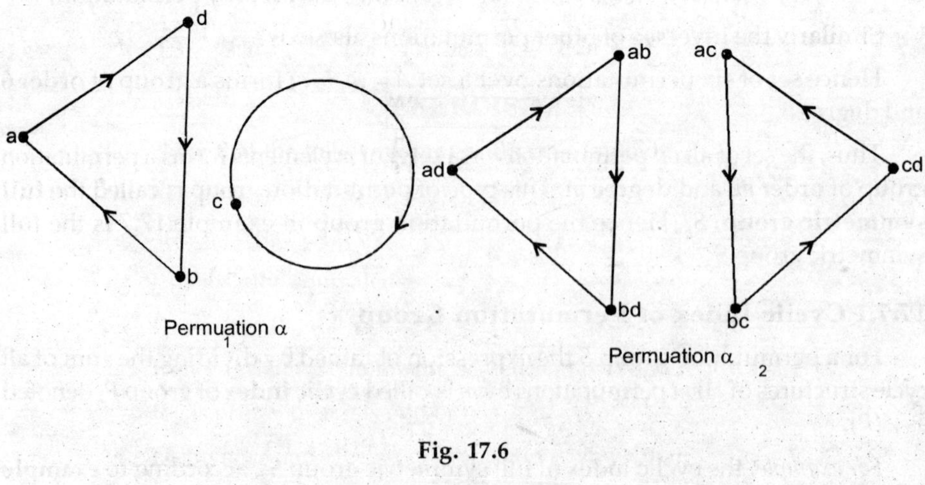

Fig. 17.6

Similarly, each of the $n!$ permutation induces a permutation on the $\frac{n(n-1)}{2}$ unordered vertex-pairs.

17.7.3 Definition

The full symmetric group S_n on n-vertices of a graph G induces a group of $n!$ permutations on the pairs of vertices. This induced group is called the **pair group** and denoted by R_n.

For example, consider the full symmetric group S_3 of 3-vertices the induced pair group R_3 has $3! = 6$ permutations as :

The identity permutation as:

$(a)(b)(c)$ in S_3 induces $(ab)(bc)(ac)$

$(a)(bc)$ in S_3 induces $(bc)((ac)(abb))$

$(ab)(c)$ in S_3 induces $((cb)(ab))(ca)$

(abc) in S_3 induces $((bc)(ca)(ab))$

(acb) in S_3 induces $((ca)(ab)(cb))$.

Table 17.1 summaries the results for R_3 as:

Table 17.1

Permutation type	*Number of permutation*	*Cycle structure*
(3, 0, 0)	1	y_1^3
(1, 1, 0)	3	$y_1 y_2$
(0, 0, 1)	2	y_3

$$\Rightarrow \qquad Z(R_3) = \frac{1}{6}[y_1^3 + 3y_1y_2 + 2y_3]$$

Similary the cycle index of group R_4 can also be obtained. **Table 17.2** gives the details about all the permutations in R_4.

Table 17.2

Terms in $Z(S_4)$	*Induced Term in $Z(R_4)$*	*No. of Permutations*
y_1^4	y_1^6	1
$y_1^2\, y_2$	$y_1^2\, y_2^2$	6
$y_1\, y_3$	y_3^2	8
y_2^2	$y_1^2\, y_2^2$	3
y_4	$y_2\, y_4$	6

Therefore the cycle index of the pair group R_4

$$Z(R_4) = \frac{1}{24}(y_1^6 + 6y_1^2\, y_2^2 + 8y_3^2 + 3y_1^2\, y_2^2 + 6y_2\, y_4)$$

$$\boxed{Z(R_4) = \frac{1}{24}(y_1^6 + 9y_1^2\, y_2^2 + 8y_3^2 + 6y_2y_4)}$$

Example 17.70. Find the cyclic index of a full symmetric group S_4.

Solution: Let the set $A = (a, b, c, d)$ has 4-elements then there will be a total of $4! = 24$.

permutations The permutations are

$$(a)(b)(c)(d), (a)(bcd), (a)(bdc), (abc)(d)$$
$$(acd)(d), (acd)(b), (adc)(b), (abd)(c)$$
$$(adb)(c), (a)(b)(cd), (a)(c)(bd), (b)(c), (ad)$$
$$(a)(d)(bc), (b)(d)(ac), (c)(d)(ab), (ab)(cd)$$
$$(ac)(bd), (ad)(bc), (abcd), (acbd)$$
$$(adbc), (abdc), (acdb), (adcb)$$

It can be seen that there is one permutation of type (4, 0, 0, 0), 6 of type (2, 1, 0, 0, 0), 8 of type (1, 0, 1, 0, 0). 3 of type (0, 2, 0, 0) and 6 of type (0, 0, 0, 1). Table 17.3 gives the cycle structures of the permutations.

Table 17.3

Permutation type	*Total Number*	*Cycle structure*
(4, 0, 0, 0)	1	y_1^4
(2, 1, 0, 0)	6	$y_1^2\ y_2$
(1, 0, 1, 0, 0)	8	$y_1 y_3$
(0, 2, 0, 0)	3	y_2^2
(0, 0, 0, 1)	6	y_4

So $\quad Z(R_4) = \frac{d}{24}(y_1^4 + 6y_1^2 y_2 + 8y_1 y_2 + 3y_2^2 + 6y_4)$ is the cycle index of group S_4.

17.8 EQUIVALENCE CLASSES OF FUNCTIONS

Before describing Polya's theorem, we want to introduce one additional concept of equivalent classes.

Let us consider two non-empty finite selts D and R, with the number of elements $|D|$ and $|R|$, respectively. Let $\oint$ be a function from D onto R *i.e.,*

$$\oint : D \to R.$$

Such that for each element of $d \in D$ there exist a unique element $\oint(d) \in R$, sets D and R called domain and range respectively.

Since each element of the $|D|$ elements can be mapped into any of the $|R|$ elements, the total number of different mappings from domain set D to range set R is $|R|^{|R|}$.

Let there be a permutation group P on the elements of set D. Then we define two mappings $\oint_1$ and $\oint_2$ on D as P-equivalent if there is some permutation f in P such that

$$\oint_1(d) = \oint_2(f(d)) \text{ for all } d \in D.$$

Since P is a permutation group:

(*i*) P contains identity permutations and thus is reflexive.

(*ii*) If P contains permutations f then it also contains f^{-1} *i.e.*, the inverse permutation. The relation is symmetric also.

(*iii*) If P has two permutation f_1 and f_2, it must also contain the $f_1 \times f_2$ permutation. Hence P-equivalence is a transitive relation also.

Polya s Counting Theorem (Definition)

Let us consider the following terms used in Poly's counting theorem:

(*a*) The two non-empty sets D and R denotes the **domain** and **range** respectively, together with permutation group P.

(*b*) For each element $r \in R$, $w(r)$ is called the **weight** or content of r. $r \in R$ are called figures.

(*c*) Defined a function f: $D \rightarrow R$ called **configurations**, is described by a sequences of $|D|$ elements of set R such that the ith element in the sequence is the image of the ith element of set D under f. Thus the **content** $W(f)$ of function f can be defined as the product of the contents of all its images.

That is $$W(f) = \prod_{d=D} w(f(d))$$

(*d*) Sometimes the weights of the elements of R can be expressed as power of some common quantity say x and the weight assignment in this case can be described by a **counting series** $A(x)$ for the elements of set R.

Thus $$A(x) = \sum_{p=0}^{\infty} a_p x^p$$

where a_p is the number of elements in set R with weight x^p.

(*e*) The number of configurations can be expressed in term of another series $B(x)$ such that

$$B(x) = \sum_{m=0}^{\infty} b_m x^m$$

where, b_m is the number of different configuration having weight x^m. B (x) is called the **configuration counting series.**

17.9 POLYA S ENUMERATION THEOREM

The configuration counting series is obtained by substituting the figure counting series into the cycle index of the permutation group P.

$$B(x) = Z(P; A(x^1), A(x^2), \dots A(x^m))$$

The proof of Polya's theorem is not in the scope of this book. We are mainly interested in the application of polya's theorem. The next few examples will illustrate the use of polya's theorem.

Example 17.71. Consider a square of 4-vertices. Count the number of different (non-isomorphic) ways to colour the 4-vertices of the square, if each vertex is to be either red or blue.

Solution: Let D and R be the two non-empty sets called domain and range respectively. $|D| = 4$ and $|R| = 2$. *i.e.*, set D has 4 elements (the 4-vertices of square) say 1, 2, 3 and 4 set R has 2 elements (red or blue) say r and b.

So figure counting series = $(r^i + b^i)$ for each y_i.

Now a clockwise rotation of square through an angle of 0°, 90°, 180° or 270° takes the square into itself. Thus, there are 4-circular or cyclic symmetries. Another 4 dihedral symmetries can be obtained by reflection of the square in the 2 diagonals and in the 2 lines bisecting opposite sides. So these symmetries of the square compose a subgroup G of order 8 of S_4; the elements of G are :

(*i*) Permutation induced by rotating the square through 0° with cycle index y_1^4.

(*ii*) Two permutations induced by rotating the square through 90° and 270° with cycle index y_4.

(*iii*) Permutation induced by rotating the square through 180° with cycle index y_2^2.

(*iv*) Two permutations induced by reflection in the diagonal joining corners (say a and b and c) with cycle index $y_1^2\, y_2$.

(*v*) Two permutations induced by reflection is the line joining the mid points of the opposite sides, with cycle index y_2^2.

So cycle index of $G = \frac{1}{8}[y_1^4 + 2y_1^2\, y_2 + 3y_2^2 + 2y_4]$ ...(1)

put $y_1 = r + b,\ y^2 = r^2 + b^2,\ y_4 = r^4 + b^4$ in equation (1) then the configuration counting series

$$B(r, b) = \frac{1}{8}[(r+b)^4 + 2(r+b)^2 \cdot (r^2 + b^2) + 3(r^2 + b^2)^2 + 2(r^4 + b^4)]$$

or
$$B(r, b) = r^4 + r^3 b + 2r^2 b^2 + rb^3 + b^4$$

Thus there exist one square in which 4 vertices are colored red, 1 square in which 3 vertices and red and one is blue; 2 squares in which 2 vertices are red and 2 are blue; one square in which one vertex is red and 3 vertices are blue and one square in which all vertices are blue (zero vertices are colored red).

These can be shown below :

(*i*)

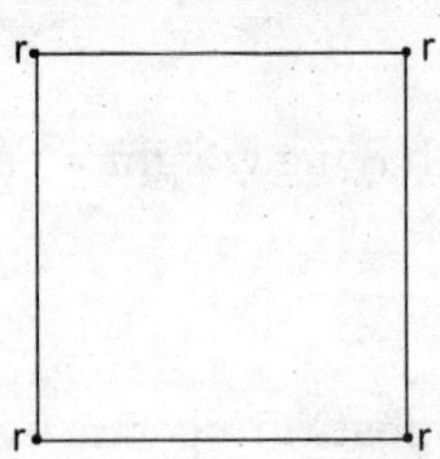

(*ii*)

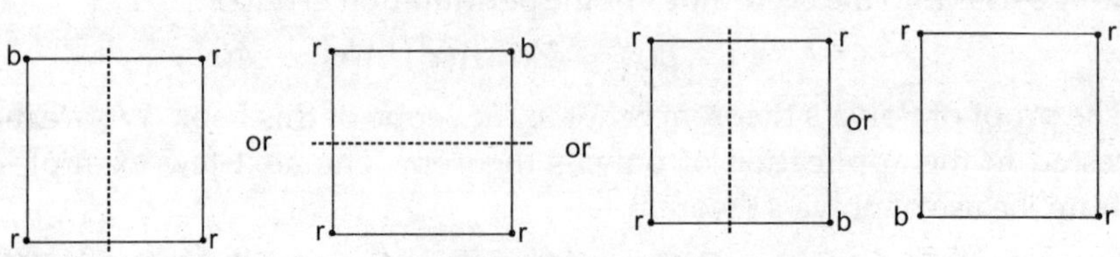

all these patterns are isomorphic

(iii)

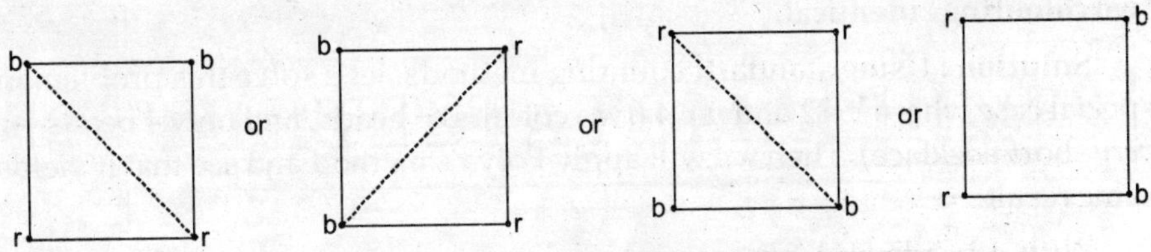

all these patterns are isomorphic patterns

(iv)

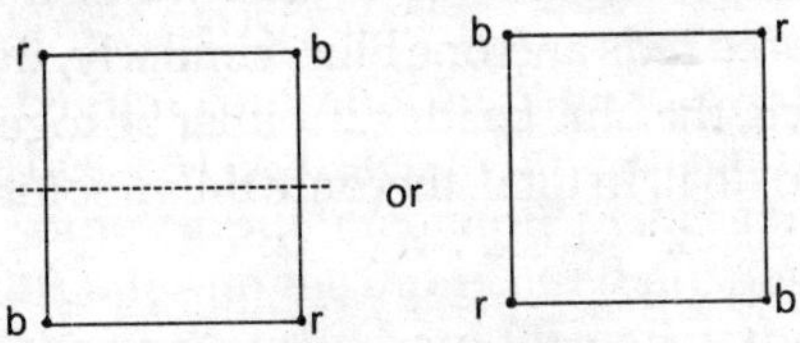

the two are isomorphic

(v)

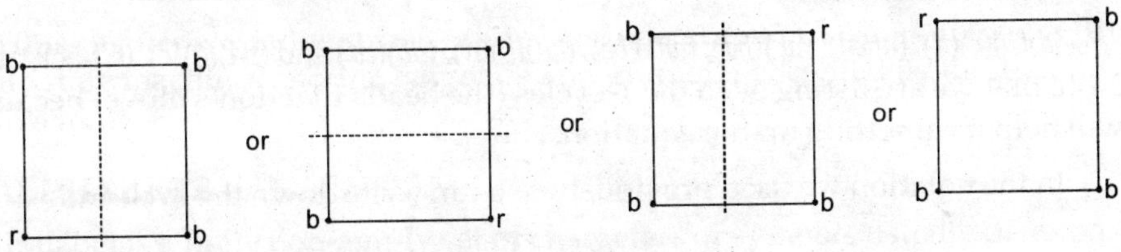

all four patterns are isomorphic

(vi)

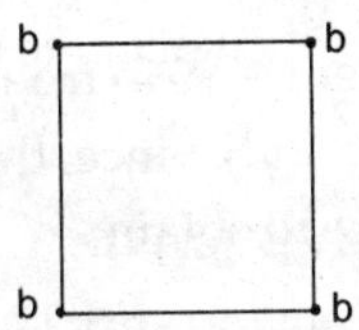

Thus the $2^4 = 16$ colourings of square falls in 6 different categories.

Example 17.72. Given a cube with 4-balls (identical). Count the number of different ways to arrange the balls on the corners of the cube. The arrangements are said to be non-isomorphic. Two arrangements are said to be same if by any rotation of the cube they can be transformed into each other.

Solution: Let D and R be the non-empty finite sets called domain and range. D consist of 8-corners of the cube R has two elements s and t with content x^1 and x^0 respectively s and t indicates the "presence of a ball" and "absence of a ball" respectively. The Fig. counting series is

Example 17.73. Count the number of ways to arrange beads on a necklace, where there are k different colours of beads, and n total beads arranged on the necklace.

With a necklace, we can obviously rotate it around, so if we number the beads in order as 1, 2, 3, 4, then for a tiny necklace with only four beads, the pattern "red, red, blue, blue" is clearly the same as "red, blue blue, red", et

cetera. Also, since the necklace is just made of beads, we can turn it over, so if there were four colours, although we cannot rotate "red, green, yellow, blue" into "blue, yellow, green, red", we can flip over the necklace and make those two colourings identical.

Solution : Using standard counting methods, let's solve this problem in the special case where $k = 2$ and $n = 4$ (two colours of beads, and only 4 beads—it's a *very* short necklace). Then we will apply Polya's method and see that it yields the same result.

With 4 beads and two colours, we can just list the possibilities. There is obviously only one way to do it with either all red beads or all blue beads. if there is one red and three blue of the reverse—three reds and one blue, similarly, there's only one way to do it. If there are two of each, the blue beads can either be together, or can be separated, so there are two ways to do it. In total, there are thus 6 solutions.

Now let's try Poly's method:

If the bead positions are called a, b, c and d, here are the permutation that map the necklace into itself.

$(a)(b)(c)(d)$, $(abcd)$, $(ac)(bd)$, $(ad)(bc)$, $(ac)(b)(d)$, $(ab)(cd)$ and $(bd)(a)(c)$. (Check these). Note that we are listing even the 1-cycles (the beads that don's move) because it will help us in setting up the equation.

In the notation we used previously, we can write down the cycle index:

$$G = \frac{1}{8}(y_1^4 + 2y_1^2 y_2 + 3y_2^2 + 2y_4)$$

Since, there are three permutations having two 2-cycles, there is a 3 in front of the term y_2^2, et cetera. Let $y_1 = (x + y)$, $y_2 = (x^2 + y^2)$, and $y_4 = (x^4 + y^4)$. Since, there are only two colours, we only need x and y. Substitute as before to obtain:

$$G = \frac{(x+y)^4 + 2(x+y)^2(x^2+y^2) + 3(x^2+y^2)^2 + 2(x^4+y^4)}{8} \quad ...(2)$$

If we expand, we obtain:

$$x^4 + x^3y + 2x^2y^2 + xy^3 + y^4$$

It's easy to check that these terms correspond to the 6 ways beads could be arranged, where there's a unique way to do it unless there are two of each colour, in which case there are two arrangements. Also notice that it gives our detailed count as well. If we think of the x as corresponding to a "red" bead, and y to a "blue" bead, the coefficient in front of the term x^4 (which is 1) corresponds to the number of ways of making a necklace with four red beads. The 2 in front of the x^2y^2 term means that there are two necklaces with two beads of each colour, et cetera.

And notice again that by substituting $x = y = 1$ into equation (2) we obtain the total count:

$$\frac{2^4 + 2\cdot 2^2 \cdot 2 + 3\cdot 2^2 + 2\cdot 2}{8} = \frac{16+16+12+4}{8} = 6$$

$$A(x) = \sum_{p=0}^{\infty} a_p x^p = a_0 x^0 + a_1 x^1 = 1 + x$$

Since a_0 is the number of Fig. with content 0, is one and a_1 is the number of Fig. with content 1 is also one.

The $2^8 = 256$ configurations are different functions (or mappings) that assign balls to the corners of the cube. The permutation group P on D is the set of all those permutations that can be produced by rotation of the cube. These permutations and their cycle structures are:

(*i*) One identity permutation with cycle structure y_1^8.

(*ii*) There 180° rotations around lines connecting the centres of opposite faces with cycle structure y_2^4.

(*iii*) Six 90° rotations both clockwise and counter-clockwise around lines connecting the centres of opposite faces with cycle structure y_4^2.

(*iv*) Six 180° rotations around lines connecting the midpoints of opposite edges with cycle structure y_2^4.

(*v*) Eight 120° rotations around lines connecting opposite centres in the cube with cycle structure $y_1^2 y_3^3$.

Thus the cycle index of this permutation group P is

$$Z(P) = \frac{1}{24}(y_1^8 + 9y_2^4 + 6y_4^2 + 8y_1^2 y_3^3)$$

Substitute $y_1 = 1 + x$, $y_2 = 1 + x^2$, $y_3 = 1 + x^3$ and the Fig. counting series thus obtained is

$$B(x) = \frac{1}{24}((1+x)^8 + 9(1+x^2)^4 + 6(1+x^2)^2 + 8(1+x)^2 + (1+x^3)^3)$$

$$B(x) = 1 + x + 3x^2 + 3x^3 + 7x^4 + 3x^5 + 3x^6 + x^7 + x^8$$

The coefficient of x^i gives the number of p-equivalent configurations of content x^i (*i.e.*, with i-balls).

The total number of p-equivalent configurations with content $x^0, x^1, ..., x^8$ is obtained by taking sum of all the coefficients in $B(x)$.

That is $$\text{sum} = 1 + 1 + 3 + 3 + 7 + 3 + 3 + 1 + 1 = 23$$

In other words one can say that the number of different ways to point the corners of a cube by two colours is 23.

Fig. 17.7 shows the 7-ways of arranging 4 balls on the corners of a cube. Because the coefficient of x^4 in $B(x)$ is 7.

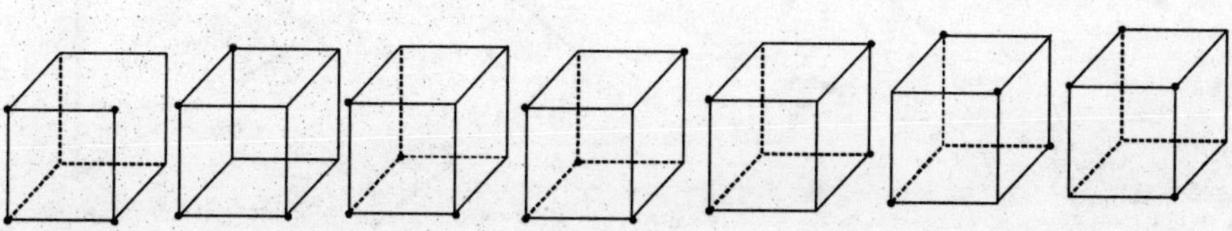

Fig. 17.7. Arranging four balls to the corners of a cube.

17.10 GRAPH EMUMERATION WITH POLYA S THEOREM

17.10.1 Enumeration of Simple Graphs

Consider the problem of enumerating all unlabeled simple graph of n-vertices. The graph G can be regarded as configuration (*i.e.*, mapping) of the elements of set D *i.e.*, all the $\frac{n(n+1)}{2}$ unordered pairs of vertices.

Range R consists of two elements s and t with contents x^1 and x^0.

An element of set D maps into s with content x^1 if a vertex pair is joined by an edge otherwise the vertex pair in D maps into t with content x^0.

$$A(x) = \sum_{p=0}^{\infty} a_p x^p = 1 + x.$$

The relevant permutation group in this case is the group of permutations on the pairs of vertices induced by S_n, *i.e.*, R_n.

Therefore, the configuration counting series $B(x)$ is obtained by putting $1 + x$ for y_1 $1 + x^2$ for y_2 and $1 + x^3$ for y^3 and so on in $Z(R_n)$.

For $n = 3$ vertices.

$$Z(R_3) = \frac{1}{6}(y_1^3 + 3y_1y_2 + 2y_3)$$

So

$$B(x) = \frac{1}{6}((1+x)^3 + 3(1+x)(1+x^2) + 2(1+x^3))$$

$$\boxed{B(x) = 1 + x + x^2 + x^3}$$

The coefficients of x^i,s in the configuration counting series $B(x)$ gives the number of configurations with content x^i.

The content of a configuration here is the number of edges in the corresponding simple graph.

Thus, the number of simple graphs of 3-vertices with 0, 1, 2 and 3 edges is 1 in each case. Because in $B(x)$ the co-efficient of x^0 is 1, x^1 is 1, x^2 is 1, and of x^3 is also 1.

All the simple graphs are shown in Fig. 17.8 (a).

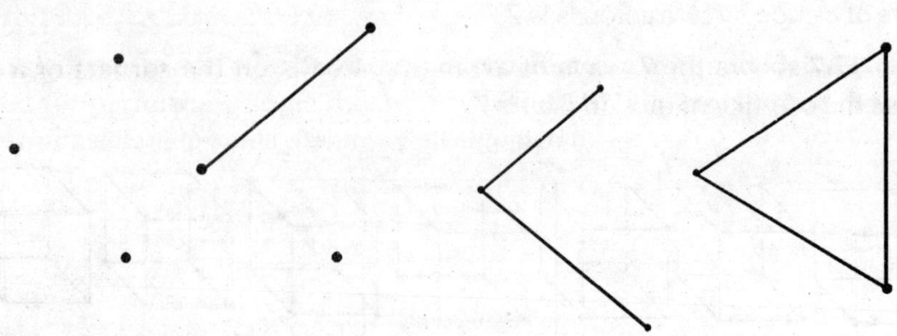

Fig. 17.8(a). Simple unlabeled graphs of 3-vertices.

Similarly, one can obtain all simple graphs of 4-vertices by substituting the Fig. counting series for each of y_i in the cycle index of the group R_4.

So $$Z(R_4) = \frac{1}{24}(y_1^{\,6} + 9y_1^{\,2}\, y_2^{\,2} + 8y_3^{\,2})$$

then $$Z(R_4) =$$

$$\frac{1}{24}((1+x)^6 + 9(1+x)^2 + (1+x^2)^2 + 8(1+x^5)^2)$$

$$= 1 + x + 2x^2 + 3x^3 + 2x^4 + x^5 + x^6.$$

i.e., one graph with no edge, 1 graph with one edge 2 graphs with 2 edges, 3 graphs with 3-edges 2 graphs with 4-edges and one graph each with 5-edges and 6-edges.

Fig. 17.8(*b*) shows all simple unlabelled graphs of 4-vertices.

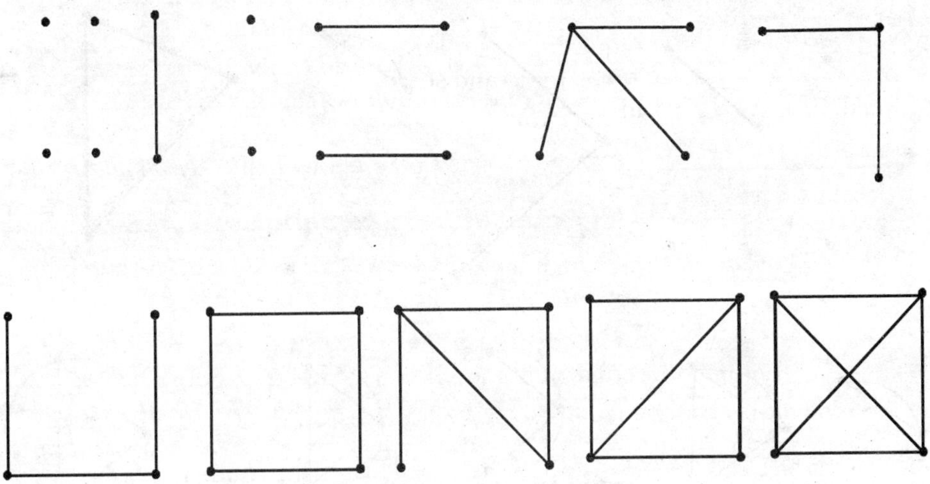

Fig. 17.8(*b*). Simple unlabelled graphs of 4-vertices.

Similarly, we can find out all the simple unlabelled graphs of 5-vertices by putting $1 + x^i$ for y^i in $Z(R_5)$.

$$Z(R_5) = \frac{1}{120}(y_1^{\,10} + 10y_1^{\,4}y_2^{\,3} + 20y_1y_3^{\,3} + 15y_1^{\,2}y_2^{\,4} + 30y_2y_4^{\,2} + 20y_1\,y_3y_6 + 24y_5^{\,2})$$

$$Z(R_5) = 1 + x + 2x^2 + 4x^3 + 6x^4 + 6x^5 + 6x^6 + 4x^7 + 2x^8 + x^9 + x^{10}$$

One can obtain the number of simple unlabelled graphs with n-vertices for any n, similarly.

17.10.2 Enumeration of Multi-graphs

Let us count all multi-graphs of n-vertices in which there exist at most 2-edges between every pair of vertices.

The domain set D will have $\frac{n(n-1)}{2}$ unordered pair of vertices as element. But range R will contain 3-elements r, s and t with contents x^0, x^1 and x^2.

An element of D goes to Ir by mapping f if there is no-edge between a vertex pair, goes to s if there is one-edge and goes to t if there are two-edges between a pair of vertices. Then the Fig. counting series.

$$A(x) = 1 + x + x^2$$

Let us calculate all multi-graphs of $n = 3$ and 4 vertices.

(*i*) $n = 3$ vertices $Z(R_3) = \frac{1}{6}(y_1^3 + 3y_1y_2 + 2y_3)$

$$Z(R_3) = \frac{1}{6}((1 + x + x^2)^3 + 3(1 + x + x^2)(1 + x^2 + x^4) + 2(1 + x^3 + x^6))$$

$$Z(R_3) = 1 + x + 2x^2 + 2x^3 + 2x^4 + x^5 + x^6.$$

Thus, the Multi-graphs of 3 vertices are shown in **Fig. 17.9 (*a*).**

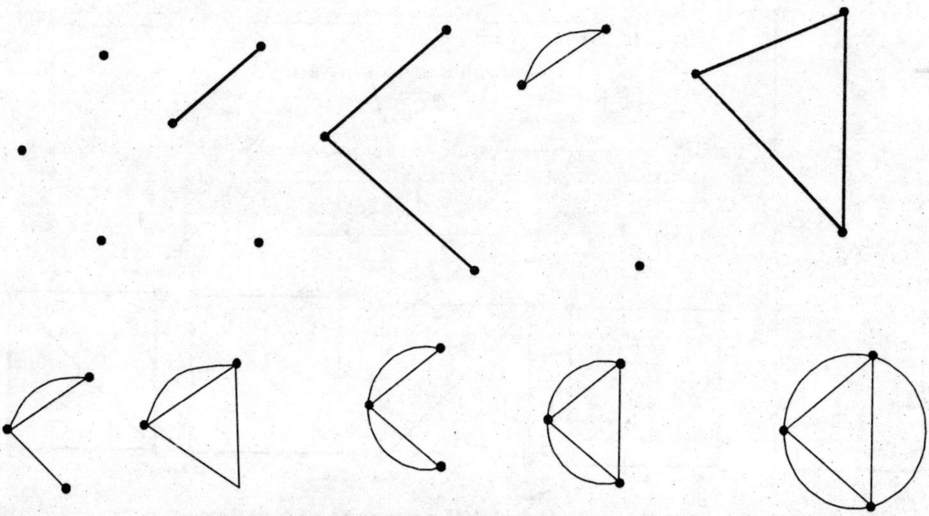

Fig. 17.9 (*a*). Multi-graphs of 3-vertices and at most 2-edges between every pair of vertices.

Similarly, Multi-graphs of 4-vertices can be obtained by substituting $1 + x^i + 2x^i$ for each y_i in $Z(R_4)$

$$Z(R_4) = \frac{1}{24}[y_1^6 + 9y_1^2y_2^2 + 8y_3^2 + 6y_2y_4]$$

$$= \frac{1}{24}[(1 + x + x^2)^6 + 9(1 + x + x^2)^2(1 + x^2 + x^4)^2 + 8(1 + x^3 + x^6)^2$$
$$+ 6(1 + x^2 + x^4)(1 + x^4 + x^8)]$$

$$= 1 + x + 3x^2 + 5x^3 + 8x^4 + 9x^5 + 12x^6 + 9x^7 + 8x^8 + 5x^9 + 3x^{10} + x^{11} + x^{12}$$

The co-efficient of x^i gives the number of multi-graphs of 4-vertices and i-edges. Each vertex pair can have at most 2-edges. (*i.e.*, parallel).

For example; the coefficient of x^4 is 8. That is there exist 8-multi-graph of 4-edges **(Fig. 17.9(*b*)).**

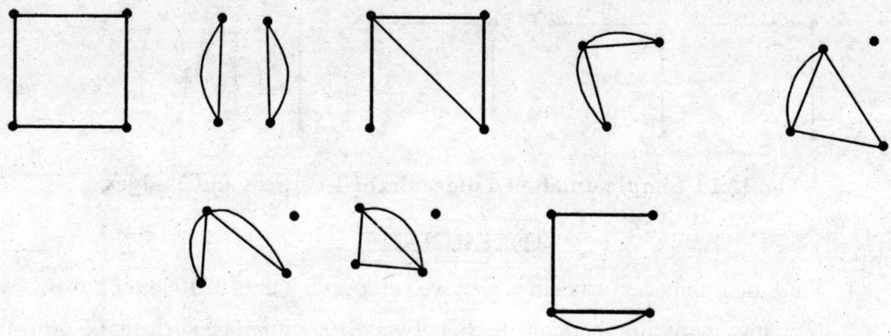

Fig. 17.9(b). Multi-graphs of 4-vertices, 4-edges and at most two parallel edges.

17.10.3 Enumeration of Digraphs

For counting digraph consider two non-empty finite sets D and R, called domain and range. D consisting of $n(n-1)$ ordered pair of vertices. The relevant permutation group will consist of permutations induced on all ordered pairs of vertices by S_n. The cyclic index of M_n, can be obtained by the same procedure as done in case of R_n. For $n = 4$ the terms in $Z(M_4)$ induced by each term in $Z(S_4)$ given in **Table 17.4.**

Table 17.4

Terms in $Z(S_4)$	*Induced term in $Z(M_4)$*
y_1^4	y_1^{12}
$y_1^2 y_2$	$y_1^2 y_2^5$
$y_1 y_3$	y_3^4
y_2^2	y_2^6
y_4	y_4^3

Therefore, the cycle index is

$$Z(194) = \frac{1}{24}(y_1^{12} + 6y_1^2 y_2^5 + 8y_3^4 + 3y_2^6 + 6y_2^6)$$

For a simple digraph, the Fig. counting series.

$$A(x) = 1 + x \text{ is applicable}$$

As for a given ordered pair (v_i, v_j) either does or does not have an edge directed from v_i to v_j. So on substituting $1 + x^i$ for each y_i in $Z(M_4)$ the configuration counting series $B(x)$ is obtained as

$$B(x) = \frac{1}{24}[(1+x)^{12} + 6(1+x)^2(1+x^2)^5 + 8(1+x^3)^4 + 3(1+x^2)^6 + 6(1+x^4)^6]$$

$$B(x) = 1 + x + 5x^2 + 13x^3 + 27x^4 + 38x^5 + 48x^6 + 38x^7 + 27x^8 + 13x^9 + 5x^{10} + x^{11} + x^{12}$$

The co-efficient of x^i in $B(x)$ gives the number of simple digraphs with 4-vertices and i-edges. For instance the five-digraphs of two edges are shown in **Fig. 17.10.**

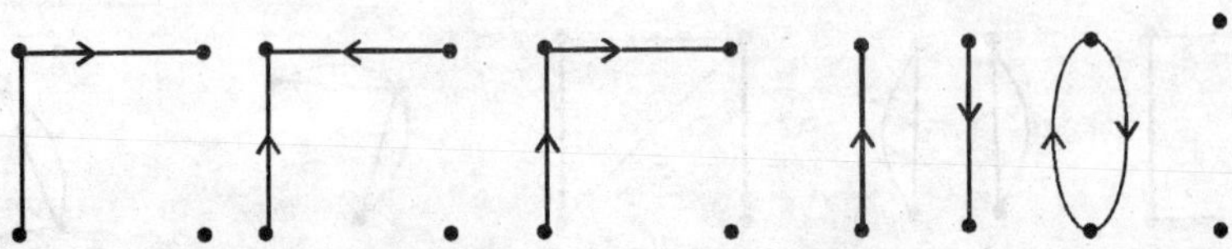

Fig. 17.10. Simple unlabeled digraphs of 4-vertices and 2-edges.

EXERCISES

1. Find the number of ways in which we can post 5 letters in 10 letter boxes.
2. (*a*) How many numbers are there between 100 and 1000 such that 5 is in the unit's place ?
   (*b*) How many numbers are there between 100 and 1000 such that at least one of their digits is 7 ?
   (*c*) How many of them have exactly one of their digits as 5 ?
   [**Hint:** (*b*) & (*c*) All numbers are of 3 digits, so there exist 3 cases, 5 is in the unit's place, 5 is in the ten's place and 5 is in the hundred's place].
3. A teacher distributes 5 different pencils among 8 students.
   (*i*) In how many ways can this be done if each student receives at most one pencil ?
   (*ii*) In how many ways can this be done if there is no restriction on the number of pencils a student can receive ?
4. A 5 person committee having members *A*, *B*, *C*, *D* and *E*, is to select a President, Vice President and Secretary.
   (*a*) In how many way can this occur if either *C* or *D* must be president ?
   (*b*) How many selections are there in which either *E* is a secretary or *E* is excluded ?
   (*c*) How many selections exclude *A* and *B* ?
5. If four dice are rolled, count the number of possible outcomes in which at least one die shows 2.
6. Find the number of signal that can be generated by using 6 differently coloured flags, when any number of them may be hoisted at a time.
7. In how many ways a five digit number divisible by 3 is to be formed using the numerals 0, 1, 2, 3, 4 and 5 without repetition ?
8. Find the sum of all the five digit numbers that can be formed using the digits 1, 2, 3, 4 and 5 without repetition.
   [**Hint:** Total numbers of 5 digit numbers are $5! = 120$. Now 1 can occur at the unit place in $4!$ ways, 2 can occur at unit place in $4!$ ways and so on. Thus, the sum due to the unit place of all the 120 numbers is $24(1 + 2 + 3 + 4 + 5) = 360$ units.
   Similarly, the sum due to the tens place = 360 tens sums for the remaining places are obtained similarly. Hence, the sum of all the 120 numbers is $360(1 + 10 + 100 + 1000 + 10000) = 360 \times 11111 = 3999960$).
9. How many number of rectangles you can find on a chessboard ?
   [**Hint:** There are 9 horizontal and 9 vertical lines on a chessboard and we require 2 horizontal and 2 vertical lie to form a rectangle].

10. Ten persons, amongst whom are X, Y and Z are to speak at a function. Find the number of ways in which it can be done if a wants to speak before B and B wants to speak before C.

11. Prove that $n\,!\,(n+2) = n\,! + (n+1)\,!$

12. Out of 7 consonants and 4 vowels, how many words can be made each containing 3 constants and 2 vowels ?

13. In how many ways can a consonant and a vowel be choosen out of the letters of the word COURAGE ?

14. In how many ways can 4 Latin and 1 English book be placed on a shelf so that the English book is always in the middle, the selection being made from 7 Latin and 3 English books ?

15. There are two works each of 3 volumes and two works each of 2 volumes. In how many ways can be 10 books be placed on a shelf so that volumes of the same work are not separated ?

17. Find the number of arrangements that can be made out of the letters of the words.

(*i*) Independence

(*ii*) Superstitious

(*iii*) Institutions.

17. A room is to be decorated with 14 flags. If 2 of them are blue, 3 red, 2 white, 3 green, 2 yellow and 2 purple, in how many ways can they be hang ?

18. (*i*) In how many ways can 7 persons form a ring ?

(*ii*) In how many ways can 7 Englishmen and 7 Americans sit down at a round table, no two American being together ?

19. In how many ways can five things be divided between two persons ?

20. There are 20 persons among whom are two brothers. In how many ways we can arrange then around a circle so that there is exactly one person between the two brothers ?

21. In an examination hall, there are four rows of chairs, each row containing 10 chairs. There are two classes sitting for the examination, each class having 20 students. In how many ways can these 40 students be seated. If in each row all students belong to the same class ?

22. Prove that

$$\binom{n+1}{r} = \binom{n}{r-1} + \binom{n}{r}$$

23. Prove that

(*i*) $\binom{n}{0}+\binom{n}{1}+\binom{n}{2}+\ldots+\binom{n}{n}=2^n$

(*ii*) $\binom{2}{1}+\binom{3}{1}+\binom{5}{1}+\binom{3}{2}+\binom{4}{2}$

(*iii*) $\binom{n}{0}-\binom{n}{1}+\binom{n}{2}-\binom{n}{3}+\ldots+\binom{n}{n}=0.$

24. A group of 8 players is composed of 5 cricketers and 3 footballs.

(*a*) In how many ways sa committee of 5 be formed ?

(*b*) In how many ways a committee of 5 be formed that has 3 cricketers and 2 footballers?

25. A committee of 5 is to be formed out of 6 men and 4 women. In how many ways can this be done if

(*a*) At least 2 women are to be included.

(*b*) At most 2 women are to be included.

26. In a box, there are 12 red balls and 8 white balls. A sample of 5 distinguishable balls is to be selected :

(*i*) How many samples of 5 balls are there?

(*ii*) How many samples contain all red balls?

(*iii*) How many samples contain all white balls?

(*iv*) How many samples are three with 3 red and 2 white balls?

(*v*) How many samples contain at least 3 red balls?

27. A person deposits Rs. 500 in a savings account at an interest rate of 5% per year compounded annually obtain the amount at the end of *r* years. Sketch the numeric function.

28. Every particle inside a nuclear reactor splits into two particles in each second. Suppose one particle is injected into the reactor every second beginning at $t = 0$. How many particles are there in the reactor at r^{th} second ?

[**Hint:** Given that at $t = 0$, number of particles are $a_0 = 1$. At $t = 1$, $a_1 = 1 \times 2 + 1 = 3$. (one particles splits into 2 and one is injected).

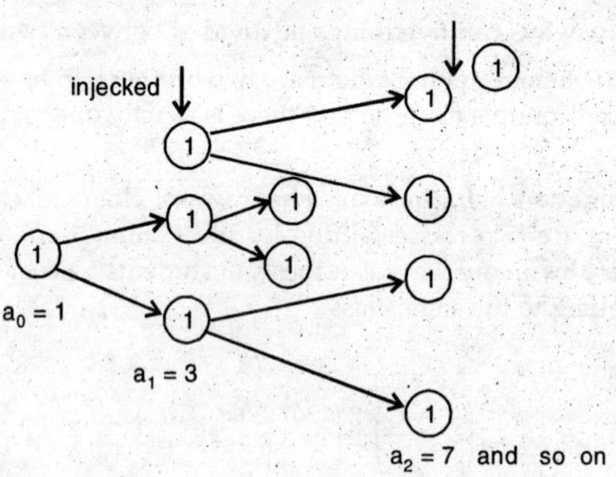

$a_0 = 1 \quad a_1 = 3 \quad a_2 = 7$ and so on

So the sum $S = 1 + 3 + 7 + ... + a_r$ is the sum of particles in *r* second. Solving this we have $a_r = 1 + 2 + 4 + 8 + ... + \infty = 2^{r+1} - 1$].

29. There are two kinds of particles inside a nuclear reactor. In every second, an α-particle will split into three β-particles and a β-particles will split into an α-particle and two β-particles. If there is a single α-particle in the reactor at $t = 0$. How many particles are there altogether at $t = 100$? **(U.P.T.U; 2005-06)**

30. Let *a* and *b* be two numeric function given by

$$a_n = \begin{cases} 0, & 0 \le n \le 4 \\ 2^{-n} + 3, & n \ge 5 \end{cases} ; \qquad b_n = \begin{cases} 1 - 2^n, & 0 \le n \le 2 \\ 2^n + 1, & r \ge 3 \end{cases}$$

Find $a + b$ and $a \,.\, b$.

31. Let $C = a \,.\, b$. Show that $\Delta C_r = a_{r+1}(\Delta b_r) + b_r(\Delta a_r)$

[**Hint:** $\Delta C_r = C_{r+1} - C_r,\ \ C_{r+1} = a_{r+1} b_{r+1},\ \ C_r = a_r \,.\, b_r$]

32. Let a be a numeric function given by

$$a_r = \begin{cases} 2 & 0 \le r \le 3 \\ 2^{-r} + 3 & r \ge 4 \end{cases}$$

(*i*) Determine $S^2 a$ and $S^{-2} a$ (*ii*) determine Δa and ∇a.

33. Determine $C = a * b$ where a and b are two numeric functions such that

$$a_r = \begin{cases} 1 & 0 \le r \le 2 \\ 0 & r \ge 3 \end{cases} ; \qquad b_r = \begin{cases} 1 & 0 \le r \le 2 \\ 0 & r \ge 3 \end{cases}$$

34. Let a, b and c be numeric functions such that $a * b = c$. Given that

$$a_r = \begin{cases} 1 & r = 0 \\ 2 & r = 1 \\ 0 & r \ge 2 \end{cases} ; \qquad c_r = \begin{cases} 1 & r = 0 \\ 0 & r \ge 1 \end{cases}$$

Find b.

[**Hint:** Use $C_r = a_r * b_r = \sum_{i=0}^{r} a_i b_{r-i} = a_o b_r + a_1 b_{r-1} + \ldots + a_r b_o$

Thus $C_0 = a_0 b_0,\ \ C_1 = a_0 b_1 + b_0 a_1$ and so on].

35. Find the generating function of the following numeric functions.

(*i*) $a_n = 2^n + 3^n;\ \ n \ge 0$ (*ii*) $a_n = {}^{n+4}C_n,\ n = 0, 1, 2 \ldots$

(*iii*) $a_n = \begin{cases} 3^n & n \text{ is even} \\ 3^{-n} & n \text{ is odd} \end{cases}$ (*iv*) $a_n = 3^{r+2},\ \ r \ge 0.$

36. Let $C = a * b$, where $a_r = 2^r$, $r \ge 0$ and $b_r = 3^r$, $r \ge 0$. Find the generating function of C.

37. Determine the numeric functions corresponding to each of the following generating functions :

(*i*) $A(x) = \dfrac{2}{1 - 4x^2}$ (*ii*) $A(x) = \dfrac{1 + x^2}{(1 - x)^4}$

(*iii*) $A(x) = \dfrac{1}{5 - 6x + x^2}$ (*iv*) $A(x) = \dfrac{x^4}{1 - 2x}$

(*v*) $A(x) = \dfrac{1}{(1 - x)(1 - x^2)(1 - x^3)}$ (*vi*) $A(x) = \dfrac{1}{1 - x^3}$

38. Use generating function to show that $\sum_{k=0}^{n} C(n, k)^2 = C(2n, n)$, where n is a positive integer. **(U.P.T.U. 2005-06)**

[**Hint:** $C(2n, n)$ *i.e.*, ${}^{2n}C_n$ is the co-efficient of x^n in the expansion of $(1 + x)^{2n}$.

So $(1 + x)^{2n} = [(1 + x)^n]^2 = ({}^nC_0 + {}^nC_1 x + {}^nC_2 + {}^nC_2 x^2 + \ldots + {}^nC_n x^n)^2$.

The co-efficient of x^n from the above expression.

$${}^{2n}C_n = {}^nC_0 \cdot {}^nC_n + {}^nC_1 \cdot {}^nC_{n-1} + {}^nC_2 \cdot {}^nC_{n-2} + \dots + {}^nC_n \cdot {}^nC_0$$

$$= ({}^nC_0)^2 + ({}^nC_1)^2 + \dots + ({}^nC_n)^2 \left[\because {}^nC_{n-k} = {}^nC_k\right]$$

$$= \sum_{k=0}^{n} ({}^nC_k)^2]$$

39. Find the number of solutions of $n_1 + n_2 + n_3 = 17$, where n_1, n_2, n_3 are non-negative integers such that
$2 \le n_1 \le 5, 3 \le n_2 \le 6, 4 \le n_3 \le 7.$ **(U.P.T.U., 2005-06)**
[**Hint:** The generating function
$A(x) = (x^2 + x^3 + x^4 + x^5) \cdot (x^3 + x^4 + x^5 + x^6) \cdot (x^4 + x^5 + x^6 + x^7).$
So the number of solutions with the given constraints is the co-efficient of x^{17} in the generating function $A(x)$.
So $\quad A(x) = x^9 (1 + x + x^2 + x^3)$ y3 $= x^4 (1 + 3x + 3x^2 + x^3) \cdot (1 + 3x^2 + 3x^4 + x^6)$
the co-efficient of $x^{17} = 3$. Thus there are 3 solutions].

40. Find the generating function for the number of integer solutions to the equation :
$$C_1 + C_2 + C_3 + C_4 = 20$$
where $C_1 \ge -3, C_2 \ge -3, -5 \le C_3 \le 5$, and $C_4 \ge 0$.
also find the number of integer solutions. **(U.P.T.U., 2005-06)**

41. Find the number of ways to select 10 marbles from a large pile of red, yellow and green marbels if
(*a*) The selection has at least 2 marbels of each colour.
(*b*) The selection has at most 2 green marbels.
(*c*) The selection has an even number of red marbels.

[**Hint:** The problem is equivalent to the number of integer solutions of the equation
$$r + y + g = 10.$$
where
(*a*) $r \ge 2, \; y \ge 2$ and $g \ge 2$.
(*b*) $r \ge 0, \; g \ge 0, \; 0 \le g \le 2$.
(*c*) $r = 0, 2, 4, 6, \dots, \; y \ge 0, \; g \ge 0$.
r is the number of red marbels, y is number of yellow marbels and g is the number of green marbels].

42. Show that (using generating function)
$$\sum_{i=0}^{n} \binom{m}{i}\binom{n}{i} = \binom{m+n}{n}, \quad m \ge n \ge 0.$$ **(U.P.T.U., 2005-06)**

43. Solve the following recurrence relations using
(*a*) characteristic root method
(*b*) method of operators
(*c*) method of generating functions
(*i*) $a_n - 2a_{n-1} + a_{n-2} = 2^n$,
(*ii*) $a_n - 5a_{n-1} + 6a_{n-2} = 2, \quad a_0 = 1, \; a_1 = -1$
(*iii*) $a_n - 9a_{n-1} + 20a_{n-2} = 2.5^n$,

(*iv*) $a_n - 3a_{n-1} + 2a_{n-2} = 5n + 3$,
(*v*) $a_n - 7a_{n-1} + 12a_{n-2} = n4^n$.

44. Solve the following recurrence relations :

(*i*) $a_{n+1} - 3a_n = 5^n \cos n \dfrac{\pi}{2}$
(*ii*) $a_{n+3} + 2a_{n+2} + a_{n+1} = 2^n + n^2 + n$
(*iii*) $a_{n+2} - 5a_{n+1} + 6a_n = \sin n$
(*iv*) $a_{n+2} - 2a_{n+1} + a_n = 2^n . n^2$
(*v*) $a_{n+2} - 7a_{n+1} + 10a_n = 12 . e^{3n} + 4^n$

45. Solve the recurrence relation $t_{r+2} - t_{r+1} + t_r = 0$.

Given that $t_0 = 1$ and $t_1 = \dfrac{1+\sqrt{3}}{2}$.

46. Find the particular solution of the following difference equations

(*i*) $a_r - 4a_{r-1} + 4a_{r-2} = 2^r + r . 2^r$
(*ii*) $a_r + 5a_{r-1} + 6a_{r-2} = 4^{r^2} - 6r + 1$

47. Solve the recurrence relation

$a_n + 3a_{n-1} + 2a_{n-2} = f(n)$ given

$$f(n) = \begin{cases} 1 & n = 5 \\ 0 & \text{otherwise} \end{cases}$$

also given that $a_0 = a_1 = 0$.

[**Hint:** Calculate $a_n^{(P)}$ $[f(n)]$ for $n = 5$ for other values of n, $f(n) = 0$].

48. Solve the recurrence relations using generating function :

(*i*) $a_r - 5a_{r-1} + 6a_{r-2} = 2^r + r$, $r \geq 2$ given that $a_0 = 1$, $a_1 = 1$.
(*ii*) $a_r + 4a_{r-1} + 4a_{r-2} = r^2 - 3r + 5$
(*iii*) $a_r - 4a_{r-2} = 0$, $n \geq 2$, $a_0 = 0$, $a_1 = 1$.
(*iv*) $a_r - 8a_{r-1} + 21a_{r-2} - 18a_{r-3} = 0$, $n \geq 3$.

49. There are n circles are drawn in a plane such that each circle intersects every other circle in exactly 2 points and no 3 circles meet in the same points. If these n circles create $f(n)$ regions, find a recurrence relation for $f(n)$.

50. Let a_n be the number of n-letter sequences that can be formed using the letters *A*, *B* and *C* such that any non terminal *A* has to be immediately followed by *B*. Find the recurrence relation for $<A_n>$ and the corresponding generating function.

ANSWERS

1. 10^5

2. (*a*) 90

(*b*) 90 + 81 + 81 = 252

(*c*) 72 + 72 + 81 = 225

4. (*a*) 24

(*b*) 36

(*c*) 48

5. 671

6. 1956

7. 216

9. $\left({}^9C_2\right) . \left({}^9C_2\right) = 1296$

10. $\dfrac{10!}{6}$

12. $\begin{pmatrix}7\\3\end{pmatrix} . \begin{pmatrix}4\\2\end{pmatrix} . 5! = 25200$

13. 12

14. 2520

15. 3456

16. (*i*) 1663200 (*ii*) 129729600, (*iii*) 3326400

17. 151351200

18. (*i*) 720, (*ii*) 3628800

19. 30

20. 17 ! (18) 2 !

21. 2 . (20 !)^2

24. (*a*) ${}^8C_5$

(*b*) ${}^5C_3 . {}^3C_2$

25. (*i*) 186, (*ii*) 126

27. $a_1 = 500\,(1.05)^r$

28. $a_r = 2^{r+1} - 1$

29. a_{100} = number of particles at $t = 100$ is $(3 + 3^1 + 3^2 + \ldots 3^{100})$

30. $a_n + b_n = \begin{cases} 1 - 2^n & 0 \le n \le 2 \\ 2^n + 1 & 3 \le n \le 4 \\ 2^n + 2^{-n} + 4 & r \ge 5 \end{cases}$ $\quad a_n . b_n = \begin{cases} 0 & 0 \le n \le 4 \\ 1 + 2^{-n} + 3 . 2^n & n \ge 5 \end{cases}$

32. (*i*) $S^2 a = \begin{cases} 0 & 0 \le r \le 1 \\ 2 & 2 \le r \le 5 \\ 2^{-(r+2)} + 3 & r \ge 6 \end{cases}$ $\quad S^2 a = \begin{cases} 2 & 0 \le r \le 1 \\ 2^{-r-2} & r \ge 2 \end{cases}$

(*ii*) $\Delta a = \begin{cases} 0 & 0 \le r \le 2 \\ 17/16 & r = 3 \\ -2^{-r-1} & r \ge 4 \end{cases}$ $\quad \nabla a = \begin{cases} 0 & 0 \le r \le 3 \\ 17/16 & r = 4 \\ -2^{-r} & r \ge 5 \end{cases}$

33. $C = a * b = \sum_{i=0}^{r} a_i b_{r-i} = a_0 b_r + a_1 b_{r-1} + a_2 b_{r-2} + \ldots + a_r b_0$

So $C = \begin{cases} 1 & r = 0 \\ 2 & r = 1 \\ 3 & r = 2 \\ 2 & r = 3 \\ 1 & r = 4 \\ 0 & r \ge 5 \end{cases}$

34. $b_r = (-2)^r$

35. (*i*) $A(x) = \dfrac{1}{1-2x} + \dfrac{1}{1-3x}$ (*ii*) $A(x) = \dfrac{1}{(1-x)^5}$, $|x| < 1$

(*ii*) $A(x) = \dfrac{1}{1+3x}$ (*iv*) $A(x) = \dfrac{9}{1-3x}$

36. $C(x) = \left(\dfrac{1}{1-2x}\right) \cdot \left(\dfrac{1}{1-3x}\right) = \left(\dfrac{1}{1-5x+6x^2}\right)$

37. (*i*) $a_n = \begin{cases} 0 & n \text{ is odd} \\ 2^{n+1} & n \text{ is even} \end{cases}$ (*ii*) $a_n = \dfrac{n+1}{3}(n^2 + 2n + 3)$

(*ii*) $a_n = \dfrac{1}{4}\left[1 - \dfrac{1}{5^{n+1}}\right]$ (*iv*) $a_n = \begin{cases} 0 & 0 \le n \le 4 \\ 2^{n-4} & n \ge 4 \end{cases}$

41. (*a*) 15 (*b*) 30 (*c*) 36

42. (*i*) $a_n = (C_1 + C_2 n) + 4.2^n$

(*ii*) $a_n = 1 - 2.3^n + 2^{n+1}$

(*v*) $a_n = (C_1\, 4^n + C_2\, 3^n) + n(A_1 + A_2\, n)\, 4^n$

43. (*i*) $a_n = C_1 \,.\, 3^n + 5^n \left(C_2 \cos\dfrac{n\pi}{2} + C_3 \sin\dfrac{n\pi}{2}\right)$

(*ii*) $a_n = (-C_1 + C_2 + C_3\, n)(-1)^n + \dfrac{2^n}{18} + \dfrac{1}{9}\left(n^2 - 3n - \dfrac{1}{2}\right)$

(*iv*) $a_n = (C_1 + C_2\, n) + 2^n (n^2 - 8n + 20)$

(*v*) $a_n = (C_1\, 2^n + C_2\, 5^n) + \dfrac{12e^{3n}}{e^6 - 7e^3 + 10} - \dfrac{4^n}{2}$

44. $t_r = \cos\dfrac{n\pi}{3} + \sin\dfrac{n\pi}{3}$

45. (*i*) $a_r^{(P)} = r^2 \,.\, 2^r$

(*ii*) $a_r^{(P)} = \dfrac{1}{3} r^2 + 8 \,.\, r + 4$

5. $a_n = + \dfrac{5}{186}(-1)^n - \dfrac{5}{186}(-2)^n + \dfrac{1}{6}$, for $n = 5$.

$= 0$; otherwise

46. (*i*) $a_n = \dfrac{5}{4} + \dfrac{1}{2}(n+1) - 3.2^n - 2^{r+1} \,.\, (r+1) + \dfrac{17}{4} 3^r$

(*iii*) $a_n = 0$ if n is even

$= 2^{n-1}$ if n is odd

(*iv*) $a_n = C_1\, 2^n + C_2\,(n+1)\, 2^n + C_3 \dfrac{(n+1)(n+2)}{2} 2^n$

47. $f(n) = f(n-1) + 2(n-1); f(1) = 2$

48. $a_n = 2a_{n-1} + a_{n-2}$; $a_1 = 3$, $a_2 = 7$ and $A(x) = \dfrac{3x + x^2}{1 - 2x - x^2}$

■■■

M.C.A.

(SEM 1) ODD SEMESTER THEORY EXAMINATIONS 2009-10 DISCRETE MATHEMATICS

Time : 3 Hours *[Max. Marks : 80]*

Note: *Question - Paper carries three sections. Read the instruction carefully and answer accordingly.*

Section A

1. Attempt all parts of this section :

(*i*) This question contains 10 multiple choice questions. Select the correct answer for each one per instruction. **10 × 1= 10**

(*a*) Consider (X, Y, f) is a morphism, where $f : X \to Y$. If image set of f is ϕ then f is

(*i*) an empty function

(*ii*) an identy function

(*iii*) a surjective function

(*iv*) an injective function.

(*b*) Let I^+ be the set of positive integers and R be the relation on I^+ defined by xRy iff $2x \le y + 1$. Then which ordered pair belongs to R:

(*i*) (2, 2) (*ii*) (3, 2)

(*iii*) (6, 15) (*iv*) (15, 6)

(*c*) Let I be the set of integers and R be the relation on I defined by xRy iff $|x - y| = 2$. Then the relation R on the set I is:

(*i*) Reflective (*ii*) Symmetric

(*iii*) Transitive (*iv*) None of above

(*d*) Let $(X, *)$ be a group, where $*$ is the multiplication operation on X. Then for any $x, y \in X$ which one is true?

(*i*) $(x^{-1}) = x$ (*ii*) $(x^{-1})^{-1} = x$

(*iii*) $(xy)^{-1} = yx$ (*iv*) None of above

(*e*) The number of minterms in the k-map of 3 variable boolen function are:

(*i*) 4 (*ii*) 6

(*iii*) 8 (*iv*) 16

(*f*) Graph shown below is an example of

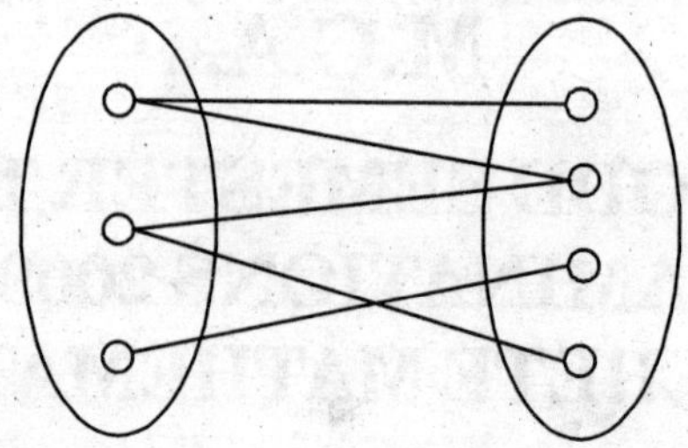

(*i*) Regular graph (*ii*) Planar graph
(*iii*) Bipartite graph (*iv*) None of above

(*g*) An undirected $G = (V, E)$ is complete iff:
(*i*) $|E| = n$ (*ii*) $|E| = n(n-1)$
(*iii*) $|E| = n(n-1)/2$ (*iv*) $|E| = \phi$.

(*h*) The truth value of the formula $(P \to Q) \vee \sim (P \leftrightarrow \sim Q)$ if P and Q are true will be
(*i*) True (*ii*) False
(*iii*) False and True both (*iv*) None of above

(*i*) Which one of the following formulas is true?
(*i*) $(P \to Q) \Leftrightarrow (\sim P \wedge Q)$
(*ii*) $(P \to Q) \vee (\sim P \vee Q)$
(*iii*) $(P \leftrightarrow Q) \Leftrightarrow (P \to Q) \vee (Q \to P)$
(*iv*) $(P \wedge Q) \wedge \sim (P \wedge Q)$

(*j*) Suppose we have n distinct letters, then total number of words can be formed with those n letters are:
(*i*) n (*ii*) $n \cdot (n-1)$
(*iii*) $n \cdot (n-1).(n-2)...3.2.1$ (*iv*) $n \cdot (n-1)/2$

2. State True/False: **5 × 1 = 5**

(*a*) For any finite set X, $\phi \in \rho(X)$ where $\rho(X)$ is the power set of X.
(*b*) A graph is bipartite iff it contains no odd cycle.
(*c*) Master theorem is used to solve the recurrences and return the solution in asymptotic bounds.
(*d*) The time complexity of the function that multiply two matrices of order n can be less than $0(n^3)$.
(*e*) Solution of the recurrence $T(n) = 2T(n/2) + c$ for $n \geq 2$ is $\theta(n \log n)$.

3. Fill in the blanks. **5 × 1 = 5**

(*a*) If S and T are two sets not necessary disjoint then $|S \cup T| = |S| + |T|$— ________ .
(*b*) A binary tree that has n nodes may need an array of size up to ________ for its representation.

(c) A regular graph with vertices of degree k is called as ________.

(d) ${}^nC_n$ = ________ .

(e) $(P \rightarrow Q) \wedge P$ = ________ .

Section B

2. Attempt *any* **three** parts: **10×3=30**

(a) Consider a relation R = $\{(x, y) \mid x, y \in I^+$ and $(x - y)$ is divisible by 3$\}$. Find the set of equivalence classes generated by the elements of set I^+.

(b) Let G be an Abelian group and N is a subgroup of G. Prove that G/N is an Abelian group.

(c) Consider a set X = $\{\alpha, \beta, \gamma\}$. Draw the Hasse diagram of the poset $(p(X), \subseteq)$, where $\rho(X)$ is the power set of X.

(d) Prove that the foumla $B \wedge (B \rightarrow C)$ is a tautology.

(e) A box is filled with three blue balls, three red balls and four yellow balls. Eight balls are taken out one at a time. In how many way can this be done. (Assume that balls of same color are indistinguishable).

Section C

Attempt any **five** parts selecting are from each questions. **10×5=50**

3. (a) For given sets $X = \{1, 2\}$, $Y = \{a, b, c\}$ and $Z = \{c, d\}$, find $(X, Y) \cap (X \times Z)$.

(b) Prove the following property of Fibonacci numbers

$$f_1^2 + f_2^2 + ... + f_n^2 = f_n f_{n+1} \quad \forall\, n \geq 1$$

where $f_0 = 0$, $f_1 = 1$ and $f_i = f_{i-1} + f_{i-2}$ for $i \geq 2$.

4. (a) Let $f: G \rightarrow H$ be a group homomorphism, prove that Ker (f) is a normal subgroup of G.

(b) Prove that $(\{0, 1, 2, 3, 4\}, +_5, *_5)$ is a finite field, whee $+_5$ is addition modulo 5 and $*_5$ is multiplication modulo 5 operators.

5. (a) For a given truth table obtain the simplified Boolean functions f_1 and f_2 is sum-of-products and products of-sum forms.

x	y	z	f_1	f_2
0	0	0	0	1
0	0	1	1	1
0	1	0	0	1
0	1	1	1	0
1	0	0	1	0
1	0	1	1	1
1	1	0	0	0
1	1	1	0	1

(*b*) Show that bipartite graph is 2-colourable.

6. (*a*) Show that $B \rightarrow E$ is a valid conclusion drawn from the following premises.

$A \wedge (B \rightarrow D), \sim C \rightarrow (D \rightarrow E). A \rightarrow C$ and $\sim C$.

(*b*) Prove that following argument is valid using predicate logic:

(*i*) All dogs are barking.

(*ii*) Some animals are dogs.

/∴ Same animals are braking.

7. (*a*) Solve the recurrence $T(n) = 2T(n/2) + n$ for $n \geq 2$ and n is a power of 2.

(*b*) Show that the generating function for the Fibonacci number sequence is $\frac{x}{1-x-x^2}$. The Fibonacci numbers are defined by the recurrence $f_n = f_{n-1} + f_{n-2}$ for $n \geq 2$ where $f_0 = 0$ and $f_1 = 1$.

B.TECH.

(SEM IV) EVEN SEMESTER THEORY EXAMINATIONS 2009-10

DISCRETE MATHEMATICS

Time : 3 Hours *[Max. Marks : 100]*

Note: *Attempt **all** the questions.*

1. Attempt **any four** parts of the following: **(4 × 5 =20)**

(*a*) Show that for positive integers *m* and *n*.

$$n\binom{m+n}{m} = (m+1)\binom{m+n}{m+1}$$

(*b*) Define a Relation R on a set $\{a, b, c, d, e\}$

(*i*) Which is only reflexive

(*ii*) Which is reflexive and symmetric.

(*iii*) Which is symmetric but not reflexive

(*c*) Let R = [(1, 2), (1, 4), (2, 3), (2, 4), (3, 4)] and S = [(1, 1), (3, 1), (4, 3), (4, 1), (4, 2)] be two relations on a set X = {1, 2, 3, 4}. Find the matrix representations M_R and M_S of R and S and show that $M_R.M_S = M_{ROS}$

(*d*) How many one to one functions are there from a set with five elements to a set with ten elements?

(*d*) Find the value (s) of n if ${}^nC_3 = {}^nP_2$.

(*f*) A computer company receives 350 applications from computer graduates for a job planning a line of new Web servers. Suppose that 220 of these graduates majored in computer science, 147 majored on business, and 51 majored both in computer science and in business both. How many of these applicants majored neither in computer science nor in business?

2. Attempt **any two** parts of the following : **(2 × 10 = 20)**

(*a*) Construct a truth table for each of the compound propositions:

(*i*) $(p \leftrightarrow q) \vee (\neg p \rightarrow r)$

(*ii*) $(\neg p \leftrightarrow \neg q) \leftrightarrow (p \leftrightarrow q)$

(*b*) State the coverse, contrapositive, and inverse of each of these conditional statements.

(*i*) If it snows tonight, then I well stay at home

(*ii*) I go to the beach whenever it is a sunny summer day.

(*iii*) When I stay up late, it is necessary that I will sleep until noon.

(*c*) Determine whether these are valid arguments,

(*i*) If x is a positive real number, the x^2 is a positive real number. Therefore, if a^2 is positive, where a is a real number, then a is a positive real number.

(*ii*) If $x^2 \neq 0$, where x real number then $x \neq 0$. Let a be a real number with, $a^2 \neq 0$; then $a \neq 0$.

3. Attempt **any four** parts of the following **4×5=20**

(*a*) Let $A\begin{pmatrix} 1 & 1 \\ 1 & 0 \end{pmatrix}$

show that $A^n = \begin{pmatrix} f_{n+1} & f_n \\ f_n & f_{n-1} \end{pmatrix}$

Where f_n is n^{th} Fibonacci number.

(*b*) Solve the recurrence relation $a_n + 6a_{n-1} + 9a_{n-2} = 0$ for $n \geq 2$ $a_0 = 3$, $a_1 = -3$.

(*c*) Use generating function, solve that $\Sigma(^nC_k)^2 = {}^{2n}C_n$.

(*d*) Using generating function, solve

$a_{n+2} - 5a_{n+1} + 6a_n = 2$

$n \geq 0$, $a_0 = 3$, $a_1 = 7$.

(*e*) Evaluate $F_{n+1}^2 - F_nF_{n+1} - F_n^2$, for $n = 0, 1, 2, 3, 4$ where F_n is n^{th} Fibonacci number.

(*f*) Show that the co-efficient $p(n)$ of x^n in the power series expansion of $1/(1-x)(1-x^3)(1-x_5)$... equals to the number of partitions of n.

4. Attempt **any four** parts of the following **4×5=20**

(*a*) Define a group. Verify whether the set of all 2 × 2 matrices of real numbers form a group with respect to matrix multiplication.

(*b*) Define an abelian group. Give two examples of non-abelian groups.

(*c*) Define a permutation group. Let S_4 be a permutation group on symbols (1, 2, 3, 4). FInd two subgroups (non-trivial) of S_4.

(*d*) Let $u(10)$ = [1, 3, 7, 9] be a group with respect to multiplication moduluo 10. Find the order of every element of $u(10)$.

(*e*) Prove that the order of a subgroup H divides the order of the group G (HCG).

(*f*) Define a field. Verify whether Zp = (0, 1, 2, ... p – 1}, p prime, with respect to addition and multiplication modulo p.

5. Attempt **any four** parts of the following **4×5=20**

(*a*) Prove that the number of vertices in a graph with even degree is always EVEN.

(*b*) Find the examples of two graphs G_1 and G_2 (with number of vertices ≥ 10 and number of edges ≥ 15) such that

(*i*) G_1 is Hamiltonian but not eulerian.
(*ii*) G_2 is eulerian but bot Hamiltonian.

(*c*) Prove that in a binary tree T the number of pendent vertices is $\left[\frac{n+1}{2}\right]$ where n is the number of vertices in T.

(*d*) Define a finite automation machine construct deterministic finite state automat that recognize each of these languages.
(*i*) The set of bit strings that begin with two 00's.
(*ii*) The set of bit strings that end with two 00's.

(*e*) Find the state diagram for the non-determinstic finite state automation with state table.

State	f	
	0 input 1	
s_0	s_0, s_1	s_3
s_1	s_0	s_1, s_3
s_2		s_0, s_1
s_3	s_0, s_1, s_3	s_1

the final states are s_2 and s_3.

B.TECH.

(SEM III) ODD SEMESTER THEORY EXAMINATIONS 2009-10

DISCRETE MATHEMATICS

Time : 3 Hours *[Max. Marks : 100]*

Note: *Attempt* ***all*** *the questions.*

1. Attempt **any four** parts of the following: **(4 × 5 =20)**

(*a*) Give a combinatorial argument to show that for integers *n*, *r* with $n \geq r \geq 2$.

$^{n+2}C_r = {}^2C_r + 2\ {}^nC_{r-1} + {}^nC_{r-2}$.

(*b*) Define a Relation R on a set X = {1, 2, 3, 4, 5}

(*i*) Which is only reflexive

(*ii*) Which is reflexive and symmetric.

(*iii*) Which is symmetric but not reflexive

(*c*) Let R = {(*a*, *b*), (*a*, *d*), (*b*, *c*), (*b*, *d*), (*c*, *d*)}

and S = {(*a*, *a*), (*c*, *a*), (*d*, *c*), (*d*, *a*), (*d*, *b*)]} be two relations on a set X = {*a*, *b*, *c*, *d*}. Find $(RoS)^5$ and R^5oS^4.

(*d*) Each user on a computer system has a password which is six to eight character long, where each character is an upper case letter or a digit. Each password must contain at least one digit. How many possible passwords are there ?

(*d*) How many positive integer between 1000 and 9999 incluse are divisible by 7 and 11 ?

(*f*) Define a one-one onto function. Show that if *f* and *g* are one-one onto then *fog* is also one-one onto.

2. Attempt **any two** parts of the following : **(2 × 10 = 20)**

(*a*) Construct a truth table for each of the compound propositions:

(*i*) $(p \rightarrow q) \vee (\neg p \rightarrow \neg r)$

(*ii*) $(p \leftrightarrow q) \leftrightarrow (\neg p \rightarrow q)$

(*b*) Show that the following premises are inconsistent :

(*i*) If Jack misses many classes through illness, then he fails high schoo.

(*ii*) If Jack fails high school, then he is uneducated.

(*iii*) If Jack reads a lot of books, then he is not uneducated.

(*iv*) Jack misses many classes through illness and reads a lot of books.

(*c*) Test the validity of the following argument:

If I study, then I will not fail mathematics; If I do not play basket ball, then I will study. **But I failed mathematics.**

Therefore, I must have played basket ball.

3. Attempt **any four** parts of the following **4×5=20**

(*a*) If n is a non-negative integer, then

$$\sum_{k=0}^{n} ({}^{n}C_k)^2 = {}^{2n}C_n.$$

(*b*) Suppose that k and n are integers with $1 \le k < n$, prove that

$${}^{n-1}C_{k-1} \cdot {}^{n}C_{k+1} \quad {}^{n+1}C_k = {}^{n-1}C_k \cdot {}^{n-1}C_{k-1} \quad {}^{n+1}C_{k+1}.$$

(*c*) Find the solution to the recurrence relation $a_n = 6a_{n-1} - 11a_{n-2} + 6a_{n-3}$ with the initial conditions $a_0 = 2$, $a_1 = 5$, $a_2 = 15$.

(*d*) Prove that for $(n + 1)$ st fibonacci number

$f_{n+1} = {}^{n}C_0 + {}^{n-1}C_1 + \ldots\ldots + {}^{(n-k)}C_k$

(*e*) Show that for positive integer n

$$\binom{-\frac{1}{2}}{n} = \binom{2n}{n} \Big/ (-4)^n$$

(*f*) Show that the co-efficient $p(n)$ of x^n in the formal power series expansion of $1/(1 - x)(1 - x^3)(1 - x^5)\ldots$ equals to the number of partitions of n.

4. Attempt **any four** parts of the following: **4×5=20**

(*a*) Define a group. Verify whether the set of integers Z forms a group with respect to difference.

(*b*) Let $M_2(R)$ be the set of all 2×2 non-singular matrices of real numbers, verify whether $M_2(R)$ forms a group with respect to matrix multiplication.

(*c*) Define a cyclic group. Prove that cyclic group is ABELIAN.

(*d*) Define a permutation group. Let S_3 be a permutation group on three letter set $\{a, b, c\}$. Find three nontrivial sub-groups of S_3.

(*e*) Let $u(8) = \{1, 3, 5, 7\}$ be a group with respect to multiplivation modulo 8. Prove that every element if $u(8)$ is its own inverse.

(*f*) Define a Ring. Verify whether $Z_p = \{0, 1, 2, 3, \ldots. (p - 1)\}$ p prime w.r.t. addition and multiplication modulo p.

5. Attempt **any four** parts of the following: **4×5=20**

(*a*) Prove that the sum of the degrees of all the vertice of a graph is even.

(*b*) Find the example an eulerian graph which is also hamiltonian.

(*c*) Define the pendent vertices of a tree. Prove that every tree $T = T(V, E)$ with V vertices and E edges, $|V| \geq 2$ has at least two pendent vertices.

(*d*) Define a finite automation machine. Construct deterministic finite state automata that recognize each of these languages :

(*i*) the set of bit strings that begin with two oo'2

(*ii*) the set of bit strings that end with two o's

(*iii*) the set of bit strings that contain at least two o's.

(*e*) Draw the transition diagram of non-deterministic finite state automata $M = (S, A, I, f, s_o)$

where

S \ I	a	b
s_0	$\{s_1\}$	$\{s_0\}$
s_1	$\{s_1\}$	$\{s_1, s_2\}$
s_2	ϕ	ϕ

$S = \{s_0, s_1, s_1\}$

$A = \{s_1\}$

INDEX

A

B

C

D

E

F

U

V

W

Z

Win Prizes!

ATTENTION READERS

We request you for your frank assessment regarding some of the aspects of the book given as under :

DISCRETE STRUCTURES

By

KALIKA PATRAI

Please fill up the given space in neat capital letters.

(i) What topics of your syllabus that are important from you examination point of view are not covered in the book ?

....

....

....

....

(ii) What are the chapters wherein the treatment of the subject matter is not systematic or organised or updated ?

....

....

....

....

(iii) Have you came across misprints/mistakes/factual inaccuracies in the book ? Please specify the chapters and the page numbers.

....

....

....

....

(iv) What other books on the same subject have you found/heart better that the present book ? Please specify in terms of price and quality.

....

....

....

....

(v) Further suggestions and comments for the improvement of the book:

....

....

....

....

(PS: If need be please attach a separate sheet to write your views.)

Other Details :

(i) Who recommended you the book ? (Please tick in the box near the option relevant to you.)

☐ Teacher ☐ Friends ☐ Bookseller

....

....

....

....

(ii) Name of the recommending teacher, his designation and address.

....

....

....

....

(iii) Name and address of the bookseller you purchased the book form :

....

....

....

....

(iv) Name and address of your institution (Please mention the University or Board, as the case may be)

....

....

....

....

(v) Your name and complete postal address :

....

....

....

....

(vi) Course you are enrolled in :

....

....

....

....

The best assessment will be awarded each month. The award will be in the form of our publications as decided by the Editorial Board.

Please mail the filled up coupon at your earliest to:

S.K. KATARIA & SONS®
4760-61/23, Ansari Road, Daryaganj, Delhi-110002
Phone : 23243489, 23269324
E-mail : katson_sanjeev@yahoo.co.in
katariabooks@yahoo.com